Materials and Processes

PART A: MATERIALS

MATERIALS ENGINEERING

A Series of Reference Books and Textbooks

SERIES EDITOR

Robert S. Shane
President
Shane Associates, Inc.
Wynnewood, Pennsylvania

1. Adhesive Bonding of Aluminum Alloys, *edited by Edward W. Thrall and Raymond W. Shannon*

2. Materials and Processes, Third Edition *(in two parts), edited by James F. Young and Robert S. Shane*

Additional Volumes in Preparation

Materials and Processes

Third Edition
(in two parts)

PART A: MATERIALS

Edited by

James F. Young
General Electric Company
Fairfield, Connecticut

Robert S. Shane
Shane Associates, Inc.
Wynnewood, Pennsylvania

MARCEL DEKKER, INC.   New York and Basel

Library of Congress Cataloging in Publication Data

Main entry under title:

Materials and processes.

    (Materials engineering; 2)
    Contents: A. Materials – B. Processes.
    Includes index.
    1. Materials – Addresses, essays, lectures.
2. Manufacturing processes – Addresses, essays, lectures.
I. Young, James F., 1917-1980. II. Shane, Robert
Samuel, [date]. III. Series: Materials
engineering; 2.
TA403.8.M375 1985   620.1'1   85-20432
ISBN 0-8247-7197-4 (pt. A)
ISBN 0-8247-7198-2 (pt. B)

First edition © 1944 by John Wiley and Sons, Inc., New York.

Second edition © 1954 by John Wiley and Sons, Inc., New York.

MARCEL DEKKER, INC.
270 Madison Avenue, New York, New York 10016

Current printing (last digit):
10 9 8 7 6 5 4 3 2

PRINTED IN THE UNITED STATES OF AMERICA

In memory of

JAMES F. YOUNG and ARTHUR M. BUECHE

whose untimely deaths during the preparation of this book
were a loss to mankind as well as to science and engineering

Robert S. Shane

James F. Young

Arthur M. Bueche

Photographs courtesy of the General Electric Company, Fairfield, Connecticut.

The scientist and the engineer. What a combination! Discovery translated into products and services for people. No two persons more fully played these roles within the General Electric Company for nearly four decades than did Arthur M. Bueche, Senior Vice-President—Corporate Technology, and James F. Young, Vice-President—Technical Resources. Powerfully motivated men, they served their nation and their company with distinction through their outstanding achievements in broad areas of science and in providing creative solutions to engineering problems.

Both men were exacting taskmasters, whose corporate and national reputations reached the highest levels. Both had an in-depth grasp of a wide range of technologies that permitted them to make significant contributions in many fields. Both played an active role in those academic, industrial, and public organizations dedicated to the advancement of the nation's technical strength. While remaining flexible and endearing personalities, they were quick to take those actions necessary to make General Electric a quality-oriented, technologically competitive, world-class industrial power.

It is significant that this two-volume book on materials and processes is dedicated to Art and Jim, considering the contributions both men made to materials science.

Fred W. Garry, Vice-President
Corporate Engineering and Manufacturing
General Electric Company
Fairfield, Connecticut
August 16, 1984

Arthur M. Bueche (1920-1981)

On October 22, 1981, Dr. Arthur M. Bueche died at the early age of 60. This tragic, untimely loss cost the world a great contributor and cost me a great friend and supporter. It is not too much to say that this book might never have been completed without his warm encouragement.

Dr. Arthur M. Bueche was the principal technical officer of the General Electric Company. He was widely recognized as a key spokesman for the technical community as it related to societal issues.

Dr. Bueche spoke often of the strength of the "technological triangle." By this he meant the nation's universities, industries, and government—working together— "each to do what it can do best." He recognized and respected those who were champions of industrial technology, those who spoke forcefully on behalf of the government's role in science, and those whose dedication was to the academic life. He took for himself the task of helping tie the entire triangle together. The remarkable scope and successes of his industrial activity, his government advisory positions, and his academic activities and interests are the benchmarks of a truly outstanding career in technological statesmanship. This book is dedicated to a man who supported its concept over many years.

Robert S. Shane

In appreciation of **James F. Young** (1917-1980)

James F. Young was a "statesman of engineering." Throughout his career his words and actions brought a basic message to all who knew him: the engineer has a special talent for serving certain needs of society, and his work output should reflect the highest standards of integrity in doing so.

In a way, this book memorializes those convictions. He discovered for himself, as a young mechanical engineer, that materials impose fundamental limits on the performance of all physical things. The first edition, published in 1944, resulted from his determination to assemble and disseminate information on materials useful to himself and other engineers involved in designing structures.

As this third significantly extended edition demonstrates, his interest in materials and processes was sustained throughout his lifetime. It is sad that he did not live to see its completion. Thanks to the partnership he arranged with Dr. Robert S. Shane, those who continue in the practice of engineering have opportunity to make use of it, and to appreciate the dedication that brought it into being.

Arthur M. Bueche

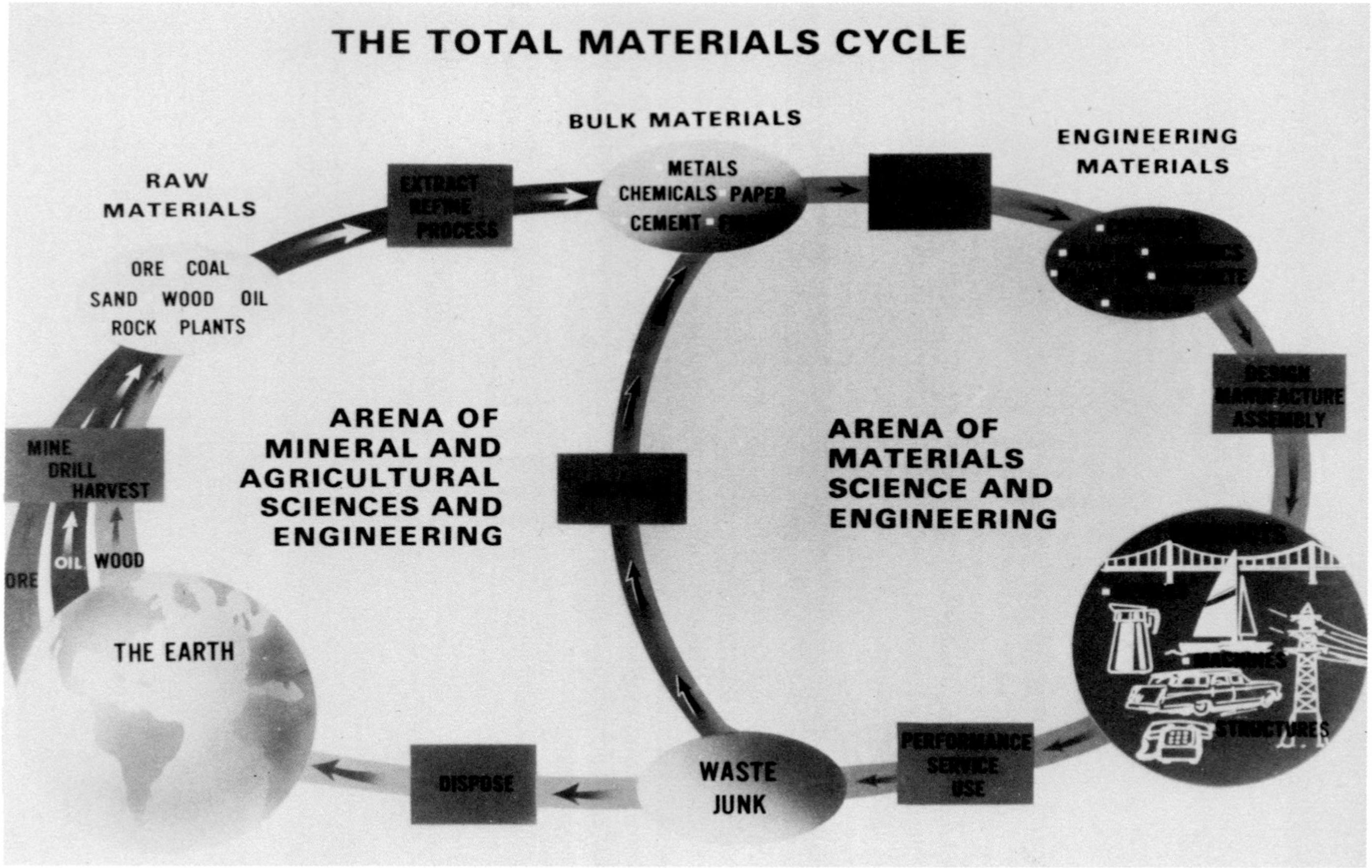

THE TOTAL MATERIALS CYCLE

BULK MATERIALS

METALS
CHEMICALS PAPER
CEMENT

ENGINEERING MATERIALS

RAW MATERIALS

ORE COAL
SAND WOOD OIL
ROCK PLANTS

EXTRACT
REFINE
PROCESS

DESIGN
MANUFACTURE
ASSEMBLY

STRUCTURES

PERFORMANCE
SERVICE
USE

WASTE
JUNK

DISPOSE

MINE
DRILL
HARVEST

OIL WOOD
ORE

THE EARTH

ARENA OF MATERIALS SCIENCE AND ENGINEERING

ARENA OF MINERAL AND AGRICULTURAL SCIENCES AND ENGINEERING

Based on "Materials and Man's Needs," National Academy of Sciences, Washington, D.C., 1974.

Preface to the Third Edition

In this third edition we have built on the second edition and consciously focused on the contributions of materials science and materials engineering to the needs of students and practitioners of engineering. Defining design as a planned use of knowledge, we present here a road map to the materials and processes by which a design engineer can find the way to contemporary knowledge of materials and processes. For detailed information special sources of knowledge must be consulted. Our purpose is to point out the existence of such knowledge.

This is a newly written and revised edition of the classic textbook on materials and processes written by James F. Young for the General Electric engineering course and adopted worldwide for the education of engineers. It is about twice the size of the second edition. This reflects the rapid growth of materials science and the addition of several topics that we consider important to the education and practice of engineers.

Review questions have been deleted from the text in keeping with our belief that earnest teachers and students can devise such questions in line with their individual interests and emphasis.

Each contributing author is eminent in his field. Some contributors have a worldwide reputation; others are only known to the cognoscenti in their discipline. However, the presence of a contributor in this book is assurance that the late Mr. Young and I believed that the reader is getting authoritative information from the best available sources. Total responsibility for the final text is assumed by me and I would be grateful for any suggestions for improvement that any reader would care to give. However, my deepest thanks go to my friend, the late Mr. Young, and all the contributors who made this work possible.

Robert S. Shane

Preface to the Second Edition

The primary objective in preparing this revision has been to maintain the approach that motivated the first edition. That is, the emphasis has been placed on presenting a broad study of engineering materials and manufacturing processes from the viewpoint of the engineer. The acceptance gained by the first work has, it is believed, confirmed the need for this approach and its usefulness. The subject matter has been organized to provide for textbook and reference use by students and practicing engineers, particularly those having product development, design, production, processing, quality, and application responsibilities.

Other objectives sought in this revision are:

1. To expand the coverage of nonmetallic materials, constructional materials, and many materials and processes used in product industries.
2. To bring the presentation up to date with latest developments in physical metallurgy and chemistry, both in theory and in available materials.
3. To reorder the presentation to assist classroom use.
4. To add typical material property data and process tolerances which should promote the book's usefulness as a "starting" reference and aid development in the student of the "sense of proportion" so necessary to engineering.

Most of the chapters have been completely rewritten to accomplish these objectives. New chapters have been added on Metallographic Examination; Structure and Properties of Nonmetallic Materials; Rubber; Ceramics, Porcelain, and Glass; Miscellaneous Nonmetallic Materials; and Statistical Methods Useful in Quality Control. The old Heat Treatment chapter has been brought up to date and consolidated with chapters on Alloys, Iron and Steel, and Heat-Treating Processes to avoid duplication. Sections have been added on Tarnishing, Electric Contacts, and Nondestructive Testing, and the coverage has been expanded on many subjects, such as bearing metals, superalloys, nonferrous elements, blow molding, shell molding, and pressure welding.

Contact with first-edition users among the engineering colleges and experience with the new chapters in training programs of the General Electric Company have been utilized as a guide to the organization. In the study of metallic materials, properties are covered first and then specific materials available are reviewed in more detail. The nonmetallic materials are similarly covered. Manufacturing processes are grouped in the order typical of most fabrication. Since many first-edition users adopted some variation in the original order to suit particular student groups and course objectives, the self-contained-chapter approach was continued to aid and encourage this practice, and textbook styling was emphasized.

Careful consideration was given to the possibility of separating the book into two volumes, because the study of materials is often separated from processing studies in engineering curricula. The single volume was retained, however, to provide for interchange in such courses; to aid reference use in more advanced courses, such as design courses; and particularly for reference use in engineering practice.

A study of the broad field covered by this book requires, of necessity, considerable condensation to achieve the present compact size. It has, therefore, been necessary, on occasion, to sacrifice preciseness for clarity of presentation and practicality. Bibliographies of many good works are listed with each chapter for those desiring more specialized treatments of the subjects. The tabular property data were selected to provide representative information on typical materials, with the hope that they may illustrate the general ranges available. This should aid the user in selecting types of materials for an application and then permit him to seek specific suppliers' data for his specification. It should also aid the student in achieving a sense of proportion for size, tolerance, and strength. Such data may not agree exactly with existing specifications or suppliers' guarantees, since the latter often deal with minimum values, and techniques among suppliers may provide materials differing from the representative values. Some trade names have been incorporated for convenience in identification. There is no intent to slight any source by its omission; any omissions have resulted from the physical impossibility of including all sources.

The book's wide scope requires dependence almost wholly on the contributions of specialists in the many fields represented. It is my hope that due acknowledgment of sources has been made with the material presented. I am also indebted to many experts in the General Electric Company, particularly the Research Laboratory, the Turbine Division Laboratory at Schenectady, N.Y., and the Household Refrigerator Laboratory at Erie, Pa., who read individual chapters and offered recommendations for technical accuracy and completeness; to contributors to individual chapters, including those who assisted in preparing portions drawn from the first edition, and, insofar as possible, their names have been incorporated with their work; to many professors who offered suggestions and criticisms; to the Metals Engineering Division Committee, ASME, for discussions and suggestions which have greatly assisted formation of the new edition; to my wife, Rita, who contributed much in companionship over a period of years, besides doing most of the typing and proofreading; and to Mrs. G. W. Snell and Miss C. Barr for assistance in typing and proofreading.

J. F. Young
December, 1953

Preface to the First Edition

This book has been written to present in one volume a broad study of the materials and manufacturing processes employed by the design engineer, and thus to provide information directly useful in the selection of materials for design. It is intended for convenient reference and for textbook use. It has therefore been organized for ease in classroom presentation and in such a manner that, it is hoped, will give the practicing engineer or designer an overall picture of the subjects discussed.

The problem of selecting the material for a piece of apparatus is no easy matter. One's first attempt at selection proves this quite convincingly. Plenty of information is available, but the real task is to find and evaluate that which has a bearing on the design. It is reasonable to assume that, if a material has the properties required in service, it is suitable, and so it may be selected. It soon becomes clear, however, that the material also must be available in the right form, and be such that it lends itself better than others to the available and desired method of processing. In addition, tne overall cost, including both of the material and of fabricating it, should represent the maximum value per dollar expended. And this combination can be obtained only when the product is both proportioned with respect to the material to be used and detailed to accommodate the method of processing.

The young engineer's training for design, which is largely obtained through working with more experienced engineers, should be preceded by some study of metallic and nonmetallic materials and of manufacturing methods. However, much of the available information on these subjects is not presented to suit the requirements of the design engineer. Data on metallic materials are presented largely from the metallurgist's viewpoint, and information on manufacturing methods usually tells how to accomplish the process and how to operate the equipment used. These approaches leave a rather broad gap in the design problem, a gap that it is hoped this book will bridge.

This textbook considers chiefly the materials and processes used in manufacturing electromechanical products. Sufficient metallurgy is included to enable the engineer to understand heat-treating practice and the effects of various processes on metallic materials. In the discussion of processes, enough detail for understanding the basic nature of each process is given; but—throughout—emphasis is laid on so designing the products that they can be easily processed. Materials and processes used only in building construction are not considered. Material-specifications systems and data have been left out also because of their changing nature and the impossibility of covering adequately all sources of supply.

Many of the chapters have been written from lectures given in a general course in materials and processes conducted in the Advanced Engineering program of the General Electric Company. I am indebted to the many engineers whose lectures and papers have been used in the preparation of this book, and, insofar as practicable, their names appear at the head of those chapters prepared from their lectures or incorporating their work. I also wish to express my appreciation to Dr. A. R. Stevenson, Jr., to whom this book is dedicated, for the opportunity of preparing the book and for his encouragement; to Mr. E. E. Parker, Mr. E. R. Boynton, and Mr. J. E. Ryan, who served as the first supervisors of the general course and laid out the early lectures; to Mr. T. S. Fuller and Mr. E. R. Parker for their helpful suggestions and consultation; and to Miss F. E. Rist for her patience and good nature in typing the manuscript and in proofreading.

J. F. Young
January, 1944

Contents of Part A

Preface to the Third Edition   *ix*
Preface to the Second Edition   *xi*
Preface to the First Edition   *xiii*
Contents of Part B   *xxi*
Contributors to Parts A and B   *xxiii*

1. Introduction, *James F. Young*   **1**

    I. Purpose   1
    II. Technology and Society   1
    III. Scope   3
    IV. Organization   3
    V. Evolution of Materials Characterization, *with Donald G. Groves*   5
    VI. Selection of Materials and Processes in Design, *Robert S. Shane* 8
    VII. Evaluation of Materials and Processes Choices   12
    VIII. Proof of Reliability   14
    IX. Use of Handbooks and Standards, *Robert S. Shane*   15
    X. Conclusion   17
    References   17

2. Structure of Matter and Introduction to Metallurgy, *Ellis D. Verink, Jr.*   **19**

    I. Atoms and Bonding   23
    II. Metallic States   25
    References   56

3. Mechanical Properties and Fracture Mechanics, *Douglas L. Jones*   **57**

    I. Static Properties   58
    II. Cyclic Loads   91
    III. Impact Loads   110
    IV. Design Applications   115
    V. Conclusion: Advice to the Engineer   122
    List of Symbols   123
    Bibliography   124

4. Chemical Properties and Corrosion, *Lewis W. Gleekman*   127

    I. Economics of Corrosion   127
   II. Corrosion, the Silent Scourge, *Elio Passaglia*   127
  III. Principles of Corrosion   136
  IV. Protecting the Metal   144
   V. Design Improvements   145
  VI. Corrosion Behavior of Common Materials   146
 VII. Nonmetallic Materials   158
VIII. Inhibitors   163
  IX. Oxidation and Hot Corrosion, *S. R. Shatynski*   164
     References   194

5. Magnetic Materials, *Daniel L. Potts*   197

    I. Why and How Magnetic Materials Are Used   197
   II. Units   198
  III. Basic Terms Used in Describing Magnetic Materials   198
  IV. Types of Materials and Range of Properties Available   201
   V. Classification of Materials   204
  VI. Permanent Magnetic Material Selection   205
 VII. Soft Magnetic Material Selection   210
VIII. Additional Information on Electrical Steel   213

6. Semiconductor Materials, *Gordon K. Teal, W. R. Runyan,*
Kenneth E. Bean, and Howard R. Huff   219

     I. Introduction   219
    II. Single-Crystal Manufacture   230
   III. Transistor Manufacture   231
   IV. Silicon Crystal Growth   235
    V. Evolving Directions   239
   VI. Semiconductor Devices: Material Considerations   241
  VII. Semiconductor Properties   242
 VIII. Analytical Procedures   253
   IX. Purification and Synthesis   256
    X. Crystal Growth   262
   XI. Behavior of Impurities During Freezing   269
  XII. Crystal Growth from the Vapor: Epitaxy   279
 XIII. Silicon Epitaxy   283
 XIV. Crystallographic Defects   291
  XV. Sawing, Cutting, and Polishing   296
     Notes and References   304

7. Radiation Effects on Materials, *S. Vaidyanathan*   313

    I. Introduction   313
   II. Microstructural Changes Resulting from Irradiation   315
  III. Irradiation Effects on Mechanical Behavior   320

IV. Design Implications 327
 V. Summary 329
   References 329

8. Iron and Steel, *Robert S. Shane* **331**

   I. Introduction 331
  II. Production of Iron and Steel 331
 III. Heat Treatments of Steel 385
  IV. Conclusion 398

9. Nonferrous Metals and Alloys **401**

      I. Introduction, *Daniel L. Potts* 401
     II. Aluminum and Aluminum Alloys, *Robert S. Shane* 406
    III. Beryllium and Beryllium Alloys, *Robert S. Shane* 414
     IV. Cobalt, *Robert S. Shane* 415
      V. Copper and Copper Alloys, *Daniel L. Potts* 418
     VI. Germanium, *Robert S. Shane* 435
    VII. Hafnium, *Miles C. Leverett* 436
   VIII. Lead and Its Alloys, *Robert S. Shane* 436
     IX. Magnesium and Its Alloys, *Robert S. Shane* 439
      X. Nickel and Its Alloys, *Robert S. Shane* 442
     XI. Silicon, *Robert S. Shane* 445
    XII. Tin, Solder, and Bearing Materials, *Joseph B. Long* 445
   XIII. Titanium, *Robert S. Shane* 460
    XIV. Tungsten, *Robert S. Shane* 468
     XV. Zinc, *Robert S. Shane* 468
    XVI. Zirconium, *Miles C. Leverett* 472
   XVII. Other Nonferrous Alloys of Commercial Importance, *Robert S. Shane* 480
  XVIII. Metal Matrix Composites, *Robert S. Shane* 490
    XIX. Liquid Metals, *Miles C. Leverett* 490
     XX. Clad Metals (Metallic Composites), *Robert S. Shane* 491
         Appendix: Spinodal Alloys 495
         References 497
         General Reading 500

10. Miscellaneous Inorganic Materials: Principles of Use and Design **501**

   I. Introduction: Properties and Structure, *Donald G. Groves* 501
  II. Nonmetallic Materials, *Robert S. Shane* 504
 III. Types of Chemical Bonding, *Robert S. Shane* 506
  IV. Structure of Inorganic Materials, *Robert S. Shane and
      Herman F. Mark with Marvin G. Britton* 511
   V. Relation of Properties to Structure (Summary), *Robert S. Shane* 559
  VI. Conclusion, *Robert S. Shane* 571
      References 573
      Additional Reading 573

11. **Plastics and Elastomers (Rubber)**, *Shalaby W. Shalaby and Barbara Greenberg Schwartz*   575

      I. Introduction   575
     II. Chemical Notations of Chain Molecules   575
   III. Interplay of Structure, Properties, Processing, and End Use   581
   IV. Material Selection for Specific Applications   630
    V. Thermoset Processing   637
   VI. Processing Thermoplastic Materials   639
  VII. Rubber, *Peter J. Larsen*   640
 VIII. Relationship of Polymer Structure to Performance, *Herman F. Mark*   654
       References   669
       Bibliography   670

12. **Adhesives**, *Raymond F. Wegman and David W. Levi*   671

      I. Nature of Adhesion   672
     II. Types of Adhesives   674
   III. Preparation of Surfaces   674
   IV. Application of Adhesives   676
    V. Durability and Reliability   678
   VI. Designing for Adhesive Bonding   678
  VII. Testing of Adhesive Properties   679
       References   680

13. **Electrical Insulation**, *Kenneth N. Mathes*   687

      I. Functional Properties and Processing Characteristics   688
     II. Insulating Materials   698
   III. Factors in Design and Application   707
       References   709
       Additional Reading   709

14. **Ceramics**   711

      I. Nonvitreous Ceramics, *R. Nathan Katz, Winston Duckworth, and David W. Richerson*   711
     II. Nature and Properties of Glass, *Marvin G. Britton, P. Bruce Adams, and J. R. Lonergan*   736
       References   781
       Recommended Readings   783

15. **Advanced Composite Materials**, *Samuel J. Dastin*   785

      I. Introduction   785
     II. Advanced Composite Material Systems   794
   III. Typical Advanced Composite Material Properties   795
   IV. Advanced Composite Material Design Considerations   795
    V. Advanced Composite Material Definitions   803
   VI. Composite Material and Labor Costs   804

VII. Composite Test Methodology   805
VIII. Typical Design of Composite Structures   808
IX. Conclusion   810
X. Composites: Challenge and Opportunity, *Robert S. Shane*   811
Bibliography   824
References   826

16. Miscellaneous Materials, *Robert S. Shane*   **829**

I. Carbon and Graphite   829
II. Cement, Mortar, and Concrete   834
III. Cork   837
IV. Felt   838
V. Fiber   840
VI. Fiberboard   842
VII. Leather   842
VIII. Lubricants   843
IX. Wood   847
X. Silicone Products   851
XI. Amorphous Metal Alloys   852
XII. Special Nonmetallic Plain Bearing Materials, *Joseph B. Long*   853
XIII. Materials for Specialized Applications, *Edwin T. Myskowski*   854

(Cumulative index at end of Part B)

Contents of Part B

17 Introduction to Processing, *James F. Young with Irving J. Gruntfest*

18 Casting Processes, *James F. Young, Edwin T. Myskowski, D. Basch, and Robert S. Shane*

19 Powder Metallurgy, *Robert S. Shane*

20 Heat Treating and Metal Strengthening Processes, *Robert S. Shane*

21 Hot and Cold Working of Metals, *Robert S. Shane*

22 Joining, *R. S. Pelton, Edwin T. Myskowski, Joseph B. Long, R. T. Gillette, R. M. Rood, James F. Young, Robert S. Shane, Raymond F. Wegman, and David W. Levi*

23 Machining, *James D. Collins*

24 Coatings, *Robert F. Brady, Jr., Howard G. Lasser, and Fred Pearlstein*

25 Electroplating, *Eric C. Svenson*

26 Interactive Process Design, Robotics, and Process Control, *Robert V. Klint, Joseph B. Gibbons, Paul F. Scott*

27 Plastics Processing, *A. M. Varner and Robert S. Shane*

28 Processing of Ceramics, *David W. Richerson*

29 Advanced Organic Matrix Composite Processing, *Samuel J. Dastin*

30 Nondestructive Evaluation, *Leonard Mordfin with David S. Dean*

31   Standards and Specifications, *William A. McAdams with J. H. Westbrook*

32   Design for Reliability and Safety, *Robert E. Warr and Robert S. Shane*

Cumulative Index, Parts A and B

Contributors to Parts A and B

P. Bruce Adams Corporate Systems Analysis, Corning Glass Works, Corning, New York

D. Basch General Electric Company, Fairfield, Connecticut

Kenneth E. Bean Central Research Laboratories, Texas Instruments Incorporated, Dallas, Texas

Robert F. Brady, Jr.* Federal Supply Service, General Services Administration, Washington, D.C.

Marvin G. Britton Education and Training Department, Corning Glass Works, Corning, New York

James D. Collins‡ Detroit Diesel Allison Division, General Motors Corporation, Indianapolis, Indiana

Samuel J. Dastin Advanced Development Department, Grumman Aerospace Corporation, Bethpage, New York

David S. Dean H. M. Propellants, Explosives, and Rocket Motor Establishment, Westcott, England

Winston Duckworth Battelle-Columbus Laboratories, Columbus, Ohio

Joseph B. Gibbons‡ Corporate Research and Development, General Electric Company, Schenectady, New York

R. T. Gillette† General Electric Company, Schenectady, New York

Lewis W. Gleekman† Materials and Corrosion Services, Southfield, Michigan

Donald G. Groves National Academy of Sciences, National Academy of Engineering–National Research Council, Washington, D.C.

Current affiliations

*U.S. Naval Research Laboratory, Washington, D.C.
‡Retired
†Deceased

Irving J. Gruntfest* Missile and Space Vehicle Department, General Electric Company, Philadelphia, Pennsylvania

Howard R. Huff† Philips Research Laboratories, Signetics Corporation, Sunnyvale, California

Douglas L. Jones Department of Civil, Mechanical, and Environmental Engineering, The George Washington University, Washington, D.C.

R. Nathan Katz Ceramics Research Division, Army Materials and Mechanics Research Center, Watertown, Massachusetts

Robert V. Klint Corporate Research and Development Center, General Electric Company, Schenectady, New York

Peter J. Larsen Chemical Products Group, Lord Corporation, Erie, Pennsylvania

Howard G. Lasser Naval Facilities Engineering Command, Alexandria, Virginia (Retired)

Miles C. Leverett‡ Nuclear Energy Business Group, General Electric Company, San Jose, California

David W. Levi§ U. S. Army Armament Research and Development Center, Dover, New Jersey

J. R. Lonergan Corning Glass Works, Corning, New York

Joseph B. Long¶ Tin Research Institute, Inc., Columbus, Ohio

William A. McAdams** International Electrotechnical Commission, Fairfield, Connecticut

Herman F. Mark Department of Chemistry, Polytechnic Institute of New York, New York, New York

Kenneth N. Mathes †† Corporate Research and Development Center, General Electric Company, Schenectady, New York

Leonard Mordfin Office of Nondestructive Evaluation, National Bureau of Standards, Gaithersburg, Maryland

Edwin T. Myskowski‡‡ Fairchild Control Systems Co., Manhattan Beach, California

Elio Passaglia§§ Center for Materials Science, National Bureau of Standards, Washington, D.C.

Fred Pearlstein U.S. Navy Aviation Supply Office, Philadelphia, Pennsylvania

---

Current affiliations

*Office of Toxic Substances, U.S. Environmental Protection Agency, Washington, D.C. (Retired)
†Market Development, Monsanto Electronic Materials Company, Palo Alto, California
‡Consultant, Monte Sereno, California
§Consultant, Succasunna, New Jersey
¶ Bohn Aluminum &Brass Division, Gulf + Western Manufacturing Co., Greensburg, Indiana
**Consultant, Fairfield, Connecticut
††Consultant, Schenectady, New York
‡‡Hughes Helicopters, Inc., Culver City, California
§§Polymers Division, National Bureau of Standards, Gaithersburg, Maryland

R. S. Pelton† General Electric Company, Schenectady, New York

Daniel L. Potts Materials Information Services, Corporate Engineering and Manufacturing, General Electric Company, Bridgeport, Connecticut

David W. Richerson* AiResearch Manufacturing Corporation, Phoenix, Arizona

R. M. Rood† General Electric Company, Fairfield, Connecticut.

W. R. Runyan Materials Division, Texas Instruments Incorporated, Dallas, Texas

Barbara Greenberg Schwartz Suture Technology Department, ETHICON, Inc., Somerville, New Jersey

Paul F. Scott General Electric Company, Schenectady, New York

Shalaby W. Shalaby Polymer Research Department, ETHICON, Inc., Somerville, New Jersey

Robert S. Shane Shane Associates, Inc., Wynnewood, Pennsylvania

S. R. Shatynski† Rensselaer Polytechnic Institute, Troy, New York

Eric C. Svenson Platers Supply Company, Twinsburg, Ohio

Gordon K. Teal‡ Texas Instruments Incorporated, Dallas, Texas

S. Vaidyanathan Advanced Nuclear Technology Operations, General Electric Company, Sunnyvale, California

A. M. Varner† General Electric Company, Fairfield, Connecticut

Ellis D. Verink, Jr. Department of Materials Science and Engineering, University of Florida, Gainesville, Florida

Robert E. Warr§ Corporate Quality Staff, General Electric Company, Fairfield, Connecticut

Raymond F. Wegman¶ U. S. Army Armament Research and Development Center, Dover, New Jersey

J. H. Westbrook* * Materials Information Services, Corporate Engineering and Manufacturing, General Electric Company, Schenectady, New York

James F. Young† General Electric Company, Fairfield, Connecticut

---

Current affiliations

†Deceased
*Ceramatec, Inc., Salt Lake City, Utah
‡Consultant, Dallas, Texas
§Corporate Engineering and Manufacturing, General Electric Company, Bridgeport, Connecticut
¶Adhesion Associates, Ledgewood, New Jersey
**Knowledge Systems, Inc., Scotia, New York

Materials and Processes

PART A: MATERIALS

1

Introduction*

I. PURPOSE

A review of current engineering curricula in U.S. colleges and universities reveals little coverage of materials or their production processes. There are, of course, presentations on specific materials that are closely associated with each engineering field, and there are topical electives available to all engineering students. But none of the curricula in the major engineering institutions offers a broad review of the subjects of materials and processes on which design engineering depends. Filling this gap is one of the major driving forces in the preparation of this third edition of *Materials and Processes*.

Another major driving force stems from the role of the experienced engineer engaged in developing and applying technology to meet perceived needs. Design problems never emerge in neat packages in a single engineering discipline. Rather, they usually require application of knowledge and understanding from many areas of applied science and from the lore of modern engineering practice. An essential foundation for all design is material selection and adaptation of processes for production and fabrication. With the rapid evolution in materials engineering and process development that has occurred in the last few decades, the designer is now faced with many new alternatives. So, a major driving force has been to provide state-of-the-art discussions, and forecasts where possible, of the materials and processes knowledge a designer needs to select, prove, and specify materials and processes for a design.

A third driving force stems from the changing societal expectations and needs for technology, as reviewed in the next section.

II. TECHNOLOGY AND SOCIETY

The term "technology" is widely used but with differing meanings. Generally, technology connotes a competence or a capability. The differences relate to scope. The purist thinks of technology simply as knowledge of the applied sciences. The more pragmatic concept is to recognize technology as a body of knowledge of the applied sciences *together* with the lore of past experience and the facilities to produce the goods and services on which society and the economy depend. In this pragmatic view, technology is the totality of

---

*Contributed by JAMES F. YOUNG.

1

the means employed to provide the objects of a material culture and, more recently, to protect the natural environment. This is the concept of technology employed in this edition.

Materials understanding, knowledge, and lore are an intrinsic part of our nation's technology. So, too, are the understanding, knowledge, lore, and capability of processes for materials production, treatment, and fabrication. Developments in either add to national technology available to meet national and, sometimes, international needs.

In earlier days, the social test for technology was pure economics. Does a product serve a useful purpose at an acceptable cost? There was concern for acute hazards affecting worker and user safety. There was concern for performance and reliability as rather obscure elements of quality. But the test was cost, so people could enjoy the fruits of their labors and raise their standard of living in terms of cars, food, clothing, shelter, leisure, recreation, and entertainment.

Under this set of expectations, technology, especially the evolving technology of materials and processes, had profound effects. Thirty million jobs were created since the mid-1950s. In the last 20 years, domestic per capita gross national product had an average annual growth rate of 2% in real terms while population increased 21%, and tax revenues of all kinds—federal, state, local, and social insurance—were increased by 450% for expanded government functions, social programs, and transfer payments. U.S. exports of high-technology products increased from $12 billion per year in the mid-1960s to well over $50 billion in 1978. Technology over a longer period helped make U.S. agriculture the acknowledged leader in feeding the world. Agricultural exports in 1978 were some $30 billion, and the United States needed less than 4% of the work force to feed itself and many others. New instruments for diagnosis, surgery, and therapy have made medical care more expensive, but care is also now more effective. Life expectancy continues to rise in the United States, up by 4 years in the last three decades.

Although it is nice to remind ourselves of these past achievements of technology, the expectations of society are changing worldwide. Underdeveloped regions still must link their technology to survival. But the developed and developing regions are reaching for a higher quality of life. The most publicized aspect of these expectations is concern for pollution of the air, the waters, and the land—in short, protection and restoration of the natural environment. A second major new concern is for the secondary effects of technology, the potential latent or chronic effects on living things that may emerge over time. Choice and control of materials and processes are profoundly affected by these changing needs and expectations.

There are also many new socioeconomic forces placing demands on technology:

1. The necessity for energy conservation and the concurrent development of new energy supplies domestically and also among developed and developing nations
2. The demand for safe, reliable systems and products for the workplace, commerce, and the home, often with an unattainable longing for zero risk
3. The expectations for high product quality, including performance, features, functions, environmental compatibility, and especially reliability and durability, as products compete with growing alternatives for discretionary expenditures
4. The expansion of health care and comfort for an aging population and the disadvantaged
5. The reduction of inflation and its erosion of living standards through technology applied to increase productivity

6. The reduction of the imbalance of national payments through improved world competitiveness, especially through reduction of cost and improvement in the availability of high-technology products
7. The expansion of employment and enhancement of goals for upward mobility through technological innovation of new products and services

In the nonfuel materials area, the United States is facing growing import dependence. The Bureau of Mines, U.S. Department of the Interior, reported that in 1978 our nation imported over half its requirements for some 19 nonfuel minerals and metals. The percentages and sources for that year are tabulated in Table 1.1. Projections to 1990 indicate this trend will continue, with potential world shortages of tantalum and tin and perhaps more local shortages of nickel and cobalt. The United States, like other developed nations, is expected to expand its exports of manufactured products to pay for the materials imports. Meanwhile, the materials-rich, less developed countries are expected to process raw materials to higher stages of added value to provide added employment and improve their balance of payments. This duality promises potential increases in materials costs that will lend added impetus for (1) conservation of materials, especially those depending on import, (2) substitution of lower cost and more available materials in a period of fluctuating materials prices, and (3) greater dependence on recycling including design of products to suit recycling processes.

All these pose new tests for technology and new dimensions for product design. They also pose added requirements and considerations for selection, proof, and specification of materials and processes that have guided preparation of this third edition.

III. SCOPE

Adaptation to these new driving forces has enlarged the content of the third edition and has led to a two-volume format, but the unique focus of earlier editions is maintained. This focus is on the information the designer needs to select, prove, and specify materials and processes for product design. It differs from other books that treat materials from the view of the applied science, although the views of the chemist, metallurgist, ceramist, and polymer scientist are by no means ignored. It also differs from books on processes that treat that subject from the view of the processor.

The objective is to familiarize the product designer with the background necessary to work effectively with specialists in metallurgy and in chemistry and related sciences and with specialists in the process fields, but it is not intended to try to make the designer a specialist in these allied fields. The treatment does not concentrate mainly on theory, as in a textbook, nor does it primarily provide extensive data, as in a handbook. Rather, the treatment can be likened to a broad road map. It is a starting reference with adequate text to explore and select routes to design application, and enough data to work out alternatives. The intent is to indicate what is available and how to get it.

IV. ORGANIZATION

The organization is intended to suit classroom use at the undergraduate level or as a basis for courses in industry. It is also intended for personal study, and especially as a starting

Table 1.1 U.S. Net Import Reliance of Selected Minerals and Metals as a Percentage of Consumption in 1978

Minerals and Metals	Net Import Reliance* as a Percent of Apparent Consumption**	Major Foreign Sources (1974-1977)
Columbium	100	Brazil, Thailand, Canada
Mica (sheet)	100	India, Brazil, Madagascar
Strontium	100	Mexico, Spain
Manganese	97	Gabon, Brazil, Australia, South Africa
Tantalum	97	Thailand, Canada, Malaysia, Brazil
Cobalt	95	Zaire, Belg.-Lux, Zambia, Finland
Bauxite & Alumina	93	Jamaica, Australia, Surinam
Chromium	91	South Africa, U.S.S.R., Turkey, Southern Rhodesia
Platinum—Group Metals	90	South Africa, U.S.S.R., United Kingdom
Asbestos	85	Canada, South Africa
Tin	79	Malaysia, Bolivia, Thailand, Indonesia
Fluorine	76	Mexico, Spain, Italy, Canada
Nickel	76	Canada, Norway, New Caledonia, Domin. Rep.
Zinc	66	Canada, Mexico, Australia, Belg.-Lux.
Potassium	64	Canada, Israel, Fed. Rep. of Germany
Mercury	64	Algeria, Canada, Spain, Mexico, Yugoslavia
Cadmium	63	Canada, Australia, Belg.-Lux., Mexico
Tungsten	56	Canada, Bolivia, Peru, Thailand
Gold	53	Canada, Switzerland, U.S.S.R.
Silver	48	Canada, Mexico, Peru, United Kingdom
Selenium	43	Canada, Japan, Yugoslavia
Titanium (ilmenite)	42	Canada, Australia
Barium	38	Peru, Ireland, Mexico
Vanadium	36	South Africa, Chile, U.S.S.R.
Gypsum	33	Canada, Mexico, Spain
Iron Ore	29	Canada, Venezuela, Brazil, Liberia
Iron & Steel Scrap	(19) NET EXPORTS	
Antimony	23	South Africa, Bolivia, China, Mexico
Copper	20	Canada, Chile, Peru, Zambia
Iron & Steel Products	12	Japan, Europe, Canada
Sulfur	12	Canada, Mexico
Aluminum	11	Canada
Salt	10	Canada, Bahamas, Mexico
Lead	9	Canada, Mexico, Peru, Australia
Cement	8	Canada, Norway, Bahamas, Mexico, United Kingdom
Pumice & Volcanic Cinder	4	Italy, Greece

*Net import reliance = imports-exports + adjustments for government and industry stock changes.
**Apparent consummation = U.S. primary + secondary production + net import reliance.
Source: Bureau of Mines, U.S. Department of the Interior, Washington, D.C. Import-export data from Bureau of the Census, Washington, D.C., revised September 1, 1979.

reference in design problem solving that involves materials and processes. Self-contained chapters permit selection for course emphasis and specialization as appropriate.

The first volume comprises the chapters on materials. Chapters providing back-grouns on the nature and properties of materials are presented first. They are followed by presentations of specific materials categories that typify current technology. Of these, metals and alloys are discussed first, and nonmetallic materials follow. Emphasis is balanced by current importance to design. The evolving field of composites is handled in two ways: (1) under the appropriate matrix and/or fiber material classifications, and (2) in a special chapter.

The second volume comprises chapters on processes. They are presented in the general sequence of production for metals, followed by the processes unique for nonmetallic materials. Concluding chapters emphasize nondestructive testing, standards, specifications, and design for reliability. Some duplication has occurred in the interest of clarity and completeness.

V. EVOLUTION OF MATERIALS CHARACTERIZATION*

The discovery of the chemical elements began some 10,000 years ago with identification of elemental materials, such as carbon, copper, iron, lead, mercury, sulfur, tin, silver, and gold. These elements are either found in a pure form in nature, or can be separated from their ores at relatively low temperatures. Much of the technological progress and earthly survival of the ancients has been largely dependent on the lore of extracting and employing these materials, a lore passed from generation to generation among the artisans of the evolving social orders.

By the middle of the nineteenth century, some 62 chemical elements had been identified. At that time, two notable events occurred. In 1868, Dmitri I. Mendeleev, a Russian chemist, published his *Elements of Chemistry* with the periodic table he had devised. It systematized the properties of the known elements, and permitted prediction of elements not yet discovered. Today, over 100 elements have been discovered, 40 since publication of Mendeleev's classic work.

The other notable event was the discovery of carbon in increasing amounts in wrought iron, steel, and cast iron. This knowledge led in turn to the discovery by Henry Clifton Sorby in 1864 of the role played by crystallinity in behavior of metals.

These events changed materials science from consideration of chemical composition alone. Atomic structure was introduced into the study and prediction of material behavior. Properties were now related to shape, size, relative distribution, and interrelationships of distinguishable microstructural features of materials.

Since 1900, many new alloys have been developed, and new and sophisticated analytical techniques for probing further into the internal structure of matter on an atomic scale have come into existence. For example, the microstructure of materials has been revealed by the microscope, crystal structure by various spectroscopies, and electronic structure by excitation techniques.

In the late 1930s to the early 1950s, pioneering work by such eminent scientists as Einstein, Fermi, and Seaborg led to discoveries regarding the nuclear structure of

---

*With contributions from DONALD G. GROVES.

matter. In turn, these led to understanding of radioactivity, the creation of isotopes through irradiation and radioactive decay, and the processes of fission and fusion.

Concurrent and continuing studies on the physics of matter have led to a growing identification of subnuclear particulates. While these developments continue the search for basic understanding of the nature of materials, they have not yet reached the stage where they play a role in the design of materials and processes. The presentations in *Materials and Processes,* Third Edition, accordingly are based on chemical composition and atomic structure and, where needed, on nuclear structure. Subnuclear particulate physics is not employed except in Chapter 7 but can be expected to add to materials understanding and prediction in the future.

For all these recent developments, we are still unable, for the most part, to relate properties of the important engineering materials to the character (composition, structure, and defect distribution) of the material. Yet this relationship and predictability is a major goal of current materials science. If properties could be predicted solely by the characterization of materials, we should be able to "tailor make" more satisfactory materials having needed reproducibility and reliability with less expenditure of time and money, with less hazard in use, and with greater alternatives in choice of ingredients and processes.

The weakest, or missing, link in the chain of materials science is knowledge that relates the significant features of a material with the properties they occasion and the processes they permit. Some properties are greatly affected by very small variations of these significant features. And in some cases, the important variations are apparently too small to be detected with present techniques. Predictive knowledge of these significant features is the goal of characterization of materials.

In 1967, the Materials Advisory Board, (MAB) of the National Research Council undertook a study of materials characterization. This study is still regarded as a landmark for its still applicable perception of characterization, as follows:

> Characterization describes those features of the composition and structure (including defects) of a material that are significant for a particular preparation, study of properties, or use, and suffice for reproduction of the material.

Pictorially, the MAB related characterization to the production and use of materials by the diagram in Fig. 1.1. Under this concept, the Materials Advisory Board study reported:

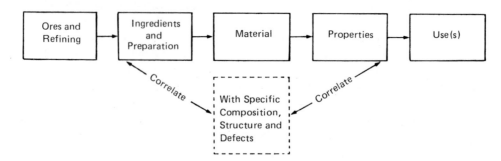

Fig. 1.1   Relation of characterization to production and use of materials. (From Ref. 1.)

Preparation is not necessarily included in the ultimate characterization, because preparation is not a quality of the final material. Therefore, we should work toward eliminating the need to describe preparation, although recording may be necessary in early and intermediate stages of work on material.

Properties are not included in characterization, because properties are determined by composition and structure, and so we should seek to correlate composition and structure with properties until eventually characterization suffices. In this connection, the more that measurements of properties are used to analyze for specific features of composition and structure, the more correlation there will be between composition and structure and properties.

There is a widespread misconception about the use of property measurements as a means of characterizing materials. True, every characterization method uses some property of the atoms; e.g., mass (mass spectrographic chemical analysis), characteristic nuclear decay (activation analysis), and x-ray scattering factor (x-ray analysis). Nevertheless, it needs to be proved, always, that the property measurement reflects directly and unambiguously the relevant compositional or structural features of a material before it can be accepted as a valid characterization method. Most property measurements do not fall in this category and require, therefore, an independent means of characterizing the structure or composition of the material.

Similarly, a description of method of preparation has often been taken to be necessary, or even sufficient, for the characterization of a material. Such a description is, however, only a substitute for the true characterization of a material in terms that are independent of its past history.

It is clear from this assessment that a true science of materials has yet to evolve. It also seems clear that materials research needs more rigor in characterization concepts and methodology to strengthen the correlation of research findings and their use in furthering basic understanding of materials behavior.

From an engineering point of view, materials cannot yet be specified without inclusion of methods of preparation, and the values of property tests along with descriptions of composition limits and structure. In this sense, materials engineering is still highly empirical (2). The designer must depend on knowledge that a particular composition, prepared in a certain way, and exhibiting certain values when specimens are prepared and tested a particular way, has been found to perform satisfactorily when used under certain environmental and physical conditions. A similar characterization (preparation, composition, and test values) is utilized for patent protection of new materials that possess novel properties.

This knowledge gives some assurance that materials for similar environmental and physical conditions will also perform well if the same compositions are prepared in the same way and have similar values from tested specimens.

It should be emphasized that the property values are not intrinsic. They are simply the test values of material specimens selected and prepared usually in a prescribed way and subjected to test procedures that are usually standardized. *Materials and Processes*, Third Edition, will refer to most of the property tests employed in materials engineering. Lesser coverage will be given to techniques that are evolving in materials characterization work, such as neutron and electron diffraction, spectroscopy (Chap. 10), wet chemistry, polarography, electron spin resonance, and cyclotron resonance. Section VIII will deal with the question of whether preselection and preproduction testing are necessary.

VI. SELECTION OF MATERIALS AND PROCESSES IN DESIGN*

There are many ways to classify products and services. The Department of Commerce uses SIC codes to classify products by industry groups, and at one time by subgroups. Finance markets use categorizations that combine nature of the product and the product market. Engineers tend to think of products as commodities, modified system designs, or custom designs. The range is enormous, and evergrowing. But all products, whether commodities, durables, or the tools of production and services, are processed from materials. All depend on materials and processes for their functional performance, their reliability, and their attractiveness. Accordingly, selection materials and processes is an essential and often determining element of product design.

A. Selection, an Aspect of Systems Engineering (3)

The considerations in materials and processes selection have often been thought of as an aspect of systems engineering. This concept can be illustrated by extending the unit process analysis often employed in chemical engineering to higher stages of product added value. Table 1.2 shows a set of selected added value stages and an example of a product at each stage. Each of the lower level products is a component for each higher stage. As many stages can be designated as may be needed for the purpose.

   When used to depict engineering in a business, and its product technology, the output stage of the business is highlighted. The stages within the business are plotted below

---

*Contributed by ROBERT S. SHANE.

Table 1.2 Stages of Product Added Value[a]

	Added value stage	Product example
↑ *Increasing added value*	Higher order system	Electric power system
	Higher order system	Electric power plant
	Higher order system	Nuclear steam supply system
	Higher order system	Reactor
	System	Recirculation loop
	Subsystem	Motor and control center
	Equipment	Motor
	Device	Stator
	Part	Insulated bar
	Piece	Bar form
	Preform	Bar
	Casting	Ingot
	Refined material	Electrolytic copper matte
	Concentration/separation	Concentration (23% Cu)
	Milling/gathering	Ground ore (0.4 to .8% Cu)
	Extraction	Ore removal
	Explore and develop resource	Copper mine

[a]The number of stages, and the size of the added value stage are arbitrary and depend on the requirements for their use.

to the point of product purchase from vendors. The stages above the business output are its applications and markets.

A technology-focused business would have the nature depicted in Fig. 1.2. The business shown is based on expertness in motor technology, and the materials and processes involved in motor design and production. This is often called a do-focus business because of the emphasis on an ability to do. Applications are shown as a broad spectrum to meet the needs of other equipment manufacturers (OEM) to whom the motor products are sold. In this business, the engineer (1) continuously seeks improvement in motor performance per unit of cost, and (2) with marketing and application engineering, continuously seeks performance and features for new market applications. Stress on the latter results from customer options to produce its own product needs, to use another technology, or to buy from competitors. To offset these possibilities, or to make them less attractive, the business objective and the engineering role is continued product improvement and adaptation to added applications. This focus is typical of materials suppliers and suppliers of component devices or equipment.

A system-focused business has just the opposite depiction, as illustrated in Fig. 1.3. This is often called a want-focus business because of its emphasis on what the customer wants or needs. The concern is for meeting the system application requirements, using any available technology. The designer's selection of materials and processes for a product or equipment takes on the focus and the options of Fig. 1.3. Serving the needs or requirements, using any available technology, is a primary aspect of selection.

This concept can also be utilized to visualize product integration of a business. Downward integration implies extending the business scope of materials and process operations to lower stages of added value and reducing dependence on suppliers. Systems-focused businesses generally avoid extensive downward integration because it tends

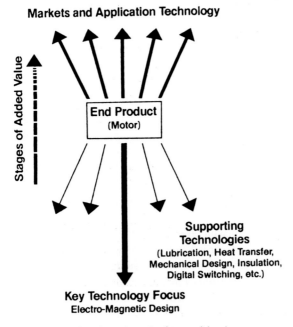

Fig. 1.2 Technology in a do-focused business.

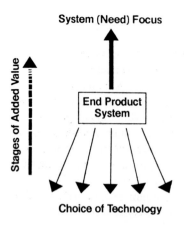

Choice of Technology   Fig. 1.3 Technology in a want-focused business.

to limit choices of alternative technologies to meet evolving systems requirements. Upward integration implies business participation at higher stages of added value, thus extending materials supply and processing to higher stage markets by providing economy through the integration. Horizontal integration implies extending business participation by adding related products, equipments, or systems at the same stage of added value. Here the advantage may be in multiple use of materials and processing technology and/or of marketing, application, and distribution capability.

There is another more prevalent aspect that likens materials and processes selection to systems engineering. This is viewing the application as a functioning system in itself. Active and passive elements of a design are thought of as parts of systems to perform a product function. Examples are structural support systems; alignment systems; corrosion protection systems; heat dissipation of cooling systems; lubrication systems; electromagnetic supply, concentration, and containment systems; torque, force, and load distribution systems; and finish systems. Segregation of a design problem or design improvement task into its functioning elements usually enhances effectiveness of materials and process selection while assuring rigor and thoroughness in considering alternatives and meeting all essential requirements.

B. Selection of Candidate Materials and Processes

A designer begins to choose materials, processes, raw and in-process inventory, and manufacturing costs with the first line in a design layout or the first display in CAD/CAM interactive graphics. The design concept therefore undergoes a substantial selection among alternatives before the layout stage is reached.

The logic of the conceptual stages usually follows a subjective character suited to the individual designer, but there are a number of key considerations that are addressed in the course of the design process, as follows.

1. A premier question is what must the product (component, equipment, or system) do—what function must be accomplished? Broad functional requirements, like weight limitations, energy sources and levels, acceptable failure modes and secondary effects, essential operating life and reliability, and allowable cost ranges, establish a preliminary sorting of alternative structures, materials, and processes.

2. A second determining question is, how many products will be produced? This question addresses the production rate (units to be produced per shift or per year) and

the design life (years in which the production process can be employed to produce the generic design).

Where few units are required, relatively cumbersome and operation-inefficient processes will be employed. Materials and designs suited to general-purpose tooling will predominate, perhaps aided by direct numerical control or other programmable processes.

As the number of required units increases in rate or total, more specialized processes will be economic. Per unit capital cost for tooling can be traded off against per unit inventory, process time, and labor. At high production levels, dedicated specialized tooling and automation will become attractive for parts fabrication and ultimately for assembly. Developments in programmable automation may make it possible to employ automated processes at lower throughput or at shorter product design life.

3. A third key question is the anticipated performance requirements. This includes not only the range of conditions under which the product must perform its intended service and retain its appearance, it also includes the conditions the product must survive during fabrication; during transportation to its point of use; during assembly, installation, and maintenance; and during extraordinary events like earthquakes, storms, and shocks. The selection process at this stage will produce a multiplicity of candidate materials or material systems and their processes. A number of noncandidate materials will also be set aside as unsuited to the performance requirements.

4. A fourth set of key questions is used to narrow and refine the range of candidate materials and processes. In familiar applications, the limiting conditions are usually well known, and quantitative information is easily obtained. There is a history of successful application and often a record of performance inadequacy where the product was misapplied or misused. Service failures, too, are often known, whether they result from new or unforeseen exposure, manufacturing errors, materials defects, or inadequate maintenance. Product improvement in such applications usually concentrates on materials and process improvements to reduce cost and upgrade performance, to reduce overdesign of material content, to consolidate or combine part functions and structures, and to structure the line of products for more commonality and longer production runs. Component portions of the product may take on basic redesign and thorough proof to take advantage of new materials and changes, like new alloys, angineered plastics, or lower cost materials substitutions.

At the other extreme is "first-of-a-kind" products. Often these will perform new functions and/or operate in unfamiliar environments. Initially a wide scope of materials and processes are among the alternatives. Tests may be required to establish property values for environmental regimes not previously documented. Depending on the mission, the design may be simulated analytically to establish requirements, means of proof, redundancy in design elements and spares to achieve required reliability, and backup system designs where feasibility is not yet proved. The range of choices converges as the design progresses, the limited options are discarded, and the final configuration reaches a design basis.

Whether a product is a modest or major improvement for a familiar application, or a more extensive first-of-a-kind innovation, there is an added set of considerations that will further narrow the range of acceptable candidate materials and processes. Primary among these are

> Possibility of environmental exposure to reactive liquids or vapors, continually, intermittently, or rarely, and its consequences.
> The consequences of failure, including avoidance of acute and immediate property

damage or bodily harm, and chronic effects of exposure to the failed product
that may appear over time.

The secondary effects of the materials and processes technology that may affect
workers, the natural environment, or third persons during the course of the
product life cycle. This set of considerations runs the gamut from materials
acquisition, through product production, product application, and use, to
product disposal.

All these are dealt with in major projects through preparation of environmental impact
statements or product safety analysis reports. This practice will probably become more
prevalent for other products with which the public comes in contact, either through
evolving regulations or as a result of the legal case history in product liability. Materials
and process considerations will play an increasingly significant role in such analyses and
documentation.

VII. EVALUATION OF MATERIALS AND PROCESSES CHOICES

The evaluation of candidate materials and processes to serve a design concept traditionally employs three criteria. They are suitability, availability, and cost. All three must be
adequately satisfied.

A. Suitability

The considerations reviewed in Sec. VI on selection of candidate materials and processes
apply to this criterion. A suitable material or material system must meet the performance
requirements through the product life cycle. For materials, this includes

Permanence or stability of dimensions and properties on which maintenance
of product output or efficiency depends.
Predictable wear or deterioration on which availability, reliability, and appearance
depend. This may rely on preventive maintenance or planned parts replacement, with attendant need for ease of serviceability.
Ability to survive the nonoperating conditions (temperature extremes, mechanical
extremes, and unusual events).
Acceptable potency of toxic material content, or means to prevent exposure or
uptake.

For processes, manufacturability or producibility is important to suitability determinations. Included are

Ease of forming or fabrication
Reproducibility of processes, including controllability of the process and the vari-
ances in parts produced
Availability of the process, including yield, downtime, and spare capacity
Effect of process on materials properties, beneficial or detrimental
Detectability of processing flaws in finished material

B. Availability

This criterion includes tactical and strategic considerations, both monetary and political.
With an increasing number of materials depending on imports, the designer is faced with a

potential for sudden price increases, political disruptions of supply, embargoes on supply, and import tariff and quota constraints. The designer is also faced with unavailability through source nation decisions to process materials to higher stages of added value, or to develop a glut on the market through oversupply based on maintaining employment or meeting needs for hard currency and balance of payments.

Multiple sourcing and investment in raw materials inventory are tactics for smoothing the uncertainties of import supply. Downward integration to domestic sources and supply from free world stable nations are other alternatives.

Another aspect of availability is the capacity of the supply industry. If there is overcapacity, production and price should be competitive although tactical steps may be needed to build inventory to ride through protracted supply disruptions for whatever reason, such as labor problems. A more difficult availability situation occurs when there is little spare industry capacity, and the designer needs a rather small volume of materials (compared with the industry production) that have special properties and require special processing. Alternatives may be multiple sourcing (maybe only two suppliers) for continuity of supply, longer term source contracting, and integration downward to more available base materials.

C. Cost

The criterion of cost is applied to the overall cost of the finished goods as well as to the best price for equivalent starting materials. Attractive price benefits from astute purchasing require good material specifications, so the advantages for bargain material are not lost in subsequent processing costs or the costs of internal and external product failures.

By concentrating at the selection stage on overall costs, the designer takes into account the trade-off of processing steps, yields, and validation versus cost of starting materials. When two acceptable materials have fluctuating market prices, it may be desirable for the designer to permit substitution. An example is the use of jute or cotton for some electrical cables. Another is the use of copper or aluminum for electrical bus bars.

Materials costs have traditionally shown real cost reduction through economy of scale. In the 1970s, the effect of scale economy waned, and materials have shown an escalating price trend. The magnitude of this trend has also been increased by the trend in energy costs since materials production is some four times as energy intensive as other manufacturing. Cost escalation of base materials is expected to continue with some extraordinary changes for individual materials.

There will be continuing need to lower costs through conservation in the use of expensive materials. For example, a hard-surface coating might be used for a part subject to wear or erosion, rather than producing the whole part from material having the properties needed at the surface. Other forms of composites can be used for structural strength, stability, and resiliency, often with substantial savings in scarce materials and weight. Still another approach to conservation is to distribute part loading, or to grade part sections to a uniform factor of safety. An example is the design of the wall sections of a die casting for near-uniform stress under peak loading.

A technique for systematic evaluation of part or device overall cost is *value analysis*. This technique evaluates value (or overall cost) to perform a part function using examples from the generic product and from a range of other products containing a similar part function. Overall savings achieved through this technique often reach 25-50% of the part cost.

Whatever the technique, approach, or analysis, the selection of materials and processes for a design must, for engineering effectiveness, assure suitability for all performance specifications, availability of the materials and processes, and acceptable overall cost.

VIII. PROOF OF RELIABILITY

The phrase "time is money" often expresses the necessity for product reliability. A product that will not perform as intended can delay start-up of a whole plant. Interest on the investment or the value of the lost production can greatly exceed the value of the faulty product. Similarly, failure of a key product in a plant in production may idle the entire plant, with losses in production output and employment that can run into hundreds of thousands to millions of dollars a day. Redundancy of key components, like pumps and valves, can be employed to help avoid plant shutdown, and an on-hand supply of spare parts of essential components can help to reduce downtime. Even so, the cost of delays or outages is one factor that demonstrates beyond question the worth of reliability.

Product reliability is also essential to product safety. Some ships are driven with a single engine (diesel or gas turbine). An engine failure, however infrequent, can put the ship, cargo, and crew in jeopardy of storms or collision with other ships or bridges. Similar critical reliability requirements apply to airplanes, bridges, elevators, safety systems, alarms and emergency controls, to name a few.

Concern for and need for product reliability have been greatly stimulated in the last two decades. The increased scale of buildings, plants, and facilities, magnified by inflation, have added to the economic worth of reliability. It can be seen that money has been a prime driver for product reliability, but concern for public safety has been even more demanding. Designers have always had concern for the acute effects of product failure. From the beginning, bicycles had twin forks, and electricity grew up with protective measures. Recently, however, the mores of U.S. society have changed and added to the expectations for safety. Society has become less daring. There is less risk taking, and some have an overwhelming desire for zero risk that is unfortunately not attainable. In addition, society has become more litigious. Civil (sometimes criminal) suits abound seeking relief for accidents that often involve misuse of a product, product modification by the user, and wornout or unmaintained products. Such suits are promoted by the contingency fee system and escalating jury awards. Liability insurance has been escalating in turn, jeopardizing the continuity of smaller businesses and professionals.

All this puts a greater burden on the designer to prove reliability for the protection of customers and users, and to avoid any basis for a claim of negligence in the design process. At the same time, the trend of products to higher levels of sophistication, to higher levels of complexity, and often to more demanding missions has increased the challenge of attaining reliability. The response to all these forces has been the development of sophisticated tools for reliability analysis and reliability proofs.

Among these are MTBF (mean time between failures), MTTR (mean time to repair), and Duane plots for failure history and failure prediction. Failure mode and effects analysis is used for appraising mitigation of the consequences of failure. Reliability data banks are being developed for component selection and for reliability prediction. Reliability analysis techniques have been devised for assessing reliability potentials in the predesign phase. These techniques are discussed in Chap. 32.

There are also growing techniques for risk analysis. At this writing, regulations are being developed relative to potential and actual carcinogens and toxic substances. The test techniques used for characterization now indicate carcinogenicity and toxicity to be rather common properties of materials, although they differ in potency by factors up to a million to one. Risk analysis is beginning to incorporate potency and exposure, including uptake, for assessment purposes.

Testing techniques have also become more sophisticated and more prevalent. Tests preceding materials selection are used for screening candidate materials. These may seek property performance under previously undocumented conditions or under boundary limit conditions.

Simulated service tests are used for design proof before mass production and sale. These tests evaluate behavior of the materials and design elements when used in a manner similar to their intended use under conditions subject to measurement and control. In more complex product systems, computer modeling and simulation are used to explore limiting conditions. *Field tests* are often made in preproduction samples of a product to be sure the laboratory-simulated service tests have not missed some conditions of actual service. *Service tests* are performed on selected samples of ongoing production to monitor product performance and assure that production tests, which are usually more limited, have not overlooked product variations affecting service.

Life tests are not always feasible, but should be done on higher volume, shorter life products. Special tests for knowledge can often be performed on modified design products to prove design procedures, and plant tests can often be made to affirm time constants, design margins, and often factors. New product introduction is often spread over a period of time. The objective is to get field performance feedback and assurance before full production is committed, and thereby give proof to all the preproduction and initial production proof of the design.

A statistically significant data base is often available for parts reliability or for material property variations or consistency for familiar products or materials applications. Preproduction and production testing can often be designed to add to these data bases. The designer is therefore pretty well prepared when designing in familiar products areas.

The proof of reliability for "first-of-a-kind" applications of necessity requires more testing. As noted above, the tests may seek knowledge of behavior of materials in new environmental regimes. The tests may seek a series of successive approximations with simulated models. Nature and configuration of the tests will depend on the experience and judgment of the designer and the departure of the first-of-a-kind product from related product experience. Replication may be a particularly difficult task to gain a sufficient data base to make the statistical analysis valid. Successive or repeat testing and comparative evaluation with known materials systems can be used to improve confidence limits.

IX. USE OF HANDBOOKS AND STANDARDS*

It is recognized that engineers require reference books and other sources of authentic information as they go through the iterations of a design. Too often design decisions are based on the input of a potential vendor's representative rather than on a carefully prepared handbook. Technical data from an advertising piece are a last resort; carefully evaluated data from handbooks are a first resort.

---

*Contributed by ROBERT S. SHANE.

A. Handbooks

A general handbook is the *Handbook of Chemistry and Physics,* published by Handbook Publishers, Inc. For metals information, the ninth edition of *The Metals Handbook,* published by the American Society for Metals, is a reliable source of evaluated data. No single handbook of evaluated data exists for polymers, but the *Modern Plastics Encyclopedia,* published annually by McGraw-Hill, is a compilation of typical values that may be used as a first step toward a design specification. The National Bureau of Standards Monograph 132, a Compilation and Evaluation of Mechanical, Thermal and Electrical Properties of Selected Polymers, published by the U.S. Department of Commerce, gives a critical review of data for six important classes of polymers: polytetrafluoroethylene and its copolymer with hexafluoropropylene, polychlorotrifluoroethylene, polyethylene terephthalate, polypyromellitimide, polyparaxylylxylene, and polycarbonate. This 1973 book is a model for what is acutely needed for the entire field of polymers.

Disciplinary handbooks have been produced and are well-known. Examples are *Marks' Standard Handbook for Mechanical Engineers,* published by McGraw-Hill, and *Engineering Manual,* published by McGraw-Hill.

An excellent example of a processing handbook is the *Forging Industry Handbook,* published by the Forging Industry Association, in 1970. Other industries and processes are represented by handbooks whose worth has been amply demonstrated.

Frequent use of the library to collect needed information is the hallmark of a competent engineer. However, besides collecting the information it is of paramount importance that the information be evaluated in terms of the design problem at hand. This means careful scanning of the data source to determine whether the data were collected under conditions relevant to the proposed use. It also means determining whether raw or evaluated data are collected and whether a sufficient confidence level has been ascertained by a statistical evaluation of the original data.

An example of proper use of a handbook is selection of a metallic conductor for a design. If the maximum temperature that the part will ever see is -50°C, mercury will serve the purpose since its resistivity is about the same as lead. But a closer examination of the handbook will reveal that the measurement of resistivity was made at -50°C. However, the melting point of mercury is -38.5°C and use above that temperature will be accompanied be melting of the mercury. Hence, data should not be carelessly lifted from a handbook without due consideration of the conditions of measurement and its applicability to the design problem in hand.

B. Standards

The definitive engineering tool of communication is a specification. This tells in sufficient detail and with sufficient rigor all that is necessary to transmit engineering information replicably. A specification that has been validated by a recognized process that is widely accepted is a standard. The subject is discussed in some detail in Chap. 31.

For this introductory chapter it is important to note that selection and use of standards is an easy way for an engineer to utilize the accumulated lore and experience of his profession. Standards are embodied in codes that directly affect the acceptability of an engineer's work. There are more than 400 standards-developing organizations in the United States. Of these two most prolific are the American Society for Testing and Materials and the Society of Automotive Engineers. The U.S. government has maintained a system of independent standards for the military and civilian departments. The

Office of Management and Budget has ordered that, beginning in 1980, voluntary standards shall be used wherever possible.

The same caveat applies to using standards for materials, processes, and test methods that applies to using handbooks. In each case the engineer must satisfy himself that the standard applies to his design problem. A mere similarity of title is not sufficient; the standard—no matter how carefully developed—must be appropriate to the task in hand.

X. CONCLUSION

This chapter has presented a host of considerations that underlie the preparation of *Materials and Processes,* Third Edition. It provides the background purposes of the new edition and the scope and organization of its contents.

The chapter has also presented the background status of materials characterization, the considerations and criteria for selection of materials and their attendant processes, the special and growing concerns for reliability, and the use of testing and the literature for selection and application of materials.

Intrinsic in the discussion has been a conviction of the essential importance of materials and processes to the selection, proof, and specifications of a design. Examination of this part of the designer's role attests to its basic contribution to business success, its excitement and challenge to creativity and innovation, and its great significance in serving heightened social concerns for protection of health, safety, and the environment. Perhaps, as never before, design engineers will be seen, and will see themselves, as key planners and key contributors to societal goals for raising the standard of living, especially among the disadvantaged, while enhancing the quality of life for all.

REFERENCES

1. "Characterization of Materials," Report MAB-229M, Materials Advisory Board, National Research Council, Washington, D.C., 1967, pp. I-8 to I-10.
2. "Organic Polymer Characterization," Report NMAB-332, Materials Advisory Board, National Research Council, Washington, D.C., 1977.
3. *Science Base for Materials Processing—Selected Topics* (NMAB-335). Committee on Science Base for Materials Processing, Materials Advisory Board; Commission on Sociotechnical Systems, National Research Council (National Materials Advisory Board, 1979; 147 pp.; available from NTIS, Springfield, Virginia; PB80-122948).

2
Structure of Matter and Introduction to Metallurgy*

One main concept has emerged from modern thought regarding the nature of materials. This is the concept of structure. It is no longer adequate to consider only the "building blocks" of which material is composed. It is also essential to recognize that the form and structure of materials determine their engineering properties. The pattern extends from the realm of the molecules, crystals, and phases through composite materials and encompasses metals, ceramics, polymers, and combinations of these classes of materials. In this chapter we will be dealing primarily with metals. There are 70 elements classified as metals. Of these, some 40 are of commercial importance today. Metals are distinguished from other elements by characteristic physical properties, so-called metallic properties, such as high melting temperature, low specific heat, good thermal and electrical conductivity, metallic luster, hardness, and the ability to be deformed permanently without fracture.

Not all the elements classified as metals possess these properties in equal measure. Some are notably lacking in one or more of the metallic properties (see Table 2.1). The classification of metals is therefore not an exact one and some exceptions are made.

Except for certain precious metals, notably gold and copper, and some iron in meterorites, all metals are found in nature chemically combined with other elements as "ores." The science of extracting them, refining them, and adapting them to use is known generally as metallurgy. Extractive metallurgy deals with the purification and reduction of ores. Process metallurgy deals with the refining, working, and heat treatment of the metallic materials obtained. Physical metallurgy covers the study of the fundamental nature of metallic materials and leads to the development of new materials and processing techniques on which engineering progress relies heavily.

All engineers in industry must have a working knowledge of physical metallurgy so that they can take into account the specific properties of the materials and thereby avoid wastage and spoilage due to misuse and optimize the usefulness of materials based on their structure and properties. Practically all useful properties of materials are strongly dependent on their internal structure. The term "internal structure" refers to the arrangement of electrons and atoms within the material. For a material of given composition, the internal structure is not necessarily constant. It can vary greatly depending on manufacturing processes and service conditions. This means that the internal structure of

---

*Contributed by ELLIS D. VERINK, JR.

19

Table 2.1 Metallic Elements of Commercial Importance[a]

Metal	Chemical Symbol	% Wt of Earth's Crust *	Specific † Gravity	Melting ‡ Pt., °F	Boiling ‡ Pt., °F	Specific ‡ Heat at 32 F vs. Water	Electrical § Resistivity, microhm-cm	Thermal ‡ Conductivity, Btu/ft²/ in./°F/ sec
Aluminum	Al	8.13	2.7	1220	3740	0.215	2.655	0.43
Antimony	Sb	..	6.71	1167	2620	0.049	39.0 (0C)	0.037
Arsenic	As	0.0000001	5.73	..	1130⁺	0.082	35 (0C)	..
Barium	Ba	0.05	3.7	1300	2980	0.068	..	..
Beryllium	Be	0.001	1.93	2340	5020	0.52	5.9 (0C)	0.31
Bismuth	Bi	..	9.92	520	2590	0.034	106.8 (0C)	0.016
Cadmium	Cd	..	8.7	610	1409	0.055	6.83 (0C)	0.18
Calcium	Ca	3.63	1.55	1560	2625	0.149	3.43 (0C)	0.24
Cerium	Ce	..	6.8	1100	2550	0.042	78	..
Chromium	Cr	0.037	7.1	3430	4500	0.11	13 (28C)	0.13
Cobalt	Co	0.001	8.92	2723	5250	0.099	6.24	0.134
Columbium	Cb	..	8.4	4380	5980	0.065	13.1	..
Copper	Cu	0.01	8.94	1980	4700	0.092	1.673	0.759
Germanium	Ge	..	5.36	1760	5070	0.073	89.000 (0C)	..
Gold	Au	0.0000001	19.37	1945	5380	0.031	2.19 (0C)	0.57
Iridium	Ir	..	22.42	4449	9600	0.031	5.3	0.11
Iron	Fe	5.01	7.88	2802	4960	0.11	9.71	0.144
Lead	Pb	0.002	11.35	621	3160	0.031	20.65	0.067
Lithium	Li	..	0.59	367	2500	0.79	8.55 (0C)	0.14
Magnesium	Mg	2.09	1.74	1202	2030	0.25	4.46	0.303
Manganese	Mn	0.1	7.39	2273	3900	0.115	185	..
Mercury	Hg	0.00001	15.63	−38	675	0.033	94.1 (0C)	0.0163
Molybdenum	Mo	0.0001	10.2	4760	8670	0.061	5.17 (0C)	0.28
Nickel	Ni	..	8.91	2651	4950	0.105	6.84	0.18
Palladium	Pd	..	12.16	2829	7200	0.058	10.8	0.135
Platinum	Pt	..	21.7	3224	7970	0.032	9.83 (0C)	0.133
Potassium	K	2.6	0.875	145	1420	0.177	6.15 (0C)	0.19
Rhodium	Rh	..	12.5	3571	8100	0.059	4.5	0.17
Selenium	Se	..	4.8	428	1260	0.084	..	..
Silicon	Si	27.72	2.34	2605	4200	0.162	100,000 (0C)	0.16
Silver	Ag	0.000001	10.75	1761	4010	0.056	1.59	0.81 (0C)
Sodium	Na	2.85	0.97	208	1638	0.295	4.2 (0C)	0.26
Strontium	Sr	..	2.54	1420	2520	0.176	23	..
Tantalum	Ta	..	16.6	5425	7590	0.036	12.4	0.11
Tin	Sn	0.0000001	7.30	449	4120	0.054	11.5	0.127
Titanium	Ti	0.63	4.5	3300	5430	0.126	80 (0C)	..
Tungsten	W	0.005	20.2	6170	10700	0.032	5.5	0.39
Uranium	U	..	18.69	2065	5250	0.028	60	0.052
Vanadium	V	..	5.7	3150	6150	0.120	26	..
Zinc	Zn	0.004	7.13	787	1663	0.0915	5.916	0.22
Zirconium	Zr	..	6.4	3200	5250	0.066	41 (0C)	..

[a] Properties are for annealed metals of high purity. Impurities and strain from working or heat treatment can alter these properties considerably.

* Layer land and ocean, 3 miles deep, according to U. S. Geological Survey (0.0000001% = 24,500,000 tons).

† S.g. × 0.0362 = lb per cu in. at room temperature. Point of sublimation at atmospheric pressure.

‡ At atmospheric pressure; from *Metals Handbook*, A.S.M., 1948.

§ At room temperature; from *Metals Handbook*, A.S.M., 1948.

a material may be controlled to provide special properties of immense usefulness. It also implies that, depending on the conditions of manufacture or use, the otherwise seemingly useful properties of a material may be destroyed if it is handled in an inappropriate fashion.

The structure of two forms of carbon, diamond and graphite, are shown in Fig. 2.1. In graphite, the carbon atoms are disposed in flat, hexagonal arrays with each carbon atom bonded to three other carbon atoms. Graphite is black and relatively soft, and because the planes of atoms can glide fairly easily over one another, graphite frequently serves as a lubricant. Graphite also is an electrical conductor. By contrast, diamonds, also composed of carbon atoms, are extremely hard and colorless and are nonconductors. Clearly, the differences do not reside in chemical composition. Instead, the differences lie in the structure and bonding of the two materials. In the diamond lattice each carbon atom is bonded to four other carbon atoms in a three-dimensional array that provides much greater strength.

Another example of the variation in properties with structure is shown in Fig. 2.2. The mechanical strength of a carbon steel (iron plus 0.8% carbon) depends on the thermal history of the material. Before testing, three specimens of this steel were heated to 900°C and cooled to room temperature at different cooling rates. The variation in yield strength (the stress needed to produce permanent deformation) with cooling rate is given in Fig. 2.2. As the cooling rate increases, yield strength increases drastically. Thus, even though the three samples have the same chemical composition and are tested at the same temperature, the mechanical strength varies with the thermal history. Physical evidence

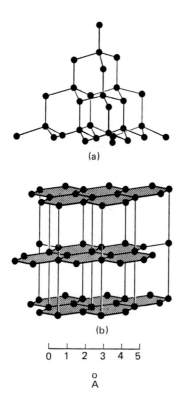

(a)

(b)

0 1 2 3 4 5

Å

Fig. 2.1 (a) The crystal structure of diamond. The carbon atoms are virtually in contact, with a distance between carbon centers of 1.54 Å. (b) The crystal structure of graphite. The fine horizontal lines join atoms of successive sheets, which are in the same vertical relationship. The carbon-carbon distance in the planes is 1.42 Å; that between planes is 3.40 Å. (From Ref. 1.)

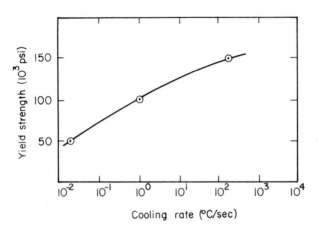

Fig. 2.2 Variation of yield strength with cooling rate for an Fe + 0.8% C steel initially held at 900°C. (Adapted from Ref. 2.)

that the structure has indeed changed as a result of variation in cooling rates is given in the photomicrographs included in Fig. 2.2.

Another example of the importance of internal structure on physical properties involves the phenomenon of superconductivity. Superconductivity occurs in certain metals and alloys at temperature near absolute zero and is characterized by a loss of all electrical resistance. Thus, an electric current in a superconducting cricuit will flow indefinitely without any resistance loss. These superconducting properties are very dependent upon the imposed current density and on applied magnetic fields. For example, superconductivity will disappear if a sufficiently strong magnetic field or current density is imposed. Figure 2.3 shows the relationship between the current density and magnetic field and indicates the regions over which a material will exhibit superconductivity and normal conductivity (with resistance losses). The maximum current density that can be tolerated for a given magnetic field in the superconductivity state is known as the "critical current density" (Fig. 2.3). Figure 2.4 shows that the critical current density is extremely sensitive to variations in internal structure. For example, the critical current density varies by three orders of magnitude for a Nb-25% Zr alloy, depending on the exact thermal-mechanical

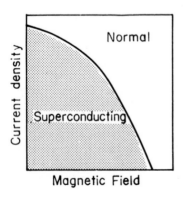

Fig. 2.3 Relation between current density, applied magnetic field, and the conductivity of a superconducting material. (From Ref. 3.)

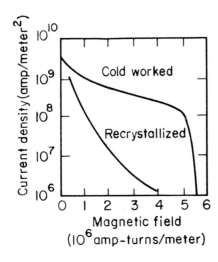

Fig. 2.4 Superconducting properties of Nb₃Zr for two different internal structures. (From Ref. 3.)

history of the wire. The terms "recrystallized" and "cold worked" will be explained in more detail later; however, it would be useful at this point to recognize that the recrystallized sample has been heated to a temperature slightly below the melting point and then cooled to room temperature, whereas the cold-worked sample was given the same recrystallization treatment but was then plastically deformed at room temperature. Apparently the plastic deformation changed the internal structure of the Nb–25% Zr alloy in such a fashion that it altered its superconducting properties. In fact, microstructural examination with the aid of an electron microscope revealed that the recrystallized and cold-worked structures were vastly different.

I. ATOMS AND BONDING

Primary bonding between atoms may be by ionic bonds, covalent bonds, or metallic bonds. In addition, there are weaker, secondary bonds called van der Waals forces. It should be remembered that although one bond type may predominate within a material, other bond types can be present also; thus, "mixed bond" types are not uncommon.

In ionic bonding, one of the atoms receives an electron from the outer shell of the other. The donor becomes a positively charged ion and the receptor a negatively charged ion. These unlike-charged ions are attracted to each other and form ionic bonded crystals like table salt (NaCl). Such bonds are commonly found in ceramics and tend to be very strong and brittle.

Polymeric materials commonly exhibit covalent bonding. Such materials have molecular structures and include plastics, paper, wood, and elastomeric materials, such as rubber. In these materials, the atoms share electrons. The atoms may be of the same kind, as in diamond, or they may be different kinds of atoms, as in hydrocarbons, where the carbon and hydrogen atoms are linked by covalent bonds. The properties of ceramics and polymers will be discussed in later chapters.

Metallic bonding, as the name implies, occurs in metals. The valence electrons in metals are not tightly bound to any particular atom but are considered to be "free" to form an "electron cloud" in the metallic structure while the metal atoms become positively charged ions. This is believed to explain a number of metallic properties, such as ductility and electrical and thermal conductivity.

A number of generalizations may be made about materials properties which relate to the atomic structure and bonding mode [4].

1. Density is controlled by atomic weight, atomic radius, and coordination number. Coordination number becomes important because it controls the packing of atoms.

2. Melting point and boiling point are relatable to the energy required to separate atoms from one another (Fig. 2.5).

3. Strength also is correlatable with the force necessary to separate atoms. Materials with high melting points, such as diamond or aluminum are stronger and harder than materials with low melting points (lead or plastics). Where mixed bonding occurs there may be exceptions, as in graphite and clay.

4. The modulus of elasticity can be calculated from the slope of the sum of attractive and repulsive forces when the atoms are at equilibrium spacing (i.e., when the sum of attractive forces equals the sum of repulsive forces; see Fig. 2.5).

5. The thermal coefficient of expansion of materials having comparable atomic packing varies inversely as the melting temperature.

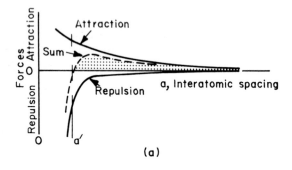

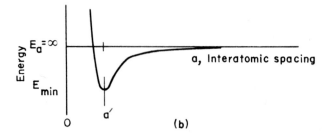

Fig. 2.5 Interatomic distances. (a) The equilibrium spacing 0-a' is the distance at which the attractive forces equal the repulsive forces. (b) The lowest potential energy occurs when 0-a' is the interatomic distance. Since E = ∫F da, the shaded area of (a) equals the depth of the energy trough in (b). (From Ref. 4.)

6. Ionically and covalently bonded materials are poor electrical and thermal conductors. Metallically bonded materials are excellent conductors of heat and electricity because of the mobility of the delocalized electrons in the electron cloud.

II. METALLIC STATES

A metal can exist as a solid, a liquid, or a vapor (gas). The state depends on the pressure exerted on the metal and on its temperature. The pressure versus temperature diagram for a pure metal has the form illustrated in Fig. 2.6. The three lines define the pressure-temperature regions for each state. These lines also indicate the pressures and temperatures at which the respective states may exist. All three states may exist together at the "triple point," where the three lines intersect.

Values for the triple point and the state curves are characteristic of each metal, but all the common metals are solid at atmospheric pressure and temperatures, except mercury, which is a liquid. All except arsenic change to liquid and thence to vapor as temperatures increase (line 1-2, Fig. 2.6).

Arsenic will change directly from solid to vapor at atmospheric pressure as its temperature is raised. This process is known as "sublimation." A common example of sublimation is the wasting away of naphthalene mothballs. All the common metals will sublime at very low pressures, such as those used in vacuum tubes, if their temperatures are raised sufficiently.

A. Vapor and Vapor Pressure

The vapor pressure of a metal is the pressure of its vapor over a solid (or a liquid) established by the equilibrium of the atoms entering and those leaving the solid (or the liquid). If the metal is placed in an evacuated space, the vapor pressure is the only pressure, but when the metal is open to the atmosphere, the vapor pressure constitutes a partial pressure of that exerted on the solid (or liquid) surface.

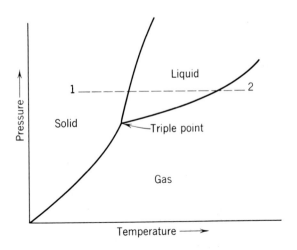

Fig. 2.6 State of a pure metal versus pressure and temperature.

The vapor pressures of metals at atmospheric temperature are not large. That for tungsten has been calculated to be 10^{-149} atm. This is equivalent to less than one atom in a volume equal to that of the universe. Although tungsten is somewhat less volatile than most metals, all are of low vapor pressure, and even mercury has only 1/10,000 the vapor pressure of water at 30°C.

Although metals volatilize at a more rapid rate at high temperatures, the vapor pressures for most common metals are not of practical importance at temperatures used in commercial metallurgical practice. Such metals as mercury, zinc, cadmium, and arsenic are exceptions. The vapor pressure of mercury is high enough at relatively low temperatures so that the properties of its vapor are successfully utilized for power generation in a vapor turbine. Zinc, arsenic, antimony, and even magnesium have high enough vapor pressures at elevated temperatures for purification by condensation on a commercial scale. Special difficulties have been experienced in outer space with certain electrical circuits resulting from the formation of metallic whiskers that short out the electrical circuits at the very low pressures of outer space.

B. Liquid Metal

The atoms within a liquid do not take fixed positions, nor is their arrangement as chaotic as in a vapor. The atoms are restrained in their motion by their neighbors and, of course, by the walls of the containing vessel. Some liquid metals show preferences for characteristic distances between their atoms.

Atoms continually leave and reenter the surface of a liquid metal, as described above. The pressure of the atoms above the liquid surface at any time is increased with increasing temperature. When the temperature has been increased until the vapor pressure equals the externally applied pressure, the passage of atoms from the liquid induces a mechanical agitation known as "boiling."

The temperatures at which common metals boil at atmospheric pressure are given in Table 2.1. If the external pressure is lowered, the boiling temperatures are lowered. Since the pressure on any part of the liquid metal is the sum of the external pressure on the surface and the weight of the liquid column it supports, the boiling action takes place almost exclusively at the surface at low pressures. Even at atmospheric pressure, boiling takes place closer to the surface for metals than for water.

C. Solidification

As a liquid metal cools, the motion of the atoms becomes slower and slower until the "freezing point" is reached. The random motion of the atoms stops at this temperature and the atoms take fixed locations with regard to one another, forming a solid metal. All the atomic motion does not cease as a result of solidification, but what motion persists is confined to vibration within these fixed locations. A further decrease in temperature will cause some modification of this atomic vibration in keeping with a lower energy state.

If heat is ideally removed at a constant rate during freezing, the cooling curve (temperature versus time) will be similar to that shown in Fig. 2.7. Cooling of the liquid is indicated by the line from point 1 to point 2. At point 2, freezing begins, and constant temperature is maintained until freezing is complete at point 3. Continuing the removal of heat from the now solid metal causes the drop in temperature from 3 to 4. During the time interval 2-3, the temperature remains constant, although heat was being removed

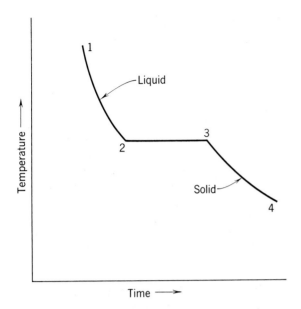

Fig. 2.7   Ideal cooling curve for freezing of a pure metal.

at a constant rate. The heat energy given up during this interval is called the "latent heat of fusion." It is supplied by the atoms as they take fixed positions and lose their kinetic energy of random motion.

D. Solid Metal

The atoms of solid metal are held together by the electrostatic attraction between the positive ions in the nucleus and the negative electron cloud. Since these forces are greater in some direction than others, the atoms arrange themselves in geometric patterns or structures characteristic of the metal. These crystalline structures are composed of three-dimensional "building blocks" called unit cells. A unit cell is constructed of atoms arranged in intersecting plane surfaces. Although there are seven principal crystal systems, or "unit cell patterns," the most commonly encountered forms of unit cells in metals are the cubic and hexagonal patterns. The cubic patterns are of two types, the body-centered cubic and the face-centered cubic. Table 2.2 lists several metals and their corresponding lattice types at room temperature, but at a certain temperature the lattice form changes to another type stable over another temperature range. Such metals are said to be "allotropic." These different lattice forms are, of course, identical by chemical analysis but they usually possess widely different physical properties.

Iron is the most familiar example of an allotropic metal. In Table 2.2, it is listed as having a body-centered cubic cell at room temperature. This type of structure is retained up to 910°C but above this temperature iron changes to face-centered cubic. The face-centered cubic structure persists up to the temperature of 1400°C at which temperature the body-centered cubic is again stable.

The unit cell of the body-centered cubic structure has an atom at each corner of a cube and one in the center, as illustrated in Fig. 2.8a. The atoms should not be pictured

Table 2.2 Lattice Types for Several Metals at Room Temperature

Body-centered cubic (BCC)	Face-centered cubic (FCC)	Simple hexagonal (close packed; HCP)
Barium	Aluminum	Berylium
Chromium	Calcium	Cadmium
Columbium	Copper	Cobalt
Iron	Gold	Magnesium
Molybdenum	Lead	Titanium
Tantalum	Nickel	Zinc
Tungsten	Platinum	Zirconium
Vanadium	Silver	

as concentrated at these positions but rather as having their centers of activity located there and their electron clouds packed close together, as illustrated by the model made from Ping-Pong balls (Fig. 2.8b).

The unit cell of the face-centered cubic structure has an atom located in the center of each face as well as one in each corner of the cube but none at the center. This arrangement and the corresponding Ping-Pong ball model are illustrated in Fig. 2.9.

The hexagonal close-packed structure is illustrated by the Ping-Pong ball model in Fig. 1.20b. The unit cell of the space lattice is simple hexagonal, as shown in Fig. 2.10a. A pair of atoms (a corner atom and an atom within the hexagon) is associated (or clustered) with each lattice point. The hexagonal close-packed structure and the face-centered cubic structure are closely related. They represent the closest packing of spheres, with each atom having 12 near neighbors. The body-centered cubic has 8 near neighbors for each atom and is a less densely packed structure.

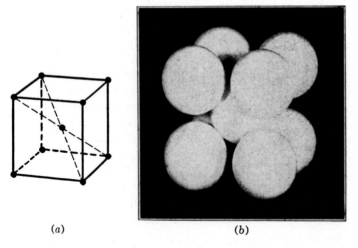

(a)   (b)

Fig. 2.8 Unit cell of the body-centered cubic structure. (a) Sketch showing location of atom centers. (b) Ping-Pong ball model.

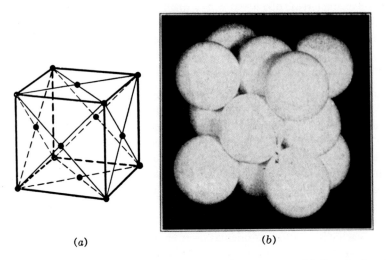

(a) (b)

Fig. 2.9 Unit cell of the face-centered cubic structure. (a) Sketch showing location of atom centers. (b) Ping-Pong ball model.

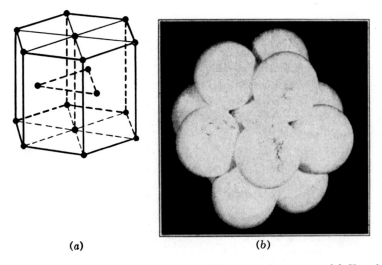

(a) (b)

Fig. 2.10 Unit cell of the close-packed hexagonal structure. (a) Sketch showing location of atom centers. (b) Ping-Pong ball model with top hexagonal plane removed and an extra cluster of three center atoms below.

The unit cell size characteristic of a metal is very small (e.g., 3.6×10^{-8} cm on an edge for copper and 4.078×10^{-8} cm for silver). A crystal of metal normally contains many, even millions, of unit cells located side by side and extending in all directions.

E. Grain Formation

At some temperature within a liquid metal the forces of attraction between atoms equal forces of repulsion, and atoms come together to form a solid nucleus. For this nucleus to remain stable and not redissolve, the latent kinetic energy given up by the atoms must equal or exceed the work done in forming the nucleus. This requires that the nucleus achieve a critical size for it to continue to grow. If no other nuclei form, the single nucleus will grow continuously as atoms of the single crystal in which all the corresponding atomic planes are parallel. Such single crystals have been produced as large as several inches in diameter and a foot long.

At a usual rate of cooling, however, the solidified metal is generally composed of thousands of crystals, each with a different orientation from its neighbors. Each of these crystals started from its own nucleus and grew until it met its neighbors, which were, of course, growing simultaneously from other nuclei.

When two adjacent crystals (more commonly called grains in metals) grow together during freezing, the atoms of the last liquid to solidify are mutually attracted to both grains. Since the orientations of the two grains differ (otherwise they would unite and become one), these atoms cannot form on either lattice, but must occupy compromised positions, as illustrated in Fig. 2.11. This transition zone between grains is called the "grain boundary." Grain boundaries interrupt the continuity of the lattice planes and increase the resistance of a metal to cold deformation.

The number of grains that form depends jointly on the rate at which nuclei form and on their rate of growth. For most metals the rate of grain growth is low enough that the nucleation has a predominant influence on the grain size. Since the amount of supercooling determines the minimum size of nuclei that are stable, rapid cooling usually

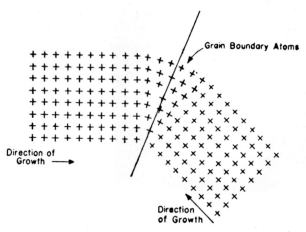

Fig. 2.11 Schematic diagram of the arrangement of atoms at a grain boundary. The crosses indicate atom positions.

results in the formation of many small grains, whereas slow cooling produces fewer but larger grains. The preparation of metallic single crystals is discussed in Chap. 6.

F. Grain Structures in Castings

Metals are refined from their ores commercially in three ways. They are refined by (1) furnace reduction involving chemical action, (2) furnace reduction involving condensation, and (3) electrolytic deposition from ore partially refined by furnace reduction. These refined metals are then put into form for use by (1) pressing and sintering of powders or (2) casting from the liquid state into a suitable solid state.

Most of the tonnage of metal produced today is refined by furnace reduction involving chemical action followed by casting. Some is cast into the final form in which it will be used, such as sand, die, or plaster castings (see Chap. 18). Pure metals are rarely used as cast shapes, however. More frequently, pure metals are cast into ingots that are later forged, drawn, extruded, or hammered into various shapes, such as large cylinders, thin sheets, plates, rods, bars, rails, and wire. But with either ingots or cast shapes, the casting operation introduces two grain structures peculiar to the process.

1. Columnar Grains

In the discussion of grain formation it was assumed that heat was removed uniformly from all the molten metal. Nuclei were then formed throughout the melt; no special grain alignment was noticed; the grains tended to be of the same size and were equiaxed (i.e., of nearly equal dimensions in all directions).

The entire melt in a casting is not cooled uniformly because heat is removed through the wall of the container or mold. This causes the metal in contact with the wall to freeze first, and solidification proceeds toward the center from nuclei formed in the vicinity of the walls.

Grains grown from nuclei along the wall impinge on neighboring grains; thus their size in this direction normally is restricted. The growth toward the center of the liquid is less restricted by impingement with other grains, because the removal of heat energy through the solid metal at the wall causes limited nucleation in the liquid. As a result, when freezing is complete, the grains are very long in a direction perpendicular to the walls of the container. Grains of this type are called "columnar grains." Their occurrence is most pronounced in chilled, thick sections. Columnar grain growth is particularly undesirable in ingots that have square corners because planes of weakness are established at the intersections of the columnar grains. These planes nearly bisect the corner angles, as shown in Fig. 2.12, and reduce the mechanical strength of the ingot making it susceptible to fracture during working.

Corrugated ingot molds with round corners are used to minimize the formation of columnar grains. They cause the grains to grow in all directions. Equiaxed grains will also form and grow if the metal is cooled in such a manner that the center reaches a temperature at which nucleation may occur.

2. Dendrites

A structure known as dendritic structure of "pine tree structure" may occur within both equiaxed and columnar grains. It occurs to some degree in most castings because of preferred directional growth. Such growth is a natural characteristic of metals since it is

Fig. 2.12 Photograph of the cross section of a copper ingot showing columnar grains (full size).

easier to add atoms to the unit cells in certain planes than in others. The more rapid growth in the preferred directions during solidification results in a skeleton crystal formation (a dendrite) that contains many interstices filled with melt. These liquid areas freeze last, and their contraction causes regions of microscopic porosity that contribute to somewhat reduced ductility in the cast metal as compared with the same metal treated mechanically and thermally for highest ductility.

The regions of microporosity, which are a result of final solidification of the melt, outline the pine tree formation of the dendrites. Dendrites are similarly outlined by insoluble impurities or by other structural constituents in the alloys, and this is known as "dendritic segregation."

Columnar growth and dendrite formation can be modified in casting shapes or ingots by choice of mold shape and control of grain size. The degree of occurrence of columnar grains in dendrites in alloy castings, however, is largely a function of composition of the alloy. Alloying and casting technique therefore render these effects subject to control.

G. Amorphous Structures

Amorphous (noncrystalline) structures differ from molecular and crystalline structures in that amorphous structures do not exhibit a long range, repetitive, structural pattern. Gases, liquids, and glasses are included among the amorphous materials. Gases and liquids are outside the scope of our discussion. On the other hand, glass may be considered a "rigid liquid." Although glass does not exhibit long-range order, it does show short-range order. As shown in Fig. 2.13, in a typical simple glass each silicon atom is bonded to three surrounding oxygen atoms and forms a strong, continuous network structure. It is possible to produce glassy (amorphous) metal structures by exceedingly rapid cooling rates from the liquid state.

H. Macrostructures

Composite materials, combinations of metals or ceramics with polymers, may produce unique engineering structures with properties not available in any single class of materials. Wood is a natural composite material. Figure 2.14 shows five types of composites based on the structure of the composite material. One of the most common procedures is to imbed the structural constituent in the matrix of the composite. However, some composite structures have no matrix but rather are constructed from one or more forms (or orientations) or different compositions of materials (e.g., laminar structures). The intermixing of different constituents introduces interfaces at common boundaries. The coatings normally applied to glass fibers in reinforced plastic and the adhesives that bond layers of laminate comprise "interphases" whose properties must also be understood in designing and using composite structures. More information is provided on ceramic and polymer structures and composite structures in later portions of this book.

I. Elastic Deformation

Any solid deforms elastically in response to an applied stress. The piece becomes slightly longer when subjected to a tensile stress. Removal of the load permits the piece to return to its original size. Response to a compressive stress is a slight shortening of the piece.

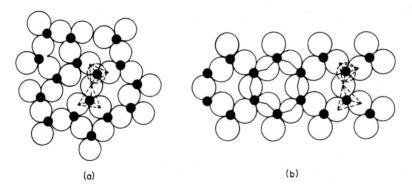

(a)   (b)

Fig. 2.13 Two-dimensional representations of (a) silica glass and (b) crystalline silica at room temperature. Each has a short-range network structure. Only the crystalline silica has a long-range network order. (The fourth oxygen above or below the silicon is not shown.) Oxygen serves as a bridge between adjacent tetrahedra. (From Ref. 4.)

Fiber composite   Particulate composite   Laminar composite

Flake composite   Filled composite

Fig. 2.14 Five types of composites based on structure. (From Ref. 5.)

When the compressive stress is relieved the part will assume its original dimensions. The elastic deformation, or strain that occurs is proportional to the applied stress in accordance with Hooke's law. The ratio of stress to strain (under conditions wherein Hooke's law applies) is the modulus of elasticity (Young's modulus).

$$E = \frac{\sigma}{e}$$

where

      E = modulus of elasticity
      σ = stress
      e = strain

The magnitude of Young's modulus may vary with crystallographic direction in a material and will be larger for metals with greater interatomic bonding forces, as shown in Table 2.3. The value of E varies with temperature, as shown in Fig. 2.15.

Because the *volume* of the unit cell remains constant, any uniaxial elongation will be accompanied by a contraction normal to the applied force. This is referred to as the "Poisson effect," and the negative ratio between the lateral strain e_y and the direct ten-

Table 2.3 Modulus of Elasticity (Young's Modulus)

Metal	Maximum		Minimum		Random	
	GPa	10^6 psi	GPa	10^6 psi	GPa	10^6 psi
Aluminum	75	11	60	9	70	10
Gold	110	16	40	6	80	12
Copper	195	28	70	10	110	16
Iron (BCC)	280	41	125	18	205	30
Tungsten	345	50	345	50	345	50

Source: From Ref. 4. Adapted from E. Schmid and W. Boas, *Plasticity in Crystals,* London, Hughes and Co.

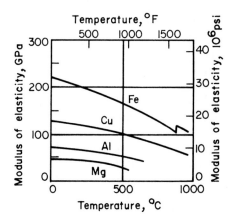

Fig. 2.15 Modulus of elasticity versus temperature. (From Ref. 4. Adapted from Ref. 5.)

sile strain e_z is called the "Poisson ratio". (Only the absolute magnitude of ν is used in calculations.)

$$e_y = e_z = -\nu e_x = -\frac{\nu \sigma_x}{E}$$

If a solid is subjected to two opposing forces that are parallel but not aligned, a shear stress will result. The shear stress τ is that force F_s divided by the sheared area,

$$\tau = \frac{F_s}{A_s}$$

A shear stress produces an angular displacement α. The shear strain γ is defined as the tangent of that angle, and the elastic shear strain is proportional to the shear stress; the equation is

$$G = \frac{\tau}{\gamma}$$

where G is the shear modulus (sometimes referred to as the modulus of rigidity). Young's modulus and the shear modulus are related by the expression

$$E = 2G(1 + \nu)$$

Since Poisson's ratio ν normally ranges between 0.25 and 0.5, the value of G is approximately 35% of E.

A third elastic modulus, the bulk modulus K, is the reciprocal of the compressibility β of the material and represents the hydrostatic pressure P_h per unit of volumetric compression $\Delta V/V$:

$$K = P_h \frac{V}{\Delta V}$$
$$= \frac{1}{\beta}$$

The Bulk modulus K is related to Young's modulus E in the following way:

$$K = \frac{1}{3}E(1 - 2\nu)$$

J. Plastic Deformation

1. Single Crystals

Cubic metals and their nonordered alloys deform plastically by slip. This is a plastic shearing process wherein one plane of atoms slides over the next adjacent plane. Hexagonal metals also exhibit this characteristic. Shear stresses develop even when tensile or compressive stresses are applied because these may be resolved into shear stresses on the active slip plane(s) (Fig. 2.16). Certain slip systems are predominant in metals (Table 2.4).

The shear stress required to produce slip on a crystal plane is the "critical shear stress" τ_C. The axial stress (either tension or compression) required to produce plastic deformation depends not only on the critical shear stress τ_C for the slip system of the material under consideration, but also on the orientation of the applied stress with respect to that slip system. Schmid's law relates the resolved shear stress τ to the axial stress S ($S = F/A$).

$$\tau = \frac{F_s}{A_s} = \frac{F}{A} \cos \lambda \cos \phi$$

where ω and θ are as defined in Fig. 2.17. Slip occurs with minimum axial force (that is, the resolved shear stress is maximized) when both ω and θ are 45°. Under these circumstances the shear stress τ is equal to ½ F/A, which is also equal to ½ the axial stress. It is noted that the resolved shear stress drops to zero as either ω or θ approaches 90°. Since plastic deformation changes the internal structure of a metal it is not surprising that it changes metal properties as well. For example, resistivity increases with deformation. Perhaps of greater engineering significance, strength and hardness also increase with increased plastic deformation while ductility tends to decrease (Fig. 2.18). An index of plastic deformation is the percentage of cold work (CW).

$$\%CW = \frac{A_o - A_f}{A_o} \times 100$$

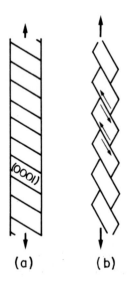

(a)   (b)

Fig. 2.16 Slip in a single crystal (hcp). Slip parallel to the (0001) plane contains the shortest slip vector. (From Ref. 7.)

Table 2.4 Predominant Slip Systems in Metals

Structure	Examples	Slip planes	Slip directions	Number of independent slip systems
BCC	α-Fe,Mo,Na,W	$\{101\}$	$\langle\bar{1}11\rangle$	12
BCC	α-Fe,Mo,Na,W	$\{211\}$	$\langle\bar{1}11\rangle$	12
FCC	Ag,Al,Cu,γ-Fe,Ni,Pb	$\{111\}$	$\langle\bar{1}10\rangle$	12
HCP	Cd,Mg,α-Ti,Zn	$\{0001\}$	$\langle11\bar{2}0\rangle$	3
HCP	α-Ti	$\{10\bar{1}0\}$	$\langle11\bar{2}0\rangle$	3

Source: From Ref. 4.

where %CW equals percentage cold work, A_o is equal to the original cross-sectional area, and A_f is the final cross-sectional area. The traces of slip planes are evident in Fig. 2.19.

Plastic deformation is caused by loads greater than the limit of elasticity of the metal. Removal of the load after plastic deformation will allow recovery of the superposed elastic deformation, but the plastic deformation is retained. The shape is changed and changes are caused in the structure and properties of the metal. The space lattice of each grain of the metal may be reoriented during plastic deformation by rotation, distortion, or both. There are three mechanisms by which the deformation takes place: slip, twinning, and formation of deformation bands. Slip deformation occurs by translation of two parts of the grain much as occurs in a deck of cards distorted into a parallelopiped. The region between slip planes extends over many atomic distances within which the lattice is distorted. There are several combinations of planes of atoms and direction of slip that can function in a crystal. The most favorable planes are those that contain the greatest number of atoms and are most widely spaced. Slip takes place along these planes of dense atomic population even though they may not be aligned with the direction of loading, as shown in Fig. 2.20b.

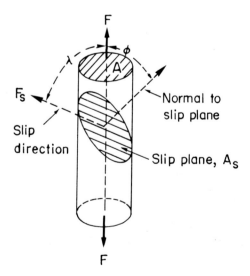

Fig. 2.17 Resolved stresses. The axial stress τ equals F_S/A_S, where $F_S = F\cos\Sigma$, and $A_S = A/\cos\phi$.

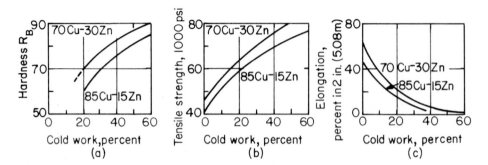

Fig. 2.18   Effect of amount of cold work on mechanical properties. For the tensile strength, multiply pounds per square inch by 6.9×10^{-3} to obtain mega-Newtons per square meter (MN/m^2). (From Ref. 4.)

In twinning deformation a thin platelike section or lamella of a crystal changes lattice orientation with respect to the rest of the crystal. The lattice within the twinned portion becomes a mirror image of the rest of the crystal, as would be obtained by rotating this portion 180°. The process of twinning is not one of rotation, however, but occurs by shearing movements of atomic planes over one another. The movement involved does not occur over a multiple of interatomic distances as does slip. Instead, the near neighbor relationship of atoms is maintained. The formation of twins takes place with an audible click; a rapid succession of such clicks is responsible for the "cry" that is heard when a bar of tin is bent. Twinning is a major mechanism of the deformation of zinc, and it is believed that bismuth and antimony deform entirely by twinning at room temperature.

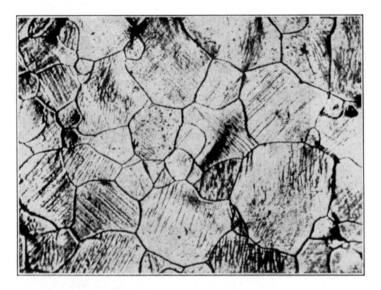

Fig. 2.19   Photomicrograph of pure iron with a slight amount of cold work. Note the slip lines on each grain. Approximately ×100.

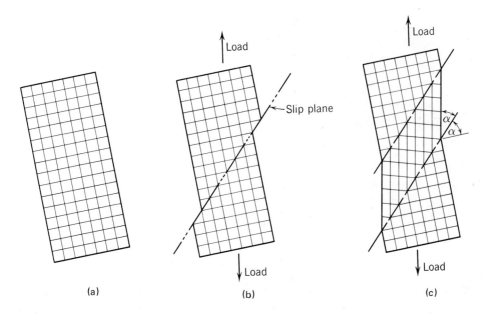

Fig. 2.20 Diagrammatic representation of slip and twinning in a single crystal. (a) Unstrained crystal. (b) Slip deformation. (c) Twinning deformation.

Deformation bands consist of lamellar regions of a grain within which portions of the lattice rotate in different directions as a result of slip on different planes. The photograph of 70% copper-30% zinc-brass (Fig. 2.21), contains many of these bands. They appear as shaded stripes across the surface of individual grains.

    Although these three mechanisms define the types of structural changes that occur during plastic deformation, there is much to be learned about the exact nature of the process in different metals. Whatever their nature, the structural changes produce changes in properties. The electrode potential and cohesive force are increased, and the coefficient of thermal expansion and compressibility are slightly increased, whereas ductility is decreased and a slight decrease occurs in density, electrical conductivity, and magnetic permeability. Such changes are of considerable importance to metals engineering.

2. Strain Hardening

Atomic planes, which are suitably oriented, slip readily as a load sufficient to cause plastic deformation is first applied. Slip generated on these planes progresses to the grain boundaries or to other imperfections, where a stress field is set up that tends to oppose additional slip. Further slip must then take place in grains with less favorably oriented slip planes, and additional load is required to orient these grains so that their slip planes may be utilized. Thus, as the deformation proceeds, the availability of slip planes decreases, and the tendency to resist deformation increases. When the resistance of a metal to deformation has been increased in this way, the metal is said to have been strain hardened or work hardened. This can be illustrated by reference to the stress-strain curve (Fig. 2.22). If a stress is applied to a test bar to point A in Fig. 2.22a (a stress considerably above the yield strength of 50,000 psi but below the tensile strength of 100,000 psi),

Fig. 2.21 Photomicrograph of annealed 70% copper, 30% zinc brass showing deformation bands or stripes across the surfaces of the grains (×75).

and then the sample is unloaded, the unloading cycle will be described by the dotted line (Fig. 2.22a). Upon reloading the sample, reloading will return up the dotted line to point B, which will be the new yield strength (Fig. 2.22b). It is observed that this is over 80,000 psi in place of the 50,000 psi originally observed. Stressing the sample in the region of plastic strain results in slip taking place on favorably oriented planes with dislocation production and movement. As more and more slip occurs, dislocations interact, pile up, and dislocation tangles form. This makes it more and more difficult for further slip to take place. As a consequence, the stress-strain curve rises, demonstrating that to produce more strain, more stress is required. In other words, when point A was reached, all the planes and dislocation sites for easy slip were used up. When the load was removed, there was not change in this situation. Therefore, when the load was reapplied, no plastic strain was possible until the stress level of point A was again reached. Thus, only elastic strain was encountered on reloading until we again reached point A, the new yield point. This phenomenon is referred to as work hardening or strain hardening or cold work. The term "cold" is a relative term and normally involves deformation that proceeds at temperatures below approximately ½ the melting temperatures on the absolute temperature scale. Figure 2.23 shows a number of methods for work hardening during processing. In each case it is observed that the mechanical working of the incoming material results in a distortion or deformation of individual grains. This in turn leads to the development of a so-called texture in which the individual grains will have properties which are different in different directions. This anisotropy is also referred to as a preferred orientation. Wrought products, such as sheet and extrusion, exhibit marked differences in mechanical properties in different directions of measurement. For example, the tensile and yield strength in the direction of rolling or extrusion is likely to be considerably higher than that measured transverse to the sheet or extrusion, which in turn will be considerably better than

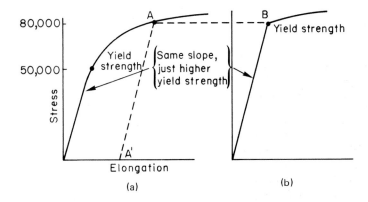

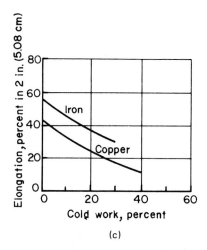

Fig. 2.22 Schematic correlation between work hardening and the stress-strain curve (stress in pounds per square inch). (c) Correlation of cold work and elongation in tensile testing. [Parts (a) and (b) from Ref. 8; (c) from Ref. 4.]

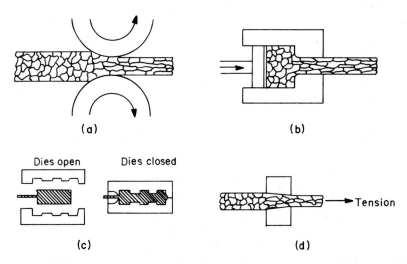

Fig. 2.23 Methods for work hardening during processing. (a) Cold rolling bar or sheet. (b) Cold extrusion. (c) Cold forming, stamping, forging. (d) Cold drawing. (From Ref. 8.)

the properties measured through the thickness (the short transverse direction). Thus, it is important in design of critical parts to specify the orientation of deformation in materials forming parts that will be highly stressed.

It is not always desired to have the maximum strength and hardness in a part because, as hardness increases, ductility decreases (Fig. 2.22c). For example, in the fabrication of a metal cup, it may be discovered that the sheet cracks at corners during fabrication. This indicates that the capacity for plastic elongation has been exhausted. In order to permit deeper drawing it is necessary to restore the original structure by eliminating the extensive slip and dislocation tangles that have developed in the material. Inasmuch as the atoms are not rigidly fixed but can diffuse from their positions, it is possible to provide the necessary thermal energy to facilitate restoration of formability of the material. The process of heat treatment is known as "annealing." The metal softens as a function of heating temperature and time at temperature. At elevated temperatures new equiaxed grains grow inside the old distorted grains and at the old grain boundaries. At still higher temperatures grain size increases. Meanwhile, hardness and strength decrease with grain size, but elongation increases. On the microscopic scale, slip on a given plane stops when we reach a grain boundary, which is an area of high dislocation density and entanglements. Consequently, the greater the number of grain boundaries, the greater the limitation of slip and the higher the strength. Elongation is lower because slip is limited by the larger grain boundary area. Additional examples of material structure will be found in Figs. 2.24-2.31.

3. Recovery, Recrystallization, and Grain Growth

It is convenient to divide the effects of temperature on cold-worked materials into three generalized regimes of temperature.

a. Recovery

Recovery occurs in a temperature range just below recrystallization. Electron microscopy reveals that in the process of recovery, stresses are relieved in the most severely slipped regions. Mechanistically, this is explained on the basis that dislocations move to lower energy positions giving rise to subgrain boundaries in the old grains. This process is called "polyganization." Hardness and strength do not change greatly during recovery, but corrosion resistance is believed to be improved somewhat. Optical microscopy does not reveal any visible change in the structure, however, in contrast to electron microscopy.

b. Recrystallization

In the recrystallization temperature range, new stress-free, equiaxed grains are formed. This leads to lower strength and higher ductility. There is a relationship between the amount of prior cold work and the grain size of recrystallized materials. The lower the amount of prior cold work, the fewer nuclei exist for formation of new grains. Consequently, under these circumstances, the recrystallized grain size will be larger than if cold work were more extensive.

4. Grain Growth

Increasing the temperature still further causes grains to continue to grow. The driving force is the reduction of surface energy since larger grains have less surface area per unit

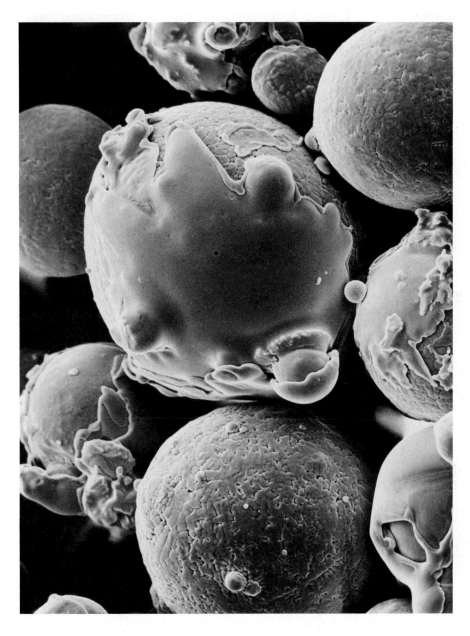

Fig. 2.24 Particles of a superalloy powder showing evidence of collision between solid and liquid particles during flame spraying operation. (Courtesy of Climax Molybdenum Company, Greenwich, Connecticut.)

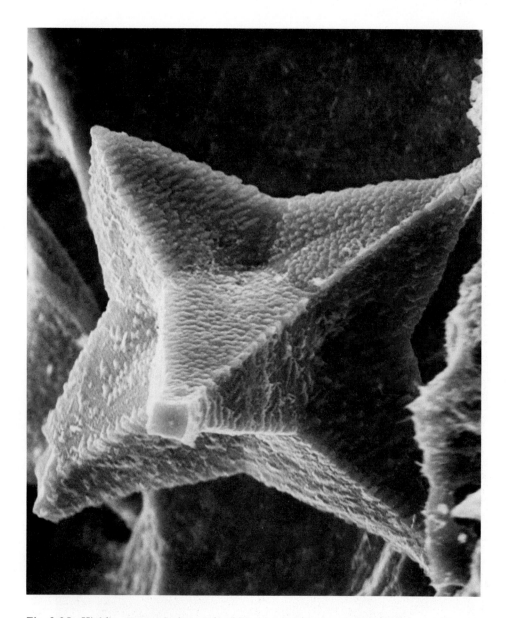

Fig. 2.25 Highly symmetrical complex MC-type carbide in a cobalt-based superalloy exposed by electrolytic dissolution of the matrix phase. (Courtesy of Climax Molybdenum Company, Greenwich, Connecticut.)

Fig. 2.26 Interconnected triangular crystal platelets of sintered molybdenum monocarbide stabilized with tungsten. (Courtesy of Climax Molybdenum Company, Greenwich, Connecticut.)

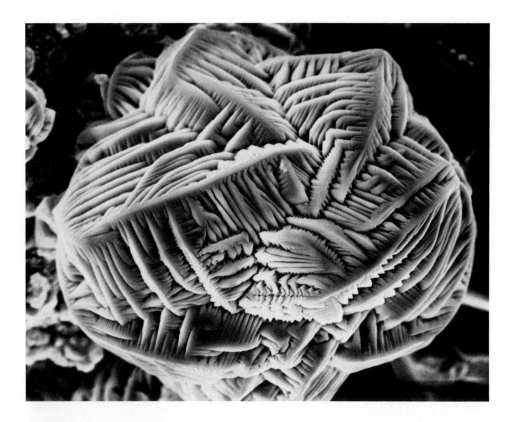

Fig. 2.27   Dendritic solidification surface texture of a nickel sulfide droplet. (Courtesy of Climax Molybdenum Company, Greenwich, Connecticut.)

of volume than do smaller grains. Large grains grow at the expense of smaller grains. Figure 2.32 summarizes the effects of annealing on cold-worked materials. The temperature required for recrystallization varies with the metal but is approximately $\frac{1}{3}$ to $\frac{1}{2}$ the melting temperature of a pure metal expressed on the absolute scale. In the case of steel, annealing temperatures are in the range of 870-980°C. By contrast, lead will recrystallize at room temperature. Cold working tends to provide good surface finish and close dimensional tolerances with no weight loss during working. For relatively ductile materials, cold working is often more economical than hot working. In addition, in those cases where the alloy cannot be strengthened by one of several different treatments (such as heat treatments) cold working may be the only feasible method by which strength can be increased.

5. Hot Working

Hot working (see also Chap. 21 and Ref. 9) combines the working and the annealing process by deforming a metal above the recrystallization temperature. Thus, ductility is constantly restored by recrystallization and grain growth during or immediately after recrystallization. Because they have relatively high recrystallization temperatures, most

Fig. 2.28 Spatial arrangement of leaflike and globular columbium carbide and acicular chromium nitride exposed by electrolysis in experimental heat-resistant ferritic Cr-Mo steel. (Courtesy of Climax Molybdenum Company, Greenwich, Connecticut.)

commercial metals must be worked at elevated temperatures. Hot working can be continued until the metal cools to the recrystallization temperature, at which point reheating is required before continuing. As with cold working, several cycles of operation may be necessary before a piece of metal is reduced to the desired shape and size.

Hot working has a number of effects on metal characteristics or properties beyond the mere change of shape. Among these are (1) general densification of the metal if blowholes or other cavities are present (providing the cavity surfaces are not oxidized), and (2) refined grain structure. Since grain size is reduced by recrystallization during hot working, the working can be continued until the metal is cooled to the lowest possible recrystallization temperature. At that temperature little grain growth is possible. (3) The metal is made more homogeneous since inclusions will tend to be broken up and, if plastic, will be elongated in the direction of metal flow. Inhomogeneities in composition will tend to be reduced through increased diffusion. (4) Hot working may develop a preferred orientation. As indicated above, for best service performance, the flow direction in hot-worked metals should coincide with the direction of maximum stress. (5) Hot-working processes require less power than cold-working operations for the same degree of reduction. (6) Exposure of hot metal to air usually causes scaling. This may not be seriously

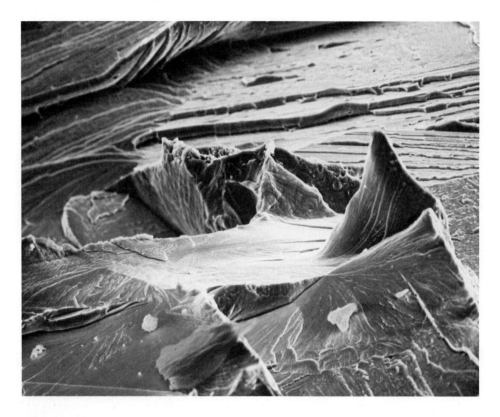

Fig. 2.29 Lunar craterlike appearance of a weld fracture detail in 18% Cr-2% Mo ferritic stainless steel. (Courtesy of Climax Molybdenum Company, Greenwich, Connecticut.)

damaging, but it is difficult to maintain tolerances as close or surface finish as good with hot working as is the case with cold working.

The upper temperature limit for hot working is determined by metal composition and is influenced by impurities that may be present. For example, iron sulfide segregates to grain boundaries and can cause hot shortness at too high forming temperatures. Care must be exercised in deformation of large parts since the too rapid application of heat in a localized area may result in large thermal gradients that may lead to cracking or cause local phase transformations.

6. Creep

At elevated temperatures (of the order of half the melting temperature in absolute terms), metals, ceramics, and polymers exhibit time-dependent plasticity. If a constant stress is applied, and maintained constant despite any changes in cross-sectional area, the material will deform over a period of time regardless of whether the stress is greater or less than the yield strength. Such deformation is called *creep* and in many cases it is the limiting factor in the selection of materials for use at high temperatures. For example, creep deformation of turbine blades in a jet engine could cause the rotating blades to strike the walls of the turbine, resulting in bending or breakage of the blades. Small creep deforma-

Fig. 2.30 Eutectic carbides having a frozen waterfall appearance exposed by electrolysis in a Ni-2.0 C-30 Cr-9 Mo base hard-facing alloy. (Courtesy of Climax Molybdenum Company, Greenwich, Connecticut.)

tion of nuclear fuel element cladding materials can lead to rupture of the cladding and result in buildup of radioactivity in the cooling system. High-pressure steam lines subject to creep can crack and cause the shutdown of a power plant. As a consequence, most high-temperature designs are governed by the creep properties of the structural materials. A normal criterion is to permit a certain amount of deformation over the desired lifetime of the structure. In the case of rocket nozzles, this may be a matter of minutes or, as for jet engine blades, it may be thousands of hours. Figure 2.33 shows a schematic creep curve for a crystalline material. There are three stages in the creep process. During the first stage, the strain rate (the slope of the strain versus time curve) decreases. During stage 2, the strain rate remains constant for an extended period of time. This is referred to as the steady-state stage of creep. Eventually, the creep rate again increases rapidly until final rupture occurs. The latter stage following steady-state creep is referred to as "tertiary" creep. Most of the lifetime of the structure is spent in the steady-state range; thus, the creep rate in stage 2 plays a dominant role in determining the lifetime of the structure.

Creep is dependent on stress and is a thermally activated process. As a consequence, the creep rate is accelerated as temperature or stress increases. Correspondingly, since

Figure 2.31 Molybdenum trioxide crystals grown on a resistance-heated molybdenum wire by evaporation and condensation. (Courtesy of Climax Molybdenum Company, Greenwich, Connecticut.)

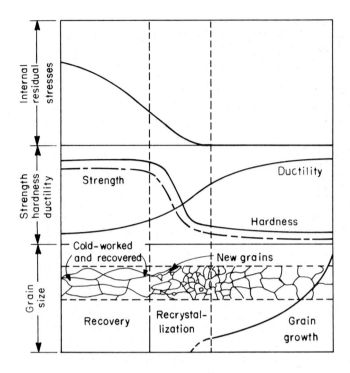

Fig. 2.32 Schematic representation of annealing effects. (From Ref. 8.)

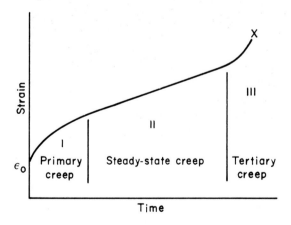

Fig. 2.33 Schematic illustration of creep curve showing time-dependent plastic strain. (From Ref. 3.)

creep rate increases with stress and/or temperature, the rupture lifetime decreases as stress and/or temperature is raised. Amorphous polymeric materials also exhibit a time-dependent form of deformation at elevated temperatures. However, in some instances, the time-dependent strain is recoverable after the load is removed. Figure 2.34 illustrates the strain-time curve for a polymer exhibiting an elastic response known as viscoelasticity. Further information with regard to polymers appears in a later chapter.

K. Impurities in Solids

We are conditioned to admire the "real thing." As a consequence, we prefer pure wool, refined sugar, 24 carat gold, and so on. Although these ideals may be noble, there are instances where, because of cost, availability, or properties, it is desirable to have impurities present. One example is sterling silver, which contains 7½% copper. We rate sterling silver quite highly, but it could be refined to well over 99% purity. However, it happens that very pure silver would not only cost more but would be inferior in use characteristics to sterling silver, since the addition of 7½% copper makes the silver stronger and harder, and consequently more durable, at a lower cost. Another example is brass. The addition of zinc to copper to form brass lowers the cost of the material as compared with pure copper. In addition, brass is harder and stronger and may be more formable than copper. On the other hand, brass also has lower electrical conductivity than copper, so if an electrical conductor is contemplated, pure copper would be superior to the stronger, less expensive alloy. Alloys are combinations of two or more metals in one material. Such combinations may be mixtures of two kinds of crystalline structures, such as body-centered cubic iron and the intermetallic compound Fe_3C in steel, or they may form solid solutions. Although the term "alloy" generally is not used specifically, various combinations of two or more nonmetallic substances, such as oxide components, may be incorporated advantageously in ceramic products, such as spark plug insulators. A number of polymeric materials take advantage of the combination of several types of molecules to provide strong structures.

L. Alloys

A metal alloy is a substance composed of two or more chemical elements such that metallic atoms predominate in composition and the metallic bond predominates. The element

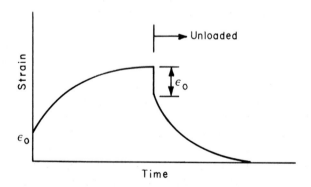

Fig. 2.34 Viscoelasticity or time-dependent elastic strain as exhibited by many polymeric materials. (From Ref. 3.)

present in the largest proportion is called the base metal, whereas all other elements are considered alloying elements. The physical, chemical, and mechanical properties of the base metal may be changed drastically by the presence of alloying elements. The type and extent of change of properties depends on whether these alloying elements dissolve in, are insoluble in, or form a new phase with the base metal.

1. Insoluble Mixtures

When the alloying atoms and the base metal atoms are relatively insoluble in the solid state, each exists almost independently in an intimate mixture of grains. Grains of each element retain their own identity, space lattice, and properties.

Impurities retained from the ore or from processing when insoluble in the base metal atoms are called "inclusions." They are usually oxides, sulfides, silicates, and similar compounds. If the impurities are gaseous they may produce "porosity." Some degree of porosity also results from imperfections in lattice formation in pure metals and alloys. Both inclusions and porosity constitute defects in the base metal structure with consequent reductions in mechanical properties. Complete insolubility is thermodynamically impossible. However, in practice, solubility can be extremely limited. For example, lead is essentially insoluble in iron; thus from a practical viewpoint, an alloy of iron and lead actually is an intimate mechanical mixture of the components where each component retains its own identity, properties, and crystal structure. Leaded bronze bearings are another example. Small globules of lead are trapped in a bronze matrix, which serves to support the bearing load while the lead acts as a lubricant.

2. Solid Solutions

Alloys containing alloying elements that are relatively soluble in the base metal in the solid state are called "solid solutions." They are the most common form of alloy. Two kinds of solid solutions are formed: (1) substitutional solutions in which some of the atoms of the base metal are replaced in their normal lattice sites by solute atoms and (2) interstitial solutions in which solute atoms are found in the "holes" or interstices between solvent atoms. Figure 2.35 shows schematic representations of substitutional and interstitial solid solutions. A further distinction is often made between solutions formed upon the adding of first solute atoms (those that retain the crystal structure of the solvent) and solutions that develop after extensive alloying (those that normally have a different crystalline structure than either component). The former are known as primary or terminal solid solutions, whereas the latter are referred to as secondary or intermediate solid solutions.

3. Substitutional Solutions

The most important considerations in forming solid solutions are related to atomic radii and chemical properties. Hume-Rothery and coworkers have pointed out that to form extensive solid solutions (greater than 10 atomic percent soluble) the following general rules must be obeyed: (1) the difference in atomic radii should be less than 15%, (2) proximity within the periodic table is important, and (3) for a complete series of solid solutions the metals must have the same crystal structure. Two metals with great chemical affinity for each other will not form an extensive solid solution. For example, strongly electronegative and strongly electropositive elements will tend to form compounds rather

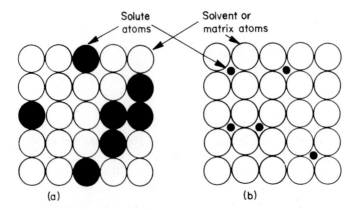

Fig. 2.35 Schematic representation of solid solutions. (a) Substitutional. (b) Interstitial. (From Ref. 10.)

than solutions. Metals of high valence can dissolve only a small amount of a lower valence metal, although the lower valence metal may have reasonable solubility for the higher valence metal.

Ideally, solute and solvent atoms are randomly distributed on lattice sites with statistically uniform distribution. In certain cases, however, substitution is regular rather than random. This can occur in binary systems with 1:1 or 1:3 atomic ratios. The 1:1 system often has a body-centered cubic structure with atoms of one type at the corners and atoms of the other type in the center. Such regular arrangements are known as "superlattices." Although they are extremely temperature sensitive, they are otherwise stable and the ordering is accompanied by changes in mechanical and electrical properties. Alloys of copper and gold possess the ability to form superlattices either as AuCu3 with gold at the corners and copper on the faces of the face-centered cubic structure or as AuCu in the tetragonal structure (see Fig. 2.36).

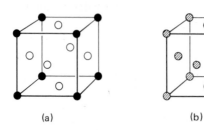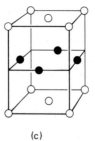

Fig. 2.36 Ordering in gold-copper alloys. (a) Unit cell of face-centered AuCu3, showing ordering with copper on faces and gold at corners at lattice. (b) Disordered space lattice with gold or copper at atom locations of face-centered cubic; proportions are 25% gold and 75% copper. (c) Unit cell of tetragonal ordered AuCu. Open circle, copper atoms; solid circle, gold atoms; dashed circle, copper or gold atoms.

4. Interstitial Solutions

Inasmuch as solute atoms in an interstitial solid solution must occupy the space between solvent atoms, the solute atoms must be small relative to the solvent atoms. The elements hydrogen, carbon, nitrogen, boron, and oxygen all have radii less than 1 Å and form interstitial solutions. There is increased interest in taking advantage of engineering changes in alloys as a result of intentional control of interstitial species. By far, the most important of these interstitial species is carbon in iron. The control of this interstitial species forms the basis for hardening of steels, which is discussed in Chaps. 8 and 21.

Iron atoms in γ-iron (austenite) are on face-centered cubic sites and the carbon atoms occupy interstitial positions. The largest interstitial positions in this lattice are at the midpoint of the structure cell edges and at the cube center. Crystallographically, these positions are equivalent. Smaller interstices exist that are surrounded by a tetrahedral arrangement of four solvent atoms. The largest interstices in iron have a radius of 0.52 Å, whereas the next smallest interstices have a radius of only 0.28 Å. A carbon atom has a radius of 0.8 Å, whereas a nitrogen atom has a radius of 0.7 Å. Thus, even the larger of the two interstitial sites can accommodate these interstitial species only by expansion of the lattice.

In α-iron (body-centered cubic) the largest interstices have a radius of 0.36 Å whereas the next smaller interstitial location has a radius of only 0.19 Å. As a consequence, interstitial solution of carbon in α-iron should be more limited than in γ-iron as will be seen in a later chapter. Although interstitial solid solutions depend on the size, valency, and chemical affinity factors in the same manner as do substitutional solid solutions, they do not require the same crystal structure in order to form solid solutions. It is possible to form interstitial solid solutions with ceramics; however, in this case, the addition of ions in interstitial sites requires some adjustment to maintain electrical neutrality. This can be accomplished by forming vacancies or substitutional solid solutions or by changing the electronic structure. For example, if YF_3 or ThF_4 is added to CaF_2, Y^{3+} or Th^{4+} substitute for Ca^{2+} and F^- ions go into interstitial positions to maintain electrical neutrality.

5. Intermetallic Compounds

Some alloys containing at least two metals possess the ability to form upon solidification an ordered lattice corresponding to simple, whole-number ratios of the combining metallic elements, for example, A_1B_1, A_2B_3. Such ordered lattices are known as "intermetallic compounds." They are also termed an "intermediate phase," that is, a phase that is not continuous with one of the pure metals in the system. Classification of an intermediate phase as an intermetallic compound is justified only if there is a narrow range of homogeneity, simple atom proportions, and atoms of identical kind occupying identical points throughout the space lattice. Intermediate phases often occur over a broad range of concentration, however. The lattice form of an intermetallic compound is characteristic of the compound; it differs from those of its components and usually is more complex.

The lower order of symmetry in the complex lattice results in low electrical conductivity and greater resistance to slip. Consequently, such compounds usually are both hard and brittle. Particles of them properly dispersed throughout a softer base metal will strengthen the latter. There are many intermetallic compounds that are of importance for this effect. Two examples are cementite, Fe_3C, which contributes to the hardening and strengthening of steel, and copper aluminide, $CuAl_2$, which hardens certain aluminum base alloys.

REFERENCES

1. W. L. Bragg, *Atomic Structure of Minerals*, Cornell University Press, Ithaca, N.Y., 1950.
2. D. S. Clark and W. R. Varney, *Physical Metallurgy for Engineers*, Litton Educational Publishing, New York, 1952.
3. C. R. Barrett, W. D. Nix, and A. S. Tetelman, *The Principles of Engineering Materials*, Prentice-Hall, Englewood Cliffs, N.J., 1973.
4. L. H. Van Vlack, *Elements of Materials Science and Engineering*, 4th ed., Addison-Wesley, Reading, Mass., 1980.
5. H. R. Clauser, *Industrial and Engineering Materials*, McGraw-Hill, New York, 1975.
6. A. G. Guy and J. J. Hren, *Elements of Physical Metallurgy*, 3rd ed., Addison-Wesley, Reading, Mass., 1974.
7. C. Elam, *Distortion of Metal Crystals*, Clarendon Press, Oxford.
8. R. A. Flynn and P. K. Trojan, *Engineering Materials and Their Applications*, Houghton Mifflin, Boston, 1975.
9. W. L. Roberts, *Hot Rolling of Steel*, Marcel Dekker, New York, 1983.
10. C. O. Smith, *The Science of Engineering Materials*, 2nd ed., Prentice-Hall, Englewood Cliffs, N.J., 1977.

3

Mechanical Properties and Fracture Mechanics*

The properties that define the behavior of a material under applied forces or loads are broadly classified as *mechanical properties.* They characterize the strength and durability of a material and are of great importance to the design engineer, who must assure adequate service life and performance in components and structures.

The mechanical properties necessary to assure that a material will sustain the conditions imposed on it are best established by monitoring similar components while they are in service, where service includes fabrication, installation, and actual use of the component. However, without the use of standard test methods to determine the appropriate mechanical properties, the adequacy of such properties for specified service conditions can only be estimated in a qualitative way. In many structures overload or proof tests may be performed to assure the adequacy of the mechanical properties. In other instances the selection of a material can only be based on the satisfactory behavior of the material under similar conditions in other applications or on simulated service tests of typical components.

The widespread use of metals for many years has been based primarily on their high strength and stiffness together with some ductility. The ductility is especially desirable for its ability to prevent or reduce the likelihood of unstable or brittle fracture. The strength and ductility of metals are due to the nature of the metallic bond between ions in the lattice. The metallic bond is characterized by the freedom of the valence electrons to move through the ion core lattice and results in high melting temperatures, thermal and electrical conductivities, and elastic moduli.

The necessity of defining, controlling, or predicting mechanical properties, such as strength, stiffness, or ductility, has led to the development of many standard laboratory tests for measuring these properties. Although many standard tests were developed on a voluntary or consensus basis by such organizations as the American Society for Testing and Materials (ASTM) and the Society of Automotive Engineers, a large number of them have also been adopted by regulatory agencies to assume the force of law. Examples of the latter are the Pressure Vessel Code of the American Society of Mechanical Engineers (ASME) or the many military standards issued by the U.S. Department of Defense.

---

*Contributed by DOUGLAS L. JONES.

The results of standard tests are employed for many purposes, including

1. Establishing the ability of a material to meet certain specifications previously determined to be adequate for a particular application
2. Controlling production of materials and fabrication of components
3. Establishing parameters for design calculations
4. Developing and evaluating new materials

Although most of the topics discussed in this chapter have been developed for application to metals and alloys, many of the concepts, test methods, and material properties will be appropriate for nonmetals as well. Since the appearance of the first edition of this book, engineering applications involving nonmetals have increased dramatically. These materials include plastics, ceramics, and advanced composite materials and have been applied to nearly all components and structures in service today.

For purposes of organization, all the tests discussed in this chapter will be arbitrarily classified according to how the loads are applied. Although some overlap exists among test methods, the mechanical properties will be identified as *static, cyclic,* or *impact* properties.

I. STATIC PROPERTIES

A. Stress and Strain

Whenever a load is applied to a machine component or assembly, reaction forces are produced throughout the component and transmitted through the component to its supports. The intensity of the reaction force at any point in the component is called the *stress.* The stress may be visualized as the force per unit area exerted by the material on one side of an imaginary plane through the component against the material on the other side of the plane. When the forces are parallel to the imaginary plane, the stress is called a *shear stress,* and when the forces are perpendicular to the plane, the stress is called either a *tensile* or *compressive stress,* depending on whether the forces tend to separate or compress the material. The dimensions of stress are force per unit area and are usually expressed in pounds per square inch (psi), ksi (1 ksi = 1000 psi), or Pascals (1 Pa = 1 newton per square meter), where 1 psi = 6.89×10^3 Pa.

To examine the behavior of materials under each of these types of stress, sample specimens are selected of such shape that the distribution of reaction forces, and therefore the stress, is known. Figure 3.1 illustrates several of the common test specimens. In tension and compression tests the stress is assumed to be uniformly distributed over the cross section perpendicular to the applied load. The stress in the bending and torsion tests is not uniformly distributed, but its variation can be predicted from the theory of elasticity and is also presented in Fig. 3.1.

When forces are applied to a component, it is deformed or strained from its initial configuration. The amount of strain in any direction is a function of the applied load and is equal to the amount of deformation per unit length. Strain is thus a dimensionless quantity, measured in inches per inch or in percentage change. The strains resulting from the loads applied to the test specimens illustrated in Fig. 3.1 are shown in Fig. 3.2.

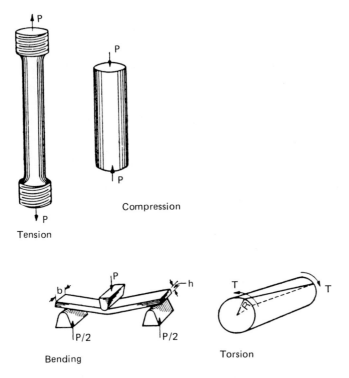

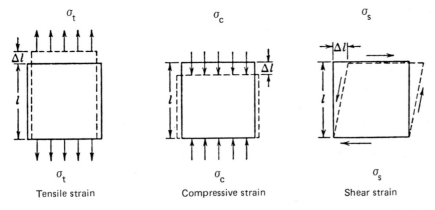

Fig. 3.1 Sketches of the common test specimens. The stress distribution in each is given by the accompanying equations. Tension: S_T = tensile stress = P/A; P = load; A = cross-sectional area of specimen. Compression: S_c = compressive stress = P/A; P = load; A = cross-sectional area of specimen. Bending: S = tensile stress (tension or compression depending on value of c) = Mc/I; M = bending moment, lb-in.; c = distance from neutral axis; I = moment of inertia of specimen = $bh^3/12$. Torsion: S_s = shear stress = Tr/J; T = torque, lb-in.; r = radius, neutral axis; J = polar moment of inertia of specimen = $\pi R^4/2$.

Fig. 3.2 Strains resulting from three types of stress. Solid line denotes unloaded shape; dotted line is the loaded shape.

B. Stress-Strain Curves

In determining static material properties, a gradually increasing load is applied to the specimen, and the elongation or strain in the direction of loading is measured until failure occurs. When the stress σ is calculated by dividing the applied load P by the original cross-sectional area, $\sigma = P/A_0$, and the strain ϵ is determined by $\epsilon = \Delta l/l_0$, a plot of stress versus strain for the entire test is called an *engineering stress-strain curve*. If, alternatively, the instantaneous stress is calculated by $\sigma_i = P/A_i$, where A_i is the instantaneous area, and the strain is calculated by

$$\epsilon_i = \int_{l_0}^{l_i} \frac{dl}{l}$$

$$= \ln\left(\frac{l_i}{l_0}\right)$$

where l_i is the instantaneous length, the resulting curve is called a *true stress-strain curve*.

Because of their widespread use in engineering applications, engineering stress and strain will be employed throughout this chapter. Tensile stress-strain curves for two different metals are presented in Figs. 3.3 and 3.4; tension and compression curves for a third metal are shown in Fig. 3.5. Figure 3.3 illustrates the typical stress-strain response of a mild steel; Fig. 3.4 is more typical of nonferrous metals, such as copper or aluminum.

The point at which the engineering stress attains its maximum value coincides with the onset of "necking" or local reduction in cross-sectional area where the fracture eventually occurs. Since the area of the necked regions A_i is significantly less than A_0, the true stress continues increasing until fracture, but the engineering stress declines beyond this point. Although the true stress represents a more accurate measure of the loading condition in the necked region, the engineering stress is more widely employed in characterizing the overall material behavior.

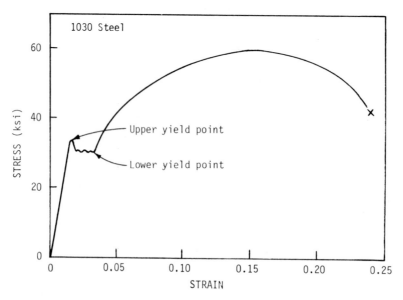

Fig. 3.3 Typical engineering stress-strain diagram for a low-carbon steel.

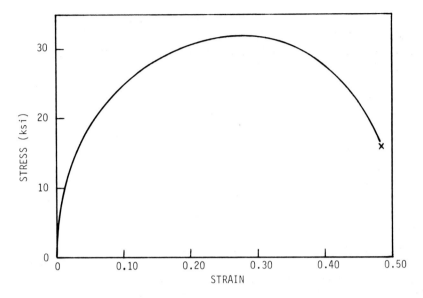

Fig. 3.4 Engineering stress-strain diagram for polycrystalline copper.

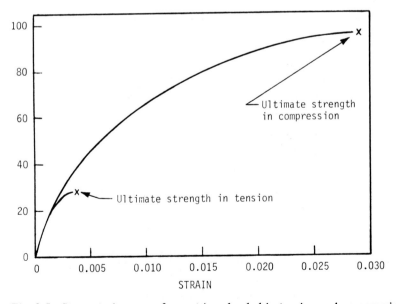

Fig. 3.5 Stress-strain curves for cast iron loaded in tension and compression.

C. Strength

Stress-strain curves are a measure of the strength of a material—the capacity of the material to support a load. Many of the various strength properties taken from these curves are listed below.

1. Proportional Limit

As long as the stress-strain curve is a straight line from the origin, the strain is proportional to the stress and *Hooke's law* of proportionality between stress and strain applies. The value of stress at which the curve first departs from linearity is called the *proportional limit.* Materials that are hard or have been cold worked significantly during the fabrication process generally exhibit a high proportional limit, whereas softer or less worked materials normally have a much lower proportional limit (e.g., Figs. 3.3 and 3.4).

2. Elastic Limit

The *elastic limit* is the maximum stress that can be applied to a material without causing permanent or plastic deformation (deformation that remains after the stress has been removed). The elastic limit cannot be uniquely identified from a standard tensile test but must be determined through successive loading and unloading of a test specimen. Since loading-unloading types of tensile tests are less frequently performed than standard tensile tests, less emphasis is placed on the elastic limit as a mechanical property than on the proportional limit and the yield strength. For ductile metals the elastic limit is normally just above the proportional limit but close enough to it that they are often considered to have the same value. By contrast, other materials, such as cast iron or rubber, often exhibit elastic limits that are considerably greater than their proportional limits and thus require the formalism of nonlinear elasticity to accurately describe their behavior.

3. Yield Point

The yield point is defined as the lowest value of stress at which an increase in strain occurs without a corresponding increase in stress. A typical stress-strain curve for mild steel (Fig. 3.3) exhibits both upper and lower yield points, where the lower yield point is determined by the minimum value of the stress when constantly increasing displacements are applied to the specimen. With the advent and widespread use of servohydraulic materials testing systems, tensile tests may be performed by controlling load, displacement, or strain. Displacement-controlled testing corresponds most closely to the performance of the older types of testing machines. When the tensile tests are performed in load control, the lower yield point cannot be observed because the rapid system response will not permit the load to decrease. Thus, identification of the upper and lower yield points is somewhat dependent upon the type of testing system and procedure employed.

4. Yield Strength

Many metals and other materials do not exhibit a sharp change in slope or "knee" in the stress-strain curve associated with the onset of yielding. To assign to such materials some practical measure of the limit to their elastic response, the concept of *yield strength* is employed. The yield strength is defined as the stress necessary to cause a certain prescribed amount of plastic strain, usually 0.2%, in the material. The extent of plastic strain is determined by constructing a line parallel to the initial slope of the stress-strain curve but offset by 0.2% on the strain axis, since this offset line generally coincides with the unloading line. The value of stress at the intersection of the stress-strain curve and the offset line defines the yield strength and is illustrated in Fig. 3.6. For some materials offsets other than 0.2% have been employed in determining the yield strength. In such situations the amount of offset should be specified whenever yield strength values are reported.

5. Ultimate Strength

The maximum stress σ_u that any material can sustain prior to fracture is called the *ultimate strength,* or the *tensile strength* if the loading is in tension. The ultimate strength is determined by dividing the ultimate load P_u by the original area, or $\sigma_u = P_u/A_o$. For brittle materials, the ultimate strength coincides with the *fracture strength,* the value of the stress at fracture. However, for ductile materials, the fracture strength may be significantly lower than the ultimate strength, depending on the amount of necking in the material. For many years it was customary to base designs on the ultimate strength reduced by certain design factors. However, more recently developed design procedures have been based on the yield strength in conjunction with other mechanical properties, such as the fracture toughness. These newer procedures have resulted in much improved accuracies in predicting, and avoiding, premature failures, while also facilitating more effective utilization of the materials.

D. Stiffness

The resistance of a material to elastic deformation is called its *stiffness.* Since the deflections produced by operating loads on structural components may have a significant

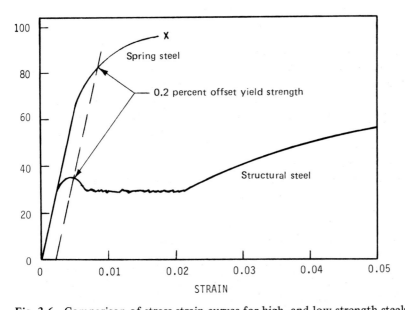

Fig. 3.6 Comparison of stress-strain curves for high- and low-strength steels.

effect on the suitability of the part, the material stiffness is often a limiting design factor. The stiffness of a material is determined primarily by the cohesive forces between the atoms. Since these forces cannot be altered without changing the basic nature of the material, it follows that the stiffness is one of the most structurally insensitive mechanical properties. The stiffness is only slightly influenced by alloying, cold working, or heat treatment. Therefore, if the stiffness of a component is seen to be inadequate, the only two choices are to increase the cross section or replace the material with another material possessing a higher stiffness. The latter alternative has become much more feasible in recent years, due to increasing applications of fiber-reinforced composite materials.

The stiffness of a material is characterized by a set of elastic constants, the number of which varies with the complexity of the material. For a homogeneous, isotropic solid only two independent constants are needed. The two most widely used constants are the modulus of elasticity (Young's modulus) and Poisson's ratio. Other important constants are the shear modulus and the bulk modulus.

1. Modulus of Elasticity

The *modulus of elasticity* or *Young's modulus* is a measure of the elastic deformation of a material when it is stressed in tension or compression within the proportional limit. Its value is equal to the ratio of stress to strain, and thus corresponds to the initial slope of the stress-strain curve. Referring to Fig. 3.2, the modulus of elasticity E is

$$E = \frac{\sigma_t}{\Delta l / l}$$

or

$$E = \frac{\sigma_c}{\Delta l / l}$$

where σ_t and σ_c are the tensile and compressive stress and $\Delta l / l$ is the strain. Although Young's modulus is usually determined from a tensile test, the same result will be obtained from a compression test, as seen in Fig. 3.5. Since strain is a dimensionless quantity, Young's modulus has the dimensions of stress (psi or Pa).

Young's modulus for steels at room temperature, regardless of condition, is 27-30 $\times 10^6$ psi (186-207 $\times 10^9$ Pa). For gray cast irons Young's modulus is 14-18 $\times 10^6$ psi (96-124 $\times 10^9$ Pa); for copper and its alloys, 15-17 $\times 10^6$ psi (104-117 $\times 10^9$ Pa); and for aluminum alloys, 9-11 $\times 10^6$ psi (62-76 $\times 10^9$ Pa).

2. Poisson's Ratio

When a material is subjected to a longitudinal tensile stress, it will elongate and, in an attempt to retain its original volume, will decrease in cross section. If the applied stress were compressive, the longitudinal shortening would be accompanied by an increase in cross section. The negative ratio of the strain in a lateral direction to the longitudinal strain under conditions of uniform uniaxial loading within the proportional limit is known as *Poisson's ratio*. This ratio is usually denoted by v and is approximately equal to 0.3 for most engineering metals and alloys. However, it does range from 0.1 to 0.2 for glass and ceramics to 0.4-0.5 for plastics and polymers. The theoretical upper limit for Poisson's ratio for an isotropic material is 0.5, which describes dilatationless (volume preserving) deformations.

3. Shear Modulus

As the name implies, the *shear modulus* represents the same basic characteristic (stiffness) of a material as Young's modulus, except that the material is loaded in shear rather than tension or compression. The shear modulus is also known as the *modulus of rigidity* and is denoted by

$$G = \frac{\sigma_S}{\Delta l/l}$$

where σ_S represents the shear stress and $\Delta l/l$ the shear strain (Fig. 3.2). For isotropic materials having only two independent constants, it can be seen that

$$G = \frac{E}{2(1 + \nu)}$$

For common structural metals and alloys having $\nu \approx 0.3$, it is evident that $G \approx 0.4E$.

4. Bulk Modulus

When an elastic body is subjected to uniform triaxial (equal in all directions) tensile or compressive stress, the ratio of the applied stress to the resulting volume change is called the *bulk modulus.* This state of stress coincides with hydrostatic tension or compression, so the bulk modulus is a basic measure of the compressibility of a material. For isotropic materials the bulk modulus can also be expressed in terms of Young's modulus and Poisson's ratio as

$$k = \frac{E}{3(1 - 2\nu)}$$

E. Resilience and Toughness

The area under the stress-strain curve represents the energy absorbed when a material is subjected to applied loads. The energy absorbed under the *elastic* portion of the curve is called the *resilience*; the total energy absorbed prior to fracture is called the *toughness* of the material.

The resilience is measured by the *modulus of resilience* U_R, which is the strain energy per unit volume stored in a body when the stress is at the yield strength. Since for most metals the stress-strain curve is linear until yielding occurs, the modulus of resilience is given by

$$U_R = \frac{1}{2} \sigma_{ys} \epsilon_{ys}$$

$$= \frac{1}{2} \sigma_{ys} \frac{\sigma_{ys}}{E}$$

$$= \frac{\sigma_{ys}^2}{E}$$

This equation shows that the modulus of resilience is greatest for materials having high values of yield strength and low values of Young's modulus. Figure 3.6 illustrates why the modulus of resilience is greater for a high-carbon spring steel than for a structural steel. It is thus apparent that, whenever it is desired to store a large amount of strain energy, such as in mechanical springs, a material having a high modulus of rigidity should be selected.

The *toughness* of a material is a measure of its ability to absorb strain energy in the plastic range. Most authorities are in agreement with this definition, but there is considerable disagreement as to how it should be measured. There is some preference for the use of impact tests (to be discussed later in this chapter) to measure the toughness, but integration of the stress-strain curve generally has more widespread acceptance. Following the latter definition, it is seen in Fig. 3.6 that the structural steel possesses the greater toughness.

The *modulus of toughness* U_T is obtained by integration of the stress-strain curve up to fracture, or

$$U_T = \int_0^{\epsilon_f} \sigma \, d\epsilon$$

where ϵ_f is the strain at fracture. Several mathematical approximations have been proposed for ease in measuring U_T. For ductile materials that have stress-strain curves like that of the structural steel in Fig. 3.6, the area under the curve can be approximated either by

$$U_T \approx \sigma_u \epsilon_f$$

or

$$U_T \approx \frac{\sigma_{ys} + \sigma_u}{2} \epsilon_f$$

For relatively brittle materials the stress-strain curve can be assumed to be a parabola, and the area is given by

$$U_T \approx \frac{2}{3} \sigma_u \epsilon_f$$

All these relations are only approximations to the area under the stress-strain curves. Also, the curves do not represent the true behavior in the plastic range, since they are based on the engineering stress and strain, rather than the true stress and strain.

F. Ductile and Brittle Deformation

Materials that exhibit a considerable amount of permanent or plastic deformation prior to fracture, such as the structural steel in Fig. 3.6, are considered to be *ductile* or *malleable*. Although the terms ductile and malleable are sometimes used interchangeably, ductility is generally associated with tensile testing, whereas malleability is considered to be a compressive quality, such as the ability of a material to be flattened by rolling or hammering. For this reason the term "malleable" is often used to describe cast irons and steels that exhibit somewhat greater ductility than others.

An important measure of the ductility in a tensile test is the *reduction in area* q' of the fracture surface in comparison with the original cross-sectional area A_0. The reduction in area is given by

$$q' = \frac{A_0 - A_f}{A_0} \, 100$$

and is expressed as a percentage. Accurate measurement of the reduction in area is difficult, thus making it less valuable as a design parameter. Although there are no widely used methods for determining a minimum reduction in area needed for a particular application, the reduction in area may serve as a qualitative indication of the formability of a metal. The reduction in area is a very structure-sensitive parameter and may thus represent a particularly good indication of material quality.

A material such as the spring steel in Fig. 3.6 exhibits much less permanent deformation prior to failure and is considered to be *brittle*. The dividing line between ductile and brittle material response is arbitrary. For example, a mild steel rod that failed after straining a few percent would be considered brittle because it normally would stretch 25-30%, whereas a normally very brittle ceramic, such as a magnesia crystal, would be considered ductile if it sustained a few percent of permanent strain before breaking.

Since brittle failures usually occur suddenly and without prior warning, they often cause an entire structure to be destroyed. By contrast, failures that exhibit some ductility often provide adequate warning signs so that corrective actions, such as replacement of the damaged component, can be taken to prevent failure of the entire structure. It has thus become very desirable to increase the ductility of structural materials without causing significant reductions in their strength properties. Considerable progress toward this goal has been made in recent years, although much more remains to be done. This progress has been aided substantially by the widespread acceptance of fracture mechanics concepts and the development of materials having increased fracture toughness, since tougher materials are more ductile.

G. Hardness

The concept of hardness of a material is both widely used and poorly defined. In general, hardness usually implies a resistance to deformation, especially a resistance to permanent penetration or indentation. There are three general categories of hardness testing depending upon the manner in which the test is performed. These are the scratch hardness, rebound or dynamic hardness, and indentation hardness. Of the three, the indentation hardness has by far the most widespread use for metals and alloys. Although the rebound and indentation hardness methods were developed for metals, their use on appropriate nonmetals is increasing.

The scratch hardness is of primary interest to mineralogists, since it is a measure of the relative ability of minerals to scratch one another. Scratch hardness is measured by the Mohs scale, which ranges from 1 to 10 in order of increasing scratch resistance, with talc equal to 1 and diamond equal to 10. The Mohs scale is not well suited for metals since the intervals are not widely spaced in the high-hardness range appropriate to most hard metals.

In rebound or dynamic hardness testing the indenter is usually dropped onto the surface being measured and the hardness is measured by the height of the indenter rebound. The difference between the two potential energies is dissipated primarily by putting an indentation in the material. Hard materials with high elastic limits and yield strengths will absorb very little energy; soft materials will cause the failing indenter to lose much of its energy on impact and rebound very little.

The use of the rebound hardness test is limited because of several hard-to-control factors that have significant influence on the test results. These factors include surface roughness and instrument alignment, so that the indenter drops vertically and impacts the surface at right angles. The Shore Scleroscope is the most widely used device for measuring rebound hardness and is comprised of a calibrated glass tube, a steel- or diamond-tipped falling indenter, and a pneumatic lifting, holding, and releasing device. The portability of this device has encouraged its use on large objects that cannot be taken to a laboratory for hardness measurement. Empirical tables for conversion of Shore hardness numbers to other measures of hardness are available, but the utilization of these tables for different types of materials often leads to inconsistent results.

Penetration hardness tests using specified penetrators and loads have attained by far the most widespread use in hardness testing. The hardness measured in this way is assigned a numerical value based either on the final contact area of indenter and metal, the projected area of the permanent indentation, or the depth of the permanent indentation. Tests in common use include the Brinell, Vickers, Knoop, and Rockwell testing devices. All these methods are employed primarily to determine the quality of a metal, rather than a quantitative design parameter.

1. Brinell Hardness

The Brinell hardness test was the first widely accepted and standardized indentation hardness test and is performed by indenting the surface with a 10 mm diameter steel ball with a load of 3000 kg for 10 sec. For soft materials the load can be reduced to 500 kg (for 30 sec), and for very hard materials a tungsten carbide or diamond ball may be used to minimize distortion of the indenter. The diameter of the indentation is measured with a low-powered microscope after the load is removed. The surface of the material being tested should be relatively smooth and free of dirt or scale. The Brinell hardness number (BHN) is expressed as the load divided by the *surface area* of the indentation. This is expressed by

$$\text{BHN} = \frac{P}{(\pi D/2)(D - \sqrt{D^2 - d^2})}$$

where

P = applied load (kg)
D = diameter of ball (mm)
d = diameter of indentation (mm)

Brinell hardness numbers using the 3000 kg load are reasonably accurate from about 130 to 750, corresponding generally to steels having tensile strengths between 60 and 350 ksi (414-2410 MPa). Although the Brinell hardness test has achieved very widespread acceptance, it has several significant disadvantages:

1. The necessity of measuring the indentation diameter is time consuming and a major source of error.
2. The large load and indenter size require the thickness of the material tested to be greater than approximately 0.1 in. (2.5 mm).
3. The test must be made far enough from the edge of the material so that no bulging results.
4. For small specimens or very soft metals, a lighter load and smaller ball having the same P/D^2 ratio as the standard test must be used.

These disadvantages have given rise to the development of several other hardness test methods that employ smaller indenters or use the depth of the indentation to determine the hardness number.

2. Vickers Hardness

The Vickers hardness test employs a pointed diamond indenter that causes fully developed plastic deformation in the metal and greatly reduced elastic deformation of the indenter. The included angle of the indenter point is $136°$, with conical and diamond-shaped indenters available. A load from 5 to 100 kg is applied and removed automatically by a weight and cam mechanism. The load is chosen in relation to the hardness and thickness of the material being tested. The hardness number, expressed as DPH (diamond-pyramid hardness), is obtained by dividing the load in kilograms by the impressed area in square millimeters, as determined by the diameter or diagonal length of the impression. Because of the shape of the indenter, the DPH is considered to be relatively independent of the applied load. There is generally good agreement between DPH and BHN values, except that elastic deformation of the Brinell ball at higher hardnesses causes the values to diverge in this range. The Vickers hardness test has received rather wide acceptance for research work because it provides a continuous scale of hardness, for a given load, from very soft metals with a DPH of 5 to extremely hard materials with a DPH of 1500. Also, the shape of the indenter makes the DPH values relatively independent of load, except at very light loads. In spite of these advantages, the Vickers hardness test has not been widely accepted for routine testing because it is slow, requires careful surface preparation of the specimen, and permits greater opportunity for operator error in determining the size of the impression.

3. Knoop Hardness

For many applications it is necessary to employ extremely small hardness impressions to obtain a meaningful result. For example, the hardness of very thin coatings or the hardness of different constituents of an alloy cannot be measured by either the Brinell or Vickers methods. Tests using very small indenters are often called microhardness tests since the applied load is very small (generally less than 1 kg). The Knoop hardness test uses a diamond-pyramid indenter with its long diagonal about seven times its short one. The depth of indentation is about $\frac{1}{30}$ the length of the longer diagonal. The Knoop hardness number (KHN) is the applied load divided by the unrecovered projected area of the indentation and is often used with the Tukon tester. Microhardness tests usually require very careful surface preparation and for that reason are not normally employed for general-purpose hardness testing.

4. Rockwell Hardness

The Rockwell hardness tester was developed primarily for production testing of large numbers of parts and has become the most widely used hardness test method. Its widespread acceptance is due to its speed, repeatability, accuracy, and small indentations. It uses the depth of penetration as its measure of hardness. To improve accuracy and minimize the effect of surface irregularities, a 10 kg minor load is applied and the depth reading is set to zero. The major load is then applied and the new depth is measured and subtracted from a constant number, with the resulting hardness number displayed on a dial.

A conical $120°$ diamond indenter with a rounded tip, called a Brale indenter, is used with a 150 kg major load for testing of hard metals (Rockwell C scale). For softer materials a $\frac{1}{16}$ in. (1.6 mm) diameter ball is used with a 100 kg major load (Rockwell B scale). The Rockwell A scale uses the diamond indenter with a 60 kg major load and represents the most extended Rockwell hardness scale. For special situations when the depth of penetration must be reduced, other Rockwell scales exist with loads down to 15 kg and a more sensitive depth-measuring system. These tests are generally called *superficial* hardness tests.

It has been determined through experimental comparisons that reasonable relations exist between the different measures of hardness discussed in this section. These relations are most reliable for steels harder than 240 BHN. Also for these materials, a good correlation between the Brinell hardness number and the tensile strength has been established as

$$\sigma_u(\text{psi}) = 500 \text{ BHN}$$

Table 3.1 presents a number of comparisons between several hardness measures and the tensile strength of many steel alloys. Considerable care must be exercised in using this table for other materials, since many materials provide results differing widely from the ones presented here.

H. Temperature Effects

For many years most material property tests were performed at room temperature, for service conditions also at room temperature, or, at worst, the normal seasonal changes in weather (bridges and other outdoor structures). However, recent engineering advances, such as for rocket development and space exploration, the discovery and use of superconductors, and the increased cost of energy transportation and use have subjected most engineering materials to a much wider range of service temperatures. These applications extend from the near absolute zero temperature and pressure that must be sustained by satellites and spacecraft to the increasing temperatures and pressures necessary for more efficient functioning of turbojet engines and power plants. Therefore, it has become increasingly important to obtain an improved understanding of the influence of extreme temperatures on the mechanical properties of materials. In response to these needs, material tests are now routinely performed at temperatures ranging from approximately $-200°C$ ($-328°F$) to $1000°C$ ($1832°F$).

1. Low Temperatures

The effect of low temperatures on the mechanical properties of metals and alloys varies widely, with the extent of the influence largely determined by the metal structure. The mechanical properties of face-centered cubic (FCC) and most hexagonal close-packed (HCP) materials exhibit a very limited temperature dependency, whereas the properties of body-centered cubic (BCC) and some polymorphic metals are dramatically influenced by the testing temperature. Most of the temperature-insensitive materials exhibit continuous moderate increases in yield strength, tensile strength, hardness, and Young's modulus and decreases in ductility as temperatures are reduced to cryogenic levels. Although these materials may possess widely varying mechanical properties, their temperature dependency is predictable and consistent.

Most BCC and some polymorphic metals and alloys exhibit dramatic changes in properties as the temperature is decreased. For example, both plain carbon and low-alloy steels become embrittled at low temperatures, sometimes at temperatures as high as 90-100°C (194-212°F). Embrittlement is indicated by a loss in toughness over a narrow or

Table 3.1 Approximate Comparisons Between Hardness Numbers and Tensile Strengths for Structural Steels

Steel	Vickers DPH	Brinell			Rockwell			Scleroscope hardness	Tensile strength ×1000 psi
		10 mm 500 kg Steel	10 mm 3000 kg Steel	10 mm 3000 kg Tungsten carbide	B 100 kg $\frac{1}{16}$ in. Ball	C 150 kg Brale	Superficial 15 kg brale		
Steel	600			564		55.2	88	74	289
	550			517		52.3	86.6	70	264
	500		465	471		49.1	85	66	240
	450		425	425		45.3	83.2	60	214
	400		379	379		40.8	80	54	190
	350		331	331		35.5	78	48	166
	300		284	284		29.8	74.9	42	141
	250		238	238		22.2	70.6	36	116
Unhardened steel, cast iron, and most nonferrous metals	175	175	210		95			35	101
	150	150	174		87.7			33	84
	125	125	142		77.5			29.5	—
	100	100	115		63.5			24	—
	75	75			41			—	—

Source: Adapted from American Society for Metals, *The Metals Handbook*, Cleveland, Ohio, 1948.

moderate temperature range, and failure is characterized by brittle, coarse-grained fracture, especially as indicated in notched-bar impact tests. The temperature at which the
impact energy absorption is reduced to 50% of the ductile or upper shelf value is generally taken as the *transition temperature* for this effect. As an example of both tempera-
ture-sensitive and insensitive material behavior, Fig. 3.7 illustrates the percentage reduction in area at fracture as a function of testing temperature for several metals. Other
properties, such as the impact energy and the fracture toughness, illustrate this behavior
even more strongly and will be discussed later in this chapter.

2. High Temperature

Numerous tests have been performed on metals at high temperatures, because of the ever-
increasing number of applications that involve high-temperature service. In general, all
metals experience a loss in static strength and stiffness and some increase in ductility as
temperatures are increased. In addition, many metals undergo metallurgical changes in
structure or phase at elevated temperatures, with consequent major changes in mechanical properties. Steel exhibits such changes to a marked degree, and as a result there are
inflections and humps in its short-time strength versus temperature curve (Fig. 3.8). The
reduction in strength is large at temperatures near the recrystallization temperature, and
the fracture generally changes from transcrystalline to intercrystalline failure as the testing temperature is increased. The elastic modulus is reduced even more abruptly above
the recrystallization temperature, and at higher temperatures the early initiation of plastic
deformation makes interpretation of the elastic modulus from the stress-strain curve a
difficult matter. Aluminum alloys exhibit the same general type of behavior, but because
of the lower melting temperature of aluminum, the changes are initiated at even lower
temperatures.

　　At elevated temperatures, the resistance to plastic deformation is more sensitive
to the rate of deformation than at low temperatures, and strain hardening effects are reduced or eliminated. Figure 3.9 illustrates how different loading rates affect the stress-
strain curve for a typical structural material tested at high temperature. Lower loading

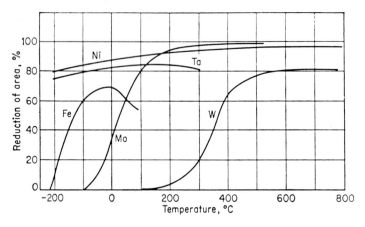

Fig. 3.7 Effect of temperature on the reduction of area of Ta, W, Mo, Fe, and Ni. (From
J. H. Bechtold, *Acta Met.,* vol. 3, p. 253, 1955.)

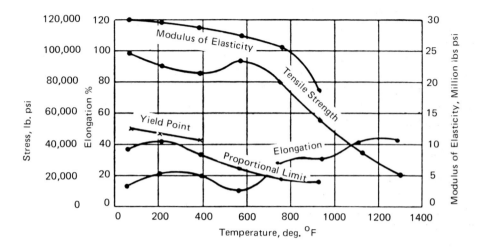

Fig. 3.8 Effect of temperature on the short-time static strength of 0.53 percent carbon steel. (From tests by H. J. Tapsell and W. J. Clenshaw, *Properties of Metals at High Temperatures II*, Engineering Special Report No. 2, Dept. of Scientific and Industrial Research, Great Britain, Sept. 1927.)

rates result in greater amounts of deformation at a given load, since microscopic plastic deformation processes have longer times in which to act.

Deformation is not exclusively elastic or plastic at high temperatures. If a load is applied to produce a small amount of strain and then is removed, the recovery of deformation will occur as illustrated in Fig. 3.10. The elastic deformation is recovered immediately as the load is removed, whereas some additional nonelastic deformation is recovered only over a period of time. This deformation is termed *anelastic* and is generally considered to be a measure of the internal damping capacity of the material. The deformation takes place primarily in the grain boundary regions of the material and is a function of the total applied load. This process is similar to a collection of hard objects held together by a viscous adhesive that takes time to stretch and time to recover. The effects of anelasticity become pronounced at temperatures above 175°C (345°F) in magnesium,

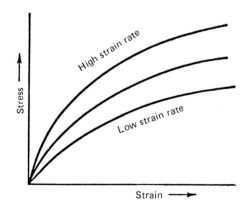

Fig. 3.9 Effect of strain rate on stress-strain curve.

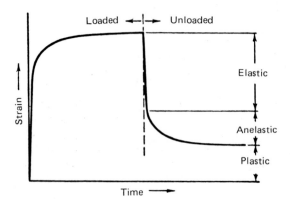

Fig. 3.10 Recovery of strain after loading to plastic range at high temperatures.

250°C (480°F) in aluminum, 300°C (570°F) in copper, 357°C (700°F) in α-brass, and 425°C (800°F) in α-iron.

I. Creep

Whenever constant loads or stresses above a certain threshold value are applied to a material or structure, the resulting deformation may not be constant but may vary with testing time, applied stress, or temperature. Such time-dependent deformation is called *creep*, and it typically becomes more severe as testing or service temperatures are increased. However, creep is not strictly a high-temperature phenomenon, since lead, rubber, and various plastics exhibit creep at room temperature. Most polycrystalline metals and alloys must be subjected to temperatures above one-half their melting temperature, $T_m/2$ measured on an absolute temperature scale, before significant creep deformation will occur.

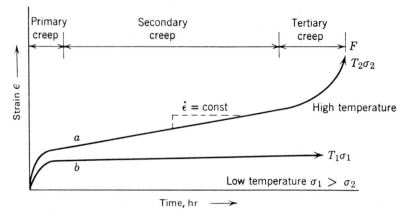

Fig. 3.11 (a) The three stages of the strain-time relation in high-temperature creep. (b) The same relation at low temperature.

The creep behavior of a polycrystalline metal at temperatures above and below $T_m/2$ is shown in Fig. 3.11. For both curves there is an initial deformation that occurs immediately upon application of the load, followed by an initial period of creep at a rapidly decreasing strain rate, usually called *primary* creep. At lower temperatures, the creep rate continues to decrease with time unless the applied stress is near the ultimate tensile strength. At higher temperatures, primary creep is followed by a period of deformation at constant strain rate, called *secondary* creep. Depending upon the conditions of temperature, applied stress, and other factors, the amount of deformation expended in secondary creep may be very small or very large. If the loads and temperatures are small, secondary creep may continue indefinitely. However, at higher loads or temperatures the creep rate may begin to increase, culminating in fracture of the specimen. The period of increasing creep rate is called *tertiary* creep. Although the separation of the creep curve into three distinct regions does not imply three completely different deformation mechanisms, generally different processes do appear to dominate the deformation in each region.

The initial instantaneous extension, which occurs prior to any time-dependent deformation, is made up of both elastic and plastic deformation. The initial deformation is not usually considered to be creep, since it is not time dependent, even though it may constitute a significant part of the total deformation. However, some researchers do consider the instantaneous extension the first stage of creep, so that with this nomenclature the creep curve is considered to have four stages.

1. Primary Creep

Primary creep is generally considered to be primarily anelastic, with a limited amount of plastic deformation. The predominant mechanisms causing primary creep, that is, the reduction of strain rate with increasing time, are typical of those associated with strain hardening in metals and alloys. Present theories explaining strain hardening are based on the motion of dislocations through individual grains of polycrystalline materials. Knowledge of the motion of individual dislocations can explain inelastic deformation processes but cannot be used to predict strain-hardening effects. However, these effects can be explained by means of interactions of dislocations with grain boundaries, crystal defects, and other dislocations. More precisely, these interactions take place between the stress or strain fields associated with each type of crystal imperfection. For example, dislocations generated during initial deformation may pile up at grain boundaries, thus creating back stresses on dislocations created during subsequent deformation. These back stresses serve to inhibit further dislocation motion, resulting in strain hardening of the material.

Some types of dislocation interactions that lead to strain hardening are

1. Interactions between dislocations on parallel slip planes (dislocation pile up)
2. Combination of dislocations
3. Nonconservative motion and climb of dislocations
4. Interactions with impurity atoms or other defects

Many theories, some incorporating specific mathematical models, have been proposed to describe primary creep. However, the number of these theories as well as the necessity of providing detailed explanations of their relative merits, preclude inclusion of them in this chapter.

2. Secondary Creep

Secondary creep is considered to be caused by a counterbalancing of *strain-hardening mechanisms* that tend to reduce or limit further straining and *recovery processes* that facilitate further straining. The strain-hardened state of a material is a condition of higher internal energy than the nondeformed metal. Therefore, there is a tendency for strain-hardened materials to revert to the strain-free condition. With increasing temperature the strain-hardened condition becomes more unstable, eventually reverting to the strain-free condition by a process known as annealing. Annealing is generally considered to be comprised of three separate processes, recovery, recrystallization, and grain growth, with each process in this sequence requiring higher temperatures for activation. Most creep behavior in metals is experienced under temperatures too low for recrystallization and grain growth to occur. Therefore, for most metals, it is the recovery process that serves to counterbalance the strain hardening and initiate secondary creep. *Recovery* is usually defined as the restoration of the physical properties of a strain-hardened material without any observable change in microstructure. The properties most influenced by recovery are those that are sensitive to point defects. The diffusion of point defects, such as vacancies and interstitial atoms, facilitates additional dislocation movement by climbing processes in which the dislocations move out of their normal planes of action into other planes more suitable for subsequent motion. Like all diffusion processes, dislocation climb is a function of time and temperature so that activation temperatures for secondary creep and diffusion should be about the same. This has been shown to be the case for a number of metals.

The average value of the creep rate during secondary creep is called the *minimum creep rate,* and is the most important design parameter derived from the creep curve. According to ASTM E139, Standard Recommended Practice for Conducting Creep, Creep-Rupture, and Stress-Rupture Tests of Metallic Materials, creep tests should be performed in such a way that the minimum creep rates range between 0.0001 and 0.00001%/hr. Minimum creep rates at the lower end of this range have typically provided data for turbojet engine applications; data from the high end have been used for steam turbines and other electric power-generating equipment.

3. Tertiary Creep

Secondary creep continues until the strain-hardening effects can no longer counterbalance the recovery processes that lead to weakening of the material. Tertiary creep is then initiated, resulting in ever-increasing creep rates until the material fractures. The cause for the initiation of tertiary creep has for many years been considered to be primarily due to necking in the specimen. However, other processes, such as the accumulation of voids, grain boundary sliding, and microcrack formation, have more recently been identified as factors causing the initiation of tertiary creep. The lack of understanding of tertiary creep suggests that a considerable amount of research remains to be done before all the critical mechanisms have been identified and their effects properly quantified.

As with most other mechanical property determinations, creep testing is subjected to several difficulties and complications. One of the major complications is that the use of constant applied loads results in increasing true stresses, since the cross-sectional area of the test section continuously decreases during the test. There are certain ranges of initial applied stress for which a constant load test will cause the material to exhibit all three stages of creep but a constant stress test will exhibit only primary creep. Outside this range of initial loadings, the behavior of both types of tests is more similar, with the primary difference a more rapid test termination for the constant load tests.

Another major difficulty with creep testing is that the combination of high temperatures and long test times usually employed in creep testing often induces metallurgical changes in the materials. The dominant effect is that of metal softening at high temperatures, but other factors, such as a combined temperature and strain energy-driven recrystallization, can also occur. In most cases the metallurgical changes induce greater strains than would otherwise be observed. Such results could then lead to nonconservative predictions of service lives and total strains in structures in actual service.

J. Interpretation of Creep Test Data

The primary objective of creep testing is to measure creep and creep rates, due to low applied stresses over long periods of time and at high temperatures, for the purpose of obtaining the creep response of materials under actual service conditions. Most data available are obtained by maintaining a standard tensile test specimen at a constant temperature under a fixed load and measuring the strain at preselected time intervals. As mentioned in the previous section, the standard recommended practice for creep testing is contained in ASTM E139. In addition, ASTM E150, Standard Recommended Practice for Conducting Creep and Creep-Rupture Tension Tests of Metallic Materials Under Conditions of Rapid Heating and Short Times, covers tests planned for faster heating rates and shorter testing times. The resulting test data may be summarized to show the effect of any two of the four parameters: creep deformation, time, stress, and temperature.

A *creep-time diagram,* such as in Fig. 3.11, is obtained by plotting creep against time for constant stress and temperature. The standard test methods cited above define creep as the time-dependent strain that occurs after the application of a load, which is thereafter kept constant. Thus the initial deformation is not used in recording creep strains.

A *creep-stress diagram* may be obtained by cross plotting the creep for a given time interval, obtained from a family of creep-time curves, as a function of the applied stress. For example, Fig. 3.12 shows a 10,000 hr creep-stress diagram for a 0.33% carbon cast steel.

A *stress-temperature diagram* shows the stress that will cause a given amount of creep or a given creep rate as a function of the testing or service temperature. A stress-temperature curve for three steels is shown in Fig. 3.13. These curves are also cross plotted from families of creep-time curves in which the stresses and temperatures are varied.

A *stress-time diagram* shows the amount of stress necessary to cause a certain amount of creep strain, as a function of the testing or service time. These diagrams are also obtained from families of creep-time curves and normally include one curve for each creep-strain increment, although temperature may also be included as a parameter.

K. Creep-Rupture and Stress-Rupture Testing

In contrast to creep-testing methods and objectives, creep-rupture tests are performed for the purpose of recording both creep and creep rates as well as the time to specimen failure (rupture). In general, both the total deformations and creep rates are considerably larger than the corresponding values developed during a creep test.

A stress-rupture test is one in which only the applied loads and times to rupture are of interest. The same test equipment may be used for all three types of tests, although

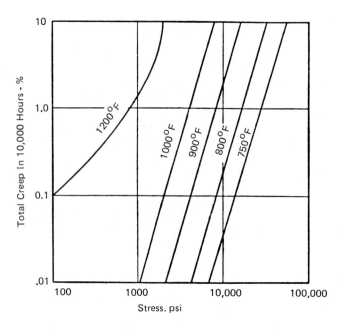

Fig. 3.12   Creep-stress curves for 0.33% carbon cast steel. (From *Symposium on the Effect of Temperature on the Properties of Metals,* courtesy of ASME and ASTM.)

equipment intended solely for stress-rupture testing is simpler to build, maintain, and operate than creep-testing equipment.

The time required to cause fracture in creep-rupture and stress-rupture testing generally varies linearly with the applied stress when the data are plotted on log-log coordinates. A stress-rupture curve for annealed Monel is presented, with testing temperature as a parameter, in Fig. 3.14. At the higher testing temperatures, this material behaves in a bilinear rather than a linear fashion, with the change in slope usually attributable to structural changes occurring in the material. These structural changes include, but are not limited to, transformation from transgranular to intergranular fracture, oxidation, recrystallization and grain growth, and subgrain formation.

Stress-rupture data are important in designing all types of high-temperature equipment, including turbojet engine components and electric power-generating equipment, as well as nuclear fission and fusion test facilities. The *rupture strength* is defined as the value of the applied stress at a given temperature necessary to cause rupture in a given time interval, such as 100, 1000, or 10,000 hr. The rupture strength data may then be used, with appropriate safety factors, to establish the maximum applied stress that could be sustained for a desired service life.

L. Stress Relaxation

If the loading applied to a structural component causes the total deformation rather than the applied stress to remain constant during long-time high-temperature service, the component may deform under the load and decrease its load-carrying capacity. This *stress relaxation* often occurs in bolts, rivets, or other fasteners and can result

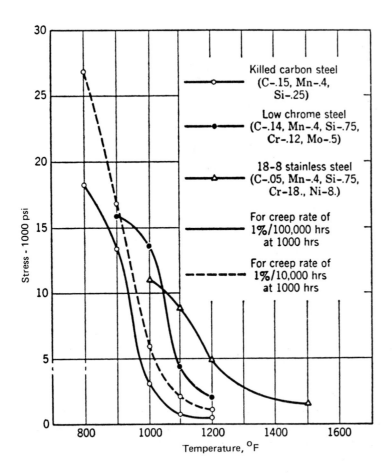

Fig. 3.13 Stress-temperature curves for three steels. (Data from *Digest of Steels for High Temperature Service*, courtesy Timken Roller Bearing Co.)

in serious problems due to the loss of beneficial residual stresses and loosening of the assembly. Although the original strain in the fastener may have been fully elastic, part of this elastic strain is gradually converted into plastic strain. The choice of suitable allowances to maintain the necessary force for the desired period requires tests to determine the relaxation of the fastener with time.

As with most other types of material testing, standards have been developed for determining the extent of stress relaxation under specified test conditions. The primary documentation for stress-relaxation testing is provided by ASTM E328, Standard Recommended Practice for Stress-Relaxation Tests for Materials and Structures, and covers a broad range of testing activities and specimen designs. Loads may be applied in either tension, compression, bending, or torsion, and the choice of test type to be utilized should be such that the applied stresses are similar to those imposed on the component. For example, tension tests are suitable for bolting applications, but bending tests should be used for leaf springs. Tension and compression tests have the advantage that the stress

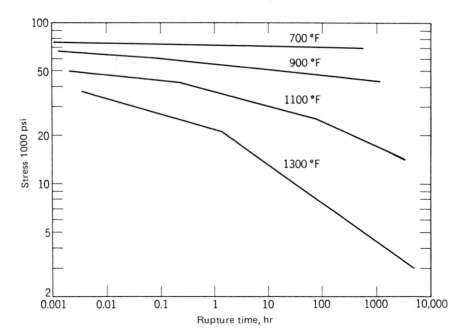

Fig. 3.14 Stress versus rupture time for annealed Monel tested over a range of temperatures. (After N. J. Grant and A. C. Bucklin, *Trans. ASM,* 1962, p. 156.)

can be reported simply and unequivocally. The state of stress is complicated in the bending tests, especially if plastic strains are generated by the loading. When the test method is not restricted by the type of stress in the component, tension testing is normally recommended.

Bending tests have the advantage of using lighter and simpler apparatus for specimens having the same cross-sectional area. Also, the strains are usually calculated from deflection or curvature measurements and are intrinsically more accurate than direct axial measurements. Due to the small required forces and test apparatus simplicity, many bending specimens may often be placed in a single oven or furnace when elevated temperature testing is required.

When the material data are to be applied to the design of a particular class of component, such as a turbojet engine or a steam turbine, the stress applied during the relaxation test should be similar to that imposed on the component. The testing temperature should also be the same so that any metallurgical changes occurring in the component will be duplicated in the test specimen.

The results of relaxation tests on several steels at fixed strain and temperature are shown in Fig. 3.15. The remnant stress is measured periodically as a function of time (usually on log-log coordinates). This type of curve permits determination of the maximum original stress that will be necessary to produce a specified remnant stress after the required duration of high-temperature service.

The shape of these curves, as with the constant load or stress-creep curves, depends primarily on whether strain hardening or annealing predominates. If the metal tends to anneal, the curve bends downward, as shown in A, B, D, and E. If it reaches a steady state

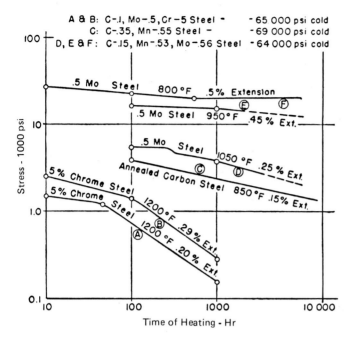

Fig. 3.15 Relaxation elastic stress-time diagram taken at constant total strain and constant temperature for three steels.

in which strain hardening dominates, the curve will bend upward and become nearly level (curve F). After 10,000 hr at 850°F, the carbon steel (curve C) has shown neither tendency.

M. Fracture Toughness

Subsequent to their initial development approximately 30 years ago, fracture mechanics concepts have been increasingly accepted and utilized by industry. The range of applications has been especially broad in the aerospace industry, because of several generally unrelated advances. These include the development of high-strength aluminum alloys and their application to new classes of aircraft powered by turbojet and turboprop engines. The desire to fully exploit these higher strength materials in new aircraft design led to unexpected structural failures resulting from the materials' increased vulnerability to stress concentrators and cracks. Fracture mechanics methodology was subsequently employed to ascertain that this problem was primarily due to decreasing fracture toughness values directly caused by the higher strength material processing methods. Further applications of fracture mechanics concepts have led to improved damage assessment methods and to modifications of the high-strength low-toughness alloys to produce significantly increased fracture toughness with only slight reduction in material strength. More recently, fracture mechanics applications have been extended to virtually all weight-sensitive structures and even to some weight-insensitive structures for which unstable structural failure cannot be tolerated. Examples of this latter category are the nuclear power-generating stations to which fracture mechanics concepts have achieved widespread application.

Fracture mechanics is defined as the study of the influence of sharp cracks on the strength of solid bodies. The fundamental fracture mechanics concepts have been developed from two distinct but compatible formulations: An energy approach first developed by Griffith in the 1920s and a stress approach developed by Westergaard in the 1940s. Both approaches will be briefly discussed in this section.

The energy approach developed by Griffith was based on an analysis of an infinite sheet containing a central crack of length 2a and subjected to equal stresses σ perpendicular and parallel to the crack (equal biaxial stresses). Griffith's basic hypothesis was that an existing crack will propagate when the associated decrease in strain energy is at least equal to the energy required to create the new crack surface. He also showed that the decrease in strain energy associated with the formation of a line crack of length 2a was

$$U = \frac{-\pi a^2 \sigma^2}{E}$$

and that the increase in surface energy was

$$U_s = 4a\gamma$$

In these equations, γ is the surface energy density, E is Young's modulus, and a unit thickness is assumed. The total change in energy due to the crack ΔU is

$$\Delta U = U_s + U$$

and the Griffith criterion stated that

$$\frac{\partial \Delta U}{\partial a} = 0 = 4\gamma - \frac{2\pi a \sigma^2}{E}$$

Thus, the stress required to propagate a crack is obtained as

$$\sigma = \left(\frac{2E\gamma}{\pi a}\right)^{1/2}$$

Subsequent work has led to the definition of the *strain energy release rate* as

$$G = \frac{\pi a \sigma^2}{E}$$

If σ is increased from zero to a critical value σ_c at which unstable fracture occurs, G assumes a limiting value,

$$G_c = \frac{\pi a \sigma_c^2}{E}$$

which is known as the *fracture toughness.* The fracture toughness is considered to be a geometry-independent material parameter that describes the strength of the material in the presence of a crack. A technique for measuring G_c was developed by Irwin and toughness values for a number of engineering materials were obtained before attention was shifted to the crack-tip stress field, or K, approach.

The stress intensity factor, K_I (opening mode), represents a one-parameter measure of the crack-tip stress field and can be best understood as a limiting case of the stress concentration factor K_t. Consider an infinite body containing (1) a round hole, (2) an elliptical hole, and (3) a line crack. All are spaced widely enough apart so that their stress fields do not interact (Fig. 3.16). The stress concentration factors ($K_t = \sigma_y/\sigma$) are

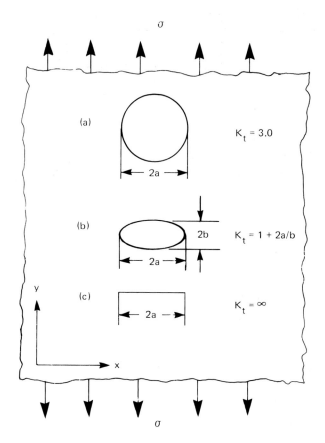

σ

(a) $K_t = 3.0$

2a

(b) 2b $K_t = 1 + 2a/b$

2a

(c) $K_t = \infty$

2a

σ

Fig. 3.16 Stress concentration factors for round and elliptical holes and a line crack.

known to be: (1) $K_t = 3.0$, (2) $K_t = 1 + 2a/b$, and (3) $K_t = \infty$. For the line crack in an elastic solid the infinite value of K_t renders it useless for practical applications. However, the dependence of the crack-tip stress field on polar coordinates (r, θ), with the origin at the crack-tip, has been established as

$$\sigma_x = \frac{K_I}{\sqrt{2\pi r}} \cos \frac{\theta}{2} \left(1 - \sin \frac{\theta}{2} \sin \frac{3\theta}{2}\right) + \cdots$$

$$\sigma_y = \frac{K_I}{\sqrt{2\pi r}} \cos \frac{\theta}{2} \left(1 + \sin \frac{\theta}{2} \sin \frac{3\theta}{2}\right) + \cdots$$

$$\tau_{xy} = \frac{K_I}{\sqrt{2\pi r}} \sin \frac{\theta}{2} \cos \frac{\theta}{2} \cos \frac{3\theta}{2} + \cdots$$

The stress field clearly incorporates the necessary behavior as $r \to 0$. In these equations, K_I is a finite real number and serves as a multiplier of the $1/\sqrt{r}$ singularity. For example, the behavior of the σ_y stress along $\theta = 0$ for three different values of K_I is illustrated in Fig. 3.17.

The opening mode (mode I) is one of three planar modes of fracture. The relative crack surface displacements associated with the opening mode, the edge-sliding mode

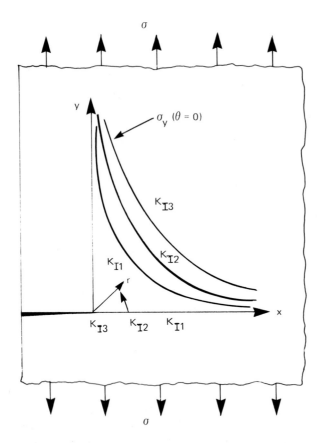

Fig. 3.17 Illustration of the effect of different stress intensity factors on the σ_y crack-tip stress ($\theta = 0°$).

(mode II), and the tearing mode (mode III) are illustrated in Fig. 3.18. The crack-tip stress fields are also different for modes II and III and are readily available in the fracture mechanics literature. Since for most fracture mechanics applications the opening mode governs the fracture process, only mode I applications will be discussed. However, when significant mode II or mode III loadings exist, these fracture modes must be taken into consideration.

Since the previous equations describe the crack-tip stress field to a reasonable degree of accuracy (the dominant first term in an infinite series expansion), knowledge of K_I and the remote stress distribution is all that is needed to perform a stress analysis of any cracked structural component. The K_I values are determined by the crack geometry, specimen or component configuration, and applied loading, so they can be uniquely established for any geometry of interest. For example, the stress intensity factor for an infinite elastic body subjected to uniaxial tension perpendicular to a central crack of length 2a (Fig. 3.16c) is

$$K_I = \sigma \sqrt{\pi a}$$

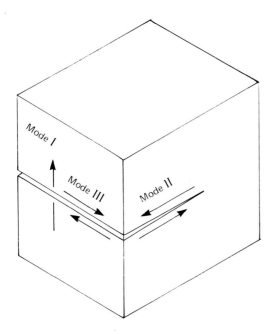

Fig. 3.18 Displacements associated with the three planar modes of fracture: opening mode (I), edge-sliding mode (II), and tearing mode (III).

This stress intensity factor is very simple in form and also establishes the units of K_I as MPa $\sqrt{m}$ or ksi $\sqrt{in}$. This expression also represents the stress intensity factor for the geometry considered by Griffith since the stress applied parallel to the crack does not affect K_I. Comparison of K_I and G for this geometry permits the establishment of a relation between them in the form

$$G_I = \frac{K_I^2}{E}$$

This expression is valid for all geometries described by plane stress conditions (materials thin enough that stresses normal to the plane of the specimen can be neglected) and establishes the mutuality between the energy and stress approaches to fracture mechanics.

The stress intensity factors for several practical geometries are presented in Table 3.2. In most cases the K_I values are presented as a polynomial function of a/w, where w is the specimen width. This factor, a/w, is used in order to permit easy determination of stress intensity factors for different crack lengths. Since practical problems rarely match the prototype solutions given in Table 3.2, it becomes necessary to obtain solutions for more practical geometric configurations. A convenient source of practical stress intensity factors is any one of several handbooks of stress intensity factors published over the last 20 years. The stress intensity factors are typically given in the form

$$K_I = \sigma \sqrt{\pi a} \ f(a, w, \ldots)$$

where $f(1, w, \ldots)$ represents a geometric correction factor and is usually given in tabular or graphic form. Stress intensity factors for many geometries included in Table 3.2 were

Table 3.2   Stress Intensity Factors for Several Important Fracture Mechanics Geometries

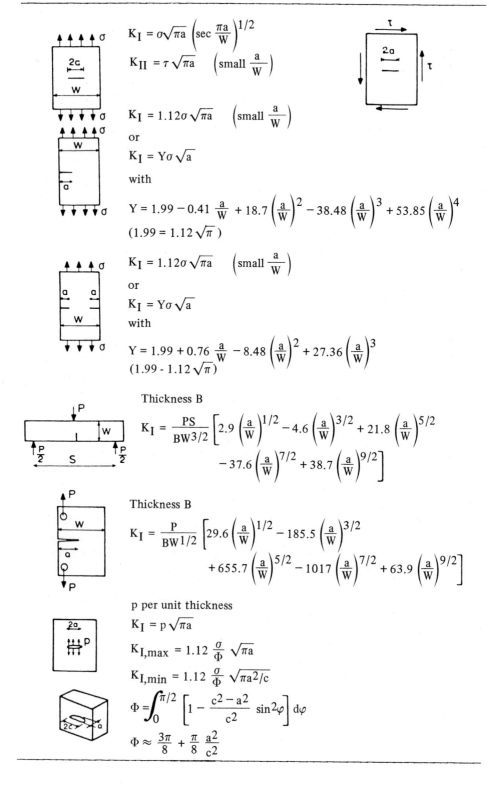

$$K_I = \sigma\sqrt{\pi a}\ \left(\sec\frac{\pi a}{W}\right)^{1/2}$$

$$K_{II} = \tau\sqrt{\pi a}\quad \left(\text{small }\frac{a}{W}\right)$$

$$K_I = 1.12\sigma\sqrt{\pi a}\quad \left(\text{small }\frac{a}{W}\right)$$

or

$$K_I = Y\sigma\sqrt{a}$$

with

$$Y = 1.99 - 0.41\frac{a}{W} + 18.7\left(\frac{a}{W}\right)^2 - 38.48\left(\frac{a}{W}\right)^3 + 53.85\left(\frac{a}{W}\right)^4$$

$$(1.99 = 1.12\sqrt{\pi}\ )$$

$$K_I = 1.12\sigma\sqrt{\pi a}\quad \left(\text{small }\frac{a}{W}\right)$$

or

$$K_I = Y\sigma\sqrt{a}$$

with

$$Y = 1.99 + 0.76\frac{a}{W} - 8.48\left(\frac{a}{W}\right)^2 + 27.36\left(\frac{a}{W}\right)^3$$

$$(1.99 - 1.12\sqrt{\pi})$$

Thickness B

$$K_I = \frac{PS}{BW^{3/2}}\left[2.9\left(\frac{a}{W}\right)^{1/2} - 4.6\left(\frac{a}{W}\right)^{3/2} + 21.8\left(\frac{a}{W}\right)^{5/2} \right.$$
$$\left. - 37.6\left(\frac{a}{W}\right)^{7/2} + 38.7\left(\frac{a}{W}\right)^{9/2}\right]$$

Thickness B

$$K_I = \frac{P}{BW^{1/2}}\left[29.6\left(\frac{a}{W}\right)^{1/2} - 185.5\left(\frac{a}{W}\right)^{3/2} \right.$$
$$\left. + 655.7\left(\frac{a}{W}\right)^{5/2} - 1017\left(\frac{a}{W}\right)^{7/2} + 63.9\left(\frac{a}{W}\right)^{9/2}\right]$$

p per unit thickness

$$K_I = p\sqrt{\pi a}$$

$$K_{I,max} = 1.12\frac{\sigma}{\Phi}\sqrt{\pi a}$$

$$K_{I,min} = 1.12\frac{\sigma}{\Phi}\sqrt{\pi a^2/c}$$

$$\Phi = \int_0^{\pi/2}\left[1 - \frac{c^2 - a^2}{c^2}\sin^2\varphi\right]d\varphi$$

$$\Phi \approx \frac{3\pi}{8} + \frac{\pi}{8}\frac{a^2}{c^2}$$

obtained numerically by the boundary collocation method. The finite-element method has also come into widespread use recently for determining stress intensity factors for various practical geometries.

When the applied stress at unstable fracture σ_c is input into the formula for the stress intensity factor, the critical value of K_I is attained and is called the plane strain fracture toughness K_{Ic}. As with G_c, the K_{Ic} value is considered to be a material constant, independent of specimen or component geometry. The relationship between K_I and K_{Ic} is directly analogous to that between the applied stress σ and the yield strength σ_{ys} as determined in a standard tensile test. For two reasons the geometry independence of K_{Ic} is of central importance in fracture mechanics applications. First, this independence is necessary in order to ensure that the same K_{Ic} value is obtained from different types of test specimens. Three specimens (compact tension, three-point bend, and C-shaped) are at present approved for K_{Ic} determination, and it has been verified that they all provide compatible results. Second, it is the geometry independence of K_{Ic} that permits its application to many different crack configurations and component geometries. As long as an appropriate stress intensity factor is known or can be determined for a particular geometry, fracture mechanics concepts can be applied when certain requirements are satisfied. These requirements have been introduced to ensure that structural components will fail in a fashion compatible with test specimens used for evaluating K_{Ic}.

N. Fracture Toughness Testing

To satisfy the objective of providing a geometry-independent material property, ASTM E399, Standard Method for Plane-Strain Fracture Toughness of Metallic Materials, has been developed. One of the major sources of variability in the test results is the tendency for increased plastic deformation and plane stress fracture as the test specimen thickness is decreased. Plane stress fracture occurs when inadequate constraint of deformation exists in the thickness direction, leading to considerable shear processes and slant fracture surfaces (Fig. 3.19). In thicker specimens, plane strain (square fracture surfaces) or transition fractures occur where the fracture surface is predominantly square but with shear lips along the lateral surfaces (Fig. 3.19). To maintain predominantly plane strain fracture, the test specimen thickness must exceed $2.5(K_{Ic}/\sigma_{ys})^2$. The characteristics of the three test specimen geometries are such that, if the thickness requirement is satisfied, all other dimensions will be adequate to assure plane strain fracture.

Since the thickness has such a strong influence on the fracturing process, particular emphasis has been placed on it during test method development. Figure 3.20 shows how the fracture toughness increases with decreasing specimen thickness. These higher toughness values are usually designated K_c and may be employed for fracture mechanics applications only if a K_c value appropriate to the particular structural application is available. However, the K_c values are usually sensitive to specimen size and shape, so it may be necessary to conduct full-scale tests to determine the proper K_c value for a particular application.

Certain additional factors capable of introducing variation in K_{Ic} values are also controlled by ASTM E399. These factors include crack-tip plasticity, loading rate, fatigue crack-tip radius and curvature, and material variability. Other conditions, such as testing

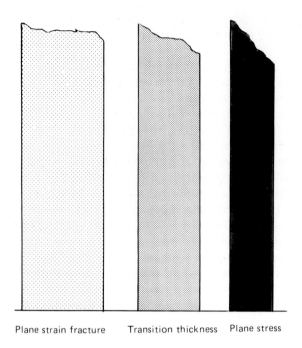

Plane strain fracture   Transition thickness   Plane stress

Fig. 3.19 Effect of thickness on fracture behavior.

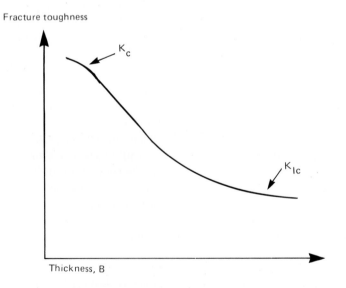

Fig. 3.20 Effect of specimen thickness on the fracture toughness.

temperature and adverse environments, may also have a strong influence on K_{Ic}. Although K_{Ic} tests may be performed at different temperatures, the temperature dependence is manifested through its influence on σ_{ys} and K_{Ic} and may result in considerable change in the minimum thickness requirement. Consideration of adverse or corrosive environments is excluded from this test method since other test methods have been developed for this purpose.

Fracture toughness values have been obtained for a wide variety of structural metals and alloys. Fracture toughness values for a number of structural materials are included in Table 3.3.

When the working stresses and material properties, including the fracture toughness, are known for a particular component and crack geometry, fracture mechanics principles can be employed to assess the tendency of the crack to initiate unstable fracture. If the combination of conditions is such that the stress intensity factor is equal to the fracture toughness, the stress and crack size have attained their critical values, denoted by σ_c and a_c, respectively.

For example, consider a center-cracked plate of 7075-T6 aluminum alloy, with the following characteristics (see Table 3.2):

$2a = 10.0$ in. (254 mm)
$w = 50.0$ in. (1.27 m)
$B = 0.50$ in. (12.7 mm)
$\sigma_{ys} = 77$ ksi (530 MPa)
$\sigma_u = 84$ ksi (580 MPa)
$K_{Ic} = 30$ ksi $\sqrt{\text{in.}}$ (33 MPa $\sqrt{\text{m}}$)

It is desired to determine whether the plate will sustain a nominal stress of 6.4 ksi (44 MPa) applied perpendicular to the crack. The minimum thickness requirement for plane strain fracture $B \geq 2.5(K_{Ic}/\sigma_{ys})^2 = 0.38$ in. (9.6 mm) is satisfied, so plane strain fracture would be expected to occur. From Table 3.2 the stress intensity factor is

$$K_I = \sigma \sqrt{\pi a} \left(\sec \frac{\pi a}{w} \right)^{1/2}$$
$$= 26 \text{ ksi } \sqrt{\text{in.}} \ (29 \text{ MPa } \sqrt{\text{m}})$$

so the stress will be sustained with a safety factor SF $= 30/26 = 1.15$. The critical crack size is determined by solving

$$K_{Ic} = \sigma_c \sqrt{\pi a_c} \left(\sec \frac{\pi a_c}{w} \right)^{1/2}$$

for a_c. Thus,

$$\pi a_c \sec \frac{\pi a_c}{w} = 22.0$$

is solved by iteration to yield a critical crack size of $a_c = 12.9$ in. (328 mm).

Although the geometry in this example was very simple, the same concepts and procedures are followed regardless of the complexity of the component or crack geometry.

Table 3.3 Typical Room Temperature K_{Ic} Data for Various Alloys

Material	Condition	σ_{ys} MPa	σ_{ys} ksi	K_{Ic} MPa $\sqrt{m}$	K_{Ic} ksi $\sqrt{in.}$	Minimum required thickness B mm	Minimum required thickness B in.
Steel							
Maraging steel							
300	900°F, 3 hr	1970	285	57	52	2.1	0.09
300	850°F, 3 hr	1670	242	93	85	7.8	0.31
250	900°F, 3 hr	1790	259	75	68	4.3	0.18
D 6 AC steel	Heat treated	1500	217	66	60	4.8	0.20
	Heat treated	1480	214	98	89	10.7	0.44
4340 steel	Forging	1480	214	56-88	51-80		
	Hardened	1830	265	47	43	1.7	0.07
A 533 B	Reactor steel	340	50	≈200	≈180	810	33
Carbon steel	Low strength	240	35	>220	>200	2150	82
Titanium							
6Al-4V	(α + β) STA	1100	160	38	35	3	0.12
13V-11Cr-3Al	STA	1130	164	27	25	1.5	0.07
6Al-2Sn-4Z-6Mo	(α + β) STA	1180	171	26	24	1.3	0.05
6Al-6V-2S	(α + β) STA	1080	157	37	34	3.0	0.12
4Al-4Mo-2Sn-0.5Si	(α + β) STA	940	137	70	64	13.6	0.55
Aluminum							
7075	T651	540	79	30	27	7.3	0.30
7079	T651	470	68	33	30	12.5	0.49
DTD 5024	Forged:						
	Longitudinal	500	72	40	36	15.9	0.65
	Short transverse	480	70	16	15	3.0	0.12
2014	T4	450	65	29	26	9.6	0.40
2024	T3	390	57	34	31	19.0	0.75
Plexiglass				1.6	1.5		

II. CYCLIC LOADS

A. Cyclic Stress and Fatigue Failures

Most structures and machine components are subjected to loads, and therefore stresses, that are not steady but vary significantly with time. The variations may be obvious, as in the case of rotating machinery, or subtle, as with traffic loads on bridges and highways. Although the magnitudes of the cyclic loads are inadequate to initiate direct failure of the components, repeated applications of these loads often lead to a different type of fracture called *fatigue*. According to ASTM E206, Standard Definition of Terms Relating to Fatigue Testing and the Statistical Analysis of Fatigue Data, fatigue is defined as "the process of progressive localized permanent structural change occurring in a material subjected to conditions which produce fluctuating stresses and strains at some point or points and which may culminate in cracks of complete fracture after a sufficient number of fluctuations." Fatigue failures have become progressively more prevalent as technology has developed more equipment, such as automobiles, aircraft and spacecraft, compressors and pumps, and turbines, which is subjected to repeated loading or vibration. This situation has led to the consensus that more than 90% of all mechanical failures are due to fatigue.

Certain standard terms used in describing cyclic loadings are illustrated in Fig. 3.21. Figure 3.21a illustrates a *completely reversed cycle of stress* of sinusoidal shape, such as those produced by standard rotating-beam fatigue-testing machines. For this type of stress cycle the maximum stress σ_{max} and minimum stress σ_{min} are numerically equal but of opposite sign. As with static testing discussed previously, tensile stress is

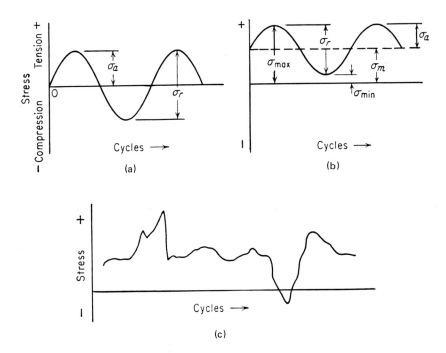

Fig. 3.21 Typical fatigue stress cycles. (a) Reversed stress; (b) repeated stress; (c) irregular or random stress cycle.

considered positive, and compressive stress is negative. Figure 3.21b illustrates a *repeated stress cycle,* in which σ_{max} and σ_{min} are not equal. In this example they are both positive (tensile), but a repeated stress cycle could just as well contain maximum and minimum stresses of opposite signs or both in compression. Figure 3.21c illustrates a complicated stress cycle that might be encountered in a component, such as an aircraft wing, subjected to variable service loads and unpredictable wind loads.

A fluctuating stress cycle may be considered to be made up of two constituents, a *mean* (steady) stress σ_m and an *alternating* (variable) stress σ_a. The *range of stress* σ_r is clearly equal to $2\sigma_a$, where

$$\sigma_r = \sigma_{max} - \sigma_{min}$$

and the mean stress is given by

$$\sigma_m = \frac{\sigma_{max} + \sigma_{min}}{2}$$

Another important quantity pertaining to cyclic stresses is the *stress ratio* R, given by

$$R = \frac{\sigma_{min}}{\sigma_{max}}$$

Fatigue failures are particularly dangerous since they usually occur without prior warning, unless careful nondestructive inspections for fatigue cracks are performed on a periodic basis. Fatigue-induced failures are usually brittle in appearance, with no gross plastic deformation in the region of the fracture surface. On a macroscopic scale, the fatigue fracture surfaces are usually perpendicular to the maximum applied tensile stress, even for materials that normally fail along planes parallel to the maximum shear stress. Fatigue loading thus serves to artificially embrittle the failure process and make impending failure more difficult to detect. A fatigue failure can usually be identified from the appearance of the fracture surface, which typically exhibits a smooth region associated with fatigue crack growth and a rough region corresponding to the failure of the remaining section during the final cycle of loading. Frequently the fatigue portion of the fracture surface exhibits *beach marks* that indicate the positions of the crack front during different portions of the cyclic life. Fatigue failures in structural components typically initiate at a point of stress concentration, such as a sharp corner, a notch or keyway, or other surface irregularities due to welding, machining, or other factors.

Three basic factors are necessary to cause fatigue failures: (1) a maximum tensile stress above a threshold value, (2) a sufficient variation or fluctuation in the applied stress, and (3) a sufficiently large number of applied load cycles. In addition, there may be many other variables, such as stress concentrations, corrosive environments, extreme temperatures, overloads, residual stresses, and structural imperfections, which tend to alter the conditions for initiating a fatigue failure.

At present there are two essentially independent approaches for analyzing and designing against fatigue failures. The first approach, which has been in use for many years, is based on a definition of the life of a smooth test specimen as the total number of cycles from the first load application until the specimen breaks. Many different specimens are tested at different cyclic load amplitudes and a relationship is obtained between the applied cyclic stress and the number of cycles to failure N. This relationship is usually plotted as a function of log N and is generally referred to as an S-N diagram. S-N diagrams have been obtained for virtually all metals and alloys of technological interest. Standard procedures have been established for correlating the life of an actual engineering structure with the S-N curve of the material employed. These procedures will be discussed in subsequent sections of this chapter. The second approach, which has been developed more recently, is based on distinguishing between the *crack initiation life* and the *crack growth life,* and subsequently adding them to obtain the total life of the component. Determination of the crack growth life is based on fracture mechanics concepts, and the resultant crack growth curve is considered to be a fundamental property of the material. The crack initiation life may also be a property of the material, but it is typically more strongly influenced by stress concentrations in the component or structure. In fact, the severity of initial stress concentrations are so great in many structures that the crack initiation life is ignored completely, and the entire structural life is assumed to be expended in fatigue crack growth. This approach to fatigue life assessment will also be presented in subsequent sections of this chapter.

B. Fatigue and Endurance Testing

The classic approach to fatigue testing is based on placing a small smooth specimen in rotating bending and either counting the number of cycles until it fails or terminating the test if a certain number of cycles is sustained (usually 10^7-10^8 cycles). Tests of this type, which result in fracture, are usually called fatigue tests, whereas those that do not cause fracture are called endurance tests. Typically a number of samples are tested under different loads and the results are summarized by plotting the maximum stress applied to each specimen against the number of cycles required to cause failure. Such a curve is called an S-N curve, and it is usually comprised of a straight line sloping downward until some point at which it curves toward the horizontal direction. The bend in the S-N curve may be sharp, as shown for the mild steel in Fig. 3.22, or it may occur gradually, as for the aluminum alloy.

Several test standards have been developed to guide in the performance of these fatigue tests and the presentation of the test results. These include three standards for definitions of terms, ASTM E206, ASTM E513, and ASTM E742, and four other stan-

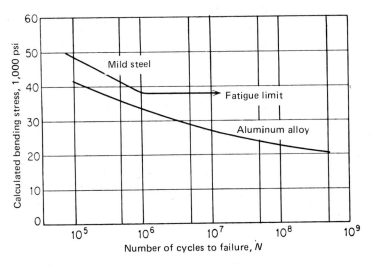

Fig. 3.22 Typical fatigue curves for ferrous and nonferrous metals.

dards controlling the test performance and data presentation, ASTM E 466, ASTM E467, ASTM E468, and ASTM E606. These standards should always be followed so the amount of uncertainty in the fatigue life data will be minimized.

1. Fatigue Strength

Any point on the S-N curve to the left of the bend of "knee," indicates a limiting stress called the *fatigue strength* associated with a selected cyclic life. For example, the fatigue strength of the mild steel in Fig. 3.22 corresponding to a life of 10^5 cycles is approximately 48 ksi.

2. Endurance Limit

As the applied loads are reduced, the metal will sustain greater numbers of cycles prior to fracture. The limiting stress below which the metal will sustain an indefinitely large number of cycles without fracturing is called the *endurance limit.* This stress corresponds to the horizontal asymptote of the S-N curve. The endurance limit can be established for most steels in about 10^7 cycles, but for many nonferrous metals, 5×10^8 cycles may be necessary since the knee in the S-N curve may be difficult to recognize or entirely absent as for the aluminum alloy data presented in Fig. 3.22. The S-N curves for several materials often used for engineering applications are presented in Fig. 3.23.

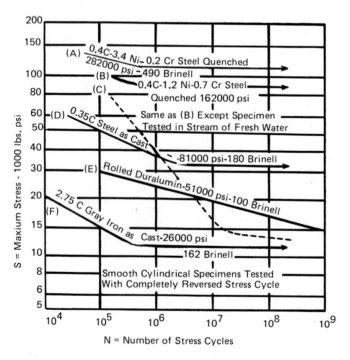

Fig. 3.23 S-N curves for some common metals. The values of stress on the curves are short-time tensile strengths. (Data from *The Fatigue of Metals,* by H. F. Moore and J. B. Kommers.)

3. Fatigue Notch Sensitivity

The effect of notches on the fatigue strength of materials is determined by comparing the S-N curves of notched and unnotched test specimens. The data for the notched specimens are usually plotted in terms of the nominal net section stress (stress concentration effect not taken into account). The effectiveness of the notch in decreasing the fatigue life is expressed by the *fatigue-strength reduction factor* or *fatigue-notch factor* K_f. This factor is simply the ratio of the fatigue strength of unnotched specimens to the fatigue strength of notched specimens. Values of K_f have been found to vary with the severity of the notch, the type of notch, the material, the type of loading, and the applied stress level. The K_f values published in the literature are subject to considerable scatter and should always be examined carefully for their limitations or restrictions. However, two general trends applicable to completely reversed loading have been observed: (1) K_f is usually less than K_t, and (2) the ratio K_f/K_t decreases as K_t increases. These observations have been confirmed by experimental results.

The fatigue notch sensitivity of a material is usually expressed in terms of the *notch sensitivity index* q, where

$$q = \frac{K_f - 1}{K_t - 1}$$

This form of the expression for q was selected so that a material that experiences no reduction in fatigue strength due to a notch has an index of zero, and a material that exhibits the full theoretical notch effect has an index of unity. However, q is not a true material constant since it varies with the severity and type of notch, the size of specimen, and the type of loading. Figure 3.24 presents q values for a family of steels and aluminum alloys subjected to reversed bending or reversed axial loads.

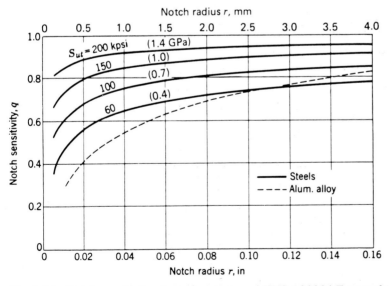

Fig. 3.24 Notch-sensitivity charts for steels and UNS A92024-T wrought aluminum alloys subjected to reversed bending or reversed axial loads. For larger notch radii use the values of q corresponding to r = 0.16 in. (4 mm). [Reproduced by permission from G. Sines and J. L. Waisman (eds.), *Metal Fatigue*, McGraw-Hill, New York, 1969, pp. 296,298.]

C. Statistical Nature of Fatigue

Although all material properties discussed in this chapter exhibit statistical variations, the dispersion in the fatigue properties is so much greater that statistical methods have been traditionally applied only to them. Since the fatigue life and endurance limit in particular are statistical quantities, considerable deviation should be expected from an average curve obtained from only a few specimens. For example, it is necessary to think in terms of the probability of a specimen attaining a certain life at a given stress or the probability of failure at a given stress close to the endurance limit. Since the fatigue life and endurance limit are statistical quantities, determination of an S-N curve requires 30-50 specimens rather than the 10 or so that would be required if the quantities were deterministic. The basic method for presenting fatigue data should thus be plotted as a three-dimensional surface representing the relationship among stress, number of cycles to failure, and probability of failure. Figure 3.25 shows how this can be presented in a two-dimensional plot, where several curves of constant failure probability are drawn. Thus, at σ_1, 1% of the specimens would be expected to fail at N_1 cycles, 50% at N_2, and so on. This figure indicates that the dispersion decreases with increasing stress, and this is usually found to be the case for actual test results. The distribution function for the endurance limit is not generally well known because of the length of time required for each test and the large number of tests required. Figure 3.26 shows the actual dispersion of test results for 39 tests on an ASTM A36 steel.

D. Applications of Fatigue Data

Since the fatigue data presented in the previous section are based on small, smooth specimens, several adjustments to these data must be made before they can be applied. The first step in making these adjustments is to calculate an endurance limit for the actual application S_e from the test specimen endurance limit S'_e, as

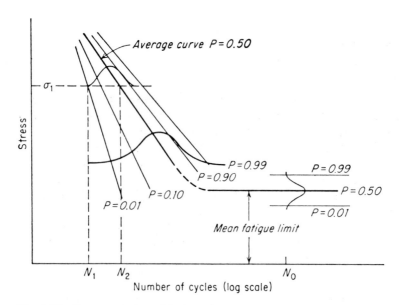

Fig. 3.25 Representation of fatigue data on a probability basis.

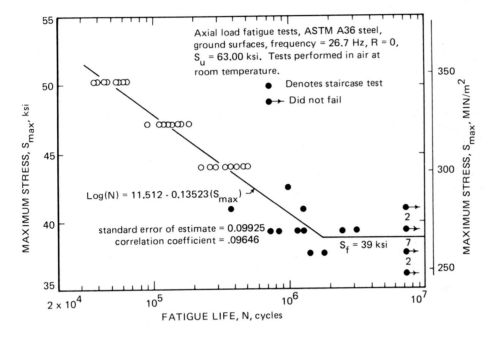

Fig. 3.26 S-N diagram for ASTM A36 steel.

$$S_e = k_a k_b k_c k_d k_e k_f S_e'$$

where

> k_a = surface factor
> k_b = size factor
> k_c = reliability factor
> k_d = temperature factor
> k_e = modifying factor for stress concentration
> k_f = miscellaneous effects factor

In order to prevent over- or underdesigning a component, it is necessary to make a careful selection of these factors. The following shows how each of them may be evaluated, although this material is included for illustrative purposes only. For actual applications, appropriate literature and test data should be used in their determination.

The surface factor is included to quantify the influence of different surface finishes employed on actual components. A series of surface finish modification factor curves for steel is shown in Fig. 3.27, in which k_a varies from 1.0 for polished components to approximately 0.2 for high yield strength forged steel.

Since smaller components are generally stronger than larger components made of the same material, a size factor is included. Although proper determination of the size effect may be complicated, a simple rule of thumb has been given as

$$k_b = \begin{cases} 1 & d \leqslant 0.30 \text{ in.} \\ 0.85 & 0.30 < d \leqslant 2.0 \text{ in.} \\ 0.75 & d > 2.0 \text{ in.} \end{cases}$$

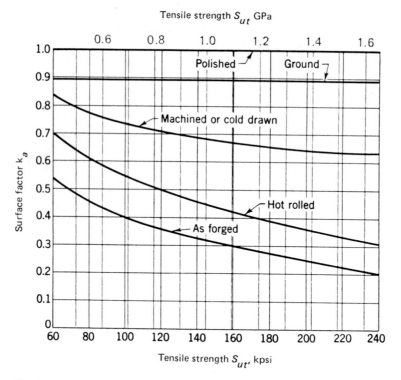

Fig. 3.27 Surface-finish modification factors for steel.

where d refers to the section depth of the component. As a first approximation, these k_f values may be suitable for components subjected to cyclic bending, torsion, or axial loads.

The reliability factor quantifies the reduction in component strength required to assure increasing probabilities of a successful design application. For a reliability of 50%, $k_c = 1.0$, and for greater reliabilities k_c has correspondingly lower values. Since tests have shown that the standard deviation of the endurance limit does not exceed 8% for a number of materials, the following expression for k_c has been proposed:

$$k_c = 1 - 0.08 Z_R$$

where Z_R is the standardized variable corresponding to the desired reliability R. Table 3.4 gives reliability factors corresponding to different desired reliabilities.

The temperature factor is included to account for the loss of strength due to high-temperature applications and may also be applied to the yield or ultimate strength. For steels the temperature factor

$$k_d = 1.0 \quad T < 160°F$$

$$k_d = \frac{620}{460 + T} \quad T > 160°F$$

may be used as a first estimate.

Table 3.4 Values of the Standardized Variable Z_R Corresponding to an 8% Standard Deviation of the Endurance Limit

Reliability R	Standardized variable Z_R	Reliability factor k_c
0.50	0	1.000
0.90	1.288	0.897
0.95	1.645	0.868
0.99	2.326	0.814
0.999	3.091	0.753
0.999 9	3.719	0.702
0.999 99	4.265	0.659
0.999 999	4.753	0.620

The stress concentration factor k_e is obtained from the fatigue strength reduction factor K_f as

$$k_e = \frac{1}{K_f}$$

where $K_f = 1 + q(K_t - 1)$. Values for q have already been presented in Fig. 3.25, and K_t values are obtained from tables or an appropriate stress analysis.

The miscellaneous effects factor is included partly to account for the influence of other factors, such as corrosions and corrosion fatigue, residual stress, and plating, on the endurance limit and partly to call attention to the importance of such effects. The design engineer should consider the effects of any such factors on the endurance limit and quantify them as well as possible for inclusion in the miscellaneous effects factor.

1. Corrosion and Corrosion Fatigue

An aggressive environment may influence a material's static properties (corrosion or stress corrosion) as well as its fatigue properties (corrosion fatigue). Except for the most aggressive environments, such as acids or highly corrosive salt solutions, the effects of environment on static properties tend to be relatively easy to diagnose and minimize. However, the environment may have a major impact on cyclic properties, such as the fatigue strength and endurance limit. In fact, environments such as brine or brine spray can cause the endurance limit of many steels, stainless steels, aluminum, and magnesium alloys to disappear completely. The influence appears to be much greater on the crack initiation portion of the fatigue life than on the crack growth portion. For this reason, the influence on the endurance limit is the most obvious evidence of the environmental effects. Because of the considerable variability in the environmental effects, resulting from the variety of material and environmental combinations available, no attempt will be made in this chapter to quantify the environmental effects. However, the extensive specialized literature that is available may be consulted to obtain estimates of such effects. The wide variety of standards to study corrosion effects is indicated by the number of ASTM standards (more than 40) related to corrosion and corrosion testing (see Chap. 4).

2. Residual Stress

Residual stresses may either improve the endurance limit or affect it adversely. Generally, if the residual stress on the surface of the component is compressive, the endurance limit will be increased. Such operations as shot peening, hammering, and cold rolling build compressive stresses into the surface layer and improve the endurance limit significantly. By contrast, residual tensile stresses tend to reduce the endurance limit. The difficulties in quantifying the residual stress effects preclude the presentation of numerical factors in this section. (See Chap. 21 on the effects of hot and cold working.)

   It should again be emphasized that, because of the brevity of this presentation, the endurance limit reduction factors presented above should be used with a good deal of caution. Whenever possible, prototype components should be tested under conditions as close as possible to the expected service conditions. If this procedure is not possible, the available specialized literature and design procedures should be used as a basis of determining the fatigue strength and endurance limit. It is often the case that certain aspects of the problem will be known in detail although very limited data are available about its other aspects. In such cases, the procedures outlined here may be employed for some of the factors and the better data used for the remainder.

   After the endurance limit (or fatigue strength corresponding to a desired cyclic life) has been obtained for a particular application, a number of different procedures are available to complete the design process. One of the more widely used approaches is to construct a *modified Goodman diagram,* which requires additional knowledge of the yield strength S_y and ultimate strength S_u of the candidate materials. Figure 3.28 shows a typical modified Goodman diagram, which uses S_u, S_y, and S_e (or S_f) in its construction. The mean and alternating components of the applied stresses σ_m and σ_a are then plotted as illustrated. All combinations of stresses that fall within the diagram would be sustained and those falling outside would be expected to cause failure of the component. When the mean stress is compressive, failure is defined by the two heavy parallel lines originating at $+S_e$ and $-S_e$ and drawn downward and to the left. When the mean stress is tensile, failure is defined either by the maximum stress line or the yield strength, depending on the relative values of σ_m and σ_a acting on the component.

E. Fatigue Crack Initiation

As mentioned in Sec. II.A, the primary emphasis of most present fatigue treatments is to consider crack initiation and growth as basically separate processes that together constitute the fatigue life of a specimen or component. The process of fatigue crack initiation in metals and alloys is caused by the accumulation of damage during each cycle of loading. For a material in which the response is perfectly elastic, the stress-strain diagram in completely reversed cyclic loading would be a straight line as shown in Fig. 3.29a. However, whenever the applied cyclic loads are large enough that some inelastic deformation occurs, the stress-strain diagram forms a closed loop, as shown in Fig. 3.29b. The area of the loop represents the irreversible work of deformation per unit volume per cycle and is called *hysteresis.*

   When the cyclic load amplitude remains constant, changes in the width of the hysteresis loop indicate whether the material response is stable or unstable. Whenever the width of the hysteresis loop remains constant or decreases, the damage accumulation processes would not be expected to cause failure of the component. This corresponds to fatigue cycling *below* the endurance limit. However, if the width of the hysteresis loop

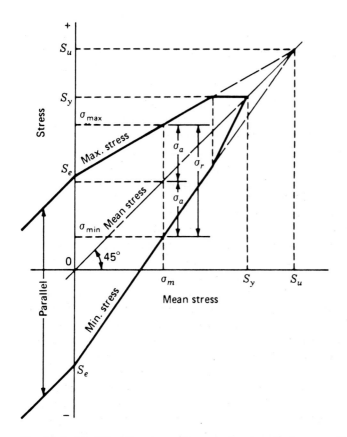

Fig. 3.28 Modified Goodman diagram showing all the strengths and the limiting values of all stress components for a particular mean stress.

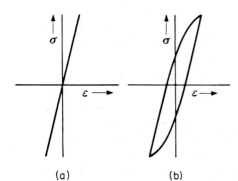

Fig. 3.29 Perfect elasticity (a) and hysteresis loop (b) exhibited by a material cyically stressed in the inelastic range.

is continuously increasing or if it starts increasing after a period of stable response, failure of the material would be expected. This corresponds to cyclically loading a test specimen or component *above* its endurance limit.

1. Hysteresis and Damping Capacity

The ability of a vibrating solid to dissipate the mechanical energy of vibration is called its *damping capacity*. If the hysteresis of the metal is large, considerable energy is absorbed during each cycle and the vibration will rapidly die out. Alternatively, if the hysteresis loop is narrow, very limited energy is absorbed in each cycle and the vibration persists over a much longer period of time.

The damping capacity is normally defined in terms of the logarithmic decrement δ:

$$\delta = \ln \frac{A_n}{A_{n+1}}$$

where A_n is the amplitude of the nth free fibration and A_{n+1} is the next successive smaller amplitude.

The damping capacity of a material is evidenced by the speed at which the sound made by a struck metal object diminishes with time. For example, a tuning fork made of a high-strength steel alloy will vibrate for a long time, but a soft piece of lead will stop vibrating almost immediately after being struck. The engine blocks of automobiles and the bases of rotating mechinery are made of high damping capacity materials to minimize the transmission of noise and vibration. On the other hand, bells are made of materials having low damping capacities.

The initiation of damage in cyclically stressed metals or alloys nearly always occurs at a free surface, although the damage mechanisms for soft, ductile materials differ considerably from those in higher strength structural metals. The emphasis of most initial efforts to identify crack initiation mechanisms was placed on pure metals, such as copper and aluminum. These metals are often called *plastic metals* since plastic deformation initiates at extremely low levels of applied stress or strain. As a result of this research, the mechanisms causing fatigue crack initition in plastic materials are reasonably well understood. However, most metallic materials employed for engineering applications are medium- to high-strength alloys having complex microstructures. These materials exhibit initially linear stress-strain curves and well-defined yield points, characteristics that lead to their being called *elastic-plastic* metals. Although many fatigue studies have been performed on these materials, most of them have not addressed the basic questions of identifying damage accumulation mechanisms. The microstructures of the elastic-plastic metals are also more complex and varied than those of the plastic metals, and as a result, there is considerably less certainty as to the basic mechanisms of fatigue crack initiation in these metals.

2. Crack Initiation in Plastic Metals

Since the plastic metals are very soft in the annealed condition, the damage processes initiated by fatigue cycling result in a significant amount of strain hardening. The extent of strain hardening is greatest during the first few cycles and is, therefore, often called the rapid hardening stage. The first few cycles of the stress-plastic strain response of a plastic metal subjected to cycling between ϵ_p and $-\epsilon_p$ is shown in Fig. 3.30. The difference in stress at the end points of adjacent loops decreases rapidly and disappears after

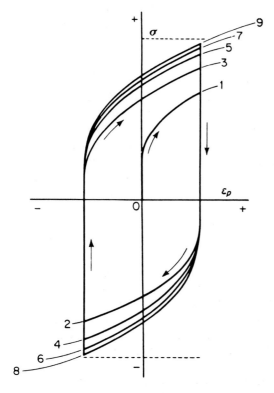

Fig. 3.30 Saturation hardening of an annealed metal cycled between fixed positive and negative strains. Dashed horizontal lines indicate the saturation stress range (schematic).

a limited number of cycles. This condition is called *saturation hardening,* and its level increases with cyclic plastic strain ϵ_p. At small alternating strains the saturation stage is approached slowly, but at large strains the steady-state condition is attained after only a few cycles.

Microscopic studies of specimens during the rapid hardening stage have shown that fine slip bands form on the metal surface in the first few cycles of loading. If the surface layer of metal containing the fine slip is removed by electropolishing before completion of the rapid hardening stage, a second rapid hardening stage similar to the first is observed. Repeated removal of metal by this procedure can significantly extend the life of the specimen. However, it is necessary to polish the surface layer before persistent slip bands occur, since they cause subsequent slip to become permanently localized in the region of these bands. The localization of plastic strain in the persistent slip bands results in the surface of the specimen becoming severely notched at the bands.

Experimental studies have shown that the persistent slip bands consist of a combination of extrusions and intrusions on the metal surface, similar to the shearing of a deck of cards back and forth, with some surfaces having less resistance to slip than others. After further cycling, microcracks will develop at the locations of the persistent slip bands. These microcracks are oriented along the persistent slip bands, in the direction of the maximum shearing stress, and are called stage I cracks. Although it is possible for stage I cracks to grow under cyclic loading until the specimen breaks, the more usual occurrence is for the stage I cracks to change direction to become normal to the direction of maximum tensile stress. These cracks are called stage II cracks and grow under cyclic loading until the specimen breaks.

3. Crack Initiation in Elastic-Plastic Metals

The microscopic changes that occur in elastic-plastic metals in response to cyclic loading are different from those in the plastic metals. In plastic metals uniform slip occurs throughout the specimen during the rapid hardening stage, followed by the formation of numerous persistent slip bands during the saturation hardening stage. In elastic-plastic metals, uniform slip does not occur, and the surface plastic strains that lead to cracking are extremely localized. Hence, the changes in surface appearance of these materials are highly localized, with certain regions exhibiting extensive change while adjacent areas appear to be unaffected. In addition, as the total amount of slip deformation decreases, a smaller percentage of the slip bands that form reach the cracking stage and fewer of the cracks grow beyond the grains in which they are initiated. Thus, although the alloying additions increase the strength, they cause fatigue to be an increasingly localized phenomenon, with this localized weakness determining the fatigue strength. In these materials, nonmetallic inclusions and intermetallic compounds often serve as sites for fatigue crack initiation.

Because of work hardening, slip does not occur under cyclic conditions in a reversible manner on the same planes. If the application of a tensile load causes slip to occur on a particular series of planes, slip in the reversed direction will occur on a different series of planes when the load is reversed. This implies that fatigue cracking is due to geometric surface effects caused by extrusions and intrusions and is not due to some internal damage process along a particular slip plane.

Depending on the irregularities in a component, either metallurgical (nonmetallic inclusions or intermetallic particles) or configurational (notches, keyways, and other stress raisers), cracks may initiate along planes of maximum shear stress (stage I) or perpendicular to planes of maximum tensile stress (stage II). However, in most cases the stage I crack growth automatically converts to stage II crack growth before the stage I cracks become very large. In most cases only one stage II crack will form, and all subsequent fatigue damage is then localized at this crack.

F. Fatigue Crack Growth

The appearance of an identifiable stage II crack in a material indicates that different damage accumulation processes are taking place in which all the damage accumulation mechanisms are concentrated at the crack tip and lead to increments of crack growth for each loading cycle. The microscopic appearance of stage II fatigue fracture surfaces shows regular markings called *fatigue striations* that indicate each stationary position of the crack front as well as the amount of crack growth during each loading cycle. A number of theories have been proposed to explain the creation of these striations, with most of the theories based on the existence of reversed crack-tip plasticity along planes of maximum shear stress.

Since the deformation processes are localized at the crack tip and there is an identifiable crack length parameter, it is natural to apply fracture mechanics concepts to stage II fatigue crack growth. Considerable research has been directed toward this matter since around 1960, and the results have been quite successful. In the analysis of fatigue crack growth data, the fatigue cycle is usually defined by $\Delta K = (K_{max} - K_{min})$, where K_{max} and K_{min} represent the maximum and minimum values of the stress intensity factor during the loading cycle. It has been shown experimentally that ΔK rather than K_{max} has the dominant influence on fatigue crack growth and that, if ΔK is constant, the fatigue crack growth rate is also constant. The use of the stress intensity factor has also proved valuable for correlating fatigue crack growth data for different geometries, such as plate bending and internally pressurized thin-walled cylinders. The overall shape of the relation between the fatigue crack growth per cycle da/dN and ΔK is sigmoidal, with a power-law region connecting the upper and lower thresholds. The general form of this curve is shown in Fig. 3.31 and is considered a basic property of the material. The power-law relationship was proposed by Paris in the form

$$\frac{da}{dN} = C(\Delta K)^m$$

where the material parameters are selected to provide the best fit with the experimental data. The lower threshold is indicative of nonpropagating fatigue crack conditions; the upper threshold describes the accelerated crack growth associated with the onset of unstable fracture.

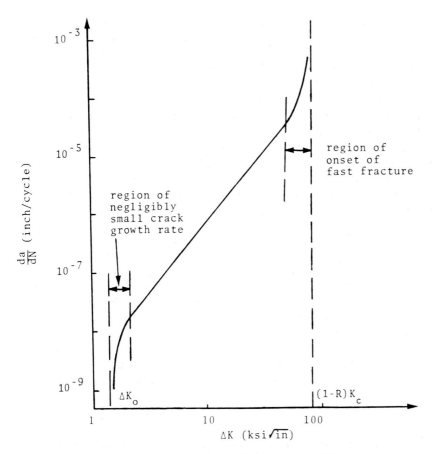

Fig. 3.31 Illustration of fatigue crack growth rate versus ΔK.

For most engineering purposes, the conditions described by the power-law region of the curve are of greatest interest since the crack growth rates in that region coincide with the growth rates encountered in many structural applications. However, a significant number of applications involve conditions for which the Paris crack growth law is inadequate. One of these is the influence of load ratio $R = K_{min}/K_{max}$ on the fatigue crack growth rate, and the other is the consideration of fatigue crack growth in the initiation threshold region. Both of these conditions have been incorporated into a fatigue crack growth law proposed by Forman, Kearney, and Engle:

$$\frac{da}{dN} = \frac{C_1(\Delta K)^m}{(1 - R)K_c - \Delta K}$$

where K_c represents the critical value of the stress intensity factor for actual conditions encountered in each application. The lower threshold values are often denoted ΔKth and range between 6 and 8 ksi $\sqrt{in.}$ (6.6-8.8 MPa $\sqrt{m}$) for steels, 2 and 6 ksi $\sqrt{in.}$ (2.2-6.6 MPa $\sqrt{m}$) for titanium alloys, and 1 and 3 ksi $\sqrt{in.}$ (1.1-3.3 MPa $\sqrt{m}$) for aluminum alloys. A typical da/dN versus ΔK curve is shown in Fig. 3.32 for types 304 and 316 stainless steel.

It should be reemphasized that, since K is geometry independent, the da/dN versus ΔK curve is also considered a basic property of a material. However, since K_{Ic} is a function of such factors as temperature, heat treatment, and orientation, it is evident that such factors will also influence the fatigue crack growth rate curve. In addition, such factors as test frequency, load ratio, periodic overloads, and environment may also have significant influence on the fatigue crack growth rates. In order to take many of these factors into account, many additional fatigue crack growth laws have been proposed but the discussion of them is beyond the scope of this chapter. Since scatter in the crack growth data also is somewhat greater than for the K_{Ic} values, additional test standards, such as ASTM E647, Standard Method for Constant-Load-Amplitude Fatigue Crack Growth Rates Above 10^{-8} m/cycle, have been developed to minimize any variations due to test procedures.

The *fatigue life* of many components and structures can be determined by combining nondestructive inspection (NDI) methods with appropriate fatigue crack growth data. In applying this method, it is normally assumed that flaws equal to the minimum detectable flaw size determined by NDI methods, a_d, exist when the component is placed into service. It is also necessary to know the material properties, especially the fracture toughness K_{Ic} and the loads applied to the component so that the stress intensity factor K_I and the critical crack size a_c may be established. Data from the fatigue crack growth rate curve and the loading history are then integrated to obtain a curve of crack size as a function of either the cycles of loading or the elapsed time, as illustrated in Fig. 3.33. The minimum detectable flaw size and the critical crack size define the limits to the number of loading cycles or service hours during which the crack may be detected.

Inspection intervals should be selected such that the structure or component will be inspected more than once during the detection period, since a crack of the minimum detectable size may just escape detection during one inspection. Since reliable inspections of structures are time consuming and expensive, it is naturally desirable to make the detection period as large as possible. The shape of the crack size curve in Fig. 3.33 illustrates that decreases in a_d by a factor of 2 increase the detection period considerably more than would be obtained by doubling a_c. Decreases in a_d can only be made through improvements in NDI techniques for a particular application. Increases in a_c can be

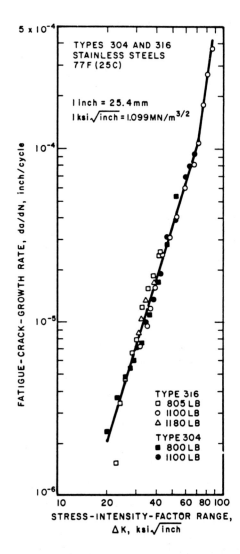

Fig. 3.32 Fatigue crack growth rates in type 304 and 316 stainless steels at room temperature as a function of range of stress intensity factor.

attained by retrofitting critical components with others manufactured from materials having higher fracture toughness values. Both types of improvements should be made if they lead to significant increases in the detection period and, therefore, in the useful life of the component.

G. Variable Amplitude Fatigue Loading

Although specimens tested in conventional fatigue machines are normally subjected to a cyclic stress of constant amplitude throughout their lives, most structural members and machine components are subjected to service loadings of varying amplitude. The variation in stress level may follow a regular pattern, which is often called block loading, or a random pattern as illustrated in Fig. 3.21c. In order to be absolutely certain that structures or components will never fail in fatigue, a designer must ensure that cyclic

Crack size

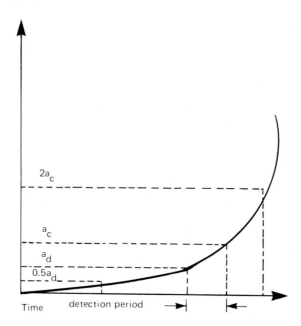

Fig. 3.33 Illustration of the relative advantage of increasing a_c and decreasing a_d on crack detection period.

stresses will never exceed the endurance limit anywhere in the body. However, this procedure will lead to overconservative designs, particularly if high service loads occur only very rarely. In order to design a component for a finite life when it is subjected to stress cycles of varying amplitude, an acceptable method of predicting life under such conditions is necessary. The investigation of fatigue under varying stress amplitudes has become known as the study of *cumulative damage,* because of the early interest in how fatigue "damage" at various stress levels is accumulated. A fatigue test in which the stress amplitude is varied in some manner is often called a cumulative damage test.

Many attempts have been made to predict the life of a specimen in a cumulative damage test, and extensive data have been generated for the purpose of establishing appropriate empirical relationships and proving or disproving the theoretical predictions. In 1924, Palmgren assumed that damage was accumulated linearly with the number of revolutions applied to ball bearings. Miner, in 1945, suggested that, in a fatigue test, damage to a specimen at a particular stress level could be considered to accumulate linearly with the number of stress cycles, with failure occurring when the accumulated damage attained some critical value.

If a specimen has a life of N_1 cycles when subjected to a cyclic stress of $\pm\sigma_1$, the portion of the life consumed after n_1 cycles will be n_1/N_1. If the same specimen is then subjected to a stress of $\pm\sigma_2$ for n_2 cycles, the total damage would be

$$D' = \frac{n_1}{N_1} + \frac{n_2}{N_2}$$

where N_2 is the cyclic life due to loading at $\pm\sigma_2$ only. This process may be continued at a number of different cyclic stress levels until the specimen fails. The Palmgren-Miner rule then prescribes that failure will occur when the sum of the portions of life consumed is equal to unity or

$$\sum_i \frac{n_i}{N_i} = 1$$

If the crack initiation and growth portions of the fatigue life can be separately identified, it is possible to rewrite the Palmgren-Miner rule in the form

$$\sum_i \frac{n_i'}{N_i'} = 1 \quad \text{and} \quad \sum_i \frac{n_i''}{N_i''} = 1$$

In these equations, n_i' and n_i'' are numbers of cycles expended in the crack nucleation and crack growth periods, respectively, and N_i' and N_i'' represent the corresponding crack nucleation and propagation lives.

The Palmgren-Miner rule is also referred to as the linear cumulative damage rule since it is assumed that the damage in a constant amplitude fatigue test is a linear function of the number of cycles. However, it has since been shown that linear damage accumulation is not necessary for the Palmgren-Miner rule to hold, although it is necessary that the material be insensitive to the sequence in which the cyclic loads are applied. As a general rule, load sequence effects are not insignificant, so great care must be exercised whenever the Palmgren-Miner rule is employed to predict the lives of structures or components.

H. Wear and Fretting

When two pieces of material are pressed together by an external static load and then subjected to relative parallel displacements, wear of the mating surfaces occurs. The amount of wear that occurs depends on the nature of the contacting materials, the sliding, rolling, or impact motion between them, the load imposed, the amount of lubrication between the surfaces, the chemical action between the surfaces, and the environment. Because of the variety of different kinds of wear mechanisms, no standards have been developed for wear tests on metals, although some standards have been developed for the testing of erosion and abrasion resistance. The tests performed in practice are designed to duplicate as nearly as possible the conditions expected to be encountered in service.

In components that have direct metal-to-metal contact, wear may be the result of one or more of the following mechanisms.

1. Tearing of particles from the surface through friction, often called *abrasive wear*.
2. The transfer of contact protuberances between members in sliding contact, often called *adhesive wear*.
3. The production of pits leading to cracks, spalling, and other surface damage.
4. The local transfer of relatively large amounts of material from one surface to another, often called *galling*. In many cases after galling has occurred between two mating surfaces, further relative motion leads to actual *seizure* between the surfaces. Seizure is a form of welding between the material of one surface and the same material already transferred to the other surface by galling.

The results of a number of experiments into the wearing characteristics of various metals has led to several conclusions that can be useful in design.

1. The wear rate (volume of wear per sliding distance) is proportional to the normal contact force.
2. The wear rate is independent of the nominal area of contact.
3. The wear rate is independent of the relative sliding speed.

If the magnitude of the relative tangential displacements are small and cyclical, the wear is called *fretting*. Fretting occurs by contacting asperities on the mating surfaces continually welding together and then breaking loose. The small fragments of metal that break off tend to oxidize, forming oxide particles that, for most engineering metals, are considerably harder than the metal itself. These particles become trapped between the mating surfaces and cause abrasive wear and scoring. In the case of steels, the oxide debris is reddish in color, but for aluminum and magnesium alloys it is black. Briefly, the characteristics of fretting are as follows.

1. Although fretting may occur in the presence of an inert gas, it is more serious when oxygen is present and the corrosion debris consists mainly of oxide particles.
2. The amount of fretting is greatest under dry conditions (no lubrication).
3. Fretting damage increases with contact load, slip amplitude, and number of oscillations.
4. Soft materials are more susceptible to fretting than harder materials of a similar type.
5. Lubricants reduce fretting damage, with solid lubricants such as molybdenum disulfide being particularly effective.

There are no completely satisfactory methods of preventing fretting. If all relative motion is eliminated, fretting will not occur. However, increasing the normal force on the surfaces increases the fretting if the relative motion is not completely eliminated. If the relative motion cannot be completely eliminated, then the reduction of friction by the use of lubricants should be attempted.

III. IMPACT LOADS

In an engineering sense, impact loading refers to any rapid application of load to a test specimen, structural member, or mechanical component. In impact testing most loads are in fact applied by the impact of some striking object on the specimen or component. Common impact test devices employ either a swinging pendulum or a drop weight to apply the impact loads. Strain rates typically associated with impact testing are in the range 10^3 in./in. per second.

Normal practice in impact testing is to utilize notches as crack initiators and apply loads of such magnitude that the specimens fracture upon the first load application. One exception to this practice is the impact testing using the split Hopkinson bar apparatus, which is used primarily to study elastic wave propagation in solids and the effect of dynamic loads on their elastic properties.

A. Brittle-Ductile Transition

It has long been known that a combination of three factors can cause many materials, which normally fracture in a ductile fashion (as discussed in Sec. I.F), to fail in brittle fracture. These three factors are

    1. A high rate of loading, such as in impact testing
    2. The existence of a stress concentrator, such as a notch or a crack
    3. Low testing or service temperatures

These factors have been identified by extensive research that has been performed to determine the cause of brittle fractures in many applications where the material would have been expected to fail in a ductile fashion. A striking example of this was the many failures of World War II Liberty Ships, which represented the first large-scale application of welding to fabricate ship structures. Normal laboratory materials testing procedures employed at that time, primarily quasi-static tensile tests and Charpy keyhole impact tests (see Sec. III.B), suggested that the brittle-ductile transition occurred in these ship steels at temperatures in the range -20 to 0°F. However, at service temperatures in the range 20-40°F many brittle fractures occurred, with eight ships having broken in two and four others abandoned after fracture occurred. These fractures were typically caused by fatigue cracks that initiated from regions fabricated with high stress concentrations, including hatch corners, flame cutting, and welding. Other Liberty Ships that were fabricated using the same methods and materials, but which were not subjected to low service temperatures, did not experience similar failures.

    One of the primary purposes of developing the various impact tests that are in use today was to simulate the worst-case conditions that would be expected to cause brittle fracture in service. By following such procedures, it is possible to identify the tendencies of otherwise ductile materials to fail in brittle fracture and to permit the utilization of materials that would not be expected to fail in brittle fracture under service conditions. This may be accomplished by determining the brittle-ductile temperature transition curves of candidate materials and then excluding their use at temperatures where they would be expected to fail in a brittle fashion. Although the brittle-ductile transition is very pronounced in many materials, such as ferritic steels, it does not exist at all in other materials, such as aluminum alloys.

B. Notched-Bar Impact Testing

A large number of notched-bar test specimens of different designs have been employed in investigations of the brittle fracture of materials. The most widely used specimen is a *Charpy specimen,* which has a square cross section and either a V-shaped or keyhole-shaped starter notch in the center. It is placed like a simply supported beam in the testing machine and struck at the midspan of the beam on the side opposite from the notch, as shown in Fig. 3.34a. The Izod specimen may be round or square and has a notch near one end, as shown in Fig. 3.34c. It is clamped like a cantilever beam and struck near the free end on the same side as the notch. In each case the notch tip is loaded in triaxial tensile stress, which strongly promotes brittle fractures.

    The amount of energy required to fracture the specimen is usually determined by the difference in potential energy of the pendulum before and after impact and is usually

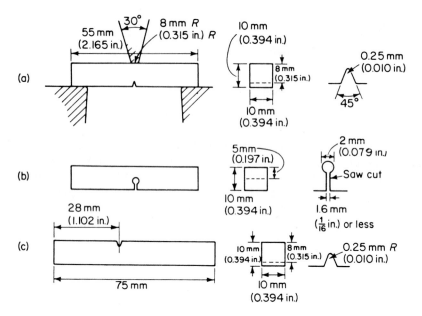

Fig. 3.34 Geometry of the Charpy V (a), Charpy keyhole (b), and Izod (c) test specimens.

referred to as the *impact energy*. The impact energy is normally recorded directly in foot-pounds by a mechanical pointer on the testing machine, although instrumented impact testers have recently been developed that record both impact force and energy as a function of time.

For more than 65 years, standards and code organizations such as ASTM and ASME have specified impact energy requirements for ferrous materials on the basis of energy absorbed in the Charpy keyhole test. More recent studies have shown that the Charpy V-notch results correlate better with actual structural failures. Therefore, the major emphasis in recent testing among these specimen types has been placed on the Charpy V-notch specimen.

Comparisons between the Charpy V-notch and Charpy keyhole results of a series of tests on a mild steel are presented in Fig. 3.35. Also included is the percentage of shear fracture and the lateral contraction, which are alternate measures of ductile fracture. These results show that the brittle-ductile transition occurs at a higher temperature for the Charpy V-notch results than for the Charpy keyhole data. Therefore, the Charpy key-hole data are less conservative and their use would not be expected to correlate as well with the actual service behavior.

Since it is difficult to use impact energy values quantitatively in actual design applications, their principal use has been to exclude materials from service when the temperatures would lead to Charpy V-notch energies of less than 15 ft-lb. However, it has also been suggested that this approach is too simplistic and that the minimum acceptable energy level should be a function of the yield strength of the material. In comparison with the fracture mechanics approach, the Charpy impact energies are much easier and cheaper to obtain, but their application is much less precise than the plane strain fracture toughness.

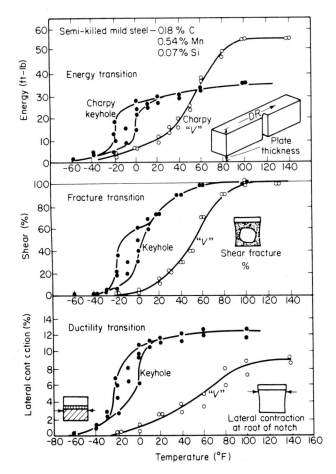

Fig. 3.35 Three methods for measuring the transition temperatures of Charpy V and keyhole specimens. (From W. S. Pellini in *Symposium on Effect of Temperature on Brittle Behavior of Metals,* ASTM STP 158, 1954.)

Several ASTM standards have been developed for the performance of Charpy and Izod impact tests. These include E23, Standard Methods for Notched Bar Impact Testing of Metallic Materials, and A370, Standard Methods and Definitions for Mechanical Testing of Steel Products.

In addition to the Charpy and Izod testing already discussed, there have been at least two additional notched-bar impact test methods that have been used widely enough to be standardized by ASTM. These are the *drop weight tear test* (DWTT) and the *dynamic tear* (DT) test, where the DWTT is used to identify the brittle-ductile transition temperature for ferritic steels and the DT test provides a fracture parameter that can be correlated with K_{Ic}.

The drop weight tear test is similar to the Charpy and Izod tests in that the primary information obtained is the breaking energy in ft-lb or joules and the fracture surface appearance (the relative amounts of brittle versus ductile fracture). The test specimen is considerably larger than the Charpy, 3 by 12 in. with full plate thickness, and it employs a pressed starter notch 0.20 in. deep. Because of the greater specimen size and the severe cold working caused by the pressed notch, the temperature transition curve occurs considerably more abruptly than in the Charpy or Izod tests. The transition in fracture surface appearance is even more abrupt, with the percentage of average shear typically decreasing from 100% to less than 20% within 20°F. The DWTT was developed primarily for ferritic steels and the testing procedure has become standardized in ASTM E436, Standard Method for Drop Weight Tear Tests for Ferritic Steels. In spite of the name, many tests are performed on a pendulum rather than a drop weight testing facility.

The dynamic tear test involves a single-edge notched beam that is impact loaded in three-point bending, and the total energy loss during fracture is recorded. The DT specimen is intermediate is size between the Charpy and the DWTT specimens, measuring 1.60 by 7.125 in. and 0.625 in. thick. The notch is prepared by machining and then pressing to obtain a repeatable profile. Tests may be performed using single-pendulum, double-pendulum, or vertical drop-weight machines. Testing procedures have also been standardized in ASTM E604, Standard Test Method for Dynamic Tear Energy of Metallic Materials. The primary emphasis of DT testing is to determine fracture energies, rather than transition temperatures, so it is applicable to a wide range of structural alloys.

Considerable emphasis in the development of DT testing has been placed on developing correlations between the DT energy and the plane strain fracture toughness K_{Ic}. An example of this relationship for several steels is presented in Fig. 3.36. Through this correlation and the use of fracture mechanics techniques, it is possible to translate the DT energy to critical crack size-stress level relationships by means of suitable charts.

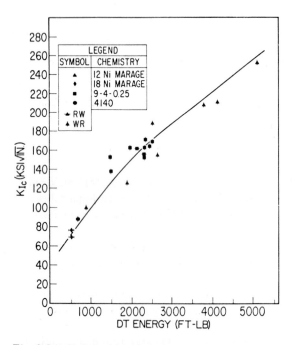

Fig. 3.36 Relationship between 1 in. dynamic tear energy and K_{Ic} values for various high-strength steels.

C. Other Impact Tests

In addition to the tests discussed in Sec. III.A, a number of additional tests have been developed to evaluate the influence of rapid loadings on material properties. For purposes of limiting the application of ferritic steels to service temperatures where their response is brittle, a test procedure has been developed to determine the *nil-ductility transition* (NDT) temperature. The NDT temperature is generally considered to be the temperature below which a material cannot deform plastically in the presence of a notch. A standard method has also been developed for measuring the NDT temperature, ASTM E208, and it is generally considered to be accurate to within 10°F.

To obtain even higher load rates the *explosive tear test* (ETT) has been developed and used to classify the fracture propagation resistance into three categories: high (approximately 15% plastic strain), intermediate (approximately 5% plastic strain), and low (no plastic strain).

Tests using the split Hopkinson bar apparatus have also been used rather widely to determine the effects of rapidly applied loads on the elastic properties of materials.

IV. DESIGN APPLICATIONS

In order to perform a successful design of a structural member or mechanical component, the choice of material, as well as the shape details and proportions, must provide for adequate performance under the anticipated service conditions. One of the necessary first steps in designing to prevent failures is to identify and characterize the many processes by which members or components may fail.

A. Causes of Failure

The most common types of mechanical failures that are encountered in practice are listed below.

1. Permanent deformation under static load, which causes misalignment or binding of components.
2. Buckling collapse due to elastic instability.
3. Creep at elevated temperatures, which causes interference between components or relaxation of fits.
4. Failure due to repeated applications of low-level service loads. Fatigue failures are usually caused by cracks that initiate at stress concentrators, such as keyways, welds, holes, or notches, and grow under cyclic loading until they reach a critical size and cause unstable fracture.
5. Wear in parts, often caused by inadequate lubrication or maintenance, which destroys fits, initiates pits and cracks, and eventually leads to failure by fracture, galling, or seizing.
6. The degradation of material properties by corrosive environments, leading to general surface degradation, pitting, crevice formation, selective dissolution, exfoliation, cavitation, stress corrosion cracking, or corrosion fatigue. Most of these corrosion mechanisms cause failure by aiding in the initiation and growth of cracks, destroying fits, or eliminating relative motion between parts (freezing or seizure).

7. Unanticipated service overloads, such as impact or fatigue loadings on a component designed for static loads only.
8. Lack of maintenance or improper maintenance, which may result in components working loose and being destroyed by excessive deformation, components in sliding contact being destroyed by lack of lubrication, major damage to machinery caused by the wearing out of a minor component, or failure to maintain protective coatings resulting in environmentally assisted failures.

Although it is very difficult to obtain accurate and comprehensive data, it appears that a great majority of service failures are caused by fatigue, wear, or lack of proper maintenance. Therefore, the possibility of service failures due to these processes should be accorded special attention during design and prototype development so that the frequency of such failures may be minimized.

It is significant that a component may fail to provide proper operation in several ways that do not involve actual breakage. This illustrates the need for determining potential failure criteria for a given design, and providing the necessary material properties or reconfiguring the component to survive the anticipated service requirements (see Chap. 32).

B. Criteria for Yield and Failure

Since either yielding or fracture may prevent the satisfactory operation of a structural member or mechanical component, it is important to establish criteria for the likelihood of either occurrence. Most of the failure criteria are similar in form to corresponding yield criteria, with the only difference the use of the ultimate strength rather than the yield strength in the formulation of the criteria.

1. Yield Criteria

Although a number of yield theories (criteria) have been proposed, only three of them have been subjected to widespread application and verification. These three are the maximum-normal-stress, maximum-shear-stress, and distortion-energy criteria.

1. The *maximum-normal-stress theory* states that yielding will be initiated in a component whenever the largest principal stress equals the yield strength of the material as measured in a uniaxial tensile test. A modification of this theory based on the maximum normal strain has also been proposed, but its use has been much more limited.
2. The *maximum-shear-stress theory* states that yielding will be initiated whenever the maximum shear stress reaches the value it has in simple tension or compression at the yield point.
3. The *distortion-energy theory* states that yielding will be initiated when the distortion energy (or shear energy) in a component attains a value equal to the distortion energy in a uniaxial tensile test at the yield point. An earlier theory similar to this predicted that yielding would occur when the total strain energy became equal to the total strain energy in a uniaxial test at the yield point. This theory is called the *maximum-strain-energy theory,* but its use has been supplanted by the distortion-energy theory.

A comparison of the maximum-normal-stress, maximum-shear-stress, and distortion-energy theories is shown in Fig. 3.37. For most structural materials the distortion-energy criterion generally provides the best agreement with experimental results.

2. Failure Criteria

The same three criteria, maximum-normal-stress, maximum-shear-stress, and distortion-energy, have also been employed as failure criteria. The application to failure is made by replacing the yield strength with the ultimate strength, or, more particularly, replacing the value of the critical quantities at yielding with those at fracture. Thus, the maximum-normal-stress-failure theory states that failure will occur in a component when the largest principal stress equals the ultimate strength of the material determined from a uniaxial tensile test. Likewise, the maximum-shear-stress and distortion-energy theories predict that failure will occur when the appropriate quantities attain the same values that they have at fracture in a uniaxial tensile test. As a general rule, the more ductile materials will fail in better agreement with the distortion-energy theory, whereas the more brittle materials fail in closer agreement with the maximum-normal-stress criterion. However, when brittle materials are loaded with the principal stress in compression, the maximum-normal-stress criterion is normally modified to take into account the greater ultimate compressive strengths that such materials exhibit. When the maximum-normal-strength criterion is modified in this manner it is often called the *Coulomb-Mohr failure theory*.

C. Stress Analysis

When the shapes of members or components are simple and the applied loads are known accurately, the stress distribution can often be calculated exactly from equations of the

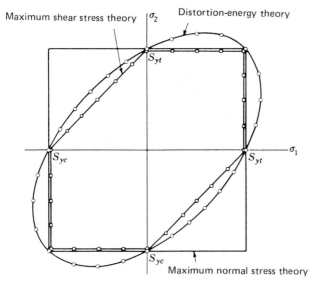

Fig. 3.37 Comparison of three yield theories under biaxial stresses.

theory of elasticity. A number of handbooks giving stresses and strains in bodies of various shapes are available for many practical geometries. However, for many more complicated shapes or for situations when the load distributions are not well known or when cracks exist, it becomes necessary to perform more detailed stress analyses. For such situations, the stresses may be determined either by numerical or experimental methods, and although they are all approximate methods, the extent of error can be controlled by the manner in which the analysis is performed. For example, numerical methods, such as the finite element, boundary collocation, and successive superposition methods, as well as many others, have come into widespread use with the concomitant increasing availability of high-speed computers. At present, by far the most work is being performed using the finite element method, in which the body is divided into a number of finite elements and a very basic numerical solution algorithm is applied to obtain nodal forces and displacements. The nodal forces are then summed and unitized for the determination of stresses.

The development of increasingly sophisticated software (computer programs or codes) for both large- and small-scale computers has made possible the widespread use of numerical stress analysis procedures. Combination of such software with interactive computer graphics routines, called *computer-aided design* (CAD), has become a very important tool for optimizing component design. Present computer systems are able to display a finite-element representation of any component geometry on a terminal screen, calculate stress distributions throughout the component, and then calculate how they change with variations in component geometry introduced by the designer. The use of such procedures permits considerably greater refinement in the stress analysis than was previously possible, with the additional advantage that a finite element analysis can often be performed in a few hours time.

In addition to the numerical stress analysis methods just discussed, several experimental methods may also be employed to obtain a detailed stress distribution in a component. At present, at least four experimental methods, brittle coatings, strain gages, photoelasticity, and moiré, have been developed to the extent that they can provide information useful to the designer.

1. Brittle Coatings

This method is based on the perfect adhesion of a brittle coating to the component being studied. When loads are applied to the component, the resulting strains are transmitted to the coating, which cracks in a direction perpendicular to the maximum tensile stresses. The first brittle coatings to be used commercially were made of shellac or lacquer, but strains beyond yields were required to initiate cracking in many of them. A number of additional brittle coatings that crack at much lower tensile strains have been developed in recent years. These include resin-based, ceramic-based, and glass lacquer coatings, which can in general provide reliable cracking at tensile strains less than 500 μin. per inch (0.05%).

The primary advantages of the brittle-coating method are as follows.

1. It provides nearly whole-field data for both magnitude and direction of principal stresses.
2. It does not require the fabrication of a special model, since it can be applied directly to a prototype or an actual working component.

3. When it is applied to the actual working component, it is not necessary to know the working loads.
4. The data obtained can be converted to stresses in a relatively simple fashion.

The primary disadvantages are as follows.

1. It is not usually practical to use brittle coatings for preliminary design because of the necessity of having a model or prototype on which to apply the coating. It is also necessary to be able to apply loads to the model that accurately represent the service loadings.
2. It is necessary to have access to any component on which brittle coatings are to be applied, since they are usually sprayed on (from aerosol cans), and they must be permitted to dry or possibly cure at elevated temperatures. For example, ceramic-based coatings must be cured at temperatures ranging from 950 to 1100°F, which precludes their use on many materials
3. There may be a loss of accuracy introduced by changes in temperature and humidity during operation.

2. Strain Gages

Another approach to the experimental stress analysis of a component is to mount strain gages at certain preselected points and then read their output when appropriate loads are applied. The type of strain gage most widely employed for this purpose is the SR-4 or resistance strain gage that is made of several types of copper-nickel-chromium alloy foils. During the early development of these gages, thin wires were used as the sensing element, but their use has been almost completely supplanted by foil. Since the foils are fragile and easy to distort, they are usually bonded to a thin plastic sheet, which is then bonded to the component or model. These gages operate on the principle that the electrical resistance of a wire or length of foil changes as it is strained in tension or compression. The changes in resistance are accurately measured, usually with a Wheatstone bridge type of circuit. To obtain greater strain sensitivity, the foil is usually fabricated in a grid pattern with the long direction of the grid aligned with the maximum tensile or compressive strain. Biaxial gages comprised of two perpendicular grids, three-element rosettes, and other special-purpose gages have been developed, along with a wide variety of electrical equipment for reading and displaying the strains. Strain gages having gage lengths from 0.008 in. (0.20 mm) to 4.0 in. (100 mm) are commercially available. Considerable care must be exercised in mounting and using these gages, but with care, strain variations from 0.001 to 20% can be reliably measured.

The principal advantages of resistance strain gages are as follows.

1. They may be quickly applied and permanently mounted on a prototype or component.
2. A wide range of gages and readout equipment is available, which permits their use over extended periods of time.
3. The individual gages are inexpensive and dependable.

The principal disadvantages are as follows.

1. A model or prototype must be available, along with a system for loading it appropriately. Therefore, it is of no value in evaluating preliminary designs,

unless the time and expense associated with fabricating a model or prototype is otherwise warranted.

2. Strains are known only where the gages are placed, so that other more critical regions of a component may be inadvertently missed.

Other strain gages than the resistance type have also been developed, including capacitance, inductance, and semiconductor strain gages. However, their applications are very limited in comparison with the resistance gage, so they will not be discussed in this chapter.

3. Photoelasticity

In performing a photoelastic stress analysis of a component, the normal procedure is to fabricate a photoelastic model, place it in a polariscope, and apply appropriate loads. The polariscope polarizes the incident light, which then passes through the model and is repolarized perpendicular to the first polarization. The transmitted light exhibits fringe patterns that illustrate the loci of points having a constant difference between the principal stresses in the plane of the model. The fringe patterns may then be photographed and measured for the determination of the strain field in the model.

The above description applies to two-dimensional photoelasticity, and the procedure is convenient only for that situation. Three-dimensional problems can be solved, however, by the use of stress freezing or other related techniques for permanently fixing the strains in the model. The model is subsequently sectioned and then layers are examined using the techniques of two-dimensional photoelasticity. Reflecting polariscopes are also available when transmission polariscopes cannot be used or when the stresses on the surface layer of an object need to be measured.

The advantages of the photoelastic method are that the strain field in the entire body can be observed and readily quantified. The disadvantages are that a carefully prepared model is necessary and that the analysis of anything but planar bodies becomes quite complicated. In addition, strain components normal to the plane of the model cannot be observed except with specialized techniques, such as the oblique incidence method.

4. Moiré Methods

The moiré methods are optical methods based on the generation of a pattern by passing light through slightly different arrays of equally spaced lines. These grids are arranged such that one array can be viewed through the other. A common example of moiré patterns can be observed by placing two layers of window screen in contact. Moiré patterns are measures of displacements that are induced in a loaded model or component. Displacements in any two perpendicular directions can be obtained by the moiré method, although two independent sets of grids are necessary. It may thus be necessary to use two models or at least two points of symmetry at which the two perpendicular sets of grids are placed.

Since the moiré approach is an optical method, it shares many of the advantages and disadvantages of the photoelastic method. The moiré method is significantly less well developed than the photoelastic method, so its limitations and capabilities are also less well known.

D. Residual Stresses

In many structures and machines, a stress analysis performed with great care and precision may provide highly erroneous results if residual stresses are not taken into account. *Residual stresses* are those stresses that exist in a body when no external forces are applied. Residual stress effects may be either favorable or unfavorable, depending upon whether they tend to subtract from or add to the applied stresses. For example, the strength of plate glass windows can be greatly increased by introducing compressive residual stresses through rapid cooling of the outer surfaces while the center is still soft. Then, when the center cools it contracts, causing residual tensile stresses in the center and residual compressive stresses on the surface. However, when the residual stresses are oriented so that they are additive to the applied stresses, the combined stresses may initiate early failures under fatigue loading or even immediate failure upon first application of the service loads. Thus, it is very important to know whether residual stresses exist in a component and whether they are additive to the service stresses.

Residual stresses in structural members or mechanical components may be evaluated by either numerical or experimental methods, or a combination of both. For example, residual tensile stresses in unannealed welds are known from both finite-element and experimental studies to approach the yield strength of the weld metal in the direction of the weld path and one-half the yield strength transverse to the weld. Most experimental methods are based on removing some of the residually stressed material and then measuring the dimensional changes induced by the removal of this material. The material removal process includes boring out the center of circular shafts, drilling holes, or machining away surface material. Another method involves applying brittle coatings and subsequently drilling holes to locally relax the residual stresses. Star-shaped relaxation patterns indicate residual tensile stresses; concentric circles indicate residual compressive stresses. X-ray diffraction methods may also be used to measure residual stresses by determining differences in the interatomic spacing.

Of the methods discussed for determining residual stresses, only the finite-element and x-ray methods are nondestructive, with only the finite-element method not requiring a model or prototype of the component.

E. Factor of Safety

Consideration of the criteria for yield and failure and the applied and residual stresses leads to the determination of a *limiting* or *damaging stress* for a member or component. This stress is the minimum stress that, if exceeded in the material, would result in failure before the end of its desired service life.

The actual stress that the component would be expected to encounter in service is called the *working stress*. The ratio of the limiting stress to the working stress is called the *factor of safety*. If the service loads and the stress distribution in the component were both known exactly, a factor of safety of unity could be used in design. However, since neither the service loads nor the stress distributions are normally known exactly, it is necessary to employ safety factors greater than unity to provide for safe operation of the structure or component. It should be evident that increases in the uncertainties of service loads and stress distributions require correspondingly greater safety factors.

Some of the conditions that influence the selection of a safety factor are

1. Variations in material properties from component to component and within one component
2. Inaccuracies in establishing the magnitude and distribution of service loads
3. Residual heat-treatment and cold-working stresses, stresses created during assembly or by thermal gradients, or stresses due to handling
4. Variations in time effects, such as strain hardening or softening under cyclic loading, creep, or corrosion degradation
5. The significance of economy of weight or material
6. The extent of risk of life or financial loss involved
7. Quality of workmanship (fits, tolerance, finish, and so on).

With all of the uncertainties indicated above, it is not surprising that safety factors ranging from 2.0 to 5.0 or even more are often employed. However, for weight-sensitive structures, such as aircraft and spacecraft, large safety factors cannot be tolerated. Thus additional stress analysis is usually performed so that safety factors less than 1.5 can be employed.

V. CONCLUSION: ADVICE TO THE ENGINEER

Determination of the mechanical properties of engineering materials encompasses a very broad range of activity, with more sophisticated and reliable testing procedures and equipment becoming available each year. Many new test methods have been developed in recent years for measuring the fracture toughness, stress corrosion cracking, and corrosion fatigue characteristics of structural metals and alloys. Improved new alloys are continually being developed by conventional processes and by entirely new processing methods, such as the rapid solidification-hot isostatic pressing process that produces material properties that are impossible to obtain by conventional methods. This diversity is enhanced even further by the increasing use of nonmetals for engineering applications. For example, the widespread utilization of composite materials has led to the development of many new test methods for determining the mechanical properties of anisotropic materials.

Associated with the development of these new materials and processes has come a wealth of data on the measurement of mechanical properties. Therefore, the user of these materials must be aware of the existence of such data and the limitations necessarily imposed by the procedures involved in their measurement. Ignorance of such data and their limitations can lead to the inefficient overdesign or dangerous underdesign of products. New design procedures, such as those incorporating fracture mechanics concepts, must also be mastered and intelligently applied.

Since new developments are certain to continue, it is important that designers keep abreast of continuing new developments in the field of mechanical property determination. This can be achieved through literature surveys, attendance at professional society technical sessions and short courses, and direct interactions with specialists in the field of mechanical property determination. By taking advantage of these professional opportunities, designers and other users of materials should be able to minimize the likelihood of unexpected failures of the materials that they are placing into service.

LIST OF SYMBOLS

a or 2a	crack length
a_c	critical crack size
a_d	minimum detectable crack size
A	test section area
A_i	instantaneous test section area
A_o	original test section area
B	fracture test specimen thickness
C	material constant
d	indentation diameter
D	penetration ball diameter
D'	damage factor
E	Young's modulus or modulus of elasticity
G	modulus of rigidity
G	strain energy release rate
G_c	fracture toughness
k	bulk modulus
k_a	surface factor
k_b	size factor
k_c	reliability factor
k_d	temperature factor
k_e	modifying factor for stress concentration
k_f	miscellaneous effects factor
K	stress intensity factor
K_c	plane stress fracture toughness
K_I	mode I stress intensity factor
K_{Ic}	mode I fracture toughness
K_f	fatigue strength reduction factor
K_t	stress concentration factor
l	gage length
l_i	instantaneous gage length
l_o	original gage length
m	material constant
n	number of cycles
N	cyclic life
P	applied load
P_u	ultimate or fracture load
q	notch sensitivity index
q'	reduction in area
(r, θ)	crack-tip coordinates
R	stress ratio
S_e	endurance limit
S_e'	test specimen endurance limit
T	temperature
T_m	melting temperature
U	strain energy
U_R	modulus of rigidity

U_s	surface energy
U_T	modulus of toughness
w	width
γ	surface energy density
δ	damping capacity
ϵ	strain
ϵ_f	fracture strain
ϵ_{ys}	strain at onset of yielding
ν	Poisson's ratio
σ	stress
σ_a	alternating stress
σ_c	critical fracture stress
σ_m	mean stress
σ_{max}	maximum cyclic stress
σ_{min}	minimum cyclic stress
σ_r	stress range
σ_{ys} or S_y	yield strength
σ_u or S_u	ultimate tensile strength
σ_x	normal stress component in x direction
σ_y	normal stress component in y direction
τ_{xy}	shear stress

BIBLIOGRAPHY

Annual Book of ASTM Standards, American Society for Testing and Materials, Philadelphia.

Broek, D., *Elementary Engineering Fracture Mechanics,* 3rd ed., Martinus Nijhoff, The
Netherlands, 1982.

Caddell, R. M., *Deformation and Fracture of Solids,* Prentice-Hall, Englewood Cliffs,
N. J., 1980.

Campbell, J. E., W. W. Gerberich, and J. H. Underwood (eds.), *Application of Fracture
Mechanics for Selection of Metallic Structural Materials,* American Society for
Metals, Metals Park, 1982.

Colangelo, V. J., and F. A. Heiser, *Analysis of Metallurgical Failures,* John Wiley & Sons,
New York, 1974.

Dally, J. W., and W. F. Riley, *Experimental Stress Analysis,* 2nd ed., McGraw-Hill, New
York, 1976.

Dieter, G. E., *Mechanical Metallurgy,* 2nd ed., McGraw-Hill, New York, 1976.

Frost, N. E., K. J. Marsh, and L. P. Pook, *Metal Fatigue,* Clarendon Press, Oxford, 1974.

Hayden, H. W., W. G. Moffatt, and J. Wulff, *Mechanical Behavior,* Vol. III, *The Structure and Properties of Materials,* J. Wulff (ed.), John Wiley & Sons, New York,
1964.

Hertzberg, R. W., *Deformation and Fracture Mechanics of Engineering Materials,* John
Wiley & Sons, New York, 1976.

Impact Testing of Metals, ASTM STP 466, American Society for Testing and Materials,
Philadelphia, 1970.

Jones, D. L., Practical Applications of Fracture Mechanics, *Metal Progress,* Part One.
Basic Concepts, Vol. 121, No. 2, pp. 26-29; Part Two. *Applications,* Vol. 121,
No. 3, pp. 64-68, 1982.

Marin, J., *Mechanical Behavior of Engineering Materials,* Prentice-Hall, Englewood Cliffs, N. J., 1962.

The Metals Handbook, 8th ed., Vol. 1-11, American Society for Metals, Metals Park, 1961-1976.

Mischke, C. R., *An Introduction to Computer-Aided Design,* Prentice-Hall, Englewood Cliffs, N. J., 1968.

Polakowski, N. H., and E. J. Ripling, *Strength and Structure of Engineering Materials,* Prentice-Hall, Englewood Cliffs, N. J., 1966.

Roark, R. J., and W. C. Young, *Formulas for Stress and Strain,* 5th ed., McGraw-Hill, New York, 1975.

Rolfe, S. T., and J. N. Barsom, *Fracture and Fatigue Control in Structures,* Prentice-Hall, Englewood Cliffs, N. J., 1977.

Shigley, J. E., *Mechanical Engineering Design,* 3rd ed., McGraw-Hill, New York, 1977.

Timoshenko, S. T., and J. N. Goodier, *Theory of Elasticity,* 3rd ed., McGraw-Hill, New York, 1970.

Zienkiewicz, O. C., *The Finite Element Method,* 3rd ed., McGraw-Hill, New York, 1977.

4

Chemical Properties and Corrosion*

Corrosion has been frequently defined as an electrochemical reaction involving a metal and an ionic environment. That definition will be followed in this chapter to differentiate corrosion from deterioration. Deterioration is the reaction of a material with its environment; the material is not necessarily a metal, the environment is not necessarily an ionic environment, and the reaction is definitely not just electrochemical. Thus, corrosion is a special form of deterioration specific to metals in contact with ionic species. For further differentiation, high-temperature oxidation of a metal would usually be considered deterioration and not corrosion because of the absence of the ionic species at high temperatures (but see Sec. IX in this chapter for a treatment of oxidation and hot corrosion). Although erosion is a mechanical action associated with the loss of materials, it is considered a special form of corrosion since the environment usually is electrically conductive and the material involved most frequently is a metal.

I. ECONOMICS OF CORROSION

Studies both in the United States and abroad (particularly the United Kingdom) have been made estimating the cost of corrosion to the nations involved. The figure arrived at for 1971 in the United Kingdom was £1365M; this was roughly equivalent to $3.3 billion that year. In 1978, the U.S. National Bureau of Standards indicated that a study carried out by Battelle Memorial Institute showed corrosion is costing $70 billion yearly in the United States. Not only is the cost of corrosion a considerable factor because of the economic waste involved, but there are also the risks to human life that occur when structures and other items deteriorate and fail.

II. CORROSION, THE SILENT SCOURGE†

In early 1976, the U.S. Senate, in its report on the appropriation bill for the National Bureau of Standards, directed that institution to embark upon a study of the economic effects of corrosion. The corrosion group at the National Bureau of Standards (NBS) is

---

*Contributed by LEWIS W. GLEEKMAN.
†Contributed by ELIO PASSAGLIA.

a technical group that had not previously conducted any formal economic studies. Hence a contract for the study was let to the Battelle Columbus Laboratories (BCL), which has both economic and corrosion expertise. The staff of BCL and NBS worked very closely together in defining the scope of the work and in the collection of data. In the fall of 1977, BCL delivered its report (1) to NBS. From this report, NBS prepared its own report for the Congress (2) and in the process extensively revised some of the figures in the BCL report. This section is a summary of the results in the NBS report.

Corrosion, defined for the purposes of this study as the degradative interaction of a metal with its environment, is only one of a number of degradative processes, such as wear, fatigue, fracture, and ultraviolet degradation, that affect materials. Corrosion, however, is unique in that a significant number of studies of the costs associated with it have been carried out (3-14). These are listed in Table 4.1. The most thorough and important of these studies is the famous Hoar report (6,7). This study came to the conclusion that the total cost of corrosion in the United Kingdom for 1970 amounted to 3.0% of the gross national product (GNP), and that 23% of the total cost (or 0.69% of the GNP) could be saved by use of currently available corrosion knowledge. In the terminology of the NBS report these latter costs are "avoidable." Where it is possible to make such an estimate, it will be seen that the total costs amount to 2-3% of the GNP, with approximately 20% of this avoidable. As will be seen below, these results are completely consistent with those obtained in the NBS-BCL study, when account is taken of the more thorough treatment by NBS-BCL, particularly with respect to the final demand sectors of the economy—personal consumption, government, and private fixed capital.

Table 4.1   Cost of Corrosion, Various Nations

Nation	Year	Cost	Avoidable	% GNP
USSR	1969	6 B Rubles		
		$6.7 B		
West Germany	1969	19 B DM	4.4 DM	3
		$6 B	$0.15 B	(0.75)
Finland	1965	150-200 M Markaa	–	–
		$47-62 M	–	–
United Kingdom	1969-1970	£1.365 B	£0.31 B	3
		$3.2 B	$0.74 B	(0.60)
Sweden	1964	0.3-0.4 B Crown	25%	
		$58-77 M		
India	1961	1.54 B Rupee		
		$320 M		
Australia	1973	$470 M	–	1.5(3)
		$550 M		
United States	1947	$5.5 B	–	2.3
	1965	$15 B	–	2.2
	1975	$9.7 B	–	–
Japan	1977	2500 B Yen		
		$9.2 B	–	(1.8)

A. Definitions and Methodology

1. Definitions

In order to carry out the directives of the Congress it was first necessary to define the "economic effects." The definition of these effects is very closely related to the costs of corrosion. The economic effects were taken to be

> Those resources (materials, energy, labor, capital) which, in the absence of corrosion, would be available for alternatives uses.

This means simply that, in the absence of corrosion, resources of materials, energy, labor, and capital that are now used because corrosion exists would become available for other uses. The yearly cost of these resources thus becomes the total cost of corrosion. This definition of the total cost of corrosion implies that the baseline for the accounting is a hypothetical corrosion-free world.

In addition, however, we need to take into account that some of these costs are *avoidable* in the sense of the Hoar study. Hence we also define the avoidable costs (or effects) as

> Those resources (materials, energy, labor, capital) that would be available for alternative uses if economic best practice were used everywhere.

Economic best practice means a minimum life-cycle cost. It therefore balances first cost against lifetime and total maintenance costs and does *not* mean the use of the most corrosion-resistant material. Indeed, in some cases, it may be economically better to use a more corrosion prone material than a more resistant one, if the first cost is sufficiently lower. Notice that the baseline for this accounting is a hypothetical *best practice* world, in which everyone uses economic best practice. This implies what is certainly known to be true, namely, that not everyone in the real world uses economic best practice.

The difference between the *total* costs and the *avoidable* costs gives what are called the *currently unavoidable* costs. The significance of these definitions is that avoidable costs can be reduced by the application of existing corrosion control technology, and hence their reduction involves *technology transfer,* but the currently unavoidable cost cannot be reduced with available corrosion control technology and hence their reduction requires *research and development.*

2. Elements of the Costs of Corrosion

The principal elements of the costs of corrosion are given in Table 4.2. These elements apply to both the total and avoidable costs. They are broken into four categories.

> *Capital costs,* which apply to consumer items as well as industrial and commercial equipment, include costs associated with shortened lifetime, any excess capacity that may be required because of corrosion, and the costs of any redundant equipment that may be necessary because of corrosion.
> *Control costs* includes costs of maintenance and repair and costs of corrosion control (cathodic protection, water treatment, and so on).
> *Design costs* includes the costs of special materials (for example, stainless steel or copper alloys) when used because of corrosion, costs of extra materials used (such as heavier sections on beams), and special processing.

Table 4.2 Some Elements of the Costs of Corrosion

Capital costs
    Replacement of equipment and buildings
    Excess capacity
    Redundant equipment

Control costs
    Maintenance and repair
    Corrosion control

Design costs
    Materials of construction
    Corrosion allowance
    Special processing

Associated costs
    Loss of product
    Technical support
    Insurance
    Parts and equipment inventory

> *Associated costs* includes such items as loss of product, research and development, insurance, and the cost of maintaining the inventory of parts and equipment needed for corrosion control. It does not include the cost of the items in the inventory since this is accounted for under control and repair costs.

3. Methodology

The methodology used for the study was input-output analysis, in which BCL has extensive expertise. For this purpose the economy is divided into 130 producing and service sectors, plus final demand. Producer capital, which is normally considered a part of final demand, was handled somewhat differently in this study, in that producer capital was considered to be an input into production (1). The economy was modeled as a steady-state economy. The year chosen was 1975 at hypothetical full employment. The study was not a "scenario" study. That is, no attempt was made to redistribute the costs; they were simply collected. It was thus a comparison of three static cases: 1975 at full employment, what 1975 would have been in the absence of corrosion, and what 1975 would have been with full best-practice use by everyone. The results will be presented as follows.

> *Final demand sectors* are personal consumption expenditures (PCE), federal government expenditures (FGE), and state and local government expenditures (SLGE). The costs for these sectors will be presented by the item of equipment and buildings for which the costs are incurred, such as automobiles. The elements included in these costs are primarily *capital costs* and *control costs* and some parts of *associated costs* where applicable. Design costs do not apply to final demand directly. They are borne in manufacture and are included in the next item.
>
> *Intermediate output (I/O)* are the costs incurred as inputs by the producing and service sectors of the economy. They include primarily *design costs* and

control costs, with *associated costs* being a small portion. They are presented by the sector bearing the cost.

Private fixed capital formation (PFCF) are the *capital* costs borne by the producing and service sectors of the economy. As previously noted, capital costs are those occurring because of shortened lifetime, excess capacity, and redundant capital. These costs are presented by the sector producing the capital.

B. Results

1. Overall Results

The total costs and the avoidable costs for all sectors are given in Table 4.3. Columns are given for the BCL results and for the NBS results derived from them as described in Ref. 2. An estimated range of uncertainty is also given.

The total costs are $69.7 billion, with a range of $52.7-86.2 billion. This amounts to 4.2% of the GNP, which is in good agreement with the studies listed in Table 4.1, when account is taken of the more thorough coverage of the BCL-NBS study. The largest costs are in intermediate output and in private fixed capital formation. The primary difference between the NBS results and the BCL results arises from the treatment of the lifetime of automobiles and other capital equipment by BCL. Reference 2 should be consulted for details.

In the avoidable costs, a significant change was made by NBS from the BCL figures, primarily because of the lifetime assessment for various items of capital equipment, notably the automobile (2). In addition, an unknown factor labeled Y is included in the NBS results. This arises because BCL estimated that best practice could be achieved without any change in input to the manufacturing sectors as compared with the present situation. That is, there would be no changes in usage of stainless steel, copper, coatings, for exam-

Table 4.3 Summary of Results

	Total costs ($B)				Avoidable costs ($B)			
Sector	BCL	BCL range	NBS	NBS range	BCL	BCL range	NBS	NBS range
PCE	22.8	–	25.8	10.3-21.3	15.9	–	4.9	3.8-15.9
FGE	8.1	–	7.9	6.2- 9.6	1.7	–	1.7	0.8- 2.5
S/LGE	2.9	–	2.4	1.2- 3.6	0.9	–	0.9	0.5- 1.4
IO	24.5	–	24.5	23.5-25.0	2.0	–	2.0-Y	-Y- 2.0
PFCF	24.1	–	19.1	11.5-26.7	12.5	–	6.2	3.0-19.1
Total	82.4	–	69.7	52.7-86.2	33.0	–	(15.7-Y)[a]	(8.1-Y)-40.9

[a]The value of Y is a matter of speculation, but assuming it costs between 10 and 70% of the expected final demand gain for best practice (extra coatings, e.g.), Y would be between $1.4 and $9.6 billion (B), and the total avoidable costs would be between $6.1 and $14.3 billion, or about 10 and 20% of the total cost. Note that these values of Y could make the avoidable 10% contribution negative. This would mean an increased cost to manufacturers in a best-practice world, to achieve a net savings to manufacturers plus final demand (life-cycle costs).

ple, and implies that first cost would not change. This view is not subscribed to by NBS, and the factor Y is listed to account for this. This factor is at present unknown, but is roughly estimated to be about $6 billion. With this value of Y, the avoidable costs are $10 billion, with the wide range of $2-40 billion. This value of Y makes the intermediate output avoidable costs negative, which merely means that the first cost of equipment would increase, although the *net* savings to the economy with best practice would be $10 billion. The avoidable costs are about 0.6% of the GNP, which is also in good agreement with the studies listed in Table 4.1.

The uncertainty in the results arises primarily from the uncertainties in estimating reliably the lifetime of equipment. These lifetimes are not well known even in the present world, and in addition estimates need to be made in both the corrosion-free and best-practice cases. This inevitably leads to a wide range of uncertainty. References 1 and 2 should be consulted for details.

2. Specific Sector Results

The results for specific sectors are given in the NBS report (2) and in more detail in the tables appended to the BCL report (1). In any study of this kind, individual sectors have more uncertainty than the overall figures, since in the overall figures, errors will tend to cancel. In this section, representative figures will be given for each of the sectors studied.

a. Personal Consumption Expenditures

The most significant of the total costs for personal consumption expenditures are given in Table 4.4 and the avoidable costs in Table 4.5. The largest item is in automobiles, which appears in two places in these tables. The entry under automobiles reflects effects of corrosion on the lifetime of automobiles, which is on the average 9.6 years, and estimated to increase to 13.6 years in the absence of corrosion (1,2). However, the distribution around this average can cause serious uncertainty (2). The entry automobile repair and service represents expenses for repair and service caused by corrosion. All other entries contain both capital-associated costs as well as maintenance and repair, except for miscellaneous chemical products, maintenance and repair construction, and personal repair service, which are solely for maintenance and repair.

Table 4.4 Total Costs, Personal Consumption (Millions of Dollars)

Misc. chem. products	159
Automobiles	8800
Trucks, buses, etc.	2090
Ships, boats	143
Household appliances	848
Radio, TV, comm. equip.	662
Maint, repair const.	992
Personal, repair serv.	331
Auto repair, service	2075

Table 4.5 Avoidable Costs, Personal
Consumption (Millions of Dollars)

Automobiles	2800
Truck, buses, etc.	500
Ships, boats	83
Household appliances	344
Maint., repair, const.	347
Personal repair serv.	115
Auto repair serv.	726

b. Government Expenditures

The most significant of the total costs for the federal government are given in Table 4.6, and the avoidable costs are given in Table 4.7. The very large bulk of these costs are borne by the Department of Defense, which has custody of by far the largest amount of equipment. The largest single cost is for aircraft, which is estimated to be the single largest item of capital equipment in the federal government. This cost arises because of redundant equipment, with the following reasoning. Approximately 30% of the time, aircraft are not deployed since they are undergoing maintenance or transfer, for example. It is estimated that 20% of this time is due to corrosion, and hence 6% of the time aircraft are not available for corrosion-related reasons. Therefore, to maintain a certain striking force, 6% more aircraft are required than would be necessary in the absence of corrosion. The yearly costs associated with the procurement of these aircraft form the bulk of these costs. The other large cost is government industry. This includes the activity of civilian employees, part of whose duties is the maintenance and repair associated with corrosion in aircraft, ships, and other machinery. For further details References 1 and 2 should be consulted. In general, the federal government has good corrosion control.

State and local governments are not presented in a separate table. The total costs from Table 4.3 are $2.4 billion, and the avoidable costs are estimated to be $0.9 billion. The largest single item is in building and repair construction, which is almost totally for highways.

Table 4.6 Total Costs, Federal Government (Millions of Dollars)

Metals	262
Gen. ind. mach.	144
Automobiles	208
Trucks, buses, etc.	309
Aircraft	2460
Ships, boats	905
Comm. equip.	1033
Sci. instr.	103
Ordnance	224
Maint., repair const.	375
Gov't. industry	1494

Table 4.7   Avoidable Costs, Federal
Government (Millions of Dollars)

Gen. ind. mach.	74
Automobiles	77
Trucks, buses, etc.	77
Aircraft	557
Ships, boats	408
Sci. instr.	53
Maint., repair const.	75
Gov't. industry	224

c. Intermediate Output

Representative total and avoidable costs for intermediate output are given in Table 4.8.
As previously noted, these costs represent added inputs (special metals, corrosion allowance, special coatings, and so on) required by the manufacturing sectors because of corrosion and corrosion maintenance and repair of the equipment and buildings in the producing and service sectors. A small amount also goes for insurance and other associated
costs.

   This sector was the most thoroughly studied and the total costs are the best known,
as is evidenced by the small uncertainty of ±3%. In many knowledgeable industries
sophisticated in corrosion control, such as petroleum refining and electric power, very
good data are available.

   With respect to the avoidable cost, however, the situation is quite different. As
previously noted, the avoidable costs include only maintenance and repair, and the figures shown in this table and in Ref. 2 are the BCL estimate of how much could be saved
by the producing and service sector by better maintenance and repair. Since the total
costs include all inputs and the avoidable costs include only maintenance and repair,
this accounts for the low value of avoidable costs. Moreover, in the capital equipment

Table 4.8   Intermediate Output Costs (Millions of Dollars)

	Total	Avoidable
Fabricated structural metal	1046	6
Screw mach. prod., stamp.	681	2
Misc. fabricated metal products	589	3
General industrial mach., equipment	668	20
Special industrial machinery	532	1
Furniture, fixtures	608	3
Petroleum refining	882	30
Steel	728	15
Automobile	1350	22
Electric power	2690	56
New construction-public utilities	1485	32
Wholesale and retail trade	606	272
Finance, etc.	617	280

producing sectors (such as fabricated structural metal or general industrial machinery), the achievement of best practice may actually increase the total cost, hence leading to a negative avoidable cost for these sectors. This leads to the factor Y in Table 4.2. No attempt was made to redistribute this factor to the relevant industries, and hence even for these industries in Table 4.8 the avoidable costs include only maintenance and repair.

d. Private Fixed Capital

Finally, Table 4.9 gives the most significant total and avoidable costs borne by the producing sectors for capital costs. These costs are caused primarily by the effects of corrosion in reducing the lifetime of capital items, but a small part is caused by redundant equipment. The listing is by the sector that produces the capital. A listing by the sectors using the equipment can be obtained from the industry indicators discussed in Ref. 2. The rather large uncertainty indicated for these costs is caused by uncertainties in the lifetime of equipment, as discussed.

C. Summary and Conclusions

It is estimated that the total costs of corrosion, and hence its economic effects, are 4.2% of the GNP, with a range of 3.2-5.2%. The avoidable costs are estimated to be about $10 billion, or 0.6% of the GNP with a very broad range of 0.2-2.4%. The range of both these estimates could be reduced by further study. By comparison of these results with previous studies, and because in such studies the errors will tend to cancel, it is believed that these results are representative of the total costs of corrosion in the economy. It is also believed the total methodology developed is better than has been used in any previous study. First, it provides a framework for determining costs, and second, it provides a sound and proven economic model for determining their economic effects. As more and better data become available, they can readily be added to the present data to make the estimate of costs more precise. However, two deficiencies still exist in the study.

First, the question of added inputs required for best practice needs to be determined in detail and the results distributed to the various final demand sectors to get a net cost for those sectors. This can be done with further study.

Table 4.9 Private Fixed Capital (Millions of Dollars)

	Total	Avoidable
Fabricated structural metal	1044	548
General hardware	282	106
Engines and turbines	334	67
General machinery	851	447
Farm machinery	1221	224
Construction machinery	1415	283
Automobiles	3640	1370
Trucks, buses	2006	401
Aircraft	227	45
Railroad equipment	267	140
Industrial controls	340	68
Service machinery	568	213

Second is the problem associated with the lifetime of equipment. We do not, unfortunately, know the actual range of lifetime of products in the economy, even in the present world. A determination of these lifetimes would be useful not only for improving the accuracy of the present study, but many other studies, such as, materials conservation, total availability of scrap, and the effects of other degradative processes, such as wear and fracture.

Despite these qualifications, it is believed that the Congress has been given a good overall picture of the economic effects of corrosion. The results in individual sectors can unquestionably be improved, but it is believed that the overall results will prove to be correct—certainly within the error limits stipulated.

III. PRINCIPLES OF CORROSION

Even with the advances in corrosion science and engineering, there is no such thing as the universal container for the universal solvent. Every material has one or more Achilles' heels, weak points subject to attack by one or more media. This is as true of stainless steel and other "corrosion-resistant" materials as it is of metal such as tantalum, titanium, gold, and platinum.

For corrosion to occur on a metal, three basic conditions must exist: (1) both the solution and the metal must be capable of conducting some electricity, (2) an electromotive force or electrical potential must set up a current flow, and (3) two or more points on the metal surface must act as electrodes for the flow of electrical current from the metal into the solution and back again. The potential that is set up occurs through the process of ionization.

When a metallic atom dissolves in an aqueous solution, it surrenders one or more electrons, with a residual positive charge. The residual part of the metallic atom carrying a positive charge is called an *ion,* and the solution in which the ion exists is called an *electrolyte.*

The driving force causing the initial tendency of a metal to corrode is the potential difference between the metal and the electrolyte into which the metal passes as ions carrying a positive charge. The charge on the remaining metal (that not going into solution) has an equivalent negative potential to that which has gone into solution. The *solution potential* is a characteristic of the metal at a specific concentration of its ions in solution. This solution potential will vary as its concentration is changed. The common reference or standard is that related to the *electropotential series,* where hydrogen is the reference point, and the potential is measured between the potential and the hydrogen. This is shown in Table 4.10. Metals above hydrogen (those having a negative potential relative to hydrogen) form positive ions and tend to displace hydrogen from solution. The positive electropotential metals tend to dissolve slowly and do not displace hydrogen from solution. The electropotential or electromotive series is similar to the galvanic series, though this latter series is specifically established to show the interrelationship of metals in seawater. When two different metals are joined one to the other and placed in an electrolyte, such as seawater, the more active metal (the one that is more negative to hydrogen) will go into solution. This is called galvanic corrosion; this is one of the several types of corrosion that will be considered. It should be noted that even a single or "pure" metal can have a nonhomogeneous structure that will cause local anodes and cathodes to form on the surface. The *anode* in corrosion terminology is the area at which

Table 4.10 Electromotive Force Series

	Element	Symbol	Standard electrode potential (volts), at 25°C
Anodic electrode	Potassium	K	-2.922
negative to hydrogen	Calcium	Ca	-2.77
	Magnesium	Mg	-2.34
	Berylium	Be	-1.70
	Aluminum	Al	-1.67
	Manganese	Mn	-1.05
	Zinc	Zn	-0.762
	Chromium	Cr	-0.71
	Iron	Fe	-0.440
	Cadmium	Cd	-0.402
	Cobalt	Co	-0.277
	Nickel	Ni	-0.250
	Tin	Sn	-0.136
	Lead	Pb	-0.126
	Hydrogen	H_2	0
Cathodic electrode	Antimony[a]	Sb	+0.1
positive to hydrogen	Bismuth[a]	+2.26	
	Copper	Cu^{2+}	+0.345
	Copper	Cu^+	+0.522
	Mercury	Hg_2^{2+}	+0.799
	Silver	Ag	+0.800
	Palladium	Pd	+0.83
	Mercury	Hg^{2+}	+0.854
	Platinum	Pt	+1.2
	Gold	Au^{3+}	+1.42
	Gold	Au^+	+1.68

[a]*International Critical Tables*, Vol. 6, McGraw-Hill, New York, 1929.
Source: H. H. Uhlig, *Corrosion Handbook*, John Wiley & Sons, New York, 1948.

the metal dissolves in the electrolyte. The *cathode* is the area which receives hydrogen, as shown by the equation

$$Fe^{2+} + 2H^+ + 2(OH)^- \rightarrow Fe(OH)_2 + 2H^+$$

The hydrogen that forms on the surface retards the reaction by forming an insulating film on the metal and by the tendency of the atomic hydrogen in the film to reenter the solution. For corrosion to proceed, the atomic hydrogen film must be removed. This removal is called depolarization; it can proceed by changing the atomic hydrogen to molecular hydrogen, which then escapes as gas bubbles or by combining the hydrogen with free oxygen in the solution to form water. Even scratches or other discontinuities on the surface of a metal can constitute inhomogeneities that allow corrosion to proceed by the formation of local anodes and cathodes.

There is more to the electrochemistry of corrosion than cited above. However, this chapter approaches corrosion from a design point of view rather than from an electrochemical standpoint (see Table 4.11).

A. Forms of Corrosion

Dr. Mars Fontana, formerly Chairman of the Department of Metallurgical Engineering at Ohio State University, categorized eight forms of corrosion as: general or uniform, pitting, stress corrosion, intergranular, galvanic, erosion, concentration cell, and selective leaching. These forms of corrosion will be subsequently discussed in relation to the design aspects of corrosion. Dr. Michael Henthorne of Carpenter Technology expanded this same concept by indicating that the forms of corrosion may be viewed as macroscopic or microscopic, as shown in Figure 4.1. Macroscopic forms are those that can be seen with the naked eye or slight magnification as opposed to those forms that require metallurgical techniques and higher magnification to delineate. Microscopic forms of corrosion include intergranular, stress, and selective leaching. The other forms of corrosion, usually visible to the naked eye, are therefore considered macroscopic.

Since, as indicated earlier, corrosion involves a reaction between a metal and its environment, then changing either or both will change the overall corrosion reaction, and if the reaction rate is retarded, the change is considered to have improved corrosion resistance. Thus, methods to improve corrosion resistance may be classed as follows: (1) change the environment, (2) isolate the metal from the environment, and (3) improve the corrosion resistance of the material. This is a very simplistic but practical way of achieving corrosion resistance by utilizing one or more of the three concepts stated above.

B. Influence of the Environment

Corrosion is an electrochemical reaction; hence, the empirical rule of physical chemistry that states that every 10 degrees Celsius increase in temperature doubles the reaction rate applies to corrosion reactions. Therefore, to slow corrosion, the temperature may be reduced. It is recognized that in many situations this is not a practical factor, since reaction of many solutions will only be completed to a desired quantitative end product and yield at a specific minimum temperature.

C. Influence of Hydrogen Ion Concentration

Electrochemical corrosion has been shown to depend upon the displacement of hydrogen ions from solution. The formation of ions is favored by high hydrogen ion concentrations; therefore, it follows that corrosion rate is increased by increase of hydrogen ion concentration. All things being equal, emphasis must be placed on the latter parts of that statement since it will be shown that some metals function best in high concentration of acidic solutions. Aqueous solutions have hydrogen (H^+) ions and hydroxyl (OH^-) ions present. Water at room temperature has approximately 10^{-14} g-mol/liter of solution as its ionization constant. Since pure water is electrochemically neutral, it has an equal number of hydrogen and hydroxyl ions, or 10^{-7} g-mol/liter of each. The hydrogen ion concentration, or pH, is expressed as follows: pH = 1 over the log (base 10) of the hydrogen ion concentration (the reciprocal of the log of the H^+ concentration). Table 4.12 shows pH as a function of the concentration in gram-moles/liter. Recognizing that a gram-mole is defined as the weight in grams equal to the molecular weight of the substance,

Table 4.11 Uses in Which Corrosion Resistance Is a Major Factor (1978)[a]

Metal	Total U.S. consumption, 1978	Type of use	Quantity	% of Total
Nickel	232,000 st	Alloying 37%, H.T. ox. res. 32%, plating 16%	197,000	85
Zirconium, metal, alloy	4,000 st	Alloying, chemical resistance	3,400	85
Titanium				
Metal	20,000 st	Metal products 65%, alloying 10%	16,000	80
Nonmetal	509,999 st	Coatings	260,000	51
Chromium	590,000 st	Alloying 58%; coatings, plating 4%	366,000	62
Cadmium	4,500 mt	Plating 40%, coatings 13%	2,400	53
Zinc	1,140,000 mt	Galvanizing 40%; coatings, sacrificial 2%	480,000	42
Tin	61,500 mt	Plating, tinning	22,500	37
Gold	4,738,000 tr oz	Plating, alloying, coating, cladding	1,660,000	35
Columbium	7,099,000 lb	Alloying, H.T. oxidation resistance	2,400,000	34
Platinum group	2,259,558 tr oz	Alloying, chemical, H.T. resistance	678,000	30
Rare earths (REO)	18,500 st	Alloying	5,500	30
Silver	160,210,000 tr oz	Alloying	41,743,000	26
Copper	2,480,000 st	Alloying and plumbing	620,000	25
Cobalt	18,870,000 lb	Alloying, H.T. oxidation resistance	4,299,000	23
Iron oxide pigments	226,500 st	Coatings (primers, undercoatings)	50,000	22
Molybdenum	67,724,000 lb	Alloying, coatings	13,500,000	20
Tantalum	1,772,000 lb	Alloying, cladding, H.T. ox. res.	320,000	18
Magnesium	109,000 st	Alloying, sacrificial	16,000	15
Hafnium	56,000 lb	Alloying	6,000	11
Thorium (ThO_2), nonenergy	35 st	Alloying	4	11
Aluminum	6,011,000 st	Alloying, coatings, cladding	600,000	10
Indium	W	Coatings	W	10
Lead	1,333,000 mt	Coatings, plating 6%, cable covering 1%	93,000	7
Beryllium	188 st	Alloying	9	5
Manganese	1,415,000 st	Alloying, cladding	55,000	4

[a]Abbreviations: st, short tons; lb, pounds; tr oz, troy ounces; met, metric tons; H.T., high temperature; W, withheld to maintain company proprietary information.

Source: Bureau of Mines, July 23, 1979.

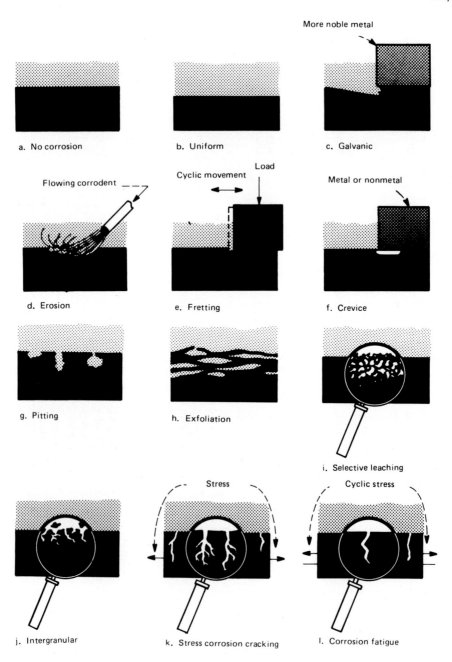

More noble metal

a. No corrosion

b. Uniform

c. Galvanic

Flowing corrodent

Cyclic movement   Load

Metal or nonmetal

d. Erosion

e. Fretting

f. Crevice

g. Pitting

h. Exfoliation

i. Selective leaching

Stress

Cyclic stress

j. Intergranular

k. Stress corrosion cracking

l. Corrosion fatigue

Fig. 4.1 Corrosion of metals and metallic alloys occurs in many different forms. (Courtesy of Dr. Michael Henthorne and *Chemical Engineering*.)

Table 4.12 Ion Concentration

Potential of hydrogen[a] on saturated platinum plate	pH	Concentration g-mol/liter	
0	0	1	
−0.058	1	0.1	
−0.116	2	0.01	
−0.174	3	0.001	H^+ ions or acid solution
−0.232	4	0.0001	
−0.290	5	0.00001	
−0.348	6	0.000001	
−0.406	7	0.0000001	H^+ and OH^- ions or neutral solution
−0.464	8	0.000001	
−0.522	9	0.00001	
−0.580	10	0.0001	
−0.638	11	0.001	OH^- ions or alkaline solution
−0.696	12	0.01	
−0.754	13	0.1	
−0.812	14	1	

[a]Voltage required to form molecular hydrogen on a blackened platinum plate. For most other metals the voltage required is greater, and the difference between the voltage required for other metals and that given above is the *hydrogen overvoltage* of the metal.

it can be seen that at pH = 1 the concentration of hydrochloric acid is equal to 26.5 g/liter, that is, a hydrochloric acid concentration of slightly greater than 3% by weight. Correspondingly, a pH of 14 has a concentration of 1 g-mol of hydroxyl ion per liter, which, in the case of sodium hydroxide, is 40 g/liter of sodium hydroxide, or 4% by weight. With these two examples, it can be seen that pH is not a good measure of indicating the relative concentration of various corrodants. Since the values of pH by definition only cover the range 0-14, concentrations of greater than 3% hydrochloric acid on the one hand and 4% caustic soda on the other hand would not be included within the range of pH values. For this reason, concentrations in either grams per liter or weight percent are used to indicate the amount of corrodant present in a solution.

D. Classes of Acids

However, it is convenient to categorize acids into three classes based on their behavior in the corrosion process.

1. Strong, nonoxidizing acids, such as hydrochloric and sulfuric acid, depend upon an external oxygen supply for depolarization. These acids in concentrated form have very low oxygen solubilities and ionize very little. Consequently, they are not as corrosive as solutions of moderate concentrations on metals which must be depolarized by oxygen. Steel, for example, is often used to hold concentrated H_2SO_4 in large tanks.
2. Oxidizing acids, such as nitric and chromic acids, supply their own oxidizing agents and therefore keep the metal surface depolarized. Metals should not be

used in contact with oxidizing acids unless a passive protective film is formed on the metal.

3. Weak acids, as found in foods or soils, corrode metals as fast as strong acids if enough oxygen is supplied. However there is usually little oxygen in foods or soils.

E. Alkaline Corrosion

Corrosion by alkalis is normally not as rapid as by neutral and acid solutions because the hydrogen ion concentration is reduced and because the hydroxides produced by corrosion are insoluble up to given values of alkalinity. Zinc and aluminum hydroxides, for example, become soluble around pH 12. Ferrous hydroxide is still insoluble in sodium hydroxide solutions at pH 14. Increasing pH on the alkaline side therefore reduces the corrosion rate unless the solution becomes alkaline enough to dissolve the protective hydroxide coating.

F. Corrosion in Salt Solutions

The action of salt solutions depends on the solubility of the corrosion products formed. Chlorides, sulfates, bromides, and fluorides are quite soluble and therefore tend to be very corrosive. The corrosion product goes into solution as fast as it is formed and therefore is not present to protect the surface. Silicates and carbonates are less soluble and protect the metal surface unless local concentration cells are set up. Since seawater contains fairly large concentrations of chloride and sulfates, particularly sodium and potassium, soluble corrosion products are formed on many materials and therefore corrosion of these materials in seawater is rapid.

G. Other Environmental Effects

Other environmental changes that can have a profound effect on the corrosion of materials include temperature, velocity, presence of gases, duration of exposure, presence of corrosion products, and trace elements. Corrosion products in a corroding medium can form a protective film of varying thickness on the surface of the metal. If the film is impermeable, then it serves to inhibit corrosion. However, if the corrosion product film is irregularly distributed and is disturbed, for example by velocity effects, then there is a possibility of a galvanic cell between the corrosion product, which is often cathodic to the metal and the metal itself. Porous coatings result in the substrate metal being more anodic by the exclusion of oxygen; an oxygen concentration cell developed; the net result is accelerated pitting (Fig. 4.2).

As indicated, high-velocity flow can not only remove the protective film that forms on the metal but under extreme conditions can result in erosion due to the turbulence, particularly at bends, changes in direction, or due to entrance or exit effects, for example in heat exchangers.

The presence of gases in a solution can have profound influences on many metals depending on the specific gases. The most common example is the presence of air (and therefore oxygen) in water with the resulting corrosion on steel. If the water is deaerated by one of several techniques, including replacing the air with nitrogen, then in most cases the same water will not cause corrosion of the steel. Conversely, if stainless steel is in contact with a liquid under reducing conditions (in particular the absence of air, along with other reducing factors), then corrosion of the stainless steel can be expected.

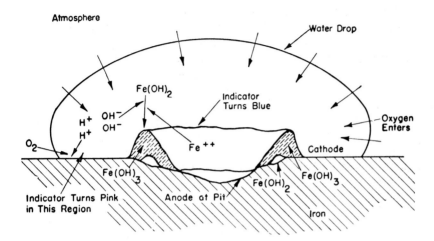

Fig. 4.2 Oxygen concentration cell set up by a drop of water on a piece of iron. The action involved is indicated by a ferroxyl indicator.

Certain materials act as inhibitors on the corrosion of specific metals. For example, the presence of cupric or ferric ions in a solution results in a relatively high degree of corrosion resistance of stainless steel due to the oxidizing conditions produced by these metal ions. As such, these ions serve as inhibitors. The field of inhibitors goes far beyond the use of metallic ions into organic complexes. Many books and articles dealing with corrosion inhibition of specific combinations of organics and metals have been written.

The duration of exposure itself has an influence on the corrosion, since the initial rate of corrosion in single-metal corrosion is usually much higher than the final rate. The rate of corrosion seldom is linear with time but frequently decreases with continued exposure due to the formation of a protective corrosion product on the surface. It is for this reason that the planned interval corrosion test has been used in many investigations. Unfortunately, with regard to pitting and other forms of localized corrosion, the predictability of the rate of pitting (depth and distribution) is not a simple matter. Due to the formation of galvanic and concentration cells, the rate of pitting in certain metals may actually increase with time.

From a design point of view, the engineer bases corrosion allowance on the assumption that the corrosion is general and that the rate of corrosion is constant. This allows a figure for the loss of metal in a particular environment to be developed, reported usually as so many inches per year (ipy) or mils per year (mpy). These figures frequently are arrived at in corrosion tests by weight loss measurements per unit area per unit time and are converted to thickness loss values by taking into account the specific gravity of the metal in question. Obviously, the validity of such corrosion allowance values falls rapidly when one has to take into account pitting, crevice corrosion, glavanic corrosion, erosion, or other localized nonuniform corrosion factors (stress, intergranular corrosion, and selective leaching).

Other influences on the environment include agitation and replacement of the solution. In the case of agitation, the oxygen supply is frequently renewed in the region of the metal, and therefore any atomic hydrogen or protecting films can be removed from the surface of the metal and the oxygen necessary for corrosion replenished. In many industrial operations (boiler water treatment), deaeration of certain solutions is carried out to remove the oxygen and thereby minimize or negate the corrosion of the base material. The oxygen removal frequently is done by the displacement of oxygen by another gas, such as nitrogen, or, more positively, by actual physical deaeration of the solution (vacuum deaeration), or by adding a reducing agent like hydrazine.

The presence of corrosion products can act as a change in the environment since these products can exist as a protective film on the surface of the metal. In certain cases of metals that form passivating films, such as stainless steel, titanium, or tantalum, the films are extremely thin whereas other coatings may have substantial thickness. If the corrosion product is porous, there is a tendency for the substrate metal to become more anodic by the exclusion of oxygen, thereby forming an oxygen concentration cell and causing accelerated localized pitting.

Another environmental factor that should not be overlooked is that of cyclic mechanical stresses resulting in a condition called corrosion fatigue. This is often related to the presence of vibrations whose fluctuations cause cyclic stresses. Fatigue, as a form of failure, is common to many materials in the absence of corrosive environments; factors, such as stress risers, that encourage fatigue in noncorrosive environments have a more pronounced effect in a corrosive environment leading to earlier corrosion fatigue.

IV. PROTECTING THE METAL

Protecting the metal to retard corrosion most often means isolating the metallic surface from the corrosive environment. This usually can be done with either organic or metallic surface layers. Organic layers or coatings can be thick or thin; they can be paint films, solid sheet linings, or nonmetallic plastics in the form of a tape, sheet, or a powder fused to the surface. Metallic coatings are often applied as electroplated materials, though some can be deposited without electricity, for example, silvering glass to make a mirror, coating by diffusion at moderate temperatures, such as application of zinc on steel in galvanizing, or aluminum on steel in aluminizing, or metallizing the surface by spray application of partially melted materials. Selection among these protective methods depends on surface factors, environmental conditions, economics, ease of application of materials, complexity of parts, desired longevity, and other factors.

In addition, a metal surface can be protected by applying an electrical potential under conditions that make the surface anodic or cathodic depending on the metal being protected and the environment. The more widely used of these two methods is that of cathodic protection, often applied to underground pipelines, tank bottoms, water boxes on exchangers, and so on. The application of cathodic protection may be done with a sacrificial anode of magnesium, zinc, or aluminum or by the application of an applied potential from a rectifier or battery where one uses a more permanent anode of graphite, Duriron, or platinized titanium. Anodic protection has been most widely applied to steel in concentrated sulfuric acid service to prevent iron pickup.

Another standard technique of protecting steel with its relatively poor corrosion resistance in many environments but taking advantage of its good load-carrying characteristics and lower cost is to have a layer of a more corrosion resistant metal over the steel. This layer of metal may be metallurgically bonded, as in rolled welding, or it may be explosively clad to the surface of the steel, or as indicated earlier, it may be a relatively thin electrodeposited metallic coating. The advantage of clad metal construction (see

Chap. 9) is that a small amount of expensive corrosion-resistant material contacts the environment while the thicker, less costly steel on the exterior carries the load or withstands the pressure. This has allowed fabrication of fairly large vessels where the chemical conditions (environment, temperature, and pressure) required the use of materials such as tantalum, zirconium, titanium, stainless steel, nickel, or other relatively expensive corrosion-resistant material. Usually the cladding thickness is 10% of the total thickness; wall thicknesses less than ¼ in. generally are not economically practical for clad construction. There are fabricating techniques that require more than passing consideration; these have been well established and therefore pose no problems. Some construction has been carried out with loose clad linings of relatively expensive corrosion-resistant materials; this has worked well provided fabricating conditions are well defined. This subject is discussed in greater depth in Chaps. 24 and 25.

V. DESIGN IMPROVEMENTS

In addition to changing the environment, protecting the metal, and, yet to be discussed, changing the metal, design improvements can be made to a material to enhance its behavior in the environment. Such design improvements require a balance of considerations for the service involved, including the strength-weight ratio, ease of fabrication and ease of joining, corrosion resistance, effect of temperature on physical, mechanical, and electrical properties, and, not least, availability and cost.

Improved design considerations include complete drainage of the metal part, ease of cleaning and inspection, and minimizing galvanic corrosion of dissimilar metals when these metals are in contact. Crevices generally should be avoided. If necessary, use continuous welds instead of skip welds. Where skip welds are required for economics or for other factors, then the use of a protective coating that is cathodic in nature should be considered. Such protective coatings generally are the zinc-rich paints, particularly those based on a high concentration of zinc in an inorganic silicate binder.

The following general rules should be considered from the standpoint of minimizing the likelihood of corrosion difficulties.

1. Continuously welded butt joints should be used whereever possible. If filleted lap welds are used, the welds should be continuous on the process side.
2. Installation of baffles, strainers, and drain nozzles and the location of valves and pumps should be such that free drainage occurs and water washing can be accomplished without any holdup.
3. Means of access and maintenance should be provided whenever possible.
4. The use of dissimilar metals in contact with each other should be avoided, particularly if they are separated in their position in the EMF or galvanic series. If they must be used together, either the anodic material should have a substantially larger area or the metals should be electrically insulated from each other.
5. Local turbulence and areas of high velocity at feed and drain connections at fittings and lines should be minimized where possible.
6. Equipment should be supported in such a way that it will not rest on a pool of liquid or damp supporting material. Porous material should be waterproofed or otherwise protected from moisture to avoid contact of the wet material with the equipment.

7. Fabrication methods used should cause a minimum of surface roughness. Inspection during fabrication and prior to acceptance is absolutely required.

VI. CORROSION BEHAVIOR OF COMMON MATERIALS

The designer in the late 1980s has available a multitude of metals, alloys, and materials to handle an equally large number of corrosive environments; this makes it virtually impossible in a limited number of pages to give recommendations for specific applications. Insofar as comparison with similar cases can be used as a basis for choosing materials, Table 4.13 and the following discussion may be used for estimating the resistance of many common metals and alloys to a number of common corrosives. Where contact between dissimilar alloys or elements is involved, Table 4.14 may be used as a guide in choosing combinations that will have little galvanic action. The reader is cautioned to temper these general considerations in response to the factors already discussed. In large installations or for large production, it would be preferable to use these data as a guide to conducting pilot tests that assure the necessary life. In addition to the common, everyday factors that influence the selection of one of several potential materials to handle a given corrosive environment, the actuality of shortage of certain materials must be considered. Most common, perhaps, is the fact that supplies of wood have reached a point where paper is not the readily consumable item it had been. Similarly, the oil shortage has a potential limitation on materials derived from oil, such as many of the plastics and solvents. When one adds the shortages of such metals as cobalt, molybdenum, and chromium, among others, due to the fluctuations of national boundaries and political upheaval, even the stainless steels may not be foreseeably freely abundant.

Steel still is the most convenient material of construction when strength and low cost are the principal considerations. Corrosion protection is achieved in most cases by coatings. However, this is only true of the more industrially advanced countries. To manufacture steel economically, not only is a sufficient and abundant supply of iron ore or steel scrap of a regular consistency needed, but also water and energy in large quantities are required. The energy can be derived from coal, oil, gas, or electricity. Due to the advances that have been made in continuous casting of metals, including continuous casting of steel, a lower capital investment is required to produce relatively large shapes and sections of steel. This is in contrast to the previous requirement of blast furnaces, open hearths, and electric furnaces, for pouring large ingots. Further, those ingots then required large breakdown mills to produce usable shapes in large volume. Continuous casting allows many of these steps to be bypassed; hence many countries with limited assets now can directly get the form of the steel that is most needed.

At one time in the late 1960s, as advanced a nation as Argentina had no major steel-producing plants for sizes greater than a 1 or 2 in. bar. As a result, the reinforced concrete construction industry in Argentina was relatively extensive since the raw materials for cement were indigenous and in large supply and the reinforcing rod could also be made locally. This situation is known to be true in many other countries throughout the world. Accepted conventional materials of construction in "advanced" countries (e.g., the United States, Great Britain, France, Germany, Scandinavia and other European countries, and the USSR, are not available from local manufacturers in many third world countries.

The arrangement of the discussion below will be in terms of the major metals used in the developed world with subsequent consideration of the major nonmetallic materials used.

A. Steel

Steel (an alloy of iron and carbon with varying amounts of other elements) is not noteworthy for its corrosion resistance in aqueous aerated environments. The most common manifestation of this lack of corrosion resistance is rust (Fig. 4.3). In spite of the wide range of physical properties, such as hardness and tensile strength, that can be achieved by alloying iron with varying amounts of carbon and other elements with or without subsequent heat treatment, the corrosion resistance of steel is basically the same irrespective of most alloying elements. With the exception of oxidation and creep resistance at high temperatures, the corrosion behavior of most steels is essentially the same in a given industrial atmosphere. Major exceptions to this are the low-alloy, high strength steels, which have excellent resistance to atmospheric environments, so much so that they are frequently called the "weathering" steels. Incorporating relatively small amounts of copper, chromium, and nickel produced these weathering steels, which are now quite common for bridge and building construction. The oxide that forms is a tenacious oxide rather than one that continues to grow and lose its attachment to the base steel, as is common in mild steels.

In spite of the relatively poor resistance of steel to most corrosive environments, it nonetheless finds widespread use in industry because of the protective treatments that can be given to it. These include, to name but a few, galvanizing (either hot dipped or electrolytic zinc deposition), aluminizing, and protective coatings, such as paints, linings, and elastomers. A common method of dealing with corrosive environments is to use a structural steel member with an elastomeric lining for corrosion resistance covered with a masonry lining, such as brick and resin mortar, for additional chemical and thermal resistance.

Cathodic protection of either bare steel or steel with a protective coating is also used in underground applications. Even a relatively mild steel corrodant as concentrated sulfuric acid (generally taken as greater than 85% by weight of sulfuric acid) has had its corrosion rate reduced by the application of anodic protection, thereby putting the corrosion rate of steel at absolutely zero (no pickup of iron). This gives steel greater corrosion resistance in areas of high velocity, such as in nozzles. The use of inhibitors allows steel to be used in environments where corrosion would otherwise be quite active.

There are beneficial metallurgical techniques that can be used in fabricating steel, such as stress relieving to overcome mechanical or thermal stresses (particularly those stresses set up by welding where contraction takes place). This allows steel, when stress is relieved, to be used in handling some alkaline solutions, such as caustic soda, which in the absence of stress relief would actively embrittle the steel. Caustic embrittlement, it should be noted, is a form of stress corrosion cracking that is unique to steel in the presence of strong caustic environments.

B. Cast Iron

Cast iron behaves in many regards similar to wrought steel in the sense that it has limited resistance in the presence of oxygen and moisture. The presence, however, of graphite as

Table 4.13 Relative Corrodibility of Uncoated Metals and Alloys[a]

Class of material	Maximum safe temp. for good service (°F)			Fruit, vegetable juices	Dairy products	Other good products	Acids, moderate concentrations, 5-15%				
	Oxidizing gases	Reducing fuel gas	Sulfur-rich gas				HCl	H₂SO₄	HNO₃	Acetic	Phosphoric
Low-carbon steel, open-hearth	800	—	—	P	P	P	P	P	P	P	P
Copper steel	800	—	—	P	P	P	PF	F	P	P	P
Cr-Cu-Si steel											
Hot galvanized iron and steel	—	—	—	P	P	P	P	P	P	P	P
Calorized iron and steel	1600	1600	1600	—	—	—	—	—	—	—	—
Gray cast iron	700	—	—	P	P	P	P	P	F	P	P
High silicon iron	—	—	—	G+	—	—	F	E	E	G	G
Nickel cast iron	—	—	—	—	—	—	P	F	F	—	—
Ni-Cr-Cu cast iron*	—	—	—	F	—	—	FG	FG	G	—	—
Chromium cast iron	1800	1800	1800								
Nickel steel											
Low nickel	840	—	—	P	P	—	P	P	P	P	—
High nickel	930	—	—	F	—	—	FG	G	F	P	—
Chromium steels											
4-6% chromium	1200	1200	1200	P			P	P	P	P	P
12-14% chromium	1400	1400	1400	G	—	G	P	P	G	G	F†
16-18% chromium	1600	1600	1500	G	G	G	P	P	G	E†	P
25-30% chromium	1900	1900	1800	E	G	E	P	P	E	F	G
Chromium nickel steels											
8-20%	1600	1600	—	—	—	—	P	G	F	G	G
18-8%	1550	1550	300-1300	G	E	E	P	F	G	G	G
18-8%, 4% Mo	—	—	—	—	—	—	P	—	G	G	G
18-12%	—	—	—	—	—	—	P	F	G	G	F
18-35%	1900	1900	1900	G	—	—	P	G	P	P	F
25-12%	2100	—	—	G	E	E	P	P	G	G	F
26-24%	2100	2000	—	G	E	E	P	F	G	G	G
Silchrome steel	1500	1500	—	F	—	—	P	P	E	G	—
	—	—	—	E	—	—	P	P	E	G	—
Stellite	2000	2000	—	E	E	E	F	G	E	G	E
Commercially pure nickel	1300	2000	P	G	GE	GE	G	G	P	G	G
Nickel alloys											
Monel metal	900	2000	P	G	FE	FG	FG	G	P	G	G
Nichrome 60-15%	1475	2100	P	G	—	—	F	F	F	G	G
Inconel, 14% chromium	2000	2000	—	E	E	E					
80-20% nickel-chromium	—	—	—	G	—	—	F	G	P	G	G
Hastelloy	—	—	—	E	G	G	G	FG	G	G	E
Commercially pure copper	—	—	—	FG	P	F	F	FG	P	FG	FG
Copper alloys											
Red brass	—	—	—	FG	P	P	P	FG	P	FG	FG
Tobin bronze	—	—	—	—	—	—	P	F	P	F	F
Phosphor bronze	—	—	—	P	P	F	P	FG	P	FG	FG
Silicon bronze	—	—	—	G	P	—	F	G	P	G	G
Aluminum bronze	—	—	—	FG	P	—	P	FG	P	FG	FG
Nickel silver	—	—	—	G	P	—	F	G	P	G	G
Admiralty metal	—	—	—	FG	—	—	P	FG	P	FG	FG

[a]P = Poor. F = Fair, only to be used in temporary construction. G = Good, will give good service. E = excellent, almost unlimited service.

*Alloys added should be increased with thickness of section.

†E at low and P at high concentrations.

	Salt solutions, moderate concentrations,			Hot sulfide liquor	Dye liquor	Refinery crudes below 400°F		Atmosphere		Water					
Alkalies, 8%	NH4Cl	MgCl2	MgSO4			Sweet	Sour	Seashore	Industrial	Domestic	Mine	Sea	Saline with H2S	Brackish with NaCl	Wet steam
E	P	F	F	P	–	F	P	P	P	P	P	P	P	F	F
E	P	F	F	P	–	F	P	F	F	P	F	F	P	F	FG
								FG	FG	F	FG	FG			
P	P	P	PF					G	G	F	F	G	F	F	
–	–	–	–	P	P	FG	FG	G	G	P	P	P			
G	P	P	P	P	P	F		F	G	P	P	P	F	F	F
E	G	G	G	P				E	E	E	G	E			
GE	–	–	–	P	P	F+	FG	F	G	F	..	F	F+	F+	
GE	F	F	F	F	F	G		GE	GE	G	..	G	G	G	G
								G	G	G	G	FG			
E	P	F	F					F	F	F	P	F	G	G	
E	F	G	G					E	E	G	FG	E	G	G	
G	P	P	F	P	P	G	G	FG	FG	FG	FG	FG	G	F	FG
E	G	G	G	P	F	E		G	G	G	FG	FG	–	FG	G
E	G	G	G	F	F	E	E	G	G	G	G	G	–	G	G
G	E	E	E	GE				G	G	E	G	G	–	G	E
G	–	–	–	P	–	G		G	G	G	G	G			
E	FG	E	E	GE	F	E	E	E	E	E	G	G	E	G	E
G	E	E	E	GE	G	E	E	E	E	E	G	G	E	G+	E
E	–	–	–	G	–	E	E	E	E	E	G	G	E		
G	–	–	–	P				E	E	E	FG	G			
G	E	E	E	GE	G	E		E	E	E	G	G	E	G	E
G	E	E	E	E	G	E		E	E	E	G	G	E	G	E
G	–	–	–	–	PG			G	G	F	F	F	F	F	F
E	–	–	–					G	E	E	E	F	E		
E	E	E	E	G	G	E	E	E	E	E	E	E	E	G	E
E	–	–	G	P	G			E	G	E	G	E	G	E	
E	E	E	G	P	E	F		E	G	E	PG	E	G	G	E
G	–	–	–	P				G	G	E	G	G			
								G	G		G	G			
E								G	G	E	G	G			
E	G	G	G	E	G	G	G	E	E	E	G	E	E	E	E
FG	–	–	–	F	FG			G	G	G	FG	G	P	–	G
P	–	–	–	F	FG			G	G	FG	FG	G	P	F	P
F	–	–	–	G	G	FG		G	FG	–	F	G	F	F	
FG	–	–	–	G	FG			G	FG	<	F	G			
FG	–	–	–	G	FG			G	G	–	FG	G			
FG	–	–	–	–	FG	G		G	G	–	FG	G	–	F	
G	–	–	–	–	FG			G	G	–	G	G			
P	–	–	–	–	FG			G	G	–	FG	G	–	F	

Table 4.14 Electromotive Force Series

Anodic end	
1 Lithium	43 Cobalt
2 Rubidium	44 Ni-resist cast iron
3 Potassium	45 50-50 lead tin solder
4 Strontium	46 17% Cr-7% Ni steel, type 301 (active)
5 Barium	47 18% Cr-8% Ni steel, types 302, 303
6 Calcium	304, 321, 347 (active)
7 Sodium	48 23% Cr-14% Ni steel, type 309
8 Magnesium and its alloys	(active)
9 Aluminum	49 25% Cr-20% Ni steel, type 310
10 Beryllium	(active)
11 Uranium	50 18% Cr-12% Ni-3% Mo steel, type
12 Manganese	316 (active)
13 Tellurium	51 Hastelloy "C" (59% Ni, 17% Mo, 5%
14 Zinc	Fe, 14% Cr, 5% W, 0.1% C)
15 Chromium	52 Lead
16 Sulfur	53 Tin
17 Gallium	54 Iron (Fe^{3+})
18 Iron (Fe^{2+})	55 Hydrogen

19 ⎫
20 ⎪
21 ⎬ A
22 ⎪
23 ⎪
24 ⎭
25

- 19 Galvanized steel[a]
- 20 Galvanized wrought iron
- 21 Al-Zn-Mg alloys (e.g., Al-75S)
- 22 Al-high Mg alloys (e.g., Al-220)
- 23 A Al-low Mg alloys (e.g., Al-4S)
- 24 Aluminum plus very low percentage of alloying constituents (e.g., Al-53S)
- 25 Alclads

26 Cadmium
27 Al-Si-Mg alloys (e.g., Al-356)
28 Al-Cu alloys (with or without small additions of Mg)(e.g., Al-24S)
29 Al-Cu alloys (with or without small additions of Zn) (e.g., Al-113)
30 Al-Cu-Si alloys (e.g., Al-108)
31 Mild steel
32 Copper steel
33 S.A.E. 4140
34 S.A.E. 3140
35 Wrought iron
36 Cast iron
37 4-6% chromium steel, type 501 or 502 (active)
38 12-14% chromium steel, types 403, 410, 416 (active)
39 16-18% chromium steel, type 440 (active)
40 23-30% chromium steel, type 446 (active)
41 Indium[c]
42 Thallium

56 Antimony
57 Bismuth
58 Arsenic

- 59 Muntz metal (60% Cu, 40% Zn)[d]
- 60 Manganese bronze (66.5% Cu, 19% Zn, 6% Al, 4% Mn)
- 61 B Naval brass (add. of ¾% Sn to Muntz metal)
- 62 Nickel (active
- 63 60% Ni-15% Cr (active)
- 64 Inconel (78% Ni, 13.5% Cr, 6% Fe) (active)
- 65 80% Ni-20% Cr (active)
- 66 Hastelloy "A" (60% Ni, 20% Ni, 20% Mo, 20% Fe, 0.1% C)
- 67 Hastelloy "B" (65% Ni, 30% Mo, 5% Fe, 0.1% C)
- 68 B Yellow brass (58-70% Cu, 0.50-1.5% Sn, 0.75-3.5% Pb, balance Zn)
- 69 Admiralty brass (71% Cu, 28% Zn, 1% Sn)
- 70 Aluminum bronze (add. of 2-2-¼% Al to 75% Cu-25% Zn alloy)
- 71 Red brass (85% Cu, 15% Zn)

72 Copper (-ic)
73 Oxygen
74 Polonium
75 Copper (-ous)

76 Iodine			87	Inconel (passive)
77 Tellurium			88	80% Ni-20% Cr (passive)
78	Silicon bronze (1.0 to 3.0% Si)[e]		89	Titanium
			90	Monel (70% Ni, 30% Cu)
79	Nickel-silver		91	12-14% Cr steel, types 403, 410, 416 (passive)
80	Ambrac (5.0% Zn, 20% Ni, balance Cu)		92	16-18% Cr steel, type 440 (passive)
81	70% Cu-30% Ni		93	17% Cr-7% Ni steel, type 301 (passive)
82 C	Comp. G-bronze (88% Cu, 2% Zn, 10% Sn)		94 C	18% Cr-8% Ni steel, types 302, 303, 304, 321, 347 (passive)
83	Comp. M-bronze (88% Cu, 3% Zn, 6.5% Sn, 1.5% Pb)		95	23% Cr-14% Ni steel, type 309 (passive)
84	Silver solder		96	23-30% chromium steel, type 446 (passive)
85	Nickel (passive)		97	25% Cr-20% Ni steel, type 310 (passive)
86	60% Ni-15% Cr (passive)		98	18% Cr-12% Ni-3% Mo steel, type 316 (passive)

99 Mercury
100 Silver
101 Lead (Pb^{4+})
102 Palladium
103 Bromine
104 Chlorine
105 Graphite
106 Gold (-ic)
107 Gold (-ous)
108 Platinum
109 Fluorine

Cathodic End

[a]Members of group A are listed relative to each other, but the position of the group with respect to the group between chromium and ferrous iron is uncertain. The group, however, is listed correctly between zinc and cadmium.

[b]The heat treatment of aluminum and its alloys has an effect upon their potentials. They fall in the aluminum group, however, regardless of the heat treatment.

[c]Indium, thallium, and cobalt are listed relative to each other, but their position relative to the stainless steels is uncertain. They are properly placed between aluminum and lead.

[d]The members of group B are listed relative to each other, but the position of the group relative to ferric iron to arsenic is uncertain. The group is, however, properly placed between tin and copper.

[e]The members of group C are listed relative to each other, but the position of this group with respect to iodine tellurium and mercury is uncertain. However, the group is properly placed between copper and silver.

[f]Graphite lies between lead and gold, but its position relative to palladium, bromine, and chlorine is uncertain.

Note: Select contacting combinations as close together as possible in this series. Keep dissimilar metals as far apart as possible and particularly avoid threaded connections between widely dissimilar metals. Avoid combinations where the area of the less noble metal is small; that is, use a more noble fastening for a less corrosion resistant part, like Monel rivets on a steel tank instead of steel rivets on a Monel tank. Insulate joints where possible. Paint both the noble and the less noble if paint insulation is used. When brazing, choose a brazing alloy more noble than at least one of the metals to be jointed.
Source: A. E. Durkin and Carroll Seversike, Thomson Laboratory, General Electric Company, Lynn, Massachusetts.

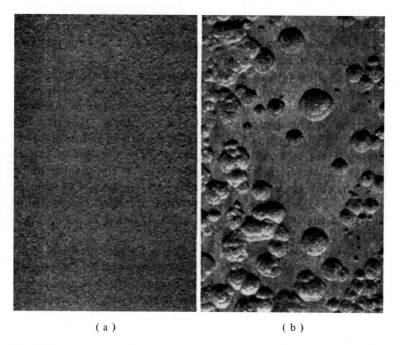

(a)   (b)

Fig. 4.3 Corrosion of plain carbon steel. (a) Uniform surface corrosion. (b) Pitting. Both about actual size. (Courtesy of Carnegie Research Laboratories, U.S. Steel, Pittsburgh, Pennsylvania.)

a distinct and separate phase in cast iron imparts a degree of internal cathodic protection to the cast iron in many environments such that the need for protection is somewhat less than for mild steel. There are, however, other factors beyond corrosion resistance that, in many cases, limit the use of cast iron. The lack of tensile strength and ductility of conventional gray and white cast iron means that their use is mainly for parts where compressive strength is the major consideration. The development of ductile iron (where the heat treatment and chemical composition results in the graphite being in a different configuration and different distribution) has contributed to the more widespread use of this form of cast iron for applications where ductility and strength must be considered. However, for many environments, ductile iron has its chemical limitations. The reason the cast irons are used in somewhat borderline applications relates most generally to the relatively heavy wall that must be used compared with the thin wall of a steel part. Allowing for a moderate corrosion rate still permits a fairly good life of a cast iron part. Wrought iron is more a historical material than a common material of construction by virtue of the advances that have been made in melting and processing cast iron and cast steel. The presence of a large amount of inclusions had at one time offered many advantages to wrought iron, but the overall corrosion resistance does not materially differ from that of mild steel or cast iron.

Cast iron has been alloyed mainly to improve corrosion resistance rather than hardness and hardenability properties, as is the case when alloying steel. Small nickel additions, about 3½%, result in so-called nickel cast iron. This is a slight improvement over the unalloyed cast irons in some environments. However, in a somewhat analogous fashion to that of mild steel, the alloys of cast iron require larger amounts of nickel (in particular) and chromium to get appreciable corrosion resistance. The series of Ni-Resist cast irons normally starts at approximately 11% nickel and then proceeds in about 10% increments to Ni-Resists II and III. These are really austenitic cast irons analogous to the austenitic stainless steels.

For environments with good ionic conductivity, cast iron frequently undergoes galvanic corrosion when coupled with another metal. This initially manifests itself by solution of the iron leaving behind a matrix of graphite. This process is known as graphitization. The graphite, being a noble material, can adversely affect what had been the earlier more noble coupled metal, such as stainless steel. Stainless steel, coupled to cast iron in most environments, is more noble than the iron in the cast iron. As the iron is depleted by corrosion, the graphite remaining in the cast iron is yet more noble than the stainless steel, ultimately causing corrosion of the stainless steel.

Because of thermal expansion and contraction properties along with ductility, ductile iron is the preferred material for pumps and valves where the danger of thermal shock may occur (e.g., rapid cooling of cast iron components by fire hoses to extinguish a fire). Ductile iron will withstand an extreme change in temperature, in contrast to conventional cast iron, which has been known to crack, releasing a previously contained flammable product (e.g., a hydrocarbon, thereby contributing to the spread of fire. Despite the need for designers to take account of its low ductility, high-silicon cast iron (with or without adding molybdenum) is widely used for handling sulfuric and certain other acids.

C. Stainless Steels

By alloying iron with chromium, to a minimum of 12% by weight, the oxidation resistance of the resulting material is tremendously changed from that of steel itself. This has given rise to the broad term, "stainless steel." However, for further improvement in resistance to chemical environments, such as nitric acid, many salts, alkalis, and organics, nickel is added to the chromium, the conventional and preferred relationship being 18% chromium and 8% nickel, minimum.

There are metallurgical changes that occur by virtue of these chromium or chromium plus nickel additions. In the case of chromium added to low-carbon steels, ferritic stainless steels result, whereas with chromium and nickel, the austenitic phase persists to room temperature and thus these materials are called austenitic stainless steels. Higher carbon content chrome steels can undergo a phase transformation after heating to form martensite. The early nomenclature of type 300 series is also applied to the austenitic stainless steels. Improvements in corrosion resistance, machining, strength, high-temperature oxidation resistance, and so on, can be made by varying the amounts of chromium and nickel as well as adding singly or in combination other alloying elements, such as molybdenum, selenium, titanium, or columbium (Fig. 4.4).

Superior as the stainless steels are in comparison with mild steels, these materials have a point of weakness, specifically in the case of the 300 series stainless steels. This is their susceptibility to chloride pitting, chloride stress corrosion attack, and/or chloride crevice corrosion. This is particularly true when the stainless steels are in their sensitized condition wherein carbon has combined with chromium, thus removing the chromium as an effective alloying element from areas such as the heat-affected zone of a weld. The result is the very common carbide precipitation in the grain boundary, particularly in the heat-affected area. This area is where the temperature (and cooling rate), specifically in the range of 800-1200°F allowed chromium to combine with the carbon to form

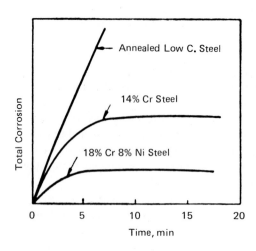

Fig. 4.4 Corrosion of two chromium-bearing steels compared with low-carbon steel. Note reduction of corrosion rate of the Cr steels with formation of a protective film.

chromium carbide. To overcome this sensitivity to carbide precipitation, the "L-grade" stainless steels have been developed. The carbon is kept at 0.03% or less. The less carbon, the less is the tendency to carbide precipitation.

Molybdenum has a substantial influence on the resistance of stainless steels to chloride attack, though even with 3-4% molybdenum, as in the case of type 317 stainless steel, there is no certainty that the chloride attack will not have an adverse effect on the stainless steel. It has been said that the minimum amount of chloride that can be handled by any austenitic stainless steel without damage is *nil*.* This has been borne out in many cases, particularly where the acidity of the solution containing the chlorides is pH 3 or less. Other proprietary stainless steels have been developed that do not fit into the AISI nomenclature, that is, compositions with higher amounts of chromium, nickel, molybdenum and copper, among others. In fact, copper in solution acts as an inhibitor in many solutions that otherwise normally attack stainless steel. Chlorides of calcium, magnesium, iron (ferric), sodium, and lithium are extremely aggressive to most stainless steels. This situation shows up in its worst form under stagnant conditions, that is, when the solution is not flowing or flowing at a low velocity (less than 1 ft/sec) or when the solution is under a crevice, such as a deposit or a gasket.

Chromium (12-17%) stainless steel with high carbon is mainly used for parts requiring a hard surface for wear resistance and for cutlery. Roughly the same chromium content with low carbon is used for resistance to oxidizing corrosion up to 1650°F and for ornamental purposes. The passive film that is formed on the surface of stainless steel is due to the presence of chromium. The film is sometimes artificailly accelerated by a passivating treatment with nitric acid and/or chromic acid. Among other things this passivating treatment removes any free iron that may be present on the surface of this stainless as a contaminant while simultaneously enhancing the protective oxide film. If the film is broken while in an oxidizing environment, it will usually heal itself; however, in the presence of a reducing environment, the break in the film will be the focus of heavy corrosion.

---

*There is evidence that chloride and oxygen content exhibit synergism. If one can be made to have a low concentration, a certain amount of the other can be tolerated.

Stainless steels as a group are resistant to corrosion by nitric acid and are widely used in the manufacture of nitric acid. Stainless steels resist most weak acids but are attacked by the strong nonoxidizing acids, as well as hydrochloric and sulfuric acids. Chlorides under acid conditions (in the presence of other acids) will break down the protective film and so cause corrosion. On the other hand, under alkaline conditions, the effect of chlorides is less critical, other than sodium hypochlorite itself. Stainless steels are used in many alkaline environments, as in caustic soda evaporators.

The relative roughness of stainless steel is a factor with regard to the durability of the protective passive film. The better the finish in terms of smoothness, the greater the resistance to corrosion. In this regard it should be noted that stainless steels, like many other metals, can be obtained in finishes as rough as rolled or pickled to as fine as mirror polished with a whole gamut of finishes between these extremes.

Cast stainless steels are available with analyses similar to the wrought forms, though slightly modified in composition for best pouring behavior, cooling rates, and so on. The material of the mold is critical since it has been well documented that cast stainless steel can pick up carbon from certain mold surfaces and thus become susceptible to surface carbide precipitation, which may diffuse into the casting. The condition of the stainless steel casting, whether it is as-cast or annealed, is also critical; this should be taken into account for preferred corrosion resistance at the time the casting is ordered. The presence of ferrite as a distinct phase in stainless steel castings (as well as in certain wrought alloys) has led to a two-phase structure that has proven very useful in many environments.

D. Copper and Copper-Based Alloys

Copper is widely used for corrosion resistance in unalloyed form. To improve its strength, copper is frequently alloyed with varying amounts of zinc to form brasses and with tin to form the generic alloy series called bronze. The nomenclature of copper alloys has been clarified recently by virtue of a numerical designation system that is part of the unified numbering system; thus, copper and its alloys are represented by UNS numbers such as C11000 to C75200 for wrought alloys.

Copper is a noble metal in the periodic table, below hydrogen in the electromotive series; it is not noble in the sense of gold and other precious metals since common acids can attack copper. Copper resists attack under most corrosive conditions, though it has limited usefulness in certain environments because of hydrogen embrittlement and stress corrosion cracking. Brass stress corrosion cracking (previously called "season cracking") is most commonly found in brasses exposed to ammonia or to ammoniacal compounds (Fig. 4.5). Brasses containing more than 15% zinc are the most susceptible to this attack. However, a small amount (1%) of tin added to the alloy markedly reduces the tendency to season crack.

The corrosion of copper generally differs from that of most other metals in that its corrosion resistance does not depend on the formation of protective films but rather on its limited solubility and reasonable hydrogen overvoltage. Since the corrosion tends to proceed by depolarization of hydrogen, the control of oxygen is important in acid corrosion as well as corrosion by neutral waters and alkalis.

For this reason, copper and its alloys (with the exception of the high-zinc brasses) are fairly resistant to nonoxidizing acids, such as sulfuric and phosphoric, and to many industrial salts and water of limited oxygen content. Similarly, copper and its alloys are generally attacked by oxidizing acids and their salts since no passivation occurs. Because of the solubility of cupric ammonia oxide, ammonia corrodes copper alloys readily

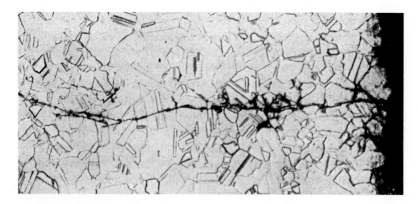

Fig. 4.5 Photomicrograph of season crack in aluminum brass (76Cu, 22Zn, 2Al). Note that the crack is transcrystalline in the body of the material and intercrystalline (the usual form) near the edge at the right.

though these alloys are resistant to other alkalis in moderate concentrations at normal temperatures. Sulfur-containing compounds, such as certain rubber insulation and tapes, have been known to cause corrosion by the formation of copper sulfide and other compounds when moisture is present. Copper is widely used in handling potable water, although at elevated temperature, the solubility of oxygen and carbon dioxide at the service temperature becomes quite critical. In the absence of a protective carbonate scale, as is common in soft waters, pitting has been observed in many copper installations handling very soft water.

E. Aluminum and Aluminum Alloys

Aluminum, which is alloyed with such metals as magnesium, zinc, silicon, manganese, and copper, singly and in various combinations, is a metal that forms a protective oxide film. Its corrosion resistance is good in neutral environments and in oxidizing acids, such as nitric acid. However, strong alkalis and strong nonoxidizing acids remove the film and cause corrosion. Chloride salts can penetrate the film, causing rapid pitting corrosion. Heavy metals, particularly mercury, are incompatible in contact with aluminum. Because aluminum is very high in the EMF and galvanic series, contact with metals lower in the series will cause rapid galvanic corrosion; aluminum acts as an anode.

The corrosion resistance of aluminum is generally best in its unalloyed form; however, the strength of this material is generally low. Thus, a clad product has been produced where the core is an aluminum alloy with sufficient strength and the cladding is a relatively pure aluminum. In addition to the nominal improvements in corrosion resistance, the Al-clad layer provides galvanic protection to the core material. The alloy of the cladding is chosen such that it is anodic to the core, and thus when the cladding is penetrated, it will sacrifice itself to protect the heavier strength-bearing core material.

Aluminum may be protected by anodizing; this is the formation of an anodic film thicker than that normally formed in air. This is done electrolytically in an oxidizing acid, producing a film up to 500 times the thickness of the film formed in air. The anodizing process also can be used to give color to the aluminum surface. Aluminum is frequently painted to offer resistance to pitting corrosion.

F. Nickel and Nickel-Based Alloys

Nickel is widely alloyed with copper (a high-nickel alloy with copper has the trade name Monel; high-copper alloys are generally called Cupronickel), as well as being alloyed with chromium, iron, and silicon, among other materials. Nickel finds great use in the handling and production of alkalis and in its resistance to some nonoxidizing acids. Low concentrations of hydrochloric acid are resisted by nickel and Monel. Hastelloy B, which is an alloy of nickel, molybdenum, and iron, is used for handling hydrochloric acid in a reducing atmosphere. In the presence of an oxidizing atmosphere, the addition of chromium to produce the alloy known as Hastelloy C is required. There are other high-nickel alloys of the Hastelloy variety that have specific resistance in other environments.

One of the dangers to nickel alloys is contamination by sulfur resulting in embrittling effects that come from nickel sulfide at elevated temperatures. Even the presence of sulfur from human perspiration has been known to cause these embrittling effects.

Nickel and nickel-based alloys are often used in clad construction to give corrosion resistance while minimizing cost.

G. Reactive Metals

Titanium, zirconium, and columbium in particular are classified as reactive metals not because of their reactivity in corrosive environments at ordinary temperatures but rather because of their reactivity with air at high temperatures. This has led to the development of special processing techniques using inert gases to get these metals to the high state of purity necessary for their successful application. Titanium is widely used in its unalloyed conditions for its resistance to wet chlorine, chlorides, nitric acid, and many oxidizing media. The danger of crevice corrosion and iron contamination must always be considered in using titanium. By alloying titanium with a small amount of nickel, the crevice corrosion and pitting tendency of titanium, particularly in brines at elevated temperatures, has been greatly decreased. This has been a great improvement over the alloying of titanium with the noble metals, such as palladium and platinum, because of the high price of these noble metals even in the small quantities (0.1%) used in alloying. For additional strength titanium is frequently alloyed with aluminum, vanadium, tin, and/or molybdenum, though this increase in strength is achieved at the expense of corrosion resistance. Titanium is frequently used as a cladding material for economical reasons. It is loose-clad, roll-clad, or explosion-clad. Hydride attack is a very serious problem for titanium, though this usually occurs only under reducing conditions. Titanium absorbs large quantities of hydrogen and in so doing becomes brittle and cracks under stress.

Zirconium has similar uses to titanium; its greatest industrial use has been in nuclear reactors because of its inherent small-neutron cross section when highly purified. Separation from hafnium is not required for nonnuclear applications. Zirconium offers resistance to dry chlorine and to many reducing environments so that it extends the resistance of many of the reactive metals to chloride-containing reducing environments. Its fabrication is quite similar to that of titanium. Titanium and zirconium are available as castings from certain well-qualified suppliers.

Columbium (niobium) has limited chemical resistance. Because of its cost columbium is mainly used for certain high-temperature applications.

VII. NONMETALLIC MATERIALS

Metallic contamination of products has become a very important consideration and in some instances is responsible for limiting the use of metals for process equipment. The use of nonmetallic material, particularly plastics and elastomers, has proved to be very successful in avoiding metallic contamination. In many instances nonmetallic materials provide greater chemical resistance at lower cost for certain aqueous environments than metals.

A. Plastics

Traditionally, the use of plastics has been limited to relatively low-temperature services and to low-pressure applications except where used as a lining bonded to or otherwise supported by a strong substrate. This is true of the nonreinforced plastics that are categorized as thermoplastic materials, specifically those materials that are softened by heat. However, the development of reinforcing materials, combined with the class of plastics known as thermoset materials (materials that are not softened by heat), and the further development of many thermoplastic materials that have relatively high thermal distortion temperatures, has meant that present-day technology uses plastics for applications requiring moderately high temperature and pressure.

 Notable among the thermoplastic materials are polyethylene, polypropylene, polyvinyl chloride, the styrene-synthetic rubber blends, the acrylics, and the fluorocarbons. Notable among the thermosetting reinforced materials are the polyesters, epoxy, and the furane resins as custom-made laid-up reinforced materials, and the phenolic and epoxy resins molded, filament wound, and/or extruded with reinforcement. All these materials are available as piping, sheet stock, and miscellaneous molded and fabricated items. The thermoplastic resins based on ethylene are shown in Fig. 4.6. They are generally not

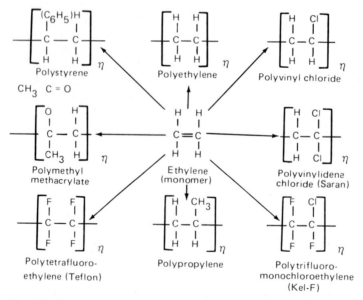

Fig. 4.6 Structural relationship of common thermoplastic compounds to that of the ethylene monomer.

subject to pitting, stress-corrosion cracking, and other forms of corrosion common to metal. However, there are design limitations, and they usually cannot be substituted for metals part for part. A point of merit that should not be overlooked is that nonmetallics do not require painting for protection against external corrosion. Plastic materials are replacing metals in many applications because of better resistance to chemical exposures and improved service life and economy.

1. Fluoropolymers

The most chemically resistant plastics commercially available today are tetrafluoroethylene (TFE) and fluorinated ethylene propylene (FEP) (Teflon or Halon). These are thermoplastic materials that are practically unaffected by all acids, alkalis, and organics at temperatures up to about 500°F. They have proved to be outstanding materials for gaskets, packing diaphragms, O-rings, seals, and other relatively small molded items. Their chemical inertness makes normal bonding and cementing operations difficult and impractical. They can be cemented to metal and other materials by using special sheets that have rough backing surfaces, which provide mechanical adherence through any one of a number of resin cements. Techniques have also been worked out that make it possible to heat-seal sheets of material together. Loose linings, including nozzle linings, may now be installed in tanks, ductwork, and other straight-sided and nonintricate equipment.

Polyvinylidene fluoride (PVF$_2$) also possesses excellent chemical resistance to almost all acids and alkalis at temperatures up to about 350°F. Although it is difficult to bond to itself and other materials, the use of tape-bonded laminated construction has widened the use of the material. It can be extruded readily and is available in the form of solid pipe and also as a lining material for steel pipe.

2. Polyethylene

Polyethylene is the lowest cost plastic that has excellent resistance to a wide variety of chemicals. Polyethylene tape with pressure-sensitive adhesive on one side is receiving increasing attention; it is used as a wrapping material to protect conduit and pipe from corrosion. Unfortunately, polyethylene's mechanical properties are relatively poor, particularly at temperatures above 120°F; hence, it must be supported for most applications. It can be readily joined to itself by heat fusing; a wide variety of equipment has been satisfactorily made using heat-sealing techniques. Weathering resistance of the unfilled grades is poor, but the carbon-filled grades have good resistance to sunlight and are satisfactory for outside use.

3. Polyvinyl chloride

Unplasticized polyvinyl chloride materials have excellent resistance to oxidizing acids, other than nitric and sulfuric, and to nonoxidizing acids in all concentrations; they are satisfactory for use at temperatures up to about 150°F. They also are resistant to both weak and strong alkaline solutions and to solutions of most chemical salts. Resistance to aromatic and aliphatic hydrocarbons is generally poor. They are not satisfactory for use with ketone or ester solvents. They are difficult to ignite but they have been known to give off large volumes of smoke when in a fire. They are resistant to sunlight and outdoor weathering. Two general types are available: regular and high-impact. The latter has appreciably better impact resistance, but somewhat lower strength and lower overall chemical resistance. Both are readily fabricated and can be joined by fusion- and solvent-welding techniques.

4. Styrene Elastomer Blends

The styrene-synthetic rubber blend materials, which are a mixture of styrene-acrylonitrile polymer with butadiene-acrylonitrile (A-B-S, for example) have good resistance to nonoxidizing weak acids but are not satisfactory for handling oxidizing acids. The upper useful temperature limit is about 150°F. Resistance to strong alkaline solutions is fair and to weak alkaline or salt solutions generally good. They are not satisfactory for use with aromatic or chlorinated hydrocarbons, and they possess only fair resistance to aliphatic hydrocarbons. They are not satisfactory for use with ketone and ester solvents. These materials normally will burn, but fire-retarded grades are commercially available. Resistance to outdoor weathering is generally good. They can be readily fabricated and can be joined by solvent-welding techniques.

5. Resin Coatings

The use of reinforced resin systems that are applied as liquids and then are converted to solids by catalytic action has increased tremendously. These include polyester resins catalyzed by peroxide, epoxy resins catalyzed by basic amines, and furane resins catalyzed by acids. The nature of furane resins is such that they are available only as a black material, which limits their application. Polyester and epoxy materials are available as translucent materials that can be pigmented to any desired color. Epoxy resins have excellent resistance to nonoxidizing and weak acids and to alkaline materials but poor resistance to strong oxidizing acids. The upper service temperature limit is generally about 200°F. The epoxy resins have good resistance to aromatic and aliphatic hydrocarbons but only fair resistance to chlorinated hydrocarbons, ketones, and esters. Polyester resins have good resistance to nonoxidizing and oxidizing acids, both weak and concentrated, with moderate resistance to many alkaline solutions and excellent resistance to salt solutions. The upper service limit is 200°F, although there is some loss of mechanical properties at this temperature. Polyester resins hve good resistance to aromatic and aliphatic hydrocarbons but only fair resistance to the strong solvents. Polyesters will burn, but fire-retarded grades, based on the addition of antimony trioxide and halogenated compounds to the polyester resins, have been developed. Resistance to sunlight and outdoor weathering is good. Overall resistance of the resin for severe chemical services is often improved by the use of a chemically resistant, synthetic-fiber cloth, such as Orlon acrylic fiber or Dacron polyester fiber. The glass-fiber laminates are not satisfactory for use in hydrofluoric acid, and it has been found that they are subject to attack and penetration in other acids and alkalis, including hydrochloric acid under certain conditions. Furane resins have broad chemical resistance to acids and alkalis as well as many solvents. Otherwise the material is similar to the epoxy and polyester laminates.

6. Polyurethane and Other Foams

Urethane resins, based on the isocyanate molecule, are finding increasing use as an insulation material in the form of a foam and as an abrasion-resistant material in the form of a compounded elastomer. Although urethane foams have been made to meet certain requirements of the various regulatory agencies with regard to flame retardancy, extreme caution is recommended in the use of these materials when the application parameters differ from the flammability parameters. The insulation properties are outstanding, as is the ease of fabrication, since urethane foam can be applied as a mixture of two solutions, each sprayed simultaneously onto a substrate. Other resin systems, including epoxy, can also be spray-applied to form a foam, although properties are somewhat different from those of the urethane foam. Top coating of the urethane foam with a flame-retardant coating is virtually a requisite, along with the additional protection given to the foam from ultraviolet degradation.

B. Rubber and Other Elastomers

1. Natural Rubber

Natural rubber has been used for many years as a material for molded and lined equipment for chemical service. It can be compounded for maximum resistance for a number of service conditions and has proved to be a very useful material for many conditions that are highly corrosive to metals. Natural rubber compounds will resist a wide variety of chemical solutions, including all concentrations of hydrochloric acid and phosphoric acid, up to about 50% concentration sulfuric acid, saturated salt solutions, such as ferric chloride, brine, bleaching solutions, and most plating solutions. Rubber compounds are readily attacked by strong oxidizing acids, such as nitric and chromic, and by aliphatic, aromatic, and chlorinated solvents. The maximum service temperature for rubber compounds varies with the chemical and its concentration. The temperature limitation for continuous service is about 140°F for most soft rubber compounds, and for hard rubber it is about 180°F. However, heat-resisting compounds are available that may be used at somewhat higher temperatures. Soft rubber, especially compounded for maximum temperature resistance, may be used for continuous service under some chemical conditions up to 200°F, and hard rubber may be compounded for service temperatures as high as 230°F.

2. Synthetic Elastomers

A number of synthetic elastomeric materials have been developed with special characteristics that extend the overall usefulness of the elastomers for corrosion-resistant equipment. Notable among these are Buna S (GR-S), Buna N (GR-A), butyl (GR-I), neoprene (GR-M), Hypalon (chlorsulfonated polyethylene), and Thiokol compounds. In addition, polymers of ethylene and propylene have been developed with elastomeric properties. Like natural rubber, each of these may be compounded in several ways to yield maximum resistance to specific chemical exposures. Natural rubber and other elastomers are frequently used in combination with brick linings for temperature conditions that are above that allowed for the elastomeric material alone; they have proved to be excellent membrane linings for such construction (see Sec. VII.C).

C. Brick Linings

Brick-lining protection can be used for many conditions that are severely corrosive even to high-alloy metals. It should be considered for tanks, vats, stacks, vessels, and similar equipment. Brick shapes commonly used for such construction are made of carbon, red shale, or acid-proof refractory materials. Carbon bricks are useful for handling alkaline conditions as well as acid; the shale and the acid-proof refractory materials are used primarily for acid solutions. Carbon can also be used where sudden temperature changes are involved that would cause spalling of the other two materials. Red shale bricks generally are not used at temperatures above 300°F because of poor spalling resistance. Acidproof refractory bricks are sometimes used at temperatures up to 1600°F.

There are a number of cement materials that are regularly used for brick-lined construction. The most commonly used are based on sulfur, silicate, or resin. The resin cements include phenolic-epoxy- and furane-resin-based materials, which are used at temperatures up to 350°F. Carbon-filled phenolic resin cements have excellent resistance to all nonoxidizing acids and alkalis, salts, and organic solvents. Silica-filled compositions are available in both types of resins; they are almost equally resistant, except to hydrofluoric acid and alkalis. Sulfur-based cements are limited to a maximum temperature of about 200°F. In general, they have excellent chemical resistance to nonoxidizing acids and salts but are not suitable for use in the presence of alkalis or organic solvents. The sodium silicate-based cements have good resistance to all inorganic acids except hydrofluoric at service temperatures up to about 750°F. The potassium silicate-based cements are useful at somewhat higher temperatures, the upper limit depending upon specific conditions and requirements.

D. Concrete

Concrete is a material of construction not usually used under severe corrosive conditions other than as a substrate. For example, there are tanks and vessels whose shape and size make concrete an economical material of construction, provided there is a barrier that separates the corrosive environment from the concrete. Such a barrier sometimes is an elastomer or plastic sheet cemented in place, and often it is a protective coating applied by spray or trowel. For weathering atmospheres, concrete is protected against abnormal deterioration by the use of either a clear penetrating coating or a protective pigmented coating. Most common among the clear penetrating coatings are the silicone resins, which combine the water repellency of the silicone and the penetrating characteristics of the vehicle to prolong the life of the concrete. Since concrete is inherently alkaline (until it has weathered and reacted with the natural acids of the environment, such as carbonic acid), it is necessary to use an alkaline-resistant protective coating on new concrete. The coating may be based on vinyl resins, chlorinated rubber, or epoxies. Particularly not to be used on fresh concrete (unless acid-etched) are the oil-based paints; the presence of free alkali in the concrete will cause the oil-based paint to saponify and possibly be removed by rain or other weathering factors. The two major reasons for protecting concrete are (1) appearance and (2) improved longevity based on the fact that most concrete structures are reinforced, usually with steel in one form or another. If the concrete is not dense or is not protected, there is the possibility that moisture or other chemicals will penetrate the surface; under conditions of severe freezing and thawing the concrete will then spall, ultimately either exposing the reinforcing steel or allowing the moisture and its contaminants to attack the reinforcing material. When steel is attacked in a crevice condition such as exists where reinforcing rod or mesh is embedded in concrete, rust is formed, which results in expansion, further lifting the concrete and further exposing the reinforcing material.

E. Protective Coatings

It is important here to emphasize that it is unwise, and generally uneconomical, to try to use steel equipment with a chemically resistant coating to contain chemicals that are quite corrosive to the steel (see also Chap. 2A). This results from the fact that it is almost impossible to avoid some pinholes or holidays in the coating. Rapid attack of the steel will occur at such points, and continued maintenance attention will be required. This is the reason for the more stringent requirements on coatings for continuous-immersion service, such as tank linings. Such requirements include thickness (sometimes minimum and sometimes maximum), number of coats, freedom from pinholes, and degree of cure.

The chemically resistant coatings, such as baked phenolics, baked epoxies, and the air-dried epoxy, vinyl, and neoprene coatings, do well in minimizing contamination of chemicals handled in steel equipment. They should not be used where 100% protection from corrosion is required. An improved material for immersion service has been developed; it consists of flakes of glass dispersed in a polyester resin. This is applied by spray to a properly prepared surface, and the wet coating is rolled with a paint roller to orient the glass flakes in a plane parallel to the substrate, thereby providing maximum resistance to chemical attack.

F. Glass Linings

Glass-lined equipment is available that is chemically resistant to all acids except hydrofluoric and concentrated phosphoric acid (at ambient and elevated temperatures). It will resist many alkaline conditions at ambient and slightly higher temperatures. Specifically, the glass lining resists any comcentration of hydrochloric acid at temperatures up to 300°F, dilute concentrations of sulfuric acid to the boiling point, concentrated solutions of sulfuric acid to about 450°F, and all concentrations of nitric acid up to the boiling point. A special acid-resistant glass with improved alkali resistance is commercially available for use with alkaline solutions up to pH 12 at temperatures of 200°F. Glass-lined equipment, such as tanks, pressure vessels, reactors, pipelines, valves, and accessory equipment, are available. Improved resistance to impact has been developed for the glass linings. Methods of field repair of glass linings have been developed; these include the use of cover plates and plugs of tantalum in combination with resin cements and Teflon.

G. Wood

All woods are affected adversely by acids, particularly the strong oxidizing acids, but they are regularly used in dilute hydrochloric acid solutions at ambient temperature. Improved corrosion resistance can be imparted to wood by pressure impregnating the wood with resin solutions, such as asphalt, phenolic, and furane. This greatly extends the utility of woods in corrosive service. Strong alkaline solutions, particularly sodium hydroxide, generally cause disintegration and cannot be used with impregnated wood. Weak alkaline solutions can be used with wood equipment with reasonably good service life.

VIII. INHIBITORS

The corrosion of iron and other metals immersed in aqueous solutions can frequently be minimized or inhibited by the addition of soluble chromates, phosphates, molybdates, silicates, amines, or other chemicals, singly or in combination. Such materials are called inhibitors and are generally attractive for use in recirculating systems or closed systems. They are also used in neutral or very slightly acid solutions. Sodium silicate has also been effective as an inhibitor for aluminum in alkaline solutions. The concentration of an inhibitor for maximum corrosion control depends upon the solution, composition, temperature, velocity, metal system, and the presence of dissimilar metals in contact in the solution. Care should be taken in the selection and application of inhibitors, since in some instances they can cause increased localized attack.

Although chromate treatment is widely used for aqueous solutions, attention to maintain the concentration above the required minimum for specific environmental conditions is required. In addition, there is the always present danger of pollution from loss of chromate to the surrounding environment.

One of the most common uses of inhibitors is in brine systems. When calcium or sodium chloride brine is used in steel equipment, it is generally recommended that sodium dichromate be used. Where chromates cannot be used, disodium phosphate is recommended for sodium chloride brines. Where aluminum equipment is used in brine service, it is recommended that sodium dichromate be used on the basis of 1% of the chloride concentration.

For recirculating cooling water steel systems, it has been found that 0.02% by weight of sodium silicate (40° Baume) is effective in inhibiting corrosion. Sodium dichromate at 0.01% concentration is also effective and can be used where toxicity effects can be disregarded.

For preventing corrosion of steel and other ferrous-based materials, particularly in protecting machine parts and equipment in storage, the use of volatile (VCI) or vapor-phase (VPI) corrosion inhibitors has been found to be effective. These materials are amine nitrite salts. They can also be used to protect steel process equipment when idle or in standby condition. These materials are also available as crystals or as impregnated paper. The inhibitors are slightly volatile at atmospheric temperature; the protection obtained results from the diffusion and condensation of the vapors on the surface of the items being protected.

IX. OXIDATION AND HOT CORROSION*

The immense progress in technology in the past few years has imposed increasingly greater demands on the mechanical and chemical properties of metallic materials; in this section, we will be chiefly concerned with only the chemical aspect, specifically the oxidation and scaling resistance of the metallic components. The terms metal oxidation, tarnishing, and scaling are used in this context to describe the attack of a metal or an alloy by oxidizing gases, such as oxygen, sulfur, the halogens, or water vapor. This attack of metallic components can take place under a variety of conditions varying from the "mild" oxidizing conditions that may exist at room temperature in air to the severe conditions imposed by hot furnace gases contacting metallic surfaces. Especially stringent requirements are placed upon the scaling resistance of surfaces where condensed molten salts and oxidizing conditions simultaneously exist. The attack under such conditions is commonly referred to as hot corrosion and is a particularly severe form of corrosion.

Industrial materials without sufficient scaling resistance frequently fail after a short period of time as a result of rapid oxidation or hot corrosion, in conjunction with severe spalling due to poor adherence of the scale to the metallic component. As a result, the

---

*Contributed by S. R. SHATYNSKI.

permissible limits of wear are often exceeded and premature replacement of parts is required. An extensive effort is being made to develop alloys that are not only heat resistant but that also retain their mechanical properties at high working temperatures.

A. Oxidation

1. Scale Morphology

Scale morphology is dependent on the conditions of reaction, the rate of oxidation, the composition of the corrosive medium, and the type and composition of the particular alloy involved. A wide variety of scale morphologies can result depending upon the above conditions (15).

The most undesirable scale morphology is the porous oxide. If a scale is porous throughout its entire cross section (as shown in Fig. 4.7) the chemical composition and structure of the reaction products are of no essential significance to the course of oxidation. The oxidation rate is determined by the chemical reaction at the metal/scale interface since the oxidizing gas penetrates sufficiently rapidly through fissures in the scale. This mode of oxidation automatically excludes the possibility of using such a material at high temperatures. Fortunately, most materials do not behave in such a manner.

Frequently two or more multilayer scales form on a surface with at least one of the layers being impervious. The division of layers may be due to (1) variation of the morphological structure of the individual layers; for example, the layers may be distinguished by grain size, such as coarse or fine grained, or (2) different chemical composition of the layers. A single-phase scale is formed when only one compound can form; an example is the oxidation of cobalt at 1050°C, as shown in Fig. 4.8, where only CoO can form a homogeneous scale. At lower temperatures the oxidation of cobalt results in the formation of CoO and Co_3O_4. A schematic illustration of the initial stage of oxidation of single-phase and two-phase oxidation is shown in Fig. 4.9. Initially, when the scale is thin it is compact throughout the cross section and is closely adherent. At a later stage, as shown in Fig. 4.10, the morphology changes and a porous layer forms with no change in composition.

The structure and phase composition of scales are more complex when alloys are considered. A number of limiting types of scale morphologies can be distinguished on binary alloys. If each of two alloying components, Me and M, can form chemical compounds under reaction conditions with oxidant X and there is complete miscibility of MeX and MX, then irrespective of alloy composition a single-phase scale is formed, as

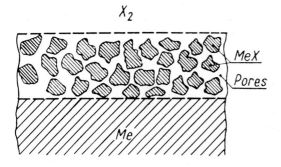

Fig. 4.7 Scheme of the structure of a porous scale. (From Ref. 15.)

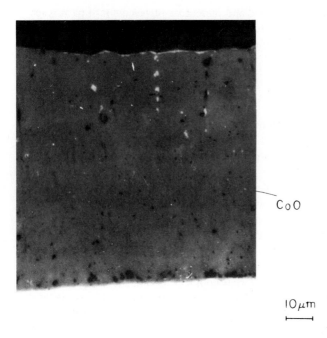

CoO

$10\,\mu m$

Fig. 4.8 Cross section of reaction zone above cobalt after 100 min at 1050°C in pure oxygen. (From Ref. 16.)

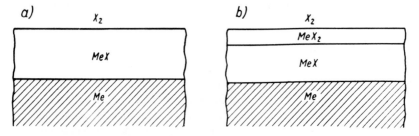

Fig. 4.9 Scheme of the structure of scale on metal in the early stage of its formation: (a) single-phase; (b) two-phase. (From Ref. 15.)

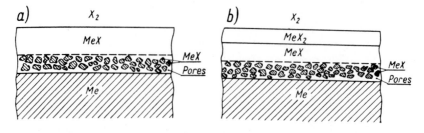

Fig. 4.10 Scheme of the structure of scale on metal in the later stage of its formation: (a) single-phase; (b) two-phase. (From Ref. 15.)

shown schematically in Fig. 4.11a. At a later stage of the oxidation a region of porosity will again develop in a way similar to previous examples; this is schematically illustrated in Fig. 4.11b. If the two oxidation products show no mutual solubility, then for alloys rich in more noble metal Me, a homogeneous scale composed of only MeX may be formed initially. On the other hand, in alloys rich in metal M, a scale consisting of MX may initially form. Schematically both are illustrated in Fig. 4.12. If, however, the alloy contains substantial amounts of both metals the scale can form a heterophasic mixture of MeX and MX in the manner shown schematically in Fig. 4.13a and b for an alloy rich in Me and rich in M, respectively. When MeX and MX display a certain mutual solubility, then the inner layer of MX saturated with Me becomes a three-layered scale with the development of a porous inner layer. If the attacking oxidant gas dissolves in the metallic phase in a concentration such that its activity in the alloy is sufficient to form a compound of the less noble metal with the oxidant, then in alloys rich in Me, under MeX islands of MX form as schematically shown in Fig. 4.14a. This region of precipitated oxide in the metal is called the internal oxidation zone. On alloys rich in M, a scale composed entirely of MX phase is formed, again with the possibility of an internal oxide zone (Fig. 4.14b). In certain specific cases, the partial pressure of the oxidant may be lower than the dissociative pressure of MeX but sufficient for creating MX. In that circumstance no scale but an internal oxide, as shown in Fig. 4.15, will be formed.

2. Thermodynamics of Oxidation

The thermodynamics of oxidation can be easily understood by considering the following generalized reaction:

$$bB + cC \rightarrow dD + eE$$

The Gibbs energy for such a reaction can be written as

$$\Delta G_I = \Delta G_I^o + RT \ln \frac{a_D^d a_E^e}{a_B^b a_C^c}$$

where $\Delta G_I^o = d \, \Delta G_D^o + e \, \Delta G_E^o - c \, \Delta G_C^o - b \, \Delta G_B^o$.

The criterion for a spontaneous reaction at a constant temperature and pressure is the reduction of ΔG. These reactions can be applicable to oxidation reactions, such as

$$Ni + \frac{1}{2} O_2 = NiO(s)$$

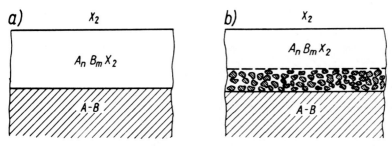

Fig. 4.11 Scheme of the structure of a single-phase scale on a binary alloy: (a) initial reaction state; (b) later stage. (From Ref. 15.)

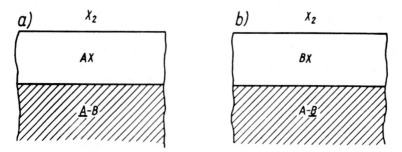

Fig. 4.12 Scheme of the structure of scale on a binary alloy in the case of selective oxidation: (a) alloy rich in metal Me; (b) alloy rich in metal M (initial reaction stage). (From Ref. 15.)

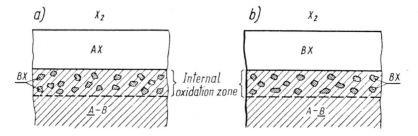

Fig. 4.13 Scheme of the structure of heterophasic scale on a binary alloy: (a) alloy rich in metal Me; (b) alloy rich in metal M. (From Ref. 15.)

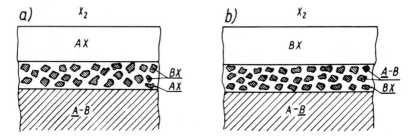

Fig. 4.14 Scheme of the structure of scale and internal oxidation zone in the simplest case of the formation of a single-phase: (a) alloy rich in metal Me; (b) alloy rich in metal M. (From Ref. 15.)

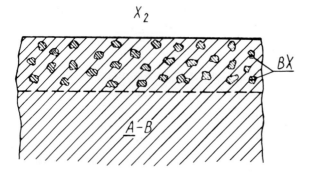

Fig. 4.15 Scheme of the formation of an internal oxidation zone in conditions in which the MeX compound is thermodynamically unstable. (From Ref. 15.)

It is possible to calculate the pressure of oxygen (P_{O_2}) at which the formation of pure NiO on Ni can just be prevented. Stating it another way, it is possible to calculate the partial pressure of oxygen necessary for the coexistence of Ni and NiO at equilibrium. At equilibrium,

$$\Delta G^o_{NiO} = -RT \ln \frac{a_{NiO}}{a_{Ni}(P_{O_2})^{\frac{1}{2}}}$$

Since both Ni and NiO are pure, then their standard states are of unit activity; therefore, rewriting the above equation

$$\Delta G^o_{NiO} = \frac{RT}{2} \ln P_{O_2}$$

or

$$P_{O_2} \text{ (Ni-NiO equilibrium)} = \exp \frac{2\Delta G^o_{NiO}}{RT}$$

$$= 7 \times 10^{-11} \text{ atm}$$

Using the above analysis it is possible to decide whether a given nonequilibrium oxygen partial pressure will result in oxidation of Ni to NiO. Writing the Gibbs energy as

$$\Delta G = +RT \ln (P_{O_2})^{\frac{1}{2}} \text{equil} - RT \ln (P_{O_2})^{\frac{1}{2}} \text{actual}$$

then $\Delta G < 0$ when PO2 (actual $>$ PO2(euil) and oxidation results. This is plotted in chart form as shown in Fig. 4.16 and is called on Ellingham diagram.

3. Kinetics of Oxidation

The process of compact scale formation can be thought of as a series of partial processes occurring in the solid phase and at phase boundaries. Thus the kinetics of scale formation can be further subdivided into individual heterogeneous surface reactions and the stepwise processes of transport of matter through the scale. The heterogeneous surface reactions are composed of

1. Adsorption and chemisorption of the oxidizing gas on the surface of the continuous layer formed during the initial stage of oxidation
2. Incorporation of the oxidizing ions originating at this surface into the crystal lattice
3. Transformation of metal from the metallic phase to the reaction product in the form of ions with electrons available to be transferred to the oxidant.

The processes of transport of matter through the scale then consists of

1. Diffusion of metal ions and the equivalent electrons from the metal/scale phase boundary to the layer surface (by means of defects)
2. Diffusion of oxidizing ions from the gas/scale boundary toward the metal surface and simultaneous electron diffusion in the opposite direction
3. A simultaneous and counterdirection diffusion of both reactants in the form of electrons and ions
4. Diffusion along grain boundaries

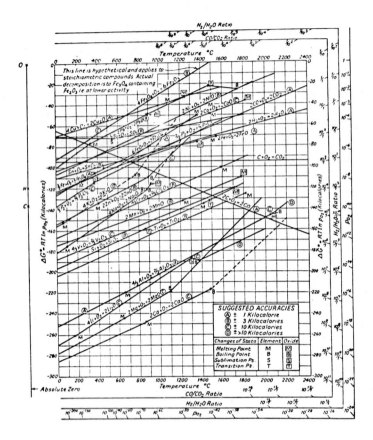

Fig. 4.16 Standard free energy of formation of oxides as a function of temperature. (From L. S. Darken and R. W. Gurry, *Physical Chemistry of Metals*, McGraw-Hill, New York, 1953. Modified from F. D. Richardson and J. H. E. Jeffes, *J. Iron Steel Inst.* 160:261, 1948.)

Depending on temperature, time, and pressure, one step from the above two sequences may be rate determining.

Kinetic measurements are generally carried out as a function of temperature and pressure and supply valuable information concerning the elementary processes that determine the overall kinetic rate. A large number of different rate laws have been observed during the oxidation of metals. The most simple is perhaps the linear rate law. The linear rate law has been observed when the product is either a gas or a liquid and thus leaves the metallic surface during the reaction, thereby allowing continuous direct contact between the metal and the oxidizing environment. In this case the reaction rate is independent of time and is determined by the chemical reaction resulting in product formation. The rate law then can be written as

$$\frac{dx_{Me}}{dt} = k_1$$

where

x_{Me} = loss of metal thickness
k_1 = linear rate constant of oxidation

On the other hand, where the reaction product is solid, the character of the oxidation process and its rate depend very much on the macro- and microstructure of the scale and the temperature. For example, if the scale is porous throughout the cross section, transport of the oxidizing gas can be rapid so that formation of the product at the metal/scale interface is the slowest process; linear kinetics results. Linear kinetics also occurs when the oxide is spalling concurrently with oxidation of the alloy.

Over a wide range of temperatures, it has long been noted that a parabolic rate law is obeyed:

$$\frac{dx}{dt} = \frac{k_p}{x}$$

where

x = thickness of the layer
k_p = parabolic rate constant

Wagner has theoretically related this rate law to the diffusion of ionic species through the scale. Thus, chemical reactions at the phase boundaries are faster than the diffusion. Hence, the slowest partial process is diffusion of one or both reactants in the reaction product phase. The diffusion of reactants in the scale takes place under the effect of the concentration gradient of lattice defects that result from a gradient of the chemical potential of the oxidant. Assuming that the transport rate of the electrons is much greater than the transport rate of the ionic species, Wagner theoretically derived the parabolic rate constant as a function of the self-diffusion coefficients of both species (D_1^* amd D_2^*), the concentration in equivalents ($\tilde{C}$), and the nonmetal activity:

$$K_p = \tilde{C} \int_{a_X'}^{a_X''} \left(\frac{Z_1}{Z_2} D_1^* + D_2^* \right) d \ln a_X$$

where a_X' and a_X'' are the nonmetal activities at the scale/metal and scale/gas interfaces, respectively.

Cubic kinetics has also been observed. Since the character of the course of oxidation is dependent on time and temperature, some metals can transform from a lower rate law to a cubic law:

$$X^3 = 3K_c t$$

Kinetic measurements are usually performed by measuring the mass of scale per unit surface area ($\Delta M/g$) as a function of time. Figure 17 illustrates the dependence of the kinetic reaction rate on temperature for titanium. Table 4.15 illustrates the effect of temperature on the kinetics of oxidation for many common metals. Experimentally, a thermogravimetric apparatus is used to measure such kinetics. This equipment, shown in Fig. 4.18, consists of a sample suspended from a sensitive balance into the hot zone of a furnace. The weight gain (which occurs during oxidation) and weight loss (during spallation and evaporation) can be monitored accurately as a function of time.

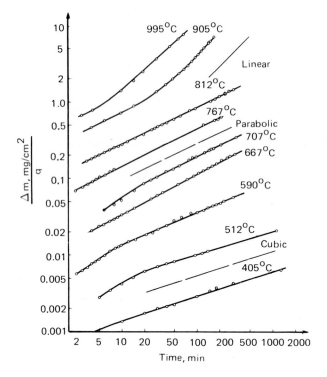

995°C 905°C

Linear

812°C

767°C

Parabolic

707°C

667°C

590°C

512°C

Cubic

405°C

$\Delta m, mg/cm^2$ / q

Time, min

Fig. 4.17 The course of titanium oxidation for a number of temperatures, in the double logarithmic plot. (From Ref. 15.)

Table 4.15 Effect of Temperature on the Course of Oxidation of Some Metals

Metal	Temperature, °C										
	100	200	300	400	500	600	700	800	900	1000	1100
Mg	log.		par.	paralin.	lin.						
Ca	log.		par.	lin.	lin.						
Ce	log.	lin.	incr.								
Th			par.		lin.	lin.					
U	par.	paralin.		incr.							
Ti			log.		cu.	cu.	paralin.		paralin.		
Zn			log. cu.			cu.			cu.	cu.	lin.
Nb			par.	par.	paralin.		lin.	lin.		incr.	
Ta	log. inv. log.			par.	paralin.		lin.		lin.		
Mo			par.	paralin.		paralin.	lin.		lin.		
W				par.		par.	paralin.		paralin.		paralin.
Fe	log.	log.	par.	par.	par.		par.		par.		par.
Ni		log.	log.	cu.	par.				par.		par.
Cu		log. cu. cu.			par.	par.		par.			
Zn		log.	log. par.								
Al	log. inv. log.		log.	par.	lin.						
Ge				par.		paralin.					

Denotations: log. — logarithmic law; inv. log. — inversely logarithmic law; cu. — cubic law; par. — parabolic law; paralin. — paralinear law; lin. — linear law, incr. — increased oxidation rate.

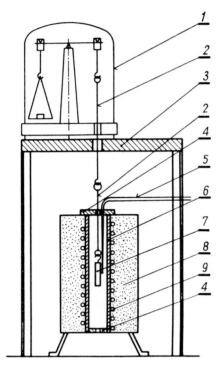

Fig. 4.18 Thermogravimetric apparatus for investigation of the oxidation kinetics of metals in air: (1) semimicroanalytical balance; (2) platinum wire; (3) heat-insulating plate; (4) ceramic plug; (5) thermoelement; (6) alundum tube; (7) sample; (8) heat-insulating mass; (9) heater. (From Ref. 15.)

4. Design Considerations

The oxidation protection of most modern alloys is dependent on the formation of a compact protective film of a slow-growing chemically stable oxide, such as Cr_2O_3, Al_2O_3, or SiO_2. The growth of these oxides is particularly slow since the native defect concentration responsible for oxidation during parabolic oxide growth is so small that it has yet to be measured. Choice of corrosion-resistant materials is based on their maintaining a compact protective scale so as to induce a slow-growing highly coherent oxide. Those alloys that depend upon the formation of a protective Cr_2O_3 layer are limited to service temperatures below 1000°C. Above this temperature volatile CrO_3 and, in the presence of water vapor, volatile $CrO_2(OH)_2$ are formed above the Cr_2O_3; rapid vaporization of the protective Cr_2O_3 results.

Few alloys use SiO_2 as the protective scale since alloys that have a high enough silicon content to allow the formation of a SiO_2 film have such poor mechanical properties that they are not of practical importance.

The formation of an Al_2O_3 protective layer appears to be the most useful of the protective oxides. But, note that in Fe-Al alloys, poor mechanical properties limit the amount of Al that can be used in such alloys. Furthermore, in the presence of a molten salt and an oxidizing atmosphere, rapid degradation may ensue in the temperature range 850-900°C.

All these protective oxides are susceptible to spallation upon thermal cycling. The great difference in the coefficient of thermal expansion between the protective oxides and the base alloys results in severe spallation of the oxide; thus, base metal is restored to contact with the oxidizing environment. Because of the parabolic nature of the oxidization kinetics, rapid oxidation ensues. After a number of thermal cycles the alloy adjacent to the protective scale can become depleted in those elements used to form protective oxides. In such a case more rapid oxidation will occur; enhanced degradation of the alloy will ensue. Despite such drawbacks, there are no useful alternatives to the formation of protective Cr_2O_3 or Al_2O_3 scales.

5. Stainless Steels

Ferritic stainless steels, high in chromium, depend upon this chromium for their high-temperature corrosion resistance. It has been noticed in practice that a Cr_2O_3 scale may form on an alloy at a temperature above 600°C when the chromium content is above 13% by weight (17,18). Such a scale has excellent protective properties and occurs in the form of a very thin layer containing up to 2% iron. The dependence of the oxidation rate on the chromium content is shown in Fig. 4.19a. It can be seen that when the alloy contains more than 18% Cr by weight the metal loss due to oxidation at 950°C is quite small. Such alloys are quite resistant to attack by water vapor at 600°C, as shown in Fig. 4.19b. Rohrig et al. (19) note that the ferritic stainless steels were the most resistant of all the steels examined. Fig. 4.20 shows the long-term isothermal oxidation resistance for some ferritic stainless steels after 10,000 hr at 815°C (20). It should be noted that types 410 and 430 behaved significantly better than 409. Type 409 has a Cr content of 11% by weight, whereas types 410 and 430 contain 11.5-13.5% Cr and 14-18% Cr, respectively. The advantage of type 410 over type 430 is due to the enhanced spallation of Cr_2O_3 on type 430 compared with type 410.

    The course of oxidation of austenitic stainless steels is in many ways analogous to that of the ferritic stainless steels. As before, it is desirable that a Cr_2O_3 scale forms as a protective layer on such alloys. Most investigators find that austenitic stainless steels exhibit better scaling resistance than ferritic stainless steels. This is attributed to the greater adhesion and spallation resistance of such scales. The influence of nickel on the oxidation resistance of chromium steels at 1000°C after 24 hr is shown in Fig. 4.21. Wood and coworkers (22-24) observed that at 1200°C in an initial period lasting approximately 15 min the reaction obeys the linear, not the parabolic, rate law and oxidation is quite high during this time. Bénard and coworkers (25) found that a relationship exists between the induction period, which is observed to be parabolic in nature, and the initial linear period. As shown in Fig. 4.22, the scale recovers after 15 min and proceeds in a parabolic manner. Because of the temperature of the oxidation reaction, it is believed that vaporization of the Cr_2O_3 plays an important role in this linear region. Figures 19 and 20 show the oxidation of some austenitic stainless steels. It is noticed that the oxidation of these steels is comparable with the ferritic stainless steels (with the exception of type 321).

6. Nickel and Cobalt Superalloys

Nickel-based superalloys, depending upon composition, can form Al_2O_3 or Cr_2O_3 surface layers. Those alloys that depend for their corrosion resistance upon the formation of Cr_2O_3 are essentially Ni-Cr alloys to which small amounts of Si, Zr, and C are added together with solid solution strengtheners W, Mo, Fe, and Co and carbide formers, such as Nb and V (26). However, one can usually consider such alloys to be essentially Ni-Cr alloys in which it is necessary to have 20-25% Cr present to maintain a coherent protective oxide. Initially in these alloys a thin layer of NiO is formed since its growth rate is

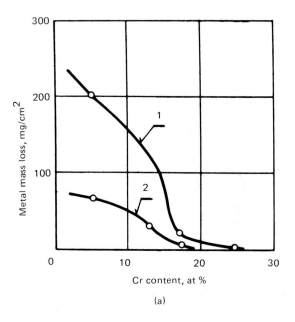

(a)

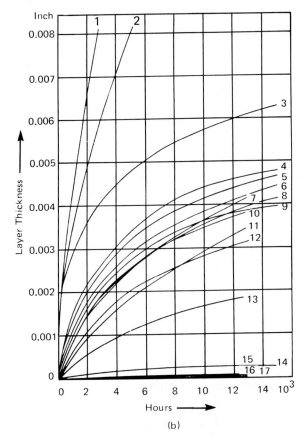

(b)

Fig. 4.19 (a) Dependence of the oxidation rate of iron-chromium alloys on alloying addition concentration after 20 hr oxidation: (1) 1100°C; (2) 950°C. (Data from Ref. 15.) (b) Corrosion with time of a few steels in water vapor at 595°C. (From Ref. 19.)

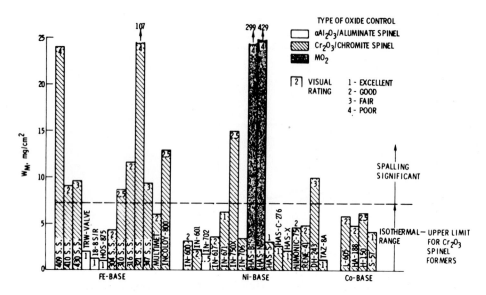

Fig. 4.20 Specific metal loss estimates W_M for 10-1000 hr exposures (10,000 hr total) at 815°C (1500°F) in static air for 33 high-temperature alloys compared to the W_M limiting value for Cr_2O_3/chromite forming alloys after 10,000 hr in isothermal oxidation. (From Ref. 20.)

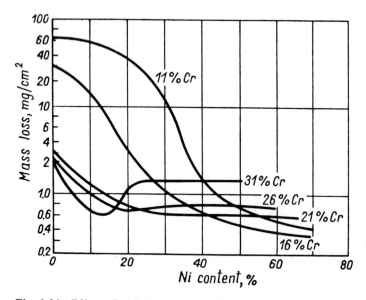

Fig. 4.21 Effect of nickel content on the oxygen corrosion rate of chromium steels at 1000°C after 24 hr oxidation. (From Ref. 21.)

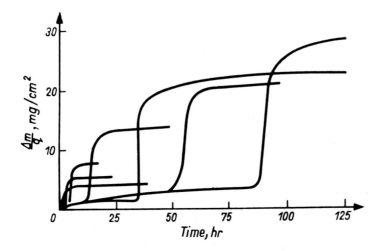

Fig. 4.22 Oxidation course of 18-8 type steel at 1050°C illustrating a relationship between the time of slow reaction course in induction stage in the following accelerated corrosion period. (From Ref. 15.)

much more rapid than Cr_2O_3. Beneath the NiO the Cr_2O_3 then begins to grow along with the chromium-nickel spinel $NiCr_2O_4$. As the Cr_2O_3 grows Cr depletion occurs, especially after successive spallation. When a significant depletion zone forms, then surface NiO formation and internal oxidation may occur.

The oxidation behavior of multicomponent γ' strengthened alloys can be estimated by considering the NiCrAl content of alloy. It is reported that initially for typical Ni(Cr,Al) alloys, such as U-500 and 713C, the spinel $Ni(Cr,Al)_2O_4$, NiO, and Cr_2O_3 initially form during transient oxidation. Thereafter, both Cr_2O_3 and Al_2O_3 are present as internal oxides. The proportions to which the elements are present in both the external and internal scales is determined by the alloy composition. Wasielewski and Rapp (26) report that the steady-state morphology can be classified into three types:

1. An external NiO scale with Cr_2O_3 and Al_2O_3 present as internal oxides for alloys that have low concentrations of both Cr and Al
2. A Cr_2O_3 scale with Al_2O_3 present as an internal oxide for high Cr (>15%) but low Al (<3%) contents
3. An exclusive α-Al_2O_3 scale for relatively high Cr (>15%) and high Al (>3%) contents

The critical parameter in such alloys is the Cr-Al ratio, which determines the composition of the protective scale. The oxidation kinetics of such alloys are generally quite complex and may depend on other elemental additions to the alloy, particularly the Nb, Mo, and W additions. These elements have an adverse effect on the oxidation resistance of the Ni-based superalloys. Barrett and Lowell (28) studied the cyclic oxidation of a series of Ni-Cr-Al alloys in still air for 500 1-hour heating cycles at 1100°C and 200 1-hour heating cycles at 1200°C. At 1100°C compositions estimated to have the best cyclic oxidation resistance were Ni containing 45% Al and Ni containing 30% Cr and 20% Al. At 1200°C the compositions estimated to have the best cyclic oxidation resistance were Ni-45% Al alloy and Ni containing 35% Cr-plus 15% Al. In general good cyclic oxidation resistance is associated with Al_2O_3 or $NiAl_2O_4$ formation. Barrett and Lowell (27,28) also investigated the isothermal and cyclic oxidation of numerous Ni-based sheet alloys. Figure 4.23 shows a comparison of the isothermal to cyclic oxidation of many of these alloys.

Cobalt-based alloys generally rely on chromium for high-temperature corrosion resistance. Therefore, most cobalt-based alloys contain at least 20-25% Cr so as to form a protective Cr_2O_3 film incorporating only minor amounts of CoO and $CoCr_2O_4$. Although chromium is generally recognized as the most useful element for high-temperature oxidation resistance, Beltran (29) has shown that yttrium significantly reduces the oxidation rate and improves scale adherence between 900 and 1200°C of a pure Co alloy containing 20% Cr. Similar results have been noticed in steels with additions of lanthanum and cerium. Work by Davin (30) demonstrated the effects of ternary alloying element additions in both Co-10Cr and Co-20Cr alloys oxidized in still air between 800 and 1200°C. The results are displayed as a bar graph in Fig. 4.24; the large difference in weight gain between the two sets of alloys should be noted. Obviously, the Co-30Cr alloys are far superior to the Co-10Cr alloys even at temperatures as high as 1200°C. The detrimental effect of Mo and Nb additions to the Co-30Cr alloys at 1200°C is evident. It should be noticed that the oxidation resistance of Co-30Cr is improved by additions of Zr, Ce, Al, and B; Ta appears to be of some benefit to Co-10Cr alloys. Wheaton (31) reported thermogravimetric studies on the oxidation of five cobalt base alloys WI-52, X-45, Mar-M-509, Mar-M-302, and

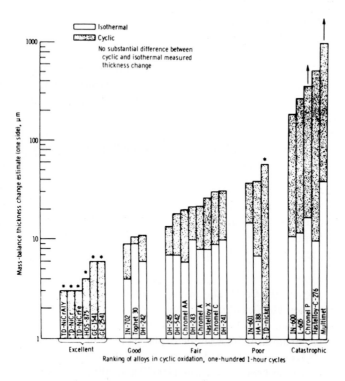

Fig. 4.23 Comparison of cyclic and isothermal oxidation for 100 hr in still air at 1150°C. Thickness change (one side) estimated from gravimetric data by mass-balance approach. (From Ref. 28.)

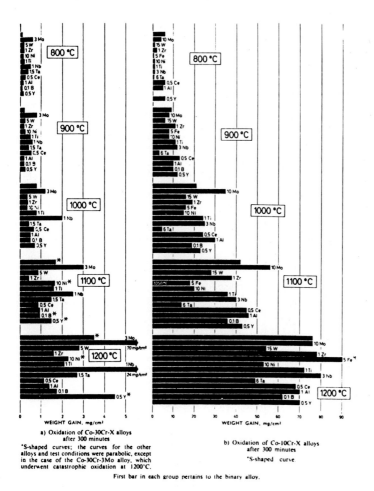

a) Oxidation of Co-30Cr-X alloys
after 300 minutes

*S-shaped curves; the curves for the other
alloys and test conditions were parabolic, except
in the case of the Co-30Cr-3Mo alloy, which
underwent catastrophic oxidation at 1200°C.

b) Oxidation of Co-10Cr-X alloys
after 300 minutes

*S-shaped curve.

First bar in each group pertains to the binary alloy.

Fig. 4.24 Effect of ternary elements on oxidation resistance of Co-10 and 30%Cr alloys in still air. (After Ref. 30.)

FSX-414 in static air at 1093°C (see Fig. 4.25). The continuous weight gain data indicated that the linear kinetic mode predominated for WI-52 and the parabolic kinetic mode for X-45 and Mar-M-509. On the other hand, FSX-414 and Mar-M-302 appear to oxidize at a rate slower than parabolic with FSX-414 the more resistant. These results suggest that the oxide formed on WI-52 is nonprotective; the rate-determining step is simply the surface reaction of metal converting to oxide. Conversely, the parabolic oxidation of X-45 and Mar-M-509 denotes a diffusion-controlled process, mainly the growth of protective Cr_2O_3. In terms of composition, the differences between the two types of alloys lend support to the above observations in that WI-52 contains 2% Nb in contrast to 3.5% Ta for Mar-M-509. The superior resistance of FSX-414 and Mar-M-302 are probably due to the 30%Cr and 9%Ta contents, respectively. Wlodek [reported by Beltran (20)], using the continuous weight gain technique, characterized the oxidation of X-40 and L-605. He concluded that the heterogeneous nature of the oxidation behavior of these alloys precluded establishment of an absolute rate behavior. Brief periods

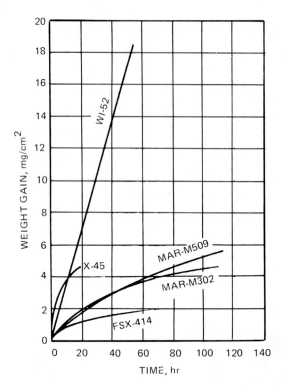

Fig. 4.25 Static oxidation of selected Co-based alloys at 2000°F (1093°C). (After Ref. 31.)

of parabolic rate kinetics were observed and appeared to be related to the occurrence of Cr_2O_3; spinel $CoCr_2O_4$ formed in all other cases examined. At higher temperatures (1204°C) catastrophic oxidation occurred with the formation of a low melting point scale containing $CoWO_4$, CoO, and $CoCr_2O_4$. Barrett (20) tested the isothermal oxidation resistance of four Co-based alloys, L-605, HA-188, H-150, and S-57, and found all of them to have adequate oxidation resistance after 10,000 hr at 815°C (as shown in Fig. 4.20); however, the cyclic oxidation of L-605 was noted as being poor, as seen in Fig. 4.23.

7. Refractory Metals

The oxidation of refractory metals is particularly severe. Molybdenum oxidizes to MoO_2 and MoO_3. The melting point of MoO_3 is 797°C, thus allowing rapid transport through the oxide (32). Furthermore, molybdenum oxides are highly volatile, forming $(MoO_3)_3$, MoO_3, and MoO_2. At 1100°C the vapor pressure of $(MoO_3)_3$ is approximately 1 atm in still air. Significant vapor pressures of the other two species are also apparent. The detrimental effect of Mo additions on the oxidation resistance of alloys has been noted by many authors. Tungsten oxides similarly have a high volatility and, as noted by Aliprando (16), linear kinetics were observed in all cases within the temperature regime of 750-1050°C. Although a large number of tungsten oxides are stable, only WO_2 and WO_3 were found in the temperature regime studied. Both Ta and Nb oxidize rapidly to form porous oxides. Only Nb_2O_5 is readily identified in scale formed in air oxidation at high temperature. The scale is porous on a very fine scale, besides having large cracks. The scale grows according to ionic diffusion-controlled growth until a certain scale thickness is reached and then the scale cracks and becomes porous. This process is then repeated many times. Cracking results from the growth mode being anion diffusion inward. In such cases oxide is formed at the metal/oxide interface. Other elements that grow by anion diffusion inward are Ti, Zr, V, and Hf. Because the ratio of the molar volume of oxide to molar volume of metal is much greater than unity and the oxide is continuously formed at the metal/oxide interface (beneath previously formed scale of given dimensions), the first formed (external) layers of scale are in tension and crack and the oxide becomes very porous. After oxide cracks and porous scale forms, molecular oxygen goes through the scale nearly to the metal. Direct reaction and rapid oxidation ensue.

B. Hot Corrosion of Superalloys

Hot corrosion is an accelerated form of oxidation that occurs due to the presence not only of an oxidizing gas, but also of a molten salt on the component surface. The molten salt interacts with the protective oxide so as to render the oxide nonprotective. Most commonly hot corrosion is associated with the catastrophic oxidation associated with the condensation of a thin molten film of Na_2SO_4 on superalloys, which are expected to withstand simple oxidation with a minimum of corrosion. Such superalloys are commonly used in components for gas turbines, particularly first-stage blades and vanes. Other examples of hot corrosion have been identified in energy conversion systems, particularly coal gasifiers and direct coal combusters. In these cases the salt originates from alkali impurities in the coal that condense on the internal components, thereby initiating the hot corrosion.

1. Reactions

The origin of the thin molten Na_2SO_4 deposition in gas turbines is believed to be related to the reaction between the residual sulfur in fuel and sodium, which may be contained either in the fuel or the intake air. The sodium in the air is frequently present as an aerosol of sea salt. An analysis of seawater indicates that the majority of sodium present is as NaCl but approximately 11% is present already as Na_2SO_4. The pertinent reaction is

$$2NaCl + SO_3 + \frac{1}{2}O_2 \rightarrow Na_2SO_4 + Cl_2$$

DeCrescente and Bornstein[33] have shown that such a reaction should go almost to completion. Bessen and Fryxell[34] determined the vapor pressure of Na_2SO_4 necessary for condensation of Na_2SO_4 on the cooler vanes and blades from the hotter gas. The dew points are shown in Fig. 4.26 for two pressures and for both complete sulfidation of NaCl and zero sulfidation of the NaCl. From such analysis it can be shown that salt concentrations of over 0.01 ppm in the intake air may be necessary to initiate hot corrosion.

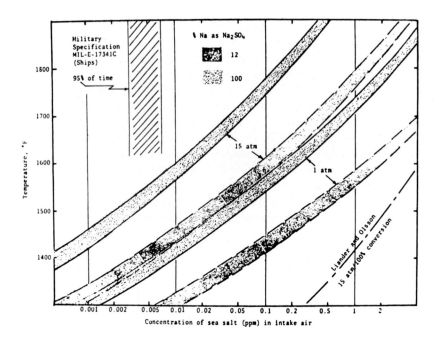

Fig. 4.26 Condensation temperature for Na_2SO_4 as a function of concentration and pressure. (From Ref. 34.)

2. Morphology of the Reaction

The morphology of the hot corrosion reaction is also different from simple oxidation in that concurrent oxidation and sulfidation commonly occur. It has been observed, as shown in Fig. 4.27, that a layer of sulfide particles forms beneath a region of porous oxide. In most service cases the sulfides are nearly always chromium rich; however, some reports of nickel sulfides have been reported. The extent of the internal sulfide layer can vary considerably. At times the sulfide layer appears to be virtually absent or consist of a very thin band of fine discrete sulfide particles; under different conditions a large band of interconnected sulfide particles may form. For the superalloys of importance the outer porous oxide consists of the simple oxide of the base metal, either NiO or CoO. Beneath this zone in some cases a zone of mixed oxide, sulfide, and pure metal may exist. In such cases the oxide may be a spinel, Cr_2O_3 (for chromia formers), or Al_2O_3(for alumina formers). As noted previously the sulfides are usually rich in chromium and the underlying metal is generally rather pure discontinuous fragments of the base metal.

3. Corrosion Resistance Tests

A number of tests have been used to test the hot corrosion resistance of alloys and to investigate the mechanism of hot corrosion. Depending on the particular objective the tests may either try to reproduce the hot corrosion kinetics and morphology or to examine one or more particular aspects of hot corrosion. One of the most common tests is the so-called crucible test. In this test an alumina crucible is filled with a salt of desired

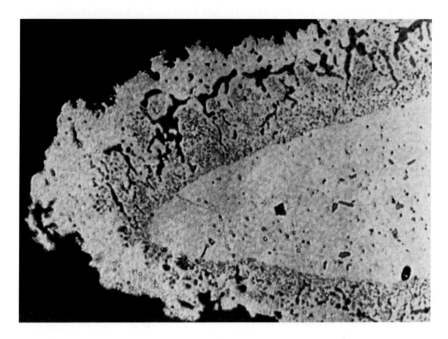

Fig. 4.27 Heavy internal sulfation at the trailing edge of an originally pack-aluminized Nimonic 105 turbine blade from a Proteus engine. (From Ref. 35. Crown Copyright Reserved.)

composition, which is melted. The test specimens are placed in the molten salt and held at a definite temperature for a prescribed length of time either in air or a controlled environment. Such a test is generally considered to be a poor representation of the hot corrosion observed during practice. The specimens are attacked in a deep salt melt differently than the attack in practice, where a thin molten salt is present on the alloy surface. In addition, the dissolution of the alumina crucible continuously changes the salt composition until a saturation level is reached. The crucible test is generally believed to be a more severe form of attack than actual practice. It is useful, however, as a means of classifying alloys and can be performed quite simply.

Another test that is widely used in the laboratory is the salt-coated test. In this case samples are heated to 150°C on a hot plate and then sprayed with an aqueous salt solution of desired composition. A number of varieties of this test exist. The samples can be periodically sprayed and, in addition, thermally cycled so as to more accurately model the thermal cycles and environmental conditions of in-service components. In many cases, coated samples are placed in alumina boats and exposed to a controlled atmosphere. The salt-coated samples can also be hung in a furnace and attached to a microbalance. The kinetics of the hot corrosion reaction can then be determined from weight change per unit area versus time measurements. Such thermogravimetric measurements may be performed in controlled environments. Using such salt-coating tests, it is possible to investigate various aspects of the mechanism of hot corrosion. By carefully controlling and modifying the conditions, variables such as temperature, time, salt, alloy, and gas composition can be separately or totally examined.

A variation of salt-coating test, the Dean-rig, has been used extensively in the United Kingdom. A crucible of salt of desired composition is placed in a hot zone of the furnace downstream from the sample. The sample is placed in a second hot zone, which is cooler than the salt crucible. Salt thus is vaporized and condenses on the samples placed at the lower temperatures. This test has been observed to be considerably milder than hot corrosion observed in practice.

Another technique that is rapidly gaining acceptance is the use of electrochemical cells to investigate the mechanism of hot corrosion. Rapp et al. (36,37) used electrochemical techniques to investigate the solubilities of Cr_2O_3, and α-Al_2O_3 in fused Na_2SO_4 at 927°C under conditions of known oxygen partial pressure and activity of Na_2O. A schematic of the experimental apparatus is shown in Fig. 4.28. The galvanic cell can be represented as

$$Na_2SO_4(1), O_2(g), Pt \quad \big| \quad Na^+\text{-} \quad \big| \quad Na_2SO_4(1), SO_2(g), O_2, Pt$$
$$(Na_2O) \text{ anode} \quad \big| \quad \text{mullite} \quad \big| \quad (Na_2O) \text{ cathode}$$

Because the Na^+ ion is the conducting species in mullite, the following half-cell reactions can be written:

Anode: $Na_2O(a) \rightleftharpoons i/2\, O_2(a) + 2e^- - 2Na^+(a)$

Mullite electrolyte: $2Na^+(a) \rightleftharpoons 2Na^+(c)$

Cathode: $2Na^+(c) + 2e^- + Na_2SO_4(c) \rightleftharpoons 2Na_2O(c) + SO_2(c)$

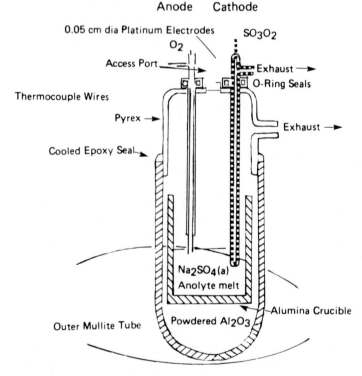

Fig. 4.28 Experimental electrolytic cell to study dissolution of Al_2O_3 in Na_2SO_4. (From Ref. 36.)

Overall: $Na_2O(a) + Na_2SO_4(c) \rightleftharpoons 2Na_2O(c) + \frac{1}{2}O_2(a) + SO_2(c)$

From this one can determine the open-circuit EMF for the reaction:

$$\epsilon = \epsilon^0 - \frac{RT}{2F}\ln\frac{a_{Na_2O}^2(c)\,P_{SO_2}(c)\cdot P_{O_2}^{1/2}(a)}{a_{Na_2O}(a)\cdot a_{Na_2SO_4}(c)}$$

where ϵ^0 is the standard cell potential. By measuring the EMF it is possible to determine the solubility of Cr_2O_3 and α-Al_2O_3 in Na_2SO_4 melt. Difficulties, however, also plague this experimental technique. Because the alumina crucible is soluble in the Na_2SO_4 melt, it is difficult to obtain reproducible results and to separate the dissolution of the container from the dissolution of the test specimen. Furthermore, no good stable reference electrodes are available for use in molten sodium sulfate. Rapp et al. (37) compared six reference electrodes in molten Na_2SO_4. Of these materials the most promising electrode appears to be the Ag/Ag^+ electrode, which consisted of a mullite tube containing a solution of 10m/o Ag_2SO_4 and 90m/o Na_2SO_4. Into this solution a 1 mm silver wire that was spot-welded to a platinum lead wire was immersed.

The test that perhaps best models the hot corrosion observed in practice is the burner rig test. Many of these rigs use a combustion chamber from a small gas turbine, using air from compressors and burning fuel in the usual way. Salt may be added either to the fuel or the air or sprayed into the hot combustion gases. The hot gases can then be led down a channel or may pass freely from the combustion chamber. They then pass over an array of rod samples, which may be in the form of miniature turbine blades. These samples are rotated in the gas stream on a carousel mount. An example of a burner rig using a dynamic combustor is shown in Fig. 4.29. In this particular rig salts were injected near the fuel nozzle. This rig also has the capability of introducing particulates so as to study the influence of erosion on the hot corrosion process.

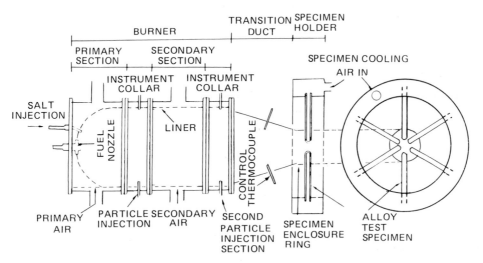

Fig. 4.29 Schematic diagram of dynamic combustor used in the hot corrosion studies. Salts were injected into the rig as an aqueous solution near the fueld nozzle. Particulates could be injected into the burner at the instrument collars to introduce an erosive component into the hot corrosion test. (From Ref. 38.)

4. Test Results

There has been great difficulty in correlating the results of various test procedures. Stringer (35) reports results of the Hot Corrosion Task Force of the Gas Turbine Panel of ASTM. They organized a round robin test of six alloys: 713C, IN100, IN738, Mar-M-421, Udimet 500, and Udimet 700. The results, shown in Table 4.16, were from 15 participants, of whom 13 used burner rigs. There are some important anomalies in these results. The test ranking U-700 as the best of the six alloys should be regarded with some suspicion, along with those that rank IN100 above 713C. In practice there is little doubt that U-500 has the best corrosion resistance, and this is borne out by many tests.

The results of these tests enable some generalizations to be made on aspects of the hot corrosion reaction. Concerning the kinetics of hot corrosion, Kaufman determined the rate of attack of several nickel-based alloys in a G.E. burner rig for up to 100 hr using 200 ppm salt. He found that the majority of the tests indicated that the samples corroded at nearly a parabolic rate with only a few alloys showing a transition to a linear rate. Hardt (39) found that coating pure Ni with Na_2SO_4 and heating to 900°C in dry oxygen resulted in rapid initial oxidation. If the Na_2SO_4 was removed by washing the specimen, the rate immediately slowed to that of simple oxidation.

5. Temperature Effects

Hot corrosion tests have a strong dependence on temperature. For most alloys, the corrosion rate is a maximum at 850-900°C, and decreases very markedly at temperatures up to 1000°C (35). Hence, it is generally thought that a molten salt is necessary in order to initiate not corrosion. Shatynski and Dannemann (40), however, have shown that it is possible to initiate hot corrosion in aligned Co-based TaC-strengthened alloys in a solid pack of $MgCo_3$ at 900°C. Since $MgCo_3$ is solid at this temperature, this indicates that, although in many cases a fused salt is present on the alloy surface during hot corrosion, hot corrosion can also be generated in packs of solid salts.

Table 4.16 Results of ASTM Round Robin Test

	Ranking of alloy by participant														
Alloy	A	B	C	D	E	F	G	H	I	J	K	L	M	N	O
U 500	1	2	2	2	1	3	1	1	3	3	3	3	1	3	1
IN 738	2	1	1	3	3	2	3	2	2	2	1	2	3	1*	2
Mar-M421	3	3	3	4	4	1	2	4	1	5	2	1	2	2.	3
U 700	4	4	4	1	2	4	4	3	4	1	4	4	4	4	4
713C	5	6	5	5	6	5	5	5	5	6	6	5	5	5	5
IN 100	6	5	6	6	5	6	6	6	6	4	5	6	6	6	6

G is a static rig. specimens coated with 50% Na_2SO_4, 50% NaCl and exposed to simulated combustion atmosphere.

H is a high pressure rig, 15 atm.

K is a crucible test with 90% Na_2SO_4, 10% NaCl.

N is a high-velocity rig, Mach 1.

The remainder are a variety of burner rigs. In most cases the test temperature was 899°C and the test duration was 100 hours.

Source: Ref. 25.

6. Effect of Salt Composition

The influence of salt composition on the hot corrosion has also been extensively studied. It has been found that additions of NaCl to Na_2SO_4 increase the severity of hot corrosion reaction. In fact, Waddams et al. (41) noted that as little as 1% NaCl in Na_2SO_4 can increase the hot corrosion dramatically. Archdale has noted that mixtures of Na_2SO_4 and carbon can also give accelerated attack. Bornstein et al. (42) examined additives to the salt and found that additions of 5.5% V_2O_5 to the Na_2SO_4 resulted in a corrosion rate representative of simple oxidation. Bornstein et al. (43,44) and Shatynski and Dannemann (40) have reported that hot corrosion can occur in salts of other compositions. Salts that have been shown to induce catastrophic oxidation are $NaNO_3$, NaCl, $MgSO_4$, $MgCO_3$, and Na_2CO_3.

7. Effect of Alloy Composition

The effect of alloy composition is still open to debate. It is generally conceded that the chromium content is the most important single factor in hot corrosion resistance. Much of the disagreement concerning the effects of other elements may be due to interactive effects within the alloy scale and the salt. Giggins and Pettit (38) have recently concluded that, in discussing the influence of alloy composition, some elements can produce beneficial effects over certain concentration ranges but can be deleterious over others. This is vividly illustrated in Fig. 4.30 for the corrosion of a Ni-20Cr-6 alloy and Ni-30Cr. Berg-

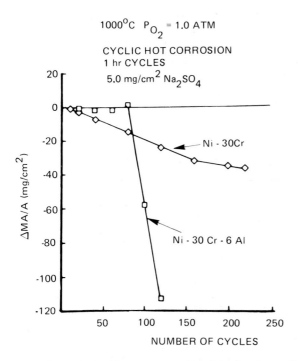

Fig. 4.30 Weight change versus time data for the cyclic hot corrosion of Ni-30Cr and Ni-30Cr-6Al specimens. The aluminum initially causes the Ni-30Cr-6Al to be more resistant than Ni-30Cr, but after longer times it causes more severe attack. (From Ref. 38.)

man et al. (45) extensively studied the influence of alloying elements on the corrosion resistance of nickel-based alloys and found chromium always to be beneficial; Co and Ta slightly improve the corrosion resistance; Ti seemed to do little; and Mo and W were seen to be detrimental especially as temperature was increased. Morrow et al. (46) dispute the detrimental effect of Mo. They found that the corrosion resistance increased for a given Al content as the Mo increased. In addition they found, using G. E. burner rig techniques, that the corrosion resistance decreased for a given Mo content with additions of Al (see also Table 4.17). Goebel et al. (47) used a salt-coating test to examine the influence of alloying elements on the hot corrosion of B-1900. They found that catastrophic corrosion was associated with the presence of Mo. Beltran (29) has reported the hot corrosion resistance of some complex cobalt-based alloys, and the results are shown in Fig. 4.31. Bourhis and St. John (48) have reported the hot corrosion of nickel-based alloys in Na_2SO_4 and NaCl using thermogravimetric techniques; their results are illustrated in Fig. 4.32.

8. Chromium Equivalent

Many investigators have attempted to predict the effect of alloying elements on hot corrosion in terms of chromium equivalent. Lewis and Smith (49) proposed an empirical equation for the chromium equivalent for a crucible test as

$$X_{LS} = \%Cr + 0.7\ (\%Al + 1.5\%Ti)$$

where X_{LS} is chromium equivalent. A more complex empirical relation was proposed by Donachie et al. (50):

$$X_R = \%Cr + 3.8(\%Al\text{-}5) + 2.0(\%W) - 12.5(\%C) - 1.4(\%Mo\text{-}1)$$

where X_R is chromium equivalent. Stetson and Moore (51) used a Solar gas turbine environmental simulator with JP-5 fuel and 35 ppm synthetic sea salt in the temperature range of 899-983°C to determine the following Cr equivalent.

$$X_{SM} = \%Cr + \tfrac{1}{2}\%Al - \%Mo$$

Table 4.17 Ranking of Nickel-Based Alloys by Various Criteria

Alloy	%Cr	%Mo	%Al	%Ti	X_{LS}	X_R	X_{SM}	C_F	log vol. loss (Ryan) 950°C
B 1900	8	6	6	1	13.2	3.6	5	2.14	− 7.743
IN 100	10	3	5.5	4.7	18.7	7.3	9.7	0.36	− 7.911
Nim 100	11	5	5	1.5	16.0	1.7	8.5	1.01	− 7.715
713C	12.5	4.2	6.1	0.8	17.6	10.7	11.3	2.18	− 7.584
U 700	15	5.2	4.2	3.5	21.6	5.9	12.1	0.31	− 7.499
Nim 105	15	5	5	1.2	19.8	8.5	12.5	0.96	− 7.444
Nim 90	20	0	1.4	2.4	23.5	6.1	20.7	0.13	− 7.860
U 500	19	4	2.9	2.9	24.1	7	16.4	0.23	− 7.448
Nim 80A	20	0	1	2.2	23.0	5.0	20.5	0.10	− 7.896
Hast X	22	9	0	0	22.0	− 6.2	13.0	—	− 7.072
Waspaloy	19.5	4.3	1.4	3	23.6	1.6	15.9	0.11	− 7.523

Source: Ref. 35.

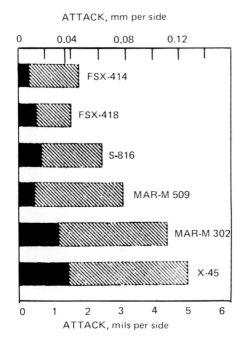

ATTACK, mm per side

FSX-414

FSX-418

S-816

MAR-M 509

MAR-M 302

X-45

ATTACK, mils per side

Fig. 4.31 Relative hot-corrosion resistance of complex cobalt-base alloys—G.E.-Schenectady burner rig data: 3%S residual oil, 325 ppm NaCl in fuel, 600 hr tests at 1600°F (871°C). Solid bar, surface loss; dashed bar, maximum penetration. (From Ref. 29.)

where X_{XM} is chromium equivalent. Ryan et al. (52) used a regression analysis to correlate volume loss due to corrosion with alloy content and temperature:

$$\log_{10} \text{(volume loss)} = 5.85 \times 10^{-9}\, T^3 - 1.34 \times 10^{-5}\, T^2 + 6.22 \times 10^{-2}\, (\%W)$$
$$+ 8.64 \times 10^{-2}\, (\%Mo) - 6.78 \times 10^{-2}\, (\%Cr)$$
$$- 8.98 \times 10^{-2}\, (\%Al) + 11.28$$

Finally, Felix (53) used a burner rig to study the corrosion of Ni-based alloys and after 300 hr with 115 ppm Na and 5 ppm V in the fuel, he reported the corrosion increased linearly:

$$C_F = \frac{\%Al}{(\%Ti)(Cr)^{1/2}}$$

where C_F is the corrosion function. All these empirical relationships have been compared. Felix overestimates the beneficial effect of Ti; Rentz overestimates the beneficial effect of Al and the harmful effect of Mo. Stetson and Moore's correlation is poor at the higher temperatures. Therefore, it appears that such empirical tabulations are only of limited value and do not universally predict the corrosion resistance of an alloy. Commercial superalloys often behave much better in laboratory tests than do model alloys with similar major element compositions. This suggests that either the minor elements or the detailed alloy structure may have a profound effect.

At present there are a number of theories that have been proposed to explain the hot corrosion of alloys. Perhaps the most popular theory is the acid-base fluxing model proposed by Bornstein and DeCrescente (54) and expanded by Goebel et al. (55). In order to simplify the fluxing processes that are relevant to hot corrosion, it is useful to present the appropriate stability diagrams; these are used as a means of predicting the

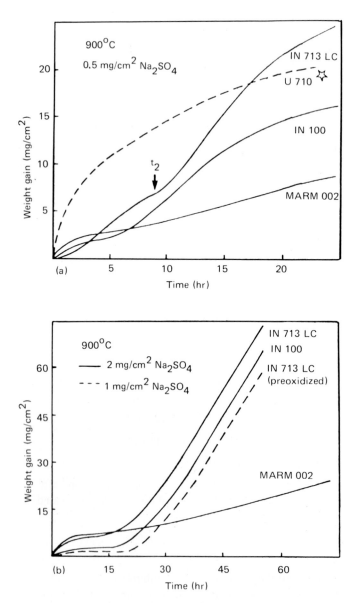

Fig. 4.32 Thermobalance curves for hot corrosion of the Al_2O_3 forming alloys with Na_2SO_4. (a) Moderate contamination. Four distinct regions are evident—an initial incubation period, a quasiparabolic region, a quasilinear region starting at t_2, and a plateau region. (b) Heavy contamination. The linear region is extended. Note that preoxidation suppresses the parabolic region only. (c) Thermobalance curves for hot corrosion with NaCl. (From Ref. 48.)

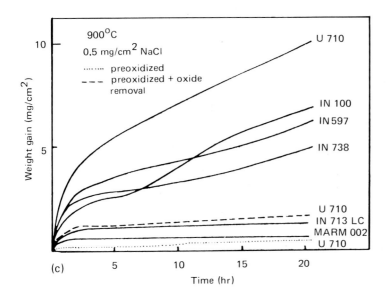

Fig. 4.32 (Continued)

fluxing reactions that are feasible. Consider the superimposed stability diagram for the phases of Ni, Al, and Cr that can exist in a Na_2SO_4 layer on a Ni-Cr-Al alloy, as shown in Fig. 4.33. Whether the oxides that are being formed on the alloy surface will make the salt basic or acidic is determined by the oxide ion concentration of the as-deposited salt; the affinities of the various relevant metals for oxide ions can be described using Fig. 4.33. This is a diagram on which are superimposed the Ni-O-S, Al-O-S, and Cr-O-S phases on the Na-O-S phase. The diagram indicates the phases that are stable in Na_2SO_4. It can be seen that there are acid melts ($\otimes$) for which NiO is more effective in developing basic conditions than Al_2O_3. On the other hand, there are basic melts ($\bullet$) for which Al_2O_3 is more effective in making acid conditions than Cr_2O_3. Basic fluxing is not self-sustaining; hence, the total amount of attack is dependent upon the amount of salt. The mechanism of basic fluxing is dependent on the ability of the salt to maintain a high concentration of oxide ions (O^{2-}) for reaction with the protective oxide layer. The possible basic salt-fluxing reactions, according to Giggins and Pettit (38), are shown in Table 4.18. In contrast to basic fluxing, acidic fluxing is self-sustaining. In this case small amounts of salt deposits can produce much more attack than when compared with basic fluxing. Salt deposits can be made acidic by two different processes. One process (alloy-induced acidity) involves the formation of oxides on the surface of alloys that have a great affinity for oxide ions, examples being MoO_3, WO_3, and Cr_2O_3. The other process occurs when there are gas species present that can make the salt acidic. The most common compounds that can make an acidic salt are SO_3 and V_2O_5. The pertinent equations for acid fluxing are shown in Table 4.18.

An electrochemical theory of hot corrosion has recently been proposed by Rapp and Goto (56). In their model chemical fluxing of the protective oxide can occur only when the solubility gradient, that is, the oxide solubility in the salt at the oxide/salt interface, is negative. In this way, the authors are able to explain the reprecipitation so

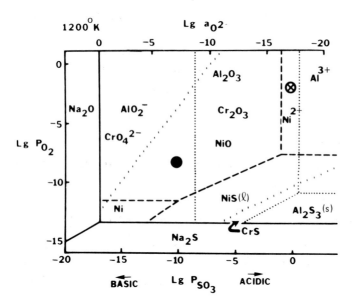

Fig. 4.33 Stability diagram to illustrate the phases of Nickel (– – –), aluminum (· · · ·), and chromium (XXX) that can exist in a Na_2SO_4 layer on a Ni-Cr-Al alloy. The Na_2SO_4 region is bounded by Na_2O and Na_2S and indicated by solid straight lines. In Na_2SO_4 of the composition ⊗ , NiO will dissolve making the Na_2SO_4 more basic whereas Al_2O_3 will not react with the melt. In Na_2SO_4 of the composition ●, Al_2O_3 will dissolve, making the melt more acidic while Cr_2O_3 will not react. (From Ref. 38.)

commonly observed during hot corrosion. This model is based on the solubility data of Stroud and Rapp (36). Based on the slopes of the solubility lines, the following acidic and dissolution reactions are substantiated:

$$Al_2O_3 \rightleftharpoons 2Al^{3+} + 3O^{2-} \qquad \text{acid dissolution}$$

$$Al_2O_3 + O^{2-} \rightleftharpoons 2AlO_2^{2-} \qquad \text{basic dissolution}$$

$$Cr_2O_3 \rightleftharpoons 2Cr^{3+} + 3O^{2-} \qquad \text{acid dissolution}$$

$$Cr_2O_3 + 2O^{2-} \rightleftharpoons 2CrO_4^{2-} \qquad \text{basic dissolution}$$

The oxide solubility gradient is established by the nature and site of the electrochemical reduction step that will generate basic ions. The interrelation of the gradient of the basic ions in the melt to the oxide solubility map determines the extent and continuance of the hot corrosion reaction. The advantage of this treatment over the foregoing fluxing model is that oxide ion production need not occur as a result of sulfur removal from Na_2SO_4.

C. Coatings

It is now common practice to apply some type of protective coating to extend the surface stability of the superalloy components. During cyclic oxidation and hot corrosion excessive spallation has been shown to occur. Coatings have been developed, particularly Co-25%Cr-6%Al-0.1%Y and Ni-15%Cr-6%Al-0.1%Y, that exhibit excellent resistance to thermal cycling. The improved resistance to spallation is generally attributed to the addi-

Table 4.18 Possible Salt Fluxing Reactions for Na_2SO_4 Deposits on Alloys

Basic processes
A. Dissolution of reaction product (i.e., AO) due to removal of sulfur and oxygen from the Na_2SO_4 by the metal or alloy:

$$SO_4{}^{2-} \text{ (sulfate} \to \frac{1}{2} S_2 \text{ (for reaction} + \frac{3}{2} O_2 \text{ (for reaction} + O^{2-} \text{ (for reaction}$$
$$\text{deposit)} \qquad \text{with alloy)} \qquad \text{with alloy)} \qquad \text{with AO)}$$

Reaction between AO and oxide ions can follow two courses:

  1. Continuous dissolution of AO

  $$A(\text{alloy}) + \frac{1}{2} O_2 + O^{2-} \to AO_2{}^{2-}$$

  Na_2SO_4 is converted to Na_2AO_2 and attack is dependent on amount of Na_2SO_4 initially present.
  2. Solution and reprecipitation

  $$A(\text{alloy}) + \frac{1}{2} O_2 + O^{2-} \to AO_2{}^{2-} \text{ (solution)} \to AO(\text{precipitate}) + O^{2-}$$

  A supply of SO_3 is required in order for attack to proceed indefinitely; otherwise attack will stop when melt becomes sufficiently basic at precipitation site.

B. Solution and precipitation of AO as a result of a negative gradient in solubility of AO in Na_2SO_4 (24).

Acidic processes

Gas Phase Induced
C. Formation of ASO_4 in Na_2SO_4:

$$A(\text{alloy}) + SO_3 + \frac{1}{2} O_2 \to A^{2+} + SO_4{}^{2-}$$

Continuous solution of ASO_4 in Na_2SO_4 requires continuous supply of SO_3 and O_2 from gas.
D. Solution and precipitation of AO in Na_2SO_4 due to reduction of SO_3:

$$A(\text{alloy}) + SO_3 \text{ (from gas)} \to A^{2+} + SO_3{}^{2-} \text{ (in melt)}$$

$$A^{2+} + SO_3{}^{2-} + \frac{1}{2} O_2 \text{ (from gas)} \to AO \text{ (precipitate)} + SO_3$$

E. Solution and precipitation of AO as a result of a negative gradient in solubility of AO in Na_2SO_4 as in B (Ref. 38).

Alloy phase induced
F. Solution of AO in Na_2SO_4 modified by second oxide from alloy (i.e., BO_3). Modification of Na_2SO_4 by BO_3:

$$B(\text{alloy}) + \frac{3}{2} O_2 + SO_4{}^{2-} \to BO_4{}^{2-} + SO_3$$

Solution reaction for AO; Na_2SO_4 becomes enriched in ABO_4:

$$A(\text{alloy}) + B(\text{alloy}) + 2O_2 \to A^{2+} + BO_4{}^{2-}$$

Solution-precipitation:

$$A(\text{alloy}) + B(\text{alloy}) + 2O_2 \to A^{2+} + BO_4{}^{2-} \to AO + BO_3$$

Precipitation of AO in Na_2SO_4 as a result of loss of BO_3 from Na_2SO_4 permits substantial attack with small amounts of Na_2SO_4.

Source: Ref. 38.

tion of yttrium. From these coatings a compact adherent scale that is either Al_2O_3/ $CoAl_2O_4$ or Al_2O_3/$NiAl_2O_4$ is formed. This has a low parabolic growth rate, thus protecting the underlying base alloy. Both $CoAl_2O_4$ and $NiAl_2O_4$ spinels have been shown to grow rather slowly compared with most oxides. Such coatings degrade after a period of time since, in addition to attack of the protective oxide by spallation, erosion, and chemical attack, inward diffusion of those critical elements (Al and Cr), which can form new scale, occurs. Thus, although coatings can improve the oxidation resistance, degradation can occur and oxidation of the alloy will ensue.

REFERENCES

1. NBS Special Publication 511-2. Economic Effects of Metallic Corrosion in the United States. Appendix B. A Report to NBS by Battelle Columbus Laboratories. SD Stock No. SN-003-01927-5. (1978). NBS GCR78-122. Battelle Columbus Laboratories Input/Output Tables. Appendix C of Economic Effects of Metallic Corrosion in the United States. PB-279-430 (1978).
2. NBS Special Publication 511-1. Economic Effects of Metallic Corrosion in the United States. A Report to Congress by the National Bureau of Standards (Including Appendix A, Estimate of Uncertainty). SD Stock No. SN-003-003-0192607 (1978).
3. Y. Kolotyrkin, quoted in *Sov. Life*, 9, 168 (1970).
4. D. Behrens, *Br. Corros. J.* 10 (3), 122 (1975).
5. V. Vlasaari, *Talouselama* (Economy), No. 14/15, 351 (1965) (quoted by Linderborg, Ref. 4).
6. S. Linderborg, *Kemian Teollusius* (Finland), 24, (3), 234 (1967).
7. F. K. Tradgaidh, *Tekn. Tedskrift* (Sweden), 95 (43), 1191 (1965) (quoted by Linderborg, Ref. 4).
8. T. P. Hoar (Chairman), Report of the Committee on Corrosion and Protection, Dept. of Trade and Industry, H.M.S.O., London (1971). An independent corrosion survey by P. Elliot, supplement to *Chem. Engr.* No. 265, Sept. (1973), substantiates the findings of the Hoar report.
9. T. P. Hoar, Information Conference, Corrosion and Protection, presented at the Instn. Mech. Engrs., April 20-21, 1971, to discuss the Hoar Committee Report.
10. K. S. Rajagopalan, Report on Metallic Corrosion and Its Prevention in India, CSIR. Summary published in *The Hindu*, English language newspaper (Madras), Nov. 12, 1973.
11. R. W. Rene and H. H. Uhlig, *J. Inst. Engr. Austral.* 46 (3-4), 3 (1974).
12. *Boshoku Gijutse* (Corrosion Engineering Journal), 26 (7), 401 (1977).
13. H. H. Uhlig, *Corrosion* 6 (1), 29 (1950).
14. NACE Committee Survey Report, Corrosion, October 1975.
15. S. Mrowec and T. Weber, *Gas Corrosion of Metals*, National Bureau of Standards TT76-54038, 1978.
16. J. J. Aliprando, Master's Thesis, Rensselaer Polytechnic Institute, 1979.
17. W. Smeltzer, *Trans. Canad. Inst. Met. Min.*, 65, 366 (1967).
18. M. Brabers and C. Birchenall, *Corrosion*, 14, 79 (1958).
19. I. A. Rohrig, R. M. van Duzer, and C. H. Fellows, *Trans. ASME*, 66, 277 (1944).
20. C. A. Barrett, in *Proceedings of Conference on Environmental Degradation of Engineering Materials*, 319 (1977).
21. A. Brasunas, J. Gow, and O. Harder, *Proc. ASM.*, 46, 870 (1946).
22. M. Hobby and G. Wood, *Oxid. Metals*, 1, 23 (1969).
23. G. Wood and M. Hobby, *J. Iron Steel Inst.*, 203, 54 (1965).
24. G. Wood, M. Hobby, and B. Vaszko, *J. Iron Steel Inst.*, 202, 685 (1964).

25. J. Bernard, J. Hertz, Y. Jeannin, and J. Moreau, *Rev. Métallurgie*, 57, 389 (1960).
26. G. E. Wasielewski and R. A. Rapp, High Temperature Oxidation, in *Superalloys*, C. T. Sims and W. C. Hagel (eds.), John Wiley & Sons, New York, 1972).
27. C. A. Barrett and C. E. Lowell, *Oxid. Metals*, 11, 199 (1977).
28. C. A. Barrett and C. E. Lowell, *Oxid. Metals*, 9, 307 (1975).
29. A. M. Beltran, *Cobalt*, 46, 3 (1970).
30. A. Davin, D. Coutsouradis, and L. Habraken, *Cobalt*, 35, 67 (1967).
31. H. L. Wheaton, *Cobalt*, 29, 163 (1965).
32. E. A. Gulbransen and S. A. Jansson, Vaporization Chemistry in the Oxidation of Carbon, Silicon, Chromium, Molybdenum and Niobium, in *Heterogeneous Kinetics at Elevated Temperatures*, G. R. Belton and W. L. Worrell (eds.), Plenum Press, New York, 1970.
33. M. A. DeCrescente and N. W. Bornstein, *Corrosion*, 24, 127 (1968).
34. I. I. Bessen and R. E. Fryxell, Proc. Gas Turbine Mat. Conf. Nav. Ship Eng. Cent., Hyattsville, Maryland, 1972, p. 73.
35. J. Stringer, *Ann. Rev. Mat. Sci.*, 7, 477 (1977).
36. W. P. Stroud and R. A. Rapp, The Solubilities of Cr_2O_3 and α-Al_2O_3 in Fused Na_2SO_4 at $1200°K$, *Proceedings of the Symposium on High Temperature Metal Halide Chemistry*, D. L. Hildenbrand and D. D. Cubicciotti (eds.), The Electrochemical Society, Inc., Princeton, N. J., 1978, pp. 574-594.
37. G. W. Watt, R. F. Andresen and R. A. Rapp, *J. Electrochem. Soc.*, to be published.
38. G. S. Giggins and F. S. Pettit, Pratt and Whitney Aircraft Report No. FR-11545, September 1978.
39. R. W. Hardt, J. R. Gambino, and P. A. Bergman, *ASTM STP* 421, 64.
40. S. R. Shatynski and K. A. Dannemann, *Oxid. Metals*, in press.
41. J. A. Waddams, J. C. Wright, and P. S. Gray, *J. Inst. Fuel*, 32, 246 (1959).
42. N. S. Bornstein, M. A. DeCrescente, and M. A. Roth, in *Deposition and Corrosion in Gas Turbines*, A. B. Hart and A. J. B. Culter (eds.), Applied Science Publishers, London, 1973, p. 70.
43. N. S. Bornstein, M. A. DeCrescente, and M. A. Roth, *Trans. AIME*, 245, 1947 (1969).
44. N. S. Bornstein and M. A. DeCrescente, *Met. Trans.* 2, 1971 (1971).
45. P. A. Bergman, C. T. Sims and A. N. Beltran, *ASTM STP* 421, 38 (1967).
46. H. Morrow, D. L. Sponseller, and E. Kahns, *Met. Trans.*, 5, 673 (1974).
47. J. A. Goebel, F. S. Pettit, and G. W. Goward, *Met. Trans.*, 4, 261 (1975).
48. Y. Bourhis and C. St. John, *Oxid. Metals*, 9, 507 (1975).
49. H. Lewis and R. A. Smith, *Proc. of 1st Int. Cong. Met. Corrosion*, 202 (1965).
50. M. J. Donachie, R. A. Spraque, R. N. Russell, K. G. Boll, and E. F. Bradley, *ASTM STP* 421, 85 (1967).
51. A. R. Stetson and V. S. Moore, *Proc. Gas Turbine Mat. Conf. Naval Ship Eng. Cent. Hyattsville, Md.*, 43 (1972).
52. K. H. Ryan, J. R. Kildsig, and P. E. Hamilton, Allison Div. Gen. Motors Tech. Rep. to Air Force Mat. Lab WPAFB AFML-TR-67-306, p. 100.
53. P. C. Felix, *Proc. of Gas Turbine Mat. Conf. Naval Ship Eng. Center, Hyattsville, Md.*, 331 (1972).
54. N. S. Bornstein and M. A. DeCrescente, *Trans. TMS-AIME*, 245, 1947 (1969).
55. J. A. Goebel, F. S. Pettit, and G. W. Goward, *Met. Trans.*, 4, 261 (1973).
56. R. A. Rapp and K. S. Goto, The Hot Corrosion of Metals by Molten Salts, in *Symposium of Fused Salts*, Pittsburgh, Pa., 1978, The Electrochemical Society, Princeton, N. J., 1979.

5

Magnetic Materials*

I. WHY AND HOW MAGNETIC MATERIALS ARE USED

A. Importance of Flux

Engineering magnetic materials are used because they provide large values of magnetic flux density B. In general they either provide B in response to small applied fields H_a or they maintain B in spite of demagnetizing fields H_d. These two functions are performed by soft and hard magnetic materials, respectively.

The flux maintained by hard magnetic materials as defined here may be used to store information and may be intentionally reversed. This type of application includes computer memory cores, recording tape, plated wire memories, and other communications-related components. Materials related to such applications, for example, recording tape, are not discussed further here.

B. Importance of Ferromagnetism

The large flux density that useful magnetic materials can provide is a consequence of their being ferromagnetic. This means that they can show large values of magnetization M even in a small field H. For any material,

$$B = \mu_0 (H + M)$$

where B is the flux density, μ_0 the space permeability, H the magnetizing force, and M the magnetization. Thus in a ferromagnetic material most of the flux can come from the material rather than the applied field. Without it, B and $\mu_0 H$ would be the same. The quantity $\mu_0 M$, the contribution of the material to B, is often called the intrinsic induction B_i.

Most substances are for practical purposes nonmagnetic. They are either paramagnetic, showing a very weak magnetization in the direction of an applied field, or diamagnetic, showing a very weak magnetization in the opposite direction. A few materials show ferromagnetism, becoming strongly magnetized in small fields. The elements iron, cobalt, and nickel are ferromagnetic at room temperature and are the bases of a

---

*Contributed by DANIEL L. POTTS.

variety of alloys used for magnetic applications. Pure manganese is paramagnetic, but many alloys of manganese show room temperature ferromagnetism. Some of the rare earth elements are ferromagnetic at low temperatures.

Ferromagnetism occurs because the individual atomic magnetic moments are coupled together by strong internal forces, which occur only in ferromagnetic materials. As the temperature is raised, thermal agitation tends to randomize these atomic moments, decreasing M. Above a temperature called the Curie temperature, thermal agitation has overwhelmed the internal coupling and the material is no longer ferromagnetic.

C. Use of Flux

The flux density B is of paramount importance in electrical engineering. It can be used either statically or dynamically, as shown in Table 5.1.

II. UNITS

Systeme Internationale (SI) units are coming into increasing use in physics and engineering and will ultimately be used as primary units throughout. The seven base quantities of this system and the corresponding units and their symbols are given in Table 5.2. Some typical derived units and their symbols are given in Table 5.3. Table 5.4 gives a list of physical quantities used in electricity and magnetism along with their symbols and with the symbol for the SI unit of each quantity. At present specifications contain a variety of metric and customary units. Table 5.5 gives conversions among them.

III. BASIC TERMS USED IN DESCRIBING MAGNETIC MATERIALS

All ferromagnetic materials show two distinctive characteristics:

1. The variation of B with H is nonlinear. For low values of H and B, the non-linearity is associated with changing physical mechanisms of magnetization reversal. At large H, it results from the material becoming magnetized to saturation.
2. If H is varied cyclically, the dependence of B on H shows hysteresis, no matter how slowly H is varied.

Table 5.1

Type of Interaction	Role of B	Typical Device
electrical	induce a voltage proportional to dB/dt	transformer, reactor
electro-mechanical	generate a force on a current-carrying conductor proportional to B	motor, loudspeaker
mechanical	create an attractive force proportional to B^2	latch, relay, separator

Table 5.2

length	meter	m
mass	kilogram	kg
time	second	s
electric current	ampere	A
thermodynamic temperature	kelvin	K
luminous intensity	candela	cd
amount of substance	mole	mol

The initial magnetization curve begins at the origin and shows the way in which B increases with H in an initially demagnetized sample. Subsequent cycling of H takes the material around a hysteresis loop. At sufficiently large H the size of the loop reaches its maximum value. Figure 5.1 shows that the initial magnetization curve is the locus of the tips of successively larger symmetrical hysteresis loops.

The area enclosed by a hysteresis loop represents irreversible energy loss, which appears as heat in the material, however slowly the loop is traversed. The limiting value for very slow changes in B and H is called the DC hysteresis loss.

The total flux density B consists of that due to the field H plus that due to the magnetization M of the material:

$$B = \mu_0 (H + M)$$

where μ_0 is the permeability of empty space, equal to $4\pi \times 10^{-7}$ H/m. The intrinsic induction B_i is the contribution $\mu_0 M$ of the material itself. In sufficiently large H this reaches a saturation value B_s.

When H = 0, the value of B on the hysteresis loop is B_r, the remanence. The field H at which B = 0 is H_c, the coercive force. The field for which M = 0 is H_{ci}, the intrinsic coercive force.

Table 5.3

Quantity	Name of SI Derived Unit	Symbol	Expressed in Terms of SI Base or Derived Units
frequency	hertz	Hz	$1 \text{ Hz} = 1 \text{ s}^{-1}$
force	newton	N	$1 \text{ N} = 1 \text{ kg} \cdot \text{m/s}^2$
pressure and stress	pascal	Pa	$1 \text{ Pa} = 1 \text{ N/m}^2$
work, energy, quantity of heat	joule	J	$1 \text{ J} = 1 \text{ N} \cdot \text{m}$
power	watt	W	$1 \text{ W} = 1 \text{ J/s}$
quantity of electricity	coulomb	C	$1 \text{ C} = 1 \text{ A} \cdot \text{s}$
electromotive force, potential difference	volt	V	$1 \text{ V} = 1 \text{ W/A}$
electric capacitance	farad	F	$1 \text{ F} = 1 \text{ A} \cdot \text{s/V}$
electric resistance	ohm	Ω	$1 \Omega = 1 \text{ V/A}$
electric conductance	siemens	S	$1 \text{ S} = 1 \Omega^{-1}$
flux of magnetic induction, magnetic flux	weber	Wb	$1 \text{ Wb} = 1 \text{ V} \cdot \text{s}$
magnetic flux density, magnetic induction	tesla	T	$1 \text{ T} = 1 \text{ Wb/m}^2$
inductance	henry	H	$1 \text{ H} = 1 \text{ V} \cdot \text{s/A}$
luminous flux	lumen	lm	$1 \text{ lm} = 1 \text{ cd} \cdot \text{sr}$
illuminance	lux	lx	$1 \text{ lx} = 1 \text{ lm/m}^2$

Table 5.4 Electricity and Magnetism

Quantity	Symbol	International Symbol for Unit
quantity of electricity	Q	C
charge density	ρ	C/m^3
surface charge density	σ	C/m^2
electric potential	V, Φ	V
electric field strength	E, E	N/C, V/m
electric displacement	D, D	C/m^2
capacitance	C	F
permittivity: $\epsilon = D/E$	ϵ	F/m
permittivity of vacuum	ϵ_0	F/m
relative permittivity: $\epsilon_r = \epsilon/\epsilon_0$	ϵ_r	–
dielectric polarization: $P = D - \epsilon_0 E$	P, P	C/m^2
electric susceptibility	χ_e	–
polarizability	α, γ	$C \cdot m^2/V$
electric dipole moment	p, p	$C \cdot m$
electric current	I	A
electric current density	j, J	A/m^2
magnetic field strength	H, H	A/m
magnetic induction	B, B	T
magnetic flux	Φ	Wb
permeability: $\mu = B/H$	μ	H/m
permeability of vacuum	μ_0	H/m
relative permeability: $\mu_r = \mu/\mu_0$	μ_r	–
magnetization: $M = \dfrac{B}{\mu_0} - H$	M, M	A/m
magnetic susceptibility	χ_m	–
electromagnetic moment	μ, μ, m, m	$A \cdot m^2$
magnetic polarization: $J = B - \mu_0 H$	J	T
magnetic dipole moment	j, j	$Wb \cdot m$
resistance	R	Ω
reactance	X	Ω
impedance: $Z = R + iX$	Z	Ω
admittance: $Y = 1/Z = G + iB$	Y	S
conductance	G	S
susceptance	B	S
resistivity	ρ	$\Omega \cdot m$
conductivity: $1/\rho$	γ, σ	S/m
self-inductance	L	H
mutual inductance	M, L_{12}	H
phase number	m	–
loss angle	δ	rad
number of turns	N	–
power	P	W
Poynting vector	S, S	W/m^2
magnetic vector potential	A	Wb/m

The permeability μ for a given value of B or H is the slope of a line through the origin and the point (B,H) on the initial magnetization curve. The limit for very small B and H is μ_i, the initial permeability. The greatest value is μ_m, the maximum permeability. If at some point H is varied slightly, the reversible permeability μ_r is not the slope of the magnetization curve at that point but a smaller quantity. This is illustrated in Fig. 5.1.

In the cgs (centimeter-gram-second) system,

$$B = H + 4\pi M$$

Table 5.5 Conversion Table: Metric Versus Customary Systems

Multiply	By	To obtain
B-Lines per sq. cm. (gausses)	6.4516	B''-Lines per sq. in.
B''-Lines per sq. in.	0.1550	B-Lines per sq. cm. (gausses)
H-Oersteds	2.0213	(A.T.'') Ampere turns per inch
(A.T.'') Ampere turns per inch	0.4949	H-Oersteds
H-Oersteds	0.7958	A.T./cm-Ampere turns per cm
A.T./cm-Ampere turns per cm	1.2566	H-Oersteds
A.T.''-Ampere turns per inch	0.3937	A.T./cm Ampere turns per cm
A.T./cm-Ampere turns per cm	2.5400	(A.T.'') Ampere turns per inch
Watts per pound	0.2834	Watts per cu in. (Sp. gr. 7.85)
Watts per pound	0.2798	Watts per cu in. (Sp. gr. 7.75)
Watts per pound	0.2762	Watts per cu in. (Sp. gr. 7.65)
Watts per pound	0.2726	Watts per cu in. (Sp. gr. 7.55)
Watts per cubic inch	3.5286	Watts per pound (Sp. gr. 7.85)
Watts per cubic inch	3.5741	Watts per pound (Sp. gr. 7.75)
Watts per cubic inch	3.6209	Watts per pound (Sp. gr. 7.65)
Watts per cubic inch	3.6689	Watts per pound (Sp. gr. 7.55)
gausses	10^{-4}	tesla (SI units)
ampere turns per cm	100	amperes per meter (SI units)

B and H having the same units. Their dimensionless ratio is the μ quoted in the specifications and widely understood as "the" permeability of a material. However, in the mks (meter-kilogram-second) system this dimensionless ratio would be $B/\mu_0 H$ or μ/μ_0 when μ is B/H in SI units. This dimensionless ratio is called the relative permeability, that is, relative to that of vacuum, and is designated μ_r in the SI system of units. However, the symbols μ and μ_r are still widely used to designate permeability and reversible permeability in cgs units.

IV. TYPES OF MATERIALS AND RANGE OF PROPERTIES AVAILABLE

The importance of flux density has already been indicated. The value of B_s, the intrinsic saturation induction, depends primarily on composition. For pure iron it is about 2.2 T, for cobalt about 1.6 T, and for nickel about 0.6 T. Fe-Co alloys can reach the highest available value, a bit over 2.4 T, at 40% cobalt. Any useful material will have a B_s that will not be less than this by much more than about a factor of 10.

The coercive force, on the other hand, is extremely structure sensitive and can vary over several orders of magnitude even for a given composition, depending on the metallurgical treatment the material has received.

Figure 5.2 shows the magnetization and coercive force of the various kinds of magnetic material. The magnetization scale is not B_s but the B_i attained in an applied field of $10H_c$. This is a more directly useful measure of the available flux density. Note that the H_c scale is logarithmic and covers a range of a factor of 10^8.

Soft magnetic materials are to the left of this diagram, hard materials to the right. The transition is in the vicinity of 10^3 A/m, and one would generally like to have H_c either as small as possible or as large as possible. Exceptions to the latter are hysteresis

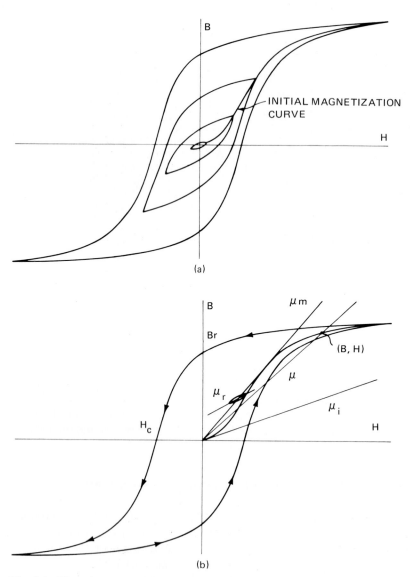

Fig. 5.1 Flux density as a function of field strength. (a) Initial magnetic curve. (b) Subsequent curve.

motor materials, memory applications, and recording tape, in all of which the usable H_c is limited by the drive available in the device. It is evident from Figure 5.2 that it is difficult to obtain materials with both high **B** and either very high or very low H_c. This is further illustrated by constant cost per pound contours, as in Fig. 5.3.

Magnetic materials can be classified by the type of application for which they are intended, as indicated in Fig. 5.4.

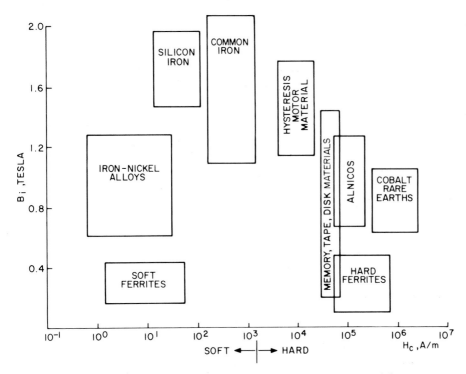

Fig. 5.2 Magnetic and coercive force of hard and soft magnetic materials.

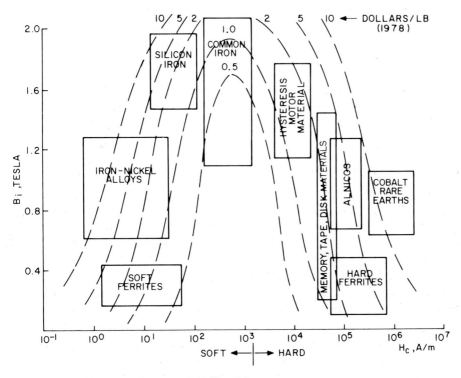

Fig. 5.3 Relative cost of materials in Fig. 5.2.

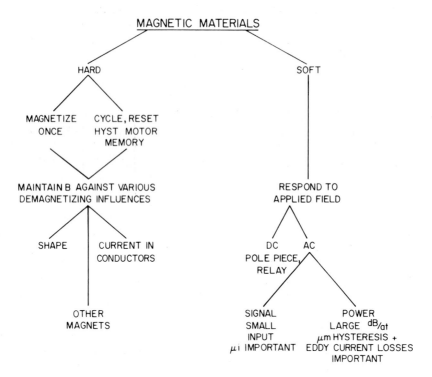

Fig. 5.4 Magnetic materials as a function of use.

V. CLASSIFICATION OF MATERIALS

Magnetic materials can be logically divided into permanent magnetic and soft magnetic materials, as discussed above. Within each of these categories, the materials can be subdivided by their metallurgical means of production, reflecting the physical forms available.

A. Bar, Rod, Plate, Slab, and Wire

Bar, rod, plate, and slab are relatively heavy, hot-rolled or hot-forged products with rough dimensional tolerances. Bar need not be round but can also be square, hexagonal, or rectangular. Thus its available sizes may overlap plate. Slab is the heaviest of all, several inches thick and several feet wide. Alloys in these forms are generally readily machinable.

   Wire is made by cold drawing, but it can be rather large and also need not be round. Rectangular or square wire is often specified.

B. Materials for Laminations

Sheet and strip are thinner than plate, and are much used in the range of about 0.010-0.040 in., in the form of laminations for power applications. Some strip materials used in continuous form for high-frequency applications may be thinner, for example, in thicknesses down to 0.001 in. Both sheet and strip are prepared either by cold rolling (CR) followed by annealing, or by hot rolling (HR). However, among the nonoriented alloys the difference in properties between CR and HR for a given composition and thickness may not be great, although the HR may be a little less expensive and looser in tolerances. The oriented alloys are cold rolled and annealed to develop a metallurgical texture that gives them superior magnetic properties.

C. Cast or Sintered Alloys

The cast permanent magnet alloys are solidified in molds, usually of sand. They have a rough surface and are hard and brittle. They can be ground (expensive) but are practically impossible to machine. Pressing and sintering of powders to final shape can produce better surface finish, tolerances, and physical strength, with some sacrifice of magnetic properties.

D. Ceramics

A few ceramic materials are included among the magnetic materials. They include the permanent magnet materials, such as ferrosoferric oxide, ferric oxide, cobaltic oxide types, and barium ferrite, and the soft ferrites, such as the sintered ferromagnetic ferrites. These are prepared by pressing and sintering oxides and have the hardness, brittleness, and high electrical resistivity characteristic of ceramics.

E. Special Alloys

The term "special alloys" often appears as a description. These materials are all sheet or strip, and include thermomagnetic materials, which are used to compensate for the temperature variation of other materials. They include the cobalt-iron-vanadium types, which are useful where high induction is at a premium, and a number of high-permeability Fe-Ni-based alloys used primarily at low and moderate flux densities.

VI. PERMANENT MAGNETIC MATERIAL SELECTION

A. Energy Product

The function of a permanent magnet is to maintain flux in spite of demagnetizing influences. In many applications the demagnetizing effect comes from the overall shape of the magnetic circuit, including the permanent magnet, as in a loudspeaker or moving-coil meter. If the magnet is used to provide the field flux in a motor, there will be additional demagnetizing fields due to rotor currents. Finally, there may be demagnetizing fields due to adjacent magnets, as in the periodic focusing ring structure of a traveling-wave tube.

For the most efficient use of a given volume of permanent magnetic material, it should operate at $(BH)_m$ on its demagnetization curve, shown in Fig. 5.5. For a typical Alnico (see Table 5.6), this occurs at B_d of 1.0 T and H_d of 39,800 A/m (B_d = 10,000 G and H_d = 500 Oe). If H_d comes entirely from an air gap of length l_g, this should be about 1/20 of l_m, the total length of the metallic part of the magnetic circuit, which may include considerable iron in addition to the permanent magnet material. This relationship follows from

$$\oint H dl = 0 \quad \text{Mks}$$

so that, neglecting leakage,

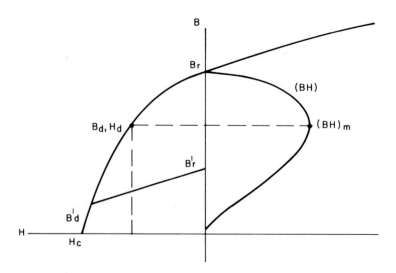

Fig. 5.5 Alnico (typical) magnetic circuit.

Table 5.6 Electrical and Magnetic Properties of Magnetically Hard Materials

FORM	MATERIAL	EMPIS SPECIFICATION	RESISTIVITY $\mu\Omega$cm	MAX. PERMEABILITY	PEAK H OERSTEDS	RESIDUAL INDUCTANCE GAUSS (MIN)	COERCIVE FORCE OERSTEDS (MIN)	PEAK INDUCTION B GAUSS	LOSS (W-SEC/CVC)/LB
BAR	Co–Cr–W STEEL	B3C6	76	35	1000	9000	910	1550	4.45
	Cr STEEL	B3C1	45	105	300	9000	63	1350	1.24
CAST	Fe–Ni–Co–Al	B3C5E (ALNICO 2)	65	11	2000	7200	540	12600	9.5
	Fe–Ni–Al	B3C2 (ALNICO 3)	60	12	2000	6700	450	12000	7.1
	Fe–Ni–Al–Co	B3C8 (ALNICO 4)	75	6	3000	5200	700	11850	10.8
	Fe–Co–Ni–Al	B3C9 (ALNICO 5)	47	18	3000	10000–13000	575–700	16700	15.3
		B3C12 (ALNICO 6)	50	11	3000	10000	750	15700	17.8
		B3C17 (ALNICO 7)	58	5.2–5.7	3000	5800–7200	950–1000	11800–12700	15.0–17.4
		B3C20 (ALNICO 8)	50	–	3000	8500	1600	–	–
		B3C21 (ALNICO 8B)	50	–	3000	7500	1850	–	–
		B3C25 (ALNICO 9)	50	–	10000	10600	1500	–	–
ROD	Ag–Mn–Al	B3C11	19	1.11	20000	550	6000	20830	–
SHEET	Cr STEEL	B3C4	50	95	300	8000	50	11500	.92
	3.5 Cr STEEL	B3C10	32	134	300	9000	48	14500	1.03
SINTERED/ POWDERS	Fe–Ni–Co–Al	B3C5G (ALNICO 2)	68	12	2000	6900	520	12400	9.3
	Fe–Ni–Al–Co	B3C8 (ALNICO 4)	68	6	3000	5200	700	11850	–
	Fe–Co–Ni–Al	B3C9 (ALNICO 5)	47	–	3000	10000	575	–	–
		B3C22 (ALNICO 8A)	50	–	3000	7600	1550	–	–
		B3C23 (ALNICO 8B)	50	–	3000	6500	1800	–	–
	Fe and Co OXIDES	B3C16	5.25×10^{12}	3.16	4000	1750	750	5800	13.2
	SAMARIUM COBALT	B3C24	25	–	60000	7500	6700	–	–
STRIP	Ag–Mn–Al	B3C11	19	1.11	20000	550	6000	–	–
	Cu–Co–Ni	B3C15	32	4	3000	3400	660	–	5.1
	Fe–Co–Ni–V	B3C18	30	285	250	14000	45	19750	1.318
WIRE	Cu–Ni–Fe	B3C14	22	7–9	2000	4800–5400	440–550	7900–7925	4.7–6.7
	Fe–Co–Ni–V	B3C19	30	285	250	14000	45	19750	1.318

$$B_d l_g = H_d l_m$$

or, in the mks system,

$$\frac{B_d}{\mu_0} l_g = H_d l_m$$

and

$$\frac{l_g}{l_m} = \frac{H_d}{B_d}$$

or, in the mks system,

$$\frac{l_g}{l_m} = \frac{\mu_0 H_d}{B_d}$$

The loudspeaker is an example of a device that contains a short piece of Alnico plus considerable iron. The magnetized Alnico by itself might be at B'_d because of its short length. Even if it were then placed in a closed iron circuit, its flux would not return to B_r but only to B'_r, along the recoil line $B'_d B'_r$. Any such structure must be magnetized after assembly, and provision must be made for this.

The volume of material needed to supply a given volume of flux in a gap is, neglecting leakage, inversely proportional to $(BH)_m$. Very high values, as in cobalt-rare-earths, are advantageous where weight and space are at a premium. The larger is $(BH)_m$, the smaller the magnet to perform a given function. Thus devices may be made possible that could not otherwise be constructed. Also, the cost per unit of $(BH)_m$ is a more accurate indicator of total cost than cost per pound. Higher $(BH)_m$ means smaller magnets and sometimes lower cost per piece. It should also be remembered that the barium ferrites are only about 2/3 as dense as metallic materials.

B. Intrinsic Coercive Force

Materials with high H_{ci} include the little-used Silmanal (silver-manganese-aluminum alloy) and Vectolite (ferrosoferric oxide-cobaltic oxide), the inexpensive and very useful barium ferrites and the cobalt-rare-earths (see Tables 5.6 and 5.7). High H_{ci} gives a demagnetization curve on which B_d and H_d are nearly equal at $(BH)_m$, corresponding to short pieces or large gaps. Since H_{ci} measures the resistance of the material itself to demagnetization, a large value is very advantageous for permanent magnet motors, in which stalled-rotor currents may cause large demagnetizing fields. It is also useful in opposed-pole structures, like microwave tube focusing arrays.

C. Types of Materials

The bulk of the permanent magnet materials now used are the Alnicos and the barium ferrites. The Alnicos are used primarily in cast form, in which they have a rough surface, are very hard and brittle, and cannot be machined. The ferrites are ceramics, also hard and brittle. Neither should be used as structural members. Sintered Alnicos are available and give smoother surfaces, better tolerances, and less brittleness, usually with reduced magnetic properties and increased cost. Barium ferrites are also available in the form of powder imbedded in a flexible plastic matrix. These materials are very useful for latching gaskets and for small permanent magnet motor stators.

Table 5.7 Mechanical Properties of Magnetically Hard Materials

FORM	MATERIAL	COMMON INDUSTRY NAME	ROCKWELL HARDNESS[a]	TENSILE STRENGTH MPa	ksi	TRANSVERSE MODULUS OF RUPTURE MPa	ksi	DENSITY lb/in³	g/cm³
BAR	Co–Cr–W STEEL		C56–64	–	–	–	–	0.296	8.2
	Cr STEEL		C45	83	12	155	22.5	0.249	6.9
CAST	Fe–Ni–Co–Al	(ALNICO 2)	C45	20	3	50	7.2	0.256	7.1
	Fe–Ni–Al	(ALNICO 3)	C45	83	12	155	22.5	0.249	6.9
	Fe–Ni–Al–Co	(ALNICO 4)	C45	62	9.1	165	24	0.253	7.0
	Fe–Ni–Al	(ALNICO 5)	C50	38	5.45	72	10.5	0.264	7.3
		(ALNICO 6)	C56	160	23	310	45	0.262	7.2
		(ALNICO 7)	C60	–	–	14	2	0.259	7.17
		(ALNICO 8)	C56	–	–	–	–	0.262	7.2
		(ALNICO 8B)	C45	83	12	155	22.5	0.249	6.9
		(ALNICO 9)	C55	35	5	55	8	0.262	7.2
ROD	Ag–Mn–Al	SILMANAL	B95	530	77	–	–	0.325	9.0
SHEET	Cr STEEL		C60–65	–	–	–	–	0.281	7.8
	3.5 Cr STEEL		–	–	–	–	–	0.280	7.73
SINTERED/ POWDERS	Fe–Ni–Co–Al	(ALNICO 2)	C43	450	65	480	70	0.245	6.8
	Fe–Ni–Al–Co	(ALNICO 4)	C42	413	60	585	85	0.232	6.4
	Fe–Co–Ni–Al	(ALNICO 5)	C50	–	–	–	–	0.241	6.7
		(ALNICO 8A)	C56	–	–	–	–	0.262	7.2
		(ALNICO 8B)	C56	–	–	–	–	0.262	7.2
	Fe and Co OXIDES	VECTOLITE	–	–	–	18	2.6	0.113	3.15
	SAMARIUM COBALT		C50	34	5	–	–	0.272	7.6
STRIP	Ag–Mn–Al	SILMANAL	B95	530	77	–	–	0.325	9.0
	Cu–Co–Ni	CUNICO	B95	585	85	–	–	0.300	8.3
	Fe–Co–Ni–V	P-6 ALLOY	–	–	–	–	–	0.285	8.16
WIRE	Cu–Ni–Fe	CUNIFE IV	B73	670–850	100–123	–	–	0.311	8.6
	Fe–Co–Ni–V	P-6 ALLOY	–	–	–	–	–	0.285	8.16

[a] HARDNESS IS NOT TO BE CONSIDERED A CRITERION OF MAGNETIC QUALITY

The bar materials of iron-chromium alloy and iron-cobalt-chromium-tungsten alloys are machinable but must then be given a postmachining heat treatment.

The punchable sheet materials*are used for hysteresis motor rotors.

D. Directionality

Many of the Alnicos are heat treated in a magnetic field by the supplier. In some cases directional solidification is also performed. These treatments improve the properties in

---

*Same nominal composition as in the bar materials.

the preferred direction at the expense of the transverse properties. Such materials must then be magnetized in their preferred direction. Cunife (copper-nickel-iron alloy) is directional, as is barium ferrite and cobalt-rare-earth.

E. Choice of Material

There are a number of Alnicos, which differ from each other in H_c, B_r, $(BH)_m$, and cost. For severe demagnetizing conditions, a high-H_c alloy would be best. Limited space and weight may dictate the use of a high-$(BH)_m$ material, perhaps at a premium. However, the Alnicos must be considered in context with other materials.

Figure 5.6 shows typical loudspeaker magnet units and illustrates that there are various ways of designing a device using permanent magnets, and that various materials may be used, resulting in devices of different shapes. The small metal-magnet loudspeaker design is widely used. It is efficient in that there is relatively little flux leakage, but the flux available in the narrow annular gap is limited by the fact that one side of the gap is permanent magnet material. In the large metal-magnet design, the gap structure is entirely iron, permitting larger gap flux, but at the expense of greater leakage and thus less efficient use of magnet material. Both of these would typically be made of an Alnico, using its good properties in the interest of size and weight, The ferrite-magnet design geometry reflects the material's low magnetization and high coercive force. This design could utilize a barium ferrite, which is less expensive and lighter in weight.

This is an example of the general principle that permanent magnet devices should be designed from the ground up with a particular material in mind. The mere replacement of one material by another in an existing device without any other change is often not effective. Suppliers are knowledgeable about these matters.

F. Hysteresis Motor Alloys

If a disk of magnetic material is rotated in an applied field H_a, the magnetization will lag the field, producing a torque. The average value of this torque over $360°$ is a measure of the rotational hysteresis W_r. The value of W_r varies with H_a and is a maximum at an H_a that nearly saturates the material, which will be in the general neighborhood of the usual H_c. At very high H_a, W_r goes to zero. Hysteresis motors start and sometimes run by virtue of the torque due to W_r. It is important that the material have its highest W_r at the magnitude of the rotating field produced by the stator. These are typically in the range of 4000-8000 A/m (50-100 Oe). Thus hysteresis motor alloys should have H_c in this range, with as high an induction as possible. Too high an H_c would be detrimental.

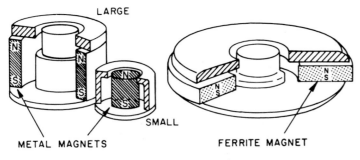

Fig. 5.6 Typical loudspeaker magnetic units.

G. Temperature Effects

Since permanent magnet materials are intended to remain magnetized once and for all, the effect of even a single temperature excursion must be carefully considered and provided for. Generally a momentary temperature rise to T_1 and return will leave the magnet at a lower M than initially. From then on, the new M is stable up to T_1, with a slight reversible temperature variation. Stabilization, when it is necessary, consists of temperature cycling the magnet after magnetization in a range somewhat larger than that expected in service

The barium ferrites lose their properties with temperature *decreasing* below room temperature. Possible low-temperature applications should be checked with the supplier.

VII. SOFT MAGNETIC MATERIAL SELECTION

A. Permeability

The function of a soft magnetic material is to provide large values of flux in response to small magnetizing fields. Its effectiveness in doing this is measured by its permeability μ. The value of this quantity is the ratio B/H in Gaussian units, or $B/\mu_0 H$ in mks units (see Fig. 5.7).

The initial permeability μ_i is a measure of the material's response to very small signals, for example in an audio transformer. This quantity is often specified, as in the Fe-Ni-based alloys, as μ for a given small ΔB, such as 0.01 T (100 G).

The maximum permeability μ_m indicates the greatest flux-multiplying power of the material. In the Fe-Ni-based alloys, this is often required to be above a specified value. The maximum permeability μ_m is important in sensitive relays, ground fault interrupters, magnetic amplifiers, magnetic shielding, pulse transformers, and other applications.

Materials for use in power handling equipment need the largest possible B for the least possible excitation. However, lamination materials for these purposes are not specified by permeability. Magnetic specifications, if any, are in terms of core loss.

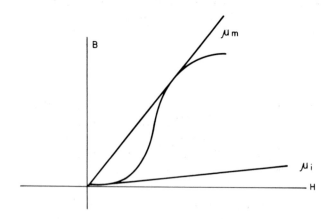

Fig. 5.7   Permeability of soft magnetic material.

B. Effect of Gaps

Permeabilities are measured in spiral-wound tape cores, rings, or with the aid of yokes to provide a magnetically completely closed circuit. If the application involves gaps in the magnetic circuit, the apparent permeability μ_a, the flux divided by the applied field, may be much less than the true material permeability.

$$\mu_a = \frac{\mu}{1 + \mu(l_g/l_m)}$$

where l_g and l_m are the lengths of the air gap and the magnetic circuit. For example, if l_g is only 10^{-3} l_m, a material with μ anywhere from 10^4 to ∞ will give an apparent permeability of 10^3. Clearly there is no point in searching for a material with $\mu > 10^4$. If the material is used in the form of a rod or as a powder, the effect may be still more pronounced. In this case the dimensional ratio of the rod affects μ in the same way as the presence of a gap. In the above expression l_g/l_m is replaced by the demagnetizing factor N, a quantity related to the shape of the sample. Then, for example, the μ_a of a rod whose length is 10 times its diameter will be about 70 for any material with μ over a few hundred.

C. Effect of Frequency

The production of a large flux in response to a small magnetizing field may be performed slowly, as in a dc relay of electromagnet pole piece, or at power or communication frequencies. If rapid response is not important, magnetic devices may be made of solid material. For rapid response or ac applications it is necessary to consider the effect of eddy currents. A changing flux induces voltages proportional to dB/dt, resulting in current flow in such a pattern that the resulting flux opposes the change. These eddy currents depend on the resistivity and permeability of the material and on the geometric path available for current flow. If a constant H is applied to a piece of material, the corresponding change in magnetization will occur rapidly on the surface and diffuse in more slowly toward the center. Under ac excitation, the surface will follow the applied field, but in the interior the flux change will be altered in phase and diminished in amplitude, because of eddy current shielding. At the same time, the eddy currents cause irreversible losses through resistive heating. The higher is the frequency, the more extreme this effect. Thus the apparent ac permeability decreases with frequency, as flux change is restricted more and more to the extreme surface. At the same time the losses increase with frequency.

D. Improvement of AC Performance

The "skin depth," the depth to which the flux approximately follows the excitation at a given frequency, is proportional to $(\rho/\mu f)^{1/2}$. Thus it becomes advantageous to increase the resistivity, which for electrical steel is done by adding silicon to the alloy, and to laminate the material in order to restrict the eddy current path. Both these procedures will improve frequency response and reduce losses. The advantage of higher resistivity is gained at the expense of B_s, which in 3.25% silicon-steel is down about 10% from that of pure iron. The presence of 3.25% silicon also alters the metallurgical phase diagram in such a way that it becomes possible to produce material with a pronounced crystal orientation, giving substantially improved properties in the long direction, at the expense of those in the transverse. This material is used in power transformers, in which the cores are built up from strips, and in distribution transformers, which have continuous spiral cores.

The advantages of thin laminations must be weighed against the cost. Most laminated materials for power applications are used in thicknesses of between 0.025 and 0.014 in. or a bit less. It becomes rapidly more expensive to make thinner material, and sometimes not even possible. Most fractional horsepower appliance motors use 0.025 in. thick punchings of a low-cost plain carbon steel with losses typically 7.6 W/lb. Larger machine sizes bring increased difficulty in cooling, less intermittent operation, and more concern for efficiency, all of which dictate the use of lower loss materials, all the way to the 0.010-0.014 in. oriented 3.25% Si sheet used in very large apparatus, with losses typically 0.6 W/lb.

Laminations must be electrically insulated from each other. This is usually done by a thin oxide layer formed during annealing. Grain-oriented silicon steel is generally coated with MgO, which reacts with silicon in the steel during annealing to form a glass.

E. Eddy Current and Hysteresis Losses

In a number of materials core loss is shown as a function of induction for a specified frequency and is divided into hysteresis and eddy current loss. This is arrived at by measuring the total loss with a watt meter, subtracting the area of the dc hysteresis loop, and identifying the difference as eddy current loss, although it may not agree well with the classically calculated eddy current loss. The reason for attempting to separate out the eddy current losses is that they presumably are the portion that will vary with resistivity, frequency, and thickness.

F. Types of Materials

Bar and plate iron and steel and lamination steels are primarily dc or low-cost ac materials. The specifications are generally limited to chemical composition, mechanical properties, and tolerances. In the silicon steel sheet materials, the better nonoriented grades and the oriented ones have in addition to chemistry and mechanical properties a core loss specification, which is in general the only required magnetic property for these power equipment materials.

The Fe-Ni-based alloys are specified by B_s, μ_m, and μ_i at various small values of ΔB. The latter may be dc or ac. The Fe-Co-V alloys can be made to have the highest available B_s of any material and are useful for applications where space and weight are at a premium.

The ferrites have a variety of quantities specified that are appropriate for their usual high-frequency applications.

The thermomagnetic materials in bar form and as strip, have magnetizations that decrease rapidly with increasing temperature around room temperature. A piece of such a material can be used, for example, as a shunt across a permanent magnet. The amount of flux shunted decreases with rising temperature, increasing the gap flux enough to compensate for its normal decrease with temperature. These materials are specified in terms of the temperature variation of their permeability.

G. Directionality

The grain-oriented silicon steels have strongly directional magnetic properties and must be used magnetized along their length, as in the legs of a power transformer core or along the spiral-wound core of a distribution transformer. Such a material would not be suitable for a complete motor punching. Nonoriented materials are used in such applications. However, oriented steels can be used in segmented cores in very large rotating machines.

The fact that "nonoriented" material is not perfectly nonoriented is reflected by specifications calling for samples that are 1/2 with-grain and 1/2 across-grain.

The iron-nickel-molybdenum alloys are available in either nonoriented or oriented form. In the latter it has an extremely high μ_m, but only in the preferred direction.

H. Processing

1. Stress Relief

Soft magnetic materials are in general stress sensitive. They must be handled with care and in many cases given a final "factory" anneal by the device manufacturer to develop their best properties.

In the specifications for the better nonoriented silicon steels, and in all the oriented grades, it is stated that stresses from punching or assembly are detrimental, and a final stress-relief anneal is specified. The same is true for all the Fe-Ni-based alloys, which must be annealed after fabrication is completed.

2. Semiprocessed Materials

The specifications for some materials begin with the statement that the material as received is not suitable for use in magnetic applications. Such materials must be given their final metallurgical processing in a specified way in the device manufacturer's factory and cannot be used unless such facilities are available.

3. Aging

Aging is the deterioration of soft magnetic properties, usually increase of losses, with time at operating temperature. For example, it is stated that the hysteresis loss of ingot iron will increase from 5 to 80%. Aging is caused by the presence of small amounts of impurities, such as carbon and nitrogen. The addition of Al and Ti to tie these up can improve aging behavior. In many of the sheet materials, aging coefficients are given.

VIII. ADDITIONAL INFORMATION ON ELECTRICAL STEEL

A. Classes and Types

Flat-rolled electrical steel is usually described by a *class* and *type number* system using the following variables as required: the degree of directionality, the method of rolling, the degree to which the magnetic properties are developed by the producer, and the maximum level of the core loss.

Table 5.8 Some Characteristics and Typical Applications for Specific Types of Flat-Rolled Electrical Steel

ORIENTED TYPES

AISI Type No.	Some Characteristics[a]	Typical Applications
M-4 M-5 M-6	Highly directional magnetic properties due to grain orientation. Very low core loss and high permeability in rolling direction.	Highest efficiency power and distribution transformers with lower weight per KVA. Large generators and power transformers.

NONORIENTED TYPES

AISI Type No.	Some Characteristics	Typical Applications
M-15	Lowest core loss, conventional grades. Excellent permeability at low inductions.	Small power transformers and rotating machines of high efficiency.
M-19 M-22 M-27	Low core loss, good permeability at low and intermediate inductions.	High-reactance cores, generators, stators of high-efficiency rotating equipment.
M-36 M-43	Good core loss, good permeability at all inductions, and low exciting current. Good stamping properties.	Small generators, high efficiency, continuous duty rotating ac and dc machines.
M-45 M-47	Ductile, good stamping properties, good permeability at high inductions.	Small motors, ballasts, and relays.

[a] Core loss is considered the principal measure of quality for flat-rolled electrical steel. Tests for other properties, such as permeability, lamination factor, electrical resistivity, interlamination resistance, aging, ductility, tensile strength and hardness are sometimes required. Methods for performing tests are described in ASTM A34.

B. Classes of Flat-Rolled Electrical Steel

Flat-rolled steel is available in the following specific classes (see Table 5.8):

    Nonoriented fully processed
    Nonoriented semiprocessed
    Nonoriented full hard
    Grain-oriented fully processed

These specific classes are described below. They can be supplied in coils or cut lengths.

    Nonoriented fully processed (NO-FP) is a class of flat-rolled electrical steel in which the magnetic properties have been developed by the producer. The Epstein specimens used in determining the magnetic properties will meet the maximum core-loss values shown in Table 5.9, when tested "as sheared."

Nonoriented semiprocessed (NO-SP) is a class of flat-rolled electrical steel in which inherent magnetic properties must be developed by the consumer with a suitable anneal. The Epstein specimens used in determining the inherent magnetic properties will meet the maximum core-loss values shown in Table 5.10, when tested after a "quality evaluation anneal" (QA) as outlined in this table.

Nonoriented full hard (NO-FH) is a class of flat-rolled electrical steel which is produced to a Rockwell hardness minimum of HRB84. Magnetic properties must be developed by the consumer with a suitable annealing treatment to promote decarburization and grain growth. Even with suitable annealing by the customer, magnetic properties are not as good as those of nonoriented semiprocessed material at comparable silicon and aluminum levels.

Grain-oriented fully processed (GO-FP) is a class of flat-rolled electrical steel in which the magnetic properties are fully developed by the producer and the properties of a sample in the direction of rolling after a suitable stress-relief anneal by the consumer will meet maximum core-loss values. Maximum core-loss values for various types and thicknesses of this class are found in Table 5.11.

Table 5.9 Maximum Core Losses,[a] Nonoriented Fully Processed Types, Flat-Rolled Electrical Steel (ASTM A677)

AISI Type	ASTM Designation	Thickness Inch	Thickness mm	Maximum Core Loss at 15 kG (1.5T) W/lb 60 Hz	Maximum Core Loss at 15 kG (1.5T) W/kg 50 Hz
M-15	36F145	0.014	0.36	1.45	2.53
M-15	47F168	0.0185	0.47	1.68	2.93
M-19	36F158	0.014	0.36	1.58	2.75
M-19	47F174	0.0185	0.47	1.74	3.03
M-19	64F208	0.025	0.64	2.08	3.62
M-22	36F168	0.014	0.36	1.68	2.93
M-22	47F185	0.0185	0.47	1.85	3.22
M-22	64F218	0.025	0.64	2.18	3.80
M-27	36F180	0.014	0.36	1.80	3.13
M-27	47F190	0.0185	0.47	1.90	3.31
M-27	64F225	0.025	0.64	2.25	3.92
M-36	36F190	0.014	0.36	1.90	3.31
M-36	47F205	0.0185	0.47	2.05	3.57
M-36	64F240	0.025	0.64	2.40	4.18
M-43	47F230	0.0185	0.47	2.30	4.01
M-43	64F270	0.025	0.64	2.70	4.70
M-45	47F305	0.0185	0.47	3.05	5.31
M-45	64F360	0.025	0.64	3.60	6.27
M-47	47F460	0.0185	0.47	4.60	8.01
M-47	64F575	0.025	0.64	5.75	10.01

[a] Tests are made on Epstein specimens not annealed after shearing, and consisting of strips of which half are cut parallel and half are cut transverse to the rolling direction.

Table 5.10 Maximum Core Losses,[a] Nonoriented Semiprocessed Types, Flat-Rolled Electrical Steel (ASTM A683)

AISI Type	ASTM Designation	Thickness		Maximum Core Loss at 15 kG (1.5T)	
		Inch	mm	W/lb 60 Hz	W/kg 50 Hz
M-27	47S178	0.0185	0.47	1.78	3.10
M-27	64S194	0.025	0.64	1.94	3.38
M-36	47S188	0.0185	0.47	1.88	3.27
M-36	64S213	0.025	0.64	2.13	3.71
M-43	47S200	0.0185	0.47	2.00	3.48
M-43	64S230	0.025	0.64	2.30	4.01
M-45	47S250	0.0185	0.47	2.50	4.35
M-45	64S280	0.025	0.64	2.80	4.88
M-47	47S350	0.0185	0.47	3.50	6.10
M-47	64S420	0.025	0.64	4.20	7.31

[a] Core loss is determined after a quality evaluation anneal. This anneal is performed on Epstein specimens consisting of strips of which half are cut parallel and half are cut transverse to the rolling direction. Customarily, the carbon level is reduced to 0.005% or less and the anneal requires using a suitable atmosphere and a soak temperature of 1550 F (844 C) for approximately one hour, except for M-47 where 1450 F (788 C) for approximately one hour normally applies.

NOTE: Above table does not apply to "full hard" material.

C. Some Characteristics and Typical Applications

To assist the user in selecting the most suitable type of flat-rolled electrical steel for a given application, some characteristics and typical applications for each type are given in Table 5.8. There are advantages inherent in each of the different types as well as limitations.

D. Magnetic and Physical Properties

Flat-rolled electrical steel of each type, regardless of the method of manufacture, is produced to meet certain magnetic requirements. Flat-rolled electrical steel is normally produced to maximum core-loss values. The maximum core-loss values corresponding to

Table 5.11 Maximum Core Losses,[a] Grain-Oriented Fully Processed Types, Flat-Rolled Electrical Steel (ASTM A665)

AISI Type	ASTM Designation	Thickness		Maximum Core Loss at 15 kG (1.5T)	
		Inch	mm	W/lb 60 Hz	W/kg 50 Hz
M-4	27G053	0.0106	0.27	0.53	0.89
M-5	30G058	0.0118	0.30	0.58	0.97
M-6	35G066	0.0138	0.35	0.66	1.11

[a] Tests are made on Epstein specimens cut parallel to the direction of rolling and stress-relief annealed in a manner to ensure magnetic characteristics are like those inherent in the materials from which the specimens were taken. Practices vary, but anneals at temperatures in the range of 1450 F (788 C) to 1550 F (843 C) for times of about one hour in atmospheres having dew points no greater than 0 F (– 18 C) and comprised to mixtures of pure nitrogen and pure hydrogen usually produce satisfactory results.

various types and thicknesses can be found in Tables 5.9 to 5.11. Nevertheless, it has other characteristics of interest to the consumer. These include permeability, exciting current, and lamination factor. The producer should be consulted for the values of these characteristics that may be expected for the material type specified.

E. Aging (Magnetic)

Aging, as it pertains to flat-rolled electrical steel, is generally used to describe the gradual impairment of magnetic properties over an extended period of time or in a relatively short period of time at elevated temperatures.

Aging can be minimized by control of chemical composition in the final processing of the material or by suitable heat treating of the laminations.

F. Punchability

In the manufacture of flat-rolled electrical steel, the prime requisite is achievement of good magnetic characteristics. The achievement of good punchability is important since a high percentage of electrical steel is stamped for fabrication into laminated-core structures. In general, punching characteristics are influenced by the silicon content and the processing by the producer. A knowledge of the application involved is generally helpful when processing material for good punchability.

G. Effect of Silicon

As the silicon content of flat-rolled electrical steel is increased in order to improve magnetic properties, the mechanical properties may be adversely affected. The increased silicon, in addition to embrittling the steel at room temperature, increases the yield point, shear strength, and ultimate strength, and generally decreases elongation.

H. Electrical Steel Density

The density of electrical steel varies according to the amounts of silicon and aluminum present in the steel. From studies conducted by ASTM, it has been determined that densities of commercially available iron-silicon or iron-silicon-aluminum alloys are given with good accuracy by the equation:

$$\text{Density } (\delta) = 7.865 - .065\,[\%\text{Si} + 1.7\,(\%\text{Al})] \text{ g/cm}^3$$

Table 5.12 Assumed Densities by Silicon and Aluminum Content (ASTM A34)

Silicon-Aluminum Factor % Si + 1.7 (% Al)	Assumed Density, grams per cubic centimeter [a]
0.00-0.65	7.85
0.66-1.40	7.80
1.41-2.15	7.75
2.16-2.95	7.70
2.96-3.70	7.65

[a] Kilograms/m^3 = (grams/cubic centimeter) x 1000.

For purposes of determining testing and design parameters, assumed densities applicable to ranges in composition provide for sufficient accuracy. Data showing standard assumed densities applicable to ranges in the factor %Si + 1.7 (% Al) are presented in Table 5.12. In practice, the applicable density is determined by the materials produced from the median or aim silicon and aluminum of the melt.

6

Semiconductor Materials*

I. INTRODUCTION

It seems reasonable in this introduction to remark on some of the more interesting aspects of the earliest germanium and silicon single-crystal experiments. This research led not only to greatly improved point-contact transistors but also to the first successful junction transistor and the first microwatt junction transistors (these were germanium devices), to the development of the first commercially feasible silicon junction transistor, to the adoption of single-crystal material broadly in the manufacture of transistors and, later, to the development of integrated circuits as well as universal adoption in their manufacture (1).

A. Early Days

A few materials have long been reported as semiconductors, primarily because their electrical properties place them midway between insulators and metals. Insulators, of course, have very low electrical conductivity, whereas the electrical conductivity of metals is high. The temperature coefficient of resistivity of semiconductors is negative over some temperature range such that increasing the ambient temperature results in increased electrical conduction. This is in contrast to the behavior of metals, whose electrical conductivity is almost invariably decreased with an increased temperature ambient.

The discovery by Willoughby Smith (2) in 1873 of the increased conductivity in selenium when illuminated by light (photoconductivity) stirred up considerable excitement. Braun (3), Schuster (4), and Siemens (5), during the period from 1874 to 1883, discovered an asymmetry in the conduction of electrical current flowing in opposing directions in selenium, oxidized copper, and certain sulfides. In 1883, Fritts (6) manufactured selenium photocells and observed rectification ratios of electrical current of 256:1 in some of the cells. Between 1915 and 1925 a variety of materials were examined by Pfund (7), Case (8), and Coblentz (9), and others for photoconductivity, positive results being found in the case of cuprous oxide made from oxidation of pure copper plates at $900°C$ (7) in the minerals argentite (Ag_2S), cuprite (Cu_2O), and molybdenite (MoS_2) (8)

---

*Contributed by GORDON K. TEAL, W. R. RUNYAN, KENNETH E. BEAN, and HOWARD R. HUFF.

and in oxidized thallous sulfide [called thalofide by its discoverer, Case (8)] and in MoS_2, Ag_2S, Bi_2S_3, and more complicated compounds (9). Thalofide, particularly, because of its very high photosensitivity and because of its being a prototype of later important commercial photocells, was the subject of extensive additional research.

During the 1920s and continuing through the 1930s, Gudden and Pohl (10) and their collaborators undertook a series of basic investigations of alkali halides, which could readily be grown into single crystals. They used an elegant and perceptively designed series of experiments and laid the groundwork for understanding the important role of trace amounts of impurities on the electrical conduction and photoconductivity in the alkali halides. These trace impurities were introduced into the crystals by special techniques the experimentalists themselves developed. The presence of these impurities caused the materials, which initially resembled insulators, to behave, in some respects, more nearly like semiconductors.

A large increase in semiconductor activity took place in 1924 when Grondahl and Geiger (11) discovered methods of forming red cuprous oxide films on copper that yielded especially good rectifiers or photocells when careful attention was given to the nature of the external electrical contact, which in most cases was a metal film. Somewhat similar developments were made by Lange (12) using selenium. The devices were useful as rectifiers, battery chargers, and photographic exposure meters, becoming widespread in their usefulness in these applications by 1935. The increasing application of these materials in devices used as modulators, nonlinear elements, and rectifiers led to the need for large numbers of devices, each with closely controlled electrical characteristics. This in turn required careful control of the processes of manufacture and a better understanding of the scientific aspects surrounding the fundamental chemical and physical phenomena involved in making and using the devices.

About 1930, McEachron (13) of the General Electric Company in Schenectady discovered that granules of the refractory semiconductor silicon carbide bonded together into a fired clay disk provided with sprayed or evaporated metal contacts on opposite surfaces of the disk had nonlinear current-voltage characteristics and could be used to make very effective lighting arresters. Although Bell Telephone Laboratories (BTL) developed an early interest in silicon carbide varistors because of their nonlinear resistance characteristics, it was much later, during the period 1946-1952, that a Bell System application of silicon carbide varistors was visualized that would exert a major impact on improved telephone communication. Within a few years BTL rapidly developed two small, thin, silicon carbide varistors that were essential to the satisfactory operation of the new telephone handset introduced by the Bell System in 1952 (14). The nonlinear resistance characteristics gave excellent equalization of subscriber handset response, independent of the distance of the subscriber from the local telephone central offices that were distributed nationwide. These silicon carbide varistors were the first close-electrical tolerance solid-state devices manufactured in large scale—the cumulative number exceeded 100 million devices within a few years.

During this same period (early to middle 1930s) the development of new thermistor materials at Bell Telephone Laboratories, usually made from sintered metal oxides, having large negative temperature coefficients of resistivity gave impetus to an increased understanding of such semiconductor materials and of their usefulness and resulted in a great expansion of the literature on the basic nature of the oxides and their applications (15). However, because of the thermistor's dependence on the relatively slow heating or cooling of the thermal mass of the semiconductor the high-frequency response was quite low. Thus, with the trend through the years toward higher frequency communication systems, thermistors were supplanted in many applications by active devices, such as silicon carbide varistors. The nonlinear current-voltage relationship of the latter results from electronic barrier layers at the contacts between the silicon carbide granules that form the semiconductor matrix of the varistor. In this case, changes of current with change of applied voltage by barrier layer transmission phenomena can occur readily and persist to very high frequencies.

The galena crystal used as the detector in early radio receivers was essentially a semiconductor device. The crystalline material normally was placed in a metal cup that acted as one electrode and then a flexible wire "cat whisker" was held in light contact with the crystal. Outstanding rectification occurred if the cat whisker by some chance settled upon a "sensitive" spot on the crystal (16). Normal experience was that frequent adjustment of the flexible contact was required as a result of accidental jarring of the detector and consequent slippage of the cat whisker from randomly located sensitive spots onto the surrounding lower sensitivity areas. The large increase in news and entertainment made available by radio caused people to become fascinated with the radio and its crystal detector. They were stimulated to acquire far more knowledge of the devices than had occurred in prior applications of semiconductors.

With the beginning of World War II and the development of radar (19), there arose, concomitantly, important requirements for the development of radar's crucial components. Since frequencies both in radar and in early independent radio research proceeded toward higher values than could be handled effectively by the vacuum tubes of the early 1930s, the search for a suitable detector returned to the consideration of the crystal detector known in the early vacuum tube era (i.e., 1914-1917). Silicon was chosen by a number of experimenters at about the same time, as a compromise directed toward achieving considerable sensitivity while retaining some circuit stability (20).

In Britain, as early as February 14, 1935, because of a perceived air threat to England, the Tizard Committee for Air Defense had been set up (21,22). It had three meetings within a month and had considered a preliminary note and a detailed proposal by Watson-Watt, of The Radio Research Station at Slough, for the development of radar for the detection of military airplanes, Within an additional 12 days a successful test was made using radio waves from a transmitter in the 49 meter band and a Heyford bomber as target flying at an altitude of 10,000 ft (21). Radar research and development for military purposes began in earnest—early enough for England to save itself in the Battle of Britain, which was at its height in September 1940 (23) and was essentially over by February 1941, although it lagged on for an additional 2 or 3 months (24).

The exchange of scientific and technical information between the United Kingdom and the United States proved invaluable for both nations and their allies (20).

With the role of radar becoming of increasing importance, further attention was given to improving crystal detector mixer diodes. Although some diodes performed well at the frequencies (3000-30,000 Mc/sec) involved, reproducibility and reliability were lower than they should have been for a radar system in which reliability was a vital concern. Several English companies, such as British Thomson-Houston, Ltd., used commercial silicon for their detectors. The Radiation Laboratory of the Massachusetts Institute of Technology also employed this semiconductor material for microwave detectors. The first steps toward the use of higher purity silicon than was commercially available were taken by the General Electric Company, Ltd., of England. Polycrystalline ingots were obtained from melts of higher purity silicon powder mixed with very small quantities of deliberately added impurities (namely, a fraction of a percent of aluminum and of beryllium (16,18). The improved performance of rectifiers made from these ingots, along with early fundamental experimental and theoretical studies, which predicted that conductivity properties would be affected by small amounts of impurities, stimulated development of materials, and devices along similar lines in the United States. E. I. duPont de Nemours and Co. developed a method of producing increased purity silicon following a request from Seitz of the University of Pennsylvania (16). The improvement was achieved by the reduction of silicon tetrachloride with zinc vapor. Ingots made from this silicon, with an appropriate dopant-impurity added, exhibited improved conductivity and rectifying properties as compared with prior commercial materials.

Lark-Horowitz of Purdue University recommended that the Eagle-Pitcher Company prepare high-purity germanium. He particularly wanted his Purdue group to undertake an investigation of the physics of germanium semiconductor material (16-18). At about the same time (February 1942), Teal of Bell Telephone Laboratories was successful in preparing germanium on tantalum filaments by low-temperature thermal decomposition of digermane gas obtained from a former colleague, Toonder at Brown University, where Teal had, during 1927-1930, prepared monogermane gas and studied some of its related germanyl compounds. Decomposition of digermane gas took place on contact with the heated tantalum filament. Decomposition occurred readily at 400-600°C and could be accomplished at temperatures as low as 200-400°C. Later, Teal found that a tungsten point placed on the tiny mounds that resulted from melting and refreezing the germanium produced rectifiers that gave high conversion in the 3000 Mc/sec region, with the tungsten points and germanium mounds acting as excellent n-type rectifiers. This was the first germanium research undertaken at Bell Telephone Laboratories and represented the first germanium devices made there. Somewhat similar exploratory experiments were done by Teal and Storks with silicon pyrolytically deposited on tantalum heated in hydrogen and silicon tetrachloride vapor. A photograph of some of the early germanium data is shown in Fig. 6.1. Lark-Horowitz also had an opportunity to study these germanium results in the spring of 1942 just as his research group was getting underway (1).

Scaff and Theuerer at BTL investigated metallurgical methods of preparing ingots of silicon and, later, germanium, for use in point-contact diode studies (16). They developed a directional freezing technique that produced polycrystalline ingots of improved purity because of the segregation of impurities that occurred between the liquid and solid phases. Both p- and n-type silicon ingots were produced, and in some cases, p and n regions occurred in contact with each other in the same ingot. In one particular ingot of special interest, although the interregional contacts and orientation of the regions occurred with considerable randomness, the barriers tended to be longer and straighter than usual and gave excellent rectification and photoelectromotive forces.

In addition to the studies just mentioned, somewhat similar research was undertaken by other companies and groups (16). Particularly active were General Electric Company (Harper North), Sylvania Corporation, Sperry Research Laboratory, and the Massachusetts Institute of Technology Radiation Laboratory. Additionally, photoelectric effects both in silicon and in germanium were studied in connection with mixer diodes by individuals in the Radiation Laboratory, University of Pennsylvania, Purdue University, and Sperry Research Laboratories.

Outstanding microwave device developments were made during World War II years. Also, the knowledge of impurity behavior continually increased. A good deal was learned about groups III and V of the periodic table of the elements as p and n dopants for both germanium and silicon. However, because of the large electrical effects induced

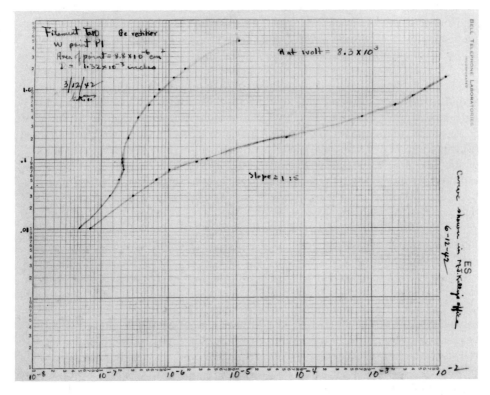

Fig. 6.1 Forward and reverse current versus voltage of early germanium rectifier made by Teal. The vertical axis is in volts. The horizontal axis is in amperes.

by extremely dilute impurities much remained confused and unknown. As late as 1948, high back-voltage germanium diodes were made with tin doping, a dopant later shown not to affect germanium conductivity. The change in conductivity that occurred probably was due to an antimony trace impurity in the tin dopant. There were many things concerning germanium and silicon that were appropriate subjects for further study (18).

B. Threshold Events

The success of silicon and germanium diodes in the gigantic World War II technical effort, the intriguing nature of the unanswered questions concerning semiconductors, and the ever-present power waste involved in heating the filament of the vacuum tube, together with the filament's limited life, stimulated scientists to study ways of making a solid-state amplifier. At the end of World War II a semiconductor group was established within BTL with Shockley in charge, to study the fundamental physics of semiconductors. He conceived the idea of an amplifier that would operate by varying the electrical field across a film of semiconductor through which electrons were flowing lengthwise. In the early experiments the magnitude of the current change resulting from a change in the applied field was always many times lower than the theoretical value that was required if the envisioned device could be said to be working in accordance with theory.

Bardeen and Brattain, who were members of the Bell Telephone Laboratories semiconductor physics group, made their earth-shaking experimental observation of the transistor effect* employing germanium on December 16, 1947 (25,26,176). The date of the invention, however, has usually been taken as December 23, 1947—which was the day the point-contact transistor was demonstrated to top executives of Bell Telephone Laboratories. The first public demonstration of their invention and the announcement of their discovery were not made until June 30, 1948. During this interval Bell Telephone Laboratories proceeded with its experimentation and rapidly expanded its research effort. Study of the point-contact transistor soon indicated that a key phenomenon facilitating the amplification of a signal was injection, into the upper plane of a germanium block, of minority carriers from one point contact (the emitter, biased in the forward direction) and flow to a nearby second point contact (the collector, biased in the reverse direction), biasing in both cases being with respect to a large-area, low-resistance contact on the lower surface, the triode base.

Polycrystalline germanium and silicon were the basic materials used in all the earlier transistor research and development. Now, however, it is well known that, with few exceptions, semiconductor devices require single-crystal starting material as well as control of impurities far beyond what was thought necessary at the time of the transistor invention (1).

Teal was involved in the development of single crystals at Bell Telephone Laboratories during the period closely following the invention of the transistor (1,37-39). He believed very strongly that a crucial need existed for high-purity, high-perfection, controlled-composition single crystals of germanium and of silicon in order to ensure the achievement of devices produced in large numbers to close performance tolerances, with high reliability and at low costs, and to achieve an optimum functional use of the devices through analysis of available electron and hole conduction processes and the device design means that might provide useful operations. All transistor research and development at Bell Laboratories and elsewhere used available polycrystalline material. No one was growing single crystals of either germanium or silicon and there were no schedules for undertaking any such programs. In the Bell Laboratories the importance of having single crystals of germanium or silicon was a very controversial matter. The opinions outside Bell Laboratories apparently were much the same; there was no effort anywhere to grow germanium and silicon single crystals. Nor was there any effort being made to obtain higher purity germanium. During 1948, Teal continued writing research program proposals pointing out the importance of undertaking research designed to develop improved germanium materials. On August 23, 1948, he wrote a memorandum proposing a program aimed at preparing germanium single crystals which, inter alia, could be used to make newly invented filamentary transistors. He added that other experimental work should be directed at the chemical treatment of the filaments to form p-n junctions along the filaments at will.

In the latter part of September 1948, Teal and Little met by chance as they were leaving Bell Laboratories for the day. Little disclosed that he needed some small germanium wafers for a new mechanical design. These might be cut from a small-diameter rod of germanium. Teal saw that in helping Little achieve his new mechanical design lay an

---

*This observation was on a point-contact arrangement. The effect was shown, according to Ref. 25, by both silicon and germanium. Only the use of the latter was described, however, in Refs. 25 and 26.

exciting opportunity to also make the germanium rod a single crystal by pulling it from a melt, as he had often thought (see p. 623 of Ref. 1), and thereby obtain a device of greatly improved electrical characteristics. Equipment for pulling single crystal germanium was quickly designed, and within 2 days a crude machine, set up in Little's laboratory, had furnished several germanium rods with some large single crystals in them (43). These experiments broadened into a single-crystal program on germanium and silicon proposed in a memorandum by Teal. The successful results obtained in the joint initial single-crystal effort with Little won immediate and, somewhat later, long-term support of their collaboration (1,40-42).

The experiments (1) were continued, and it was soon found that the single crystals of germanium exhibited strikingly new and different properties compared with polycrystalline ingots. Minority carrier lifetimes up to 140 μsec (20-300 times greater than those in polycrystalline materials) and mobilities three to four times higher due to the high perfection and purity of the single crystals were observed (27-30).

After it was known that longer minority carrier lifetimes and higher mobilities had been achieved in the single-crystal materials pulled in the Teal-Little experiments, in marked contrast to lifetimes and mobilities observed in polycrystalline germanium, Olsen and Theuerer undertook to grow germanium single crystals by other techniques than those used by Teal and Little. Some of their results were encouraging, but improved minority carrier lifetimes were sporadic and control of resistivity in three dimensions was poor. By early October 1949, all investigators of the semiconducting properties of germanium preferred to use pulled single crystals.

The orientation of the crystal was selected to give the best growth, (100 or 111) and the initial materials were repeatedly recrystallized to improve purity. By mid-1949, bottom slices* of polycrystalline germanium ingots were used to provide ultra-high-purity materials as free of deleterious impurities as possible in the early stages of the research being undertaken on the growth of high-purity germanium single crystals. Specific resistivities of 45 Ω-cm were obtained by this means (1).

The early aims of the single crystal research were stated by Teal (37,38,42) as follows.

> I reasoned that polycrystalline germanium, with its variations in resistivity and its randomly occurring grain boundaries, twins, and lattice defects that acted as uncontrolled resistances, electron or hole emitters, and traps would affect transistor operation in uncontrolled ways. Additionally, it seemed to me that use of this material to produce many complex units meant to be identical, with close performance tolerances, would be inconsistent with high yields and, therefore, also with low costs. Even in developing complex transistor devices, it seemed to me essential to have a high-perfection, high-purity controlled composition semiconductor in order to achieve a separation of various available electron and hole conduction processes in order to analyze and understand the operation of the devices and thus to finally achieve an optimum functional use of them.
>
> My general aims for the single crystal research were as follows.
> 1) to produce a conducting medium in which a high degree of lattice perfection, of uniformity of structure, and of chemical purity is attained;

---

*Slices cut from the first portion of the ingot to solidify. This ensured that, for the desired high purity single crystals the starting material was approximately 2-3% of the volume of germanium in an ingot usable for making acceptable microwave diodes.

2) to build into this highly perfect medium in a controlled way the required resistivities and electrical boundaries to give a variety of device possibilities by control of the chemical composition (i.e., donor and acceptor concentration) along the direction of single crystal growth.

The germanium purification experiments were continued in view of their proven value. On December 28, 1949, Teal urged starting a group effort to undertake research aimed at developing a method that would extend the purification process with the objective of producing germanium single crystals of known composition within the range that donors and acceptors affect the properties of the single crystals.

Teal continued to champion growing single crystals, justifying it as essential for (1) elucidating the science surrounding semiconductor devices, (2) facilitating the design and interpretation of experiments connected with the achievement of new types of transistors, and (3) firmly establishing highly pure and highly perfect single crystals as the technology of mass-produced devices (1).

C. Relation of Semiconductor Properties to Composition and Structure

To discuss further insights into the potential of germanium and silicon single crystals it is essential to consider some of the fundamental scientific aspects that relate some of the properties of elemental semiconductors to their purity, structure, donor-acceptor impurity composition, and a number of electronic conduction phenomena as viewed by theory (1,17,29,31-35).

It should be noted that both germanium and silicon crystallize in the diamond structure. They, like carbon, are elements of group IVA of the periodic table of the elements, and each elementary atom is surrounded by four loosely held valence electrons that facilitate the formation of electron pair (covalent) bonds between like atoms, thus binding each atom to four nearest neighbor atoms located at equidistances from the center atom, all making equal angles with each other and thereby resulting in the diamond structure for all group IVA elements. This structure is shown in Fig. 6.2.

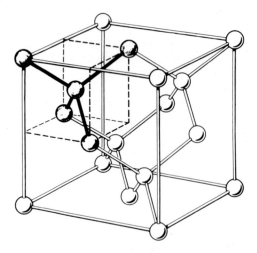

Fig. 6.2   The diamond structure of germanium and silicon, with detailed illustration of neighboring atoms, equidistances, and equal angles between nearest neighbors.

Figure 6.3 is a different two-dimensional representation, showing the electron-hole pair liberated by light or thermal vibration of the lattice with sufficient energy to leave the two carriers to drift in the solid.

n-Type germanium can be formed by addition of about 1 atom of arsenic to about 10^8 atoms of germanium. The electron can diffuse away, leaving the arsenic bound as a positively charged donor ion in the lattice. Under the influence of an electrical field the electron drifts, gives an easily detectable n-type conductivity to the germanium. Similarly, p-type germanium can be created by addition of about 1 atom of gallium to 10^8 atoms of germanium. The gallium remains fixed in the lattice as an acceptor ion, and the hole can drift away, contributing a detectable p-type conductivity to the germanium.

A further part of the physical picture is shown in Fig. 6.4, where the electronic bands of germanium, silicon, and so on, are shown. The states in the valence bond band are nearly fully occupied, and the states in the conduction band are practically unoccupied. The amount of energy needed to extract an electron from the covalent band to give a free electron-hole pair to the lattice corresponds to the energy ΔE_1 required to raise an electron from the top of the valence bond band to the bottom of the conduction band, thus leaving a free hole in the valence bond band. The activation energy for this process differs in diamond, silicon, and germanium, as indicated in the first column. Of primary interest for the present instance are the values 1.12 and 0.72 eV for silicon and germanium, respectively. The energy ΔE_2 to extract an electron from a donor atom and place it in the conduction band, and the energy ΔE_3 to create a free hole by raising an electron from the valence bond band to the acceptor level, are indicated in the second

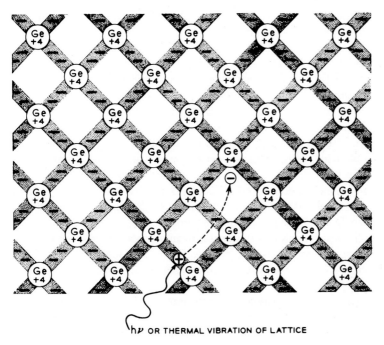

$h\nu$ OR THERMAL VIBRATION OF LATTICE

Fig. 6.3 Two-dimensional representation of intrinsic germanium illustrating creation of an electron-hole pair by an incidence photon or by a thermal vibration of the crystal lattice with sufficient energy.

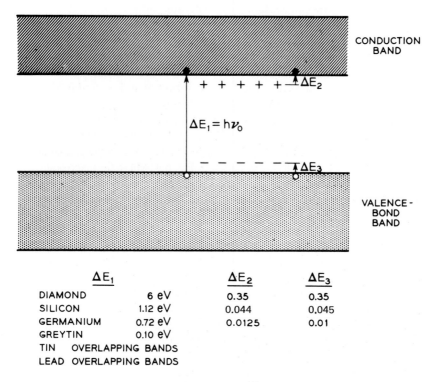

	ΔE_1	ΔE_2	ΔE_3
DIAMOND	6 eV	0.35	0.35
SILICON	1.12 eV	0.044	0.045
GERMANIUM	0.72 eV	0.0125	0.01
GREYTIN	0.10 eV		
TIN OVERLAPPING BANDS			
LEAD OVERLAPPING BANDS			

Fig. 6.4 Electron activation energies of group IV elements. ΔE_2 and ΔE_3 depend on the specific impurity. The values shown are typical.

and third columns, respectively. The somewhat lower ΔE_2 and ΔE_3 values of germanium (about 0.01 eV), as compared with silicon, account for the almost complete ionization of donor or acceptors for germanium at room temperature in contrast to a lower degree of ionization in silicon (due to the larger thermal band gap in silicon).

It should be emphasized that the early physical pictures upon which theories had to be built were highly idealized. They assumed a perfection in structure and control of purity and composition of germanium and silicon that were not present in the actual materials until the single-crystal research described here was done. This is perhaps made more evident by the photographs of Fig. 6.5 and 6.6. In Fig. 6.5 is a view of a specially treated cross section of polycrystalline germanium, the cross section having been cut perpendicular to the solidification axis of a standard germanium ingot. In Fig. 6.6 the photograph to the left is also a specially treated cross section of a standard polycrystalline germanium ingot, the cross section in this case, however, having been cut parallel to the solidification axis. The obvious imperfections that can be seen are accompanied by more subtle ones that are not visible, such as lattice vacancies, interstitial atoms, and unidentified foreign atoms. All these may act at traps for the carriers of electrical charge (34,35). Moreover, uncontrolled variation in donor and acceptor concentration within the polycrystalline ingots occurred in sufficient degree to be disturbing. These were, however, the best materials available to anyone until the research on the preparation of germanium single crystals and a study of their properties, had been done. The germanium single crys-

Fig. 6.5 View of cross section of polycrystalline germanium standard ingot, cut perpendicular to the solidification axis of the ingot.

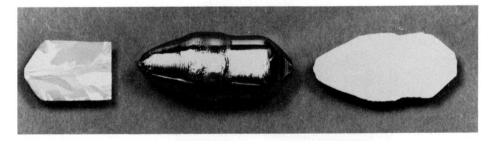

Fig. 6.6 Comparison of cross sections of a germanium single crystal (on the right) and of a polycrystalline ingot cut parallel to the solidification axis (on the left). The polycrystalline cross section reveals many grain boundaries, whereas the single-crystal cross section shows the absence of grain boundaries. An uncut single crystal of germanium is shown in the middle for comparison with the sectioned materials.

tal in Fig. 6.6 was grown using the type II crystal puller shown in Fig. 6.7. On the right is shown another germanium single crystal sectioned parallel to the growth axis of the pulled crystal. Sand blasting of the surface followed by suitable etching showed the complete absence of grain boundaries formed during solidification. If present, these grain boundaries act as traps for atoms and/or free carriers that could adversely affect properties desired in the semiconductor.

II. SINGLE-CRYSTAL MANUFACTURE

As this single-crystal research progressed, further evidence accumulated indicating that the overall purity of polycrystalline germanium was well below the level needed in

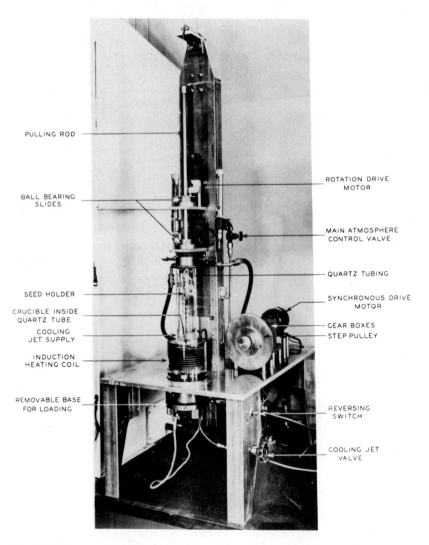

Fig. 6.7 The type II crystal-pulling machine designed and made by Teal and Little.

transistors (1). Thus, during the first half of 1950, Teal and Little designed and built the just mentioned type II crystal puller, which was a more sophisticated machine.

In this model the inert gas atmosphere was now more positively contained within an enclosure of quartz; the cooled high-frequency heater coil was placed outside the quartz tube for maintenance of added cleanliness of surfaces near the semiconductor melt, and the single-crystal seed was attached to the lower end of a rod with a well-designed mechanical means for rotation and easy vertical movement of the crystal as desired.

The germanium single crystals prepared by Teal and Little were especially effective in stimulating scientific interest since in addition to providing absence of grain boundaries and improved uniformity of distribution of added impurities, numerous intriguing questions were raised by their different properties from those found in polycrystalline germanium.

Moreover, the availability of highly pure single crystals contributed greatly to the study of injection phenomena and the determination of carrier mobilities using the method of Haynes and Shockley (30) in which the time taken for an injected minority carrier to move down a filament, as measured by a pulse technique (28), gave credibility to the concept that injection of excess minority carriers that then drift in an electrical field (as well as simultaneously diffusing) gives a very different type of conduction from that normally thought of as metallic conduction. The high purity, uniformity of distribution of added impurity atoms, and perfection of structure of single crystals permitted lengthening of the distance traveled by injected minority carriers prior to their detection at the collector, reduced spurious carrier generators, and provided bars of uniform resistivity.

The higher purity of the germanium, obtained by repeated recrystallization prior to growing the crystal of interest for specific study, led to a higher perfection of the crystal structure and greater uniformity of composition. These attributes resulted in higher mobilities; the values of 1100 cm^2/V-sec (44) measured for electrons in germanium in 1945 gave way to 3800 (45) in 1952.

The first announcement that germanium single crystals had been grown was the publication by Bell Telephone Laboratories of the abstract Growth of Germanium Single Crystals, by Teal and Little (43), Feb. 6, 1950. The brief abstract stated that germanium single crystals had been developed to control the relevant properties of the semiconductor and that single-crystal rods of up to 8 in. length and 3/4 in. diameter had been grown, and these had a high degree of lattice perfection as well as a strikingly new property, such that the bulk lifetimes of injected carriers in these materials have been shown to be greater than 200 μsec.

The oral presentation and discussion were delayed until the historic Bell Telephone Laboratories and Western Electric Company Transistor Symposium, April 21-29, 1952. Over 100 individuals representing 35 companies licensed by Bell Laboratories and Western Electric to manufacture transistors were present. The information presented by Bell Laboratories and Western Electric in the symposium was published in September 1952 in a two-volume set for licensees and certain government officials. The books were later made available to the public. A revised edition of the books was published in 1958.

III. TRANSISTOR MANUFACTURE

The longer lifetimes of minority carriers found in germanium single crystals were of vital importance in the design of transistors. In addition, longer lifetimes of minority carriers greatly facilitated basic studies of surface effects on hole and electron recombination rates and of the influence of various surface treatments of germanium on the rates.

The availability of single crystals was crucial in subsequently realizing the first junction transistor (1). Shockley, in considering several possible theories to account for the operation of the point-contact transistor, conceived of a model, which although not applicable for the point-contact transistor, offered the possibility that it might provide amplification in a solid-state device—with some distinct advantages of its own, particularly, greater resistance to mechanical shock and greater electrical stability. The model was visualized as a rod of germanium with a thin n-type layer sandwiched between two long n-type end regions.* The rod was thus divided into three zones, with p-n junctions occurring at the boundaries between the zones. Ohmic electrical contacts were firmly attached to the two ends of the rod and to the edge of the p section. An input signal applied between the emitter (one of the n layers) and the base (the p layer), suitably biased with a battery, should, according to theory, inject minority carriers into the p layer. This allows the flow of an amplified current from end to end through the rod and through an external load.

About the time, July, 1949, that Shockley published his definitive theory of the junction transistor (50), Teal suggested to Sparks, who was working with Shockley, attempting to make a junction transistor, that they collaborate (1). Within a few months they had prepared the first high purity, high-perfection single-crystal p-n junction (46,47), the first junction transistor and microwatt junction transistors (48). The key technology used in these achievements was the "double-doping" technique generally known as the grown junction method of making junction structures.† The first experiments were accomplished in the movable hydrogen-filled bell jar single-crystal puller type I, which Teal and Little had used in their initial single-crystal experiments. Slight modifications were added to the equipment to facilitate the timed addition of impurity pellets to the melt during the growth of a single crystal of germanium. During the growth of an n-type germanium single crystal from a n-type germanium melt, addition of a tiny pellet of an acceptor (p-type) element in a suitable amount causes that portion of the subsequently grown crystal to be p type. This results in a single crystal with a p-n junction perpendicular to the direction of crystal growth. The amount of impurity required to be added is so minute that the thermal equilibrium between the melt and the crystal growing from it is not disturbed, and the crystalline perfection of the crystal containing the p-n junction is preserved throughout. To achieve the n-p-n structure, a second tiny pellet (this time a donor element such as arsenic or antimony) is added with the timing of total growth and pellet dropping set as required to give a thin p-type layer situated between two much thicker n-type zones. The new junction transistors were remarkably good amplifiers for the signals found in the modern telephone electronic equipment in use since the early 1950s. The quiescent (standby) power of a junction transistor amplifier could be made nearly as low as the signal to be amplified (e.g., microwatts), unlike a vacuum tube, which might use watts in filament power alone. Power gains were also substantial, in the range of 40-50 db per stage. Another salient characteristic of the new junction-type transistor was its low noise figure (1000 times less than its predecessor, the point-contact transistor). Stability of the operating characteristics with respect to mechanical shock of the device was also excellent.

---

*Either p-n-p or an n-p-n structures will work; although the theory was originally developed for p-n-p, the first junction transistors were actually n-p-n.
†The method is described in U.S. Patent 2,727,840 issued to Teal.

The revolutionary nature of the germanium-grown junction transistor amplifier with its reduction in size and operating power as compared with previous amplifiers was a most striking aspect of the research. Additionally, however, the close agreement obtained between the experiments on high-purity, high-perfection single-crystal germanium p-n junction (46,47) and p-n junction transistor amplifier (48) and the theoretically predicted characteristics for these devices (49,50) was astounding. It was a great success for theory, and certainly no less so for the new single-crystal materials and technology employed. However, the results of the Bell transistor research were not made public as soon as laboratory successes were achieved since it was necessary to assess the adequacy of preparative procedures employed for ultimate reproducibility and reliability of the devices in a stepped-up scale of manufacture that would be needed if extensive use were to be made of the devices.

Following the success in making the first grown junction single-crystal transistor in April 1950, the type II crystal puller and, somewhat later, the improved type II crystal puller were installed for the much heavier load of experimentation and were depended on more and more in making a variety of different kinds of materials, including many n-p-n structures. The temperature of the graphite crucible from which germanium single crystals were pulled was, for a period, controlled with the best equipment that could be devised from items that could be obtained from commercial manufacturers of temperature-control equipment (1).

Figure 6.8 shows some early results of Teal and Buehler (1) obtained by controlling impurity distribution coefficients by changing the growth rate during the growth of a germanium single crystal to achieve a "hand tailoring" of the resistivity along the crystal growth direction using empirical programs and the Type II crystal puller. Such programs facilitated improved control of single crystal resistivity and lifetimes of carriers in various parts of the semiconducting solid in which transistor structures centrally located along the growth direction of the single crystal were an important feature. W. P. Slichter was

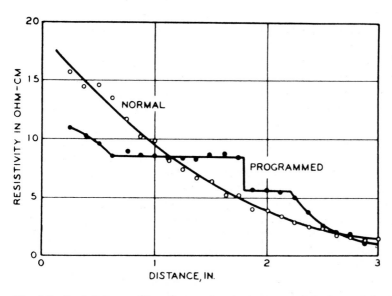

Fig. 6.8 Resistivity profiles of normal and programmed single crystals of germanium.

later able to achieve excellent results with theoretically calculated programs based on concepts of crystal growth developed with J. A. Burton (1,52,54,55).

In February 1951, Bell Laboratories management interest in the transistor project rapidly reached a peak because of the success of single crystals in providing not only greatly improved performance in point-contact transistors with respect to frequency and temperature but also improved reproducibility and reliability as well as providing, in addition, outstanding microwatt junction transistors. The microwatt junction transistors were first prepared in 1950, but their public announcement was not made until July 5, 1951 (48). At the same time Bell Laboratories issued a news release on the revolutionary operating characteristics achieved in the new junction transistor contrasted to those previously available in vacuum tubes and in the junction transistor's solid-state predecessor, the point-contact transistor. The news release also announced achievement of new understanding of reliability and reproducibility leading to the expectation that regular production of transistors could be started. It was stated that transistors had been produced that withstood shock better than any vacuum tube and that they were expected to have a service life considerably longer than that of the commercial vacuum tubes then in common use.

Shockley, Sparks, and Teal (48) disclosed that the junction transistor employed single-crystal germanium in which a thin p-type layer is interposed between two n-type sections and that the lifetime of minority carriers in the three sections were all approximately 300-400 μsec. The approximate values of the resistivity in the three sections were also disclosed. Further discussion of the technology of the semiconductor was delayed until the previously mentioned symposium (April 21-29, 1952) for representatives of companies licensed by Bell Laboratories and Western Electric to manufacture transistors.

Returning, however, to February 1951, there had been a rapid expansion of the research effort on pulled single crystals. The original cadre of chemists was given assignments in a new organization—The Chemical-Physics Laboratory, headed by Leland A. Wooten and Addison H. White, with increased authority for coordinating the transistor materials research. To Teal and two or three associates in this new laboratory went added responsibility for coordinating the transistor materials research in addition to carrying on their own research programs. To achieve suitable materials for a wide range of devices, extensive work on specifying the crystal-growing processes was undertaken by Bradley (56,57).

To prevent melt-back of just formed p layers and deterioration of the production yield of transistor structures, Bell Laboratories undertook development of a temperature controller, which for the first time achieved for the short periods necessary, control to $\pm 0.2°C$ temperatures at about $1000°C$. The circuits were developed largely by Lozier and Weller (59,60), who worked closely with members of the semiconductor crystal-growing group headed by Teal.

A good deal of interest in the single-crystal work resulted from recognition by device development engineers of their need for improved materials. An extensive program was thus carried out in the development area of Bell Laboratories to determine the suitability of single-crystal materials that were being made for new devices under development. Here again, single crystals proved themselves in the achievement of better reliability and reproducibility in point-contact transistor devices (53,58). A key factor in this and in the development of a number of p-n junction devices was the use of single crystals with a high degree of lattice perfection and composition control in three dimensions.

The scientific advances made possible by the preparation of a great variety of highly pure and perfect single crystals of germanium and silicon were subsequently summarized by Teal (39). An important conclusion of this presentation was ". . . these crystals have provided us with semiconducting media having a degree of perfection sufficient that a more powerful type of experimentation has become possible. Avoidance of complicating features in the material has permitted direct observations of theoretical parameters to be made and certain parts of a picture of electronic conduction to be established that might otherwise have remained a matter of speculation."

Hall and Dunlap at the General Electric Company succeeded in preparing single p-n junctions by an alloying technique (61). Later, Saby prepared p-n-p junctions by an alloy technique in which small indium dots were placed in contact with the center of a thin n-type germanium single-crystal wafer on both sides and the whole system was heated until the indium dots melted, dissolved enough germanium that cooling of the assembly resulted in supersaturation of the alloy by indium and growth by recrystallization of a layer of p-type germanium upon the wafer. The center n-type layer of the germanium wafer was used as the base and the two p-type recrystallized layers were used as the emitter and collector, respectively (62). The n-p-n grown junction (48) and the p-n-p alloy transistor (62) were, interestingly, discussed publicly for the first time at the same informal meeting (63).

An important development in germanium was the rate-growing technique of Hall (64) (General Electric) in which, by suitably controlled change of the rate of crystal growth and the proper selection of the donors and acceptors and their concentrations in the germanium melt, many layers of n-p-n transistor structures were grown in a given single crystal. The technique offered thinner p layers and produced transistors with good yields and performance at intermediate radio frequencies (65,66). Further extension of the utilization of two different impurities in various structures within a single crystal were carried out by both Hall and Bridgers and Kolb of Bell Laboratories (64,65,67,68). Bridgers and Kolb succeeded in making germanium single crystals having several n-p-n transistors in them by rate growing with boron and antimony as the acceptor and donor impurities, respectively. Tetrodes made from these structures generally oscillated in excess of 500 Mc, and several units exceeded frequencies above 1000 Mc (69).

Pfann (70,71) devised a simple method for repeating the action of normal melting and freezing, which avoided handling the material between each recrystallization. This resulted in extremely high purity material, which was then grown into single crystals by the pulling technique. He adapted seeding techniques to his zone-leveling equipment, thereby combining zone leveling and horizontal growth of single crystals. The combination of zone leveling and horizontal growth was later used in germanium transistor manufacturing but could not be applied to silicon because of container problems (see Sec. IV). Germanium crystal growing and associated experiments were also discussed by Pfann. The information presented was written up in the two-volume set of books later published in 1958.

IV. SILICON CRYSTAL GROWTH

Silicon single-crystal growth by the pulling technique was tried in the movable open-bottom bell-jar equipment by Teal and Little but failed because of excessive oxidation of the silicon at the elevated temperatures. Utilizing the type II puller, however, (i.e., improved atmosphere control) and adding a quartz liner to the graphite crucible, Teal and Buehler pulled the first silicon rods. The rods contained encouragingly large volumes of single-crystal material. With increasing experience and added purification of initial

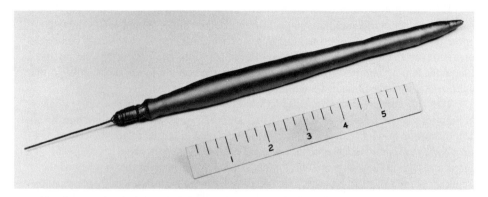

Fig. 6.9 Single crystal of silicon, length 9-10 in. Grown by Teal and Buehler in Bell Telephone Laboratories.

materials single crystals similar to the one shown in Fig. 6.9 were obtained. This crystal was sandblasted and etched to reveal the absence of twinning or grain boundaries. It was a great improvement over the best polycrystalline silicon (Fig. 6.10), which exhibited many grain boundaries and was typical of the period. Haynes and Westphal measured the mobility of electrons in p-type single-crystal silicon, obtaining a value of 1200 cm^2/V-sec, in contrast to 300 cm^2/V-sec, the highest value previously obtained for polycrystalline silicon (73). The mobility of holes in n-type single-crystal silicon was 250 cm^2/V-sec, substantially higher than for polycrystalline materials. These values were later confirmed by Pearson, using Hall effect measurements.

Some of the silicon single crystals exhibited lifetimes of holes as high as 200 μsec, compared with less than 1 μsec for polycrystalline silicon (72). Diodes made from grown silicon p-n junctions by McAfee and Pearson (74) had low reverse saturation currents, as expected on the basis of the band gap in silicon. One can understand the reverse current being lower by a factor of 10^4-10^5 obtained in the case of single-crystal silicon p-n junctions as compared with single-crystal germanium junctions based on a theoretical analysis by Shockley. The low reverse currents, good saturation of the reverse current, and sharp Zener voltage breaks were due to the silicon single crystals having a high degree of lattice perfection and chemical purity. In addition, n-p-n junction structures were also made in grown silicon crystals.

It was recognized by some scientists that silicon with a separation of the conduction and filled bands of 1.12 eV, as compared with 0.72 eV for germanium, should give a transistor that would operate at a higher ambient temperature than germanium. Opinions of the technology to be employed in attempting to make the device, however, varied. Most companies went the alloying route. Some experimenters, moreover, were recommending "leap-frogging" silicon by attempting to make a high-temperature transistor by using the III-V compound aluminum antimonide or a related compound with an even wider band separation than that of silicon. These recently discovered compounds, while offering important new and exciting research avenues, did not appear attractive for providing a high-temperature transistor. This was because of both the added complexity of the intermetallics and the superior chemical and physical stability of silicon; therefore, high-temperature operation favored silicon. Silicon seemed the better choice, too, for highest reliability in consideration of its readily formed protective oxide film which is highly resistive to aqueous or other chemical attack.

Fig. 6.10 Polycrystalline silicon ingot cross section revealing a large number of grain boundaries.

A commercially feasible grown junction silicon transistor was developed by Texas Instruments by April 1954. Without any warning to the audience the achievement was announced at an IRE National Conference on Airborne Electronics in Dayton, Ohio, on May 10, 1954, under rather dramatic circumstances. One after another, the speakers remarked how hopeless it was to expect the development of a silicon transistor in less than several years. They advised industry to be satisfied with germanium transistors for the present. Texas Instruments speaker, Teal, director of its transistor research program, was the last speaker of the day. As he neared the end of his 31-page prepared speech he announced that TI had in commercial production three types of silicon transistors and spoke of having a handful of excellent silicon transistors in his pocket. He also demonstrated that a silicon transistor operated at a higher temperature than a germanium transistor (75). When copies of Teal's paper were made available to the audience there was a rush to obtain copies.

A second paper (76) describing the silicon transistor, by Adcock, Jones, Thornhill, and Jackson, was forwarded to the Proceedings of the IRE for early publication. Teal's announcement was a bombshell to the profession and the industry. In addition to those already mentioned, Boyd Cornelison of the Semiconductor Division of TI made an important contribution by newly designing and constructing a crystal-growing machine that saw extensive use.

The ready availability of silicon transistors immediately raised the power outputs and doubled the maximum operating temperatures previously allowed by germanium transistors. These higher operating temperatures made it possible to meet military electronic specifications and vastly expanded the types of equipment in which transistors could be used.

Shortly thereafter silicon n-p-n junction transistors were made at Bell Laboratories from rate-grown single crystals. Tanenbaum et al., using gallium and antimony as doping agents, produced single crystals of silicon containing up to five n-p-n regions suitable for production of transistors (79).

Within a few years of the development of the transistor it became evident that more sophisticated devices would be developed and that these structures would require doping profile and conductivity change control beyond that attainable by standard grown junction and diffusion control. Because of these requirements, almost simultaneously several people at several different laboratories began to look at vapor phase deposition for the development of thin well-controlled overgrowth films on single-crystal substrates. The first known work published in the open literature was in October 1954 when U.S. Patent No. 2,692,839 was issued to Teal and Christensen (135). This patent proved basic to epitaxial devices. It was later developed and was very broad in its coverage of the details of the best germanium and silicon halide and hydrogen pyrolytic deposition technology—including controlled impurity additions. Essentially all high performance transistors and most transistors incorporated into integrated circuits are of this type. In September 1960 Theurer, Kleimack, Loar, and Christensen of Bell Telephone Laboratories presented a paper entitled Epitaxial Diffused Transistors. Diffused transistors of both germanium and silicon were made in a thin high-resistivity substrate. This arrangement substantially improved the transistor performance and was the first practical application of vapor growth (or "epitaxy") to semiconductor manufacturing.

The work of Frosch at the Bell Telephone Laboratories in developing the use of silicon oxide films as a diffusion mask was crucial to the development of planar devices and is noted by Petritz (18).

Because of the anticipated new demands for high-purity silicon, the development of a chemical reduction method for producing high-purity silicon was begun in 1954. The process employing high-temperature reduction of silicon tetrachloride by hydrogen was selected by Teal and Adcock because of the very high purity of the product, as reported by Teal, Fisher, and Treptow (*J. Appl. Phys., 17,* 879-886, Nov. 1946, "A New Bridge Photocell Employing a Photoconducting Effect in Silicon. Some Properties of High Purity Silicon.") The first examples of silicon photoconductive cells described in this paper were made by this method in 1943. Many people contributed over the years to the success experienced, the earliest experimenters were Adcock, Sangster, Ross, and Fischer who first established this new activity as a technically sound business for Texas Instruments. For a number of years the Texas Instrument Corp. has furnished industry with more than half of the total supply of high-purity silicon (1).

It is hard to overestimate the impact of the commercial lead of Texas Instruments in silicon transistors; certainly there were many benefits for the company in addition to the money made from the sale of devices. It contributed to a broadening of the company's military electronics, where high-temperature performance of the silicon transistors was essential to meeting military equipment specifications. The unique expertise in silicon was also important in the successful negotiations for setting up fully owned TI subsidiaries in England and France. Of greatest importance, perhaps, was that this silicon technology led to the development of integrated circuits.

The availability of high-perfection single-crystal silicon and germanium and the successful commercial introduction of the silicon transistor opened the floodgates to new technologies (41,42) making use of single-crystal materials. Diffusion, mesa, planar, and epitaxial techniques produced useful new discrete devices. A specially significant advance was made in 1958, when Kilby of Texas Instruments fabricated the first working integrated circuit, employing a concept that has made possible the implementation of many functions on a single chip of single-crystal silicon. A large-scale MOS-integrated circuit on a chip of single-crystal silicon approximately 0.86 cm long by 0.45 cm wide is shown in Fig. 6.11. It is capable of storing 262,144 bits of data in a random access memory.

With the rapidly increased use of silicon in a great variety of applications to which it is well adapted, the silicon single crystals are still basically the same as those made about 25 years ago, although they are usually larger and the details of processing may differ.*

In addition to chemical purification for silicon as just discussed, metallurgical purification has also been achieved by float-zone refining developed independently by Keck (80), Emeis (81), and Theurer (82). These experimenters contrived to form a stable molten zone in a vertical rod of silicon such that zone refining could be extended to silicon.

Although float zoning avoids the contact of any external material with the feedstock silicon and produces the highest purity silicon available (with a resistivity of up to several thousand ohm-cm) it is at present used only for specialized applications. This circumstance has developed primarily because of the relative cost of the two processes. It is also quite difficult to produce large-diameter float-zone crystals, and only recently has a 12.5 mm diameter capability been achieved. The oxygen content in float-zoned crystals is very very low, and for cases in which a higher oxygen concentration is required, pulled crystals are much more appropriate.

Integrated circuits are very demanding with respect to low cost of the material and the utmost in performance and reliability. To keep the costs of processing low, the trend has been to enlarge the silicon single crystals from 3 to 6 in. diameter by 1985. It is estimated that more than 90% of the market is presently served by silicon single crystals pulled from quartz crucibles.

V. EVOLVING DIRECTIONS

The next major advance in material technology developed to precisely control impurities in single-crystal materials to form devices was the use of the diffusion process (more correctly, the use of a variety of diffusion processes). Research by Fuller at Bell Laboratories and by Dunlap at General Electric company "laid the foundation for transistor fabrication using diffusion as a key process step" (18,78). Bell Laboratories was the first

---

*Additionally, with successful introduction of dislocation free crystals, the influence of residual point defects, such as silicon self-interstitials, and/or oxygen precipitates has come to the fore.

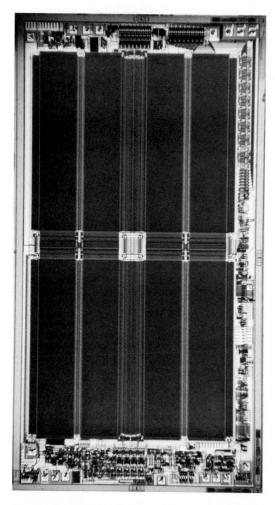

Fig. 6.11 VLSI single crystal silicon chip, 256K DRAM. Individual memory cells not visible at this magnification due to fine geometries.

to use the diffusion techniques to exploit their full potential. A symposium on the technology of diffusion devices was held at Bell Laboratories (Murray Hill, New Jersey, January 16-17, 1956). At this meeting Bell Telephone Laboratories had its technical staff detail for Western Electric Company licensees the various diffusion processes that had been developed by the Bell System and announced the development of "a high-frequency diffused base germanium transistor" by Lee (83) and "diffused emitter and base silicon transistors" by Tanenbaum and Thomas (84). Diffusion technology was an entirely new technique for transistor fabrication; in addition, it was adaptable to various existing processes (85). Cornelison and Adcock (77) combined the principles used in making both grown junctions and impurity diffusion into a semiconductor. The technique resulted in a thinner base layer of impurities in the germanium single crystal, thus enabling higher frequency performance. Grown-diffused germanium junction tetrode transistors were the first transistors in production that could handle 250 MHz for amplification and generate radio frequencies as high as 400 MHz. Grown-diffused silicon junction tetrode transistors

Table 6.1 Commercial Applications of Semiconductor Materials

Material	Applications
Germanium (Ge)	Tunnel diodes, transistors
Silicon (Si)	Diodes, photodiodes, rectifiers, solar cells, SCR, triacs, transistors, integrated circuits
Gallium arsenide (GaAs)	Light-emitting diodes, solar cells, transistors, integrated circuits
Selenium (Se)	Photoresistors, rectifiers, xerography
Silicon carbide (SiC)	Varactors, light emitting diodes
Cadmium sulfide (CdS)	Photoresistors, solar cells
Indium antimonide (InSb)	Infrared detectors (photodiodes), Hall effect devices
Lead sulfide (Pbs)	Infrared detectors (photoresistors)
Bismuth telluride (Bi_2Te_3)	Thermoelectric elements
Lead telluride (PbTe)	Thermoelectric elements
Mercury cadmium telluride (HgCdTe)	Infrared detectors (photodiodes)

were improved eightfold in their frequency range. Power gains of 15 db at 30 MHz were typical.

Although as discussed a number of semiconductors have been used for over 50 years, the semiconductor industry was not really born until the transistor was invented in 1947 and the extreme significance of reproducibly producing single crystals was understood. The first transistors were of germanium. Very remarkable amplification of low current with low power consumption and a number of p-n junction devices were achieved for the first time with germanium devices. However, their temperature range was somewhat restricted, and high reliability was difficult to achieve because of surface problems. After a few years, except for some high current applications, germanium was replaced by silicon because of the higher operating temperatures achieved and because of the great chemical and physical stability of silicon and its surface oxide. Besides providing superior reliability for the silicon devices the surface oxide also acts as a useful diffusion mask and is an insulator permitting the overlay of metallic conductors on it.

Simultaneously, a wide variety of compound semiconductors have been synthesized for special applications requiring different electronic properties. Typical materials and applications are summarized in Table 6.1.

VI. SEMICONDUCTOR DEVICES: MATERIAL CONSIDERATIONS

As discussed earlier, with few exceptions, semiconductor devices require single-crystal starting material as well as carefully controlled quantities of specific impurities. By ordinary analytical standards, these quantities are very small, often in the parts per billion range. The kinds of impurities involved fall into three categories. The first includes those that directly affect the electrical conductivity of the semiconductor at room temperature. Over a rather wide range, the conductivity is directly proportional to the atomic percentage of impurity; for example, the typical concentration of this type of impurity would be a few hundred parts per billion. The second kind of impurity in small quantities does not affect the conductivity but radically affects performance of the finished semiconductor device. These are referred to as "lifetime killers" and in silicon consist of such elements as gold, copper, and iron. The third set includes those elements that may be present in substantial quantities but to first order are electrically inert. Examples include carbon and oxygen* in silicon.

The initial synthesis of the semiconductor material concentrates on providing stock material whose impurity levels are well below those needed in the finished device. This operation usually results in needles, pellets, nodules, or rods of polycrystalline material. To provide the necessary physical shape, crystallinity, and final impurity level, a crystal-growing operation is the next step.

In most cases, not only are single crystals required, but slices cut from the crystal must have a particular crystallographic orientation. This in turn entails growing crystals in a predetermined orientation, usually in one of the two low-index directions, $\langle 100 \rangle$ or $\langle 111 \rangle$. In principle, crystal growing can be from the melt, from various solutions, or by deposition from a vapor. The chemistry of the common semiconductors is such that, except when using some molten metal solvents, growth from solutions is extremely difficult. Vapor growth is widely used, but ordinarily only for adding very thin layers to a substrate prepared from a melt-grown crystal. In practice, the dominant growth mechanism is from the melt. Under some conditions of manufacture, various semiconductors ordinarily considered crystalline can be prepared amorphous or glassy and may have use in special applications. Examples are amorphous selenium for joining optical elements, photoconductive amorphous silicon for light sensors, and glassy chalcogenides for switches. Furthermore, there are classes of organic materials that can be broadly described as semiconductors, and indeed even some liquids fall into this classification. None of these latter materials, however, have any commercial semiconductor-related applications at this time.

VII. SEMICONDUCTOR PROPERTIES

A. Appearance

The appearance of semiconductors varies from a distinctive metallic texture for germanium and indium antimonide through dark gray for silicon and gallium arsenide to transparent for gallium phosphide. In addition, silicon that has been subjected to high temperatures may be covered with a thin layer of native oxide and appear iridescent because of interference effects. Most semiconductors are hard, brittle, and tend to shatter like glass. Large pieces exhibit conchoidal fracture, although thin sections will usually break along well-defined crystallographic planes. Organic and liquid semiconductors are two exceptions to this general brittleness characteristic.

B. Crystallographic Orientation Effects

Many of the important material properties depend on crystallographic orientation. Since most semiconductor materials are used in single-crystal form, these dependencies must be considered (87). In many cases it is possible to ascertain directly from the tensor properties of the crystal whether anisotropy is to be expected (88). However, without actually making the measurements it is not possible to determine the extent. Further,

---

*Depending on the thermal history of the material, oxygen may be either active or inactive (see, for example, J. R. Patel, Oxygen in Silicon, in *Semiconductor Silicon 1977,* H. R. Huff and E. Sirtl, Editors, Electrochemical Soc., 1977, and J. R. Patel in *Semiconductor Silicon 1981,* H. R. Huff, R. J. Kriegler, and Y. Taleishi, Eds., Electrochemical Soc., 1981.)

Table 6.2 Properties that Are Anistropic in Cubic Crystals

Elastic constants
Piezoelectric coefficients
Piezoresistance coefficients
Stress-optical coefficients
Hardness
Breaking strength (cleavage)
Most chemical etch rates
Thermal oxidation rate
Diffusion depth from a free surface (note that the bulk diffusion coefficients are independent of orientation in cubic crystals
Ion implant depth

the properties may depend on such a complex set of interactions that no definitive predictions are currently possible. The only properties that will be isotropic in all crystal classes are those of tensor rank zero (those relating scalar to scalars). Examples are density, which relates mass to volume, and heat capacity, which relates the amount of heat required to increase the temperature. In addition, properties that are second-rank tensors are isotropic in cubic crystals. Examples are the thermal expansion coefficient and the electrical conductivity. Table 6.2 lists a variety of the more complex properties that are anisotropic in all crystal classes.

C. Chemical Properties

Because of the diverse nature of the various semiconductor materials, involving as they do combinations of nearly all the elements in groups IIB, IIIA, IVA, VA, and VIA of the periodic table, it is difficult to make general statements about the chemical properties. Figure 6.12 is a section of the periodic table showing these elements, along with their atomic numbers. Chemical reactions that are important are those that produce surface etching or oxidation and those that are involved in the synthesis and/or purification of the material.

The kinds of etching that are used can be broadly defined as follows.

1. Polishing yields a flat featureless surface.
2. Orientation dependent: The etch rate is substantially different from one crystallographic orientation to another.
3. Resistivity and/or conduction type dependent: The etch rate is dependent on the resistivity and whether the material is n- or p-type.
4. Defect sensitive: Etching is enhanced at structural defects, such as dislocations and stacking faults.

Both liquid and vapor-phase (including plasma) etching are used. Liquid etchants usually are aqueous solutions, but they may be molten fluxes or metals. Most aqueous etchants for silicon, germanium, and III-V compounds have three components involving an oxidizing agent, an etchant for the oxide, and a diluent to modulate the etch rate. Diamond requires molten fluxes, such as potassium nitrate. Vapor-phase etching proceeds by producing a reaction product that is volatile at the temperature used. Many vapor-phase etchants require temperatures of several hundred degrees, but some proceed by

IIIA	IVA	VA	VIA
5 **B** Boron	6 **C** Carbon	7 **N** Nitrogen	8 **O** Oxygen
13 **Al** Aluminum	14 **Si** Silicon	15 **P** Phosphorus	16 **S** Sulfur

IIB	30 **Zn** Zinc	31 **Ga** Gallium	32 **Ge** Germanium	33 **As** Arsenic	34 **Se** Selenium
	48 **Cd** Cadmium	49 **In** Indium	50 **Sn** Tin	51 **Sb** Antimony	52 **Te** Tellurium
	80 **Hg** Mercury	81 **Tl** Thallium	82 **Pb** Lead	83 **Bi** Bismuth	84 **Po** Polonium

Fig. 6.12 Section of the periodic chart showing the elements useful for semiconductor synthesis.

using a plasma near room temperature to produce free radicals. Table 6.3 lists representative etchants for various semiconductors.

Silicon oxidation reactions, apart from those required in etching, are quite important since the dioxide of silicon is an integral part of all silicon devices. Generally, the required oxide is "grown" in situ by the high-temperature reaction of the silicon surface with oxygen or steam (89). Reactions involving synthesis will be discussed in a later section.

D. Crystal Structure

Elemental semiconductors have a diamond structure (two interpenetrating face-centered cubes). Compound semiconductors formed from group III and group V elements have the zincblende form (like diamond except that there is one face-centered cube for each element). The majority of the other semiconductors are either some form of cubic or hexagonal structure. Lattice spacings and structure for the more common semiconductors are given in Table 6.4. Information regarding the structure of liquid* and amorphous semiconductors is quite sketchy although it is clear that there is substantial short-range order (90,91). For example, molten selenium appears to be a mixture of chain molecules and 8-membered rings, and amorphous silicon and germanium are thought to have a random, tetrahedral network. The "organic semiconductors" are either molecular crystals or polymers (92).

E. Mechanical Properties (34,93-95)

Hardness, breaking strength, the elastic constants, and surface tension of various semiconductors are given in Table 6.5. At room temperature most semiconductors are quite

---

*It should be noted that melting a semiconductor may not necessarily result in a liquid semiconductor. For example, in the case of group IV and III-V, it does not. However, mixtures involving tellurium or selenium often do.

Table 6.3 Semiconductor Etchants[a]

Material	Etch composition by volume	Use
Silicon	2 HF, 15 HNO_3, 5 acetic (planar etch)	Polishing
	3 HF, 5 HNO_3, 3 acetic, 0.06 Br (CP4)	Polishing
	2% HCl in H_2 at 1150	Vapor etch for removing damage before epitaxial deposition (gives polished surface)
	CF_4 with 4% O_2	Plasma etch for removing thin polycrystalline silicon layers.
Ge	3 HF, 5 HNO_3, 3 acetic, 0.06 Br (CP4)	Polishing
	1 HF, 1 H_2O_2, 4 H_2O	Polishing
GaAs	1 HF, 3 HNO_3, 2 H_2O	Polishing
InAs	99.6 ml acetic, 4 g bromine	Polishing
GaSb	1 HF, 9 HNO_3	Polishing

[a]For a wider selection of aqueous etchants see: W. R. Runyan, *Semiconductor Measurements and Instrumentation,* McGraw-Hill, New York, 1975; K. E. Bean and W. R. Runyan, Dielectric isolation: Comprehensive, current, and future, *Journal Electrochemical Society,* 124, pp. 5C-12C, 1977; and J. W. Faust, Jr., Etching of the III-V intermetallic compounds, in *Compound Semiconductors,* Vol. 1, R. K. Willardson and H. L. Goering, Reinhold, New York, 1962. For a review of plasma etching, see D. L. Flam and V. M. Donnelly, The design of plasma etchants, *Plasmachemistry and Plasma Processing,* 1, 317, 1981.

brittle, with little evidence of plastic flow. Hence, under the conditions of typical stress-strain measurements, the stress-strain curve is linear until fracture occurs. However, if the material is deformed on a microscopic scale, flow occurs. This can be observed by noting that small indentations into the surface can be made during hardness measurements without fracturing the material. Despite being brittle, semiconductors can be quite strong in tension, but that property may only be measured or utilized if precautions are taken to prevent the application of a substantial bending moment.

The ability to be cleaved along a well-defined crystal plane is moderately strong in semiconductor materials, but generally is not as pronounced as in ionic crystals. It is not precise enough to be a practical method of shaping material, except for breaking thin slices into smaller pieces (see Sec. XV) and occasionally for rough shaping diamonds.

When subjected to very high pressures, Si, Ge, and other semiconductors undergo phase transformations into the metallic state. These phases are unstable at atmospheric pressure so that when the pressure is released, they may revert to the semiconductor form (96,97).

Densities are shown in Table 6.6. Although silicon is about 10% lighter than aluminum, the specific gravity of most semiconductors is in excess of 5, and some are above 8.

Table 6.4   Crystal Structure and Spacing[a]

Material	Group	Structure	Spacing (Å)
Si	IV	Cubic diamond	5.4307
Ge	IV	Cubic diamond	5.6575
Diamond	IV	Cubic diamond	3.5668
GaAs	III-V	Cubic zincblende	5.6533
GaP	III-V	Cubic zincblende	5.4504
GaSb	III-V	Cubic zincblende	6.0961
InAs	III-V	Cubic zincblende	6.0584
InP	III-V	Cubic zincblende	5.8687
InSb	III-V	Cubic zincblende	6.4788
βSiC	IV-IV	Cubic zincblende	4.3596
βZns	II-VI	Cubic zincblende	5.4039
CdTe	II-VI	Cubic zincblende	6.481
PbS	IV-VI	Cubic rocksalt	5.935
PbSe	IV-VI	Cubic rocksalt	6.122
PbTe	IV-VI	Cubic rocksalt	6.460
Bi_2Te_3	V-VI	Hexagonal	$\alpha = 4.3835$, c = 30.487
CdSe	II-VI	Hexagonal wurtzite	$\alpha = 4.2985$, c = 7.0150
CdS	II-VI	Hexagonal wurtzite	$\alpha = 4.1368$, c = 6.7163
αSiC	IV-IV	Hexagonal	Polytype 4H, $\alpha = 3.073$, c = 10.053
			Polytype 6H, $\alpha = 3.073$, c = 15.079
αZnS	II-VI	Hexagonal	Polytype 2H, $\alpha = 3.819$, c = 246
			Polytype 4H, $\alpha = 3.814$, c = 12.46
Se	VI	Monoclinic hexagonal	

[a]Spacings from W. B. Pearson, *Handbook of Lattice Spacings and Structures of Metals and Alloys,* Vol. 2, Pergamon Press, New York, 1979, and L. V. Azaroff, *Introduction to Solids,* McGraw-Hill, New York, 1960.

F.   Thermal Properties (34,93-95)

Table 6.7 summarizes the thermal properties of the more common semiconductors. The thermal conductivity is mainly by phonons, rather than by free electrons as in metals, and may be higher than for many metals. The thermal conductivity of diamond is particularly high, and small diamond chips are sometimes used as heat sinks when electrical isolation is simultaneously required. The good thermal conductivity of silicon is a major factor in its utility for very large integrated circuits since the amount of heat that can be removed from active elements often limits the size of an integrated circuit. Thermal expansivity below the melting point is somewhat less than that of most metals. There are few data above the melting point. Some semiconductors expand on melting, but like ice, the more common ones expand when they freeze. This last feature complicates crystal growth since provisions must be made to eliminate any stress that might be caused by the expansion. It is of interest to note the high value of the latent heat of fusion for silicon (12.1 kcal/mol, or 432 cal/g compared with 80 cal/g for water). Because of this property, silicon has been proposed as a heat-storage material, but the problems in handling a 1400°C melt that is quite reactive and expands on freezing have not been solved.

Table 6.5 Mechanical Properties

Material	Hardness Mohs	Hardness Knoop (kg/mm^2)	Elastic constants (dynes/cm^2) $C_{11} \times 10^{11}$	$C_{12} \times 10^{11}$	$C_{44} \times 10^{11}$
Si[a,b]	6.5	950	16.740	6.523	7.959
Ge[c,d]	6	750	12.98	4.88	6.73
Diamond[c,e]	10	8,800	107.6	12.5	57
GaAs[f,g]	4.5	750	11.88	5.38	5.94
GaSb[f,g]	4.5	450	8.85	4.04	4.33
InAs[f,g]	3.8	380	8.32	4.53	3.96
InSb[f,g]	3.8	220	6.72	3.67	3.02
SiC[c]	9	2,880			
PbS[h]	2.5		12.7	2.98	2.48
PbTe[h]	3		13.9	−0.8	1.31

[a] A. A. Giardini, A study of directional hardness in silicon, *Am. Mineral,* 43, pp. 957-969, 1958.

[b] H. J. McSkimin, W. L. Bond, E. Buehler, and G. K. Teal, Measurement of the elastic constants of silicon single crystals and their thermal constants, *Phys. Rev.,* 83, pp. 1080 (L), 1951.

[c] G. A. Wolff, L. Tolman, N. J. Field, and J. C. Clark, in *Proc. Int. Coloq-Semicond. and Phosphors,* Interscience, New York, 1958.

[d] W. L. Bond, W. P. Mason, H. J. McSkimin, K. M. Olsen, and G. K. Teal, The elastic constants of germanium single crystals, *Phys. Rev.,* 78, pp. 176, 1950.

[e] H. J. McSkimin and W. L. Bond, Elastic Moduli of Diamond, *Phys. Rev.,* 105, pp. 116-121, 1957.

[f] N. A. Goryunoua, A. S. Borscheuskii, and D. N. Tretiakov, Hardness, in *Semiconductors and Semimetals,* Vol. 4, R. K. Willardson and A. C. Beer, Editors, Academic Press, New York, 1966.

[g] J. R. Drabble, Elastic properties, in *Semiconductor and Semimetals,* Vol. 2, R. K. Willardson and A. C. Beer, Editors, Academic Press, New York, 1966.

[h] E. H. Putley, Lead sulphide, selenide, and telluride, in *Materials Used in Semiconductor Devices,* C. A. Hogarth, Editors, Interscience, New York, 1965.

Table 6.6 Densities

Material	Density (g/cm^3)	Material	Density (g/cm^3)
Si	2.329	CdSe	5.81
Ge	5.327	CdTe	6.06
Diamond	3.51	Se (hex)	4.80
GaAs	5.307	PbS	7.60
GaP	4.13	PbSe	8.15
GaSb	5.619	PbTe	8.16
InAs	5.7	ZnS	4.1
InP	4.787	Bi$_2$Te$_3$	7.859
InSb	5.775	SiC	3.21
CdS	4.820		

Table 6.7 Thermal Properties[a]

Material	Melting point (°C)	Boiling point (°C)	Thermal conductivity[b]	Thermal expansivity (per °C)	Contraction on melting (%)	Latent heat of fusion (kcal/mol)	Specific heat (cal/g °C)
Si[d]	1412	2878	.35	2.3×10^{-6}	9.6	12.1	0.18 (18-100°C)
Ge[a]	937	2830	.14	5.7	5.4	7.6	0.074 (0-100°C)
Diamond[a]	Sublimes at atmospheric pressure		5	1.2	—		
GaAs[d]	1237	*	.09		19e		
GaP[d]	1465	*			23e		
GaSb[d]	712	*	.08	5.3	7.0	12e	
InAs[d]	942	*	.05	4.5		15e	
InP[d]	1010	*	.16			16e	
InSb[d]	525	*	.04	5.0	11.4	10e	
Se[f]	219	685	.007	27	-15.2	1.2	0.075-0.085
	(Hexagonal)			(polycrystal)			(0-100°C)
PbS[d,f]	1114	*	.007	26	-8.8		
PbSe[d,f]	1065	*	.004	20	-6.2		
PbTe[d,f]	917	*	.004	27	-3.1		
Bi$_2$Te$_3$[d,f]	585	*		12.9d		29.0	37c
				22.2d			
SiC[g]	Decomposes at atmospheric pressure			4.5-5 (200-700°C)		—	—

[a]It should be noted that there is often a divergence of values in the literature and that new determinations are made from time to time. For compounds marked with an asterisk, there is no defined boiling point for the compound. The individual constituents vaporize separately.

[b]Cal/(sec) (cm) (°C).

[c]Near room temperature.

[d]W. R. Runyan and S. B. Watelski, Semiconductor materials, in *Handbook of Materials and Processes for Electronics*, C. A. Harper, Editor, McGraw-Hill, New York, 1970.

[e]N. N. Sirota, Heats of formation and temperatures and heats of fusion of compounds A^{III}B^V, in *Semiconductors and Semimetals*, Vol. 4, R. K. Willardson and A. C. Beer, Editors.

[f]C. A. Hogarth, Ed., *Materials Used in Semiconductor Devices*, Interscience Publishers, New York, 1965.

[g]A. Taylor and R. M. Jones, The crystal structure and thermal expansion of cubic and hexagonal silicon carbide, in *Silicon Carbide*, J. R. O'Connor and J. Smittens, Editors, Pergamon Press, New York, 1960.

G. Electrical Properties (98,99)

The resistivity of semiconductors is intermediate between metals (10^{-6} ohm-cm) and insulators (10^{12} ohm-cm) and generally is in the range of 10^{-3}-10^8 ohm-cm. Further, it can be tailored to a given value within a broad range covering many orders of magnitude. Silicon, for example, can be prepared to have a resistivity anywhere between 10^{-3} and 10^5 ohm-cm. In contrast, the resistivity of a metal will usually vary by only a few percent as a function of processing, and the resistivity of insulating oxides, with few exceptions, cannot be reproducibly altered, although variations of an order of magnitude may occur as a function of stress, stoichiometry, and other factors.

Figure 6.13a shows the resistivity as a function of absolute temperature, plotted in a way to exaggerate the low-temperature behavior, and Figure 6.13b presents the same data linearly with temperature. The decrease of resistivity with increasing temperature in the low-temperature range is caused by the thermal activation of electrons or holes from the doping impurities. The increase in the midrange is caused by the decrease of the carrier mobility with increasing temperature. The rapid drop at high temperatures is caused by the thermally generated hole-electron pairs dominating the conductivity (intrinsic conduction). The temperature at which intrinsic conduction begins will increase as the semiconductor band gap increases. Table 6.8 lists semiconductor thermal band gaps at room temperature and illustrates the wide range available to the designer.

The midtemperature (extrinsic) range, after the impurities are all ionized and before intrinsic conduction begins, is the region where most semiconductor devices operate; it is the region where the resistivity can be readily changed by the introduction of a properly selected impurity. The impurities suitable for doping vary with the semiconductor, but

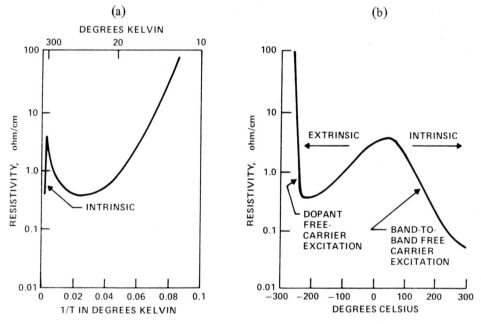

Fig. 6.13 The effect of temperature on resistivity. The portion of the curve between about $-200°C$ and $50°C$ will shift up and down with the concentration of impurities in the semiconductor. (These data are for Ge.)

Table 6.8 Thermal Band Gaps of Various Materials (in Electron Volts, at 300 K)

InSb	0.17	InP	1.27
Bi_2Te_3	0.20	GaAs	1.47
PbSe	0.26	CdTe	1.50
PbTe	0.29	CdSe	1.7
InAs	0.36	Se	1.8
PbS	0.37	GaP	2.25
Ge	0.66	CdS	2.59
GaSb	0.68	SiC	2.8
Si	1.12	ZnS	3.58

Source: Data from E. L. Kern and E. Earleywine, New developments in semiconductor materials, *Solid State Technology,* Oct. 1965.

in general, group IIIA and VA elements are used for group IV semiconductors and group IIB and VIA elements are used for III-V compound semiconductors. Over a wide range of concentration (expressed in atoms per cubic centimeter) the conductivity σ is given by

$$\sigma = q\mu_h N_A \quad \text{or} \quad q\mu_e N_D$$

where q is the absolute value of the electronic charge, μ_h the hole mobility for p-type material, μ_e the electron mobility for n-type material, N_A the concentration of acceptor (p-type) impurities,* and N_D the concentration of donor (n-type) impurities. Table 6.9 shows typical values of majority carrier mobility for several materials in the lightly doped limit. However, to take into account the mobility dependence on concentration and the amount of ionization as a function of concentration, experimental curves as shown in Fig. 6.14 are ordinarily used to relate the impurity level to the resistivity. (In Fig. 6.14 the p-type material is boron doped and the n-type materials is phosphorus doped.)

The Hall voltage is inversely proportional to the number of current carriers and thus is much larger in semiconductors than in metals with comparable dimensions. Advantage is taken of this property in a variety of devices, such as Hall multipliers and calculator keyboard sensors.

The thermoelectric power of some semiconductors is quite high. This has led to their use both as thermoelectric power generators and as thermoelectric coolers. The efficiency is poor, but for small quantities of electric power or thermal cooling, this approach is quite practical. For example, small refrigerators for recreational vehicles use thermoelectric cooling and some satellite power sources have used a combination of nuclear heating and thermoelectric current generators. Compounds of bismuth, selenium, or tellurium, such as Bi_2Te_3, are commonly used.

The resistivity is quite sensitive to stress on the semiconductor (piezo resistivity), and diffused strain gages are used for pressure sensors.

---

*If both p- and n-type impurities are present, the material is "compensated," and

$$\sigma = q\mu_x |N_A = N_D|$$

To a first approximation, μ_x is the mobility appropriate for the majority impurity at the net doping density present.

Table 6.9 Majority Carrier Mobilities of Various Materials at 300 K

Group	Material	μ_e (cm²/V-sec)	μ_h (cm²/V-sec)
IV	GE	3,640	1,900
	Si	1,400	380
IV-IV	SiC	4,690	
III-V	GaAs	7,200	200
	GaP	300	400
	GaSb	5,000	1,000
	InAs	33,000	450
	InP	4,600	100
	InSb	80,000	450
II-VI	ZnS	120	
	Cds	340	18
	CdSe	600	
	CdTe	700	65
IV-VI	PbS	600	200
	PbSe	1,400	1,400
	PbTe	6,000	4,000
V-VI	Bi₂Te₃	10,000	400
VI	Se	2	17

Source: Data from E. L. Kern and E. Earleywine, New developments in semiconductor materials, *Solid State Technology,* Oct. 1965.

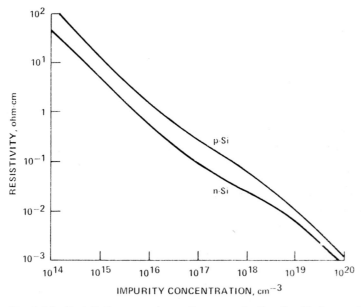

Fig. 6.14 Resistivity versus impurity concentration for Si at room temperature. (Adapted from J. C. Irving, *Bell System Tech. J.* 41:387, 1962. Used with permission from the *Bell System Technical Journal.* Copyright 1960 AT&T.)

H. Optical Properties (100,101)

In cubic materials, the optical properties are independent of the direction of light relative to the crystal axes. However, both noncubic and cubic materials, subjected to mechanical strain, may become anisotropic. These effects are generally quite small in semiconductors; nevertheless, they have been used both to measure strains and to produce light modulators.

The optical properties of semiconductors have been widely studied, particularly in conjunction with the electrical properties, since such studies offer great insight into semiconductor behavior. The optical absorption coefficient and index of refraction are the properties of most practical interest.

The optical absorption coefficient as a function of wavelength for semiconductors has the general characteristics shown in Fig. 6.15. The rather steep change at wavelength A corresponds to the optical band gap of the material. For longer wavelengths the photons do not have the energy to produce electrons and hole pairs and the absorption coefficient decreases abruptly. Figure 6.16 shows absorption coefficients in the vicinity of A for several materials as a function not of wavelength but of photon energy.*

The property of converting photons to electrons and holes allows semiconductors to be used as light sensors and solar cells. The more abrupt change of absorption coefficient as a function of wavelength, exhibited by materials such as GaAs, allows thinner pieces of material to be used to absorb the incident light. By choosing the material, maximum sensitivity can be adjusted to a given wavelength. The high absorption coefficient of silicon in the visible and shorter wavelengths, coupled with its durability, has prompted a limited use as a thin film absorber on solar collectors. For wavelengths

---

*The wavelength corresponding to a given photon energy can be calculated from λ (micrometers = 1.237/photon energy in electron volts.

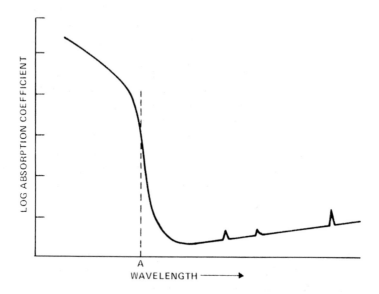

Fig. 6.15 Absorption coefficient versus wavelength. The long wavelength tail is due to free carrier absorption; the small peaks are due to specific impurity absorptions.

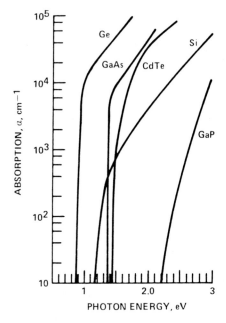

ABSORPTION, α, cm^{-1}

PHOTON ENERGY, eV

Fig. 6.16 Absorption coefficients versus photon energy. (Adapted from E. B. Bylander, *Materials for Semiconductor Functions*, Hayden Book Co., New York, 1971.)

longer than A many of the semiconductors transmit light very well and are often used for making optical elements for infrared imaging systems. Silicon and germanium are used, both singly and together, as achromats, and a variety of amorphous infrared glasses can be formulated from such components as As, Ge, and Te. The sharp peaks at the longer wavelengths are ordinarily due to specific impurity absorption in the lattice. The background absorption, which increases as the square of the wavelength, is due to free carrier absorption peaks, and thus is proportional to the doping level. Both the specific absorption peaks and the background are often used for impurity analysis. In the former case, the background (free carrier absorption) can be reduced and sensitivity improved by operating at liquid nitrogen temperature or below.

Figure 6.17 shows the index of refraction for germanium and illustrates the general behavior of other materials. Table 6.10 gives data for other materials. For wavelengths well beyond point A of Fig. 6.15 the index is relatively constant if the material is very pure. If there are substantial free carriers, plasma resonance will produce an observable perturbation in the vicinity of the plasma wavelength. The plasma wavelength is inversely proportional to the free carrier concentration to the one-half power and is sometimes used as a measure of impurity level at high doping level. Peaks in the index at wavelengths below A occur whenever there is a rapid and large change in the absorption coefficient.

VIII. ANALYTICAL PROCEDURES (102,103)

Semiconductor material analytical procedures can be roughly broken into two completely separate classes. One involves analyzing for the materials themselves, such as Ge, in various ores. The other requires determining the amount of impurities, often in the parts per billion range, in the semiconductor. For the first set of procedures, the reader may consult the standard texts of chemical analysis. For investigating low levels of impurities,

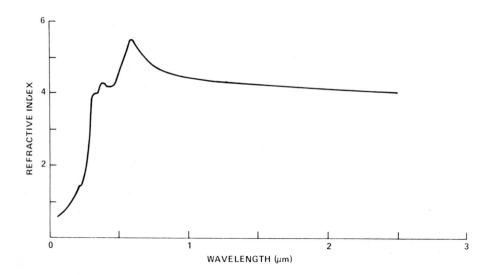

Fig. 6.17   Refractive index of germanium at room temperature.

there are a bewildering array of techniques already available (14,16), with more being continually introduced. Table 6.11 summarizes some of them and briefly discussed their range of applicability. The electrical measurements are generally nonspecific. That is, certain classes of impurities can be detected with good precision over a tremendous range of levels. Unfortunately, it is difficult to determine which impurity element is involved.

All analytical methods depend on standards for calibration. It is very easy to make mistakes in this calibration and thereby have an error by a factor of 10 in the level determination. Caution must be the watchword for the analyst.

Table 6.10   Index of Refraction

Material	Wavelength (μm)	Index
Germanium	6.0	4.01
Silicon	.6	4.0
	1.0	3.5
	8.0	3.42
Diamond	.6	2.41
GaAs	.6	3.6
	2.0	3.6
InSb	8.0	4.0
	18.0	3.86
PbS	3	4.1
PbSe	3	4.6
PbTe	3	5.3
Se (amorphous)	2–10	2.4

Table 6.11 Analytical Methods

Method	Comments

Electrical
  Resistivity
  Hall coefficient
	These measurements are primarily nonspecific. The electrically active impurities can be detected with good precision over a tremendous range of levels, but it is difficult to determine the specific impurity. The equipment is relatively inexpensive.

Optical
  Emission spectroscopy[a]
  Infrared absorption
    spectroscopy
	Emission sensitivity is ordinarily not sensitive enough for evaluating high-purity semiconductor impurities. Infrared optical spectroscopy requires samples up to a few millimeters thick, and may require the sample to be cooled to liquid nitrogen temperatures or below. For applicable elements (e.g., oxygen in silicon), sensitivity is in the parts per million range.

Atomic absorption[a]	

Mass spectroscopy
  Conventional
  Ion microprobe[a]
	The ion microprobe locally removes material by sputtering and examines it with a mass spectrometer. This provides not only spatial resolution, but also resolution normal to the surface. The latter is very important since the impurity level may change several orders of magnitude in a few micrometers.

Ion back scattering	Ion back scattering is used for separating interstitial versus substitutional location of impurities, and for analyzing surface contamination.

Radioactivity
  Neutron activation[a]
    analysis
  Charged particle
    activation analysis
	Neutron activation analysis is widely used for detecting low levels of heavy metals.
	Charged particle analysis is more applicable for detecting light elements, like C, O, N, and B.

X-ray
  Electron microprobe[a]
  X-ray fluorescence[a]
	The electron microprobe can detect only in the parts per thousand range, but can scan a surface and provide spatial resolution of a few micrometers. Useful for inclusions and precipitates.

Electron spectroscopy	Electron spectroscopy, such as Auger, allows the top few atomic layers to be investigated for specific elements, quite independently of what is underneath. The concentration in those layers must be relatively high, but the detection limit in terms of absolute magnitude is very small.

[a]For sensitivity limits, see charts in R. D. Dobrott, Chemical characterization techniques, *Proceedings of the National Workshop on Low Cost Polycrystalline Solar Cells,* T. L. Chu and S. S. Chu, Editors, SMU, 1976.

IX. PURIFICATION AND SYNTHESIS

A. Introduction

The very high purity requirements for semiconductors put demands on production techniques that, in most cases, are orders of magnitude more stringent than those required
for any other industry. For example, instead of parts per thousand, or perhaps parts
per million, impurity levels must be in the parts per billion range. Nevertheless, the purification procedures used are essentially the same as those previously available. The difference is that many must be used in series, and indeed material may be recycled through
the same process several times (104). Further, preference is given to reactions that use the
most easily purified feedstock rather than those that give the greatest conversion efficiency.

Historically, purification has proceeded by such routes as precipitating the desired
compound while leaving the undesirable ones in solution and repeated distillations. The
latter is the method most widely used in silicon manufacturing. There are, however, a
number of other approaches. Electrolysis, applicable to indium, for example, can be
rather selective. By choosing the current density and cathode potential, various impurities can first be plated out of solution, then the potential is changed and the indium is
extracted. Vacuum sublimation is useful for materials having a high vapor pressure that
are contaminated with low vapor pressure materials. Arsenic is often purified in this
manner.

Liquid extraction, in which some ions are considerably more soluble in organic
liquids than in aqueous solutions, has application. Again, using indium as an example,
In^{3+} is four times more soluble in a tributylphosphate-benzene mixture than in a 10
normal HCl solution. Cd^{2+} is only one-half as soluble, so that a substantial reduction in
cadmium concentration can be accomplished.

Selective adsorption, in which some ions become tightly bound to various resins
or natural zeolites, is used to prepare water of unparalleled purity for the semiconductor
industry and is one way of removing contaminants from silane, sometimes used in silicon
manufacture.

Recrystallization, during which impurities are generally rejected as a crystal form, is
very important and will be discussed in more detail later. In many cases, the only place
the rejected impurities can accumulate is at grain boundaries. If such material is subsequently crushed and leached with appropriate acids, many of the impurities can then be
permanently removed. This approach was taken in the 1940s to provide silicon for microwave diodes (16); it has been recently resurrected as one approach to low-cost silicon
solar cell material (105).

The volume of semiconductor-grade silicon produced now completely overshadows
that of all other semiconductors combined. In the early years of the industry, germanium
dominated, but its usage has drastically decreased. For comparative purposes, Table 6.12
gives some historical rates.

B. Silicon (16,34,106)

In the early 1800s, Berzelius heated a mixture of silica, carbon, and iron and obtained
what he called "silicon," although it was apparently an iron silicide, or what would now
be called "ferrosilicon." Soon thereafter, Gay-Lussac and Thénard reduced silicon tetra-
fluoride with potassium and formed "amorphous" silicon. In the following years a wide
variety of other reactions were studied. However, the most common method of producing

Table 6.12 Estimated Semiconductor Usage (Tons)

	1955	1960	1965	1970	1975	1980
Si	5-10	30	50		1500	3000
Ge[a]	40[b]	30[b]	45[b]	80	150	

[a]Includes optical elements.
[b]Includes resmelted scrap.
Source: Based on *Mineral Yearbook* and industry estimates.

commercial-grade silicon today is similar to the first one used. Quartzite rock is reduced by coke in an electric furnace. The resulting low-purity silicon can then be used as raw material for further processing. The order and number of those subsequent steps determine the final purity. The usual sequence is to chlorinate the impure silicon, carefully purify the resulting halides, which will be predominantly $SiCl_4$ and $SiHCl_3$, and then reduce them. If a compound other than a chloride, such as a bromide, is required, it may either be prepared directly or by further reactions involving the tetrachloride. Large quantities of silicon tetrachloride ($SiCl_4$) and trichlorosilane ($SiHCl_3$) are produced annually for use in silicone manufacturing, so a ready source of feedstock for the semiconductor industry is available.

Halides are most often purified by direct distillation. There is difficulty in separating some of the group IIIA and VA halides from those of silicon; the addition of complexing agents prior to distillation is sometimes helpful. Since many of the halides of major impurity elements can either be reduced or thermally decomposed at lower temperatures than those of silicon, various predeposition furnace arrangements are sometimes used for feedstock purifications. There are also problems in removing all traces of hydrocarbons. For example, around 1955, silicon from $SiHCl_3$ was plagued by a silicon carbide dross that floated on the melt at the crystal-growing step. This material was traced to contamination of the $SiHCl_3$ by normal heptane, which has a boiling point of 33°C and is not efficiently removed during the distillation of $SiHCl_3$, which boils at 31.8°C. It may however be removed by complexing.

1. Halide Reduction

The first commercial process for the production of semiconductor-grade silicon used zinc to reduce silicon tetrachloride by the reaction $SiCl_4 + 2Zn \rightarrow Si + 2ZnCl_2$. Zinc was an acceptable reducing agent since it is available in high purity, does not form a silicide, is relatively noncorrosive, and is not explosive in the presence of air. In this process, the zinc and silicon tetrachloride were separately vaporized and metered and were then introduced into a 950°C reaction chamber in approximately stoichiometric proportions. The silicon nucleated and grew out in dendrites (needlelike structures) from the walls of reaction chambers. In the 1960s the zinc process was completely replaced by hydrogen reduction of the halides.

Trichlorosilane will thermally decompose at a slow rate to give silicon at temperatures above 850°C. Silicon tetrachloride is thermally stable up to at least 1200°C. With hydrogen, each reacts and produces silicon with reasonable efficiencies and rates. Reactions involving Si, H_2, and the various halides can be quite complicated. Even though only one halide is introduced into the system, a wide variety of products will be present in the reaction chamber. Some of the reactions are:

$$SiCl_4 + 2H_2 \rightleftharpoons Si + 4HCl$$
$$SiHCl_3 + H_2 \rightleftharpoons Si + 3HCl$$
$$Si + 2HCl \rightleftharpoons SiH_2Cl_2$$
$$Si + HCl + H_2 \rightleftharpoons SiH_3Cl$$
$$Si + 2HCl \rightleftharpoons SiCl_2 + H2$$

If these equations are combined with the appropriate equilibrium constants, the partial pressure of the various species can be predicted as a function of input concentration and temperature (107).

In either hydrogen reduction or thermal decomposition, a lower temperature favors the production of very small particles. For example, at 800°C most of the $SiCl_4$ reduction output will be powder. $SiHCl_3$ decomposition at 830°C gives aggregates of fine powder. At 1230°C the silicon begins to have a nodular appearance. Hydrogen reduction of either $SiCl_4$ or $SiHCl_3$ may vary from dendrites to large, well-defined crystals. The use of trichlorosilane is favored over silicon tetrachloride because of faster deposition rates at lower temperature and the ease of removing phosphorous and boron compounds from it. Figure 6.18 is a schematic illustration of a typical silicon-production process based on $SiCl_4$ or $SiHCl_3$. Early manufacturing processes allowed the silicon to deposit directly on the inside walls of the hot reaction chamber (a quartz tube). This required that the tubes be broken up and etched away from the silicon at the end of each run. Today deposition proceeds on thin resistance-heated silicon rods, and typical runs will produce a final rod of perhaps 15 cm in diameter and 2 meters long (or hairpin rods of 4 m in total length). The deposition rate for this kind of process varies with the reaction chamber geometry, temperature, and feed stream composition, and semiempirical equations are generally used to predict the growth rate.

2. Silane

The thermal decomposition of silane (SiH_4) above 400°C to give silicon and hydrogen is, in principle, a very simple process. Further, silane can be produced with the very high purity required. The process was not widely used in the early days because of the difficulty of making silane, and because of its instability. Silane will ignite or explode on exposure to air, is decomposed by water containing traces of alkali, and reacts explosively with halogens to produce silicon halides. Silane is now used in moderate quantities for producing thin layers of polycrystalline silicon directly on oxide-covered slices that are required in some MOS-integrated circuit processes. Because of the low-temperature deposition characteristics, silane is also useful for low-temperature crystal-growing operations (see section on crystal growth from the vapor).

The desire to provide low-cost silicon of a quality useful for solar cell manufacture has led to a renewed study of many of the old processes, as well as a variety of new approaches (108). For example, a new process eliminates the requirement for long-term storage of the silane. Hydrogen and $SiCl_4$ are fed into a hot fluidized bed of metallurgical-grade silicon. The temperature and concentrations are adjusted so that approximately 15% of the exit gas is $SiHCl_3$. The $SiHCl_3$ is then collected, distilled, and partially converted to dichlorosilane (SiH_2Cl_2). By use of a catalyst, a portion of the SiH_2Cl_2 is con-

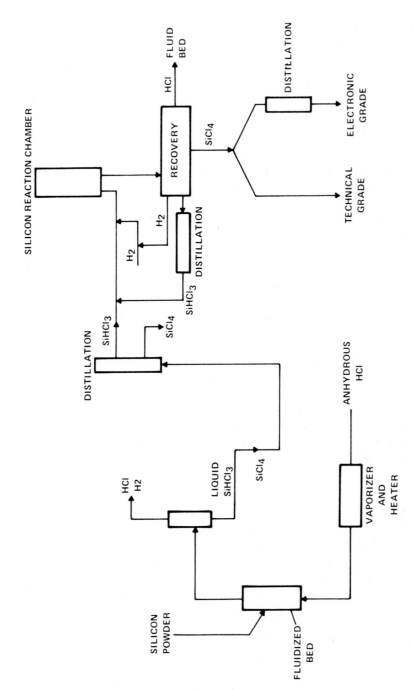

Fig. 6.18 Silicon manufacturing process flow. (Adapted from L. D. Crossman and J. A. Baker, Polysilicon technology, in *Semiconductor Silicon 1977*, H. R. Huff and E. Sirtl, eds., The Electrochemical Society, Princeton, New Jersey, 1977.)

verted to silane. The silane is further purified and decomposed to form silicon, and the other reaction products are recycled to produce more $SiHCl_3$ and SiH_2Cl_2.

3. Iodide

Silicon tetraiodide will thermally decompose to form silicon. Because the iodide is easily purified, it will produce extremely high purity silicon. In order to obtain reasonable deposition rates, low pressures are required, along with a combination of vacuum pumps and iodine traps. SiI_4 freezes at 124°C, so all transfer lines must be kept heated (as was also required in the Zn-$SiCl_4$ method). Further, because of the high cost of iodine, a recovery and recycling process is necessary.

C. Germanium (106,109)

Germanium is widely distributed in the earth's crust at the 6-7 ppm level. It is found as the sulfide, and is associated with the sulfide ores of other elements, such as copper, zinc, tin, lead, and antimony. Winkler isolated germanium from the mineral argyrodite ($4Ag_2S \cdot GeS_2$) in 1886 and named it after his native country, Germany. Other minerals containing germanium are germanite ($7CuS \cdot Fe \cdot GeS_2$), renierite ($(Cu, Fe, Ge, As)_xS_y$, and ultrabasite ($(Pb, Ag, Ge, Sb)_xS_y$. Since germanium does not occur in sufficient concentration to justify extraction by itself, it is obtained as a by-product from other operations. Coal contains small amounts of germanium, which can be recovered from the flue gases; this has been used as a major source in England. In the United States the primary source was the lead and zinc ores of the Joplin, Missouri (tristate), area. However, that supply is essentially exhausted, and most germanium ore now comes from Africa and the USSR. It should also be noted that the semiconductor requirement for germanium has dropped markedly, and only a small part of production is now used for that purpose.
A typical reclamation flow is shown in Fig. 6.19. Specific waste residues are treated with HCl to give crude germanium tetrachloride ($GeCl_4$). The tetrachloride is then purified by distillation, and perhaps by solvent extraction. Very pure (filtered and deionized) water is then reacted with the $GeCl_4$ to give HCl and GeO_2. After filtering and drying, the GeO_2 is reduced in pure hydrogen at 650°C to give germanium. The Ge powder is then melted in graphite boats and cast to give ingots. Finally, the germanium ingot is purified by zone refining (see Sec. X) and is ready for crystal growth.

D. Compound Semiconductor (110)

The problems of producing compound semiconductors are at least three times as difficult as are those of the elemental semiconductors since each constituent must be separately purified and then reacted to give the desired compound (Fig. 6.20). Indium antimonide can be compounded by directly melting the two elements together. Gallium arsenide (and similar materials), because of the high vapor pressure of arsenic, must be approached differently. In this case, gallium can be melted and a stream of arsenic vapor passed over it. If only small quantities are desired, as in the case of thin crystalline layers, the simultaneous reduction of two gaseous species containing, respectively, gallium and arsenic, allows direct combination to form GaAs. An example is the hydrogen reduction of a mixed vapor of gallium monochloride (formed by passing HCl over molten gallium) and arsenic trichloride.

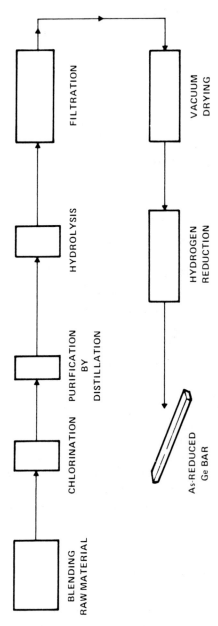

Fig. 6.19 Flowchart for germanium production.

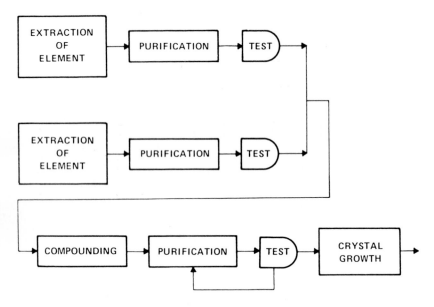

Fig. 6.20 Material flow for compound semiconductors. (Adapted from E. B. Bylander, *Materials for Semiconductor Functions,* Hayden Book Co., New York, 1971.)

X. CRYSTAL GROWTH

A. Growth from the Melt (34,111,112)

The Teal-Little method (commonly known as "Czochralski") is widely used for Si, Ge, and GaAs (34,51,52,54,55). Of all the crystal-growing methods available, it is the most widely used and least expensive. Growth proceeds by allowing the melt to slowly freeze onto a single-crystal seed, which is simultaneously rotated and withdrawn (pulled) from the melt (Fig. 6.21). The freezing (growth) rate is controlled by the temperature of the melt at the growing interface and the amount of heat lost from the crystal by radiation from its surface and conduction up the crystal to the seed. The relation among these variables may be expressed by the heat-balance equation:

> Latent heat + heat transferred to crystal from melt
> = heat conducted away from the interface by the crystal

Stirring via crystal rotation minimizes unsymmetrical crystal growth arising from uneven heating of the melt; in general, it reduces thermal gradients in the melt. As the diameter of the crystal increases, the maximum possible growth (pull) rate decreases. The relation is a rather complex one that depends on the machine construction, properties of the material being grown, and the ambient gas (113). Silicon crystals as small as 3 mm in diameter and as large as 200 mm in diameter have been grown by this process. Figure 6.22 shows a machine capable of holding an initial charge of 20 kg silicon and producing a crystal 100 mm in diameter and 1 m long. In 1984 similar machines use a 40 Kg charge and can produce 100 mm crystals 2 meters long. This is to be contrasted with early machines that only had the capability for crystals 20-30 mm in diameter and 10 or 20 cm in length. The increase in diameter of silicon crystal production with time is shown in Table 6.13. It was not, however, growing technology that limited production (150 mm diameter

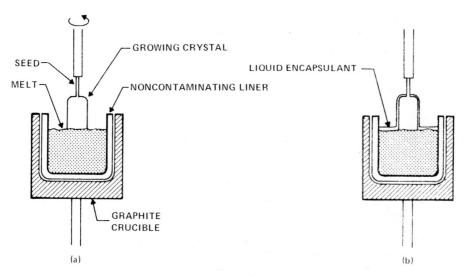

SEED

MELT

GROWING CRYSTAL

NONCONTAMINATING LINER

GRAPHITE
CRUCIBLE

(a)

LIQUID ENCAPSULANT

(b)

Fig. 6.21 Crystal pulling from the melt. (a) Normal arrangement: A chamber to provide an inert atmosphere will surround this portion of the puller. (b) Liquid encapsulant: normally used in conjunction with a pressurized chamber. Counter rotation of the seed and crucible, as indicated in (a), is often used.

germanium crystals were grown in 1957) (114), but rather the state of semiconductor processing technology.

The primary advantage of pulling is that it puts few physical constraints on the growing crystal. This is quite important since most semiconductors expand several percent when they freeze. Silicon poses an additional problem, which is solved by pulling from the melt. It bonds to almost all container walls during freezing so that differential expansion on further cooling causes fracture of both crystal and container. A disadvantage of this method, or any other using a container, is possible contamination from the container. Since molten silicon attacks all materials investigated to some degree, progress in preparing highly pure crystals has been dependent first on careful selection of the container material and ultimately in obtaining very high purity levels in the container material.

The best compromise for high-purity Si production is to use a fused quartz container, even though quartz is slowly dissolved. The quartz container in turn sits in a graphite crucible. The use of these two materials had led to some significant problems for the semiconductor industry, and well illustrates that one should never underestimate the subtleties associated with small quantities of impurities. It was observed as early as 1955 that oxygen incorporated into a growing silicon crystal at levels of 1 to 2×10^{18} atoms cm^{-3} often resulted in a significant resistivity shift (115). This was quite unexpected since simple theory indicated that oxygen should be electrically inactive in silicon. Subsequent studies identified the mechanism to be that of electrically active SiO_4 complexes forming during the prolonged heating of the upper portion of the crystal as the remainder is grown. These complexes are electrical donors and form most rapidly near 450°C. Although they usually disappear during subsequent high-temperature device-manufacturing steps, their presence makes it difficult to accurately evaluate crystal

Fig. 6.22 Commercial silicon crystal puller. (Courtesy of Hamco Inc., Dallas, Texas.)

Table 6.13 Silicon Crystal Diameter Trend with Time (Volume Production)

Year	1955	1960	1965	1970	1975	1980	1985
Diameters (mm)	25	25	37	50	75	100	150

resistivity before processing. Consequently, semiconductor material manufacturers developed methods to annihilate the donors before shipment to the users. Not until the 1977-1980 period was it recognized that these anneals, typically at 600-850°C for 1-4 h followed by a quench cool, produced micro defects in the silicon material (116). Subsequent growth of these defects, perhaps combined with low levels of carbon from the graphite crucibles and heaters, can cause warpage of slices during device processing (117). On the positive side, these same defects, when confined to the middle region of a slice, can gather unwanted impurities ("lifetime killers") away from the surface where the active devices are located (118,119).

Heating of the melt is either by inductive coupling into the graphite crucible or by a separate resistive winding (also usually of graphite). The temperature is sensed and controlled automatically. The diameter of the crystal can be optically monitored and controlled through variations in the temperature and pull rate. Small-diameter seeds are very desirable and minimize the chance of imperfections in the growing crystal. However, extreme care must be taken to prevent a heavy crystal from exerting enough bending movement on the small brittle seed to break it. This is now accomplished by attaching the seed chuck to a bead chain or a cable rather than to a rigid shaft.

For crystal growing from melts with high vapor pressures (particularly those that dissociate near the melting point) the whole furnace assembly can be enclosed in a pressure vessel. If a lower melting point inert material is available, the crystal can be grown using it as a mantle, as shown in Fig. 6.21. Not only does this reduce vaporization, it allows an inert gas to be used for pressurizing (12).

B. Zone Leveling (121)

A single-crystal seed is placed in one end of a horizontal boat, and the remainder of the boat is filled with polycrystalline material. A relatively narrow molten zone is then formed across the single-polycrystal boundary and moved along the boat until all of the polycrystalline material has been melted and grown as a single crystal (Fig. 6.23b). The

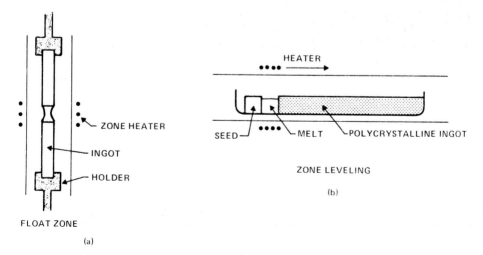

Fig. 6.23 Growing crystals from moving zones.

equipment is simple and works well for materials such as germanium and InSb, which are relatively easy to grow and do not react with nor stick to the boat.

C. Float Zoning (18,80,82)

In order to minimize the purity problems associated with the melt interacting with a container, float zoning is often used. A polycrystalline rod is held vertically with a seed crystal adjacent to one end, as shown in Fig. 6.23. A molten zone is then initiated partially in the seed and partially in the polycrystalline material and caused to traverse the length of the rod. Both ends of the rod can be rotated, and provisions are made to vary the vertical spacing between the end jaws so that the crystal may be "stretched" if desired. Heating is usually by radiofrequency (RF) induction. The maximum diameter rod that can be float zoned depends on the amount of levitation produced by the interaction of the RF currents induced in the melt with those doing the heating, the surface tension of the melt, and the length of the zone. For silicon, the present (1980) limit is about 100 mm. This method is the only one capable of producing silicon crystals with a resistivity greater than a few hundred ohm-cm.

D. Bridgman/Stockbarger

If the material to be grown is not reactive and contracts upon freezing, a vertical tubular container may be used to hold a melt that is allowed to freeze from the bottom up by passing the container through a rather abrupt temperature gradient. Either a single-crystal seed or a sharp-pointed container can be used to promote a single-crystal growth. This method is very simple and is widely used for growing metal crystals. It will also work for low-melting-point compound semiconductors, such as indium antimonide (even though they expand upon freezing), and was used in some of the crystal-growing experiments aboard spacecraft in the late 1970s.

E. Ribbon Growth (122)

Most semiconductor manufacturing processes require thin, flat large-area single-crystal sections with damage-free surfaces as starting material for subsequent processing. The methods discussed thus far generate long ingots from which slices must be sawed, lapped, and polished. To circumvent these steps, various attempts have been made to grow single-crystal ribbon directly. None of these have been commercially successful. The simplest one used a series of vertical spacers running the length of a zone-leveling boat. Another approach used the basic Teal-Little process without crystal rotation; thermal gradients formed by an orifice above the melt forces rectangular cross-sectioned growth (Fig. 6.24). To further assist the crystal growth, an orientation was chosen so that the sides of the crystal were bounded by the slowest growing planes—(111) for Si and Ge.

Instead of defining the crystal size by thermal gradients it is more feasible to define the source size, for example, by the methods shown in Fig. 6.24. The "wick" in the edge-defined growth of Fig. 6.24 is comprised of a pair of closely spaced plates wetted by the melt and arranged so that surface tension pulls the melt up from the main reservoir to the top of the plates. Growth is then confined to the cross section of the top. When used for silicon, contamination from the wick is a major problem.

If thermal gradients in the melt are such that a supercooled melt results, transmission of the latent heat up the growing crystal is no longer the growth-limiting step, and

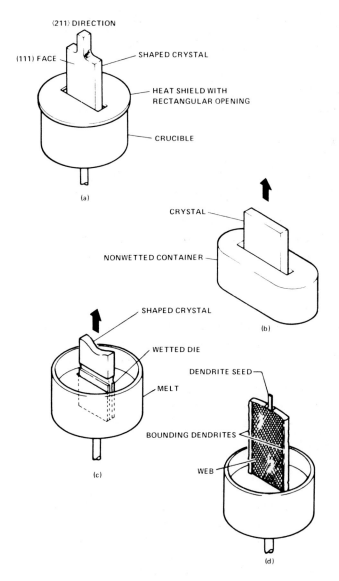

Fig. 6.24 Melt-confined crystal growth. (a) By using thermal gradients alone. (b) From a nonwetted rectangular reservoir generally fed from a larger one. The cross section is determined by the size of the rectangle. (c) From a wetted die. The crystal cross section is determined by the area of the top of the die. (d) Using dendrites. The ribbon width is determined by the separation of the two side dendrites, which are grown out from the seed dendrite.

high-velocity dendritic growth can occur. For diamond lattice materials the growth is in a [112] direction with (111) exposed faces. Besides being difficult to control, these dendrites are very narrow, and all have several twin planes running the length of the ribbon. To extend the width, two dendrites originating from the same seed but separated by up to 1.5 in. can be simultaneously pulled from the melt. As these two dendrites are withdrawn, they drag up a thin web of molten semiconductor between them (Fig. 6.24). This web freezes and sometimes produces good quality very thin single-crystal ribbon.

F. Growth from Low-Temperature Melts

If the desired semiconductor is soluble in a lower melting point material (usually a metal), the molten metal can be saturated and then, as the temperature is lowered, the solubility will decrease and the semiconductor will freeze out. The grown crystal will be saturated with solvent atoms as it freezes, so any application must take this into account. This approach (referred to as liquid-phase epitaxy)* is widely used for growing compound semiconductor layers (Fig. 6.25) (123). Ordinarily, growth will be from one of the consti-

*Epitaxy means oriented overgrowth and is generally applied to the growth of thin layers over a large-area substrate.

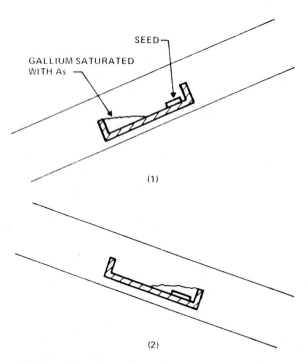

Fig. 6.25 Liquid-phase epitaxy. The gallium is saturated with As in position 1. The whole furnace assembly is then tilted to position 2 until growth is completed. Next, the furnace is tilted back to quickly remove the melt from the GaAs slice. There are also methods for removing the slice that involve sliding a reservoir of melt onto a horizontal slice line and then off when growth is to be terminated.

tuents of the compound, such as gallium for GaAs, and thus there are no contamination problems. One major difficulty with this method is ensuring that the surface of the seed is free of foreign material so that single-crystal growth can occur. For example, the difficulty in removing SiO_2 from silicon at such low temperatures has precluded successful silicon liquid-phase epitaxy.

G. Verneuil or Flame Fusion (112)

In this system, the top of the seed crystal is melted and powdered feedstock is slowly added. Simultaneously, the crystal is lowered so that freezing occurs on the bottom of the molten zone. Thus it is not unlike float zoning except that the zone is replenished, not by melting the upstream part of a bar, but by melting a fine stream of powder. Its original use (1890s) was for growing sapphire and other refractory oxides. For this purpose heating could be by flame (hence, flame fusion). Verneuil growth has not proved feasible for semiconductor materials.

H. Heat Exchange Method (124)

By controlling the heat removal rate from a seed in the bottom of a melt, freezing can occur out into the melt, but not at the walls of the container until the very end, and not even then, if the residual melt can be satisfactorily removed. This method has thus far not been successfully applied to silicon, primarily because of container problems.

I. Electrolysis (125,126)

Germanium can be electrolytically deposited from aqueous solutions, and thin, single-crystal overgrowth has been reported. However, more promising, and applicable to a wider range of materials, is electrolysis from various molten salts. Silicon, CdTe, InP, and SiC have all been produced in this fashion.

J. Casting

For some applications, for example, silicon infrared optics and low-cost solar cells, polycrystalline material may be acceptable. In such cases, casting becomes an alternative to crystal growth. Most semiconductors expand several percent upon freezing, and that, combined with the reactivity and tendency to stick makes casting much more difficult than the casting of metals. There are, however, now several commercially available processes (127).

XI. BEHAVIOR OF IMPURITIES DURING FREEZING (51,52,54,55,68,113,121)

A. Segregation Coefficient

When a multicomponent melt freezes, the composition of the portion just frozen almost never matches the composition of the melt. For two components (i.e., the semiconductor and one impurity) this behavior can most conveniently be depicted by a standard phase diagram such as is shown in Fig. 6.26. If the melt is initially at temperature T_0, its composition will remain constant at x_0 as the temperature is decreased, until the liquidus line is reached. At that point, material of composition Y_1 will begin freezing. If the

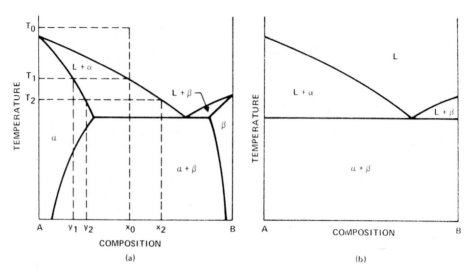

Fig. 6.26 Binary phase diagram. Usually, composition is measured in atomic percentage from A to B. The y_1 means y_1 percent B, the amount of component A is given by 100 − B.

temperature is then further decreased to T_2, a new composition Y_2 freezes. Thus, as freezing progresses, the ratio of component A to component B in both the solid and the liquid continually changes.

    In most cases, the desire is to grow a crystal from a melt that is nearly all one component, such as Si, with a small amount of impurity, so that only the region very near one end of the diagram is of interest. The terminal phase is material A (or B) that has some B atoms in it (or A), but not enough to change the crystal structure or precipitate as a separate phase. The demarcation line between the α and ($\alpha + \beta$) phases therefore maps the solubility limit as a function of temperature. If the solubility of one phase in the other is very low, the diagram of Fig. 6.26a collapses into that of Fig. 6.26b and no solubility data can be obtained from a diagram drawn to that scale, Instead, curves of solubility versus temperature are commonly used, as shown in Fig. 6.27 for Si.

    The ratio k_0 of the amount of component A freezing out in the solid to that in the melt is called the equilibrium segregation coefficient and can, in principle, be determined from examining an expanded portion of the liquidus and solidus curves near the freezing point of B. These curves will appear as in Fig. 6.28, and k_0 is given by $(100 - Y)/(100 - X)$. In practice, it is generally more appropriate to determine the impurity concentration in a melt independently, grow a section of crystal, determine its impurity content, and then calculate a segregation coefficient. Values for selected impurities in Si and Ge are given in Table 6.14.

B. Effective Segregation Coefficient

It should be remembered that, although phase diagrams are constructed from equilibrium data, in practice, freezing will take place too rapidly for equilibrium to be reached. Thus, in many cases the observed values will deviate markedly from those predicted from the phase diagram. Typical of this behavior is the segregation coefficient. Under steady-state

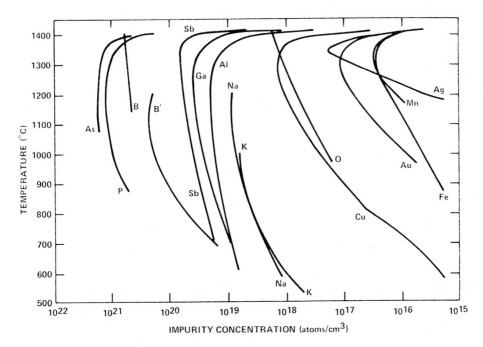

Fig. 6.27 Solid solubilities of impurity elements in silicon. (Adapted from F. A. Trumbore, Solid solubilities of impurity elements in germanium and silicon, *Bell System Tech. J.*, vol. 39, pp. 205-233, 1960. Low-temperature boron data (B′) from G. L. Vick and K. M. Whittle, *J. Elec. Chem. Soc.*, vol. 116, pp. 1142-1144, 1969.)

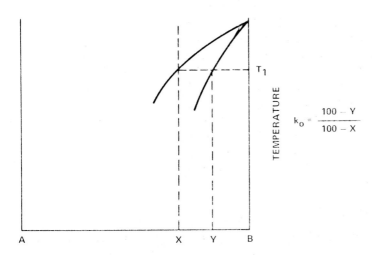

Fig. 6.28 Determination of the equilibrium segregation coefficient k_0 from a phase diagram.

Table 6.14 "Equilibrium" Segregation Coefficients

Element	Si	Ge
B	0.80	17
Al	0.0020	0.073
Ga	0.0080	0.87
In	4×10^{-4}	0.001
P	0.35	0.08
As	0.3	0.02
Sb	0.023	0.003
Bi	7×10^{-4}	
Cu	4×10^{-4}	
Au	2.5×10^{-5}	
Fe	8×10^{-6}	
O	1.25	
C	0.07	

conditions, a plot of concentration of the impurity (component A in the earlier discussion) versus distance away from the freezing interface would be as in Fig. 6.29a, where N_0 is the concentration in the melt and N_s that in the solid. Since the level is lower on the solid side than in the liquid in the examples shown, there is a rejection of impurities as the freezing interface moves. Unless growth is very slow, so that the impurities can diffuse back into all of the remaining melt, there will be a buildup in concentration at the interface as in Fig. 6.29b.* The level N_0' at the interface thus becomes a function of the freezing rate and is greater than the N_0 of Fig. 6.29a. The N_s in the solid is still k_0 times the concentration in the melt at the interface. However, since that value is greater than N_0, the new N_s is greater than the old one, and any calculations based on the impurity concentration N_0 in the body of the melt and an equilibrium k_0 will be in error. The interface concentration in the melt is also experimentally difficult to obtain. Therefore, a new (effective) distribution coefficient is defined as

---

*Stirring will also help, but there will still be a thin stagnant layer adjacent to the crystal interface.

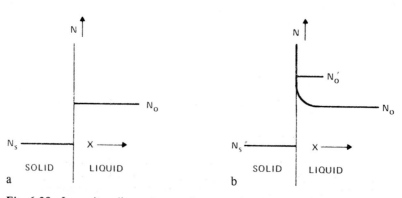

Fig. 6.29 Impurity pileup at a moving crystal-melt interface.

$$K_e = \frac{N_s}{N_0}$$

where K_e takes these concerns into consideration. In principle, this K_e can be calculated from the Burton-Prim-Slichter expression (179)

$$K_e = \frac{k_0}{[1 - (1 - k_0) \exp(-r\delta/D)]} \qquad (1)$$

where r = the microscopic rate of growth, δ = width of the boundary layer adjacent to the growing interface, and D = the diffusivity of the impurity in the melt. However, it is generally determined experimentally. Figure 6.30 shows typical data for Si and Ge.

If conditions are maintained so that K_e does not vary during the growing cycle, and if no additional impurities are added, the impurity concentration at any time is given by

$$N_s(l) = N_0 k_e (1 - l)^{k_e - 1} \qquad (2)$$

where l is the fraction of the original melt that has solidified, N_s is the concentration in the solid, and N_0 is the original melt concentration. This behavior is shown in Fig. 6.31 for several values of k. If k is very much less than 1,

$$N_s = \frac{N_0 k_e}{1 - l} \qquad (3)$$

and the functional dependence of N_s on the amount grown is independent of k, although the absolute value is directly proportional to k_e [cf. Eq. (2)].

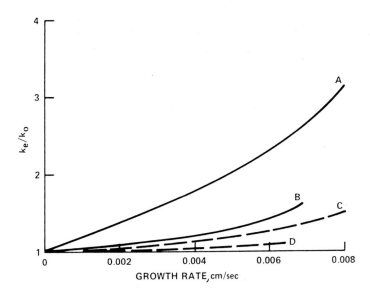

Fig. 6.30 Effect of growth rate on effective segregation coefficient. The value also depends on crystal rotation rate. Thus, the curves cannot be directly compared. A, Sb in Ge, 144 rpm; B, Sb in Si, 200 rpm; C, Ga in Ge, 144 rpm; D, Ga in Si, 200 rpm. (Curves A and C from Ref. 67, B and D from H. Kodera, *J. Appl. Phys. (Japan)* 2:527-534, 1963.)

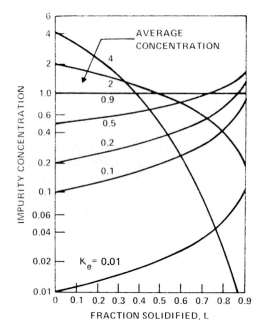

Fig. 6.31 Curves for normal freezing, showing impurity concentration in the solid as a function of the fraction solidified. (Adapted from Ref. 121.)

C. Crystal Doping (51,52,54,55)

When crystals are grown by the Teal-Little (Czochralski) method, the impurity concentration (doping) is described by Eq. (2), and varies with the amount grown, as shown in Fig. 6.31. Since the requirement is usually for a rather narrow resistivity (and thus impurity) spread, efforts are made to level out the profile. One obvious way is to use impurities with segregation coefficients near unity. Unfortunately, except for boron in silicon, most coefficients are much lower. The two remaining ways of controlling the profile are by changing the melt impurity concentration or varying the growth conditions to change k_e. Impurities with high vapor pressures over the melt will slowly evaporate and help hold the profile flat. Changing the effective segregation coefficient can be accomplished by varying the spin rate or growth rate, and is the procedure most used (34,51,52,54,55).

If the melt is continuously replenished during the growth so that larger crystals can grow from a given size of crucible, the impurity concentration can be held quite constant if the concentration of impurity in the incoming feed is uniform. The general expression for concentration $N_s(x)$ under continuous growth is

$$\frac{N_s(x)}{N_s(0)} = \frac{N_r - (N_r - k_e N_i) \exp(V_x/V_i)}{k_e N_0} \tag{4}$$

where

$\quad N_i$ = initial impurity concentration in the melt
$\quad V(i)$ = melt volume
$\quad V(x)$ = volume of crystal grown

$N_s(0)$ = concentration in the first part of the grown crystal
N_r = concentration in the replenishing material

Thus, if $N_r = k_e N_i$, then N(x) is constant.

As an alternative to doping during growth from a large melt, doping can be done from the molten zone of zone leveling or float zoning. All the dope is added in the zone before it is moved along, and the concentration $N_s(x)$ is given by

$$N_s(x) = k_e N_L e^{-(L/kx)} \tag{5}$$

where N_L is the initial impurity concentration in the zone, L is the zone length, and x the length frozen. This procedure is typically used for Ge and InSb crystals, and for those silicon crystals grown by float zoning. Unlike the previous behavior, the most uniform doping occurs when k_e is very small, so that the exponential term approaches 1. An impurity profile other than that defined by Eqs. (2), (3), (4), or (5), is difficult to achieve unless it is based on discrete additions. Provisions can be made to periodically add impurities and produce a profile as shown in Fig. 6.32a, but impurities cannot then be removed in any practical fashion to allow the profile to continue as shown by the dotted line of Fig. 6.32b. The resistivity of the crystal can, however, be changed by taking advantage of compensation so that with impurity level changes as shown in Fig. 6.33a, the resistivity changes as in Fig. 6.33b. If enough compensating impurity is added, the type can be changed as in Fig. 6.33c. This particular profile is the one required for an n-p-n bipolar transistor and is the method used most to dope the crystals for grown junction transistors. (The grown junction transistors were the first commercially successful silicon transistors and were principally produced between 1954 and 1960.) An alternative doping scheme, which leaves the impurity concentration in the melt virtually unchanged, but can vary the level in the crystal enough to substantially affect resistivity, is shown in Fig. 6.34. It takes advantage of the fact that the k_e values for different impurities may

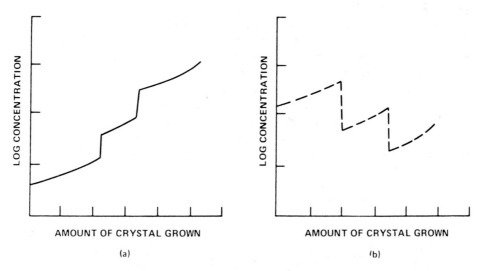

(a) (b)

Fig. 6.32 (a) A melt-grown doping profile easily realizable. (b) A melt-grown profile virtually impossible to obtain.

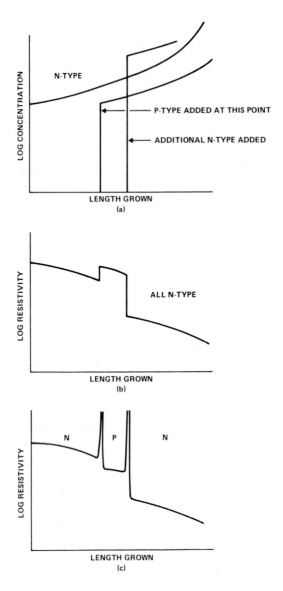

LOG CONCENTRATION

N-TYPE

P-TYPE ADDED AT THIS POINT

ADDITIONAL N-TYPE ADDED

LENGTH GROWN
(a)

LOG RESISTIVITY

ALL N-TYPE

LENGTH GROWN
(b)

LOG RESISTIVITY

N   P   N

LENGTH GROWN
(c)

Fig. 6.33 Effect of compensation in (a) impurity profile, (b) resisitivity, (c) resistivity if sufficient p-type was added in (a).

have different dependencies on growth rate (64,67). For those cases where the impurity level itself must change substantially during growth, vapor-phase methods, discussed later, must be used.

In addition to the rather smooth increase in impurity concentration predicted by Eq. (2), there are a variety of phenomena that produce additional radial and longitudinal variations. The more common ones are melt stirring by the rotating crystal, convection currents, and nucleation effects (51,52,54,55,127,128). Both stirring and convection

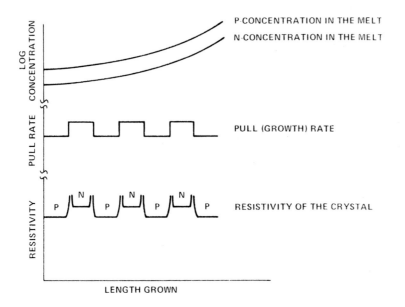

Fig. 6.34 Use of changes in growth rate to produce alternating p and n regions. This procedure was used in the 1950s to manufacture "rate-grown" germanium transistors. In such applications, the growth rate cycle would be asymmetric so that the base regions would be much narrower than the emitter and collector regions. From a practical stand-point, the rate cannot be changed as abruptly as shown. Thus the resistivity will actually change more gradually than indicated.

currents cause rather severe short-term temperature fluctuations to occur in the melt and produce similar fluctuations in the growth rate. These in turn may lead to closely spaced unwanted impurity concentration variations. The growth interface will ordinarily be slightly curved, but if the growth direction is close to a slow-growing plane, facets may develop as indicated in Fig. 6.35. Because the curved interface has an abundance of steps for nucleation, lateral growth can proceed uniformly. However, substantial undercooling may be required to nucleate each new layer on the facet, so that, when it does nucleate, growth across it will be very rapid. The net result is that k_e will be higher in the faceted region and produce impurity coring.

D. Purification During Growth

Reference to Fig. 6.31 shows that, even though there is a substantial increase in impurity concentration in the solid as freezing progresses, there is a reduction of impurity in the first part to freeze so that purification occurs (i.e., for $k_e \leqslant 1$). If the last portion to freeze is discarded, and the remainder remelted and resolidified, continued purification can be obtained. Such a procedure is, however, time consuming as well as wasteful of material. If the material is in the form of a long rod, and a short zone melted and then moved the length of the rod as shown in Fig. 6.36a, the impurity concentration freezing out is still given by

$$N_s = k_e N_0 \tag{6}$$

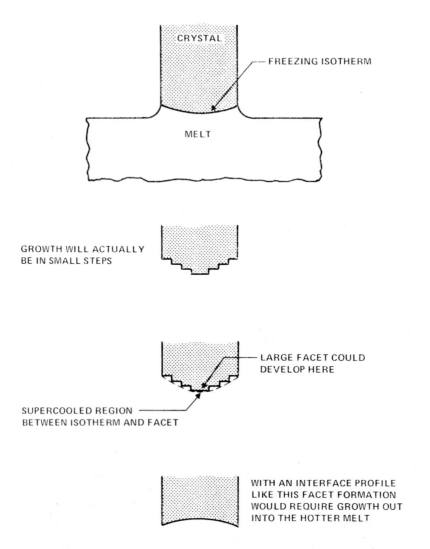

Fig. 6.35 Mechanics of facet formation. (Adapted from J. B. Mullin in Ref. 110.)

at any given instant. However, because the whole volume is never all melted at the same time, impurities cannot redistribute and the concentration versus length, analogous to Eq. (2), is now given by

$$N_s(x) = N_0 \left[1 - (1 - k_e) \exp\left(-k_e \frac{x}{L}\right) \right] \qquad (7)$$

where L is the length of the zone and x is the distance from the starting end. This expression is plotted in Fig. 6.36b. It can be seen that some purification occurs in the front part of the crystal, although not nearly as much as occurred in Fig. 6.31. Since only a zone is melted at any given time, the zone can be repeatedly passed along the ingot, and additional purification achieved without resorting to cutting off the end each time. Eventually,

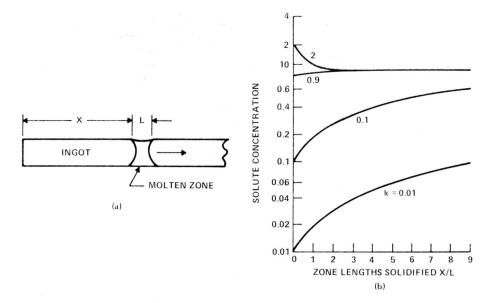

Fig. 6.36 Zone refining. (a) A zone is formed as in zone leveling and slowly swept the length of the ingot. (b) Curves for zone melting showing the impurity concentration in the solid as a function of x/L. (From Ref. 121.)

though, a limiting impurity profile will be reached that will depend on k_e and the ratio of L to total length. This process is called zone purification and, when semiconductor synthesis technologies were less advanced, it was an important part of most purification procedures. Zone refining, in which a horizontal boat is used (Fig. 6.23b), was well suited to materials such as Ge, InSb, and GaAs since they could be easily confined in noncontaminating boats. A single crystal need not be grown with each pass, although it is preferable since grain boundaries can trap impurities. To speed up the process (the zone rate of travel is of the order of centimeters per hour), spatially separated zones can be initiated and moved along the ingot simultaneously.

To eliminate container contamination, silicon must use the float-zone configuration of Fig. 6.23a, and since there is no horizontal support, only one zone at a time can be used. Because of this restriction, zone refining of silicon is particularly expensive and is seldom used except to produce material for a few specialized devices, such as particle detectors, large-area photodetectors, and very high voltage power control devices.

XII.   CRYSTAL GROWTH FROM THE VAPOR: EPITAXY (129)

A. Introduction

Growth from the vapor involves more variables and in general is less applicable to large-volume, high-perfection crystals than is growth from the melt. However, there are some materials, such as CdS, that sublime rather than melt and are virtually impossible to grow in any other manner. In addition, because of the minimal interaction between the vapor stream and the solid, it is relatively easy to overgrow thin crystalline layers onto a single-crystal substrates that have impurity concentrations differing widely from the layers. The growth of such layers is referred to as epitaxy. The word *epitaxy* comes from the Greek words *epi,* meaning upon, and *taxis,* meaning arrangement or order. It was apparently coined by Royer (130) to describe the orientation of a crystal layer of a water-soluble salt grown by evaporation onto a naturally occuring mineral face. Webster's Collegiate Dictionary, which did not add the word until the early 1970s, defines it as, "The growth on a crystalline substrate of a crystalline substance that mimics the orientation of the substrate." In the semiconductor industry epitaxy is used to specifically denote over-growth onto a substrate of the same material, such as Si on Si or Ge on Ge. That special-ized usage has become so deeply entrenched that the original context, silicon overgrown on sapphire, is now described as heteroepitaxy. Further, epitaxy now always denotes growth from the vapor. Growth from a melt is called "liquid-phase epitaxy," or LPE (discussed in an earlier section).

Vapor-phase grown single crystals of silicon and germanium were first grown short-ly after melt-grown crystals became available (135), but the process received little atten-tion until a problem arose that melt growth could not solve. Vapor-phase growth is more expensive than melt growing, and it is difficult to grow thick (greater than a few tenths of a millimeter) high-perfection layers. However, shortly after the advent of planar transis-tors, a transistor structure requiring a thick slice of low resistivity, overlaid with a thin high-resistivity layer, was developed (86). Most transistors and some MOS integrated cir-cuits still require such a configuration, and bipolar integrated circuits use a high-resis-tivity starting slice ("substrate") with a low-resistivity epitaxial layer. Due to the indus-trial importance of silicon epitaxy, Sec. XIII is devoted exclusively to it.

The first requirement for vapor-phase growth is a means of transporting material to the growing surface. This can be by a vapor of the same composition as the crystal to be grown, or by a compound or compounds that react at the growing surface to pro-duce the desired material. The second requirement is that, if a chemical reaction is in-volved, it must be surface catalyzed, so that the material is produced only at the surface of the growing crystal and not in the gas stream. Third, if single crystals are to be grown, the surface mobility of the impinging atoms must be high enough for them to migrate to the correct position on the surface before they become bonded in place.

B. Sublimation

Sublimation is one of the simpler approaches. For materials with low vapor pressures, a very low chamber pressure must be used to obtain practical growth rates. Silicon and germanium fall in this category and have been grown from evaporation in a bell jar, but residual gases usually cause excessive crystallographic defects. However, by using a very clean system with the pressure held to less than 10^{-8} torr during deposition, satisfactory quality layers of silicon and most other materials can be grown. This procedure is com-monly referred to as "molecular beam epitaxy" (131).

C. Closed-Tube Transport (132)

Instead of sublimation, material can be transported from a region at one temperature to a region of another temperature via a series of reactions involving a transport interme-diary. Although suitable agents can be found for transporting almost all semiconductor materials, the inconvenience of such a system precludes its use if other means are avail-able. Two of the more common types of reactions are disproportionation and halide transport. A typical disproportionation reaction would proceed as follows. An evacuated tube would have polycrystalline silicon for source at one end, a single-crystal seed at the other, and a charge of iodine. To transport silicon from the source to the seed, the seed would be held at 900°C and the source at 1100°C. The I_2 would react with Si to give SiI_4. The SiI_4 would further react with Si to give $2SiI_2$. At the cool end of the tube (seed end) the SiI_2 disproportionates to Si + SiI_4, the SiI_4 diffuses back to the hot end, reacts with more Si to give $2SiI_2$, and the process repeats.

Halide transport involves the reaction of a halide at one end of the tube with the feedstock and hydrogen reduction at the other end to deposit single-crystal material on the seed. For example, HI will react with germanium at the hot end of a tube to form GeI_2 and H_2. At the cold end, the hydrogen reduces the germanium iodide to germanium and produces HI again.

D. Open-Tube Transport

Despite all the other approaches available, most semiconductor vapor-phase crystal growth now uses an open-tube transport system. It allows many more choices of transporting agent and a wider range of deposition rates and impurity concentration than any of the other methods.

A carrier gas is commonly used to transport the various components required. Some source materials dissociate at high temperature and require a carrier gas only for dilution. Examples are

$$SiH_4 + heat + H_2 \text{ (carrier)} \rightarrow Si + 2H_2 + H_2 \text{ (carrier)}$$

$$GeH_4 + heat + H_2 \text{ (carrier)} \rightarrow Ge + 2H_2 + H_2 \text{ (carrier)}$$

$$SiH_2Cl_2 + heat + H_2 \text{ (carrier)} \rightarrow Si + 2HCl + H_2 \text{ (carrier)}$$

The last example is really more complex than indicated because additional reactions take place to produce a whole series of silicon-bearing compounds. They range from very simple gases to long-chain oily and explosive liquids.

Hydrogen reduction reactions, such as

$$SiCl_4 + heat + 2H_2 \rightarrow 4HCl + Si$$

are also widely used. In this case the hydrogen is both a reagent and a carrier. It is present in excess without interfering with the reaction.

For III-V compounds, the transport system becomes more elaborate since two components are involved. Often, even though an open tube is used, one component will be deposited by a disproportionation reaction similar to that used in a closed tube. Simultaneously, the other component might be deposited by sublimation, pyrolysis, or hydrogen reduction of a halide. Typically, the group III elements are transported as subchlorides (disproportionation) manufactured in the same reactor, upstream from the deposition. The group V elements are transported as hydrides, halides, or pure vapor.

E. Substrate Etching (34)

During the early and mid-1960s silicon and germanium substrates were mechanically polished, and considerable crystal damage remained in the polished surface. This damage gave rise to epitaxial films having a high density of stacking faults. This effect was recognized in 1963 and the problem eliminated by in situ HCl vapor etching in the epitaxial reactor.

F. Orientation Effects (87)

The deposition rate is usually dependent on the orientation of the substrate. Such effects are more noticeable at lower deposition temperatures since high-temperature rates tend to be limited by the availability of reactants at the surface (i.e., diffusion kinetics), and not by growth kinetics. In the case of silicon, the deposition rate is minimum on (111) faces, and since most bipolar structures use (111) orientation, a slight misorientation is commonly used to both speed up deposition and minimize the growth impact of structural defects.

G. Doping

Impurity control during vapor-phase growth is accomplished by transporting and depositing the dopant along with the semiconductor. For silicon and germanium, hydrides, such as diborane, phosphine, and arsine, are appropriate. For III-V compounds, hydrides can be used if available, but in some cases, it may be necessary to use metal vapor, such as zinc, or a halide.

H. Diffusion and Autodoping Effects (133)

In normal CVD epitaxy abrupt changes in impurity concentration at the substrate-epitaxial layer interface cannot really be obtained because of two separate effects. The first is that the growth operation must be done at high temperatures; hence, some impurity redistribution by diffusion will occur. Second, for most reactions there is some etching as well as deposition, so any impurities in that portion of the substrate that is etched away can be subsequently incorporated into the growing layer. The development of molecular beam epitaxy overcomes this problem and atomically sharp junctions can be formed.

I. Heteroepitaxy (134)

The requirement to grow materials onto foreign substrates has arisen because of the desire to combine specialized properties of the layer and substrate. As examples, devices built in silicon overgrown on an insulating oxide or on a sapphire substrate can be truly electrically isolated, and relatively expensive III-V compounds can be overgrown on relatively inexpensive germanium substrates. The problems of such growth arise from (1) lattice mismatch between the two materials, (2) thermal expansivity mismatches, and (3) incompatibility of the substrate with the layer deposition conditions. Lattice mismatch is the major problem. Nucleation will begin at many sites over the surface, and if the two lattices do not exactly match, there will be defects between the various regions when they grow together. Further, bonding forces between the overgrowth atoms and the substrate may be so weak that either no orientation occurs, or a substantial number of those nucleation sites are unoriented. However, despite these difficulties, some combinations, such as Si on sapphire and GaAs on germanium, have proven commercially feasible, and silicon on insulator (SOI) looks promising for VLSI MOS devices (134).

XIII. SILICON EPITAXY

A. Historical Introduction

As noted in Sec. IV, within a few years of the development of the transistor, it became evident that more sophisticated devices would be developed and that these structures would require doping profile and conductivity change control beyond that attainable by standard grown junction and diffusion control. Because of these requirements, people at several different laboratories began to look at vapor-phase deposition for the development of thin well-controlled overgrowth films on single-crystal silicon substrates. The first known work published in the open literature was in October 1954, when Patent No. 2,692,839 was issued to Teal and Christensen (135). This patent was filed in April 1951, based on work done in 1950, and describes the deposition of a layer or successive layers of different conductivity types. The patent describes both germanium and silicon deposition. In May 1957, a paper titled Growth of Silicon Crystals by a Vapor Phase Pyrolytic Deposition Method (136) was published by Sangster et al. of Hughes Aircraft Company. They used the silicon tetrabromide ($SiBr_4$) hydrogen reduction process in the temperature range of 1100-1350°C. They also recognized the importance of a clean, oxide-free, substrate at this early date: "Proper pretreatment of the Si seed surface is necessary if oriented growth is to be obtained. Chemical etching is desirable for producing a superficially clean surface, but all such techniques apparently leave an oxide film which must be destroyed before deposition is begun. For this purpose the only successful technique that has been found is preheating in a stream of pure H_2 for 30-60 minutes at 1275°C."

Approximately 3 years passed before the next publication on silicon overgrowths appeared. In June 1960, Mark (137) of the U.S. Army Signal Research and Development Laboratory, Fort Monmouth, New Jersey, reported that: "A Single Crystal Silicon Overgrowth has been produced on a silicon substrate. This phenomenon confirms a fundamental principle for silicon showing that it is possible to have a vapor phase chemical reaction between the solid and vapor of one of the compounds. The silicon monolayers grow by accretion of atoms from the generating medium and assimilate as a single crystal overgrowth to the (100) orientation of the silicon substrate." This work was carried out using the hydrogen reduction of silicon tetrachloride ($SiCl_4$). In July 1960, Wajda et al. (138) and Glang et al. (139) published papers on the subject of epitaxial silicon, as did Marinace (140) (who also gave the definition for the word epitaxial) on the subject of epitaxial germanium. All these papers related to disproportionation of the iodide, SiI_2 or GeI_2, to transport the silicon or germanium from a high-temperature source zone to a lower temperature substrate zone.

At the Boston, Massachusetts meeting of the AIME Conference, August 20-31, 1960, Sivalda and Glang of the Federal System Division, International Business Machines, presented a paper titled Epitaxial Growth of Silicon p-n Layers from the Vapor Phase. This work made use of the silicon-iodine system. "Through thermodynamic analysis of the silicon-iodine system, it has been demonstrated that the SiI_2 disproportionating reaction can be used for silicon transport. This has been verified experimentally in the 1100 to 850°C temperature region. By placing a suitably prepared silicon single crystal, as a substrate, in the lower temperature region, expitaxial growth was achieved. Controlled impurity introduction, for growing n- or p-type layers, has been demonstrated in the range of 5×10^{16} to 2×10^{20} atoms/cm^3. Multiple layers of p, n, p+, and n+ conductivity have been grown and show characteristics similar to junctions fabricated by diffusion techniques."

"The vapor growth process can be used to fabricate novel device structures and promises to play an increasing role in semiconductor technology" (141).

Allegretti et al. of Mercke, Sharp, and Dohme Research Laboratories, Rahway, New Jersey (142), reported on "Vapor Deposited Silicon Single Crystal Layers" at this same conference. This work made use of SiH_4 with emphasis on epitaxial deposition on single-crystal rods and the study of the effect of substrate orientation on single-crystal growth. It was attempted to obtain single crystals with growth planes that would have maximum surface area free of imperfections. The crystal substrates (rods) that were used in the studies were [111], [211], and [110] oriented. The substrate orientation that produced the highest yield of usable single-crystal area was the [111] oriented substrate. Good control of alternate p-n layers was demonstrated on the rods. In September 1960, Theurer et al. of Bell Telephone Labs, Murray Hill, New Jersey, presented a paper entitled Epitaxial Diffused Transistors (86) that mentions both silicon and germanium. Transistors of both germanium and silicon have been made on epitaxial material. In the case of silicon, a low-resistivity (0.002 ohm-cm) n-type crystal was used as a substrate. Epitaxial n-type layers several micrometers thick were produced by the decomposition of $SiCl_4$. The base and emitter layers were then diffused using conventional boron and phosphorus processes. Although we have emphasized the compatibility of epitaxial films with conventional diffusion and alloying processes, the use of the material is by no means restricted to these techniques. For example, the need for a diffused base layer may be eliminated by doping the epitaxially grown film to an appropriate resistivity. The use of epitaxial material, however, is not restricted to conventional devices. Complex structures, which are now difficult to fabricate or are not possible with conventional technology, now become feasible. These techniques will open many new horizons in the semiconductor device area. "The thin high resistivity layer can be produced by thermal decomposition of silicon tetrachloride or germanium tetrachloride on a low-resistivity substrate of the same material." Theuerer (143), July 1961, published Epitaxial Silicon Films by the Hydrogen Reduction of $SiCl_4$. In this work a deposition temperature of 1270°C and (111) oriented substates were used. At the same time, work was being conducted at Ferranti Ltd. (144) under a U.S. government contract, AD 268 314, at Texas Instruments by Sigler and Watelski (145) and by Schnable and Hirshon of Philco Corporation (146). In Hirshon's work as well as in Schnable's work, they extended the technology of silicon epitaxy to that of selective epitaxial deposition on a SiO_2 masked substrate. Schnable described HCl vapor etching through the oxide mask and epitaxial growth from $SiCl_4$ + H_2. This type of preferential deposition work was again reported by Joyce and Baldrey (147) of Texas Instruments. These authors reported minimization of polycrystalline silicon deposition on the oxide mask by using low flow rates of $SiCl_4$ at 1200°C in hydrogen. Also in that year, Deal (148) of the Raytheon Company reported on the deposition of silicon on several slices simultaneously, and Bylander (149) of Texas Instruments reported on the kinetics of silicon crystal growth from $SiCl_4$ decomposition. These references are the first known publications or reports to appear in the open literature referring to "epitaxially deposited layers."

B. Silicon Source for Epitaxy

During the early and mid-1960s the primary production deposition process used throughout the industry was the hydrogen (H_2) reduction of silicon tetrachloride ($SiCl_4$) at approximately 1250°C:

$$\overset{\sim 1250°C}{SiCl_4 + 2H_2 + \Delta H \to Si + 4HCl} \quad \text{(simplified reaction equation)}$$

Most of this material was used in bipolar discrete and integrated circuit processing, and (111) oriented substrates were used. The silicon substrates were, in general, mechanically polished, and considerable crystal damage remained in the substrate polished surface. This damage gave rise to imperfect epitaxial films having a high defect (especially stacking fault) density. This effect was recognized in the early 1960s and the problem was eliminated by in situ HCl vapor etching in the epitaxial reactor (150). This work was reported in 1963 by Bean and Gleim (151). This paper was followed almost immediately by a publication by Lang and Stavish (152), entitled Chemical Polishing of Silicon with Anhydrous Hydrogen Chloride. (See also Refs. 153 and 154, as well as Vapor Phase Deposition and Etching of Silicon, by Shepard, Ref. 155.)

From the above simplified chemical reaction it is evident that hydrogen chloride is one of the primary end by-products of the reaction (150,158). Four volumes of HCl gas are produced for each volume of silicon produced. From this observation and the data from Fig. 6.37, it is evident that high, >0.2 mol, ratio of $SiCl_4$ in hydrogen may also be used to etch silicon.

The etching rate of silicon falls off very rapidly at temperatures below 1100°C and becomes preferential at mole ratios less than 0.02 and greater than 0.06 in hydrogen. Because of this mole ratio limitation on producing smooth noncrystallographically preferential etching, and because of the high cost of the silicon halides, such as $SiCl_4$ or $SiHCl_3$, high-purity anhydrous HCl is commonly used for in situ vapor etching today (Fig. 6.38). (The words "high purity" cannot be overemphasized for all materials in epitaxial processing: H_2, HCl, Si halides, and so on.) During the late 1960s and 1970s other silicon sources became commonly used as the demands on epitaxial film thickness control, junction abruptness, autodoping, and pattern shift or washout control over buried layers, became dictatorial. The silicon halides $SiCl_4$, $SiHCl_3$, and SiH_2Cl_2 and the hydride SiH_4 are all commonly used today depending upon the requirements of the epitaxial film, polycrystal film, and economical factors. The epitaxial reaction temperatures are given in Table 6.15 for the hydrogen reduction, chemical vapor deposition (C.V.D.), of high-quality epitaxy silicon films from the indicated sources. The bromides can be reduced at ~50°C lower temperature than the chlorides; however, good epitaxial film growth at below 100°C is very difficult at atmospheric pressure.

The hydrogen reduction reaction and equilibrium constant for the commonly used silicon halides are listed in Table 6.16 (see also Ref. 157). Silicon tetrachloride, $SiCl_4$, does not thermally decompose as does dichlorosilane, SiH_2Cl_2, or trichlorosilane, $SiHCl_3$, but silane, SiH_4, thermally decomposes quite readily at low temperature (see Table 6.17).

The reaction rates are a function of temperature, and the equilibrium constant can be expressed in terms of thermodynamic quantities through the relationship

$$K \sim \exp\left(-\frac{\Delta G}{RT}\right)$$

where ΔG is the change in Gibbs free energy, R is the universal gas constant, and T is the absolute temperature. For reactions at constant temperature and pressure, the Gibbs free energy change may be written in terms of ΔH, the enthalpy of formation of the reaction, and ΔS the change in entropy.

$$\Delta G = \Delta H - T\,\Delta S$$

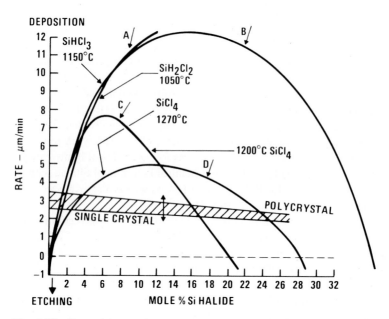

Fig. 6.37 Deposition and etch rate versus mole percent silicon halide. The deposition rate is also affected by reactor design. Curves A and B are from a vertical reactor as in Fig. 6.39C. Curve C is from a multiple-slice vertical reactor in which each slice rotates on its own susceptor. Curve D is the data in Ref. 143, using a single-slice vertical reactor as in Fig. 6.39A. Above 2 or 3 μm/min growth rate, the material is usually polycrystalline. The cross-hatched region shows the variability observed in the position of the line of demarcation.

where $T \, \Delta S$ is the amount of work done in a reversible process (see Table 6.17).

Silicon tetrachloride (SiCl$_4$ and trichlorosilane (SiHCl$_3$) are liquids at room temperature, therefore requiring a bubbler or volatilizer and a metering system for controlled use in epitaxial systems. Dichlorosilane (SiH$_2$Cl$_2$) and silane (SiH$_4$) are gaseous at room temperature, supplied in pressurized cylinders, and require only the proper metering systems for control in the epitaxial systems. In general, the silicon-bearing component of the reaction contains about 1-2% hydrogen to produce high-quality epitaxial films. Higher mole ratios lead to the deposition of polycrystalline films.

Table 6.15 Epitaxial Silicon Deposition Temperature for Silicon Halides and Hydride at 1 atm Pressure[a]

Si source	Temperature required (°C)
SiCl$_4$	1250 ± 25
SiHCl$_3$	1150 ± 25
SiH$_2$Cl$_2$	1050 ± 25
SiH$_4$	1000 ± 25

[a]In high-purity hydrogen.

Table 6.16 Hydrogen Reduction Reaction and Equilibrium Constant for Commonly Used Silicon Chlorohalides

$SiCl_4 + 2H_2 \rightarrow Si + 4HCl$	$K(SiCl_4) = \dfrac{p^4HCl^a}{p\,SiCl_4\,p^2H_2}$
$SiHCl_3 + H_2 \rightarrow Si + 3HCl$	$K(SiHCl_3) = \dfrac{p^3HCl}{p\,SiHCl_3\,pH_2}$
$SiH_2Cl_2 + H_2 \rightarrow Si + 2HCl$	$K(SiH_2Cl_2) = \dfrac{p^2HCl}{p\,SiH_2Cl_2\,pH_2}$

[a]p = partial pressure.

The use of silicon halides as the silicon source has dominated the industry's epitaxial processing. However, the high temperatures required for H_2 reduction cause redistribution of impurities from the substrate into the epitaxial layer. Additionally, the HCl etching of the substrate both prior to and simultaneously with deposition add undesirable doping agents to the epitaxial film (autodoping). This would favor the use of silane, as it can be reduced at lower temperatures (see Table 6.15), thus reducing outdiffusion of the substrate, and there are no halides produced that may cause undesirable autodoping from the substrate or from buried layers (diffusion under the film, DUF) in the substrate (Fig. 6.38). Due to these factors, SiH_4 is becoming more commonly used to produce thin abrupt junction epitaxial films. A difficulty does exist, however, in depositing thick ($>8\,\mu m$) films from SiH_4. This problem arises due to the relatively easy thermal reduction of SiH_4 in the gas phase, above the silicon substrate, as well as at the substrate surface. This gas-phase reduction produces small silicon particles that may fall onto the silicon substrate, thus producing spurious nucleation sites on the epitaxially growing surface, which may produce epitaxial defects called epi spikes. This problem can be minimized by the use of epitaxial reactors in which the silicon substrates stand in a near vertical position, contrasted to that of a horizontal position, and/or by the use of cold wall reactors.

C. Substrate Orientation

The deposition rate of epitaxial silicon is also dependent upon the orientation of the substrate (87,160). The deposition rate increases as the misorientation from the (111)

Table 6.17 Values of ΔS, ΔH, and ΔG at $1500°K$ for the Reactions in Table 6.15

Silicon source	ΔS (cal/mol-K)	ΔH (kcal/mol)	ΔG (cal/mol)
$SiCl_4$	35.9	59.8	5,950
$SiHCl_3$	29.9	49.3	4,450
SiH_2Cl_2	26.0	31.1	−7,900
SiH_4	23.6	5.2	−30,200

EPITAXY

(a)

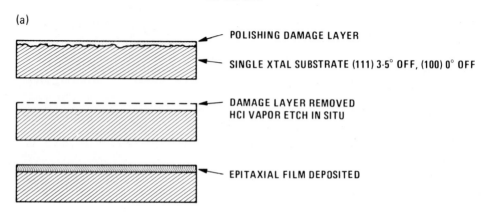

POLISHING DAMAGE LAYER

SINGLE XTAL SUBSTRATE (111) 3-5° OFF, (100) 0° OFF

DAMAGE LAYER REMOVED
HCl VAPOR ETCH IN SITU

EPITAXIAL FILM DEPOSITED

"DUF" EPITAXY

(b)

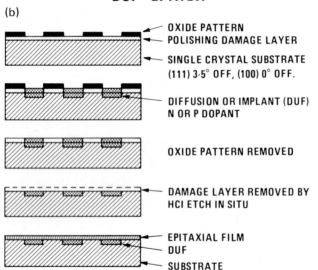

OXIDE PATTERN
POLISHING DAMAGE LAYER
SINGLE CRYSTAL SUBSTRATE
(111) 3-5° OFF, (100) 0° OFF.

DIFFUSION OR IMPLANT (DUF)
N OR P DOPANT

OXIDE PATTERN REMOVED

DAMAGE LAYER REMOVED BY
HCl ETCH IN SITU

EPITAXIAL FILM
DUF
SUBSTRATE

Fig. 6.38 (a) Steps in producing a Si epitaxial layer. (b) Steps to produce localized low resistivity regions buried under the epi layer. Referred to as "buried layer epi" or "DUF" (diffusion under the film).

plane toward the (110) plane increases. With r near (111) = 1.5, r (100) = 1.7, r (110) = 1.6, where r = deposition rate in micrometers per minute for $SiCl_4$ at 1% in H_2 at 1200°C. The r near (111) is actually for 1 to 3° off the (111) toward the nearest (110) due to the extreme difficulty of depositing high-quality epitaxial films directly on the (111) crystal plane. This effect, as well as pattern shift and/or washout, over buried layers, was observed in the early 1960s by Bean and Runyan (see U.S. Patents No. 3,379,584 and 3,534,236 filed September 4, 1964). Practically all epitaxial deposition on (111) silicon since that time has made use of this off-orientation effect. During the

1970s the use of (100) oriented silicon, due to its low surface state density, became the industry standard for MOS (metal oxide silicon) field effect structures, but (111) off-oriented silicon retained its popularity for bipolar structures. Most of the early MOS circuits were built on bulk silicon substrates, and epitaxial films were not required. In the late 1970s and early 1980s, some of the more sophisticated, high-density, MOS and CMOS structures require the use of epitaxial films and the (100) orientation has been used.

D. Doping of Epitaxial Silicon Films

As in crystal growth from the melt, doping agents may be incorporated into the epitaxial film deposited from the vapor (139,160). p-Type dopants for silicon are selected; from Group III boron is the predominant choice. Diborane (B_2H_6) is commercially available in high purity for epitaxial doping purposes. If n-type doped films are desired, phosphorus, arsenic, and antimony are commonly used n-type dopants: arsine and phosphine are commercially available for this purpose.

E. Autodoping and Diffusion During Epitaxy

Due to the relatively high temperature required for conventional silicon epitaxy (>1000°C), solid-state diffusion of impurities, and/or doping elements, from the substrate to the epitaxial junction or interface results in some impurity redistribution. This problem alone can be calculated and to some extent negated by selection of deposition source, temperature, and rate. A more troublesome problem in producing abrupt profiles arises due to "autodoping" (162-165) during the epitaxial process. This phenomenon arises due to a combination of chemical and vapor phase side reactions, particularly the chemical vapor-phase etching of dopants and impurities from the reactor components, as well as top, edge, and back surfaces of the silicon substrate, and subsequent redeposition in the film. This is caused by the reaction products, such as chlorine (Cl) and hydrogen chloride (HCl), which result from the hydrogen reduction of the silicon halides. Some doping agents, such as arsenic, also exert considerable vapor pressure and outgas during epitaxy, thus adding to the autodoping problem. These problems must be considered and can partially be eliminated by (1) lower deposition temperatures, (2) use of SiH_4 (166) or lower halogen containing silyl halide (such as SiH_2Cl_2, rather than $SiCl_4$), or (3) by sealing the back and edge surfaces of the substrate with an oxide (SiO_2), nitride (Si_3N_4), or polysilicon film prior to expitaxial deposition.

Two advances in epitaxial systems during the late 1970s make abrupt junction or interface epitaxial processing available: molecular beam epitaxy (167) and low-pressure epitaxy (168). These systems use lower deposition temperature as well as other effects that reduce or eliminate autodoping. Also, interest has developed in liquid-phase silicon epitaxy (169), which also reduced the autodoping effects.

F. Epitaxial Systems

Epitaxial reactor development has evolved from a rather simple (by today's standard) one-slice system used in the early 1960s, as shown in Fig. 6.39A to the very sophisticated multislice systems used today, as shown in Fig. 6.39C and D and Fig. 6.40. The reactor shown in Fig. 6.39A has some very desirable characteristics for today's research and was excellent in the early days. The reactant gases enter at the top of the reactor, pass down over the single slice, and exit at the bottom of the reactor, thus eliminating the slice-to-

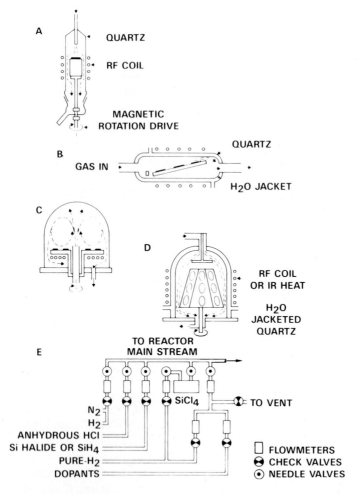

Fig. 6.39 Chemical vapor deposition gas manifold and reactor types. (Adapted from V. S. Ban, *J. Crystall. Growth* 45:97, 1978.)

slice autodoping effect of reactor B. In most cases the slice and RF susceptor were rotated, which provided good deposition uniformity of thickness and resistivity across the slice. The one-slice reactors of the early 1960s soon gave way, due to demands of production, to multislice reactors, such as that of Fig. 6.39B and C. The "horizontal reactor" of Fig. 6.39B has the inherent characteristic that the reaction gases and the by-product gases, pass from each of the leading slices to the subsequent slices located downstream, thus adding contamination (autodoping), progressively across each slice, and from slice to slice through the reactor. The slices were heated by RF induction to the graphite susceptor or by the electrical resistance of the susceptor. The "pancake" reactors, as in Fig. 6.39C, were developed in the early and mid-1960s and were widely used through the 1970s. The RF susceptor rotates, and the gases in general make one pass over the slice, thus to a large extent eliminating the nonuniformity of film thickness and doping control across the slice and from slice to slice. The "barrel" reactor concept of Fig. 6.39D was

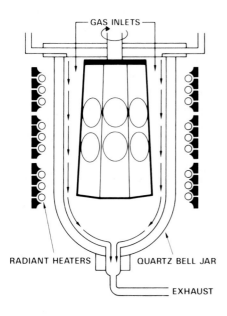

GAS INLETS

RADIANT HEATERS   QUARTZ BELL JAR

EXHAUST

Fig. 6.40 Schematic of high-volume vertical epitaxial reactor. (Courtesy of Applied Materials Inc., Santa Clara, California.)

first developed during the late 1960s using RF heating and redeveloped in the mid-1970s and late 1970s using infrared heating. This reactor has the advantage of near vertical slice positioning, which greatly reduces the probability of particle contaminants falling onto it and developing into defects during epitaxial deposition. These reactors usually have two or three slices mounted one above the other. This again produces to some extent the autodoping problems as described in the discussion on the horizontal reactor. The use of radiant heated reactor systems as well as shaped susceptors have reduced the tendency for dislocation formation during epitaxial processing. The above-discussed reactor types are those predominantly used in industry; however, there are many more designs used in research and development, which are too numerous and too varied in design to discuss in this text.

The epitaxial reactor gas distribution flow system, shown in Fig. 6.39E, is a typical, but simplified, version of what is commonly used. The extreme importance of high purity of materials must be reemphasized. Properly cleaned stainless steel plumbing is standard in these systems. The degree of sophistication ranges from simple hand-operated switches and valves to very sophisticated computer-controlled switches, timers, and mass flow controllers.

Highly automated continuous epitaxial systems are in the evaluation stages for production at this time, as are molecular beam (167,168) low-pressure, low-temperature epitaxial reactors for research and development for tomorrow's needs.

XIV. CRYSTALLOGRAPHIC DEFECTS (103)

So far, single-crystal growth has been discussed without discussing the perfection of the material produced. There are, in fact, a wide variety of defects that can occur, ranging from missing single atoms to massive errors in the lattice.

The most common kinds of defects are summarized in Table 6.18. Some of them are unavoidable, but most can be either completely eliminated or reduced to a usable level by careful attention to the crystal-growing process. For most applications, grain boundaries, lineage, and twins are rejected (but they seldom occur). Dislocation-free Si and Ge can be routinely grown from the melt, although dislocation densities of a few hundred per square centimeter can usually be tolerated.* Stacking faults are sometimes found in epitaxial growth and occur in densities of a few per square centimeter. Fortunately, the most troublesome defects can be detected rather easily by etching a polished surface (e.g., a slice) and examining it with a metallurgical microscope. Figure 6.41

---

*Densities of defects involving lines (dislocations) and planes (stacking faults) are generally measured in terms of the number that intersect the surface on which the measurement is being made.

Table 6.18   Crystal Defects

Vacancy: An atom missing from the otherwise regular lattice	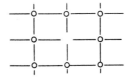	These will always be found since small numbers are thermodynamically stable.
Interstitial: An extra atom in the lattice	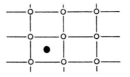	Occurs during oxidation, diffusion, and ion implanation. Small numbers are thermodynamically stable.
Edge dislocation: Appears as though there is an extra sheet of atoms in some portions of the crystal	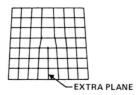 EXTRA PLANE	Observed in melt-grown crystals.
Stacking fault: Occurs when a crystal lattice has a sequence of layers e.g., ABC ABC ABC, and some region gets out of order (appears as a missing layer), e.g., ABC BC ABC.	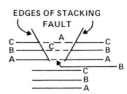 EDGES OF STACKING FAULT	Observed in vapor-grown crystals with diamond or zinc-blende structures

Table 6.18 (Continued)

Twin: Occurs when two regions of different orientation share a common plane and all nearest neighbors are in the correct position	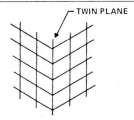	Often found in melt-grown material. The twin plane is a function of the crystal structure. For diamond and zinc-blende it is a (111).
Linage: Occurs where two regions with slightly differing orientation meet	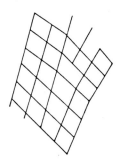	Often occurs in zone-leveled germanium.
Grain boundary: Occurs where two regions with gross differences in orientation meet		Occurs during casting, seldom during seeded crystal growth.
Hillocks and depressions		Observed in vapor-grown crystals. May not be errors in the lattice, but physical outline may cause later processing problems.

shows the way some of these defects in silicon appear after etching. Other materials will appear similarly, although the etchants will be different. Table 6.19 lists some selected defect etchants for silicon. Transmission electron microscopy can be used in most cases instead of etching, but has the disadvantage of requiring more extensive sample preparation and more elaborate equipment. Point-defects such as vacancies, interstitials, and foreign atoms and antistructure (found only in multicomponent crystals) are more difficult to observe and may require various electrical or optical measurements.

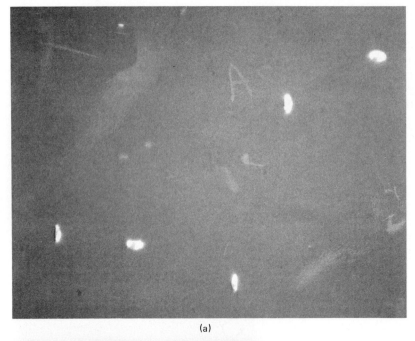

(a)

(b)

Fig. 6.41 Crystallographic defects delineated by chemical etching. (a) (100) dislocation loops under phase contrast. 846X. (b) Crystal growth twin. The straight vertical line occurs at the boundary between two orientations in twin relationship. (c) (111) stacking faults. 175X (d) (100) stacking faults and loop dislocations.

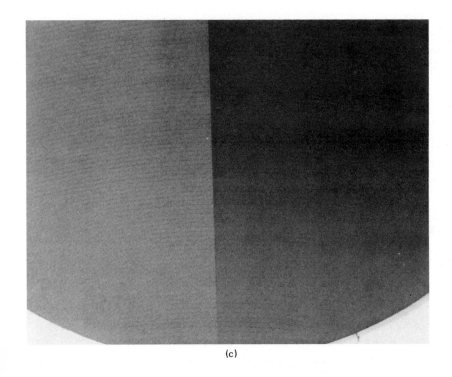

(c)

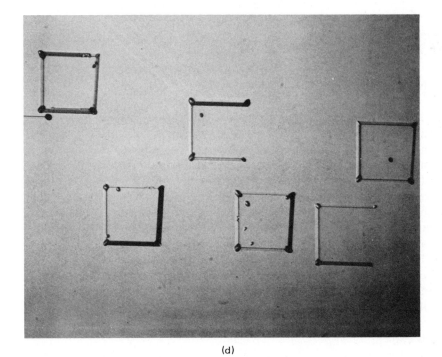

(d)

Fig. 6.41 (Continued)

Table 6.19 Etches for Defect Delineation in Silicon Crystals

Etch solution	Composition	Characteristics
Dash	$1HF:3HNO_3:10HAC^a$	Delineates defects in (111) silicon. Requires long etch times, concentration-dependent.
Sirtl	$1HF:1(5M \cdot CrO_3)$	Delineates defects in (111). Needs agitation. Does not reveal etch pits in (100) very well.
Secco	$2HF:1(0.15M \; K_2Cr_2O_7)$	Delineates oxidation stacking faults in (100) silicon very well. Agitation reduces etch times.
Wright-Jenkins	60 ml HF:30 ml HNO_3: 30 ml $(5M \; CrO_3)$:2 gm $Cu \; (NO_3)_2$:60 ml HAC^a: 60 ml H_2O	Delineates defects in (100) and (111) silicon. Requires agitation.
Schimmel	$2HF:1(1M \; CrO_3)$	Delineates defects in (100) silicon without agitation. Works well on resistivities 0.6-15.0 Ohm-cm n and p types.
Modified Schimmel	$2HF:1(1M \; CrO_3:1.5H_2O$	Works well on heavily doped (100) silicon.
Yang	$1HF:1(1M \; CrO_3)$	Delineates defects on (111), (100), and (110) silicon without agitation.

Note: Agitation is ultrasonic.
[a]Acetic acid.

XV. SAWING, CUTTING, AND POLISHING (93,170)

The shaping operations consist of grinding to reduce the grown crystal to an ingot of uniform diameter; sawing the ingot into slices of uniform thickness; sequentially lapping and polishing the slices so that the surface is damage free; and then cutting or breaking the slice into individual transistors or IC (integrated circuits) after they have been fabricated on the slice (1). These steps are shown schematically in Fig. 6.42.

Silicon, germanium, and gallium arsenide (and most other semiconductor materials as well) are brittle and have hardnesses similar to quartz. Thus the methods for working of semiconductors are more akin to those used for glass than to metal-working procedures. Material removal is by hard abrasive grit, which may either be bonded into cutters, as shown in Fig. 6.43, or used in a slurry. Figure 6.44 gives the comparative hardness of various semiconductors and abrasives. Most commonly used abrasives are harder than the common semiconductors and are thus usable. Alumina, diamond, silicon carbide, and boron carbide are all used in slurries, but only diamond is commonly found in blades and cutters.

A. Abrasives

The particle size, size distribution, shape, and fracture characteristics are all very important with regard to cutting speed, depth of damage, and type of damage. The most com-

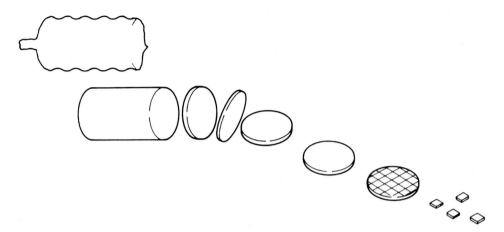

Fig. 6.42 The slicing, polishing, and dicing sequence. The crystal as grown will have surface undulations that are ground away before being sliced. The final step of dicing is done after all the semiconductor processing steps are completed.

mon way of expressing particle size is by the size of sieve the particle will pass through. Table 6.20 relates sieve number and opening size. Table 6.21 gives some examples of the kinds of abrasives used in various applications and typical range of particle sizes.

The distribution of sizes can best be described on a cumulative size distribution chart, as shown in Fig. 6.45. Large grit produces more damage, and small grit cuts more slowly, so in general, a narrow distribution about the chosen size is desirable. Further, since operating time is often used to estimate the amount of material removed, and subsequent processing steps are designed to cope with some maximum amount of damage,

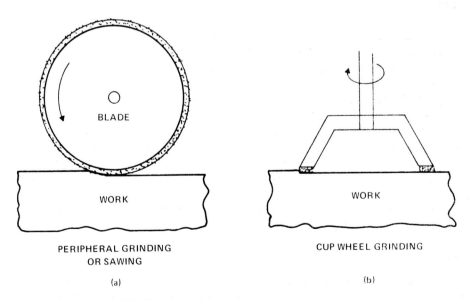

PERIPHERAL GRINDING
OR SAWING

(a)

CUP WHEEL GRINDING

(b)

Fig. 6.43 Cutting with diamond wheels.

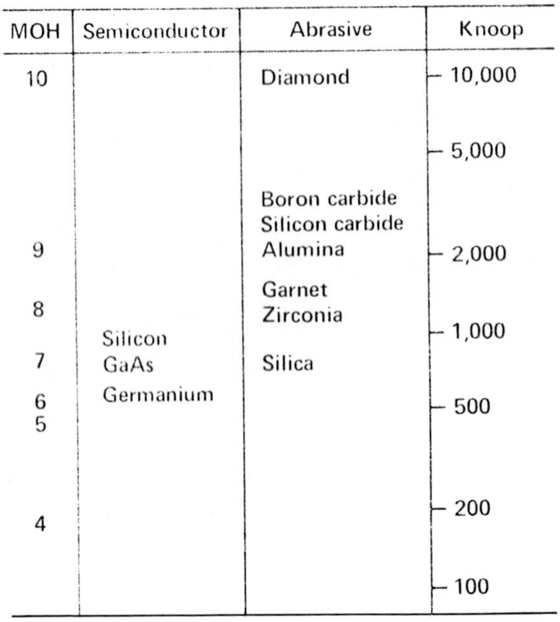

MOH	Semiconductor	Abrasive	Knoop
10		Diamond	— 10,000
			— 5,000
9		Boron carbide Silicon carbide Alumina	— 2,000
8		Garnet Zirconia	— 1,000
7	Silicon GaAs	Silica	
6	Germanium		— 500
5			
4			— 200
			— 100

Fig. 6.44 Comparative hardness of various semiconductors and abrasives. (Adapted from Geoscience Report TRIOA-1965.)

reproducibility of the particle size distribution from lot to lot or wheel to wheel is important.

There is no standard way of measuring particle shape, but it can be roughly categorized as varying from "block" to "needle" (also referred to as increasing angularity). Cutting speed is generally increased with the use of sharper particles, although a limit is reached when the particles become so fragile that they break up too rapidly.

Fracture characteristics that describe the form of the resultant particles as the abrasive breaks up are related to the shape of the original particle, the mechanical prop-

Table 6.20 Sieve Sizes

Sieve	Particle size (μm)
60	250
80	177
100	149
200	74
220	68
240	62
325	44
400	37
600	25
800	19
1200	12
3200	5

Table 6.21 Particle Size Versus Usage

Use	Abrasive	Particle Size (μm)
Sawing	Diamond	50-100
Surface grinding	Diamond	10-40
Rough lap	Diamond SiC Al_2O_3 BC	10-100
Fine lap	Diamond SiC Al_2O_3 ZrO_2	1-10
Polishing	Diamond Al_2O_3 ZrO_2	0.01-1

erties of the material, and the amount and kind of cementing material that is between the grains of the original particle. If needlelike chips occur, scratching is likely. Silicon carbide, in particular, fractures in this manner, and though it has good hardness and cutting speed, should only be used for rough grinding. If the edges wear away and round off instead of continuously fracturing and exposing new sharp cutting edges, the abrasive will have a short life. If it easily fractures into small particles, cutting speed will quickly decrease since the material will then have a finer grit. Most of the naturally occurring abrasives suffer from this defect because of the presence of weak cementing materials between grains.

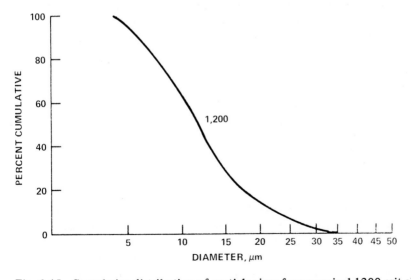

Fig. 6.45 Cumulative distribution of particle sizes for a nominal 1200 grit abrasive.

When grinding wheels or saw blades are being considered, the behavior of the bonding material used to attach the grit to the blade must be considered in much the same light as the fracturing just described. For a wheel to wear well, it must be able to drop off the worn grit; otherwise the wheel will cut more slowly, build up more heat from friction, and perhaps fail catastrophically. However, too fast a drop-off will also reduce life, so compromises affecting both the kind of bonding used and the concentration of grit in bond must be made. Characteristics for each class of bonding material can be summarized as follows.

1. Metal—most wear resistant, almost always used for semiconductor cutting
2. Vitrified—compromise between metal and resin
3. Resin—free cutting, but the least wear resistant

The concentration of diamond in the bonding matrix is typically expressed in terms of the diamond "concentration." The concentration is defined such that 100 equals 4.4 carats/cm^3 of bonding material (i.e., 880 mg/cm^3).

B. Cutting Operation

Grinding to reduce the crystal to a uniform diameter can be done by a centerless grinder, or by a conventional grinder using temporary machining centers to the ends of the crystal. In cases where slices are to be substantially reduced in thickness after integrated circuit processing, for example, before breaking the slice into individual chips, large surface grinders, such as a Blanchard, are often used.

The main criteria for the choice of a slicing method are the width of the cut (kerf), since that represents a loss of expensive material; the depth of damage induced during sawing; the flatness and thickness control possible; and the economy of operation (171). By far the most common slicing method is the use of rotating diamond blades. The cutting edge may be on the periphery of a blade (O/D), as is common in most saws used for cutting brittle materials, or an annular blade may be supported by its outer edge and have a large hole in it with the cutting being done by the inside edge of the annulus. The latter is referred to as I/D cutting, and is illustrated schematically in Fig. 6.46. A typical machine is shown in Fig. 6.47. If conventional O/D blades are used, they must be thick enough to prevent excessive wobble and flexure. For example, a 5 in. diameter blade that will only cut a 2 in. diameter crystal must be at least 0.015 in. thick. However, an I/D blade, supported by a massive hub and stretched tight like a drum, can be only 0.006 to 0.008 in. thick and have sufficient stability to cut 8 in. diameter crystals. In that case, the outer diameter of the I/D blade will generally be between 35 and 40 in. Such a large, thin diaphragm is subject to vibration, which must be minimized to prevent slice breakage. Fortunately, the natural frequency of vibration is usually well above the rotational frequency (172). Coolant introduction during I/D sawing is more difficult than in conventional sawing because instead of feeding the liquid onto the blade near the hub and letting centrifugal force sweep it to the edge and into the work, it must be sprayed directly onto the cut. The diamond-impregnated cutting edge of the blade (whether O/D or I/D) is shown in Fig. 6.48. Each of the indicated dimensions must be specified, as well as blade diameter, the type of matrix bond used to hold the diamonds, the concentration of diamonds, and their size.

The slices, as cut, have edges that are both chipped and subject to more chipping. To smooth the edge and minimize further damage, a contoured diamond cutting wheel can be used (173), as shown in Fig. 6.49.

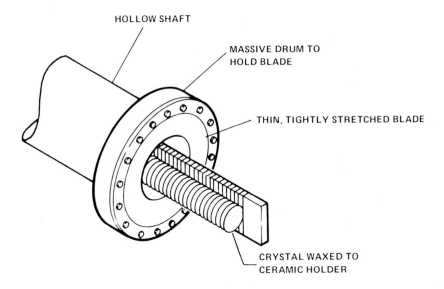

HOLLOW SHAFT

MASSIVE DRUM TO
HOLD BLADE

THIN, TIGHTLY STRETCHED BLADE

CRYSTAL WAXED TO
CERAMIC HOLDER

Fig. 6.46 I/D sawing.

Fig. 6.47 I/D saw for cutting large-diameter silicon crystals at Meyer & Burger AG.

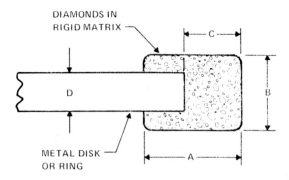

Fig. 6.48 Section of diamond saw blade.

C. Lapping and Polishing (93,173 to 176)

After the slices are sawn, they require lapping and polishing to provide a smooth, flat
surface and to remove saw damage. Lapping may be done by free floating the slices,
in which the slices are held between two oppositely rotating cast iron lapping plates and
simultaneously lapped on both sides by an abrasive slurry, or by attaching the slices to
one plate by either wax or vacuum hold-down. Lapping starts with a coarse grit to rapidly
remove saw marks and thickness nonuniformities. However, since crystal damage penetrates to a depth approximately equal to the particle size, an intermediate-size grit should
be used to reduce the depth of damage before polishing is started. This also shortens the
polishing time.

The requirements for semiconductor polishing are considerably more stringent in
some respects than those for optical element preparation. In the latter, emphasis is on
following a specified surface contour and on having an extraordinarily smooth surface.
The semiconductor process, on the other hand, may be tolerant of small surface undulations, but all vestiges of underlying damage must be removed and none introduced during
polishing. Polishing equipment (Fig. 6.50) looks much like the lapping equipment, except
pitch or cloth will be used to cover the cast iron plate.

For optical polishing (e.g., for infrared optical components), pitches such as coal
tar and burgundy used with cerium oxide or alumina are appropriate. Semiconductor

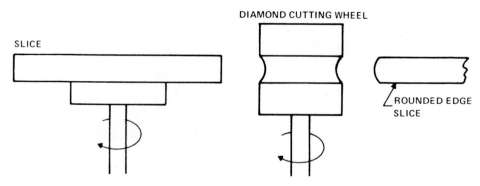

Fig. 6.49 Edge beveling.

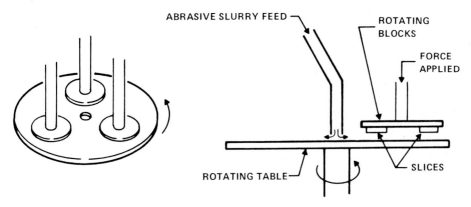

Fig. 6.50 Schematic of polishing operation.

finishing ordinarily uses cloth instead of pitch, and the performance of the various polishing agents is quite dependent on the type of cloth used. For example, harder cloths are more prone to scratching but produce flatter surfaces; nonwoven cloth must be "broken in" by polishing scrap slices before the removal rate becomes constant. Attention must also be given to suspending the polishing compound in its liquid carrier (dry polishing produces much more damage than wet polishing). Diamond is used with oil, and most other polishing agents with water. Diamond can be used for polishing semiconductors, but it generally produces a high incidence of scratches. Chromic oxide is suitable for GaAs, alumina is good for all semiconductors, and cubic zirconia is particularly good for silicon. However, the best silicon polish is a mixture of very fine silica and sodium hydroxide solution (chemical-mechanical polish). This results in locally oxidized silicon by sodium hydroxide and its subsequent mechanical removal by the silica. The high spots become hot and rapidly oxidize while simultaneously being mechanically polished. The low spots become covered with a layer of reaction products and are attacked very slowly.

D. Dicing

After a slice has been processed through the semiconductor fabrication line, there are a large number of identical components, such as integrated circuit or transistor chips, covering the entire slice in a regular pattern. It is necessary to separate slices into these individual components (174). Historically, the first method used was "scribe and break," in which a diamond was used to scribe small grooves in the slice in spaces deliberately left for that purpose (scribe lines). If the lines are oriented along traces of ⟨110⟩ or (111) planes, cleavage occurs rather easily, and if the slices are flexed, the individual chips can be broken out with minimal damage. The diamond has now been largely supplanted by either laser scribing or sawing. Laser scribing uses a finely focused beam to vaporize grooves in the slice; sawing with a very thin diamond blade cuts the grooves. The main advantage of both the laser and the saw is a more reproducible groove with cleaner breaks and fewer broken chips. The advantage of the saw over the laser is in the relative simplicity of the tool.

NOTES AND REFERENCES

1. G. K. Teal, Single Crystals of Germanium and Silicon—Basic to the Transistor and Integrated Circuit, *Special Bicentennial Issue: IEEE Transactions on Electron Devices,* vol. ED-23, pp. 621-639, July 1976.
2. W. Smith, The Action of Light on Selenium, *J. Soc. Telegr. Eng.,* vol. 2, no. 1, pp. 31-33, 1873.
3. F. Braun, Ueber die Stromleitung durch Schwelmetall, *Ann. Phys. Chem.,* vol. 153, no. 4, pp. 556-563, 1874.
4. A. Schuster, On Unilateral Conductivity, *Philos. Mag.,* vol. 48, pp. 251-257, October 1874.
5. W. Siemens, *Ann. Phys. Wied.,* vol. 2, p. 521, 1877.
6. C. F. Fritts, A New Form of Selenium Cell, *Am. J. Sci.,* vol. 26, pp. 465-472, December 1883.
7. A. H. Pfund, *Phys. Rev.,* vol. 7, p. 289, 1916.
8. T. W. Case, *Phys. Rev.,* vol. 9, p. 305, 1917; vol. 15, p. 289, 1920.
9. W. W. Coblentz, *Phys. Rev.,* vol. 13, p. 154, 1919; vol. 14, p. 523, 1920.
10. R. W. Pohl, Electron Conductivity and Photochemical Processes in Alkali Halide Crystals, *Proc. Phys. Soc.,* vol. 49, pp. 3-31, Extra Part, 1937.
11. L. O. Grondahl and P. H. Geiger, A New Electronic Rectifier, *Trans. Am. Inst. Elec. Eng.,* vol. 46, pp. 357-366, February, 1927.
12. B. Lange, Die Photoelemente und Ihre Anwendung. Leipzig, Johann Ambrosius Barth, 1936.
13. K. B. McEachron, Thyrite, A New Material for Lightning Arresters. *General Electric Rev.,* vol. 33, pp. 92-99, February 1930.
14. G. K. Teal was in charge of the materials aspects of the development of the silicon carbide varistors for the new telephone handset. This involved him in device design and device technology and in methods of device preparation best suited to achievement of close control of the final electrical characteristics of the devices. For further information concerning this silicon carbide research and project associates with whom he worked most closely the reader is referred to page 623 of reference 1 given above. Additional information on this SiC research and a comparison of it and the early years of Bell transistor research is given in a paper, Integrating Diversified Technology into Profitable Programs, presented on January 21, 1960, by G. K. Teal as part of a Research and Development Conference of an American Management Association meeting in New York City.
15. J. A. Becker, C. B. Green, and G. L. Pearson, Properties and Uses of Thermistors— Thermally Sensitive Resistors, *Elec. Eng.,* Nov., 1946; Bell System Tech. J., p. 170, 1947.
16. For general information up to the end of World War II the reader is referred to H. C. Torrey and C. A. Whitmer, Editors, *Crystal Rectifiers,* McGraw-Hill, New York, 1948. This book describes the contributions of Lark-Horowitz and his associates at Purdue University, Scaff, Schumacher, and associates at Bell Laboratories, North at the General Electric Company, Seitz and his associates at the University of Pennsylvania, and of others carrying on investigations elsewhere. References 16-18, 29, and 31-35 are primarily concerned with general information on semiconductor materials and/or semiconductor devices. Reference 36 summarizes some of Bell's World War II work. References 19-24 are especially good sources of information concerning the evolving technical, strategic, planning and operations aspects of electronic/aeronautical warfare in the early and continuing years of World War II—from about 1933 to 1945.

17. K. Lark-Horowitz, The New Electronics, from *The Present State of Physics,* pp. 57-127, Publication of the AAAS, Washington, D.C., 1954. This review is not only an excellent summary of the numerous studies of germanium made by Lark-Horowitz and his associates at Purdue University but also of the research of many others in the semiconductor field.

18. R. L. Petritz, Contributions of materials technology to semiconductor devices, 50th Anniversary Issue, *Proc. IRE,* vol. 50, pp. 1025-1038, May 1962.

19. R. Watson-Watt, *The Pulse of Radar* (The autobiography of the author), pp. 1-438, The Dial Press, New York, 1959.

20. G. C. Southworth, *Forty Years of Radio Research* (autobiography), pp. 1-274, Gordon and Breach, New York, 1962.

21. R. V. Jones, *The Wizard War—British Scientific Intelligence, 1939-1945,* pp. 1-556, Coward, McCann and Geoghegan, Inc., New York, 1978.

22. C. P. Snow, *Science and Government* (The Godkin Lectures at Harvard University, 1960), pp. 1-128, New American Library, New York, 1962.

23. Ref. 20, p. 172.

24. Ref. 21, p. 179.

25. J. Bardeen and W. H. Brattain, The transistor, a semiconductor triode, *Phys. Rev.,* vol. 74, pp. 230-231, July 15, 1948.

26. J. Bardeen and W. H. Brattain, The Nature of the Current in Germanium Point Contacts, *Phys. Rev.,* vol. 74, pp. 231-232, July 15, 1948.

27. Early in 1949, J. R. Haynes, in a personal communication to G. K. Teal, indicated a lifetime of minority carriers in single crystals being grown as being 20-100 times greater than obtained in polycrystalline germanium materials. The higher values of the lifetimes in this paragraph and the increased mobilities and the 45 ohm-cm specific resistivity for germanium single crystals followed some months later, as will become evident in the text below.

28. W. Shockley, G. L. Pearson, and J. R. Haynes, Hole Injection in Germanium— Quantitative Studies and Filamentary Transistors, *Bell System Tech. J.,* vol. 28, pp. 344-366, July 1949.

29. W. Shockley, *Electrons and Holes in Semiconductors,* Van Nostrand, New York, 1950.

30. J. R. Haynes and W. Shockley, The mobility and life of injected holes and electrons in germanium, *Phys. Rev.,* vol. 81, pp. 835-843, March 1951.

31. G. L. Pearson and W. H. Brattain, History of semiconductor research, *Proc. IRE,* vol. 43, pp. 1794-1806, December 1955.

32. E. Spenke, *Elektronische Halbleiter,* Springer, Berlin, 1955.

33. N. B. Hannay, Editor, Semiconductors, Reinhold, New York, 1959.

34. W. R. Runyan, *Silicon Semiconductor Technology,* McGraw-Hill, New York, 1965.

35. Special Issue on Transistors, *Proc. IRE,* vol. 40, pp. 1-1283, November 1952.

36. J. H. Scaff, H. C. Theuerer, and E. Schumacher, p-type and n-type silicon and the formation of the photovoltaic barrier in silicon ingots, Trans. Met. Soc. AIME, vol. 185, pp. 383-388, June 1949.

37. G. K. Teal, Chapter 4, Germanium Single Crystals: Introduction, pp. 69-78, *Transistor Technology,* vol. 1, Bell Telephone Laboratories and Western Electric Company, September 1952.

38. G. K. Teal, Chapter 4, Preparation of Germanium Single Crystals by the Pulling Method: Introduction, pp. 69-78, *Transistor Technology,* vol. Edited by H. E. Bridgers, J. H. Scaff, and J. N. Shive, Van Nostrand, New York, 1958.

39. G. K. Teal, Germanium and Silicon Single Crystals, invited paper, *Phys. Rev.,* vol. 87, p. 221, 1952. Presented at meeting of the American Physical Society, May 3, 1952, in Washington, D.C.

40. G. K. Teal, Roots of creative research, *IDEA,* vol. 11, pp. 1-6, Spring 1967 (published by the Patent, Trademark and Copyright Institute of the George Washington University).

41. G. K. Teal, Roots of creative research, *Nurturing New Ideas: Legal Rights and Economic Roles,* pp. 18-22, J. L. Harris (ed.), Bureau of National Affairs, Inc., Washington, D.C., 1969).

42. G. K. Teal, Reflections on early germanium and silicon single crystal research, invited paper, IEEE 1968 International Electron Devices Meeting, Washington, D.C., October 23, 1968.

43. G. K. Teal and J. B. Little, Growth of germanium single crystals. *Phys. Rev.,* vol. 78, p. 647, 1950. Abstract read by title only at the APS meeting in Oak Ridge, Tennessee, on March 18, 1950. The abstract gave the lifetimes of injected carriers in these single crystal materials as greater than 200 μsec but did not publicly reveal a definite comparison with lifetimes found in polycrystalline germanium. In the invited paper of reference 39 Teal gave 1-5 μsec to be lifetimes of carriers in polycrystalline germanium.

44. A mobility of 1156 cm^2/volt-sec for electrons in germanium was a measurement of K. Lark-Horowitz and V. A. Johnson in 1946. At about the same time their Hall mobility measurements for p-type germanium gave a mobility of 385 cm^2/volt-seconds for the carrier. This information comes from Ref. 16, Torrey and Whitmer.

45. This value was obtained by Hall measurements by P. Debye after special purification and by achieving great donor and acceptor uniformity in germanium. Using these materials and adding donors to lower resistivities, the mobility was found to decrease in quantitative agreement with the theoretical values relating mobility and resistivity in germanium. See P. Debye and E. Conwell, Mobility of electrons in germanium, *Phys. Rev.,* vol. 87, pp. 1131-1132, L, 1952.

46. G. K. Teal, M. Sparks, and E. Buehler, Growth of germanium single crystals containing p-n junctions, *Phys. Rev.,* vol. 81, p. 637, Feb. 1951.

47. F. S. Goucher, G. L. Pearson, M. Sparks, G. K. Teal, and W. Shockley, Theory and experiment for a germanium p-n junction, *Phys. Rev.,* vol. 81, p. 638, February 15, 1951.

48. W. Shockley, M. Sparks, and G. K. Teal, p-n junction transistors, *Phys. Rev.,* vol. 83, pp. 151-162, July 1, 1951.

49. C. Wagner, On the theory of rectification, *Physikalische Zeitschrift,* vol. 32, pp. 641-645, 1931.

50. W. Shockley, The theory of p-n junctions in semiconductors and p-n junction transistors, *Bell System Tech. J.,* vol. 28, pp. 435-489, July 1949.

51. J. A. Burton and W. P. Slichter, Chapter 5, The distribution of solute elements: Steady state growth, pp. 79-118. *Transistor Technology,* vol. 1, Bell Telephone Laboratories and Western Electric Company, September 1952.

52. W. P. Slichter and J. A. Burton, Chapter 6, The distribution of solute elements; Transient conditions, *Transistor Technology,* vol. 1, pp. 119-142, Bell Telephone Laboratories and Western Electric Company, September 1952.

53. G. K. Teal, M. Sparks, and E. Buehler, Single crystal germanium, *Proc. IRE,* vol. 40, pp. 906-909, August 1952.

54. W. P. Slichter and J. A. Burton, Chapter 6, The distribution of solute elements: Transient conditions, *Transistor Technology,* vol. 1, pp. 107-129, H. E. Bridgers, J. A. Scaff, and J. N. Shive (eds.), Van Nostrand Co., New York, 1958.

55. J. A. Burton and W. P. Slichter, Chapter 5, The distribution of solute elements: Steady state growth, pp. 79-107, *Transistor Technology,* Vol. 1, H. E. Bridgers, J. A. Scaff, and J. N. Shive (eds.), Van Nostrand, New York, 1958.

56. W. W. Bradley, Chapter 7, Preparation of germanium single crystals, pp. 143-170, *Transistor Technology*, Vol. 1, Bell Telephone Laboratories and Western Electric Co., September 1952.

57. W. W. Bradley, Chapter 7, Preparation of germanium single crystals, pp. 130-154, *Transistor Technology*, Vol. 1, Van Nostrand, New York, 1958.

58. J. A. Morton, Present status of transistor development, *BSTJ,* vol. 31, pp. 411-442, 1952.

59. J. C. Lozier, Chapter 8, Controls for the n-p-n crystal growing machines, pp. 171-182, *Transistor Technology*, Vol. 1, Bell Telephone Laboratories and Western Electric Co., September 1952.

60. J. C. Lozier, Chapter 8, Controls for the n-p-n crystal growing machine, pp. 155-164, *Transistor Technology*, Van Nostrand, New York, 1958.

61. R. N. Hall and W. C. Dunlap, p-n Junctions prepared by impurity diffusion, *Phys. Rev.,* vol. 80, pp. 467-468, November 1950.

62. J. S. Saby, Recent developments in transistors and related devices, *Tele-Tech.,* vol. 32, p. 10, 1951. (This is the first published account of Saby's research. A more complete report is: J. S. Saby, Fused impurity n-p-n junction transistors, *Proc. IRE,* vol. 40, pp. 1358-1360, November 1952.)

63. IRE-AIEE Electronic Devices Research Conference, Durham, New Hampshire, June 1951.

64. R. N. Hall, p-n Junctions produced by rate growth variation, *Phys. Rev.,* vol. 88, p. 139, October 1952.

65. R. N. Hall, Segregation of impurities during the growth of germanium and silicon crystals, *J. Phys. Chem.,* vol. 57, 836-839, November 1953.

66. R. N. Hall, Fabrication Techniques for High-Frequency Transistors (written in English), *Fortschritte Hochfrequenztechnik,* Bd. 4, pp. 129-155, 1961. (This contains an excellent list of references relating to the subject.)

67. H. E. Bridgers, Formation of p-n junctions in semiconductors by the variation of crystal growth parameters, *J. Appl. Phys.* 746-751, 1956.

68. C. D. Thurmond, Control of composition in semiconductors by freezing methods, Chapter 4 of *Semiconductors,* N. B. Hannay (ed.), Reinhold, New York, 1959. (This chapter has an excellent bibliography for the subject.)

69. H. E. Bridgers and E. D. Kolb, Rate-grown germanium crystals for high-frequency transistors, *J. Appl. Phys.* vol. 26, pp. 1188-1189, 1955.

70. W. G. Pfann, Segregation of two solutes, with particular reference to semiconductors, *J. Metals,* vol. 4, pp. 861-865, August 1952.

71. W. G. Pfann, Techniques of zone melting and crystal growing, *Solid State Physics,* F. Seitz and D. Turnbull (eds.), Academic Press, New York, vol. 4, pp. 423-521, 1957.

72. G. K. Teal and E. Buehler, Growth of silicon single crystals and of single crystal silicon p-n junctions, *Phys. Rev.,* vol. 87, p. 190, 1952; presented at the APS meeting in Washington, DC, May 1, 1952.

73. J. R. Haynes and W. C. Westphal, The drift mobility of electrons in silicon, *Phys. Rev.,* vol. 85, p. 680, February 15, 1952.

74. K. B. McAfee and G. L. Pearson, The electrical properties of silicon p-n junctions, grown from the melt, *Phys. Rev.,* vol. 87, 190, 1952; presented at the APS meeting in Washington, DC, May 1, 1952.

75. G. K. Teal, Some recent developments in silicon and germanium materials and devices, presented at National IRE Conference meeting in Dayton, Ohio, on May 10, 1954, Conference sponsored by IRE Navigational and Airborne Electronics Group (31 pages).

76. W. A. Adcock, M. E. Jones, J. W. Thornhill, and E. D. Jackson, Silicon transistor, *Proc. IRE,* vol. 42, p. 1192, July 1954.

77. Boyd Cornelison and W. A. Adcock, Transistors by grown-diffused technique, *IRE Wescon Convention Record,* vol. 1, pt. 3, pp. 22-27, San Francisco, August 20-23, 1957.

78. H. S. Reiss and C. S. Fuller, Diffusion processes in germanium and silicon, Chapter 6 of *Semiconductors,* N. B. Hannay (ed.), Reinhold Publishing, New York, 1959. (This chapter has an excellent bibliography for diffusion.)

79. M. Tanenbaum, L. B. Valdes, E. Buehler, and N. B. Hannay, Silicon n-p-n grown junction transistors, *J. Appl. Phys.,* vol. 26, 686-692, June 1955.

80. P. H. Keck and M. J. E. Golay, Crystallization of silicon from a floating liquid zone, *Phys. Rev.,* vol. 89, p. 1297, March 1953.

81. R. Emeis, Growing silicon single crystals without a crucible, *Naturforschung,* vol. 9a, p. 67, January 1954.

82. H. C. Theuerer, Removal of boron from silicon by hydrogen-water vapor treatment, *J. Metals,* vol. 8, pp. 1316-1319, October 1956.

83. Charles A. Lee, A high-frequency diffused base germanium transistor, *Bell System Tech. J.,* vol. 35, pp. 23-34, January 1956.

84. M. Tanenbaum and D. E. Thomas, Diffused emitter and base silicon transistors, *BSTJ,* vol. 35, pp. 1-22, January 1956.

85. F. M. Smits, Formation of junction structures by solid state diffusion, *Proc. IRE,* vol. 46, pp. 1049-1061, June 1958.

86. H. C. Theurer, J. J. Kleimack, H. H. Loar, and H. Christensen, Epitaxial diffused transistors, *Proc. IRE,* vol. 48, 1642-1643, September 1960.

87. K. E. Bean and P. S. Gleim, The influence of crystal orientation on silicon semiconductor processing, *Proc. IEEE,* 57, pp. 1469-1476, 1969.

88. J. F. Nye, *Physical Properties of Crystals,* Oxford University Press, Fairlawn, New Jersey, 1960.

89. B. E. Deal and J. D. Plummer, Thermal oxidation and chemical vapor deposition of insulators, in *Integrated Circuit Process Models,* J. D. Meindl (ed.),

90. M. Cutler, *Liquid Semiconductors,* Academic Press, New York, 1977.

91. W. E. Spear (ed.), *Proceedings of the Seventh International Conference on Amorphous and Liquid Semiconductors,* University of Edinburgh, 1977. Also see others in the series.

92. J. J. Brophy and J. W. Buttrey (eds.), *Organic Semiconductors,* Macmillan, New York, 1962.

93. W. R. Runyan and Stacy B. Watelski, Semiconductor materials, in *Handbook of Materials and Processes for Electronics,* Charles A. Harper (ed.), McGraw-Hill, New York, 1970.

94. M. Neuberger, *Germanium Data Sheets,* AD 610 828, Feb. 1965, and *Silicon,* AD 698 342, Oct. 1969, NTIS, Springfield, Va. 22161.

95. J. W. Harrison, *Gallium Arsenide Technology,* Research Triangle Institute, Tech. Report AFAL-TR-72-312, Jan, 1973.

96. R. W. Keyes, The effects of hydrostatic pressure on the properties of III-V semiconductors, in *Semiconductors and Semimetals,* R. K. Willardson and A. C. Beer (eds.), Academic Press, New York, 1968.

97. G. J. Piermarina and S. Block, Ultrahigh pressure diamond-anvil cell and several semiconductor phase transition pressures in relation to the fixed point pressures scale, *Rev. Sci. Instrum.* 46, pp. 973-979, 1975. See also the included references.

98. R. A. Smith, *Semiconductors,* Cambridge, 1961.

99. D. H. Navon, *Electronic Materials and Devices,* Houghton Mifflin, Boston, 1975. For later developments, see, for example, B. L. H. Wilson (ed.), *Physics of Semiconductors,* 1978, The Institute of Physics, London, 1979, and others of the series.

100. J. L. Pankove, *Optical Processing in Semiconductors,* Prentice-Hall, New Jersey, 1971; Dover, New York, 1975.
101. T. S. Moss, *Optical Properties of Semiconductors,* Butterworths, London, 1959.
102. P. F. Kane and G. B. Larrabee, *Characterization of Semiconductor Materials,* McGraw-Hill, New York, 1970.
103. W. R. Runyon, *Semiconductor Measurements and Instrumentation,* McGraw-Hill, New York, 1975.
104. M. S. Brooks and J. K. Kennedy (eds.), *Ultrapurification of Semiconductor Materials,* Macmillan, New York, 1962.
105. T. N. Chu, Thin films of silicon on metallurgical silicon substrates, Proceedings of the Photovoltaic Advanced Materials Review Meeting, SERI/TP-49-105, pp. 137-147, October 24-26, 1978.
106. J. W. Mellor, *A Comprehensive Treatise on Inorganic and Theoretical Chemistry,* Longmans, Green, New York, 1957.
107. R. F. Lever, The equilibrium behavior of the silicon-hydrogen chlorine system, *IBM J. Research and Dev.,* 8, pp. 460-465, 1964; E. Sirtl, Thermodynamics and silicon vapor deposition, in *Semiconductor Silicon,* R. R. Haberecht and E. L. Kern (eds.), Electrochemical Society, Princeton, New Jersey, 1969.
108. L. P. Hunt, Low-Cost, Low-Energy Processes for Producing Silicon, in *Semiconductor Silicon 1977,* H. R. Huff and E. Sirtl, (eds.), Electrochemical Society, Princeton, New Jersey, 1977. See also the various proceedings of the ERDA Semiannual Solar Photovoltaic Program Review Meetings after 1976.
109. F. I. Metz (ed.), *New Uses for Germanium,* Midwest Research Institute, 1974.
110. R. K. Willardson and H. L. Goering (Editors), *Preparation of III-V Compounds,* Reinhold, New York, 1962.
111. J. C. Brice, *Growth of Crystals from Liquids,* North-Holland Publishing Co., Amsterdam, 1965.
112. R. A. Laudise, *The Growth of Single Crystals,* Prentice-Hall, Englewood Cliffs, New Jersey, 1970.
113. S. N. Rea and G. F. Wakefield, Large area Czochralski silicon for solar cells, Proceedings of the International Solar Energy Conf., Winnipeg, Canada, 1976.
114. W. R. Runyan, Growth of large diameter silicon and germanium single crystals, *J. Appl. Phys.* 27, p. 1562, 1956.
115. C. S. Fuller, J. A. Ditzenberger, N. B. Hannay, and E. Buehler, Resistivity changes in silicon single crystals induced by heat treatment, *Acta Metallurgica,* 3, pp. 97-99, 1955; W. Kaiser, P. H. Keck, and C. F. Lange, Infrared absorption and oxygen content in silicon and germanium, *Phys. Rev.,* 101, pp. 1264-1268, 1956; W. Kaiser, Electrical and optical properties of heat-treated silicon, *Phys. Rev.,* 105, pp. 1751-1756, 1957. The first reference reported the change of resistivity with heat treatment. The second described the finding of oxygen in silicon crystals. The last one related the resistivity change to the oxygen.
116. J. R. Patel, *Semiconductor Silicon/1977,* H. R. Huff and E. Sirtl (eds.), The Electrochemical Society Soft Bound Symposium Series, Princeton, New Jersey, pp. 521-545, 1977.
117. B. Leroy and C. Plougonven, Warpage of silicon wafers, *J. Electrochm. Soc.,* 127, pp. 961-970, 1980.
118. T. Y. Tan, E. E. Gardner, and W. K. Tice, Intrinsic gettering by oxide precipitate induced dislocations in Czochralski Si, *Appl. Phys. Lett.,* 30, 175-176, 1977.
119. S. C. Baber, H. R. Huff, J. T. Robinson, H. Schaake, and D. Wong, Some observations of bulk oxygen precipitation gettering in device processed Czochralski silicon, Electrochemical Soc. Extended Abstract 79-2, pp. 1262-1265, 1979.

120. J. B. Mullin, R. J. Heritage, C. H. Holliday, and B. W. Straughon, Liquid encapsulation crystal pulling at high pressures, *Journal of Crystal Growth*, 3/4, pp. 281-285, 1968.

121. W. G. Pfann, *Zone Melting*, 2nd Edition, John Wiley and Sons, New York, 1958.

122. For early work on semiconductor dendrites and crystal shaping by various other means, see papers in R. O. Grubel (Editor), *Metallurgy of Elemental and Compound Semiconductors*, Interscience Publishers, New York, 1961. The September 1980 special issue of the *Journal of Crystal Growth* is completely dedicated to shaped crystal growth and covers the 1980 state of understanding and application.

123. H. Nelson, Epitaxial growth from the liquid state and its application to the fabrication of tunnel and laser diodes, *RCA Rev.*, 24, pp. 603-615, 1963; and C. S. Kang and P. E. Greene, Preparation and properties of high-purity epitaxial GaAs grown from Ga solution, *Appl. Phys. Lett.*, 11, pp. 171-173, 1967.

124. D. Viechnicki and F. Schmid, Crystal growth using the heat exchanger method (HEM), *Journal of Crystal Growth*, 26, pp. 162-164, 1974. See also discussions of similar approaches in H. E. Buckley, *Crystal Growth*, John Wiley & Sons, New York, 1951.

125. See references in W. R. Runyan, Melt and solution growth of silicon, pp. 26-39, in Proceedings of the National Work Shop on Low Cost Polycrystalline Silicon Solar Cells, T. L. Chu and S. S. Chu (Editors), Southern Methodist University, 1976.

126. U. Cohen and R. Huggins, Silicon epitaxial growth from molten salts, *Recent Newspaper 159*, Fall Electrochemical Society Meeting, Dallas, 1975; and in the *Proceedings of the Fifth Conference on Crystal Growth*, American Association for Crystal Growth, 1980.

127. E. Sirtl, Current aspects of silicon material processing for solar cell applications, in *Semiconductor Silicon 81*, ed. by Huff, Kiegler, and Takeishi, Electrochemical Society, Princeton, N.J.

128. F. Rosenberger, *Fundamentals of Crystal Growth I*, Vol. 5, Springer, Series in Solid State Sciences, New York, 1979.

129. J. J. Gilman (Editor), *The Art and Science of Growing Crystals*, John Wiley and Sons, New York, 1963.

130. M. L. Royer, Recherches experimentales sur-1-epitaxie ou orientation mutelle de Cristaux d'especes differentes, *Bull. Soc. Franc, Mineral*, vol. 51, 7-159, 1928.

131. P. E. Luscher, Crystal growth by molecular beam epitaxy, *Solid State Technology*, Dec., 1977.

132. H. Schäfer, *Chemical Transport Reactions*, Academic Press, New York, 1964.

133. G. R. Srinivasan, Autodoping effects in silicon epitaxy, *J. Electrochem. Soc.*, 127, pp. 1334-1342, 1980.

134. H. W. Lam, A. F. Tasch, Jr., and R. F. Pinizzotto, Silicon-on-insulator for VLSI and VHSIC, *VLSI Electronics Microstructure, Science*, Vol. 4, Edited by N. G. Einspruch, Academic Press, New York, 1982.

135. G. K. Teal and H. Christensen, Method of fabricating germanium bodies, U.S. Patent No. 2,692,839, Oct. 26, 1954.

136. R. C. Sangster, E. F. Maverick, and M. L. Croutch, Growth of silicon crystals by a vapor phase pyrolytic deposition method, *J. Electrochem. Soc.*, 104, pp. 317-319, 1957.

137. A. Mark, Growth of single crystal silicon over-growth on silicon substrates, *J. Electrochem. Soc.*, 107, pp. 568-569, 1960.

138. E. S. Wajda, B. W. Kippenhan, and W. H. White, Epitaxial growth of silicon, *IBM J.*, pp. 288-295, 1960.

139. R. Glang and B. W. Kippenhan, Impurity introduction during epitaxial growth of silicon, *IBM J.*, pp. 299-301, 1960.

140. J. C. Marinace, Epitaxial vapor growth of Ge single crystals in a closed-cycle process, *IBM J.*, pp. 248-255, 1960.

141. E. S. Wajda and R. Glang, Epitaxial growth of silicon layers from the vapor phase, *Mettalurgy of Elemental and Compound Semiconductors*, pp. 229-253, Edited by R. O. Grubel, in Metallurgical Society Conferences, vol. 12, John Wiley & Sons, New York, 1960.

142. J. E. Allegretti, D. J. Shombet, E. Schaarschmidt, and J. Waldman, Vapor deposited silicon single crystal layers, *Metallurgy of Elemental and Compound Semiconductors*, pp. 255-270, Edited by R. O. Grubel, in Metallurgical Society Conferences, John Wiley & Sons, New York, 1960.

143. H. C. Theuerer, Epitaxial silicon films by the hydrogen reduction of $SiCl_4$, *J. Electrochem. Soc.*, 108, pp. 651-653, 1961.

144. Study of the epitaxial growth process in silicon, Ferranti Ltd., Wythenshawe, Manchester, 22, 10th Nov. 1961 Annual Report Project RP1-21.

145. J. Sigler and S. B. Watelski, Epitaxial techniques in semiconductor devices, *Solid State J.*, pp. 33-37, 1961.

146. G. Schnable, Epitaxial overgrowth structures in silicon, Final Progress Report, April 1961-June 1963, Contract No. DA-36-SC-89070. See also J. M. Hirshon, Silicon epitaxial junctions with compatible masking. Abstract No. 101, Electrochem. Society, Extended Abstracts, May 1962.

147. B. Joyce and J. Baldrey, *Nature*, No. 4840, pp. 485-486, 1962.

148. B. E. Deal, Epitaxial deposition of silicon in a hot-tube furnace, *J. Electrochem. Soc.*, 109, pp. 514-517, 1962.

149. E. G. Bylander, Kinetics of silicon crystal growth from $SiCl_4$ decomposition, *J. Electrochem. Soc.*, 109, pp. 1171-1175, 1962.

150. Motorola Inc., Epitaxial Film Growth Report, No. 2, Dec. 1962-March 1963. ASD Project No. 7-995, Contract AF 33 (657)-9678.

151. K. E. Bean and P. S. Gleim, Vapor etching prior to epitaxial deposition of silicon, *Late Newspaper*. Fall Meeting Electrochemical Society, New York, 1963.

152. G. A. Lang and T. Stavish, Chemical polishing of silicon with anhydrous hydrogen chloride, *RCA Rev.*, 24, pp. 488-498, 1963.

153. D. M. Jackson, Jr., Epitaxial control system, Report No. 5, Contract AF 19 (604)-8351, Dec. 1963.

154. P. Wang, B. Selikson, V. Sils, and R. Berkstresser, Development of epitaxial technology for microelectronics and large area device applications, 7th Quarterly Report on Contract No. 85431, Index No. SR0080302, ST-9348 Nov. 1963 through Jan. 1964.

155. W. H. Shephard, Vapor phase deposition and etching of silicon, *J. Electrochem. Soc.*, 112, pp. 988-993, 1965.

156. S. K. Tung, The influence of process parameters on the growth of epitaxial silicon, *Metallurgy of Semiconductor Materials*, 115, pp. 87. AIME, Interscience Publishers, Los Angeles Meeting, Aug. 30, 1961.

157. T. O. Sedgwick, Analysis of the hydrogen reduction of silicon tetrachloride process on the basis of a quasi-equilibrium model, *J. Electrochem. Soc.*, 111, pp. 1381-1383, 1964.

158. J. Bloem and W. A. P. Classen, Rate determining reactions and surface species in CVD silicon, *J. Crystal Growth 49*, 435-444 (1980), and subsequent articles in volumes 50, 51, 52 (1980, 1981, 1982).

159. L. P. Hunt and E. Sirtl, A thorough thermodynamic evaluation of the silicon-hydrogen-chlorine system, *J. Electrochem. Soc. 119*, pp. 1741-1745, 1972.

160. S. K. Tung, The effect of substrate orientation on epitaxial growth, *J. Electrochem. Soc.*, 112, pp. 436-438, 1965.

161. W. J. Corrigan, Doping of silicon epitaxial layers, *Metallurgy of Semiconductor Material,* 15, p. 103, AIME Interscience Publishers, Los Angeles Meeting, Aug. 30, 1961.

162. R. Nuttall, The dependence on deposition conditions of the dopant concentration of epitaxial silicon layers, *J. Electrochem. Soc.,* 111, pp. 317-323, 1964.

163. D. Kahng, R. C. Manz, M. M. Atalla, and C. O. Thomas, Paper 135 presented at Electrochemical Society Meeting, Detroit, Oct., 1961.

164. C. O. Thomas, D. Kahng, and R. C. Manz, Impurity distribution in epitaxial silicon films, *J. Electrochem. Soc.,* 109, p. 1055, 1962.

165. G. R. Strinivasan, Autodoping effects in silicon epitaxy, *J. Electrochem. Soc.,* 127, pp. 1334-1341, 1980.

166. A. L. Armirotto, Silane, Review and applications, *Solid State Tech.,* pp. 43-47, 1968.

167. G. E. Becker and J. C. Bean, Acceptor dopants in silicon molecule-beam epitaxy, *J. Appl. Phys.,* 48, pp. 3395-3399 (1977). J. C. Bean and S. R. McAfee, Silicon Molecular Beam Epitaxy. A Comprehensive Bibliography, 1962-1982. Proc. Intl. Mtg. on the Relationship Between Epitaxial Growth Conditions and the Properties of Semiconducting Epitaxial Layers, Perpignan, France, p. 795.

168. R. B. Herring, Advances in reduced pressure silicon epitaxy, *Solid State Tech.,* pp. 75-84, 1979. D. Bellavance and R. N. Anderson. Continuous CVD: A new type of silicon reactor, ECS Extended Abstracts 83. Electrochemical Society, Princeton, N.J., 1983.

169. B. J. Baliga, Dopant distribution in silicon liquid phase epitaxial layers: Meltback effects, *J. Electrochem. Soc.,* 126, pp. 138-142, 1979.

170. A. C. Bonora, Silicon wafer processing technology: Slicing, etching, polishing, in *Semiconductor Silicon 1977,* Edited by H. R. Huff and E. Sirtl, Electrochemical Society, Princeton, New Jersey, 1977.

171. G. S. Kachajian, A systems approach to semiconductor slicing to improve wafer quality and productivity, *Solid State Technology,* pp. 59-65, 1972.

172. S. E. Forman and W. J. Rhines, Vibration characteristics of crystal slicing ID saw blades, *J. Electrochem. Soc.,* 119, pp. 686, 690, 1972.

173. Anon. Silicon wafers with optimum edge rounding, *Solid State Technology,* pp. 16-17, May 1976.

174. A. C. Bonora, Flex-mount polishing of silicon wafers, *Solid State Technology,* pp. 55-58, Oct. 1977.

175. T. M. Buck and R. L. Meek, Crystallographic damage due to silicon by typically slicing, lapping, and polishing operations, in *Silicon Device Processing, NBS Special Publication 337,* Charles P. Marsden (Editor), 1970.

176. E. W. Jensen, Polishing compound semiconductors, *Solid State Technology,* pp. 55-58, 1977.

177. J. F. Marshall, Scribing and breaking of semiconductor wafers, *Solid State Technology,* pp. 70-75, Sept., 1974.

178. J. Bardeen and W. H. Brattain, Physical principles involved in transistor action, *Physical Review,* vol. 75, pp. 1208-1225, April 1949.

7

Radiation Effects on Materials*

I. INTRODUCTION

The effects of radiation on the behavior of materials have received considerable attention in recent years, stimulated by the needs associated with the development of nuclear energy. Although neutron irradiation is the primary focus of interest to the reactor designer, other types of radiation are also of interest, particularly from the point of view of shielding. Design considerations arising from advanced reactor types, such as liquid metal fast breeder reactors, high-temperature gas-cooled reactors, and controlled thermonuclear reactors, indicate that a better understanding of material behavior under irradiation and development of new materials to withstand radiation damage are necessary prerequisites for their successful application. In this chapter we will be concerned primarily with particulate irradiation: protons, deuterons, α-particles, β-rays (electrons or positrons), γ-rays (photons), neutrons, and other heavy ions. Irradiation with these particles generally results in atomic displacements and net energy transfer. In contrast, the phenomena of diffraction, refraction, and scattering are best understood within a theoretical framework embodying the wavelike nature of radiation and matter.

A. Microstructural Changes

The microstructures of irradiated material are affected by several aspects of irradiation. First, depending upon the kinetic energy and mass of the incident particle as well as the characteristics of the target material, physical displacement of atoms could take place. These displacements give rise to point and line defects in a crystalline solid. Two consequences of major importance to design that result from the generation of these defects are void swelling and irradiation creep. Second, solute atoms in alloys used as structural materials can react at the site of irradiation-induced defects to produce microchemical segregation. In turn, the composition of the matrix often changes as a function of irradiation. Third, in fissile materials, transmutations could take place leading to both chemical and physical changes. Even in nominally nonfissile materials, such as structural steels, transmutations occur, such as the interaction between the boron present in the steel and

*Contributed by S. VAIDYANATHAN.

neutrons, leading to production of helium. Fourth, irradiation could produce highly localized temperature gradients that could alter the microstructure.

B. Effect on Mechanical Properties

The microstructural changes produced during irradiation alter the mechanical behavior
of the material. A variety of conventional properties, such as fatigue strength, creep resistance, fracture toughness, yield strength and ultimate strength are changed by irradiation. In addition, entirely new forms of behavior come into play. As examples, metal
swelling and irradiation creep occur only as a result of irradiation. An understanding of
the microstructural changes induced by irradiation and their subsequent effect on mechanical behavior are therefore necessary for the design of components intended for use
in a radiation environment.

C. Units

In considering particulate irradiation, a major factor of interest is the *flux*. The flux of
radiation impinging on a material is usually defined by the radiation type and by the density and energy range of the particles. The *total flux* Φ is given by

$$\Phi = \int \phi(E) \, dE \quad \text{particles cm}^{-2} \text{ sec}^{-1}$$

where $\phi(E)$ is the flux of particles that have a kinetic energy between E and E + dE.
Only a portion of the flux with energy greater than a specific minimum value can be considered to give rise to atomic displacements. With neutron irradiation, particularly when
applied to fast reactor designs, it is customary to specify the flux above the 0.1 MeV
energy level. A related factor of interest in design is the fluence, which is the integrated
total radiation seen by the material over a period of time:

$$\text{Fluence} = \int_0^t \Phi(t) \, dt \quad \text{particles cm}^{-2}$$

where t is the time. It is also customary to specify the fluence above the 0.1 MeV energy level
when dealing with neutron irradiation as applied to fast reactor design. In many instances
the selection of a particular material for a given radiation environment is dictated by the
physical characteristics of the major atom species rather than its mechanical behavior.
For instance, the need to moderate the neutrons might specifically require the use of certain materials, such as graphite. The property of importance here is the microscopic
cross section σ, which represents the probability of a certain reaction occurring. It is
expressed as an area through which the particulate radiation must pass to effect a certain
reaction in the target material. The rate at which reactions occur is given by

$$\text{Rate} = \Phi \sigma N \quad \text{reactions sec}^{-1}$$

where N is the number density of the particles in the material that are involved in the
particular reaction and σ is the area of interaction between the radiation and an atom in
the target.

In comparing the displacements produced by different types of particles, a basis
for comparison is necessary. This can be done by comparing the number of displaced

atoms calculated on the basis of the displacement cross section. With this normalization, the unit for fluence would be displacements per atom (dpa). Insofar as possible, however, extrapolation of the results of particulate radiation of one kind, such as electrons, to another kind, like neutrons, should be avoided except for making qualitative evaluations.

II. MICROSTRUCTURAL CHANGES RESULTING FROM IRRADIATION

A. Mechanisms of Single-Particle Interaction

When an energetic neutron or charged particle enters a solid, it interacts with the electrons and atomic nuclei. Given sufficient kinetic energy this interaction could result in atomic displacements. Another type of reaction with which we will not be concerned here relates to ionization events, which are more dependent upon the charge of the particle. It is the atomic displacements that for the most part produce the microstructural changes that in turn result in changes in the mechanical behavior of the solid.

An isolated atomic displacement produces an interstitial-vacancy pair in a crystalline solid. It should be noted that in this case the interstitials refer not to typical interstitial solute atoms, such as carbon or nitrogen, but to lattice atoms displaced to interstitial sites. The production of interstitial-vacancy pairs is the simplest type of radiation damage and leads to the generation of point defects. Even if this were to be the only type of damage, it is clear that the equilibrium concentration of vacancies has been affected, and given thermal activation, the associated diffusion phenomena will also be affected. When more than a single isolated atom is displaced, the directionality of the displacements has to be considered. In amorphous materials the displacements could occur in any direction due to the implicit isotropy of the solid, but the regular spatial arrangement of atoms in a crystalline solid leads to a more focused displacement effect along specific directions. Thus the energy transfer takes place preferentially along close-packed rows of atoms by focusing the collision sequences [1,2]. In addition to the propagation of transferred energy, mass transport could result directly from the collision[3,4], as when a displaced atom moves along a row of atoms of through a channel of interstitial sites.

The scenario of isolated interstitial-vacancy pair production has to be modified when considering highly energetic neutrons or charged particles. The atoms that are directly displaced can in these instances move through the solid with sufficient energy to produce a collision cascade. During a collision cascade the energy is dissipated rapidly, possibly resulting in a highly localized and sudden temperature rise, which leads to a final rearrangement of atoms that is different from the case of isolated interstitial-vacancy pairs. Theoretical simulation models of the collision cascade process[5] indicate that, after all the atoms have come to rest, a vacancy-rich zone exists with interstitials in the immediate neighborhood. In turn, not only the equilibrium concentration of vacancies has been disturbed as a result of irradiation, but the local concentration gradients have been disturbed as well. In the presence of temperature activation the diffusive processes will be modified not only by the overall change in concentration but also by potentials arising from local concentration gradients. The formation of depleted zones or vacancy clusters resulting from collision cascades has been observed in irradiated tungsten using a field ion microscope [6]. Since the formation of collision cascades is highly dependent upon the mass and velocity of the incident particle, electron irradiation may not produce the type of cascade effect observed with neutron or heavy ion bombardment.

This is one of the reasons the effects of one type of particle radiation cannot be easily extrapolated quantitatively to another type.

At higher temperatures of irradiation the nature of the microstructure changes due to the mobility of the defect population. The end result of a condensation/collapse process of the defect structures described thus far is the formation of interstitial or dislocation loops and void embryos. The atom movements that must accompany the collapse of radiation-induced point defect structures to form dislocation loops are not well known at the present time. The dislocation loops so formed are generally sessile, enclosing a stacking fault, but could be unfaulted, that is, made glissile, at a later time by a number of processes. The sizes and the number densities of the dislocation/interstitial loops would depend on the material, the irradiation temperature, the fluence, and possibly the flux. The generation and movement of these dislocations have important consequences to the phenomenon of irradiation creep.

B. Physical Effects of Fission Fragments

The physical effects arising from impact due to fission fragments differ in many respects from the particulate radiation damage considered so far. Because of their large masses, high kinetic energies, and highly ionized conditions, fission fragments produce a more or less continuous track in nonmetallic materials [7,8]. In metals, though fission fragments produce an array of defects [9], no continuous tracks are to be seen. The continuous track has been attributed to electron interactions and the consequent thermal spike [4] or from coulombic interactions arising from ionization due to the passage of the fission fragment [10]. It is only near the ends of such a track that the displacement type of damage takes place as the velocity of the fission fragment is attenuated sufficiently.

C. Microchemical Evolution

The creation of mobile point defects and the local rearrangements of atoms resulting from collision cascades produce under thermal activation microchemical changes in the material that are important to an understanding of mechanical behavior. Changes to the microchemistry of the solid, that is, the types and distribution of various phases on a microscale within the solid, may be expected as a result of the changes in the diffusion phenomena arising from the generation of irradiation-induced point and line defects. In the simplest case the vacancy concentration is increased as a result of radiation; this may be considered to lead to an irradiation-induced increase in diffusivity. This enhanced diffusivity could in turn promote precipitation, as many precipitation reactions in solids are diffusion controlled. As an example, the precipitation of copper from a supersaturated solution of Al-2 atom% Cu was accelerated by electron irradiation [11]. In this example irradiation has promoted the transformation of a metastable phase to one that is more stable, and in this respect the effect of irradiation is similar to a thermal treatment. In other cases phases have been noted in irradiated material that are not normally induced by thermomechanical treatments [12]. In view of the fact that microchemical changes take place continuously as a function of irradiation and in turn affect the microstructural and microchemical changes during subsequent irradiation, it is more appropriate to speak of microchemical evolution rather than specific changes.

In considering microchemical evolution the major area of interest relates to alloys in which complex phases may be present as a result of specific solutes added to enhance the mechanical properties of the material. These solute atoms can trap either vacancies or interstitials and provide preferred sites for recombination for the opposite type of incoming defect. More importantly, the solute atoms may be mobile at the working temperature. In this case, the flux of point defects can become coupled with the net fluxes of solute atoms toward or away from sinks [13], such as grain boundaries, dislocations, or voids. This coupling introduces concentration gradients that could lead to solute enrichment or depletion in the vicinity of the sinks. Under conditions favoring strong segregation, the local solubility limit may be exceeded, leading to radiation-induced precipitation. The type of segregation brought about by irradiation will depend upon the radiation temperature, the type of solute, the extent of coupling between the interstitial/vacancy and the solute atom, and the type of sink. The extent and type of microchemical segregation has a large influence on the subsequent mechanical behavior of the material.

The microchemical changes arising from irradiation may be best illustrated for the case of solid solution alloys by the behavior of 300 series austenitic stainless steels [14]. The element nickel, an important constituent in the 300 series alloys, tends to segregate at defect sinks, such as voids, grain boundaries, and dislocations. In turn, the matrix nickel content is reduced. This is particularly the case if the alloy initially contains carbon or silicon. These elements coprecipitate with nickel at the defect sinks. In type 316 stainless steel, for instance, the matrix nickel content could be reduced from an initial value of approximately 13.5% to a final value of 9% after prolonged irradiation. In contrast, chromium, another important consituent of the 300 series alloys, is depleted near the sink surfaces with a corresponding enrichment in the matrix.

The type of precipitates that develop from the matrix in 300 series stainless steel alloys depends upon the time-temperature history, minor variations in composition, and preirradiation thermomechanical treatment. These precipitate phases are generally nitrides, carbides, or intermetallic compounds such as Laves, sigma and eta phases. In addition, silicides of nickel of the composition Ni_3Si corresponding to a γ precipitate structure have been noted [12] under fast neutron irradiation. This phase is not seen in unirradiated 300 series alloys and appears to require irradiation for its formation. Once formed, the nickel silicide phase Ni_3Si requires continued irradiation to be stable. An out-of-reactor anneal of the irradiated specimen redissolved the Ni_3Si phase [15]. Direct evidence for the segregation of nickel, chromium, carbon, and silicon has been obtained using energy dispersive x-ray analysis techniques [16].

The consequences of removing the nickel from the matrix could be large, particularly in terms of the subsequent behavior of material under irradiation as it relates to void swelling. In austenitic steels of a lower initial nickel content, such as type 304, the depletion of nickel from the matrix is large enough to transform isolated areas of the matrix to ferrite [14].

The foregoing discussions relative to the effect of irradiation on solid solution-type alloys can be extended to precipitation-strengthened alloys. In general, the precipitation-strengthened alloys used in the reactor environment belong to the high-nickel austenitic series containing either γ' or γ'-γ'' precipitates. The strong affinity of silicon, titanium, and aluminum for nickel would indicate that, if these alloys were irradiated in an initially solution-treated condition, that is, without the necessary heat treatment to form the precipitates, precipitation of these phases will occur in the radiation environment. In the case of an initially solution treated and aged material where the precipitates exist at the start of irradiation, the effect of irradiation would be similar to prolonged heat treatment at higher temperatures, in that coarsening of the precipitates would occur.

Such a coarsening of γ' precipitates in Nimonic PE16 has been noted [17]. In Inconel 706, a γ'-γ''-strengthened alloy, neutron irradiation appears to lead to dissolution of γ'' and coarsening of γ' in the temperature range of 400-500°C, but at higher temperatures overaging predominates and several types of carbide phases and γ' are formed near the grain boundaries. Based on the considerations outlined earlier, the spatial distribution of the precipitates could be expected to change depending upon the initial condition of the material. For instance, the availability of a large network of dislocations introduced as a result of initial cold work would facilitate the occurrence of irradiation-induced precipitation on a finely distributed scale over the preexisting dislocation network. Therefore, the final properties of the material will depend upon whether the material was in the solution treated or solution treated and aged or cold-worked condition prior to irradiation. The microstructural stability during irradiation will largely determine the final mechanical behavior of the alloy, especially as regards ductility and hardness.

The underlying reasons for the importance of microchemical segregation and phase stability under irradiation in the design of structures intended for use in a radiation environment will be brought out more clearly in later discussions on void swelling, irradiation creep, and conventional mechanical behavior.

D. Transmutation Effects

In designing components for use in a radiation environment, the effects of transmutation reactions have to be taken into account. The transmutation rate depends upon the energy spectrum and on the atom species being irradiated as characterized by the cross section for the reaction.

1. Fuel Element Sheaths

Chemical and physical effects arising from transmutation reactions are particularly important when considering sheath material for fuel or absorber elements in a reactor. Transmutation of the uranium in uranium oxide, a principal form of fuel used in power reactors, produces a variety of fission products [18]. In addition, the fission products themselves can undergo transmutation reactions by further interactions with neutrons or other particulate irradiation. The chemical and physical state of these fission products depends upon the temperature and the ability to form compounds within themselves or with the uranium oxide fuel. In terms of the design of the sheath material, three important factors have to be taken into account. First, one must consider the major atom species of the sheath material and whether the cross section of this atom species is favorable for overall neutron economy. Second, one must consider if the sheath can withstand the chemical environment posed by the fuel, fission products, and coolant and third, if it can withstand the loading imposed by the fuel at the operating temperatures without failure. Thus in considering transmutation effects, both physical and chemical effects have to be taken into account. A more complete description of irradiation effects on nuclear fuel material and sheath design can be found in Ref. 19. In the rest of this section, we will be concerned primarily with transmutation effects in structural materials that are considered to be nominally nonfissile.

2. Structural Materials

Transmutation effects that occur in nominally nonfissile material, such as structural steels, can often lead to changes in the mechanical behavior of the material. Virtually all commercial steels contain trace amounts of boron, of which some 20% is the ^{10}B isotope. A thermal neutron reacts with this isotope of boron to produce helium by the reaction

$$^{10}B + n \rightarrow {^7}Li + {^4}He + 2.76 \text{ MeV}$$

The relatively large cross section for this reaction means that a considerable amount of helium would be generated. The generation of helium has several consequences. First, though the initial thermal neutron may not have sufficient energy to produce displacement damage, the large energy associated with the helium atoms produced as a result of the transmutation could result in displacement damage. Second, helium atoms can stabilize the initial void nucleus and thereby permit void growth and swelling. Third, helium has a tendency to migrate to the grain boundaries resulting in embrittlement of steel. Although the production of helium with thermal neutrons is associated with boron, similar transmutation reactions can occur between fast neutrons and virtually all elemental constituents of steel. The implications of helium embrittlement for design will be considered more fully later.

E. Surface Effects

The earlier discussions on displacement damage, microstructural changes, and microchemical evolution focused on irradiation effects in the matrix of the solid rather than its surface. A number of physical effects associated with the surface of the solid take place as a result of irradiation. These include processes such as sputtering and blistering.

Sputtering denotes material removal by the ejection of atoms from a target surface by particulate bombardment. It is commercially important in ion etching, surface cleaning, and in the production of thin films in the electronics industry. The process of sputtering is governed essentially by momentum transfer and requires a certain minimum threshold energy level of the incident ions to eject surface atoms. This level depends upon the material properties of the ion and the target. At high levels of incident energy the surface action is replaced by the displacement type of damage described earlier so that sputtering efficiency decreases beyond this value. The surface of the target after sputtering is generally not uniform but presents a geometric etched pattern [20]. The type of pattern reflects the underlying crystallographic symmetry of the crystal surface. The rate of removal depends upon the crystallographic orientation, and for this reason adjoining grains with different orientations are sputtered at different rates. This property can be used to obtain a relief etch of the grain structure of the material.

When the incident radiation is highly energetic, for example, neutrons with energy levels above the 10 MeV range, more extensive surface damage occurs in the form of blistering and exfoliation [21]. This phenomenon is of importance in the design of the first wall of a controlled thermonuclear reactor that will be exposed to such high-energy neutrons. In most instances the re-emission of the implanted atoms or accumulation of transmutation-induced helium close to the surface is the principal cause of exfoliation or bursting of blisters. The tendency to exfoliate is a function of the solubility and diffusivity of the atom species being re-emitted. When this atom species is more soluble or diffuses rapidly from the surface into the matrix of the solid, there is less tendency to blister.

So far, irradiation-induced changes of the structure, both in the matrix and on the surface, have been described. In applications involving structural materials used in design, it is the displacement damage and associated changes to the microstructure and microchemistry that are vital areas of concern. The specific changes to the mechanical behavior of the solid are considered in the next section.

III. IRRADIATION EFFECTS ON MECHANICAL BEHAVIOR

A. Introduction

In considering the effects of particulate irradiation on the mechanical behavior of materials, the primary emphasis will be on structural materials. Changes to mechanical behavior in materials specifically used as a result of their desirable transmutation reaction cross sections, such as nuclear fuel materials and absorber materials, are treated in more detail in Refs. 18 and 19, for example. In these instances, the mechanical behavior is secondary to considerations on the achievement of specific reaction rates in the total system. Two major aspects of mechanical behavior, void swelling and irradiation creep, which are present only as a result of irradiation, will be discussed first, followed by the effects on more conventional forms of mechanical behavior.

B. Void Swelling

An aspect of irradiation-induced changes in mechanical behavior that has major consequences to design, particularly under fast neutron irradiation, is the swelling or volume change in the material during irradiation. This volume change is a consequence of the formation and growth of voids. Figure 7.1 shows typical void images produced by nickel ion bombardment of cold-worked type 316 stainless steel, as revealed by transmission electron microscopy. An appreciation for the magnitude of this problem may be gained from the fact that for solution-annealed type 316 stainless steel, which was the reference core component structural material until the early 1970s for fast breeder reactors, the volumetric swelling at goal fluences of about 1×10^{23} neutrons cm^{-2} ($E > 0.1$ MeV) and at an irradiation temperature of 500°C was found to be greater than 10%. The design evaluations are complicated by the fact that this volume increase is a sensitive function of temperature so that a component subjected to a temperature differential would encounter differential swelling resulting in high internal stresses. The bowing of structural members under such differential swelling and the extent to which the internal stresses are relaxed by irradiation creep have to calculated in order to design these components.

Until the late 1960s it was believed that the radiation-induced production of defects would reach a steady state whereby equal amounts of defects will be created and annihilated beyond a certain fluence level. The vacancy-interstitial pairs produced by irradiation-induced atomic displacements, for instance, could be lost by recombination if the defects are sufficiently mobile. In such a case there will be no net change to the defect population or the density of the material. The discovery of metal swelling in the observation that the metal contained a profusion of voids that were several tens of nanometers in diameter, which could not be accounted for by transmutation reactions, such as helium production, led to a more detailed study of the phenomenon of void swelling. A number of mechanisms have been postulated whereby the movement of the interstitials is biased, leaving an excess flux of vacancies, toward neutral sinks. Dislocations, for instance, act as preferential absorption sites for interstitials. In this case, provided an initial void embryo has been formed and stabilized from redissolution, the excess vacancy flux could move to this neutral sink and result in void growth. The net effect of the growth of voids is an associated volume increase in the material. The fact that the

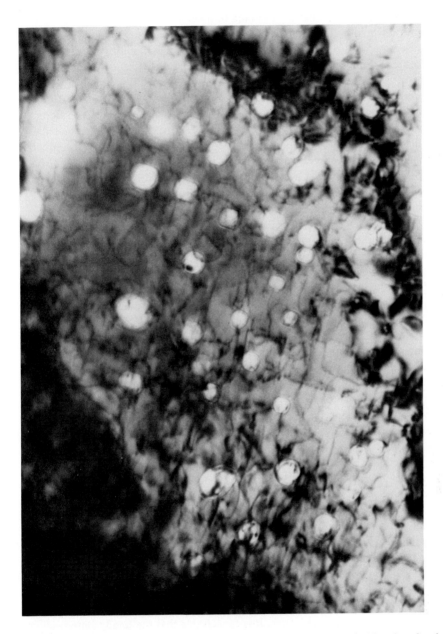

Fig. 7.1 Transmission electron micrograph showing voids in nickel ion bombarded cold-worked type 316 stainless steel. Magnification 125,000X. (Courtesy of T. Lauritzen, General Electric Company, Sunnyvale, California.)

vacancies should be mobile and that there should be a net excess flux of vacancies without recombination would indicate that at high temperatures a lower volume swelling will result as there is more chance for recombination due to the high mobility of both interstitials and vacancies. Similarly, at low temperatures, the mobility is insufficient for void growth. Unfortunately, the temperature regime where void growth is predominant in structural materials such as austenitic stainless steels, that is, between 350 and 650°C, is also the temperature regime of interest to the designer of commercial breeder reactors.

Experimental evaluations of the swelling resistance of materials were conducted extensively during the 1970s using various ion bombardment techniques applying heavy metal ions, carbon ions, electrons, deuterons and protons.* In one of the earliest studies [22] simple ternary alloys of Fe, Cr, and Ni were bombarded with 5 MeV nickel ions. Maximum swelling was observed in the ternary composition range with 15% nickel and 10-30% chromium. Unfortunately this composition is similar to that of the major alloying elements in the 300 series austenitic stainless steels. In compositions of high nickel contents representative of precipitation-strengthened superalloys, such as Nimonic PE16, the swelling was extremely low. In the low nickel range representative of the ferritic class of alloys, once again the swelling was found to be low. In the same series of experiments, the effect of minor element additions to an Fe-15Cr-20Ni ternary was evaluated with systematic additions of silicon, titanium, niobium, zirconium, manganese, phosphorus, boron, aluminum, and carbon. Of these, the first four were found to reduce the swelling considerably but the rest had no significant effect. The highly significant effect of minor element additions could be appreciated from the fact that at 675°C after nickel ion bombardment to 166 dpa (displacements per atom), the swelling of the ternary was reduced from about 60% in the unalloyed condition to below 5% with the addition of 3 atom% silicon.

Recent work [23-25] on theoretical models for swelling has clarified the mechanisms involved in void swelling, particularly on the effect of minor element additions. Initial modeling in this area was based on a rate theory approach that takes into account the relative bias strengths of sinks for vacancies. Dislocation networks, grain boundaries, and precipitates all act as preferential sinks and influence the void swelling phenomenon. This model has to be modified when solute atoms can interact with vacancy/interstitial defects. One possibility is that solute atoms can trap either vacancies or interstitials to form point defect/solute atom complexes. These complexes could act as sites for recombination for the opposite type of incoming defect thereby reducing the excess vacancy flux and in turn the void growth rate. Another possibility relates to the effect of microchemical evolution on swelling behavior. As discussed earlier, binding interactions between mobile solute atoms and point defects result in segregation and precipitation reactions. The exact mechanism whereby such segregation processes affect the swelling process is not well understood at this time. It is conceivable that the steady-state swelling of complex alloys, such as type 316 stainless steel, is a function of the matrix composition. As nickel, silicon, and carbon are preferentially removed to sink surfaces, the final matrix composition reaches some steady state. Based on the work on Fe-Ni-Cr ternaries

---

*Ion bombardment techniques permit the rapid qualitative assessment of the swelling resistance of various alloys, though extrapolation to other particles such as neutrons is not straightforward. In particular, these techniques do not provide the detailed variation of the swelling behavior with temperature (because of a substantial temperature shift) required for design but are useful screening techniques for the development of swelling resistant materials.

cited earlier, it would appear that the swelling rate would now correspond to the new composition of the matrix with the lower nickel content. This is a somewhat qualitative picture for describing the effects of microchemical segregation on the swelling phenomenon. In order to more accurately describe the effect of alloying elements on swelling it is necessary to consider the extent of binding between solute atoms and point defects, the change in solute and defect diffusivity with temperature, and the extent to which the sink capture rates are affected due to the local segregation near the sink. In addition, the irradiation-induced precipitates themselves can in some instances act to reduce swelling by acting as preferential recombination sites. Although most of the discussion on solute atoms here applies to interstitial solutes, similar effects arising from the interaction of the interstitial defect and substitutional atoms in the alloy have also been postulated.

The change in swelling properties as a function of alloy composition has led to an intensive development effort in many countries to design specific materials suitable for use in fast neutron environments. Perhaps one of the best illustrations for swelling variation as a function of composition was illustrated by the microstructure of IN-744 [26]. This alloy has a duplex microstructure consisting of austenite and ferrite grains. After neutron irradiation the microstructure revealed no voids to be present in the ferrite grains but voids were evident in the austenite grains. New materials for better swelling resistance include the higher nickel alloys and ferritic alloys, as well as austenitic alloys of 316-type composition with the addition of small amounts of minor elements, such as silicon and titanium, which suppress swelling. It should be clear that in applying the constitutive relations for swelling in design care should be taken to use well-characterized data specific to the particular heat in question. Differences in swelling behavior between materials of the same nominal 316 stainless steel composition have been noted [27] when one heat was procured commercially with trace impurities and another heat was made from high-purity base material. As might be expected from earlier discussions, it was the commercial purity material with trace impurities that showed better swelling resistance.

In addition to compositional effects, thermomechanical treatments, such as cold working, also affect the swelling behavior. Considerations relative to the initial critical stable void size necessary for growth indicate that the initial dislocation density plays an important role in the nucleation process. The presence of a large initial dislocation density or impurity concentration delays the appearance of voids. This is more easily understood by considering the swelling dependence on fluence. Experimental results from neutron irradiation [28] indicate that, for most alloys, this dependence follows a three-stage process as illustrated by the example of 20% cold-worked type 316 stainless steel shown in Fig. 7.2: an initial incubation period over which there is little or no swelling of the material, a transition stage during which the rate of swelling increases with fluence, and a final steady-state behavior during which the swelling rate remains constant. The incubation fluence, based on the premise that the initial dislocation density affects the void nucleation process, could be expected to depend upon the initial level of cold work. This hypothesis is supported by data on type 316 stainless steel. The incubation fluence for 20% cold-worked material was in the range of 7-8 $\times$ 10^{22} neutrons cm^{-2} (E $>$ 0.1 MeV); the incubation fluence for the annealed material was as low as 3 $\times$ 10^{22} neutrons cm^{-2} (E $>$ 0.1 MeV). The transition stage over which the swelling rate increases with fluence generally occurs over a typical fluence interval of 1 $\times$ 10^{22} neutrons cm^{-2}. Limited experimental data would indicate that the steady-state swelling rate is not highly influenced by the initial level of cold work. The three-stage swelling dependence on fluence can also be interpreted in terms of the microchemical evolution process: during the incubation period the solute additions intended to suppress swelling increase the recombination rate and quite possibly also affect the nucleation process resulting in little or no swelling. Beyond the transition stage, the microchemical evolution reaches some steady state for the given temperature resulting in a steady excess flux of vacancies to the voids giving rise to the third stage of swelling.

The variation of swelling with irradiation temperature is another key design issue. As seen from Fig. 7.2, at a constant fluence of about 15×10^{22} neutrons cm^{-2} (E > 0.1 MeV), peak swelling in 20% cold-worked type 316 material occurs at a temperature of between 577 and 605°C. Irradiation temperatures both above and below this level result in lower swelling. The peak swelling temperature is affected by minor alloying additions. The temperature variation of swelling is thought to arise as a result of the variation in diffusivity of the solute atoms and the differences in the types of precipitates formed at different irradiation temperatures. The large variation in swelling with temperature could promote high internal stresses in structural components subject to a differential temperature.

Three additional effects of void swelling are less well known but have important consequences to design. The first relates to the effect of stress on swelling. Theoretical studies [29,30] indicate that a tensile hydrostatic stress would increase the net flow of vacancies to voids, thereby increasing the swelling rate. Limited experimental evidence

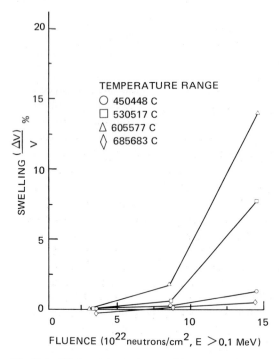

Fig. 7.2 Fluence dependence of swelling in 20% cold-worked type 316 stainless steel. (Adapted from Ref. 28.)

for stress-affected swelling is available [31] , but a complete description needs further testing. A second major effect poorly understood at this time relates to the change in swelling behavior with temperature changes. Limited experimental data [32] relative to the effect of temperature changes reveal extremely complex behavior. For instance, in type 316 material, subsequent to irradiation at the peak swelling temperature, a temperature decrease was found to result in a swelling rate higher than that corresponding to the new temperature. Since many reactor components are subject to a variable temperature history as a result of imposed duty cycle events, there is a need for further development in this area. Finally, changes to the elastic moduli of the material subsequent to radiation have been reported [33] that can be attributed to the formation of voids in the material.

C. Irradiation Creep

Time-dependent deformation at elevated temperatures at constant imposed loads is termed creep. The associated deformation rates are dependent upon the particular diffusive phenomenon dominant at the imposed temperature. When specimens were subjected to particulate radiation, deformation was noted in temperature regimes where control thermal tests exhibit little or no deformation. This deformation process, termed irradiation creep, has to be taken into account when designing a component for use in a radiation environment.

A number of aspects of irradiation creep are important to design and differ significantly from thermal creep. First, there is considerable evidence that irradiation creep is superplastic, that is, thermal creep is ductility limited, but irradiation creep is not. Second, the irradiation creep deformation appears to vary linearly with fluence and stress. In contrast, stress exponents in thermal creep are generally higher than 4. Third, the deformation rate does not appear to be as highly sensitive to temperature for irradiation creep as compared with thermal creep. Finally, materials that are swelling resistant also appear to be resistant to irradiation creep.

Several mechanisms for irradiation creep have been suggested in the literature [34,35] . Most of these mechanisms involve a preference for vacancies and interstitials to agglomerate on different crystallographic planes depending upon the direction of applied stress. A simplified interpretation would be that vacancies are more easily compressed than interstitials. On the application of a stress field, there is a net difference in the flow or collapse of vacancies and interstitials, which is directional. This difference produces growth along the tensile axis and a corresponding compression on the orthogonal axes. The preferred directionality could be explained in terms of either stress-induced preferential nucleation or stress-induced preferential absorption of the defect loops.

Irradiation creep data are generally obtained from pressurized tube tests, although deformation modes such as bending and extension have also been successfully employed. It is difficult to separate the components due to thermal and irradiation creep. This is not a shortcoming, however, since the designer would prefer to have a single constitutive relation for describing the deformation in the radiation environment. An appreciation for the extent of deformation may be gained from considering the results on 20% cold-worked type 316 stainless steel [36] . In this material, at an irradiation temperature of 475°C and at a stress level of 200 MPa, hoop strains of the order of 0.3% were found to result in a pressurized tube test with an accumulated fluence of 2.5×10^{22} neutrons cm^{-2} (E > 0.1 MeV). In the control thermal creep test for the same conditions, no observable creep deformation was noted. Though irradiation creep strains of 2 to 3% can occur without leading to rupture, it is necessary to evaluate such deformations to preclude creep collapse and maintain the deformation limits for structural adequacy. Currently available data would suggest that the deformation rate stays constant with fluence up to the incubation fluence for swelling. Beyond the incubation fluence for swelling, the irradiation creep rate increases in the transition region for swelling. In the region where the swelling rate becomes constant, irradiation creep rate also becomes constant as a function of fluence but at a higher level than that noted below the incubation fluence. Based on these experimental observations, irradiation creep rates are generally prescribed in terms of the swelling rate of the material with additional terms to describe the deformation process in the absence of swelling. Some implications of irradiation creep for design are discussed in more detail in a later section.

D. Conventional Mechanical Properties

The changes in conventional mechanical properties and behavior, such as yield strength, ultimate strength, flow behavior, thermal creep subsequent to irradiation, and fatigue as the result of irradiation, received considerably more attention before swelling and irradiation creep became the dominant issues. Irradiation-induced changes to conventional mechanical behavior, however, continue to be of major importance to design. Many of these changes are consistent with the microstructural and microchemical changes discussed earlier.

The introduction of point and line defects produced by irradiation increases the defect density in the solid. In this respect, the effect of irradiation is similar to that of cold work. This effect is dramatically illustrated in the case of annealed type 304 stainless steel [37], whose room temperature yield strength increases from the unirradiated value of about 200 to 460 MPa following irradiation to a fluence of 1.4×10^{22} neutrons cm^{-2} ($E > 0.18$ MeV) at 530°C. Although the increase in yield strength is relatively large, there is only a small increase in the ultimate strength of the material resulting in a decrease in the work-hardening capacity. The effect of irradiation saturates quickly, however, as at higher fluences there is no additional increase to either the yield or ultimate strength levels. At higher temperatures of irradiation, the effect is less dramatic as partial recovery of the irradiation-induced hardening takes place. At still higher temperatures, typically above 700°C for austenitic stainless steels of the 300 series, recovery is rapid enough that very little radiation hardening takes place and the properties subsequent to irradiation are similar to that for the annealed unirradiated material.

In contrast to the increase in yield strength, ductility decreases with fluence at all temperatures [38]. The reduced ductility at lower temperatures could be attributed to irradiation hardening similar to the decrease in ductility due to cold working. At higher temperatures the decrease is due to helium embrittlement. The loss of ductility at lower temperature irradiations can be recovered by postirradiation annealing. In contrast, the ductility loss related to helium embrittlement cannot be recovered by thermal treatments. Helium is insoluble in steel and segregates at the grain boundary promoting premature intergranular failure.

The strength and ductility changes discussed for solid solution strengthened stainless steels must be modified when considering precipitation-strengthened steels. When high-nickel alloys are irradiated in the solution-treated condition, precipitation of γ' occurs provided the initial composition is favorable to its formation. The effect of this reaction dominates the effect of irradiation hardening so that an increase in yield and ultimate strength may be expected even in the temperature range where recovery mechanisms could become operative. A pronounced loss of ductility has been reported with these materials [39]. The mechanisms causing this ductility loss are thought to be associated with the precipitation of brittle phases at grain boundaries.

Irradiation-induced changes in fatigue and fracture properties have also been reported [40,41]. Since fatigue crack propagation and fracture toughness are intimately tied to the capability for plastic deformation at the crack tip, significant reductions in these properties are to be expected. Limited available data on elevated temperature strain-controlled fatigue for irradiated types 304 and 316 steels and Incoloy 800 show that fatigue life may be reduced by as much a factor of 30. It is believed that the loss of ductility due to irradiation mainly affects low-cycle fatigue life and that irradiation effects on high-cycle fatigue life are minimal. The subject of fracture behavior subsequent to low irradiation levels in pressure vessel steels has received considerable attention [42,43]. In low-alloy ferritic steels, irradiation to low doses such as 1×10^{21} neutrons cm^{-21} results in a pronounced shift in the ductility transition temperature of the material. For instance, A302-B, an alloy steel containing 1.5% Mn, 0.5% Mo, 0.3% Si, and 0.25% C, exhibited a shift of nearly $100°C$ in the ductile-to-brittle transition temperature after irradiation at $288°C$ to a dose of 3×10^{19} neutrons cm^{-2} ($E > 1$ MeV). Control of residual elements, particularly copper and phosphorus, has been found to be beneficial in reducing the susceptibility to radiation embrittlement.

Another area of conventional mechanical behavior that has received considerable attention is postirradiation thermal creep. Although it is desirable to conduct creep experiments under the radiation environment, the facilities for performing such experiments were limited in the 1960s. Most of the thermal creep experiments during this time were conducted on specimens that were previously irradiated to a specific fluence without stress. Pronounced loss of creep ductility and reduction of up to an order of magnitude in creep rupture times were observed in these postirradiation tests compared with unirradiated material [44]. Due to irradiation hardening the postirradiation creep rates were generally lower than the unirradiated material. With high irradiation temperatures, where recovery effects dominate over radiation hardening, the postirradiation creep rates increased to more nearly the values of unirradiated material. However, the rupture times were found to be extremely low due to helium embrittlement. In recent years postirradiation tests have been de-emphasized in favor of in-reactor creep experiments. The in-reactor creep rupture time of the material where the specimen is stressed while being irradiated has been found to be superior to that observed in postirradiation tests. Limited in-reactor data indicate the stress rupture properties to be comparable or superior to postirradiation creep rupture properties [45]. The reasons for the difference between in-reactor and postirradiation behavior are not well understood at this time, but it is believed that irradiation creep may relax stresses at crack tips and sliding boundaries, thus retarding the initiation of growth of cracks.

IV. DESIGN IMPLICATIONS

In the preceding sections irradiation-inducted changes to microstructure and their effect on mechanical behavior have been outlined, with particular reference to structural materials. In considering the design implications of irradiation effects, it is useful to distinguish among four major areas of application. First, there are materials specifically chosen and used as a result of their desirable transmutation reaction properties. This would include nuclear fuel materials, such as uranium oxide, plutonium oxide, and others, as well as absorber materials, such as boron carbide. In these instances mechanical behavior is less important than obtaining the desired transmutation rates. The behavior of these materials is a strong function of the fission products produced as a result of the transmutation and their effect on the physical and chemical structure of the material.

Second, there are materials that are used to physically isolate materials of the first category. They should exhibit the fine balance of properties in terms of desirable reaction rates, high resistance to radiation damage, and resistance to the chemical environment produced by the fission products resulting from transmutation reactions of the first type of material. A nuclear fuel pin sheath would belong to this category. It should have a low neutron absorption cross section and should not be subject to significant corrosion due to the generation of fission products in the fuel. It should also have desirable mechanical properties in the radiation environment, such as a low swelling rate to avoid constrictions in the coolant flow channel and a high in-reactor creep rupture strength.

Third, there are materials that are used as structural members in a high-radiation dose environment. These, for example, would be duct structures that enclose an assembly of fuel pins. Mechanical behavior under radiation is the dominant variable in their selection. The swelling and irradiation creep of these structures are important to avoid duct-to-duct contact and to ensure proper design of the restraint to the overall core, which contains many duct assemblies. These ducts are subject to temperature gradients that might induce differential swelling. In turn, this could lead to duct bowing, or if constrained from bowing, to the development of high internal stresses. Irradiation creep could relax these stresses if the creep rates are sufficiently high. The interaction between swelling and irradiation creep is therefore a major design issue [46]. Duct materials should also have a high fracture toughness subsequent to irradiation to avoid unstable crack growth and fracture when occasionally subject to high loads, such as during fuel-loading operations.

Finally, a fourth class of materials is used nominally well outside the high radiation environment but is still subject to low levels of radiation. Examples include the core support structure and the pressure vessel. They are designed for the lifetime of the plant and have to satisfy codes such as the American Society for Mechanical Engineers Pressure Vessel Codes. In these instances the designer should make sure that sufficient shielding is in place that the fluence seen by these structures at design lifetime will not significantly affect the underlying mechanical behavior modes and properties implicit in the code rules. Specific areas requiring attention in this case are the ductility of the material and the shift in transition temperature at design fluences if the material exhibits a ductile-to-brittle transition, such as is associated with the low-alloy ferritic class of steels.

The evaluation procedures and analytical tools required for design in a radiation environment are similar to those used for unirradiated structural design. Thus, once irradiation creep constitutive relations become available, the deformations in the structure are predicted by methods similar to those already in use for thermal creep. It is only important to note that irradiation creep occurs in a temperature range where no thermal creep would be expected. An externally pressurized tube can collapse by irradiation creep, whereas thermal creep may not produce collapse under the same conditions. Similarly, the volume swelling of the material due to irradiation can be treated in a manner analogous to thermal expansion. A major difference is that the imposition of opposite temperature gradients common to on-off operations will result in alternate cycles of thermal expansion and contraction, but swelling-induced deformations are permanent and do not disappear when the radiation field is removed.

V. SUMMARY

In this chapter a brief description of the major irradiation-induced changes to the mechanical behavior of the material has been presented. It was shown that many of these changes could be related to the displacement of atoms by the incident radiation and the consequent production of point defects. The movement of these defects as affected by both atomistic and structural features under the application of thermal and strain fields is seen to lead to irradiation creep and metal swelling. Conventional modes of mechanical behavior are also affected due to changes in defect density and precipitate structures. An understanding of the essential features of these microstructural and microchemical changes is important to predicting the behavior under irradiated conditions.

Many of the investigations discussed in this chapter are quite recent. The discovery of swelling, for instance, is generally dated from the time of Cawthorne and Fulton's article in *Nature* [47] in 1967. For this reason, the reader should consult more specialized publications in the field for more recent data and interpretations. The biennial symposia held under the auspices of the American Society for Testing and Materials are a good source. A number of reports related to neutron irradiation effects cited in the list of references refer to reports of the Hanford Engineering Development Laboratory which can be obtained by writing to the National Technical Information Service of the U.S. Department of Commerce (Springfield, VA 22161).

REFERENCES

1. G. K. Wehner, *Phys. Rev.*, 102:690 (1956).
2. R. H. Silsbee, *J. Appl. Phys.*, 28:1246 (1957).
3. G. H. Vineyard, in *Physics of Solids* (D.S. Billington, ed.), Academic Press, New York, 1963, p. 291.
4. L. T. Chadderton, *Radiation Damage in Crystals*, Methuen, London, 1965.
5. A. Seeger, in *Radiation Damage in Solids*, Vol. 1, IAEA, Vienna, 1962.
6. L. A. Beavan, R. M. Scanlan, and D. N. Seidman, *Acta Met.*, 19:1339 (1971).
7. P. B. Price and R. M. Walker, *J. Appl. Phys.*, 33:3400 (1962).
8. T. G. Knorr, *J. Appl. Phys.*, 35:2753 (1964).
9. K. L. Merkle, L. R. Singer, and R. K. Hart, *J. Appl. Phys.*, 34:2800 (1963).
10. R. L. Fleischer, P. B. Price, and R. M. Walker, *J. Appl. Phys.*, 36:3645 (1965).
11. C. W. Tucker and N. B. Webb, *Acta. Met.*, 7:187 (1959).
12. C. Cawthorne and C. Brown, *J. Nucl. Mater.*, 66:201 (1977).
13. R. A. Johnson and N. Q. Lam, *Physical Review B*, 13:4364 (1976).
14. F. A. Garner, *HEDL-SA-2159*, Hanford Engineering Development Laboratory, Richland, Washington, 1980.
15. H. R. Brager and F. A. Garner, *J. Nucl. Mater.*, 73:9 (1978).
16. H. R. Brager and F. A. Garner, *HEDL-SA-1883*, Hanford Engineering Development Laboratory, Richland, Washington, 1980.
17. P. S. Sklad, R. E. Clausing, and E. E. Bloom, in *Irradiation Effects on Microstructure and Properties of Metals*, STP-611, Amer. Soc. Test. Mater., Philadelphia, 1976, p. 139.

18. S. Glasstone and A. Sesonske, *Nuclear Reactor Engineering,* Van Nostrand-Reinhold, Princeton, New Jersey, 1963.
19. D. R. Olander, *Fundamental Aspects of Nuclear Reactor Fuel Elements,* TID-26711-P1, Technical Information Center, Energy Res. Dev. Admn., Springfield, Virgina, 1976.
20. N. Hermanne and A. Art, *Radiat. Eff.,* 5:203 (1970).
21. R. Behrisch, *J. Nucl. Mater.,* 86:1047 (1979).
22. W. G. Johnston, T. Lauritzen, J. H. Rosolowski, and A. M. Turkalo, Report No. 76CRD019, General Electric Company, Schenectady, New York, 1976.
23. R. M. Mayer, L. M. Brown, and U. Gösele, *J. Nucl. Mater.,* 95:44 (1980).
24. R. Bullough and A. D. Brailsford, *J. Nucl. Mater.,* 44:121 (1972).
25. F. A. Garner and W. G. Wolfer, *HEDL-SA-2155,* Hanford Engineering Development Laboratory, Richland, Washington, 1980.
26. S. D. Harkness, B. J. Kestel, and P. Okamoto, in *Radiation-Induced Voids in Metals,* CONF-71061, U.S. Atomic Energy Commission, 1972, p. 334.
27. L. M. Leitnaker, E. E. Bloom, and J. O. Stiegler, *J. Nucl. Mater.,* 49:57 (1974).
28. T. A. Kenfield, W. K. Appleby, H. J. Busboom, and W. L. Bell, *J. Nucl. Mater.,* 75:85 (1978).
29. W. G. Wolfer and M. Ashkin, *J. Appl. Phys.,* 46:547 (1975).
30. J. F. Bates and E. R. Gilbert, *J. Nucl. Mater.,* 71:286 (1978).
31. F. A. Garner, E. R. Gilbert, and D. L. Porter, *HEDL-SA-2004,* Hanford Engineering Development Laboratory, Richland, Washington, 1980.
32. B. A. Chin and J. L. Straalsund, *J. Nucl. Mater.,* 74:260 (1978).
33. J. L. Straalsund and C. K. Day, *Nucl. Tech.,* 20:27 (1973).
34. W. G. Wolfer and M. Ashkin, *J. Appl. Phys.,* 47:791 (1976).
35. R. V. Hesketh, *Phil. Mag.,* 7:1417 (1962).
36. E. R. Gilbert and J. F. Bates, *J. Nucl. Mater.,* 65:204 (1977).
37. J. J. Holmes, R. E. Robbins, and A. J. Lovell, in *Irradiation Effects in Structural Alloys for Fast and Thermal Reactors,* STP-457, Amer. Soc. Test. Mater., Philadelphia, 1969, p. 371.
38. J. J. Holmes, A. J. Lovell, and R. L. Fish, in *Effects of Radiation on Substructure and Mechanical Properties of Metals and Alloys,* STP-529, Amer. Soc. Test. Mater., Philadelphia, 1973, p. 383.
39. J. Barnaby, P. J. Barton, R. M. Boothby, A. S. Fraser, and G. F. Slattery, in *Radiation Effects in Breeder Reactor Structural Materials* (M. L. Bleiberg and J. W. Bennett, eds.), The Metallurgical Society of AIME, New York, 1977, p. 159.
40. J. M. Beeston and C. R. Brinkman, in *Irradiation Effects on Structural Alloys for Nuclear Reactor Applications,* STP-484, Amer. Soc. Test. Mater., Phildelphia, 1970, p. 419.
41. C. R. Brinkman, G. E. Korth, and J. M. Beeston, in *Effect of Radiation on Substructure and Mechanical Properties of Metals and Alloys,* STP-529, Amer. Soc. Test. Mater., Philadelphia, 1972, p. 473.
42. L. E. Steele, *Neutron Irradiation Embrittlement of Reactor Pressure Vessel Steels,* IAEA Report 163, IAEA, Vienna, 1975.
43. J. R. Hawthorne, in *Irradiation Effects on Structural Alloys for Nuclear Reactor Applications,* STP-484, Amer. Soc. Test. Mater., Philadelphia, 1970, p. 96.
44. E. E. Bloom and J. R. Weir, Jr., *Nucl. Tech.,* 16:45 (1972).
45. E. E. Bloom and W. G. Wolfer, in *Effects of Radiation on Structural Materials,* STP-683, Amer. Soc. Test. Mater., Philadelphia, 1979, p. 656.
46. R. A. Weiner, J. P. Foster, and A. Boltax, in *Radiation Effects in Breeder Reactor Structural Materials,* The Metallurgical Society of AIME, New York, 1977, p. 865.
47. C. Cawthorne and E. J. Fulton, *Nature,* 216:575 (1967).

8

Iron and Steel*

I. INTRODUCTION

In the development of contemporary civilization an event that ranks with the discovery of the wheel and the lever in mechanics is the development of the art of producing iron and steel for use as a structural material. There is evidence that a few iron tools were in use in India, China, and Egypt more than 4000 years ago. These early implements were probably produced by subjecting lumps of iron ore, either intentionally or accidentally, to long exposure to an ordinary campfire and then beating the extracted metallic iron into the desired shape, forming a material resembling wrought iron. However, until the invention of the blast furnace, making large quantities of the metal available, iron was a very costly material. After the invention of the Bessemer and open-hearth processes, steel became economically available. Lately, the basic oxygen furnace and the electric furnace have lowered the cost and improved the quality of steel. This process has been accelerated by the introduction of the induction and consumable electrode furnaces and by electron beam melting techniques. For small-scale controlled melting the laser beam has been used. (See Chap. 2.)

On the basis of weight, ferrous alloys are the predominant structural material at the present time. On the basis of volume, polymers and aluminum compete with the ferrous alloys. A major reason for the predominance of ferrous materials is the enormous variety of different alloys and grades that provides a range of properties not found in any other family of materials. For example, the strength of steel can be varied by an order of magnitude; corrosion resistance can be tailored to the environment; ductility and toughness can be adjusted to design needs; and composition is no longer subject to the tyranny of the equilibrium diagram of the iron-carbon alloy system as a result of rapid solidification technology.

II. PRODUCTION OF IRON AND STEEL

Iron is the name given to pure ferrite, the element Fe, as well as to fused mixtures of this ferrite with large amounts (above 2%) of carbon and substantial amounts (1-3%) of

---

*Compiled by ROBERT S. SHANE from the 2nd Edition, the American Society for Metals, *Metals Handbook* (9th Edition), and many industrial sources.

silicon. Other elements are added to alter the microstructure and properties of iron, as will be noted later. The iron mixtures may be either *pig iron* or *cast iron.* Intermediate between pure iron and cast iron is *steel,* in which none of the carbon is in elemental form. Most of the carbon in steel is combined in the intermetallic compound, iron carbide, or as a carbide of the various alloying elements, such as chromium carbide.

A. Pig Iron

In a blast furnace (Fig. 8.1) ore is smelted by adding limestone and coke and blowing heated air through the mixture. The ore is reduced by the carbon and carbon monoxide from the reaction of the air with the coke. The limestone unites with the impurities in the ore and the ash from the coke to form a molten slag that floats on the surface of the molten iron at the bottom of the furnace. The slag and molten iron are drawn off from the bottom of the furnace, which is operated continuously by addition of ore, limestone, and coke at the top. Coke helps reduce the iron oxide, furnishes carbon to saturate the iron, and ultimately burns in the lower part of the furnace supplying heat to melt the iron and slag. The temperatures shown in Fig. 8.1 are approximate. Normally, the pig iron tapped from the furnace is low in oxides and sulfur but its phosphorus content will be the same as the ore. At least 3-4% carbon is dissolved in the iron; the remainder of the carbon goes off as a gaseous oxide.

Pig iron is weak and brittle and is not used for structural purposes. It is used for making cast iron or steel. If it cannot be used immediately it is cast into blocks, or "pigs" (hence, its name), which can be remelted later. Different ores and different blast-furnace

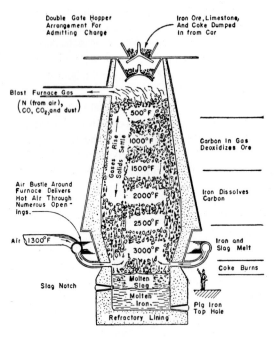

Fig. 8.1 Elementary cross section of blast furnace ready for tapping. Slag notch will be opened and slag drawn off, then iron will be drawn from the taphole. The supports and the cooling systems have been omitted.

procedures produce many grades of pig iron, each one with a special use in the foundry or steel mill (Fig. 8.2). The composition of the various pig irons and those of some of the products made from them are shown in Fig. 8.3.

B. Cast Iron

A large family of ferrous alloys is called cast iron. They are used because of low cost, good casting characteristics, high compressive strength, wear resistance, and good damping qualities. The mechanical properties, such as strength, ductility, and modulus of elasticity, and the physical properties, such as thermal conductivity and damping capacity, are strongly dependent on the structure and distribution of microstructural constituents of which the most important is free graphite.

When cast iron solidifies, the last liquid to freeze is of the eutectic composition. As can be seen in Fig. 8.4 the eutectic contains 4.3% carbon. At least 1.7% carbon must be present if no other alloying element is present for the eutectic transformation to occur. If there is less carbon, there will be no final solidification of eutectic liquid and the solid will not be cast iron. However, impurities and alloys change this value considerably; thus, 2% silicon lowers the possible carbon limit from 1.7 to 1.1%. A good rule of thumb is to estimate the carbon effect as the sum of the total carbon plus one-third of the silicon and phosphorus content.

Gray cast iron contains a substantial portion of the carbon as graphitic flakes (see Fig. 8.5). The usual microstructure is a matrix of pearlite with graphite flakes dispersed throughout. Foundry practice can be varied to yield enhanced properties. Thus the cooling rate can be varied to produce total graphitic carbon or a mixture of iron carbide (cementite) and graphite in the iron. A complete range of properties is possible, as is illustrated in Table 8.1, where the classes of cast iron are given according to ASTM standard A-48.

Completely graphitic gray iron is a soft, readily machinable metal insensitive to notches and high in damping capacity and compressive strength. Tensile strength, ductility, and impact strength are much inferior to steel, however, and there is also an absence of well-defined yield limit and modulus of elasticity, which are limitations for some uses. Irons with a high proportion of the carbon occurring as carbide are hard, brittle, unmachinable, and have good wear resistance. A close-grained iron containing graphite and pearlite is generally the strongest, toughest, and best finished type of cast iron. Medium gray irons contain some ferrite, with graphite and pearlite, resulting in less strength and poorer finish. Open gray iron has low strength and machines soft but is the best of these grades in wear resistance.

The relative amount of free and combined carbon is controlled by variations in composition, melting practice, and casting practice. The most potent single factor is the presence of silicon, which promotes the decomposition of cementite into free iron and graphite. Another important variable is the cooling rate of the iron in the mold. Slow cooling aids the formation of graphite; rapid cooling aids the formation of cementite. Iron, which would be gray throughout if cast in sand, may be given a "white" surface by casting it against chills, which cool the surface rapidly. Ordinarily, it is desired that the amount of combined carbon in gray cast iron be less, than the eutectoid percentage. For this reason, the silicon content is kept up to about 2% and the metal is cooled with chilling.

Sulfur has an effect opposite that of silicon; it stabilizes the carbide and thereby tends to "chill" the iron. This action is prevented if manganese is present in sufficient

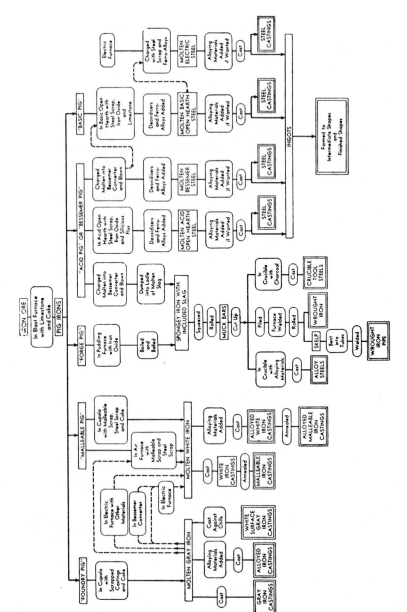

Fig. 8.2 Diagram indicating typical foundry and mill applications for different grades of pig iron. Finishes products are shown in double outline.

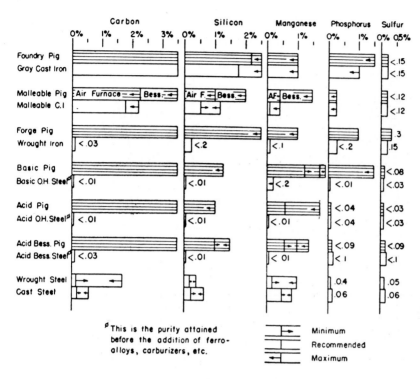

Fig. 8.3 Compositions of various pig irons and the cast irons and steels derived from them. Shaded bars represent pig irons. Unshaded bars represent refined product.

amounts (2 × S% + 0.2%) to combine with all the sulfur to form manganese sulfide. Phosphorus in cast iron occurs almost entirely combined with iron and carbon to form an eutectic called steadite. It forms an embrittling network but does not materially decrease the strength.

White iron has a characteristic white color because it contains no graphite. All the carbon is in the form of cementite, either free or in lamellar pearlite (see Fig. 8.6). White iron may be produced by two methods: (1) by casting gray iron against chills to cool it rapidly and give it a white surface layer, or (2) by adjustment of composition through keeping carbon and silicon low so that there is no free carbon throughout.

This adjustment is made in an air furnace where low-phosphorus pig iron and steel scrap are melted together. The air furnace heats the charge from above in a manner similar to the open hearth, and is used because the entire charge can be dumped in, heated, and carefully controlled. Cupola furnaces alone are sometimes used for making low-quality white castings, but better castings can be made by "duplexing" or "triplexing" processing, which combine the cupola, air furnace, Bessemer converter, and/or electric furnace (see Fig. 8.2).

White cast iron is a very hard, brittle, wear-resistant material. If the silicon is kept below 1% and the carbon is adjusted to about 2% in common cast iron, a white iron of about 400 Brinell will usually result.

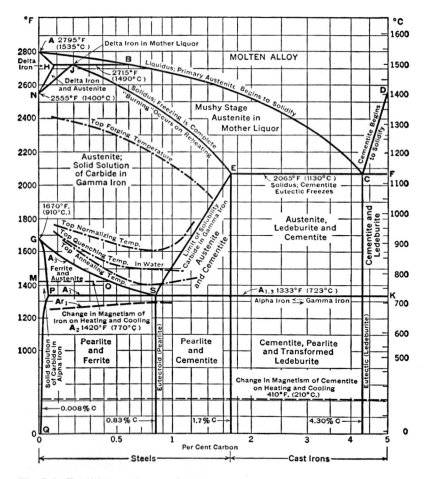

Fig. 8.4 Equilibrium diagram for the iron carbon alloy system.

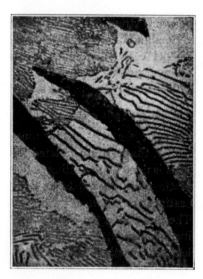

Fig. 8.5 Photomicrograph of gray cast iron, ×1000. White areas are ferrite; mottled areas pearlite; and black areas graphite. (Courtesy of the Turbine Division, Laboratory of the General Electric Company, Schenectady, New York.

Table 8.1 Properties of Typical Gray Cast Irons

Property	ASTM Class					
	20	25	30	35	40	45
Tensile strength (psi)	20,000	25,000	30,000	35,000	40,000	45,000
Compressive strength (psi)	80,000	90,000	100,000	110,000	125,000	135,000
Hardness (Brinell)	110	140	170	200	230	265
Permanent set (psi)	5,000	7,000	9,000	11,000	13,000	15,000
Endurance limit (psi)	8,000–10,000	10,000–12,000	12,000–15,000	14,000–16,000	15,000–20,000	18,000–22,000
Mod. of elasticity (psi)	11,000,000	12,000,000	13,000,000	14,000,000	15,000,000	16,000,000
Torsion modulus (psi)	4,000,000	4,500,000	5,000,000	5,500,000	6,000,000	6,500,000
Toughness (Izod impact)		Less than 1			—	—
Creep, 0.1%/1000 hr						
840°F	—	—	8,000	8,000	—	—
1000°F	—	—	0	0	—	—
Machinability	Excellent	Excellent	Excellent	Excellent	Good	Fair
Wear resistance	Good	Good	Good to exc.	Excellent	Excellent	Good to exc.
Corrosion resistance	Fair	Fair	Fair	Fair	Fair	Fair
Vibration damping capacity	Excellent	Excellent	Excellent	Excellent	Good to exc.	Good
Specific gravity	7.0	7.0	7.1	7.2	7.3	7.4
Melting point (°F)		2,150–2,300				
Thermal exp. ($\times 10^{-6}/°F$)	6.7	6.7	6.7	6.7	6.7	6.7
Thermal conductivity (cgs units)	0.11	0.11	0.11	0.11	0.11	0.11
Electrical res. (microhm-cm)	80–100	90–100	80–100	80–100	80–100	80–100
Magnetic permeability (gauses with H at 100)	9,000	9,000	9,000	9,000	9,000	9,000

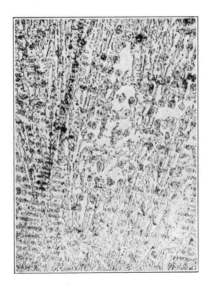

Fig. 8.6 Photomicrograph of white cast iron ×250. White areas are cementite (Fe_3C), and mottled areas are pearlite. (Courtesy of the Turbine Division, Laboratory of the General Electric Company, Schenectady, New York.)

In amounts greater than 3% chromium usually prevents the formation of graphite. The resultant iron not only has the common properties of other white irons, but also has better high-temperature strength, grain-growth resistance, and corrosion resistance.

Small additions of nickel and chromium, such as 4.5% Ni and 1.5% Cr, double the strength and significantly increase the toughness of white castings. They are used when an extremely hard (700 Brinell), tough, and strong material is desired.

Malleable cast iron is made from white cast iron annealed to dissociate the cementite to yield a product that consists almost entirely of ferrite and irregular aggregates of graphite (temper carbon) (Fig. 8.7). The annealing process consists of slow heating to 1600°F, holding at temperature for 25-60 hr according to size, and slow cooling at 10°F/hr.

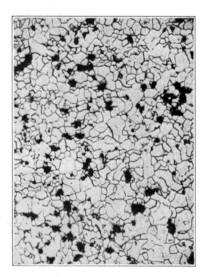

Fig. 8.7 Photomicrograph of malleable cast iron. ×125. Black "rosettes" are temper carbon in a matrix of almost pure iron (white). (Courtesy of the Turbine Division, Laboratory of the General Electric Company, Schenectady, New York.)

Cupola white iron for malleablizing (ASTM A197) results in higher bursting strength under pressure but lower strength than air-furnace, open-hearth, electric furnace, or duplexed irons. The cupola product is therefore often used for valves and pipe fittings. The higher grade products are usually ASTM-A47, grades 32510 and 35018. These grades have, for ½ in. sections, tensile strength over 50,000 psi, yield strength over 32,000 psi, 2 in. elongation of 10% or more, 7-9 ft-lb Izod impact resistance, modulus of elasticity of 25 million psi, an endurance ratio of 0.5, good moldability, machinability equal to gray irons, poor wear resistance, and sensitivity to notches. Thus, they are more like low-carbon steels than cast irons. Their cost is usually higher than gray irons because of the time required for annealing, but they are lower in cost than the softer steels.

Both short-cycle annealing, which interrupts graphitization, and addition of elements that retard graphitization result in malleable irons containing both temper carbon and 0.2-0.6% carbon combined as pearlite or sorbite. These are called *pearlitic malleable irons*. They possess higher strength and wear resistance than grade 35018, at some sacrifice in elongation and machinability. Grades of this type are often sold under special trade names.

Ductile cast iron (nodular iron, spheroidal-graphite cast iron) is cast iron in which the graphite is present as tiny spheres instead of flakes. This is accomplished by small additions (0.03-0.05%) of magnesium and/or cerium (0.005-0.02%) to gray iron. Like gray iron, ductile cast iron can be melted in cupola, electric-arc, or induction furnaces.

Ductile iron is not a single material, but is, instead, a family of materials. Its structure may be modified by alloying heat treatment, as in steel, to produce austenitic, acicular, martensitic, pearlitic, and ferritic structures. Interest to date has centered on the pearlitic-ferritic grades, the four principal types being those listed in Table 8.2. Compositions currently used range as follows: 3.2-4.2% C; 1.0-4.0% Si; 0.1-0.8% Mn; 0.1% P; 0-3.5% Ni; and 0.05-0.10% Mg. The magnesium controls the graphite form but has little influence on the matrix structure. Nickel and manganese add strength at the sacrifice of ductility. Less manganese is used than is in typical gray irons, because sulfur is very low in ductile irons. Silicon is used as an alloying element since it has no effect on size and distribution of the carbon content.

Advantages of this iron are high fluidity, excellent castability, high strength, high toughness, excellent wear resistance, pressure tightness, weldability, higher machinability than gray iron, and an elastic modulus of 25 million psi.

Alloyed cast iron is produced to increase strength and improve corrosion resistance. Since alloying is a process requiring special melting practice, alloying elements are added to good irons to make them better, never to poor irons to make them good.

Nickel in amounts up to 5% is added to cast iron primarily for its effect on machinability; it may be added either to increase the hardness and strength without appreciably decreasing the machinability, or to improve machinability without decreasing hardness and strength. The nickel also promotes corrosion resistance and uniformity of properties throughout variations in casting thickness.

Chromium, up to 3%, has the opposite effect of silicon and nickel; it inhibits the formation of graphite, promotes the formation of carbides, and, in addition, increases the corrosion resistance. Higher percentages of chromium therefore harden the iron by increasing the percentage of combined carbon.

Nickel and chromium added in a 3:1 ratio to 4% total have their graphite- and carbide-forming tendencies neutralize each other, and this results in iron with improved grain refinement, hardness, and strength, and with no impairment of the machinability.

Table 8.2 Representative Mechanical Properties of Commercial Heats of Ductile Iron, 1 in. Bars

Grade	Tensile strength (psi)	Yield strength (psi)	Elongation (%)	Bhn	Usual condition	Remarks
90-65-02	95/105,000	70/75,000	2.5/5.5	115/165	As cast	Good wear resistance
80-60-05	85/95,000	65/70,000	5.5/10.0	195/225	As cast	Combined strength and toughness
60-45-15	65/75,000	50/60,000	17.0/23.0	140/180	Annealed	Max toughness and optimum machinability
80-60-00	85/95,000	65/75,000	1.0/3.0	230/290	As cast	More Mn and P; high strength and stiffness; low shock resistance

Source: Reprinted from The industrial status of ductile iron, *Mech. Eng.,* February 1951, pp. 101-1-8, with permission.

Molybdenum, up to 1.5%, is the most effective alloying element for improving strength; wear resistance is also increased, with a consequent decrease in machinability. Molybdenum improves the uniformity of structure in heavy sections by slowing graphitization and retarding the critical transformation.

Vanadium, up to 0.5%, is a very powerful carbide former and increases the strength and hardness of cast iron considerably, even in the small amount used.

Properties typical of some alloy gray irons are listed in Table 8.3. The lower grades are used for such parts as cylinders, brake drums, and sprockets. Higher grades are used where high impact, wear, and fatigue resistance are required, as in dies, machine-tool castings, and critical machinery components.

1. Austenitic Gray Cast Iron

If iron is alloyed in such a way that the critical transformation temperature is lowered below room temperature, the iron will remain in the form of austenite (γ iron) rather than transforming to ferrite (α iron) as it cools to room temperature.

Austenitic cast iron has excellent corrosion and erosion resistance and good wearing qualities, strength, and hardness.

Nickel, 10-20%, is the only alloying element known that alone causes iron to remain in the austenitic form (at room temperature) without causing its carbide content to increase. Martensitic nickel irons are also available.

2. Inoculated Irons

High-strength irons of such composition that they would ordinarily be white as cast are often *inoculated* in the ladle with a silicon compound to cause graphitization. Typical agents used are ferrosilicon, calcium silicide, Si-Mn-Zr, or Ca-Mn-Si in crushed form. Trade names are used to designate the inoculated products, and the practice may be employed with plain carbon or alloy cast irons. The resulting product is a uniformly dense and machinable casting that might have been "white" in thin sections and soft and porous in heavy sections.

a. Stress Relief

This treatment is accomplished by heating to 800-1100°F, holding at temperature for 30 min to 5 hr, and cooling slowly in a furnace. Along with the relief of internal stress, only slight decrease in hardness or strength occurs at room temperature.

b. Annealing

Iron castings are sometimes softened to facilitate machining, and this is accomplished by annealing. The material is generally heated to 1400-1500°F, and longer time is required than with steel because of the extra carbon to be absorbed and of the retarding effect of silicon. Higher temperatures, up to 1800°F, are needed with the highly alloyed irons. The annealing operation generally increases the free carbon and decreases the strength, although the strength reduction is less for the alloyed irons.

Table 8.3 Properties Typical of Some Alloy Gray Cast Irons

Property	ASTM Class				
	30	35	40	50	60
Tensile strength (psi)	30,000	35,000	45,000	50,000	60,000
Compressive strength (psi)	100,000	110,000	135,000	150,000	175,000
Hardness (Brinell)	170	190	230	250	275
Permanent set (psi)	9,000	11,000	15,000	17,000	19,000
Endurance limit (psi)	15,000	17,000	22,000	25,000	30,000
Mod. of elasticity (psi)	14,000,000	15,000,000	17,000,000	18,000,000	20,000,000
Torsion modulus (psi)	5,500,000	6,000,000	7,000,000	8,000,000	9,000,000
Toughness (Izod impact)	—	—	Less than 1	Up to 2	Up to 2
Creep, 0.1%/1000 Hr					
840°F	—	9,500			
1000°F	—	0			
Machinability	Excellent	Excellent	Excellent	Good to exc.	Good to exc.
Wear resistance	Excellent	Excellent	Excellent	Good to exc.	Good to exc.
Corrosion resistance	Fair to good	Fair to good	Fair to good	Fair to good	Fair to good
Vibration damping capacity	Excellent	Excellent	Excellent	Good to exc.	Good to exc.
Specific gravity	7.1	7.1	7.2	7.3	7.3
Melting point (°F)			2150–2300		
Thermal exp. (×10−6/°F)	6.7	6.7	6.7	6.7	6.7
Thermal conductivity (cgs units)	0.11	0.12	0.12	0.12	0.12
Electrical res. (microhm-cm)	80–100	80–100	80–100	80–100	80–100
Magnetic permeability (gauses with H at 100)	9,000	9,000	10,000	10,000	10,000

c. Hardening

The alloyed irons are often quenched from above the transformation temperature (1500-1600°F) and then tempered to improve their hardness and resistance to wear. About 0.5-0.8% combined carbon is obtained in a pearlitic or sorbitic structure. Quenching usually is performed in oil, although water and air quenching are used for some grades.

The soft gray irons can develop a martensitic structure when quenched, but these irons are not commonly used when hardening is desired. They require a long time at austenitizing temperature to absorb the free graphite, especially if silicon content is high. Austempering, a patented process of quenching and holding at an intermediate temperature, is sometimes used for improving strength and wear resistance of gray iron. The temperature of the salt bath is maintained at 500-555°F, and the casting is held in the bath for ¼-1 hr, depending on size and composition.

Castings of eutectoid percentage combined carbon may be hardened satisfactorily in localized areas by flame and induction methods.

Alloyed irons of special composition can be nitrided to obtain high surface hardness and wear resistance. The process is performed at 950-1100°F in contact with anhydrous ammonia gas, and takes from 20 to 90 hr, depending on the size and depth of hardening desired.

Other properties of cast iron as a class are as follows.

(1) Machinability. The cast irons range in machinability from very good to the most unmachinable of the ferrous alloys, whether their rating be based on tool life, finish, or power required. Annealed permanent-mold iron is the most easily machined because it contains no combined carbon, has finely divided and dispersed graphite flakes, and is free of burned-in sand at the surface. Ductile iron also rates high. In increasing order of difficulty in machining are pearlitic-ferritic irons, pearlitic irons, mottled iron with pearlite and massive cementite, and white iron. The latter is especially difficult to machine because its structure is largely massive carbide, almost unbroken with free graphite flakes. Burned-in sand from the casting process may condemn an otherwise machinable iron, if the machining cuts are not deep enough to get below the sand-included surface.

(2) Weldability. The weldability of cast irons of all types is considered poor. Special processes are required, making it economical only to weld components as a method of salvage or repair. Forge and submerged-melt welding cannot be used. Gas and arc welding are satisfactory, particularly for sections over ¼ in. thick with special rods, if the casting is heated to a red heat before welding and cooled slowly to room temperature thereafter. Bronze welding is satisfactory both for gray irons and for white irons before malleablizing and without preheat, if temperatures of 1500-1600°F are obtained in the process. This method is probably the most used for repair.

(3) Corrosion Resistance. Though not immune to rust, the irons form rust slowly and penetration is quite slow in comparison with the low-alloy steels. High-silicon grades are suited to acid environments as are the high-chromium grades, but both, as well as the unalloyed grades, have poor resistance to caustics. High-nickel austenitic irons, like ni-resist, are resistant to some acids, except nitric, and to stress corrosion in hot, strong caustics if stresses are low.

(4) High Temperature. For pressure vessels a practical limit for gray cast iron service is 650°F. Mechanical properties for other applications are suitable to 800°F, with marked creep resistance to 600°F, and good creep and fatigue resistance to 800°F. Repeated heating of gray cast iron to temperatures above 800°F causes grain growth, distortion, and brittleness. Excessive scaling occurs above 1100°F. Lower carbon, lower silicon, and more chromium increase the permissible temperatures. High-silicon irons, high-chrome irons, and the austenitic nickel cast irons resist grain growth to still higher temperatures.

Wrought iron has been largely supplanted by mild steel. It is a pure iron containing streaks (1-2%) of slag included by the method of preparation.

C. Steel Making

After iron is won from the ore, as noted earlier, the product is a pig whose composition has to be brought to that of the desired end product, a steel. This is done in a steel-making furnace. The principal types now in use are the basic oxygen furnace, the open hearth (acid or basic) furnace, and electric furnace (electric-arc, induction, consumable electrode).

The basic oxygen furnace is a large tiltable vessel lined with basic refractory material, fitted with the accessories to bring in oxygen and charge raw materials (principally molten pig iron, scrap, and fluxes). The oxygen is brought in as a high-velocity jet either from top or bottom. It reacts with carbon and other impurities in the iron to form liquid compounds that escape either in the slag or as evolved gases. A heat of steel can be produced in less than an hour. The basic slag removes some of the sulfur and most of the silicon, manganese, and phosphorus in the iron. Almost any grade of carbon or low-alloy steel can be made in this furnace.

The open-hearth furnace dates to 1856. As in the basic oxygen furnace almost any grade of carbon or low-alloy steel can be made. The outstanding feature of this furnace is the intense heat obtainable by its regenerative process (see Fig. 8.8). The flame burns above a shallow vessel containing the charge of pig iron, steel scrap, iron ore, and flux. As the charge melts, the flux forms a slag, which is raked off into slag pockets at the side of the furnace. Figure 8.3 indicates graphically the purification that takes place in the furnace. Recarburizers and ferro-alloys (high-alloy-content iron alloys) are added after refining to bring the steel to the desired composition, since the carbon, silicon, and manganese contents may be lower than desired. This apparently wasteful procedure of eliminating the impurities and then adding them again in the desired amounts is, in reality, more efficient than trying to decrease them to just the right amounts.

At the high temperatures encountered in the furnace, the furnace lining becomes chemically active. Silica becomes an acid; the other common lining materials, calcium and magnesium oxides, become bases. The slag is also chemically active, but its acidity or alkalinity can be controlled by selection of flux. If the lining is basic, the flux must also be basic, or the slag would destroy the lining; the converse is, of course, true. The basic slag of a *basic open-hearth* furnace will remove most of the silicon, manganese, and phosphorus in the iron, as well as some of the sulfur, but the *acid open-hearth* will remove only the silicon and manganese (phosphorus and sulfur remain). The basic furnace can refine dirtier pig iron and is by far the more popular, even though the lining is more expensive than the acid lining.

Each heat (complete refining cycle) required about 10 hr, and furnaces producing as much as 250 tons per heat are in use.

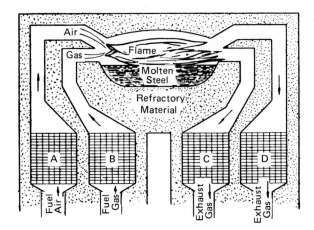

Fig. 8.8 Elementary cross section of the open-hearth furnace. A, B, C, and D are checkerworks for storing heat. When the flame is in the direction shown, C and D are being heated by the burned gases. Every 15 min the direction is reversed; that is air and gas will enter through C and D, which are hot; the flame will burn from the right side, and A and B will be heated by the burned gases. As time goes on, the system becomes hotter and the flame gradually grows more intense. This is known as the regenerative principle.

Electric furnaces (Figs. 8.9 and 8.10) are a boon to the steel industry because higher and more accurately controlled temperatures are obtainable, and because the charge may be exposed to an atmosphere that will not contaminate the metal. The high cost of electric power is the limiting factor and accounts for electric furnaces being used only for high-quality steel production, usually tool steels and alloy steels. Two types of furnaces are in general use for melting and smelting: the arc furnace and the induction furnace.

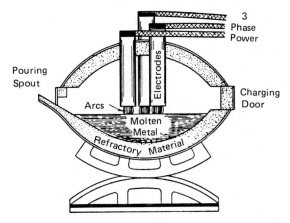

Fig. 8.9 Elementary cross section of three-phase electric-arc furnace. Electricity flows through the metal because the total resistance of this path is smallest. Mechanisms on the electrodes automatically regulate the length of the arcs.

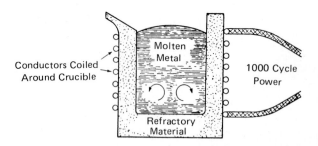

Fig. 8.10 Elementary cross section of induction furnace. Current in the coil induces currents in the metal ás shown by arrows), causing heat. Conductors are water cooled.

Some arc furnaces have been built for single-phase and for dc operation in which the molten metal is one electrode and a carbon stick is the other. Most furnaces, however, are of the three-phase type shown in Fig. 8.9. In these, the metal is heated by current flowing through it and by radiated heat from the arcs. The molten metal is not an electrode; the current flows through this metal only because the resistance of the iron path is much less than the resistance of the air between the electrodes. Suitable regulators keep the electrodes at the right distance above the molten metal.

In the induction furnace the metal is also heated by a current within itself, but the entire circuit of the current is within the metallic charge, being induced by means of a strong magnetic field. Figure 8.10 shows a simple induction furnace in which the magnetic field is set up by a coil of wire around the crucible containing the metal. The higher the frequency of the alternating current in this coil (voltage remaining constant), the better the heating effect. Frequencies of about 1000 cycles are in common use because they are readily obtainable, whereas higher frequencies require expensive generators. The heating effect of the induction furnace is analogous to the heat generated in the secondary of a transformer. The coil on the furnace is usually a copper tube, and water is circulated through it to keep it cool.

The electric furnaces can produce steel of almost any desired purity and composition because various fluxes may be used and the operator can take samples of the melt from time to time, analyze them, and add whatever is needed to bring the steel to the desired composition.

The importance of the electric furnace lies in its convenient use on a small scale; it does not require pig iron in the charge and, hence, can be located away from the blast furnaces; it is less costly to install and operate than basic oxygen or open-hearth furnaces; it can be used for producing specialty steels, such as tool steels, stainless steels, and aircraft quality steels, as well as making the common grades of carbon and alloy steels.

Note should be taken of the importance of the ladle in steel making. Besides conveying the steel from the furnace to the ingot mold or continuous casting machine, it is the place where several alloying elements are added because they would react with the slag or the oxygen in the furnace and would be lost. Additionally, deoxidizers (e.g., silicon and aluminum) are added to the ladle to achieve maximum effectiveness. Finally, the ladle is the place where final analysis of the steel composition is done and any necessary additions to achieve the desired formula can be made.

Deoxidation, already touched upon, is the removal of gases from the steel. This may be done by deoxidizers which end up in the slag, the evolution of gases while the steel is still molten, and by reducing the atmospheric pressure over the steel. Consumable electrode remelted steels usually have exceptionally low levels of dissolved gases as well as nonmetallic inclusions; therefore, they are often specified for critical applications.

The molten steel coming from the refining furnaces may either be cast into the desired shape in which it will be used, it may be cast into an *ingot* and then rolled, forged, hammered, pressed, or machined into the desired shape, or it may be poured into a continuous casting machine.

Most steel mills are arranged in such a way that the refining furnaces take molten pig from the blast furnaces, refine it, and cast it into ingots that are kept hot until ready for the blooming or rolling mill. In this way the iron is not cold from the time it enters the blast furnace as iron ore until it is a finished steel shape. See Fig. 8.11 for a diagrammatic sketch illustrating basic processing steps in a modern mill producing hot- and cold-rolled steel. (Also see Figs. 8.12 and 8.13.)

After steel has been refined in an open hearth or Bessemer, it is apt to contain oxygen in the undesirable forms of iron oxide and dissolved gas. The carburizer added to bring the carbon content of the steel up to the desired amount tends to deoxidize the iron, forming CO_2, which bubbles up and causes "boiling" in the ingot before the steel solidifies. Some of the gas becomes trapped during solidification and forms *blowholes* or voids in the ingot. Steel cast in this way has a characteristic solid rim of metal next to the mold, and is called *rimmed steel.* It has a smooth surface, and products rolled from these ingots (strip, sheet, plate) have an excellent surface finish. Blowholes deep in an ingot may cause no harm if the ingot is to be rolled, because the rolling takes place at welding heat, and the holes are effectively welded together. Blowholes near the surface, however, may break out in cracks which no amount of mechanical work will eliminate.

To minimize the occurrence of blowholes the steel may be *killed* by adding other deoxidizing materials to the ladle just before casting. Aluminum, vanadium, silicon, and manganese effectively deoxidize iron and carbon and are most commonly used.

Steel ingots are cast in a cast-iron mold. The molten metal is introduced into the mold either from the top or from the bottom. The liquid next to the mold walls solidifies first, and the solidification progresses inward. There are definite directional tendencies in the grain growth, and, in addition to coring within the individual grains, the grains differ somewhat in both composition and structure from the surface to the center of the ingot. The contraction of the metal during cooling causes a cavity, or pipe, at the top of the ingot, which extends downward at the center. Usually the ingots are provided with a "hot top," or small reservoir of metal, which feeds the ingot as it solidifies and thus prevents a large pipe from forming in the ingot.

Segregation also takes place as the ingots freeze. Dissolved elements, such as carbon, phosphorus, and sulfur, are less soluble in the solid iron than in the molten iron, and they tend to migrate from the solidifying steel into the neighboring molten steel. The molten steel therefore becomes higher and higher in impurities, and often the metal near the pipe is very "dirty" because it has solidified last. When this may cause difficulty in use, the steel is often purchased under an "ARR" (additional restrictive requirement), which calls for cropping the pipe section more than usual before rolling. Control of segregation is difficult, but it is generally true that the more impure a steel is, the greater the tendency it has to segregate. The carbon-tool steels and spring steels must be cast in small ingots to lessen the chance for carbon migration, before the whole mass solidifies. Some alloy steels are also handled in this way.

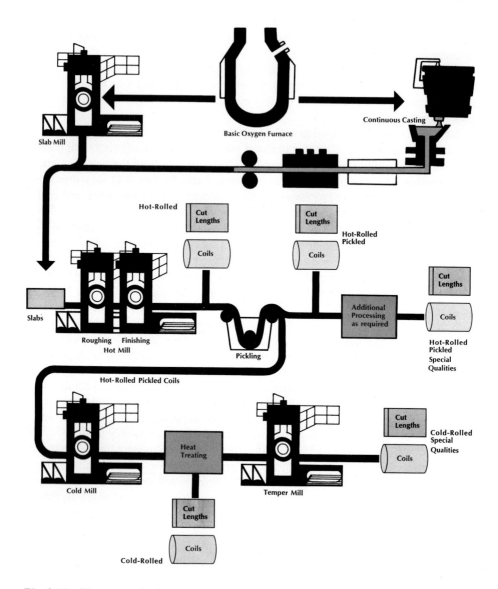

Fig. 8.11 Diagrammatic sketch illustrating basic processing steps in producing Armco hot-rolled and cold-rolled steels. (Courtesy of Armco Inc., Middletown, Ohio.)

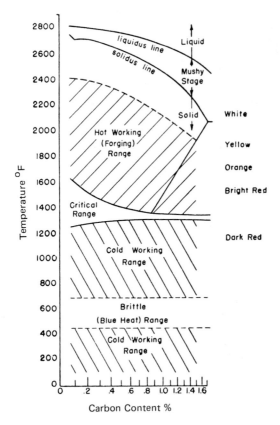

Fig. 8.12 Working temperature ranges for steel.

D. Alloying Elements

Modern steels contain alloying elements added to effect specific changes in the properties of steel. Table 8.4 summarizes the effect of alloying elements, but it must be remembered that interactive effects exist and are complex. They may be beneficial, detrimental, or neutral and should be carefully considered. The effect of alloying elements on the iron-carbon constitution diagram (Fig. 8.4), does not, in general, alter the characteristic features of the diagram. Some shifting of position of the boundary lines occurs.

By combining experimental and theoretical results it has been determined that nitrogen, manganese, nickel, zinc, and copper, in order of decreasing effect, shift line $GS(A_3)$ to the left; chromium and cobalt have little effect; and tungsten, molybdenum, silicon, vanadium, tin, aluminum, beryllium, phosphorus, and titanium, in order of increasing effect by weight percent, shift A_3 to the right. The shift is approximatel pro-portional to the percentage of each element, and is algebraically additive for several elements. Similarly, the line $SE(A_{cm})$ is altered by presence of alloying elements, although the effects are somewhat more complicated. Small amounts of manganese, silicon, chromium, copper, molybdenum, and titanium, in order of increasing effect, and of tungsten, vanadium, and probably nickel move the A_{cm} line uniformly to the left. This continues to high percentages for silicon, manganese, copper, and probably nickel, but the other

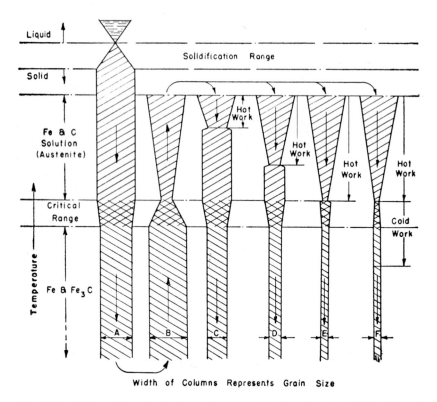

Fig. 8.13   Effect of mechanical working on grain size of steel: (a) liquid steel cooled to room temperature (grains form upon solidification, re-form to smaller grains in critical range); (b) steel heated to high temperature (reduction of grain size in critical range and then growth above this range); (c) small amount of hot working (grains are refined, but grow again to size characteristic of temperature at which hot work is stopped); (d) larger amount of hot working (grains have less chance to grow); (e) hot working to critical temperature (grains cannot grow; maximum refinement is attained); (f) working below critical (the cold work causes no further refinement; only distortion of grains results).

elements tend to introduce new carbide phases. Sufficient alloying element of the latter type causes equilibrium of lower carbon austenite with the new carbide, restricting the region at low temperatures to lower carbon content. The effects of several alloying elements superpose on A_{cm}, but will not add linearly. Effect on the line PS(A_1) has been little studied, but it is known that the line resolves into a three-phase region whose upper and lower boundaries are not necessarily horizontal. Thus, alloying elements in steel displace the phase boundaries shown in Fig. 8.4 but its features may be employed in studying the phase transformations involved.

Steel as received from the mill usually consists of ferrite and cementite. Solid solutions of which α-iron is the solvent may be called *ferrite* (see Fig. 8.14). It forms from the γ (austenitic) phase in slow-cooled alloys, within the field bounded by GSP (Fig. 8.4). Unless it has been hardened by cold working, ferrite is soft (50-100 Brinell) and is usually ductile. It may contain in solid solution manganese, silicon, chromium, nickel, and numerous other elements, but very little carbon.

Table 8.4 Effect of Alloying Elements in Steels

Element	Effects
Carbon	Provides ingredients of the microstructure that directly affect the mechanical and fabrication properties; many other alloying elements are considered because of their effect on the iron-carbon system (see Fig. 8.4)
Manganese	Deoxidizes the melt, reduces hot shortness, combines with sulfur to improve machinability, greatly increases hardenability, strength, and abrasion resistance
Silicon	A principal deoxidizer, it increases the strength without decreasing ductility, increases corrosion resistance; decreases the magnetic hysteresis loss (when used in larger amounts)
Chromium	Increases corrosion oxidation resistance, high-temperature strength, hardenability, and abrasion resistance; a high austenitizing temperature is needed for dissolution of chromium carbides; straight chromium steels tend to be brittle
Nickel	Improves toughness and shock resistance, strengthens, renders high-chromium steels austenitic, and improves resistance to heat and corrosion; nickel is very effective in combination with chromium and molybdenum in achieving high strength, toughness, and hardenability
Molybdenum	Increases hardenability, minimizes temper embrittlement, increases high-temperature tensile and creep strength, fosters formation of bainite (dispersion of iron carbide in ferrite) from austenite by slow continuous cooling
Copper	Improves corrosion resistance, detrimental to hot working
Vanadium	Inhibits grain growth during heat treatment, increases strength, toughness, hardenability (in small amounts. 0.05% V)
Niobium (columbium)	Lowers the transition temperature, raises strength, imparts fine grain size, retards tempering, increases elevated temperature strength, forms very stable carbides thereby decreasing hardenability and tendency to stress corrosion cracking
Tantalum	See niobium
Titanium	Is a powerful deoxidizer and is particularly useful in boron steels
Zirconium	Controls shape of inclusions, thereby increasing toughness; is a powerful deoxidizer
Cerium	See zirconium
Magnesium	See zirconium
Boron	Increases hardenability without loss of ductility, formability, or machinability of steel in the annealed condition
Lead	Increases machinability
Aluminum	Deoxidizer, controls grain size growth thereby promoting toughness
Calcium	Is a deoxidizer that promotes toughness and improves machinability over steels deoxidized with silicon or aluminum
Phosphorus	Increases strength and hardenability but severely decreases ductility and toughness; increases machinability and corrosion resistance; use limited to special cases
Selenium	Increases machinability
Sulfur	Increases machinability when manganese is added; otherwise its presence is very detrimental to strength and impact resistance
Nitrogen	Increases strength, hardness, machinability but it decreases ductility and toughness; in aluminum-deoxidized steels, nitrogen forms aluminum nitride particles that control size of inclusion particles thereby improving strength and toughness
Oxygen	Seriously reduces toughness
Hydrogen	Embrittles steel to produce flaking during cooling. Heating during fabrication drives out most of the hydrogen.

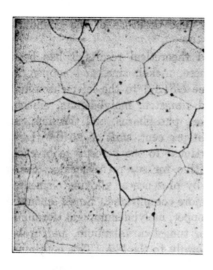

Fig. 8.14 Ferrite in commercially pure (Armco) iron ×100. (By E. D. Reilly. Courtesy of the Turbine Division, Laboratory of the General Electric Company, Schenectady, New York.)

Cementite is the intermetallic compound of carbon and iron, Fe_3C. It is very hard (approximately 1400 Brinell) and brittle, and appears in the annealed steel as parallel plates (lamellar layers), as rounded particles (spheroids), or as envelopes around the pearlite grains (see Fig. 8.15). This phase is formed in slowly cooled solid alloys within the field ESK. In alloy steels, it may contain, besides iron and carbon, one or more of the added "carbide-forming" elements.

Annealing, normalizing, and hardening of steel require formation of austenite, the solid solution of which γ-iron is the solvent. This is accomplished by heating into the range above GSE, where austenite is a stable phase. The ferrite and carbide react at their interface to form nuclei of austenite, which grow, absorbing the ferrite and cementite. Nucleation and growth require time, and proceed more quickly at higher temperatures. An increase of 50 to 100 F appears to speed the reaction tenfold. Since carbides, particularly of alloying elements dissolve more slowly than ferrite, higher temperatures are employed for reasonable austenization time when alloy carbides are present in moderate quantities.

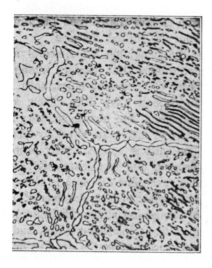

Fig. 8.15 Spheroidal, lamellar, and boundary cementite in annealed high-carbon steel ×1000. (By E. D. Reilly, Courtesy of the Turbine Division, Laboratory of the General Electric Company, Schenectady, New York.)

1. Structures Resulting from Continuous Quenching and Tempering

Nonequilibrium cooling of austenite retards the normal phase transformation indicated by the constitution diagram. The principal structures resulting from various rates of continuous cooling and from subsequent reheating, or tempering, treatments are indicated in Fig. 8.16. The temperature at which the transformation of austenite occurs is displaced from A_1 during cooling rates used in commercial heat treating. Lines indicating the displaced transformation temperature are designated by subscripts r* for cooling (and c for heating). A_{r1} is included in Fig. 8.4.

Austenite may exist at room temperature only when the normal $\gamma \rightarrow \alpha$ transformation has been fully suppressed. Even when plain high-carbon alloys are quenched very drastically, the resulting structure contains only a small fraction of austenite. Therefore austenite rarely appears in any quantity in plain or low-alloy steels. Manganese, nickel, and certain other elements are very effective in suppressing the $\gamma \rightarrow \alpha$ transformation, and even slow-cooled alloys containing certain relatively high percentages of these elements are completely austenitic at room temperature (see Fig. 8.17). Austenite is generally soft and quite ductile, unless cold worked. At elevated temperatures the austenitic steels are stronger and less ductile than the ferritic. Austenite is also more dense than ferrite, its electrical resistance and thermal coefficient of expansion are higher, and it is practically nonmagnetic (paramagnetic). Cold plastic working or chilling in liquid air will often cause "retained" austenite to transform spontaneously to martensite.

Martensite, the principal structure in fully hardened steels, is formed by a shearing mechanism (rather than nucleation and growth through diffusion, as austenite is formed) in carbon steels by fast continuous cooling of austenite to temperatures usually of 400-

---

*The subscripts r and c come from the French words *refroidir* and *chauffer,* which mean recooling and heating, respectively.

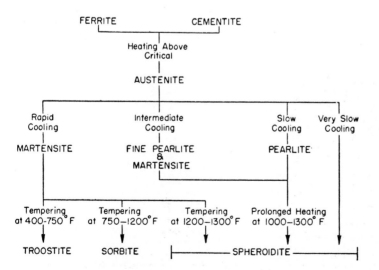

Fig. 8.16 Diagram showing the formation of the various structures of plain carbon steels. (Adapted from *The Working, Heat Treating, and Welding of Steel,* by H. L. Campbell, John Wiley & Sons, New York, 1941, p. 61.)

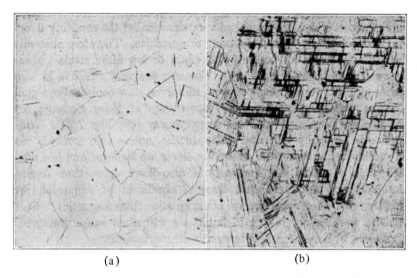

(a)   (b)

Fig. 8.17 (a) Austenite in 18 Cr-8 Ni stainless steel ×250 (by E. D. Reilly). (b) Austenite in "nonmagnetic" (16% Mn) cast steel ×100. (By Joyce Thomas, Courtesy of the Turbine Division, Laboratory of the General Electric Company, Schenectady, New York.)

600°F or lower. It has a tetragonal crystal structure, may be considered as tetragonal ferrite greatly supersaturated with carbon. Some of the carbon may have precipitated during cooling as finely dispersed cementite among the "feathers" or "needles" of the tetragonal structure. The martensitic areas are very hard, 500-1000 Brinell, according to carbon content and fineness of structure. They may contain various percentages of "retained" (untransformed) austenite and also "excess" ferrite or cementite, according to the composition of the alloy and its treatment. Several photomicrographs of these combinations are illustrated in Fig. 8.18. "Retained" austenite in martensitic structure is not easily distinguished microscopically, but its presence may be established by hardness, density, or magnetic tests. Martensite is less dense than pearlite or spheroidite, owing

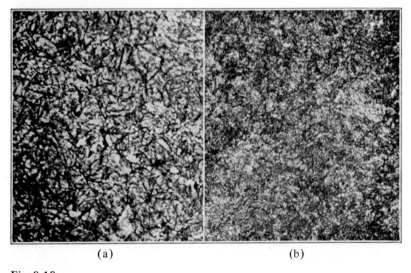

(a)   (b)

Fig. 8.18

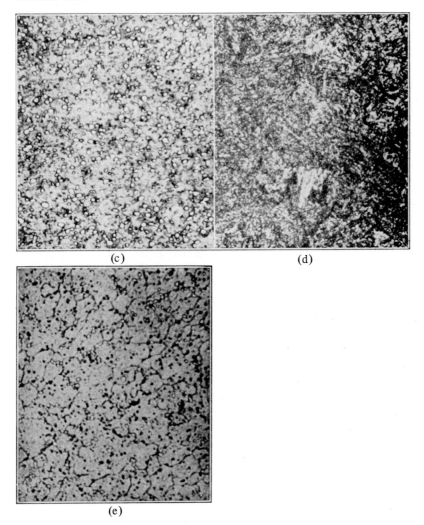

(c)   (d)

(e)

Fig. 8.18   (a) Martensite in 0.50% C steel, 700 Brinell ×500. (b) Martensite with excess cementite in high-speed steel, hardened and tempered ×500. The white areas are cementite. (c) Martensite with excess cementite in 130% C steel safety razor blade, 800 Brinell ×1000. The white areas are cementite. (d) Martensite with excess ferrite, 500 Brinell ×500. (e) Martensite and retained austenite in high-speed steel, hardened but not tempered 800 Brinell ×500. (a, b, and e by E. D. Reilly. c by R. N. Gillmor. d by J. J. Vrooman. Courtesy of the Turbine Division, Laboratory of the General Electric Company, Schenectady, New York.)

to the metastable tetragonal lattice. Therefore, an increase in volume takes place in steel when it is fully hardened.

   Pearlite consists of alternate layers of ferrite and cementite in the proportion 87:13 by weight. The areas of pearlite are formed spontaneously from slow-cooling austenite through the temperature of the line PSK (Fig. 8.4). A photomicrograph of pearlite is illustrated in Fig. 8.19. Note that pearlite is the *eutectoid structure* of two phases in iron-carbon alloys.

Fig. 8.19 Pearlite in eutectoid (0.83 C) steel ×1000. (By D. B. Blackwood. Courtesy of the Turbine Division, Laboratory of the General Electric Company, Schenectady, New York.)

Steels containing less than 0.83% carbon are called *hypoeutectoid* steels, and those that contain more are called *hypereutectoid*. This terminology applies only to plain and low-alloy steels. With high-alloy steels the eutectoid composition is altered and the structure may not even exist. To distinguish between the ferrite and cementite phases of the pearlite structure, the separate ferrite in hypoeutectoid steels and the corresponding cementite of hypereutectoid steels are known as "free" or "excess" ferrite and cementite, respectively. Photomicrographs of pearlitic steels with excess ferrite and cementite are illustrated in Fig. 8.20. An electron micrograph of pearlite with excess cementite is illustrated in Fig. 8.21. The properties of pearlite are, of course, a composite of those of ferrite and cementite.

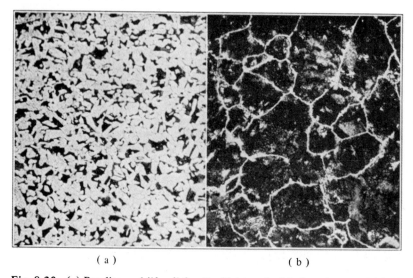

(a)   (b)

Fig. 8.20 (a) Pearlite and "free" ferrite (light grains) in hypoeutectoid (0.30% C) steel ×100. (b) pearlite and "free" cementite (light areas) in hypereutectoid (1.0% C) steel ×100. (a by W. G. Conant. b by D. B. Blackwood. Courtesy of the Turbine Division, Laboratory of the General Electric Company, Schenectady, New York.)

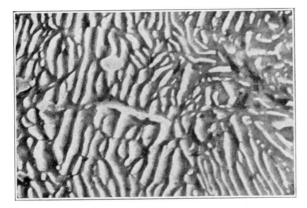

Fig. 8.21 Electron micrograph of 1.1% C steel showing excess cementite and pearlite. The light regions are cementite. ×10,000 (original ×5000). (Courtesy of Research Laboratory, General Electric Company, Schenectady, New York.)

With faster cooling, pearlitic structure may continue to form down to temperatures of 900°F, the spacing between layers becoming finer at the lower temperatures of formation (see Fig. 8.22). The hardness increases with the fineness of this spacing, ranging from 170 to perhaps 400 Brinell.

Spheroidite is the structure in which cementite takes the form of rounded particles, or spheroids, instead of plates. Just as the spacing varies in pearlite so does the size of the spheroids, as illustrated in Fig. 8.23, the finer-grained material being the harder and stronger. Spheroidite is softer and more ductile than pearlite, but not as freely machinable. It may form directly from slow-cooling hypereutectoid austenite, which contains undissolved excess cementite, present because the alloy has not been heated above the

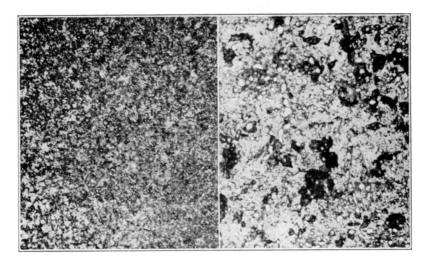

Fig. 8.22 *Left,* fine pearlite (dark) and martensite (light) in incompletely hardened high carbon steel ×100. *Right,* fine pearlite under high magnification. This structure is sometimes called "primary troostite." ×1000. (By E. D. Reilly. Courtesy of the Turbine Division, Laboratory of the General Electric Company, Schenectady, New York.)

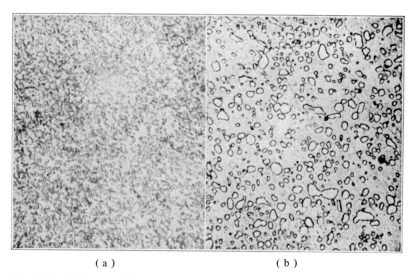

(a)   (b)

Fig. 8.23 (a) Spheroidite, fine spheroids—C 0.9, Mn 2.0, Mo 1.0. Brinell 230 X 1000. (b) spheroidite, large spheroids—C 1.60, Cr 12, Mo 1.0, Va 0.25. 216 Brinell. X1000. (By R. N. Gillmor. Courtesy of the Turbine Division, Laboratory of the General Electric Company, Schenectady, New York.)

line SE, Fig. 8.4. Also, lamellar pearlite may be spheroidized by heating the steel for relatively long periods of time just below the eutectoid transformation (Fig. 8.24). Spheroidite may also be produced by heating martensite or the temper structures of martensite in the range just below the eutectoid transformation.

If martensite is reheated (tempered) at temperatures (in plain steels) between about 400 and 750°F, a softer, more ductile structure known as *troostite* is formed. This structure is dark etching (see Fig. 8.25) and apparently consists of submicroscopic particles of cementite in ferrite.

The name "primary troostite" may be applied to the fine pearlite structure obtained on direct cooling and illustrated in Fig. 8.22, but the name, troostite, is more

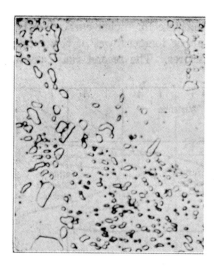

Fig. 8.24 Spheroidite in hypoeutectoid (0.20% C) steel formed from pearlite after 500 hr at 1250°F. X1500. (By E. M. Eoff. Courtesy of the Turbine Division, Laboratory of the General Electric Company, Schenectady, New York.)

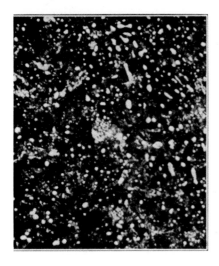

Fig. 8.25 Troostite in plain (1.10% C) steel hardened at 1475°F, quenched in brine, and tempered 1 hr at 550°F. ×1000. (By E. D. Reilly. Courtesy of the Turbine Division, Laboratory of the General Electric Company, Schenectady, New York.)

generally applied to this "secondary," or tempered, product. Some prefer to think of the structure as tempered martensite of a measured hardness because of the otherwise arbitrary degree of grain growth and agglomeration that may characterize the same structure.

When troostite (in plain carbon steel) is heated in the temperature range 750-1100°F, it changes by indistinguishable degrees into a structure known as *sorbite,* in which the cementite has grown until it has a distinctly granular appearance (see Fig. 8.26). It is softer and more ductile than troostite. Tempering at still higher temperatures causes continued grain growth, until the cementite forms the larger spheroids characterized by spheroidite.

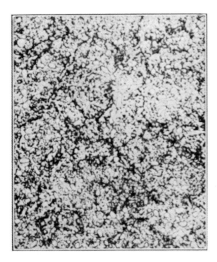

Fig. 8.26 Sorbite in plain (1.10% C) steel hardened at 1475°F, quenched in brine, and tempered 1 hr at 1000°F ×1000. (By E. D. Reilly. Courtesy of the Turbine Division, Laboratory of the General Electric Company, Schenectady, New York.)

2. Structures Resulting from Isothermal Quenching

In 1930, E. C. Bain and E. S. Davenport* published a paper on the first important investigations of constant temperature (isothermal) transformation of austenite. The method used consisted of quenching a large number of specimens of a steel from above the transformation temperature into a bath of molten lead, lead alloy, or molten salt maintained at a constant subcritical temperature. Specimens were withdrawn after various time intervals, and quenched in iced brine to prevent further change. The structures were then examined microscopically to determine the start and completion of transformation at the given temperature. Similar tests at other temperatures enabled completion of diagrams like Fig. 8.27. Because of their shape they were called s curves, but the designations T-T-T curves or time-temperature-transformation curves are more often used.

Figure 8.27 illustrates the times at which isothermal decomposition begins and ends in a steel containing a moderate percentage of carbide-forming elements. There are two ranges of rapid isothermal decomposition of austenite. The first occurs at about

*Trans. Am. Inst. Mining Met. Engrs., Vol. 90, 1930, pp. 117-154. See also *Atlas of Isothermal Diagrams,* Research Laboratory, United States Steel Corp., Kearny, New Jersey, 1943.

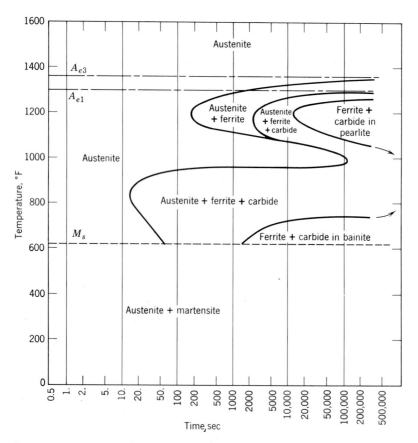

Fig. 8.27 Typical isothermal transformation diagram for steel containing carbide-forming elements (AISI 4340).

1150-1250°F and is associated with decomposition of austenite into lamellar pearlite and proeutectoid ferrite (or cementite for higher carbon steels). Coarse pearlite is formed at the upper temperatures of this range; fine pearlite at the lower temperatures. The second range occurs at 750-950°F and is related to decomposition of austenite into *bainite*. The bainite structure was named after E. C. Bain; it is formed only by isothermal transformation. It is composed of ferrite and cementite in a dispersion too fine to be resolved by the microscope. The structure produced at the lower temperatures of this range has greater hardness and strength than results at the higher temperatures of the range.

A third range is indicated on Fig. 8.27 below the M_s line at 620°F. This range is associated with the decomposition of austenite into martensite. The M_s line indicates the start of this transformation. It is influenced greatly by composition and varies from 1000°F to below room temperature for different steels. The martensite reaction starts, on cooling, at the M_s temperature independent of cooling rate. Once partial transformation has occurred, some stresses are set up in the martensite and the austenite. Further transformation involves establishing additional stresses and hence the energy associated with them. This can occur only by lowering the temperature below the point where transformation began; thus martensite forms *only* under *continuous cooling.*

In steels not containing carbide-forming elements, such as plain carbon steels, the pearlite and bainite ranges have not been segregated. It is possible that these ranges overlap for such steels, or that experimental difficulties have failed to show a region of slow transformation between them. Fig. 8.28 illustrates the T-T-T curve for such a steel. The absence of separate pearlite and bainite "noses" is to be noted.

Tempering of bainite beyond the stage of the nonuniformity of the original precipitate results in structures difficult to distinguish from tempered martensite. The relation of temperature and time for spheroidization and growth in martensite seems to apply also to tempered bainite.

E. Steel Castings

These castings are generally divided into five classifications:

1. Low-carbon steels (carbon content below 0.20%)
2. Medium-carbon steels (carbon content between 0.20 and 0.50%)
3. High-carbon steels (carbon content above 0.50%)
4. Low-alloy steels (alloy content less than 8%)
5. High-alloy steels (alloy content above 8%)

All the carbon steels contain less than 1.7% C, along with other elements, usually in the range of 0.50-1.00% Mn, 0.20-0.70% Si, 0.05% maximum P, and 0.06% maximum S, plus small amounts of other elements from the scrap used. Most of the castings produced are of the medium-carbon grade. Lower and higher carbon content grades are considered only for special product applications.

Steel castings are considered of alloy grade if the residual or added alloying elements exceed 1% Mn, 0.70% Si, 0.50% Ni, 0.50% Cu, 0.25% Cr, 0.10% Mo, and 0.05% V, W, Al, or Ti. Phosphorus and sulfur are limited as for the carbon steel casting grades. The low-alloy class is the second largest production of all the steel-casting groups. High-alloy castings are utilized for special heat and corrosion resistance, and they find limited specialized applications.

At least 75 different compositions are used commercially in these classifications. The railroad and transportation industries use 35% of the castings produced. Other large

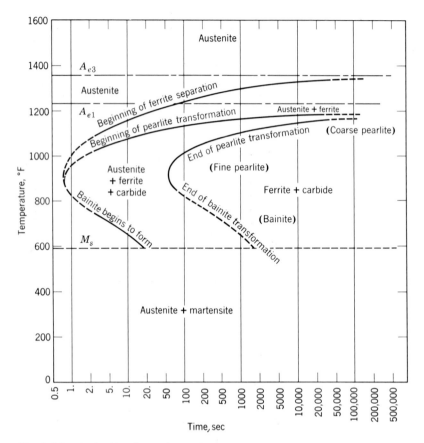

Fig. 8.28 Typical isothermal transformation diagram for steel not containing carbide-forming elements (AISI 2340).

applications are machinery parts, 25%, rolling-mill components, 15%, material-handling equipment, 8%, valves and pressure fittings, 7%, and road and building construction, 5%.

The mechanical properties of the cast steels are best summarized in terms of properties of the wrought steels. At equal hardness the cast grades have equal strength to wrought or welded grades. Ductility is often lower than for wrought material tested in the direction of working, but is higher than for transverse directions. The same differences, although usually more marked, occur with impact strength. The reversed bend endurance ratio to tensile strength varies from 0.42 to 0.50 for the cast steels, depending on the composition and heat treatment. The range of mechanical properties available is indicated by Table 8.5, taken from two ASTM minimum specifications.

Other properties, too, are similar to the wrought steels of the same composition, structure, and hardness. For wear resistance, cast steels of 0.50% C have given excellent service, as have low-alloy cast steels with Cr, Cr-Mo, Ni-Cr, Cr-V, and medium Mn, all with 0.40% C or more. For corrosion resistance, copper additions improve resistance to atmospheric attack, and high-alloy cast steels of the Cr or Cr-Ni type are used for the more corrosive conditions. Alloys with 4-6.5% Cr, especially with additions of 0.75-1.25% W or 0.40-0.70% Mo and 0.75-1% Ti, have good strength and considerable resistance to scaling

Table 8.5 ASTM Requirements for Steel Castings, Mechanical Properties

Grade	Tensile strength, minimum (psi)	Yield point, minimum (psi)	Elongation in 2 in., minimum (%)	Reduction of area, minimum (%)
	ASTM, A 27[a]			
60-30	60,000	30,000	24	35
65-35	65,000	35,000	24	35
70-36	70,000	36,000	22	30
	ASTM, A 148[b]			
80-40	80,000	40,000	18	30
80-50	80,000	50,000	22	35
90-60	90,000	60,000	20	40
105-85	105,000	85,000	17	35
120-100	120,000	100,000	14	30
150-125	150,000	125,000	9	22
175-145	175,000	145,000	6	12

[a]Carbon steel for miscellaneous industrial uses.

[b]Alloy steel for structural purposes. See other ASTM specifications for other grades and uses.

to 1000°F, but they are not comparable to the high-alloy nickel-chrome types. Weldability, hardenability, and machinability all compare with wrought steels of similar composition and strength, although the cast surface skin is itself difficult to machine unless a deep, or hogging, first cut is used.

Heat treatments of steel castings are quite similar to those to be discussed later for wrought steels of similar shape and composition. The operations performed are stress relieving, annealing, normalizing, quenching, tempering, and case hardening.

1. Stress Relieving

For relief of internal stress temperatures from 900 to 1000°F are usually used, and the castings are slowly cooled after uniform heating to this value. Stresses are usually reduced by 90% or more, and there is little effect on strength or ductility.

2. Annealing

In this operation castings are heated at a rate of 150-200°F/hr to about 200°F above the transformation temperature. After soaking at this temperature, the castings are slowly furnace cooled to 1000°F, after which the rate of cooling is accelerated. The annealing process has a twofold purpose: (1) to break up large, brittle, dendritic structures formed during solidification and slow cooling, and (2) to promote diffusion of segregated phases. Good ductility and lower tensile, yield, and impact strength result. Usually only carbon steels are given this treatment.

3. Normalizing

Castings heated to above the transformation temperature, as in annealing, and then cooled in air are said to be normalized. This operation produces somewhat higher strength properties than in annealing. Light sections have higher values than the heavier sections, which cool more slowly because of their mass. On higher alloy cast steels this operation is known as homogenizing. It is employed for these materials to promote diffusion of carbon and to break up the cast structure prior to quenching operations for hardening.

4. Quenching and Tempering

In quenching cast steel it is usually heated to 100°F above the transformation temperature, held at this temperature for ½-1 hr to assure complete austenitization, and then quenched directly in water or oil, whichever is needed to develop martensite structure.

Tempering consists of reheating to temperatures of 400-1275°F, holding at temperature for a 1 hr/in. thickness or 2 hr minimum (more for low temperatures), followed usually by furnace cooling the castings to some intermediate temperature and air cooling to room temperature. The tempering temperatures chosen depend on the hardness and strength required. Lower temperatures develop the maximum strength; high temperatures the maximum ductility. Stresses caused by nonuniform cooling in the quench are relieved by this operation.

When the size and/or design are such that direct quenching in a room-temperature bath might cause severe stresses and cracking, the castings may be given an interrupted, or timed, quench. This quench involves removing the castings from the bath when heavier sections have cooled to 500°F. The castings are then allowed to equalize temperature in air (with thinner sections which may be as cool as 250-300°F) and are then further cooled to assure completeness of the martensite transformation before tempering. Quenching into a salt bath at 500°F may also be used in this way.

Case-hardening operations may be employed for cast steels in the same manner as mentioned in discussion of wrought steels.

F. Wrought Steel

Wrought steel will be considered here in three categories: (1) plain carbon steels, (2) low-alloy steels, and (3) high-alloy steels. Plain carbon steels contain carbon as the most important alloying element. Low or mild carbon designates grades generally with 0.30% or less carbon. Medium-carbon grades contain 0.30-0.70% carbon, and high-carbon grades contain from 0.70 to 1.7% carbon. These steels also contain small amounts, usually less than 0.5%, of impurities like sulfur and phosphorus, and of deoxidizing agents, such as magnesium, silicon, and aluminum. Smaller percentages of oxygen, nitrogen, copper, nickel, chromium, molybdenum, lead, and arsenic may be present because of the raw materials utilized.

The alloy grades are thought of as low-alloy steels if one or more of the alloying elements present are within the percentages listed in Table 8.6. Higher alloy steels contain, of course, more alloying element or elements than one or more of the limits in Table 8.6.

Many of the grades in these three categories are produced according to chemical and physical specifications prepared by such groups as the American Iron and Steel Institute (AISI), the American Society of Testing and Materials (ASTM), and the Society of Automotive Engineers (SAE). Often the producers have their own identification systems or trade names for their products, and many of the large industrial consumers have their own specification systems for purchasing steels. As a result, there are many identification systems in use in engineering practice. The AISI system is probably the most widely

Table 8.6 Alloying Limits for Low-Alloy Steels

Alloying element	%
Chromium	0.15 -3.50
Copper	0.15 -1.50
Manganese	0.70 -2.0
Molybdenum	0.10 -0.50
Nickel	0.4 -5.5
Phosphorus	0.065-0.15
Silicon	0.50 -5.0
Vanadium	0.15 -0.25

adopted. It covers the plain carbon and low-alloy steel categories and will be used for convenience in identification of illustrative data to be included with succeeding paragraphs.

The AISI numbering system for steels is essentially the same as the old SAE numbering system, and the numerals used are standardized for both. The first digit of the usual four-numeral designation indicates the type to which the steel belongs. The second digit for the simple low-alloy steels indicates the approximate percentage of the predominant alloying element. The third and fourth digits (and the fifth for one group) usually indicate the mean percentage carbon content. Thus the designation 2330 indicates a nickel steel, approximately 3% (3.25-3.75%) nickel, and an average of 0.30% (0.28-0.33%) carbon content. The basic numbers for the various grades of carbon and alloy steels are given in Table 8.7.

A letter prefix is added to these numerals to indicate the manufacturing process employed in producing the steel. The prefixes used and the processes indicated are:

B: denotes acid Bessemer carbon steel.
C: denotes basic open-hearth carbon steel.
CB: denotes steel that may be either Bessemer or open hearth at the mill's option.
E: denotes electric-furnace alloy steel.
No prefix is used for open-hearth alloy steels.

A number of triple-alloy grades were designed in 1943 to alleviate shortages of nickel, chromium, and molybdenum. They were known as the National Emergency, or N.E., steels. Some of them have continued in the post-World War II period, and it is to be expected that others or modifications will continue to be important in the years ahead.

Subsequently, steel specifications have been developed that designate hardenability limits and allow composition variations (rather than grain-size variations) to achieve the specified hardenability. These steels are known as "H" steels, the H serving as a suffix to AISI numbers. Only a few of these specifications are available at present, but this practice may also be expected to expand for selected low-alloy steels.

The Unified Numbering System for steels designates chemical composition. It consists of a letter and five digits. The letter K designates wrought carbon steels. The next four digits are the AISI-SAE grade, and the final digit defines a specific alloy.

Table 8.7 Combined AISI and SAE Standard Designations for Carbon and Alloy Steels

Carbon steels

10xx	Nonresulfurized basic open-hearth and acid Bessemer carbon steel
11xx	Resulfurized basic open-hearth and acid Bessemer carbon steel

Low-alloy steels

13xx	Manganese (1.60-1.90%) steels
23xx	Nickel (3.25-3.75%) steels
25xx	Nickel (4.75-5.25%) steels
31xx	Nickel (1.10-1.40%)–chromium (0.55-0.75 or 0.70-0.90%) steels
33xx	Nickel (3.25-3.75%)–chromium (1.40-1.75%) steels
40xx	Molybdenum (0.20-0.30%) steels
41xx	Chromium (0.80-1.10%)–molybdenum (0.15 or 0.18-0.25%) steels
43xx	Nickel (1.65-2.0%)–chromium (0.40-0.60 or 0.70-0.90%) steels–molybdenum (0.20-0.30%) steels
46xx	Nickel (1.65-2.0%)–molybdenum (0.15 or 0.20-0.27 or 0.30%) steels
48xx	Nickel (3.25-3.75%)–molybdenum (0.20-0.30%) steels
50xx	Chromium (0.20-0.35 or 0.55-0.75%) steels
51xx	Chromium (mean % 0.80, 0.90 or 1.05) steels
5xxxx	Chromium steel (mean % 0.50, 1.00, or 1.45) with carbon 0.95-1.10%
61xx	Chromium (mean % 0.80 or 0.95)–vanadium (0.10 or 0.15% min) steel
86xx	Nickel (0.40-0.70%)–chromium (0.40-0.60%)–molybdenum (0.15-0.25%) steels
87xx	Nickel (0.40-0.70%)–chromium (0.40-0.60%)–molybdenum (0.20-0.30%) steels
92xx	Manganese (0.85% mean)–silicon (1.8-2.2%) steels
93xx	Nickel (3.0-3.50%)–chromium (1.0-1.4%)–molybdenum (0.08-0.15%) steels
94xx	Manganese (1.0% mean)–nickel (0.30-0.60%)–chromium (0.3-0.5%)–molybdenum (0.08-0.15%) steels (formerly N.E. type)
97xx	Nickel (0.40-0.70%)–chromium (0.1-0.25%)–molybdenum (0.15-0.25%) steels (formerly N.E. type)
98xx	Nickel (0.85-1.15%)–chromium (0.70-0.90%)–molybdenum (0.20-0.30%) steels (formerly N.E. type)
99xx	Nickel (1.00-1.30%)–chromium (0.40-0.60%)–molybdenum (0.20-0.30%) steels (formerly N.E. type)

1. Carbon Steel

Plain carbon steels constitute about four-fifths of the steel produced. They are cheaper than the alloy steels and consequently are used wherever their properties are adequate. The properties range from very soft to very hard, and these steels are suitable for a large proportion of industry's needs.

It is convenient to divide the carbon steels into two groups. One is the so-called free-cutting steels. The second is the remaining plain carbon steels not especially selected for their machinability.

The free-cutting steels consist of the three Bessemer grades B1111, B1112, and B1113, which differ mainly in sulfur content, and of the resulfurized open-hearth steels from C1108 to C1151. It is the high sulfur in these grades that promotes machinability in bar or plate and aids wear resistance in some applications. Where loading is not heavy or rapidly repeated, Bessemer or resulfurized open-hearth steel is satisfactory. Screw stock, sheets, tin plate, tubing, pipe, concrete reinforcing, and lightweight rail are commonly made of these steels. The high sulfur and phosphorus make these steels unsatisfactory for many applications, however. They have poor welding characteristics, have low ductility and malleability, are poor in fatigue, and tend to be both hot short and cold short, making them brittle under impact loads.

The second group consists of the basic open-hearth steels from C1006 to C1095. They have higher purity and are preferred for heavy rails, structural shapes for buildings and bridges, boiler plate, drawing sheet, rivets, springs, shafts, axles, gears, and high-carbon tools. Typical uses for wire, bar, plate, and sheet, or strip of these grades depends on the carbon content. Table 8.8 lists typical applications for the various levels on carbon content.

With the exception of carburizing and intermediate annealing of cold-worked material, low-carbon (under 0.30% C) steels are seldom heat treated. Medium-carbon steels (0.30-0.70% C) are frequently heat treated for hardening, improving ductility, or machinability. The high-carbon steels (0.70% C and above) are almost always quenched and tempered, because typical applications for these grades require high hardness. Both the medium-carbon and high-carbon steels are shallow hardening. That is, martensite forms only to the depth of the section that can be quenched past the "gate" of the T-T-T curve in a second or less. The larger pieces may develop only mixed ferrite and cementite, even at the surface, because of the low rate at which the large amount of heat in the volume of material can be removed from the surface.

2. Low-Alloy Steels

The first investigations on the effect of alloying elements in steel were made from 1875 to 1890, but the use of alloyed steels found little application until 1910, when reduced costs of alloys made their use practicable. The stimulus of creating materials for military requirements in the two world wars that followed greatly aided their development. Even more important, perhaps, has been the development and use of these steels for automobile applications and later for airplanes, locomotives, machine tools, and other high-efficiency products.

Today there are hundreds of varieties of low-alloy steels. Compositions occasionally involve only one or two alloying elements other than carbon, but varieties with five or more alloying elements are not uncommon. The alloying elements, when present, are limited to the ranges shown in Table 8.6 for steels of this class.

a. Effect of Alloying Elements

The alloying elements are usually added for the purposes listed in Table 8.9. Note that each element does not make a steel better in every respect. Each tends to improve some characteristics at the expense of others. The elements utilized are therefore balanced in the composition to achieve a desired combination of properties. Since the alloying elements cost considerably more than the basic steel, alloyed steels are used where they provide one or more clear-cut advantages over plain carbon steels.

There are two broad classes of these steels. One class is known as low-alloy structural steel. The other is the AISI grades already mentioned.

b. Low-Alloy Structural Steel

There are about 30 of these grades. As a group, they contain 0.30% or less carbon, 1.7% or less manganese, 1% or less silicon, 0.3-2.0% nickel, 0.1-1.5% copper with less than 0.4% molybdenum, 1.5% chromium, and 0.05-0.2% phosphorus if the latter three are

Table 8.8 Typical Uses for Plain Carbon Steels

Carbon level (%)	Typical uses
0.05-0.07	High-ductility wire.
0.70-0.15	Rimmed steel is used for sheet, strip, rod, and wire where excellent surface finish is required, such as body and fender stock, panels, deep-drawing strip, steel for lamps, hoods, sectors, pawls, clutch and transmission covers, oil pans, and a multitude of deep-drawn and formed products. It is also used for cold heading wire for tacks, rivets, and low-carbon wire products.

Killed steel should be used in preference to rimmed steel for carburized parts, expecially where both the rim and the core of rimmed steel are involved in the heat treatment. In the process of "rimming in," practically all the carbon is transferred from the outer part of the ingot for a depth of several inches, to the inner part. A cross section of the ingot or of any shape rolled or forged from it, will have an outer layer of almost pure iron and a core in which carbon has concentrated correspondingly. If the part to be carburized has been forged or machined in such a manner as to bring both the rim and the core metal into the case, irregular distortion may be expected.

These steels are of low tensile values and should not be selected where much strength is desired. Cold drawing or rolling improves their hardness and strength about 20% over the properties in the hot-rolled condition. All the properties acquired by cold working are, however, lost when these steels are heated to temperatures of $1000°F$ or higher.

These steels, being ferritic in structure, do not machine freely and should be avoided for nuts, cut screws, and operations requiring broaching or smooth finish on turning. Cold drawing, however, improves their machinability.

The higher manganese varieties have improved machinability and hardening properties.

0.08-0.18	Boiler plate, seamless, weldable boiler tubes, and ship plate.
0.15-0.20	Standard carburizing grades for wrist pins, camshafts, drag links, clutch fingers, sheet and strip for fan blades and welded tubing, and numerous forged parts where high strength is not essential.

This steel may be brazed, butt welded, and drawn into various shapes but is not as desirable for deep-drawing operations as 1008 and 1010. It shows some improvement over SAE 1010 in machining but is not recommended for smooth threading, turning, or broaching.

The higher manganese variants improve machinability and hardening properties. These steels carburize and harden freer from soft spots than 1020.

0.20-0.30	Small forgings, crank pins, gears, valves, crankshafts (0.20-0.26), railway axles (0.23-0.30), cross heads, connecting rods, rims for turbine gears, armature shafts in general (0.24-0.32), and fishplates. They have fair machining properties for threading, broaching, and turning. Forgings usually machine better without annealing or in the normalized condition.
0.35-0.45	Axles, special-duty shafts, connecting rods, small and medium forgings, cold upset wire and rod, machinery steel, spring clips, solid turbine rotors, rotor and gear shafts, armatures for turbo-generators, key stock,

Carbon level (%)	Typical uses
	shift and brake levers, forks, and anchor bolts. They possess fair machining properties and deep-hardening characteristics.
	The higher manganese variants are used for larger sections or where higher properties are desired.
0.45-0.55	Parts to be subjected to shock and heavy reversals of stress. Railway coach axles, crank pins on heavy machines, larger size forgings, such as crankshafts, starter ring gears, axles, spline shafts, and for hard-drawn wire for tempered and patented springs. Caution should be used in water quenching this steel in parts of small diameter or thin sections.
0.60-0.70	Drop-forging dies, die blocks, bolt-heading dies, plate punches, set screws, self-tapping screws, snap rings, valve springs, cushion springs, clutch springs, lock washers, spring clips, clutch disks, thrust washers, and parts for agricultural purposes, such as frogs and standards.
0.70-0.80	Cold chisels, pick axes, wrenches, jaws for vises, shear blades, hack saws, pneumatic drill bits, wheels for railway service, wire for structural work, automatic clutch disks, mower sections, plow beams, and so on.
0.80-0.90	Railway rails, plow shares, rock drills, circular saws, machine chisels, punches and dies, lock pins, clutch disks, leaf springs, music wire, mower knives.
0.90-1.00	Punches and dies, springs, balls, keys, pins, leaf and coil springs, harrow and seed disks.
1.00-1.1	Railway springs, machine tools, mandrels, springs, taps.
1.1 -1.2	Taps, tools, thread metal dies, twist drills, knives.
1.2 -1.3	Files.
1.3 -1.5	Dies for wire drawing, paper knives, and tools for turning chilled iron.
1.5 -1.6	Saws for cutting steel and dies for wire drawing.

present. They were developed to provide a high yield-strength low-cost steel that had good weldability, ductility, and impact strength, and better corrosion resistance than low-carbon steels. The low carbon aids weldability and assures that these steels will not air harden after welding or hot rolling. They are usually employed without heat treatment. Typical mechanical properties are: minimum hot-rolled strength of 50,000 psi, tensile strength 70,000-90,000 psi, and 18-30% elongation in 8 in. Although the first application was for bridges, other uses have developed for railway rolling stock, trucks, buses, cranes, shovels, and similar structures.

c. AISI Low-Alloy Steels

These steels are widely used in automobile, machine tool, and aircraft construction, especially for moving parts subject to high stresses. They are more costly than the structural class and are therefore almost always used in the quenched and tempered condition to take full advantage of the alloying additions. In this condition they have, as a class, 30-40% higher yield strength, 10-20% higher elongation in 2 in., 30-40% higher reduction in area, and often twice the impact strength of plain carbon steels of the same tensile strength and hardness. Since they are also deeper hardening than the plain carbon steels, they can be hardened through thicker sections; alternatively, less drastic quenches, producing less distortion and stress, may be employed for equivalent sections.

Table 8.9 Effects of Alloying Elements upon Properties of Steel[a]

	Carbon, 0.10-0.30	Carbon, 0.31-eutectoid	Effect of single alloying element upon low-carbon steel (0.1-0.3°C)								
			Mn, 0.25-2.00	P, 0.0-0.15	S, 0.0-0.3	Si, 0.0-2.0	Cr, 0.0-1.1	Ni, 0.0-5.0	Mo, 0.0-0.75	V, 0.0-0.25	Cu, 0.0-1.1
Hot working	+2P	− 7P	+ 7S	0	−10S	− 5P	0	− 6P	− 3P	0	−10S
Cold drawing	−4P	−10P	−10P	−10P	−10P	−10S	−10S	−10S	−10P	x	−10P
Cold bending	−2L	−10P	+ 2I	−10L	− 7P	− 6P	−10P	+ 5S	+ 3S	+3P	+ 2P
Machinability[b]	+2P	− 2P	− 6P	+ 6P	+10P	− 2P	− 2P	−10P	− 5P	0	x
Weldability	−2L	−10P	−10L	− 7L	− 3P	− 5L	−10L	−10L	−10L	+1P	− 4L
Hardenability (depth)	+1P	+ 2P	+10P	+ 3P	− 2P	+ 2S	+ 7S	+10L	+ 7S	+5I	xx
Strength[b]	+4P	+10P	+ 5P	+ 5P	− 2P	+ 7P	+ 5P	+ 5P	+ 5P	0	+ 1P
Creep resistance	+2I	− 3P	0	0	0	0	+ 1P	0	+10P	?	?
Toughness[b]	−3P	− 8P	+ 2I	− 8P	− 6P	− 2P	− 8P	+10P	+ 3S	+2P	+ 1L
Low temp., toughness[b]	−3P	−10P	+ 2I	−20S	− 6P	− 1P	−10P	+10S	+ 2S	+3P	x
Corrosion resistance	−3P	− 6P	0	+10P	−10P	+ 2P	0	+ 6L	+ 2P	0	+10S
Wear resistance[c]	0	+10P	+ 6P	+ 1P	0	+ 2P	+ 8P	+ 3L	+ 5P	+2P	xx

[a]Number indicates relative magnitude of effect: +, improvement in property; −, reduction in property; P, effect is proportional to quantity of alloying agent S, small quantities most effective; L, large quantities most effective; I, intermediate quantities most effective; x, probably negative; xx, probably positive.

[b]As rolled.

[c]Fully hardened.

Source: From data by John Mitchell in *Contributions to the Metallurgy of Steel*, Am. Iron & Steel Inst., by permission of the author and publisher.

When given the same heat treatment, tensile strength may differ by as much as 50,000 psi for these grades, but when heat treated to produce the same tensile strength and hardness, all steels of this class have approximately the same yield point, reduction of area, and elongation (see Fig. 8.29), provided tensile strength is below 200,000 psi. Above this value there is some scatter, particularly in endurance ratio and in impact strength, which may vary more than 20% probably because of stressed induced by the drastic quench required for some grades.

The manganese steels of lower carbon content are particularly adapted to carburizing and direct quenching from the furnace. The higher carbon contents are used interchangeably with equivalent plain carbon steels with advantages in lower quenching rates for the same properties.

Nickel is usually added to a low-carbon steel, because it is more easily dissolved by the ferrite than cementite. The dissolved nickel increases toughness, hardness, resistance to galling, checking, and fatigue failure, the elastic limit, and the elastic ratio. In addition, the grain structure is refined, creep at high temperature is reduced, heat treatment and casting are aided, and case hardening is not impaired.

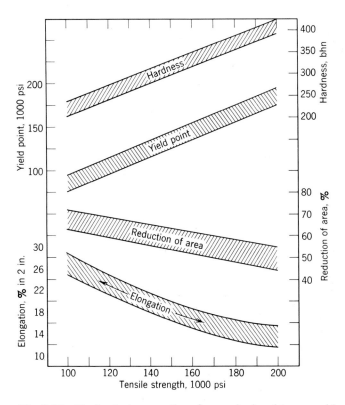

Fig. 8.29 Mechanical properties of quenched and tempered low-alloy steels containing 0.30-0.50% carbon. Bands indicate normal variations in indicated properties. Note that these properties are equivalent for various grades if they are heat treated to the same tensile strength.

Nickel-chromium steels are used where greater strength, toughness, and hardenability are desired than can be obtained with lower alloy contents. The low-carbon grades are suited to carburizing for parts, such as gears, pinions, crankshafts, and piston pins. The medium-carbon grades are used for shafts, links, studs, and bolts where properties are obtained by water quenching. The high-carbon grades of this class (0.40-0.50% C) are used for forgings and machined parts where properties are obtained by oil quenching. Typical grades and uses are given in Table 8.10.

The HSLA steels are duplex martensite-ferrite composites in which the strong martensite is incorporated into the ductile ferrite matrix by thermal control of solid-state phase transformations without the need of thermomechanical processing. A typical formula would contain 2% manganese and 1.5% silicon, with 0.1% each of molybdenum and vanadium. It is important to keep sulfur very low.

3. High-Alloy Steels—Tools and Dies

These steels are used for cutting tools, shearing tools, forming dies, and battering tools. Based on data from the Committee of Tool Steel Producers, American Iron and Steel Institute, Washington, D.C. (Sept. 1982), the following tool steels exemplify the class. More detailed information is available in the ASM *Source Book on Industrial Alloy and Engineering Data* (Feb. 1978), pp. 248-259.

For nontool purposes, type H 11 is the grade most often selected to meet the higher requirements that necessitate the quality and performance inherent in tool steels. This particularly applies to fatigue properties and toughness.

Some general remarks on the tool and die steels are given here in order to indicate the variety of tool applications and available choices of alloys (See Tables 8.11 and 8.12).

Plain carbon steels harden with a hard case and a softer core, except in small sizes, and these properties are suited to a number of applications. Hot-forming and heading dies, hammerheads, and rivet sets are made from grades with 0.60-0.75% carbon; chisels,

Table 8.10 Typical Uses for Nickel-Chromium Low-Alloy Steels

Nickel (%)	Carbon (%)	Typical uses
0.4 -0.6	0.10-0.20	Pinion gears, rear-axle differentials, case-hardened parts with little distortion
1.0	0.40	Good forging material for automobile front axles, etc.
1.5	0.15	Engine bolts, rivets
2.0	0.20	Large-frame castings, steel-mill machinery, ship castings, and castings subjected to shock and fatigue; boiler plate for locomotives, and fire box plate and tubes
2.75	0.15	Locomotive forgings, piston rods, side rods, main rods, and axles
3.25-3.75	0.15	"King" pins, piston pins, roller-bearing races, universal joints, and spring clips
3.25-3.75	0.40	Drive shafts, and other heavy splined shafting
3.25-3.75	0.50	Connecting rods; tube in this class is used for bearing races, collars, valve seats, cylinder liners, stressed rings, etc.
4.5 -5	0.50	Transmission gears on trucks and heavy apparatus

shear blades, smithing tools, swages, and flatteners from 0.75 to 0.90% C; hard chisels, large taps, trimming dies, drills, milling tools, threading dies, blanking, forming dies, and punches from grades having 0.90-1.1% C; and small taps, reamers, planing tools, woodworking tools, and razors from grades having 1.10-1.40% C.

Pneumatic chisels, rivet busters, concrete breakers, and heavy-duty punches, and so on, are applications involving higher shock resistance than is available in the 0.75-0.90% carbon steels. Steels used for these applications are of the medium-alloy type and include chrome-vanadium, low-silicon-manganese, high-silicon-manganese, and chromium-tungsten "shock-resisting" steels.

So-called high-speed steels are of compositions that have very good wear resistance and retain their hardness at high temperatures. They are used in simple shapes, where their distortion and moderate susceptibility to cracking during heat treatment are not limiting. The general-purpose grades are either of the high-tungsten or high-molybdenum type. The former contains 18% tungsten, 4% chrome, and 1% vanadium. Greater vanadium content to 3% improves abrasion resistance and makes these steels suited for tools for fine cuts, like broaches and reamers. The 4.5% molybdenum, 6% tungsten, 4% vanadium, and 4% chrome grades have similar uses. Cobalt additions from 5 to 12% in both types increase high-temperature hardness and abrasion resistance. The cobalt-tungsten grades are used for heavy-duty tools for heavy cuts. The cobalt-molybdenum grades are especially used for cutting gritty, scaly material, such as sand-cast alloys or heat-treated steels. Since these tool steels are more brittle than the other grades, they are not recommended for tools for finishing cuts.

Die operations, such as forging, coining, sheet and wire drawing, cold heading, die and permanent-mold casting, and plastic molding, make use of a number of different steels. Choice depends on the temperatures involved, need for abrasion resistance, and the amount of production expected from the die set. Some of the tool steels already mentioned are used where their strength, toughness, and "red" hardness suffice. For higher red hardness, a group of steels known as *hot-working tool* steels are available. They are medium to high C, medium Cr steels with vanadium, tungsten, and molybdenum additions. Another group of steels, known as the "nondeforming" tool steels, are used where design prevents use of water-hardening steels because of the hazard of distortion and cracking during hardening. These nondeforming grades include manganese oil hardening, tungsten oil hardening, chromium oil hardening, manganese air hardening, and high-carbon high-chromium steels.

4. High-Alloy Steels—Corrosion-Resistant Steels

The corrosion-resistant steels are frequently called the "stainless steels." Depending on their composition and history they fall into four major groups: the austenitic, which is hardenable by cold working but not by heat treatment, the ferritic, which is not hardenable by either cold working or heat treatment, the martensitic, which is hardenable by heat treatment, and the precipitation-hardening group, which is related to martensitic steels.

The ASM *Source Book on Industrial Alloy and Engineering Data* (pp. 188-195) gives the principal necessary information for selection among this important group of alloys. The substitution of nickel by manganese in austenitic alloys is an economical procedure that should be considered. Tables 8.13A and B give comparative properties of some typical heat- and corrosion-resistant steels.

Table 8.11 Tool Steels, AISI Type and Mechanical Properties

AISI type	Annealing data, temperature, °F (°C)	HRB	Room-temperature properties as annealed[a]			Heat-treat[b] austeni- tizing, °F (°C)	Tempering, °F (°C)
			Yield, 10^3 psi (MPa)	Tensile, 10^3 psi (MPa)	Elong- ation (%)		
A-8							
0.55 C	1550 (840)	97	65 (448)	103 (710)	24	1850 (1010),	800 (425)
5.00 Cr						air	1050 (565)
1.25 W						quench	1100 (590)
1.25 Mo							1200 (650)
H-11							
0.35 C	1600 (870)	96	53 (365)	100 (689)	25	1850 (1010),	1000 (540)
5.00 Cr						air	1100 (590)
1.50 Mo						quench	1200 (650)
0.4 V							
L-2							
0.50 C	1425 (775)	93	74 (510)	103 (710)	25	1575 (855),	400 (200)
1.0 Cr						oil	600 (315)
0.2 V						quench	800 (425)
							1000 (540)
							1200 (650)
L-6							
0.70 C	1425 (775)	93	55 (379)	95 (655)	25	1550 (840),	600 (315)
0.75 Cr						oil	800 (425)
0.25 Mo						quench	1000 (540)
1.50 Ni							1200 (650)
P-20							
0.35 C	1425 (775)	97	75 (517)	100 (689)	17	1575 (855),	400 (200)
1.70 Cr						oil	600 (315)
0.40 Mo						quench	800 (425)
							1000 (540)
							1200 (650)
S-1							
0.50 C	1475 (800)	96	60 (414)	100 (689)	24	1700 (925),	400 (200)
1.50 Cr						oil	600 (315)
2.50 W						quench	800 (425)
							1000 (540)
							1200 (650)
S-5							
0.55 C	1450 (790)	96	64 (444)	105 (724)	25	1600 (870),	400 (200)
0.80 Mn						oil	600 (315)
2.00 Si						quench	800 (425)
0.40 Mo							1000 (540)
							1200 (650)
S-7							
0.5 C	1525 (830)	95	55 (379)	93 (641)	25	1725 (940),	400 (200)
3.25 Cr						air	600 (315)
1.40 Mo						quench	800 (425)
							1000 (540)
							1200 (650)

[a]Tests made on longitudinal specimens of small cross-sectional bar stock.

[b]Consult steel producer for specific heat treatments; these are nominal.

[c]Single temper except H-11 and A-8 (double temper)

[d]Charpy V-notch for A-8, H-11, L-2, L-6, and P-20; others are charpy unnotched.

Room-temperature properties[a] after hardening and tempering

Yield, 10^3 psi (MPa)	Tensile, 10^3 psi (MPa)	Elongation (%)	Hardness (HRC)	Impact energy,[c] lb(J)d
225 (1551)	265 (1827)	9	52	5 (6.8)
219 (1510)	255 (1758)	10	50	
165 (1138)	184 (1269)	12	39.5	
250 (1724)	295 (2034)	9	55	10 (14)
190 (1310)	220 (1517)	12	46	20 (27)
130 (896)	160 (1103)	18	36	25 (34)
260 (1793)	290 (1999)	5	54	21 (28)
240 (1655)	260 (1793)	10	52	14 (19)
200 (1379)	225 (1551)	12	47	19 (26)
170 (1172)	185 (1276)	15	41	29 (39)
110 (758)	135 (931)	25	30	92 (125)
260 (1793)	290 (1999)	4	54	9 (12)
200 (1379)	230 (1586)	8	46	13 (18)
160 (1103)	195 (1344)	12	42	17 (23)
120 (827)	140 (965)	20	32	60 (81)
205 (1413)	270 (1862)	10	52	15 (20)
210 (1448)	250 (1724)	11	49	18 (24)
200 (1379)	225 (1551)	12	46	21 (28)
175 (1207)	200 (1379)	15	42	40 (54)
115 (793)	140 (965)	20	29	46 (62)
275 (1896)	300 (2068)	–	57.5	184 (249)
270 (1862)	294 (2027)	4	54	172 (233)
245 (1689)	260 (1793)	5	50.5	150 (203)
221 (1524)	244 (1682)	9	47.5	170 (230)
180 (1241)	195 (1344)	12	42.0	–
280 (1931)	340 (2344)	5	59	152 (206)
270 (1862)	340 (2241)	7	58	171 (232)
245 (1689)	275 (1896)	9	52	179 (243)
200 (1379)	220 (1517)	10	48	139 (188)
150 (1034)	170 (1172)	15	37	–
210 (1448)	315 (2172)	7	58	180 (244)
230 (1586)	285 (1965)	9	55	228 (309)
205 (1413)	275 (1896)	10	53	179 (243)
200 (1379)	264 (1820)	10	51	239 (324)
150 (1034)	180 (1241)	14	39	264 (358)

Table 8.12   General Characteristics of Various Tool Steels

Type	Depth of Hardening	Distortion in heat treat	Toughness	Wear resistance	Decarb. resistance	Machinability
H-11	Deep	Very low	High	Medium	Medium	Medium to high
L-6	Medium	Low	Very high	Medium	High	Medium
M-2	Deep	Low	Low	Very high	Medium	Medium
O-1	Medium	Low	Medium	Medium	High	High
S-2	Medium	High	Highest	Low to medium	Low	Medium to high
S-7	Medium to deep	Air quench lowest	Very high	Low to medium	Low	Medium to high

Table 8.13A   Comparative Properties of Some Typical Corrosion and Heat-Resistant Steels[a]

Grades and AISI Types

Austenitic Stainless Steels	UNS Numbers	Alloys for Severe Corrosion	UNS Numbers
301	S30100	20Cb-3* Stainless	N08020
304	S30400	Carpenter 7-Mo (Type 329)	S32900
304H	S30452	HASTELLOY** Alloy B-2	N10665
304L	S30403	HASTELLOY** Alloy C-4	N06455
309S	S30908	HASTELLOY** Alloy C-276	N10276
309S CbTa	—	HASTELLOY** Alloy G	N06007
310S	S31008	INCOLOY*** Alloy 825	N08825
310S Cb	—	Pyromet* Alloy 600	N06600
316	S31600	Pyromet* Alloy 625	N06625
316L	S31603	Pyromet* Alloy 800	N08800
317	S31700		
317L	S31703	**High Temperature Alloys**	
318 (316CbTa)	—	Pyromet* Alloy 102	N06102
321	S32100	Pyromet* Alloy 600	N06600
329 (Carpenter 7-Mo)	S32900	Pyromet* Alloy 625	N06625
330	—	Pyromet* Alloy 800	N08800
347	S34700	HASTELLOY** Alloy X	N06002
347H	S34709	HAYNES** Alloy 25 (L-605)	R30605
348	S34800	Type 330 Stainless	
20Cb-3* Stainless	N08020		
21Cr-6Ni-9Mn	S21904		
Ferritic Stainless Steels			
405	S40500		
409	S40900		
430	S43000		
XM-8 (439)	S43035		
18 Cr-2Mo	S18200		
Precipitation Hardening Stainless Steels			
Custom 450* Alloy	S45000		
Custom 455* Alloy	S45500		
17-7PH*	S17700		

* Registered trademark of Carpenter Technology Corporation
* Registered trademark of Armco Steel Corporation
** Registered trademark of Cabot Corporation
*** Registered trademark of International Nickel Company

Table 8.13A (Continued)

Comparative Properties of Stainless Steel (Selected Alloys)

Chemical Analysis		304	304L	309S
Chemical composition by percent	Carbon	0.08 max.	0.03 max.	0.08 max.
	Chromium	18.00/20.00	18.00/20.00	22.00/24.00
	Nickel	8.00/12.00	8.00/11.00	12.00/15.00
	Manganese	2.00 max.	2.00 max.	2.00 max.
	Phosphorous	0.040 max.	0.040 max.	0.040 max.
	Sulfur	0.030 max.	0.030 max.	0.030 max.
	Silicon	0.75 max.	0.75 max.	0.75 max.
	Molybdenum	—	—.	—
	Other Elements	Fe-bal	Fe-bal	Fe-bal

Physical Constants

Structure		Austenitic	Austenitic	Austenitic
Is it Magnetic?		No	No	No
Density — lbs.-mass/in^3 *(Mg/m³)*		0.29 *(8.0)*	0.29 *(8.0)*	0.29 *(8.0)*
Specific Heat Capacity — Btu (Int)/lbm • °F from 32° to 212°F *(kJ/kg • K from 0° to 100°C)*		0.12 *(0.50)*	0.12 *(0.50)*	0.12 *(0.50)*
Thermal Conductivity Btu (Int)• ft/hr • f² • °F	(W/m • K)			
212°F	*(100°C)*	9.4 *(16.3)*	9.4 *(16.3)*	8.0 *(13.9)*
932°F	*(500°C)*	12.4 *(21.5)*	12.4 *(21.5)*	10.8 *(18.7)*
Mean Coefficient of Thermal Expansion 10^{-6}/°F	*(10^{-6}/°C)*			
32°/212°F	*(0°/100°C)*	9.6 *(17.3)*	9.6 *(17.3)*	8.3 *(14.9)*
32°/600°F	*(0°/316°C)*	9.9 *(17.8)*	9.9 *(17.8)*	9.3 *(16.7)*
32°/1000°F	*(0°/538°C)*	10.2 *(18.4)*	10.2 *(18.4)*	9.6 *(17.3)*
32°/1200°F	*(0°/649°C)*	10.4 *(18.7)*	10.4 *(18.7)*	10.0 *(18.0)*
32°/1500°F	*(0°/816°C)*	—	—	—
32°/1800°F	*(0°/982°C)*	—	—	11.5 *(20.7)*
Electrical Resistivity at Room Temperature ohm/circular mil foot *(microhm-centimeter)*		433 *(73.6)*	455 *(77.4)*	469 *(79.7)*
Tensile Modulus of Elasticity psi x 10^6	*(GPa)*	28.0 *(193)*	28.0 *(193)*	29.0 *(200)*
Torsional Modulus of Elasticity psi x 10^6	*(GPa)*	12.5 *(86.2)*	12.5 *(86.2)*	10.5 *(72.4)*

Typical Mechanical Properties (annealed condition)

Ultimate Strength ksi	*(MPa)*	85 *(586)*	80 *(552)*	90 *(621)*
Yield Strength ksi	*(MPa)*	35 *(241)*	30 *(207)*	45 *(310)*
Elongation in 2", %		50	60	45
Izod Impact Strength foot-pound-force	*(J)*	110 *(149)*	110 *(149)*	110 *(149)*
Rockwell Hardness		B-80	B-80	B-85

Table 8.13A   (Continued)

Comparative Properties of Stainless Steel (Continued)

Chemical Analysis		310S	316	316L
Chemical composition by percent	Carbon	0.08 max.	0.08 max.	0.03 max.
	Chromium	24.00/26.00	16.00/18.00	16.00/18.00
	Nickel	19.00/22.00	11.00/14.00	11.00/14.00
	Manganese	2.00 max.	2.00 max.	2.00 max.
	Phosphorous	0.040 max.	0.040 max.	0.040 max.
	Sulfur	0.030 max.	0.030 max.	0.030 max.
	Silicon	0.75 max.	0.75 max.	0.75 max.
	Molybdenum	—	2.00/3.00	2.00/3.00
	Other Elements	Fe-bal	Fe-bal	Fe-bal

Physical Constants

Structure		Austenitic	Austenitic	Austenitic
Is it Magnetic?		No	No	No
Density—lbs.-mass/in^3 *(Mg/m^3)*		0.29 *(8.0)*	0.29 *(8.0)*	0.29 *(8.0)*
Specific Heat Capacity—Btu (Int)/lbm • °F from 32° to 212°F *(kJ/kg • K from 0° to 100°C)*		0.12 *(0.50)*	0.12 *(0.50)*	0.12 *(0.50)*
Thermal Conductivity Btu (Int) • ft/hr • f^2 • °F	*(W/m • K)*			
212°F	*(100°C)*	8.0 *(13.9)*	9.4 *(16.3)*	9.4 *(16.3)*
932°F	*(500°C)*	10.8 *(18.7)*	12.4 *(21.5)*	12.4 *(21.5)*
Mean Coefficient of Thermal Expansion 10^{-6}/°F	*(10^{-6}/°C)*			
32°/212°F	*(0°/100°C)*	8.0 *(14.4)*	8.9 *(16.0)*	8.9 *(16.0)*
32°/600°F	*(0°/316°C)*	9.0 *(16.2)*	9.0 *(16.2)*	9.0 *(16.2)*
32°/1000°F	*(0°/538°C)*	9.4 *(16.9)*	9.7 *(17.5)*	9.7 *(17.5)*
32°/1200°F	*(0°/649°C)*	9.7 *(17.5)*	10.3 *(18.5)*	10.3 *(18.5)*
32°/1500°F	*(0°/816°C)*	—	11.1 *(20.0)*	11.1 *(20.0)*
32°/1800°F	*(0°/982°C)*	10.6 *(19.1)*	—	—
Electrical Resistivity at Room Temperature ohm/circular mil foot *(microhm-centimeter)*		469 *(79.7)*	445 *(77.4)*	445 *(77.4)*
Tensile Modulus of Elasticity psi x 10^6	*(GPa)*	29.0 *(200)*	28.0 *(193)*	28.0 *(193)*
Torsional Modulus of Elasticity psi x 10^6	*(GPa)*	11.2 *(77.2)*	10.4 *(71.7)*	10.4 *(71.7)*

Typical Mechanical Properties (annealed condition)

Ultimate Strength ksi	*(MPa)*	95 *(655)*	90 *(621)*	80 *(552)*
Yield Strength ksi	*(MPa)*	45 *(310)*	40 *(276)*	35 *(241)*
Elongation in 2", % • *(.05 m)*		45	50	55
Izod Impact Strength foot-pound-force	*(J)*	90 *(122)*	110 *(149)*	110 *(149)*
Rockwell Hardness		B-85	B-85	B-80

317	317L	321	347	405	430
0.08 max.	0.03 max.	0.08 max.	0.08 max.	0.08 max.	0.12 max.
18.00/20.00	18.00/20.00	17.00/20.00	17.00/20.00	11.50/13.50	16.00/18.00
11.00/14.00	11.00/14.00	9.00/11.00	9.00/12.00	0.50 max.	0.50 max.
2.00 max.	2.00 max.	2.00 max.	2.00 max.	1.00 max.	1.00 max.
0.040 max.	0.040 max.	0.040 max.	0.040 max.	0.03 max.	0.040 max.
0.030 max.	0.030 max.	0.030 max.	0.030 max.	0.030 max.	0.030 max.
0.75 max.	0.75 max.	0.75 max.	0.75 max.	0.75 max.	0.75 max.
3.00/4.00	3.00/4.00	—	—	—	—
Fe-bal	**Fe-bal**	Ti 6 x C min. 0.60 max. **Fe-bal**	Cb + Ta = 10 x C min. 1.00 max. **Fe-bal**	Al 0.10/0.30 **Fe-bal**	**Fe-bal**
Austenitic	Austenitic	Austenitic	Austenitic	Ferritic	Ferritic
No	No	No	No	Yes	Yes
0.29 (8.0)	0.29 (8.0)	0.29 (8.0)	0.29 (8.0)	0.28 (7.8)	0.28 (7.8)
0.12 (0.50)	0.12 (0.50)	0.12 (0.50)	0.12 (0.50)	0.11 (0.46)	0.11 (0.46)
9.4 (16.3)	9.4 (16.3)	9.3 (16.1)	9.3 (16.1)	—	15.1 (26.1)
12.4 (21.5)	12.4 (21.5)	12.8 (22.2)	12.8 (22.2)	—	15.2 (26.3)
8.9 (16.0)	8.9 (16.0)	9.3 (16.7)	9.3 (16.7)	6.0 (10.8)	5.8 (10.4) (77°/212°F)
9.0 (16.2)	9.0 (16.2)	9.5 (17.1)	9.5 (17.1)	6.4 (11.5)	6.1 (11.0) (77°/842°F)
9.7 (17.5)	9.7 (17.5)	10.3 (18.5)	10.3 (18.5)	6.7 (12.1)	6.3 (11.3)
10.3 (18.5)	10.3 (18.5)	10.7 (19.3)	10.6 (19.1)	—	6.6 (11.9)
11.1 (20.0)	11.1 (20.0)	11.2 (20.2)	11.1 (20.0)	—	6.9 (12.4) (77°/1652°F)
—	—	—	—	—	—
445 (77.4)	445 (77.4)	433 (73.6)	439 (74.6)	—	361 (61.4)
28.0 (193)	28.0 (193)	28.0 (193)	28.0 (193)	29.0 (200)	29.0 (200)
10.4 (71.7)	10.4 (71.7)	10.5 (72.4)	10.5 (72.4)	—	10.5 (72.4)
90 (621)	80 (552)	90 (621)	95 (655)	65 (448)	75 (517)
40 (276)	35 (241)	35 (241)	40 (276)	40 (276)	45 (310)
45	50	45	45	30	30
110 (149)	110 (149)	100 (135)	110 (149)	25 (34)	20/50 (27/68)
B-85	B-80	B-85	B-85	B-85	B-85

Table 8.13A (Continued)

Comparative Properties of Stainless Steel (Continued)

Chemical Analysis		XM-8 (439)	18Cr-2 Mo	20Cb-3 Stainless	INCOLOY Alloy 825
Chemical composition by percent	Carbon	0.07 max.	0.25 max.	0.07 max.	0.05 max.
	Chromium	17.00/19.00	17.50/19.50	19.00/21.00	19.50/23.50
	Nickel	0.50 max.	1.00 max.	32.50/35.00	38.00/46.00
	Manganese	1.00 max.	1.00 max.	2.00 max.	1.00 max.
	Phosphorous	0.40 max.	0.40 max.	0.035 max.	—
	Sulfur	0.30 max.	0.30 max.	0.035 max.	0.30 max.
	Silicon	1.00 max.	1.00 max.	1.00 max.	0.50 max.
	Molybdenum	—	1.75/2.50	2.00/3.00	2.50/3.50
	Other Elements	Ti = 12xC min; 1.10 max. **Fe-bal**	Ti + Cb = 0.20 + 4 (C+N) min; 0.80 max. **Fe-bal**	Cu 3.00/4.00 Cb + Ta = 8 x C min. 1.00 max. **Fe-bal**	Cu 1.50/3.00 Al 0.20 max. Ti 0.6/1.2 **Fe-bal**

Physical Constants

Structure		Ferritic	Ferritic	Austenitic	Austenitic
Is it Magnetic?		Yes	Yes	No	No
Density—lbs-mass/in³ (Mg/m³)		0.28 (7.8)	0.28 (7.8)	0.29 (8.0)	0.29 (8.0)
Specific Heat Capacity—Btu (Int)/lbm • °F from 32°/212°F (kJ/kg • K from 0° to 100°C)		0.11 (0.46)	0.11 (0.46)	0.12 (0.50)	—
Thermal Conductivity Btu (Int) • ft/hr • f² • °F	(W/m • K)				
212°F	(100°C)	14.0 (24.2)	15.5 (26.8)	7.6 (13.1)	—
932°F	(500°C)	—	—	10.5[1] (18.1)	10.9 (18.8)
Mean Coefficient of Thermal Expansion 10⁻⁶/°F	(10⁻⁶/°C)				
32°/212°F	(0°/100°C)	5.6 (10.1)	—	8.31 (15.0) (68°/200°F)	7.8 (14.0)
32°/600°F	(0°/316°C)	—	5.8 (10.4) (73°/400°F)	9.43 (17.0) (68°/700°F)	8.4[2] (15.1)
32°/1000°F	(0°/538°C)	6.4 (11.5) (68°/932°F)	6.4 (11.5) (73°/800°F)	—	8.8 (15.8)
32°/1200°F	(0°/649°C)	—	—	—	—
32°/1500°F	(0°/816°C)	6.9 (17.3) (68°/1472°F)	—	9.97 (17.9) (68°/1500°F)	9.6 (17.3)
32°/1800°F	(0°/982°C)	—	7.3 (13.1) (73°/1800°F)	—	—
Electrical Resistivity at Room Temperature ohm/circular mil foot (microhm-centimeter)		—	—	625 (106.3)	678 (115.3)
Tensile Modulus of Elasticity psi x 10⁶	(GPa)	29.0 (200)	29.0 (200)	28.0 (193)	28.0 (193)
Torsional Modulus of Elasticity psi x 10⁶	(GPa)	—	—	11.0 (76.0)	—
Typical Mechanical Properties (annealed condition)					
Ultimate Strength ksi	(MPa)	70 (483)	80 (552)	95 (655)	85/105 (586/724)
Yield Strength ksi	(MPa)	45 (310)	50 (345)	55 (379)	35/65 (241/448)
Elongation in 2", % (· c 5 m)		40	35	35	50/30
Izod Impact Strength foot-pound-force	(J)	—	—	110 (149)	—
Rockwell Hardness		B-85	B-90	B-90	B-120/180

(1) 752°F (400°C)   (2) 500°F (260°C)

HASTELLOY Alloy G	Carpenter 7-Mo (329)	Custom 450 Alloy	Custom 455 Alloy	HASTELLOY Alloy B-2	HASTELLOY Alloy C-276
0.05 max.	0.08 max.	0.05 max.	0.05 max.	0.02 max.	0.02 max.
21.00/23.50	23.00/28.00	14.50/16.50	11.00/12.50	1.00 max.	14.50/16.50
Balance	2.50/5.00	5.50/7.00	7.50/9.50	Balance	Balance
1.00/2.00	1.00 max.	0.50 max.	0.50 max.	1.00 max.	1.00 max.
0.04 max.	0.04 max.	0.03 max.	0.040 max.	0.040 max.	0.040 max.
0.03 max.	0.030 max.	0.03 max.	0.030 max.	0.030 max.	0.030 max.
1.00 max.	0.75 max.	0.50 max.	0.50 max.	0.10 max.	0.08 max.
5.50/7.50	1.00/2.00	0.50/1.00	0.50 max.	26.0/30.0	15.00/17.00
Co 2.50 max. W 1.00 max. Fe 18.00/21.00 Cu 1.50/2.50 Cb + Ta = 1.75/2.50	Fe-bal	Cu 1.25/1.75 Cb 8 x C min. Fe-bal	Cu 1.50/2.50 Ti 0.80/1.40 Cb✦Ta 0.10/0.50 Fe-bal	Fe 2.00 max. Co 1.00 max.	W 3.00/4.50 Fe 4.00/7.00 V 0.35 max.
Austenitic	Austenitic/Ferritic	Martensitic	Martensitic	Austenitic	Austenitic
No	No	Yes	Yes	No	No
0.30 (8.3)	0.28 (7.8)	0.28 (7.8)	0.28 (7.8)	0.33 (9.4)	0.32 (9.1)
0.11 (0.46)	0.11 (0.46)	—	—	0.091 (0.3810)	0.102 (0.4270)
6.5 (11.2)	8.8 (15.2)	—	10.4 (18.0)	7.1 (12.3)	6.4 (11.1)
10.0 (17.3)	12.5[(2)] (21.6)	—	14.3 (24.8)	10.0 (17.3)	11.0 (19.0)
—	5.60 (10.1) (75°/200°F)	5.88 (10.6) (72°/200°F)	5.90 (10.6)	5.7 (10.3) (68°/200°F)	—
—	6.50 (11.7) (75°/600°F)	5.91 (10.6) (72°/600°F)	6.31 (11.3)	6.2 (11.2) (68°/600°F)	—
—	—	6.08 (10.9) (72°/900°F)	6.68 (12.0)	6.5 (11.7) (68°/1000°F)	—
—	—	6.17 (11.1)	—	—	—
—	8.0 (14.4) (68°/1500°F)	—	—	—	—
—	—	—	—	—	—
—	451 (76.7)	—	—	824 (140.1)	779 (132.4)
—	29.0 (200)	—	29.0 (200)	—	29.8 (205.6)
—	—	—	11.0 (76.0)	—	—
102 (703)	105 (724)	143 (986)	160 (1103)	130 (896)	115 (793)
46 (317)	80 (552)	118 (814)	135 (930)	60 (414)	52 (359)
61	25	13.3 (in 4D)	18 (in 1")	61	61
—	40 (54)	—	—	—	—
B-84	B-99	C-28	C-33	B-94	B-90

Table 8.13A (Continued)

Comparative Properties of High-Temperature Alloy Steels

Chemical Analysis			Pyromet Alloy 600	Pyromet Alloy 625	Pyromet Alloy 800
Chemical composition by	Carbon		0.10 max.	0.10 max.	0.10
percent	Chromium		14.00/17.00	20.00/23.00	19.00/23.00
	Nickel		72.0 min.	Balance	30.00/35.00
	Manganese		1.0 max.	0.50 max.	1.5
	Phosphorous		—	0.015 max.	—
	Sulfur		0.015 max.	0.015 max.	0.015
	Silicon		0.50 max.	0.50 max.	1.0
	Molybdenum		—	8.00/10.00	
	Other Elements		Cu 0.50 max. Fe 6.0/10.0	Fe 5.00 max. Co 1.00 max. Al 0.40 max. Ti 0.40 max. Cb + Ta 3.15/4.15	Al 0.15/0.60 Ti 0.15/0.60 Cu 0.75 max. Fe Balance

Physical Constants

Density-lbs-mass/in³		*(Mg/m³)*	0.307 *(8.498)*	0.305 *(8.440)*	0.290 *(8.027)*
Specific Heat Capacity— Btu (Int)/lbm • °F *(kJ/kg • K)*			0.109 *(4.564)*	0.098 *(4.103)*	—
Electrical Resistivity at 70°F *(21°C)* ohm/circular mil foot *(microhm-centimeter)*			620 *(102.9)*	776 *(129.0)*	559 *(93.0)*
Coefficient of Thermal Expansion 10⁻⁶/°F *(10⁻⁶/°C)* 68°/1200°F *(20°/649°C)*			8.6 *(15.5)*	8.20 *(14.8)*	10.2 *(18.0)*

Room Temperature Mechanical Properties

Tensile Strength ksi		*(MPa)*	80/100 *(552/690)*	120/140 *(827/965)*	75/100 *(517/690)*
Yield Strength 0.02% ksi		*(MPa)*	30/45 *(207/310)*	60/75 *(414/517)*	25/50 *(172/345)*
Elongation in 2", % **(0.5m)**			55/35	30/55	50/30
Rockwell Hardness			B84 max.	—	B95 max.

Typical Elevated Temperature Tensile Strength—ksi *(MPa)*

	°F	*(°C)*			
	700°F	*(371°C)*	—	—	—
	800°F	*(427°C)*	88.5 *(610)*	131.5 *(907)*	74.6 *(514.3)*
	1000°F	*(538°C)*	84.0 *(579)*	130 *(896)*	72.0 *(496)*
	1200°F	*(649°C)*	65.0 *(448)*	119 *(820)*	54.0 *(372)*
	1300°F	*(704°C)*	—	—	—
	1400°F	*(760°C)*	27.5 *(190)*	78 *(538)*	32.1 *(221.3)*
	1500°F	*(816°C)*	—	—	24.8 *(171)*
	1600°F	*(871°C)*	15.0 *(103)*	40 *(276)*	—

Typical Stress for Rupture—ksi *(MPa)*

In 10 Hours at	1200°F	*(649°C)*	34 *(234)*	—	—
	1300°F	*(704°C)*	—	64 *(441)*	—
	1350°F	*(732°C)*	—	—	—
	1400°F	*(760°C)*	13 *(90)*	40 *(276)*	—
	1500°F	*(816°C)*	—	25 *(172)*	—
In 100 Hours at	1200°F	*(649°C)*	23 *(159)*	75 *(517)*	28.0 *(193)*
	1300°F	*(704°C)*	—	47.5 *(328)*	—
	1350°F	*(732°C)*	—	—	—
	1400°F	*(760°C)*	8.4 *(58)*	28 *(193)*	12.5 *(86.2)*
	1500°F	*(816°C)*	—	15 *(103)*	—
In 1000 Hours at	1200°F	*(649°C)*	14.5 *(100)*	58.5 *(403)*	19.5 *(135)*
	1300°F	*(704°C)*	—	34.5 *(238)*	—
	1350°F	*(732°C)*	—	—	—
	1400°F	*(760°C)*	5.6 *(39)*	18 *(124)*	8.5 *(58.6)*
	1500°F	*(816°C)*	—	—	—
	(1) 100/1600°F *(38/871°C)*				

HAYNES Alloy No. 25 (L-605)	HASTELLOY Alloy X
0.10	0.05/0.15
20.00	20.50/23.60
10.00	Balance
—	1.00 max.
—	0.04 max.
—	0.03 max.
—	1.00 max.
—	8.00/10.00
Fe 3.00 Co Balance W 15.00 Other 3.00	Fe 17.00/20.00 Co 0.50/2.50 W 0.20/1.00 B 0.01 max.
0.330 (9.130)	0.297 (8.221)
0.092 (3.850)	0.116 (4.856)
531 (88.6)	—
8.20 (14.8)	—
146 (1007)	109.5 (755)
67 (462)	55.9 (385)
64	45 (310)
C24	
—	—
—	99.7 (687)
—	94.0 (648)
103 (710)	83.0 (572)
—	—
—	63.1 (435)
50 (345)	—
—	36.5 (252)
—	67 (462)
—	—
—	—
—	32 (221)
30 (207)	—
—	48 (331)
—	—
—	—
—	22.5 (155)
22 (152)	—
—	34 (234)
—	—
—	—
—	15.8 (109)
17.5 (121)	—

Table 8.13B Typical Mechanical Properties of Stainless Steel at Cryogenic Temperatures[a]

	Test Temperature		Yield Strength 0.2% Offset		Tensile Strength		Elongation in 2"	Izod Impact	
	°F	°C	psi	MPa	psi	MPa	%	ft. lbs.	J
Type 304	−40	−40	34,000	234	155,000	1,069	47	110	149
	−80	−62	34,000	234	170,000	1,172	39	110	149
	−320	−196	39,000	269	221,000	1,524	40	110	149
	−423	−252	50,000	344	243,000	1,675	40	110	149
Type 310	−40	−40	39,000	269	95,000	655	57	110	149
	−80	−62	40,000	276	100,000	689	55	110	149
	−320	−196	74,000	510	152,000	1,048	54	85	115
	−423	−252	108,000	745	176,000	1,213	56	—	—
Type 316	−40	−40	41,000	283	104,000	717	59	110	149
	−80	−62	46,000	317	118,000	814	57	110	149
	−320	−196	75,000	517	185,000	1,276	59	—	—
	−423	−252	84,000	579	210,000	1,448	52	—	—
Type 347	−40	−40	44,000	303	117,000	807	63	110	149
	−80	−62	45,000	310	130,000	896	57	110	149
	−320	−196	47,000	324	200,000	1,379	43	95	129
	−423	−252	55,000	379	228,000	1,572	39	60	81
Type 430	−40	−40	41,000	283	76,000	524	36	10	14
	−80	−62	44,000	303	81,000	558	36	8	11
	−320	−196	88,000	607	92,000	634	2	2	3

[a]Data courtesy of Carpenter Steel Div., Carpenter Technology Corporation, Reading, Pennsylvania.

5. High-Alloy Steels—Heat-Resistant Steels: Superalloys

Heat-resisting steels (superalloys) have as their primary properties good resistance to high-temperature creep and rupture, good oxidation and corrosion resistance, and sufficient ductility for fabrication operations. In addition to the alloys produced by conventional methods, the new rapid solidification rate (cooling at up to 106 °C/sec) process yields powders that can be formed by powder metallurgy techniques to have the following advantages.

> Unusual structures can be obtained.
> Dispersions of fine intermetallics can be obtained.
> Large amounts of elements can be held in supersaturated solution, thereby permitting production of new alloys by altering precipitation sequences and aging kinetics. These new alloys have raised the allowable design temperature limit by more than 200°C over conventional alloys, and the end is not in sight.

Superalloys of a more conventional sort are alloys based on iron, cobalt, and nickel that contain chromium for resistance to oxidation and hot corrosion and other elements for strength at elevated temperatures (see Table 8.13). In iron-based superalloys (either solid-solution or precipitation-hardening) the effect of the added elements is as follows.

> Aluminum forms γ' $Ni_3(Al,Ti)$, precipitated in the face-centered cubic alloy.
> Niobium (columbium) and tantalum form ordered body-centered tetragonal γ'' and metal carbides that strengthen the alloy.

Titanium forms γ' Ni$_3$(Al,Ti) and metal carbides that strengthen the face-centered cubic alloy.

Carbon forms reinforcing metal carbides and stabilizes the desirable face-centered cubic structure.

Phosphorus promotes general precipitation of carbides.

Nitrogen forms strengthening metal carbonitrides.

Chromium provides oxidation resistance and solid solution strengthening.

Molybdenum and tungsten provide solid solution strengthening and provide metal carbides for strengthening by precipitation.

Nickel stabilizes the desirable face-centered cubic matrix, forms the desirable γ' strengthening agent, and inhibits formation of deleterious phases.

Boron and zirconium improve the creep properties and inhibit undesirable grain boundary constituents.

Lanthanum improves oxidation resistance.

Iron-based superalloys have found considerable usage in gas-turbine engines as blades, disks, casings, and fasteners. Additional information is presented in Volume 3, *Metals Handbook* (Ninth Edition) published by the American Society for Metals.

III. HEAT TREATMENTS OF STEEL

The various forms of steel were described in Sec. II.F. The purpose of heat treatment is to achieve under controlled conditions a desired microstructure of the finished steel. These heat treatments are variously called stress-relief annealing, grain refining and machinability improvement annealing, normalizing, spheroidizing, quenching, tempering, and surface hardening. In each case the time-temperature-transformation (T-T-T) curves become of great interest since, as shown in Figs. 8.27 and 8.28, the thermal history dictates the phase composition and hence properties of the alloy.

A. Stress-Relief Heat Treatment

A heat treatment used to relieve stresses that remain locked in a structure as a consequence of manufacturing procedures is called stress relief. It is generally the uniform heating of a structure to a suitable temperature below the transformation range, holding at this temperature to ensure uniformity, and then cooling at constant rate, to relieve stresses without introducing new stresses. The benefits are reduction of distortion, stress corrosion cracking, and creep. In fact, stress relief may be looked at as a microcreep step at the stress-relief temperature. The degree of stress relief at a given temperature is given by the Larson-Miller relationship: Effect = T($^\circ$Rankin) [log t (hr) + 20] (10^{-3}). Thus, holding a piece at 1100°F for 6 hr provides the same degree of residual stress relief as 1200°F for 1 hr.

B. Annealing

This is a generic term for a heat-treat cycle (heating and cooling) that is primarily used to soften metallic materials. Both stress relief and phase transformation are contemplated by annealing. The critical temperatures for annealing of steel are those that define the onset and completion of the transformation to or from austenite. This depends on whether the steel is being heated or cooled. Figures 8.27 and 8.28 illustrate the point and also show the effect of carbide-forming alloying elements.

The events of annealing are austenization at temperatures above Ae_3 followed by slow cooling. The grain of hot-worked products is refined, and machinability is improved. For cold-worked products a temperature somewhat below Ae_3 is used to improve ductility for further working. Annealing produces lowest hardness, strength, and wear resistance.

C. Normalizing

In steels containing 1% or more carbon, previous processing treatments involving slow cooling from a high temperature frequently may cause the excess cementite to form a network around the grains. This network is very stable and it makes further heat-treating operations more difficult because it is unaffected by ordinary annealing temperatures, and it does not go entirely into solution at the regular hardening temperatures. The treatment commonly employed to break up this network and to keep it from reforming is known as *normalizing*. It consists of heating above the A_{cm} temperature line (Fig. 8.4) to dissolve the excess cementite in austenite, and then of cooling at an intermediate rate, such as air cooling, so that sufficient time is not allowed for the excess cementite to reform a network or large plates. Instead, the structure is a mixture of fine cementite and pearlite, and is a little harder, stronger, and less ductile than annealed steel of the same composition.

Normalizing is also employed to *homogenize* structures in mild carbon steels, especially heavy forgings. Owing to unequal deformation in hot working, and possible thermal gradients during working, the structure of such parts may not be uniform. The normalizing treatment is performed at a high enough temperature (farther above A_{c3} temperature line than for full annealing) to permit the formation of a uniform austenite solid solution. The normalizing treatment is usually followed by heating to below the transformation temperature to promote softness and improve ductility. Steels given this treatment after normalizing generally have equal or slightly lower strength but greater ductility than before treatment; the two treatments are known collectively as *homogenization*.

D. Spheroidizing

If a normalized high-carbon steel is reheated to just below the A_1 line, the fine particles of carbide coalesce into spheres, hence the name spheroidizing. In this condition the steel is soft and readily machinable. More particularly, it is well suited to absorption of carbides during subsequent hardening, which improves the potential hardness. The process is often used as a treatment preliminary to hardening of spring steel.

E. Quenching

The T-T-T curves are a very useful guide to the quenching rates required to develop desired structure in a particular kind of steel. They also explain why very fast quenching is required for hardening plain carbon steels, whereas alloy steels may be given equal hardness with slower cooling rates. Perhaps the easiest way to utilize the T-T-T curves for these purposes is to superpose on them the cooling curve obtained in quenching.

Since the T-T-T curves are obtained by isothermal quenching, some justification is required for superposing on them the continuous cooling curves obtained by immersion in room-temperature water, brine, oil, or air. Studies* have shown that the T-T-T curves are displaced to some extent under continuous cooling conditions. For example, the start of transformation for eutectoid plain carbon steel is lowered and moved to the right—the gate being increased from 1 to 2 sec at 800°F instead of 1000-1100°F. Such small changes are important for very small parts, like razor blades, but would have no influence on heat treatment of most components. With the exception of this effect, the continuous cooling curve may be interpreted on the T-T-T curve as a series of short isothermal quenches, each at a lower temperature. Intersection of the cooling curve with the T-T-T start curve will indicate start of transformation. Time at this temperature, in proportion to the total time for transformation, may be used to indicate the degree to which the indicated transformation has progressed.

In Fig. 8.30 cooling curve A might represent slow cooling in a furnace, yielding coarse pearlite as shown. Curve B could represent the temperature at the center of a thick bar during water quenching, and curve C could represent the surface temperature for water quenching the same part. Curve C has a cooling rate fast enough to avoid any proeutectoid ferrite or cementite, pearlite, or bainite. The transformation on continuous cooling is therefore martensite at the surface. Curve B indicates the center cooled more

---

*R. A. Grange and J. M. Kiefer, *Trans. Am. Soc. Metals,* Vol. 29, 1941, pp. 85-116.

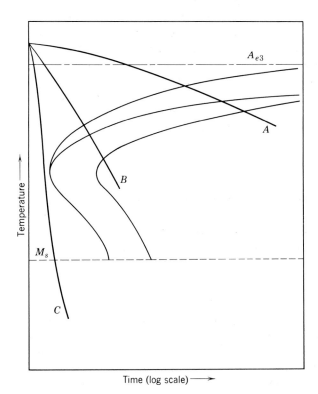

Fig. 8.30 T-T-T curve and three cooling curves.

slowly, and that fine pearlitic structure was formed. This part would therefore show full hardness at the surface grading to fine pearlite at the center for the cooling rate shown.

A cooling curve, which enters the knee and recrosses the transformation start curve before crossing the transformation end curve, causes a split transformation. A portion of the material transforms where the first crossing occurs, and the remainder where the cooling curve later crosses the bainite or martensite regions. The term *slack quench* is often used to denote this rate of cooling.

The space between the T-T-T curve knee and the vertical axis may be looked upon as a "gate"* through which the cooling curve of the steel must pass for full hardening. The faster cooling rates are hazardous to components because of the stresses introduced. These stresses are partly due to thermal gradients but mainly are the result of the $\gamma \to \alpha$ transformation. The transformation from γ to α is accompanied by a decrease in density because of the "closer packing" of atoms in the face-centered lattice. At 1670°F, the densities of the phases are $\alpha = 7.54$ g/cm^3, and $\gamma = 7.63$ g/cm^3. The volume changes may be indicated by an instrument called a *dilatometer,* which measures change in length. The dilatometer curves in Fig. 8.31 show that the volume changes and transformations

---

*A term coined by H. C. McQuaid.

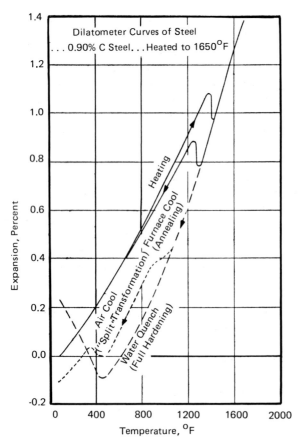

Fig. 8.31 Dilatometer curves of 0.9% C steel, heated to 1650°F and cooled at three different rates.

occur at lower temperature for faster cooling rates. Since temperature gradients are produced during the quenching operations, the transformations in near adjacent material may take place at different times, producing considerable stress. Design can help mitigate such stresses.

An *interrupted quench* may be used to reduce the stress and distortion produced. The heat treater may quench a part to below the knee in water, and then take advantage of the bay opposite the bainite region to remove the part from the water and quench it from just above the M_s in oil. In some high-carbon and alloy steels the bainite bay is so wide that time is available to straighten or perform other work on the steel while it is still in the ductile austenitic state, without affecting its capacity to harden to martensite.

A process termed *martempering* developed by B. F. Shepherd* makes use of this technique. The steel is quenched into a molten bath held slightly above the M_s point. The bath is chosen so the quench will exceed the critical rate at the gate. The part thus is quenched just fast enough to avoid transformation to any higher temperature products. It is maintained in the bath until temperatures are equalized but not long enough for bainite to form. Then it is allowed to transform to martensite by quenching in *air* (see Fig. 8.32). It is claimed this treatment improves ductility and toughness, for the same hardness, possibly because stress gradients are lower than for continuously quenched steels.

The name given to isothermal quenching of steels in the manner used by Bain and Davenport in developing the T-T-T curves is *austempering.* It is used for high-carbon and low-alloy steels of suitable composition and small cross section (½ in. thick or less). Such parts are quenched in a bath at the temperature that produces the desired structures. After holding in the bath for the time required for the transformation to end, the part is removed and cooled to room temperature in air (see Fig. 8.32). The process cannot be used to develop a very hard martensitic structure, such as is obtained by water quenching and tempering at 200 or 300°F. However, bainite structure corresponding to tempered martensite of any required degree of fineness may be produced. Steels so treated are more ductile and tough than similar steels quenched to room temperature and tempered to produce corresponding structure with equal hardness and tensile strength. Freedom from submicroscopic cracks resulting from the drastic quench to martensite is advanced as the explanation.

F. Tempering

This is a process to increase ductility and toughness as well as to relieve stresses and reduce hardness developed by previous operations. Previously hardened or normalized steel is heated to a temperature below the transformation range and cooled at a suitable rate. The principal variables that affect the results are temperature, time at temperature, cooling rate, and composition of the steel. As the temperature of tempering is raised, room-temperature hardness and strength decrease. The time at temperature directly affects the hardness; after an initial precipitous decrease, the hardness decreases on a straight line as a function of the logarithm of time. The cooling rate affects toughness; a slow cooling rate produces a decrease in toughness. (This is called "temper embrittlement.") Temper embrittlement seems to occur at grain boundaries and may be decreased by minimizing

---

*See *Product Eng.,* July 1945, p. 438, and August 1945, p. 515.

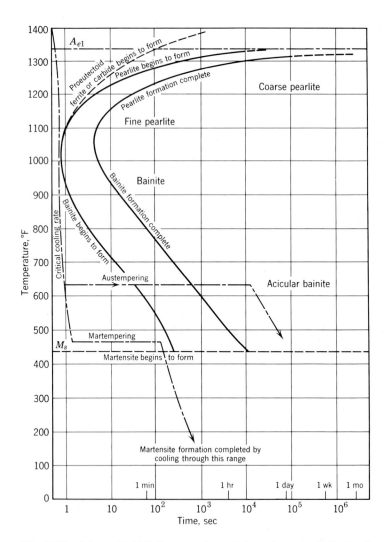

Fig. 8.32 Schematic T-T-T curves for a plain carbon steel. Cycles for martempering and austempering are indicated.

the content of trace elements, like tin, arsenic, antimony, and phosphorus. Addition of small amounts of molybdenum seem to retard embrittlement.

The effect of the steel composition on tempering varies with the additive (Figs. 8.33 and 8.34). Carbon has a marked effect; it produces a large as quenched increase in hardness, and this is retained after tempering. Alloying elements that form little or no carbides (e.g., nickel, silicon, aluminum, and manganese) remain in solution and have only a minor effect on tempered hardness. The carbide-forming elements (e.g., chromium, molybdenum, tungsten, vanadium, tantalum, niobium, and titanium) retard the softening process by formation of alloy carbides. The effect is minimal at low tempering temperatures; at high temperatures alloy carbides form, and as with highly alloyed steels, hardness may actually increase.

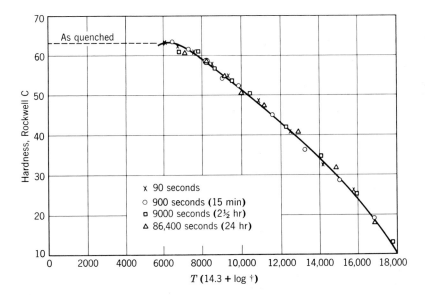

Fig. 8.33 Temperability of plain carbon steel. T = absolute temperature, t = time. (From J. H. Hollomon, and L. D. Jaffe, *Ferrous Metallurgical Design,* John Wiley & Sons, New York, 1947.)

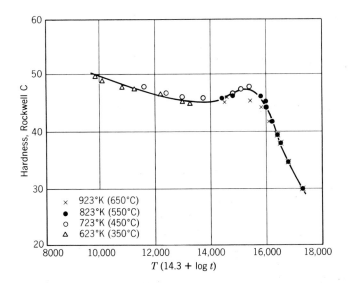

Fig. 8.34 Temperability of a carbide-forming steel. Secondary hardening is evident at 15,000 = T(14.3 + log t). (From J. H. Hollomon, and L. D. Jaffe, *Ferrous Metallurgical Design,* John Wiley & Sons, New York, 1947.)

G. Martempering

Martempering is a term to describe an elevated-temperature quenching procedure aimed at reducing cracks, distortion, and residual stresses. The process produces fairly uniform martensite through the structure. After cooling the primary martensite is brittle and must be tempered. The product has lower residual stresses than those developed by conventional quenching, and crack susceptibility is markedly reduced.

The steps of martempering are (1) quenching from the austenitizing temperature into a hot fluid medium kept above the martenizing range, (2) holding in the medium until the temperature is uniform throughout the steel, and (3) cooling to room temperature at a moderate rate to prevent steep temperature gradients between the outside and center of the piece.

H. Austempering

This is the isothermal transformation of a ferrous alloy at a temperature below that of pearlite formation and above that of martensite formation (see Fig. 8.4). The advantages of this process are increased ductility and notch toughness at a given hardness, reduced distortion, and shortest overall time cycle to through harden within hardness range of 35-55 (Rockwell C).

The details of the process comprise (1) heating steel to an austenitizing temperature (790-870°C); (2) quenching the steel in a constant temperature bath (260-400°C); (3) allowing steel to transform to bainite (dispersion of Fe_3C in Ferrite) in this bath; and (4) slowly cool to room temperature in still air (U.S. Patent 1924 099).

I. Surface Hardening of Steel

Many parts used in industry must have a hard, wear-resisting surface and a soft, ductile core. Sometimes this soft core is desired to facilitate forming or machining operations, sometimes for impact resistance during service. For such a combination of properties, it is not feasible to harden and quench a high-carbon steel to obtain surface hardness, particularly when the hardness may be required on only a small area. Two other methods make possible such surface hardening. When the full surface is to be hardened, a method known generally as *case hardening* is usually employed. In it, the surface is impregnated with carbon or some other hardening agent to give, through quenching, a hard surface. The hardenability of the core remains unchanged, and it retains its softness after quenching. Cyaniding, carburizing, and nitriding are operations of this type.

The second method, using laser, flame, and induction hardening, was developed especially for hardening small areas. A steel having the necessary hardenability is used, and it is hardened locally. Use of these methods to produce surface hardness is increasing, because of the short time in which the operation may be performed, the low costs, and the little distortion produced as compared with the furnace heating of pieces in the case-hardening methods. These advantages are especially important on large parts, such as gears, or on parts processed on an assembly line.

1. Cyaniding

The cyanide process is particularly applicable to small parts requiring a thin, hard, wear-resisting surface. Typical parts cyanided are small gears, ratchets, pawls, pins, bushings, screws, and small hand tools. The process consists in heating the steel in a molten bath of sodium cyanide and then quenching in water or oil. The time of heating will vary from several minutes to 4 hr, and a case depth up to 0.025 in. is obtainable.

2. Carburizing

Carburizing is used for heavier parts or for those that require a deeper case (up to 0.70 in.) than is normally developed in cyaniding. Two general methods, pack carburizing and gas carburizing, are in use. In the former the steel pieces, either stamped or fully machined and containing only a small allowance for finish grinding, are packed in solid "carburizing compound" (mixture of crushed wood or bone charcoal and an energizer, usually a carbonate), and heated for periods ranging from 4 to 72 hr. The heating causes the formation of carbon monoxide gas, which reacts with the iron to form carbon and CO_2. The carbon dissolves in austenite and penetrates below the surface by diffusion. Case depths of less than 0.025 in. should not be specified for pack-carburizing work.

In the gas method the pieces are heated in a furnace having an excess atmosphere of hydrocarbon gas and carbon monoxide. The gas can also be obtained by employing a volatile hydrocarbon liquid. The same action takes place as in the pack method, but is faster, cleaner, and permits better control of the depth of the hardened area. With either method the pieces have a surface carbon content ranging from 0.8 to 1.4% C. They are reheated to a temperature somewhat lower than that used in carburizing, then quenched in oil or brine to develop the required hardness. A preliminary higher temperature quench may be used to refine the grain of the core if it has grown too much during the carburizing cycle.

Surfaces can also be decarburized if they are heated in a hydrogen atmosphere. Where this treatment is purposely employed, promotion of surface softness is usually not the direct purpose. Decarburization is sometimes used to decrease carbon content of silicon steel sheet for motor applications.

3. Nitriding

Whereas almost any type of plain, low-carbon steel, or low-carbon alloy steel may be carburized or cyanided, a steel containing a nitride-forming element (e.g., Cr, Al, Mo, or V) is required for nitriding. The process consists in heating the parts in a sealed container, through which is passed a stream of ammonia gas. The usual temperature is 1000°F, and the time cycle may range from 8 to 96 hr. The nitrided case may be up to 0.030 in. depth, longer time being necessary for the heavier case. It is by far the hardest case produced by any of these processes. An advantage of nitriding is that distortion is reduced to a minimum. This is because of the low temperature used, and because the parts are cooled slowly, no quenching is needed.

Besides the hardness and excellent wear resistance obtained, the surface has good resistance to corrosion by alkalies, crude oil, tap water, salt water (not in motion), ethyl gasoline, and other media, and retains its hardness at elevated temperatures. The process is relatively costly because a special steel must be used, the equipment is expensive, and a long time cycle is required.

4. Case Hardening Small Areas

When it is desired to case harden only part of the surface, those areas to be soft may be protected by a metal plating. Copper plating is the correct protection for the cyanide and carburizing processes, and tinning or nickel plating should be used to oppose nitriding.

5. Case Depth

In the literature, case depth is usually reported as the *total* observed by metallographic methods. This often leads the designer or shopman to believe that he can grind to almost the full case depth and still have uniform hardness or wear resistance. Such is emphatically not true. For example, cyanided parts should not be ground over 0.003 in. per side, no matter how deep the case is. On carburized parts the depth of "effective," or useful, case will average about half the total. Nitrided parts should be finish ground, polished, or lapped only deep enough to remove the "blush" or frosty deposit on the surface.

Boron and silicon can be impregnated in low-carbon steel surfaces by processes similar to carburizing. The processes are known, respectively, as boriding and siliconizing (Ihrigizing), and both result in very hard wear-resistant surfaces.

J. Hardenability

Hardenability of steel is the property that determines the depth and distribution of hardness induced by quenching. Hardenability is usually the most important factor in the selection of steel for heat-treated parts. Note the distinction between hardness and hardenability. The maximum attainable hardness depends solely on the carbon content of a steel; hardenability is governed almost entirely by the chemical composition (carbon and alloying elements) at the austenitizing temperature and the austenite grain size at the instant of quenching. However, ingredients that are not dissolved at the austenitizing temperature do not contribute to the hardenability and may even decrease hardenability.

Hardenability is measured by making hardness measurements across the diameter of a bar subjected to a given quench. Figure 8.35 shows hardness versus cross section for hardened bars of various diameters. Such curves clearly show that the 4140 steel is *deeper hardening* (or *hardens through* to a greater extent), under a given quench, than

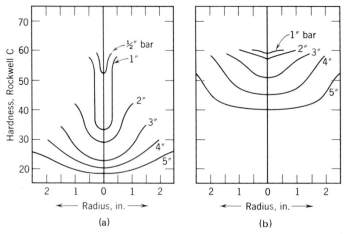

Fig. 8.35 Hardness versus cross section for hardened bars of various diameters. (a) SAE 1045 quenched in water. (b) SAE 4140 quenched in water.

1045 steel. The latter hardens through for a ½ in. bar but is *shallow hardening* on thicker bars. Since this method involves considerable testing time, and must be repeated for each different quench, it is giving way to the Jominy method.

In the Jominy test for hardenability of steel, a sample of approximately 1 in. diameter and 4 in. long is heated to the proper temperature and then quenched on one end by means of a standardized "fountain" of water, as illustrated in Fig. 8.36. This produces very rapid cooling on the quenched end, with progressively slower cooling along the bar. The rates of cooling at different points have been carefully determined and are given in Fig. 8.37.

After quenching, two flats are ground lengthwise on the bar, and indentation hardness readings taken at measured distances from the quenched end (see Fig. 8.38). The hardenability specifications mentioned in Sec. III.I are of this form, two curves being used. One establishes the high values, the other the low, thus forming a band of hardness to be achieved in the Jominy test.

Comparison of hardness readings on the Jominy bar with the rates of cooling given in Fig. 8.37 makes it possible to ascertain the hardness that would be obtained in a section of a part made of the same material and cooled at some specified rate. Conversely, if the hardness of the material of some part is known, the cooling time can be determined by reference to Jominy tests for the material. It is also possible, on the basis of these tests, to select a material that will have the proper gradient of hardness for the cooling times given by a quench, such as in water or oil, by plotting the desired Jominy hardness curve and comparing it with the curves for the available alloys.

The Jominy test can be made relatively quickly and cheaply, and agreement between various laboratories has been good. In addition, the agreement between results on cast and rolled sections has been good for plain carbon and low-alloy steels, and is about 5 Rockwell C lower for cast deep-hardening steels than for wrought. This enables use of the test prior to rolling a batch of steel at the mill to determine ability to meet hardenability specifications.

There are some disadvantages. Presence of pearlite is indicated by hardness measurements, but presence of bainite is not. Therefore, bainite structures may be present with consequent influence on properties. Steels that are segregated or otherwise not homogeneous will have large deviations from a Jominy bar result. Steels subjected to a large amount of work may differ from cast bar Jominy test. And finally, a great deal of internal stress may affect transformation rates and introduce error in interpreting a Jominy

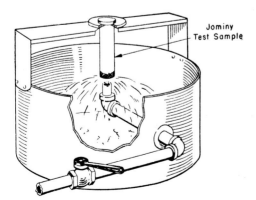

Fig. 8.36 Sketch of Jominy end-quenching operation. (Courtesy Joseph T. Ryerson & Sons.)

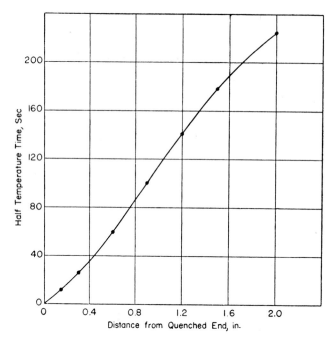

Fig. 8.37 Cooling rates on a Jominy test bar expressed as half-temperature. (From *Correlation between Jominy Test and Quenched Round Bars*, M. A. Grossman, *Trans. Am. Inst. Mining, Met. Engrs.*, Vol. 150, 1942, pp. 337-259.)

bar test for properties in parts of, say, 2 in. diameter. These factors must be considered if Jominy tests are to be applied accurately.

The variables that affect hardenability are the alloying elements present, the grain size of the steel, the homogeneity of the starting steel, and the homogeneity obtained in the austenite before quenching.

Carbon is the primary hardening element in steels. The "potential" maximum hardness that can be developed is practically unaffected by alloys, and depends almost entirely on carbon. Figure 8.39 shows the relation of carbon to the hardness of steel. The upper curve indicates hardness for 100% martensite. This is seldom achieved in commercial

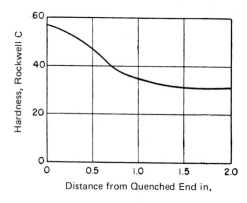

Fig. 8.38 Jominy hardness for NE 9440 alloy steel. (Courtesy of Joseph T. Ryerson & Sons.)

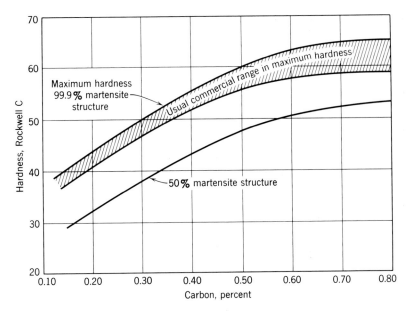

Fig. 8.39 Relation of carbon to the hardness of plain carbon steel. Data apply approximately to low- and medium-alloy steels. (*Metals Handbook,* 1948 edition, p. 497.)

practice, and values of 5-10 Rockwell C less are generally considered full hardness. The median line is for a hardness with 50% martensite and 50% pearlite, which is often called half-hardening. It applies, of course, only to steels that do not develop bainite in transformation.

Grain size, too, has an important influence on hardenability. Since the transformation of austenite begins principally at the grain boundaries, small grain size increases the likelihood of early transformation. Thus, small grain size decreases hardenability. In general, the effect of grain size is most pronounced just below the equilibrium-transformation temperature, and therefore influences pearlite hardenability to a greater extent than bainite. The ease with which grain size can be controlled suggests that large grain size would be advantageous in improving hardenability of shallow-hardening steels. The disadvantage of this practice is that the coarse-grained steel is liable to cracking on quenching, and has low impact strength at room temperature. For highly stressed parts, steels are maintained of fine grain size, and deep hardening is secured by alloying additions.

Lack of homogeneity exerts an influence on hardenability through premature transformation in segregated areas according to local composition, and through spread of the transformation over the whole section. Thus the effect of segregation and incomplete austenitization is to produce reduced hardenability.

K. Hydrogen Embrittlement and Control

Many forms of embrittlement can lead to brittle fracture of steel parts. In every case a mechanism that decreases ductility is at work. Generally the appearance of a phase that enhances stress risers in the body of the metal, thus promoting cracking and carck propagation, is characteristic of embrittlement.

Although only one type of embrittlement will be discussed here, the reader should consult the American Society for Metals *Metals Handbook,* (Ninth Edition, Volume 1, 1978, pp. 683 ff) for a fuller discussion of two major classes of embrittlement: (1) thermal treatment or elevated temperature service embrittlement and (2) embrittlement by environmental conditions. In the latter case is found hydrogen embrittlement.

Hydrogen embrittlement is a long-standing problem in the use of steel. Absorption of hydrogen results in a general loss of ductility as evidenced by bending, fatigue, reduction in area, and elongation tests. Impact tests are not generally regarded as a good method for detecting hydrogen embrittlement.

Hydrogen is an extremely mobile element in steel and other metals. In the atomic form hydrogen is present interstitially, that is, between the grains. In the molecular form hydrogen is present in microcracks and may be present at dislocations. Hydrogen comes from many sources: water vapor, pickling, and electrolysis. The solubility of hydrogen in steel increases as the temperature increases, with a great increase at the melting point. The more hydrogen is present in the steel, the greater is the chance of porosity, flakes, or cooling cracks in the solid steel. The austenitic form of steel will dissolve more hydrogen than either of the ferritic forms. Carbon and aluminum additions to steel decrease the solubility of hydrogen in steel; manganese and nickel increases its solubility.

The degree of hydrogen embrittlement is a function of the strength level of the steel. Higher strength steels are more susceptible, particularly those with pearlitic, martensitic, and ferritic structures, although austenitic steels are not totally immune.

Hydrogen embrittlement may be countered by aging at room or elevated temperature. Hydrogen embrittlement during pickling can be lessened by use of passivating agents, such as organic cyclic amines. Hydrogen embrittlement in electroplating can be minimized by choice of plating chemicals, use of a passive underlay (e.g., electroless nickel), or by use of suitable passivating agents. Hydrogen embrittlement during welding can be inhibited by taking pains to avoid moisture in electrodes, fluxes, or ambient atmosphere. Hydrogen embrittlement during sour oil well drilling can be lessened by raising the well temperature above $150°C$ or increasing the pH to 6.0 or higher. Alternatively austenitic steels, nickel alloys, or cobalt alloys may be used, but even these are not totally immune to hydrogen embrittlement. Cold working increases the susceptibility of alloy steels to hydrogen embrittlement.

IV. CONCLUSION

It is evident that the irons and steels available to the design engineer constitute a broad and versatile group of low-cost high-reliability construction materials. Properties can be found over the spectrum of soft iron to hard steels from easily corrodible sheet to forms highly resistant to corrosion and high-temperature oxidation, and from brittle to tough over a wide range of temperatures. Size and commercial forms range from small rods and wire to heavy plate and castings. The designer's ingenuity is challenged by the abundance of choices to select the material that not only meets the design specification at lowest cost and with highest reliability but also meets society's need for energy conservation and environmental preservation.

Since work on this chapter began, the American Society for Metals has published the first five volumes of the monumental 9th Edition of its *Metals Handbook*. These, with the ASM sourcebooks on *Industrial Alloy and Engineering Data* and *Selection and Fabrication of Aluminum Alloys,* should be available to all designers. In addition the technical data and advice disseminated by the Cabot Corporation, International Nickel Co., Carpenter Technology Co., Armco Corp., and the American Iron and Steel Institute (with apologies to all contributors not named here) should be sought by all those with serious decisions to make in the selection and fabrication of metals.

9

Nonferrous Metals and Alloys

I. INTRODUCTION*

A. Nonferrous Metals and Alloys in General

Nonferrous metals and alloys include all the metals in which iron is not present in large quantities. Because of their number and because the properties of individual metals vary widely, they provide an almost limitless range of properties for the design engineer. The range of available melting points is shown in Fig. 9.1.

   The nonferrous metals are generally used for parts requiring considerable fabrication and where the advantages gained from proper use of their characteristic properties justify their higher cost relative to iron-based (ferrous) alloys. No single property or group of properties will be the basis for every choice. But, one or more of the following criteria are often involved:

   1. Ease of fabrication
   2. Resistance to corrosion
   3. High electrical and thermal conductivity
   4. Light weight
   5. Reasonably high strength in the strain-hardened or heat-treated condition
   6. Wide range of modulus of elasticity (Fig. 9.2)
   7. Color

Nearly all the nonferrous alloys possess at least two of the qualities listed above in superior relationship to ferrous alloys, and some possess five or more. Fabrication of nonferrous alloys is usually much easier than for ferrous alloys except for the one fabrication area in which nonferrous alloys are somewhat inferior to ferrous alloys: weldability. Weldability is, however, not a major problem when considering the overall advantages in fabrication by forming nonferrous alloys, which typically have high ductility coupled with low yield points.

   Although it is true that corrosion resistance can be obtained in certain ferrous alloys, several of the nonferrous alloys possess this property without requiring special and expensive alloying. There are nonferrous alloys, or combinations of them, that will resist nearly any kind of corrosive attack.

---

*Contributed by DANIEL L. POTTS.

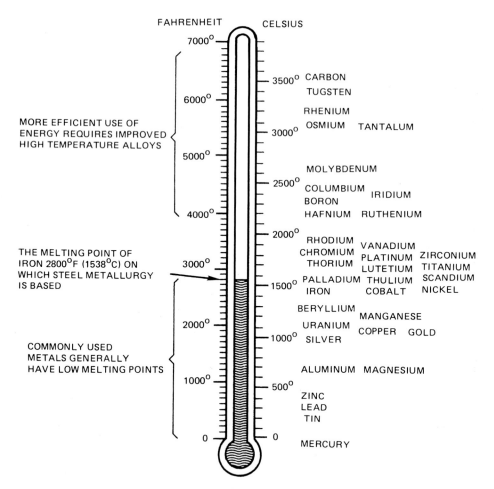

Fig. 9.1 Melting points. This chart shows all known elements with melting points above iron and major metals at lower temperatures. (Courtesy of the U.S. Department of the Interior, Bureau of Mines, Washington, D.C.)

The electrical and thermal conductivities of the nonferrous alloys vary widely from very much above the ferrous alloys to well below them. Copper is a relatively inexpensive metal and is widely used in electrical and electronic applications.

On a volume basis, the strength of the nonferrous alloys is generally below that of strong ferrous alloys. Compared on a weight basis, however, the specific stiffness (E/σ) of most of the engineering nonferrous alloys are interchangeable with the more popular ferrous alloys. (See Fig. 9.3.)

Several of the nonferrous alloys are used for their smooth finish and color variation. Aluminum and nickel and their alloys are noted for their bright silvery color. Copper and its alloys form a whole set of red, gold, and yellow finishes. Besides the generally good characteristics cited, nonferrous alloys are available in a wide variety of forms, finishes, and strain-hardened and/or heat-treated conditions.

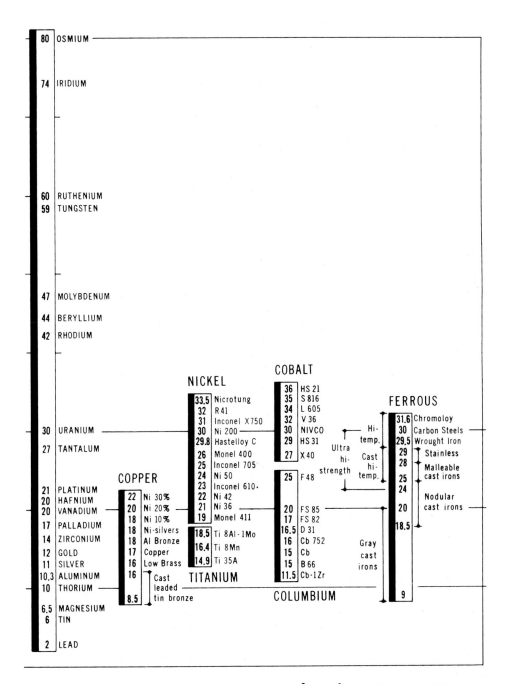

Fig. 9.2 Tensile modulus of elasticity of metals at 70°F (10⁶ psi). (Courtesy of Brooks & Perkins, Inc., Southfield, Michigan.)

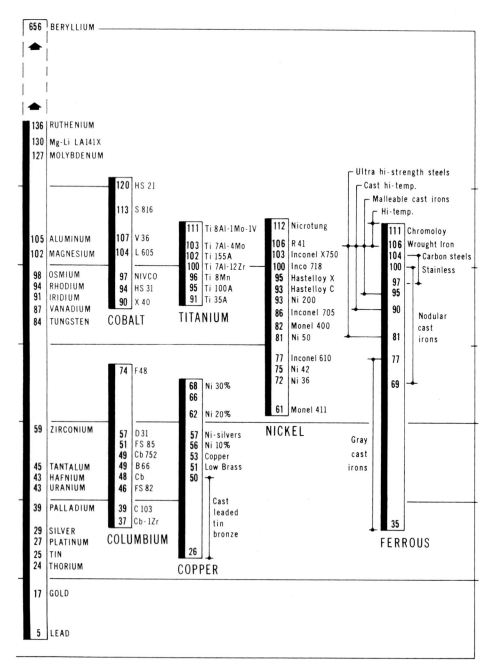

Fig. 9.3 Specific stiffness of metals at 70°F. Modulus of elasticity over density, 10^6 in. (Courtesy of Brooks & Perkins, Inc., Detroit, Michigan.)

B. Designation System

Major metal industries have organizations that establish coordinated systems of practice
in standards, specifications, tolerances, chemical compositions, and any other activity
regarding the business of selling the particular industry. Gone are the "good old days"
when competing companies within a metal industry touted alloys of identical composition as somehow better than those of the competition. The organizations referred to are
loosely termed "trade associations," such as the American Iron and Steel Institute (AISI),
the Aluminum Association (AA), the Copper Development Association (CDA), and so
forth.

   The Copper Development Association published an agreed-upon series of numerical
designations for copper and copper alloys in the 1960s. These designations take the form
of three digits after the CDA prefix, for example, CDA 110 for electrolytic tough pitch
copper.

   More recently, the American Society for Testing and Materials (ASTM), in conjunction with the Society of Automotive Engineers (SAE), initiated a unified numbering system (UNS) for metals and metallic alloys and published them in 1975. In this system
CDA 110 appears as UNS C11000. The first three numbers are the same as in the CDA
designations. The last two numbers are generally applied for variations within the chemical composition of the basic alloy, as in UNS C17500 for beryllium-copper alloy with
cobalt and UNS C17510 for beryllium-copper alloy with nickel replacing the cobalt.

   Similarly, the Aluminum Association designations fit nicely into the UNS designations; thus, AA2024 is the same as UNS A92024. Continuation of this rationale throughout yields a six-character-space alphanumeric designation of UNS numbers of national
practice in the United States.

   The unified numbering system is not in itself a specification since it establishes no
requirements for form, condition, quality, and so on. It is a unified identification of
metals and metallic alloys for which controlling limits have been established in specifications elsewhere. The unified numbering system provides the United States with a means
of correlating many nationally used numbering systems currently administered by socie-
ties, trade associations, and individual users and producers of metals and metallic alloys,
thereby avoiding confusion caused by the use of more than one identification number for
the same material, and by the opposite situation of having the same number assigned to
two or more entirely different materials.

   Since the unified numbering system is the preferred U.S. practice, UNS designations
are used throughout this chapter.

   A listing of UNS numbers assigned in alpha character followed by five digits for
major metals and metallic alloys follows.

   Axxxxx: Aluminum and aluminum alloys
   Cxxxxx: Copper and copper alloys
   Exxxxx: Rare earth and RE-like metals and alloys
   Fxxxxx: Cast irons
   Gxxxxx: AISI and SAE carbon and alloy steels
   Hxxxxx: AISI and SAE H-steels
   Jxxxxx: Cast steels (except tool steels)
   Kxxxxx: Miscellaneous steels and ferrous alloys
   Lxxxxx: Low-melting metals and alloys
   Mxxxxx: Miscellaneous nonferrous metals and alloys

Nxxxxx: Nickel and nickel alloys
Pxxxxx: Precious metals and alloys
Rxxxxx: Reactive and refractory metals and alloys
Sxxxxx: Heat; and corrosion-resistant (stainless steels)
Txxxxx: Tool steels, wrought and cast
Zxxxxx: Zinc and zinc alloys

II. ALUMINUM AND ALUMINUM ALLOYS*

Aluminum is the most abundant metallic element in the earth's crust. The principal ore
of aluminum is bauxite. Aluminum oxide derived from bauxite is dissolved in a molten
electrolyte and reduced electrolytically. This "commercially pure" aluminum is 99.5-
99.7% aluminum and is used for many decorative and electrical applications. The largest
use of commercial purity aluminum is to produce alloys of remarkable strength without
appreciable increase in density.

A. Aluminum Alloys

The principal alloying elements are copper, manganese, silicon, and magnesium; magnesium in combination with silicon, zinc, and other elements is also used. The unified numbering system identifies these alloys as follows. Wrought aluminum alloys are A9 plus
a four-digit number whose first digit indicates the alloy group, the second digit indicates
modifications of the original alloy or impurity limits, and the last two digits identify the
aluminum alloy or indicate the aluminum purity. Thus, A91xxx is aluminum with a minimum purity of 99.00%, and A91060 is a 99.60% minimum aluminum. A92xxx has copper, A93xxx has manganese, A94xxx has silicon, A95xxx has magnesium, A96xxx has
magnesium plus silicon, A97xxx has zinc, and A98xxx has other elements, as the major
alloying element. Cast aluminum alloys are similarly identified with A0 used in front of a
four-digit number whose last digit is prefaced by a decimal point. The last digit is 1 for
an ingot and 0 for a casting.

Alloying elements are used to change pure aluminum, which is not heat treatable
but work hardens, into a material with specially desirable properties. The effect of alloying is shown in Table 9.1.

B. Strengthening Processes

Non-heat-treatable alloys depend on the effect of alloying elements to harden and
strengthen the aluminum. (See Table 9.1.) In addition, aluminum and its alloys may be
cold worked for increased strength and stiffness. Such cold work must by done below the
recrystallization range of the particular alloy (650-800°F).

Dispersion hardening results from addition of elements that provide uniformly distributed microscopic particles that distort the metallurgical structure and thereby prevent
the slippage of crystal faces past each other.

Heat-treatable aluminum alloys contain elements that are more soluble at elevated
temperatures than at room temperatures. When the solid solution is rapidly quenched, a

---

*Contributed by ROBERT S. SHANE, based on material from the Second Edition using information
from *Metalworking with Aluminum,* 2nd ed., Aluminum Association, Inc., and *Sourcebook on Selection and Fabrication of Aluminum Alloys,* American Society for Metals, Metals Park, Ohio, 1978.

Table 9.1   Effect of Alloying on Wrought Aluminum Alloys

UNS	Major alloying element and effect	Principal uses and comments
A91xxx	None	Decorative, thermal and electrical conductor, excellent workability, low mechanical properties, excellent corrosion resistance; moderate increase in strength by strain hardening.
A92xxx	Copper (strengthens and hardens)	Heat treating can increase strength to that of mild steel; artificial aging will further increase yield strength with some loss of ductility; corrosion resistance is not as good as other aluminum alloys, and intergranular corrosion is a risk. These alloys are frequently clad with aluminum to protect against galvanic corrosion. A92024 is a very widely used aircraft alloy.
A93xxx	Manganese (increases strength of both dispersion and age hardening alloys)	These are generally non-heat treatable but have good workability. A93003 is a general-purpose alloy that is widely used.
A94xxx	Silicon (lowers melting point without embrittlement)	Melting point is substantially lowered without damaging ductility. Used in welding wire and as brazing alloy. Most of these alloys are non-heat treatable unless alloying constituents of the parent metal diffuse in.
A95xxx	Magnesium (increases strength and work hardening without heat treatment; improves weldability)	When magnesium alone or coupled with manganese is used, the result is a moderate to high strength non-heat-treatable alloy. The A95xxx alloys have good welding characteristics and good resistance to marine atmosphere corrosion. Special care with temper and operating conditions is required for high (>3.5%) magnesium alloys to avoid susceptibility to stress corrosion.
A96xxx	Magnesium and silicon (improves formability and corrosion resistance)	These are heat treatable. A96061 is one of the most versatile aluminum alloys. These alloys have good formability and corrosion resistance with moderate strength.
A97xxx	Zinc (increases strength markedly)	Usually coupled with a small percentage of magnesium and small quantities of other elements like copper and chromium. The alloys are heat treatable and have very high strength. An outstanding member of this series is A97075, which is one of the strongest alloys available and is used in air-frame structures and for highly stressed parts.

supersaturated condition is produced. As alloying elements precipitate out of the solution with the passage of time, the strength of the alloy increases. This is called precipitation or age hardening. Artificial aging (precipitation heat treatment at higher temperatures) is employed to develop maximum strength as quickly as possible.

C. Heat Treatments

The following classes of commonly designated non-heat-treatable aluminum alloys are, in fact, given heat treatments aimed at removing cold-working effects, residual thermal stress built up during cooling after quenching, welding, or casting (Tables 9.2 and 9.3).

D. Temper

The temper designation consists of a letter and numbers following the alloy and separated by a hyphen. The letter H is used for non-heat treatable alloys; the letter T is used for heat-treatable alloys. The letters F and O are used for all alloys. See Tables 9.4 and 9.5 for a synopsis of meanings.

E. Thermal and Electrical Conductivity

Commercially pure aluminum conducts heat and electricity about 60% as well as copper (Fig. 9.4 and Table 9.6). The lower density of aluminum (2.7 versus 8.3 for copper) results in a 50% weight saving for equal conductivity when aluminum is substituted for copper. The need for structural support of aluminum conductors prevents full realization

Table 9.2 Typical Results of Heat Treat on Tensile and Yield Strength (0.063 in. sheet)

Alloy	Yield strength (ksi)	Ultimate strength (ksi)
A91100-0	3	10
A91100-H14	14	16
A91100-H18	19	22
A95052-0	10	25
A95052-H34	26	35
A95052-H38	33	38
A95083-0	18	40
A95083-H343	38	50
A96061-0	8	15
A96061-T4	15	30
A96061-T6	35	42
A92024-0	10	25
A92024-T3	42	65
A92024-T86	65	72
A97075-0	13	30
A97075-T6	66	78
A97178-0	13	30
A97178-T6	73	85

Table 9.3 Non-Heat-Treatable Wrought Aluminum Alloys

Series	Example	Major alloying element	Typical %	Typical uses
A91xxx	A91060			Chemical equipment
	A91100			Deep-drawn parts
A93xxx	A93003	Mn	1.2	General
	A93004	Mn,Mg	1.2, 1.0	Pressure vessels, can stock
A94xxx	A94032	Si	12.0	Pistons
	A94043	Si	5.0	Brazing sheet
A95xxx	A95005	Mg	0.8	Anodized parts
	A95052	Mg	2.5	Appliances, bus, truck
	A95083	Mg	4.5	Boats, armor
A98xxx	A98001	Ni	1.0	Nuclear
	A98280	Sn	6.5	Bearings

of potential economies, but a substantial saving is achievable depending on the relative costs of each metal.

Aluminum alloys have substantially greater electrical resistivity than pure aluminum; thus A91100-0 has an electrical resistivity of 2.92 μohms-cm and A95056-0 has an electrical resistivity of 5.9. Hardening produces an appreciable increase in both thermal and electrical resistivity.

Table 9.4 Temper Designations of Non-Heat-Treatable Wrought Alloys

F	As fabricated
O	Fully annealed wrought products
H	Strain hardened
H1	Strain hardened only
H2	Strain hardened and partially annealed
H3	Strain hardened and stabilized by a low-temperature thermal treatment to improve ductility
H18	Strain hardened by a cold reduction of approximately 75% (lower digits are intermediate between O and H18)
H111	Strain hardened less than controlled H11 temper
H112	Strain hardened by shaping without close control other than by mechanical property limits
H311	Strain-hardened alloys containing more than 4% magnesium but less than H31
H321	Strain-hardened alloys containing more than 4% magnesium but less than H32
H323, H343	Strain-hardened alloys containing more than 4% magnesium with acceptable resistance to stress corrosion cracking

Table 9.5   Temper Designations of Heat-Treatable Alloys

F	As fabricated
O	Fully annealed wrought products
H	Strain hardened
W	Solution heat-treated alloy that spontaneously ages at room temperature over a long period of time
T	Thermally treated to produce stable tempers other than F, O, H; T is always followed by one or more digits
T1	Cooled from elevated shaping (e.g., casting or extrusion) and strength increases by room temperature aging
T2	Annealed cast products only, to improve ductility and dimensional stability
T3	Solution heat treated and cold worked
T4	Solution heat treated and naturally aged to a substantially stable condition
T5	Cooled from an elevated shaping temperature and then artificially aged to improve mechanical properties and/or dimensional stability
T6	Solution heat treated and then artificially aged
T7	Solution heat treated and then stabilized at elevated temperature to accelerate softening to ultimate limit
T8	Solution heat treated, cold worked, artificially aged
T9	Solution heat treated, artificially aged, and then cold worked
T10	Cooled from an elevated temperature shaping process, artificially aged, and then cold worked
	Second digits for each temper have the following meaning
T-51	Stress-relieved by stretching
T-510	Products that receive no further straightening after stretching.
T-511	Products that receive no minor straightening after stretching.
T-52	Stress relieved by compressing (after solution heat treatment
T-54	Stress relieved by combined stretching and compressing

F. Weight and Strength

Aluminum and aluminum-based alloys have densities of about 0.10 lb/in.^{3}, compared with 0.32 for copper, 0.28 for steel, and 0.065 for magnesium. By volume, aluminum is less than half as expensive as copper; it is the cheapest material for nonmagnetic parts of a given volume.

Commercial aluminum in the annealed or cast condition has a tensile strength about one-fifth that of structural steel. The strength may be more than doubled by cold-working, and by alloying aluminum with various other metals the strength may be even further increased. The ratios of strength to weight for some of the hardened aluminum alloys are as high as for any other structural material. Although some steels have comparable strength-weight ratios, the structural aluminum alloys have proved themselves superior in many applications, particularly in lightweight transportation equipment. Their superiority can be explained as follows.

Structural members are often column loaded and fail by buckling. The load sufficient to cause failure of the entire column depends upon the amount of inertia of the column cross section. In order to increase the load capacity for a given amount of material, special shapes, such as I beams or tubes, are used. The effective increase is limited by the thinnest part of the column. A material like aluminum alloy, which has thick sections

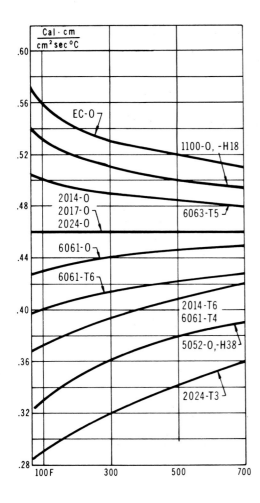

Fig. 9.4 Thermal conductivity of wrought aluminum alloys. (Courtesy of Brooks & Perkins, Inc., Detroit, Michigan; data courtesy of Alcoa Research Laboratory.)

in proportion to its weight, is ideal for lightweight columns. A typical application is in a stressed airplane skin, where compressive stresses between points on the skin may cause it to wrinkle. An aluminum alloy skin 1/16 in. thick would stand a great deal more column loading than a steel sheet 0.02 in. thick, even though they are of the same weight and their strengths in direct tension or compression would be about equal. This comparison holds true for the structural members within the airplane, and for small, light planes, wood may even be preferred to aluminum.

G. Fabrication

The melting point of pure aluminum is 1215°F, and the alloys usually melt at lower temperatures. These low melting points make them unsuitable for high-temperature applications, but do facilitate casting in sand, permanent molds, or under pressure in dies. The physical properties of aluminum alloys cast in metal molds are, in general, better than those of the same alloys cast in sand. The metal mold chills the molten metal far more rapidly, and the grains are given only a short time to grow. In a casting that has a varying cross section, the thinnest sections will cool first and have the finest grains.

Table 9.6 Thermal Conductivity of
Wrought Aluminum Alloys (cgs Units)

Alloy and Temper	Current Value	Previous Listed Value
2011-T3	.36	.34
2014-T4	.32	.29
2017-O	.46	.42
2017-T4	.32	.29
2024-T6	.36	.35
3003	.42	.46–.37
5005	.49	.48
5056	.28	.26–.28
5457	.42	.40
6053-T6	.39	.37
6061-O	.43	.41
6061-T6	.40	.37
6062-O	.47	.46
6062-T4	.41	.37
6062-T6	.43	.42
6151-O	.49	.46
6151-T4	.39	.37
6151-T6	.41	.42
7075-O	.41	.40
7075-T6	.31	.29
7079-T6	.30	.29
7178-T6	.30	.29

Source: Courtesy of Brooks & Perkins,
Inc., Detroit, Michigan.

Pure aluminum is seldom used for castings because of its hot shortness and high shrinkage. It is used only where high conductivity is demanded, for example, in cast squirrel-cage induction motor rotors.

First forming is usually done at a plastic temperature of 750-900°F, because aluminum and aluminum alloys are relatively brittle in ingot form. The hot forming may be continued at 500-900°F, depending on the type of alloy and the process. Many forms, such as the larger sizes of tubing, rod, or plate, are extruded or hot rolled to their final forms. The smaller tubing and other forms, such as wire and sheet, are finished by cold working.

A relatively new development in aluminum products is the use of pressings, which are made from suitably shaped slugs of alloys and squeezed into shape in a press. Pressings may be made hot or cold, using any alloy that can be extruded or forged.

Aluminum is inherently very reactive, and it spontaneously forms a thin oxide film when exposed to air. This oxide film must be removed and prevented from re-forming while soldering; otherwise, a very weak joint will result. Hence, ultrasonic or rubbing bonding under a film of solder is sometimes used. Even with proper precautions, the soldering of aluminum alloys is frequently unsatisfactory, because the joint between aluminum and any solder thus far developed may, in time, decompose electrolytically, except as noted in Chap. 22 covering specialized pretreatment, fluxes, or solderable coatings. Aluminum alloys containing more than 1% magnesium or 4% silicon are particularly difficult to solder. The entire subject of soldering aluminum is covered extensively in Section XII of the *Source Book on Selection and Fabrication of Aluminum Alloys.* Brazing with alloys having melting points near that of A91100 and A93003 aluminum is satisfactory, but control is difficult and brazing fluxes must be removed.

The oxygen coating offers some difficulty to welding. In nonpressure welding, the oxide must be removed by fluxes, and for resistance welding it may be necessary to brush off the oxide coat or remove it by etching. If these precautions are taken, good welds can be made by any of the usual welding methods.

Aluminum has a coefficient of thermal expansion 50% greater than copper and twice that of steel and cast iron. Where large temperature changes are encountered in an assembly of aluminum parts with those of other metals, allowance must be made for this difference in expansion.

Ease of machining aluminum varies widely from poor for annealed pure metal to good for hardened or cold-worked alloys. Although special tools and techniques have been developed, the tools used for the harder metals can be used with satisfactory results if production quantities are small.

H. Corrosion Resistance

The thin oxide coating that forms on aluminum and its alloys serves as protection against corrosion. Erosion or abrasion will, of course, remove this oxide and will accelerate corrosion. Constant rubbing of parts may make this particularly severe.

Commercially pure aluminum (A91100) is often used for such shapes as sheet, tubing, and rod where good general corrosion resistance is required. An alloy with approximately equal corrosion resistance and considerably higher strength contains 1-1.5% manganese, the remainder being aluminum (A93003). Alloys of aluminum with silicon and magnesium also have good resistance to corrosion, but zinc and iron alloys appear to have less corrosion resistance.

The alloys of aluminum with copper, tin, or zinc are attacked to various degrees by local electrolytic corrosion. The local cells may be broken up, however, by periodic cleaning of dirt and moisture from the surface. Oil or wax is good for temporary protection.

1. Oxide Coatings

A heavy surface oxide may be artifically produced by either of two methods. One method, which produces a film with considerably greater protective power than the natural oxide coating, involves boiling the metal in a chromate or carbonate solution. A second method, called *anodizing,* is more expensive, but causes formation of an even better protective coating. In this method the article is made the anode in an electrolyte, such as sulfuric acid, chromic acid, or oxalic acid. The color of the anodic coating may be controlled by surface condition and composition of the alloy, by the type of electrolyte, or by dyeing the anodic coat. Both the chemically and anodically produced oxide films are improved by "sealing" with dichromate solutions followed by a hot water rinse. In severely corrosive atmospheres anodic coatings must be kept clean to prevent breakdown of the coating and pitting of the aluminum beneath.

2. Painting

A suitable primer or treatment must be used before painting to make the surface chemically inert. An anodic coating is an ideal paint base, although often an anodic coating is painted only for appearance, not to improve its corrosion resistance. The surface can be prepared more cheaply by one of several chemical immersion treatments, but none of these gives corrosion protection equal to anodizing. For best corrosion resistance, the surface of the paint should be impermeable to moisture, and for this purpose aluminum paint with a synthetic resin is excellent.

3. Metallic Coatings

Nearly all the alloys have poorer corrosion resistance than commercial aluminum, and when corrosion protection is needed, some alloys are coated with commercially pure aluminum. The coating oxidizes and gives high electrolytic corrosion resistance, but in addition its electrolytic properties (see Chap. 4) are such that it prevents the corrosion of the base metal at the sheared edges or where the coating is deeply scratched. The entire coating will be corroded away before the base metal in the vicinity will be disturbed. The usual thickness of the coat is about 5% of the total thickness of the part, and it is applied during the hot rolling of the shape.

Rolled shapes, castings, and assemblies that cannot be readily clad may be protected by a sprayed coating of aluminum, zinc, or cadmium. These deposits are porous, and their protective action is caused only by their electrolytic properties.

Aluminum can be plated with zinc, cadmium, nickel, and chromium, all of which offer mechanical protection. Zinc and cadmium also provide electrolytic protection against corrosion, but nickel and chromium platings provide their respective corrosion resistances only if their coating is continuous. The difference in elastic modulus between aluminum and the platings limits the distortion that may be placed on the plated products.

III. BERYLLIUM AND BERYLLIUM ALLOYS*

Beryllium has recently been used in space and missile structural designs to take advantage of its low density, 1.85 (Table 9.7), its excellent mechanical properties, including high strength-weight and stiffness-weight ratios (about six times that of ultra-high-strength steel) (Figs. 9.5 and 9.6 and Table 9.8), and its superior thermal properties, especially a high melting point, 1200°C (Fig. 9.7 and Table 9.9). As a structural material it suffers from low ductility. Its electrical resistivity places it among the better conductors (Fig. 9.8).

Beryllium and its oxide are toxic when inhaled. Fabrication modes must take account of this property. The largest usage of beryllium is as an α emitter, as a moderator for nuclear reactors, and as a window for x-ray tubes. The oxide is used as a refractory with outstanding thermal conductivity and in fluorescent lights.

Beryllium shapes and parts are generally produced from the hot-pressed powder. Sheet, extrusions, bar, and rod are available. Beryllium foil is used in vacuum tubes, and wire may be drawn as fine as 25 μm in diameter. Beryllium wire, because of its great

---

*Contributed by ROBERT S. SHANE.

Table 9.7   Mass-Volume Relationships for Beryllium and Three Other Light Metals

	Be	Mg	Al	Ti
ATOMIC NUMBER	4	12	13	22
ATOMIC WEIGHT	9.013	24.32	26.98	47.90
ATOMIC VOLUME, cm³/mole	4.877	14.0	9.996	10.63
CRYSTAL STRUCTURE	hex. c.p.	hex. c.p.	f.c.c.	α = hex. c.p.* β = b.c.c.**
LATTICE CONSTANTS IN A				
				α β
a_o	2.2863	3.203	4.0491 2.95030	3.32†
c_o	3.5830	5.199	4.68312	
c/a	1.5671	1.624	1.5873	

*Below 1620°F   **From 1620° to melt. pt.   †At 1650°F
Source: Courtesy of Brooks & Perkins, Inc., Detroit, Michigan

stiffness, is used to reinforce metallic and nonmetallic composites. Thus, one such composite with 67% beryllium particles embedded in a ductile aluminum matrix has a tensile strength of 61,000 psi and a modulus of elasticity of 29×10^6 psi. Another beryllium wire-reinforced aluminum sheet has a tensile strength of 85,000 psi, a modulus of elasticity of 25×10^6 psi, and 5% elongation. Beryllium-reinforced titanium alloy composites have strengths of 140,000 psi and a modulus of elasticity of 27×10^6 psi (data from Clauser, *Industrial and Engineering Materials,* McGraw-Hill, New York, 1975, p. 173).

IV.   COBALT*

Cobalt usually occurs with nickel in arsenic ores and is obtained by reduction with aluminum. The pure metal is used as a target in x-ray tubes and for other special components;

---

*Contributed by ROBERT S. SHANE.

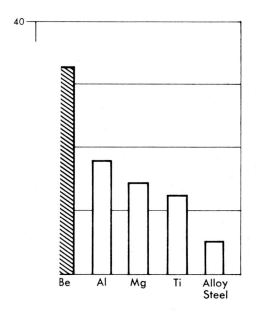

Fig. 9.5 Ratio of tensile strength to density for beryllium compared with other metals. (Courtesy of Brooks & Perkins, Inc., Detroit, Michigan.)

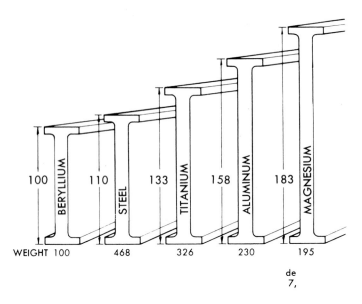

WEIGHT 100   468   326   230   195

de
7,

Fig. 9.6 Comparison of beams of the same rigidity made out of different materials. (From *The New Scientist*, vol. 7, pp. 519-521.)

the overwhelming number of applications is in alloys. Cobalt alloys are used as matrices for cutting tools, in iron permanent-magnet alloys, as a base metal in superalloys for very high temperature service, in dental and surgical applications because they resist attack by body fluids, for glass-metal seals, for hard-facing coatings, for a very low coefficient of thermal expansion, for a constant modulus of elasticity over a wide range of temperature, and for high-speed steels.

Typical properties for a cobalt superalloy (UNS 30031) that has been cast and aged 50 hr at 730°C are as follows.

Tensile strength is 75000 psi at 730°C and 100,000 psi at 425°C.
Elongation is 15% at 730°C and 12% at 425°C.
Modulus of elasticity is 22×10^6 psi at 730°C and 24×10^6 psi at 425°C.
Linear coefficient of expansion is 16×10^{-6} at 730°C and 14×10^{-6} at 425°C.
Fatigue strength (10^8 cycles) is 40,000 psi at 730°C.

Table 9.8 Mechanical Properties of Beryllium and Other Metals

	Be	Mg	Al	Ti	Alloy Steel*
DENSITY, g/cc	1.85	1.74	2.699	4.54	7.4
MOD. OF ELASTICITY IN TENSION, 10^6 psi	44	6.5	10.6	15.5	28
ULT. TENSILE STRENGTH, 10^3 psi	25-95	12-36	20-70	50-80	85-95
YIELD STRENGTH, 10^3 psi	20-45	20-30	15-60	40-70	30-35
ELONGATION, %	2-20	10-20	30	20-60	50-60

*Approx. average data.
Source: Courtesy of Brooks & Perkins, Inc., Detroit, Michigan.

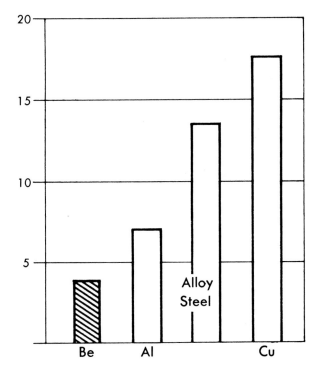

Pounds of material for equal heat absorption

Fig. 9.7 Amount of material (in pounds) for equal heat absorption. (Courtesy of Brooks & Perkins, Inc., Detroit, Michigan.)

Table 9.9 Thermal Properties of Beryllium and Other Metals

	Be	Mg	Al	Ti	Fe	Cu
MELTING TEMP., °C	1277	650	660	1668	1536	1083
°F	2332	1202	1220	3035	2797	1981
BOILING TEMP., °C	2770	1107	2450	3260	3000	2595
°F	5020	2025	4442	5900	5430	4703
THERMAL CONDUCTIVITY cal-cm²/cm-sec-°C	0.35	0.367	0.53	0.41	0.18	0.94
LINEAR THERMAL EXPANSION in micro-in./in./°C	11.6	27.1	23.6	8.41	11.7	16.5
in micro-in./in./°F	6.4	15.05	13.1	4.67	6.5	9.2
SPECIFIC HEAT, cal/g/°C	0.45	0.25	0.215	0.124	0.11	0.092
HEAT OF FUSION, cal/g	260	88	94.5	104*	65	50.6

*Estimated

Source: Courtesy of Brooks & Perkins, Inc., Detroit, Michigan.

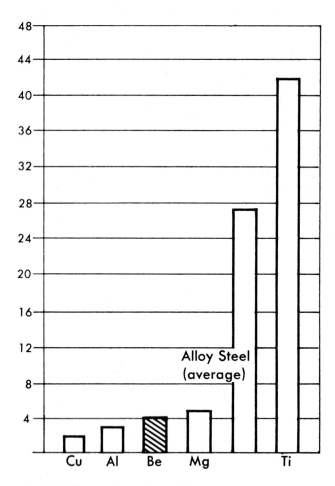

Fig. 9.8 Electrical resistivity of beryllium compared with other metals. (Courtesy of Brooks & Perkins, Inc., Detroit, Michigan.)

Cobalt alloys are characterized, beside high-temperature strength, by wear resistance, ability to take a high finish, and good machinability.

V. COPPER AND COPPER ALLOYS*

A. Production and Processing of Copper and Copper Alloys

The ores of copper are classified into three groups: (1) sulfide ores, (2) oxide ores, and (3) native ores (uncombined copper). The most important are the sulfide ores which may be very rich, containing up to 15% copper, or a very low grade containing as little as 0.9% copper. The most important sulfide minerals of copper are chalcocite (Cu_2S), chalcopyrite ($CuFeS_2$), covellite (CuS), and bornite (Cu_5FeS_4). Oxide ores are the re-

---

*Contributed by DANIEL L. POTTS.

sult of the alteration and decomposition of sulfide minerals. Large deposits of ores are located in Chile, Zaire, Poland, the Southwest and Rocky Mountains of the United States. The only important U.S. deposit of native copper is in the upper peninsula of Michigan. The product is known as "lake copper."

Copper ores are concentrated, processed into suitable shapes of exact chemical composition, and further processed into forms having specific properties and dimensional tolerances (see Fig. 9.9). The resulting forms are fabricated into the desired end product by using one or more of the metal-processing techniques discussed in Vol. 2.

B. Classification of Copper and Copper Alloys

In the designation system, numbers UNS C10000 through UNS C79999 describe wrought alloys. Cast alloys are numbered from UNS C80000 through UNS C9999. Within these two categories, the system groups the compositions into the families of coppers and copper alloys listed in Table 9.10.

C. Temper Designations of Copper and Copper Alloys

A sophisticated system of designating the "condition" or the "temper" of a wrought or cast copper or copper alloy was developed in order to assure design engineers of the

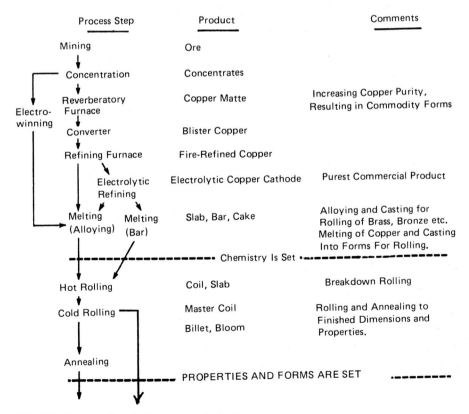

Fig. 9.9 Process flow in the copper industry.

Table 9.10   Designation by Class of Copper Alloys

Classification	UNS Designations
Coppers: Metal that has a designated mininum copper content of 99.3% or higher	C10000-C15599 C80000-C81299
High-copper alloys: For the *wrought products,* these are alloys with designated copper contents less than 99.3% but more than 96%, which do not fall into the other copper alloy groups. The *cast high-copper alloys* have a designated copper content in excess of 94%, to which silver may be added for special properties.	C16200-C19999 C81300-C83299
Brasses: These alloys contain zinc as the principal alloying element with or without other designated alloying elements, such as iron, aluminum, nickel, and silicon. The *wrought alloys* comprise three main families of brasses: copper-zinc alloys; copper-zinc-lead alloys (leaded brasses); and copper-zinc-tin alloys (tin brasses).	C20000-C29999 C30000-C39999 C40000-C49999
The *cast alloys* comprise five main families of brasses; copper-tin-zinc alloys (red, semired and yellow brasses); copper-tin-zinc-lead alloys (leaded red, semired and yellow brasses); "manganese bronze" alloys (high-strength yellow brasses); leaded "manganese bronze" alloys (leaded high-strength yellow brasses); and copper-zinc-silicon alloys (silicon brasses and bronzes).	C83300-C83899 C84200-C84899 C85200-C85899 C86100-C86899 C87200-C87999
Bronzes: Broadly speaking, bronzes are copper alloys in which the major alloying element is one other than zinc or nickel. Originally "bronze" described alloys with tin as the only or principal alloying element. Today, the term "bronze" is seldom used by itself in a technical sense. Rather it is used with a modifying adjective. For *wrought alloys* there are four main families of bronzes: copper-tin-phosphorus alloys (phosphor bronzes); copper-tin-lead-phosphorus alloys (leaded phosphor bronzes); copper-aluminum alloys (aluminum bronzes); and copper-silicon alloys (silicon bronzes).	C50000-C52999 C53200-C54899 C60000-C64299 C64700-C66199 C66700-C69999
The *cast alloys* have four main families of bronzes; copper-tin alloys (tin bronzes); copper-tin-lead alloys (leaded and high leaded tin bronzes); copper-tin-nickel alloys (nickel-tin bronzes); and copper-aluminum alloys (aluminum bronzes). The family of alloys known as "manganese bronzes" are not included here but in the brasses, above	C90200-C91799 C92200-C92999 C93200-C94599 C94700-C94999 C95200-C95899
Copper-nickels: Alloys with nickel as the principal alloying element, with or without other designated alloying elements	C70000-C72900 C96200-C96699
Copper-nickel-zinc alloys: Commonly known as "nickel silvers," these are alloys that contain zinc and nickel as the principal and secondary alloying elements, with or without other designated elements.	C73200-C79999 C97300-C97899
Leaded coppers: A series of *cast alloys* of copper with 20% or more lead and usually a small amount of silver present, but without tin or zinc.	C98200-C98899
Special alloys: Alloys whose chemical compositions do not fall into any of the above categories.	C99300-C99799

exact physical and mechanical properties of the form selected. The temper designation codes of ASTM B601 have been coordinated in both North and South America as national standard designations (Table 9.11). Representative properties of copper and copper alloys may be found in Tables 9.12 and 9.13.

D. Refractoriness

Sometimes copper and copper alloys are classified according to refractoriness, which is a term that grew out of the mills to indicate difficulty of processing. Those copper alloys that, because of their hardness or abrasiveness, require dimensional tolerances greater than those established for nonrefractory alloys are called "refractory alloys," and vice versa (see Table 9.14).

E. Conductivity of Copper

The high electrical conductivity of copper (second only to silver) has led to the use of large quantities in electrical wire and electrical machines. Conductivity is important in such applications, but unfortunately, the usual alloying agents and impurities, except for silver, reduce it. The International Standard resistance for annealed copper is 0.15328 ohm-g/m^2 at 20°C. The conductivity of a copper sample is often expressed as a percentage of the conductivity of the standard sample, such as 102% IACS (International annealed copper standard). Since the establishment of this standard, refining methods have been improved so much that copper of higher conductivity than the standard is now fairly common. Some samples run as high as 102% IACS conductivity, and a minimum standard of 99.3% IACS is frequently used. Electrolytic ingots usually have above 97.6% IACS conductivity.

F. Hydrogen Embrittlement

Oxygen-free high-conductivity copper is produced by the conversion of select refinery cathodes into continuous cast shapes. It is melted and cast under strictly controlled conditions that yield a consistently homogeneous product, virtually free of oxygen. No deoxidizing elements, such as phosphorus, are used in the production of oxygen-free copper, and there are no residual deoxidants in the finished product.

The essential difference between oxygen-free copper with or without phosphorus deoxidant and electrolytic tough pitch copper lies in the oxygen content. Oxygen-free copper normally contains less than 3 ppm (0.0003%) oxygen; tough pitch copper normally contains from 250 to 500 ppm (0.025-0.050%) oxygen.

When oxygen-bearing copper is heated to a relatively high temperature in a reducing atmosphere containing hydrogen, the hydrogen diffuses into the copper and reacts with the oxygen to form water vapor. The water vapor develops sufficiently high pressure to disrupt the structure, causing the metal to become brittle.

There are many applications where the presence of oxygen is detrimental, such as brazing or welding, or where the component must serve under high-vacuum conditions. The superior electrical and thermal conductivities exhibited by oxygen-free copper over deoxidized varieties, as well as its superior welding properties, lead to its selection for many of these applications. (See Figs. 9.10 and 9.11.)

Oxygen-free copper is subjected to an embrittlement test as a regular control procedure. A sample is cut from a continuous-cast shape, hot rolled, and cold drawn to wire. Samples of the wire are then heated in a hydrogen atmosphere at 850°C for 30 min,

Table 9.11   Temper Designation Codes

1.0   Annealed tempers, O
    1.1   Annealed to meet mechanical properties, O

Annealed tempers, O	Temper names
O10	Cast and annealed (homogenized)
O11	As cast and precipitation heat treated
O20	Hot forged and annealed
O25	Hot rolled and annealed
O30	Hot extruded and annealed
O31	Extruded and precipitation heat treated
O40	Hot pierced and annealed
O50	Light anneal
O60	Soft anneal
O61	Annealed
O65	Drawing anneal
O68	Deep-drawing anneal
O70	Dead-soft anneal
O80	Annealed to temper, 1/8 hard
O81	Annealed to temper, 1/4 hard
O82	Annealed to temper, 1/2 hard

    1.2   Annealed to meet nominal average grain size, OS:

Annealed tempers, with grain size prescribed, OS	Temper designations nominal average grains size, mm
OS005	0.005
OS010	0.010
OS015	0.015
OS025	0.025
OS035	0.035
OS050	0.050
OS060	0.060
OS070	0.070
OS100	0.100
OS120	0.120
OS150	0.150
OS200	0.200

2.0   Cold-worked tempers, H
    2.1   Cold-worked tempers to meet standard requirements based on cold rolling or cold drawing, H

Cold-worked tempers, H	Temper names
H00	1/8 Hard
H01	1/4 Hard
H02	1/2 Hard
H03	3/4 Hard
H04	Hard
H06	Extrahard
H08	Spring
H10	Extraspring
H12	Special spring
H13	Ultraspring
H14	Superspring

Table 9.11 (continued)

2.2 Cold-worked tempers to meet standard requirements based on temper names applicable to particular products, H

Cold-worked tempers, H	Temper names
H50	Extruded and drawn
H52	Pierced and drawn
H55	Light drawn, light cold rolled
H58	Drawn general purpose
H60	Cold heading, forming
H63	Rivet
H64	Screw
H66	Bolt
H70	Bending
H80	Hard drawn
H85	Medium hard-drawn electrical wire
H86	Hard-drawn electrical wire

3.0 Cold-worked tempers with added treatments
   3.1 Cold-worked and stress relieved, HR

	Temper names
HR01	1/4 Hard and stress-relieved
HR02	1/2 Hard and stress relieved
HR04	Hard and stress relieved
HR08	Spring and stress relieved
HR10	Extraspring and stress relieved

   3.2 Drawn and stress-relieved, HR

	Temper names
HR50	Drawn and stress relieved

   3.3 Cold-rolled and order strengthened, HT

	Temper names
HT04	Hard temper and treated
HT08	Spring temper and treated

4.0 As-manufactured tempers, M

	Temper names
M01	As sand cast
M02	As centrifugal cast
M03	As plaster cast
M04	As pressure die cast
M05	As permanent mold cast
M06	As investment cast
M07	As continuous cast
M10	As hot forged-air cooled
M11	As forged-quenched
M20	As hot rolled
M30	As hot extruded
M40	As hot pierced
M45	As hot pierced and rerolled

Table 9.11   (continued)

5.0   Heat-treated tempers, T

  5.1   Quenched hardened, TQ

Temper names

TQ00	Quench hardened
TQ50	Quench hardened and temper annealed
TQ55	Quench hardened and temper annealed, cold drawn and stress relieved
TQ75	Interrupted quench

  5.2   Solution heat-treated, TB

TB00   Temper name: Solution heat treated (a)

  5.3   Solution heat-treated and cold worked, TD

Temper names

TD00	Solution heat treated and cold worked: 1/8 hard
TD01	Solution heat treated and cold worked: 1/4 hard (1/4 H)
TD02	Solution heat treated and cold worked: 1/2 hard (1/2 H)
TD03	Solution heat treated and cold worked: 3/4 hard (3/4 H)

  5.4   Solution heat treated and precipitation heat treated, TF

TF00   Temper name: Precipitation hardened (AT)

  5.5   Solution heat treated and spinodal heat treated, TX

TX00   Temper name: Spinodal hardened (AT)

  5.6   Solution heat treated, cold worked, and precipitation heat treated, TH

Temper names

TH01	1/4 Hard and precipitation heat treated (1/4 HT)
TH02	1/2 Hard and precipitation heat treated (1/2 HT)
TH03	3/4 Hard and precipitation heat treated (3/4 HT)
TH04	Hard and precipitation heat treated (HT)

  5.7   Cold worked tempers and spinodal heat treated to meet standard requirements based on cold rolling or cold drawing, TS

Temper names

TS00	1/8 Hard and spinodal hardened (1/8 TS)
TS01	1/4 Hard and spinodal hardened (1/4 TS)
TS02	1/2 Hard and spinodal hardened (1/2 TS)
TS03	3/4 Hard and spinodal hardened (3/4 TS)
TS04	Hard and spinodal hardened
TS06	Extra hard and spinodal hardened
TS08	Spring and spinodal hardened
TS10	Extra spring and spinodal hardened
TS12	Special spring and spinodal hardened
TS13	Ultra spring and spinodal hardened
TS14	Super spring and spinodal hardened

  5.8   Mill hardened, TM

Manufacturing designations

TM00	HM
TM01	1/4 HM
TM02	1/2 HM
TM04	HM
TM06	XHM
TM08	XHMS

Table 9.11 (continued)

5.9 Precipitation heat treated and cold worked, TL

	Temper names
TL00	Precipitation or spinodal heat treated and 1/8 hard
TL01	Precipitation or spinodal heat treated and 1/4 hard
TL02	Precipitation or spinodal heat treated and 1/2 hard
TL04	Precipitation or spinodal heat treated and hard
TL08	Precipitation or spinodal heat treated and spring
TL10	Precipitation or spinodal heat treated, extra spring

5.10 Precipitation heat treated, cold worked, and thermal stress relieved, TR

	Temper names
TR01	Precipitation heat treated, 1/4 hard, and stress relieved
TR02	Precipitation heat treated, 1/2 hard, and stress relieved
TR04	Precipitation heat treated, hard, and stress relieved

6.0 Tempers of welded tube, W
  6.1 As-welded, WM

	Temper names
WM50	As welded from annealed strip
WM00	As welded from 1/8 hard strip
WM01	As welded from 1/4 hard strip
WM02	As welded from 1/2 hard strip
WM03	As welded from 3/4 hard strip
WM04	As welded from hard strip
WM06	As welded from extrahard strip
WM08	As welded from spring strip
WM10	As welded from extraspring strip
WM15	As welded from annealed strip, thermal stress relieved
WM20	As welded from 1/8 hard strip, thermal stress relieved
WM21	As welded from 1/4 hard strip, thermal stress relieved
WM22	As welded from 1/2 hard strip, thermal stress relieved

  6.2 Welded tube and annealed
    W050   Temper name: Welded and light annealed

  6.3 Welded tube and cold drawn, WH

	Temper names
WH00	Welded and drawn: 1/2 hard
WH01	Welded and drawn: 1/4 hard

  6.4 Welded tube, cold drawn, and stress relieved, WR

	Temper names
WR00	Welded, drawn, and stress relieved from 1/8 hard
WR01	Welded, drawn, and stress relieved from 1/4 hard

  6.5 Welded tube, fully finished, O, OS, H
    6.5.1 Fully finished tube, annealed to meet property requirements:

	Temper names
O, OS	Use appropriate designation for property or grain size requirements; see 1.1 or 1.2.
H	Use appropriate designation for property or grain size requirements; see 2.1.

Source: General Electric Company, Schenectady, New York.

Table 9.12   Typical Properties of Representative Wrought Copper Alloys

Commercial Designation*	Tensile Strength, M psi		Yield Strength ½% Offset, M psi		Elongation, % in 2 in.		Shear Strength, M psi		End. Limit, M psi 20×10⁶ cycles		Mod. of Elast. E ×10⁶ psi	Hardness, Rockwell		M.P. or Solidus, °F	Density, lb/cu in.	Volume Electrical Conduct., % of IACS	Uses	Nominal Composition, %
	Hard	Soft	Hard	Soft	Hard	Soft	Hard	Soft	Hard	Soft		Hard	Soft					
Not hardenable by heat treatment																		
Pure copper		32				45					16			1981.4	0.324	103.6		
Electrolytic T.P. copper	50	33	45	10	6–12	50	28	22	13	11	16	F90	F40		0.3217	101	Sheet (also rod and wire) for electrical parts	99.9+ Cu and Ag min
Deoxidized copper	55	32	50	10	8	45	28	22	19	11	17	F95	F40	1981	0.323	85	Sheet (also rod, wire, and pipe) for hot- and cold-worked parts, suited to welding	99.9 Cu and low P
Gilding brass (95% Cu)	56	34	50	10	5	45	37	28			17	B64	F46	1920	0.320	56	Sheet (also wire) for jewelry trade—hot and cold forming	94/96 Cu, bal. Zn
Commercial bronze (90% Cu)	61	37	47	12	5	45	38	28	21		17	B70	F53	1870	0.318	44	Sheet (also wire) for screening, radiators, bullets	89/91 Cu, bal. Zn
Red brass (85%)	70	40	57	12	5–8	47–55	42	31		20	17	B77	F59	1810	0.316	37	Sheet (also rod, tube, and pipe) of good corrosion resistance	84/86 Cu, bal. Zn
Low brass (80%)	74	44	59	14	7	50	43	32	23		16	B82	F61	1770	0.313	32	Sheet (also rod) of excellent hot-working properties	78.5/81.5 Cu, bal. Zn
Cartridge brass (70%)	77	47	64	15	8	62	44	33	21	13	16	B82	F64	1680	0.308	28	Sheet (also rod, wire, and tube) for cold deep drawing	68.5/71.5 Cu, bal. Zn
Yellow brass	74	49	60	17	8	57	43	34	14	12	15	B80	F68	1660	0.306	27	Sheet (also rod and wire) good for cold-work parts	64/67.5 Cu, bal. Zn
High-leaded brass	74	49	60	17	7	52	43	34			15	B80	F68	1630	0.306	26	Sheet, particularly machined clock parts	62.25 Cu, 35.75 Zn 2 Pb nominal
Munts metal (annealed and ½ hard)	70	54	50	21	10	45	44	40			15	B75	F80	1650	0.303	28	Sheet (also rod) good hot-working properties, forgings, trim, bolts, etc.	59/63 Cu, bal. Zn
Naval brass (0.250 annealed sheet and ½ hd. rod)	80	60	57	28	20	45	45	41	15		15	B85	B58	1630	0.304	26	Sheet (rod and tube) for hot and cold forming, and salt exposure	59/62 Cu, 0.5/1 Sn
Phosphor bronze (1.25%) (E)	65	40	50	14	8	48					17	B75	F60	1900	0.321	43	Sheet (also wire) suited to many cold-forming operations	98.5 Cu min, 1/1.5 Sn
Phosphor bronze (5%) (A)	81	47	75	19	10	64		37			15	B87	B26	1750	0.320	18	Sheet (also wire and rod) good for cold forming, but is hot short	3.5/5.8 Sn, 0.03/0.35 P, bal. Cu
Phosphor bronze (8%) (B)	93	60	72	24	10	65			22		16	B93	F75	1620	0.318	13	Sheet (also wire and rod) same as 5% but stronger	7/9 Sn, 0.03/0.35 P, bal. Cu
Nickel silver (A)	88	58	74	25	3	40					18	B87	B40	1960	0.316	6	Sheet (also rod and wire) suited to cold forming. Good tarnish and corrosion resistance for tableware, instruments	63/66.5 Cu, 17/19.5 Ni, bal. Zn
Nickel silver (B)	100	60	85	27	3	40					18	B91	B55	1930	0.314	5.5	Sheet (also rod and wire) less cold forming but stronger than A	53.5/56.5 Cu, 16.5/19.5 Ni, bal. Zn
Low-silicon bronze (A)	94	60	58	25	8	60	57	43	23	16	15	B93	B62	1780	0.308	7	Sheet (also rod, wire, and tube) suited to hot and cold forming with good mech. prop. and corrosion resistance	94.8 min Cu, 2.75/3.5 Si, 1.5 Mn or Zn
High-silicon bronze (B)	70	40	55	15	15	50	40	28			15	B80	F55	1890	0.316	12	Rod (also sheet, wire, and tube) similar to A. Used for hardware, etc.	96 min Cu, 0.75/2.00 Si, 0.75 Mn max
Manganese bronze (annealed and ½ hard)	84	65	60	30	19	33	48	42			15	B90	B65	1590	0.308	24	Rod, forgings upsettings, hot-headed and punched parts	57/60 Cu, 0.80/0.20 Fe, 0.5/1.5 Sn, 0.5 Mn, bal. Zn
Aluminum bronze (5%)	92	55	65	22	7	65					15	B92	B35	1920	0.295	17.5	Strip (also rod, forgings, tube) good hot and cold forming but limited drawing	92/96 Cu, 4/7 Al, 0.5 Fe max
Aluminum bronze (8%)	105	65	64	26	7	60					17.5–19	B96	B50	1890	0.274	14.8	Sheet (bar, forgings, extrusions) good resis. to corrosion, tarnishing, wear and heat, with good mech. properties	92 Cu, 8 Al nominal
Cupro nickel (30%)	75	55	70	20	15	45					22	B80	B35	2140	0.323	4.6	Sheet, tube and rod (rod ½ hard and tube annealed) very good corrosion resistance, forming and mech. properties for condenser tubes, etc.	29/33 Ni, bal. Cu
Hardenable by heat treating																		
Beryllium copper (sol. treated and precip. hardened)	190	72	110	25	3	50			41	32	20–17	C42.5	B60	1587	0.297	21–17	Strip (and rod). Good formability, strength, conductivity, corr. and wear resis. For many uses, even at high price	1.9/2.15 Be, 0.15/0.50 Ni, bal. Cu
Chromium copper (precip. hardened)	72–63		61–45		25							B77–65		1975	0.32	80	Rod. Moderate strength and fatigue resistance with excellent conductivity	0.85 Cr, 0.05 Si, 99.1 Cu nominal

*UNS: SAE and ASTM designations or suppliers' identifications are often used in place of these terms.

quenched in water, and clamped in a special jig that holds the wire under constant tension while it is bent back and forth over a mandrel until it breaks. Duplicate wires of each sample must achieve a minimum of 10 bends in order to be acceptable. This embrittlement test can also be used to distinguish oxygen-free copper from tough pitch copper. The latter fails with a brittle fracture in less than 1 bend.

G.   Recrystallization of Copper

Copper is usually cold worked to its final size or shape; that is, wire is drawn through dies, and bar and sheet are rolled. This working, through the mechanism of work hardening, increases the strength and hardness but decreases the ductility.

If copper that has been cold worked is heated above the recrystallization temperature (200-260°C), it will return to its original soft, weak, and ductile condition. The more it has been cold worked, the lower the temperature at which it will soften.

Table 9.13 Typical Properties of Representative Cast Copper Alloys

Designation *†	Tensile Strength, M psi	Yield Strength ½% Offset, M psi	Elongation in 2 in., %	Shear Strength 0.1% Set, M psi	Compress. Strength 0.001% Set, M psi	Mod. of Elasticity, 10^6 ps	Hardness, Brinell, 500/10	Density, lb/cu in. (68 F)	Electrical Conductivity, % IACS (77 F)	Uses	Composition, %
Leaded tin bronze (Navy M)	38	16	35	34	13	13	66	0.315	14	High-grade steam or valve bronze to 550 F, moderate pressures	86/90 Cu, 5.5/6.5 Sn, 1/2 Pb, 3/5 Zn
Leaded tin bronze (G)	36	18	30	43	13	14	68	0.318	11	High-duty, wear-resistant bearing bronze	85/89 Cu, 7.5/9 Sn, 3/5 Zn, 1 Pb
Leaded tin bronze, high lead, tin	32	17	12	43	14.5	11	65	0.32	..	Bearings for high speeds and pressures	78/82 Cu, 9/11 Sn, 8/11 Pb
Leaded brass, red	32	15	24	29	11.5	13	55	0.312	..	General-purpose free machining for plumbing and other fittings	78/82 Cu, 3.25/4 Sn, 5/7 Pb, 5/8 Zn
Leaded brass, semired	32	15	22	..	12	13	55	0.314	18	Ornamental and low-pressure valves and fittings	78/82 Cu, 2.25/3 Sn, 6/8 Pb, 7/10 Zn
Leaded brass, yellow	35	12	35	30	9	13	48	0.307	25	General-purpose, hardware, trim, fittings and ornamental	70/74 Cu, 0.75/2 Sn, 1.5/3.75 Pb
Leaded brass, yellow	40	14	25	..	..	14	65	0.304	20-26	General-purpose valves, hardware, and ornamental trim	60/65 Cu, 0.5/1.5 Sn, 0.75/1.5 Pb
Leaded brass, nickel silver	34	15	20	..	..	..	60	0.323	..	Hardware, plumbing fixtures, statuary, ornaments	53/58 Cu, 1.5/3 Sn, 8/11 Pb, 11/14 Ni, bal. Zn
Manganese bronze, standard	70	28	30	87	24	15	125	0.296	18	Tough and strong grade for propeller blades, and parts contacting water, levers, stems, etc.	58 Cu, 39.25 Zn, 1.25 Fe, 1.25 Al, 0.25 Mn nominal
Manganese bronze, high strength	115	70	15	100	60	15.5	210	0.285	12	Toughness and hardness for extra-heavy-duty gears, covers, valve stems, slow-speed bearings, etc.	60/68 Cu, 3/7 Al, 2.5/5 Mn, 2/4 Fe, bal. Zn
Silicon bronze	45-55	18-30	20-30	..	15-22	..	80-130	..	..	High strength, toughness, and corrosion resistance	90/95 Cu, 4 Si, Fe, Al, Zn nominal
Aluminum bronze (9% Sn) as cast	75	27	35	67	29	17	120	0.267	13	High strength, toughness, hardness, and corrosion resistance for marine equipment, acid pump parts, bearings, gears, valve seats, guides, etc.	88 Cu, 3 Fe, 9 Al
Aluminum bronze (11% Sn)	110	65	5	..	..	18	250	0.271	13		89 Cu, 4 Fe, 11 Al nominal
Solution treat 1600-1650, water quench and age, 1 hr at 1000 F	92	50	15	..	..	..	195	..	14		89 Cu, 1 Fe, 10 Al nominal

*UNS, SAE, ASTM, and suppliers' identifications are also used in place of these terms.

†General properties: Thermal conductivity, % of Cu approx. equals electrical volume conductivity. Coefficient thermal expansion is approx. 9-12 $\times 10^{-6}$/°F for 70-400 F. Melting points range from 1570-1800 F.

Table 9.14 Refractoriness Classification of Copper and Copper Alloys

Group classification	Basic composition ranges of nonrefractory alloys (all others are refractory)	Nonrefractory		Refractory	
		UNS designation	Previous trade name	UNS designation	Previous trade name
Copper	All tough pitch, oxygen-free, or deoxidized coppers with or without silver lead	C10200	Oxygen-free copper		—
		C11000	Electrolytic tough pitch copper		
		C11300	Silver-bearing		
		C11500	Tough pitch coppers		
		C11600			
		C12200	Phosphorus deoxidized copper, high residual phosphorus		
Copper-zinc	Containing a minimum of 61.5% copper	C21000	Gilding, 95%	C28000	Muntz metal
		C22000	Commercial bronze, 90%		
		C22600	Jewelry bronze, 87.5%		
		C23000	Red brass, 85%		
		C24000	Low brass, 80%		
		C26000	Cartridge brass, 70%		
		C26800	Yellow brass		
Copper-zinc-lead	Containing a minimum of 61.5% copper	C33500	Low-leaded brass	C36500	Leaded Muntz metal
		C34000	Medium-leaded brass		
		C34200	High-leaded brass		
		C35600	Extra-high-leaded brass		
		C36000	Free-cutting brass		

				Admiralty
Special copper-zinc	Containing a minimum of 61.5% copper and only one of the following elements in not more than the indicated amounts; aluminum, 0.5%; beryllium, 0.4%; cadmium, 0.4%; chromium, 0.3%; iron, 0.4%; silicon, 0.3%; tin, 0.5%; manganese, 1.25%	—	C44300 to C44500 C46400 C67500	Inclusive Naval brass Manganese bronze (A)
Copper-aluminum	Containing up to 1.0% aluminum, inclusive	—		—
Copper-beryllium	Containing up to 0.6% beryllium plus cobalt, inclusive	— —	C17000 C17200	Beryllium copper —
Copper-cadmium	Containing up to 0.4% cadmium, inclusive	—		—
Copper-nickel	Containing up to 5.0% nickel, inclusive	—	C70600 C71500	Copper nickel, 10% Copper nickel, 30%
Copper nickel-zinc	Containing up to 5.0% nickel, inclusive	—	C74500 C75200 C75400 C75900 C77000	Nickel silver, 65-10 Nickel silver, 65-18 Nickel silver, 65-15 Nickel silver, 65-12 Nickel silver, 55-18
Copper-silicon	Containing up to 1.25% silicon, inclusive	—	C65100 C65500	Low-silicon bronze (B) High-silicon bronze (a)
Copper-tin	Containing up to 0.5% tin, inclusive	—	C50200 C51000 C52100 C52400 C54400	Phosphor bronze, 1.25% (E) Phosphor bronze, 5% (A) Phosphor bronze, 8% (C) Phosphor bronze, 10% (D) Free-cutting phosphor bronze

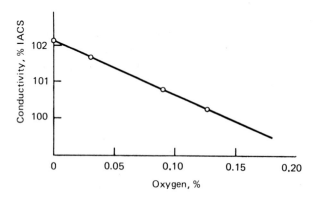

Fig. 9.10 Effect of oxygen on conductivity of copper.

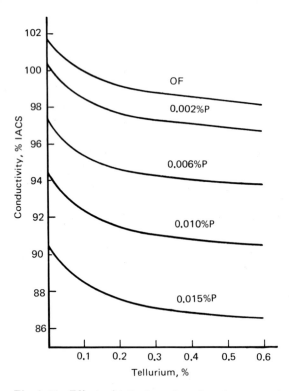

Fig. 9.11 Effect of tellurium plus phosphorus on electrical conductivity of oxygen-free copper.

Copper alloys of high recrystallization temperatures are used to avoid a loss in strength by recrystallization of cold-worked parts exposed to somewhat elevated temperatures. Several alloying agents may be used, but silver has the greatest effect. A few ounces of silver per ton raises the recrystallization temperature as much as 90°C. If cold-worked copper wires or parts are to be soldered and must retain their cold-work strength after soldering, they should be made of silver-bearing copper or of lake copper.

H. Fabrication

Tests have shown conclusively that thoroughly deoxidized copper should always be specified for copper parts that are to be gas welded. This type of welding should not be used for tough pitch copper because the copper would become spongy and porous at the weld. Spot or seam welding of high-conductivity copper is not done commercially unless the surfaces have been tin plated. Neither deoxidized nor electrolytic grades can be cut by an oxidizing gas flame because of their high thermal conductivities. They are, however, easily butt welded, silver brazed, and soldered.

Copper castings are made almost entirely in sand molds. The technique of casting high-conductivity copper is especially difficult since considerable skill is required in handling the reducing agents. The molten metal is viscous and will not fill a mold of great intricacy. Copper is hot short, and the hot casting will break as it cools if the mold or the design prevents shrinkage. If very high conductivity is not required, one of many casting alloys may be used. They are stronger and more easily machined and handled in the foundry, and as a result the finished casting will be less costly than if made of high-conductivity copper.

I. Copper-Zinc Alloys

The most widely used copper alloys are *brasses*. They are fundamentally a binary alloy of copper with as much as 50% zinc, but often their properties are modified by addition of other elements in small amounts. The brasses are stronger than copper and are used in structural applications, but the increase in strength sacrifices both electrical and thermal conductivity.

Brasses containing more than 64% copper are structurally a single-phase solid solution of zinc and copper, termed the α phase (see Fig. 9.12). They are easily cold worked, and the ductility increases with increase in zinc to a maximum at 36% zinc, 64% copper. With more than 36% zinc, the brittle β phase appears along with the α, making the alloy increasingly difficult to cold work. The high-zinc brasses are therefore most easily *hot* worked, whereas the lower zinc brasses (α brasses) are formed more easily while *cold.*

All the brasses containing up to 40% zinc are useful, and there are dozens of varying compositions. Slight modifications of composition usually do not produce important changes in properties. For the purpose of discussion, the common alloys are divided into groups. The names of the alloys in each group are generally unrelated, but they are used by custom throughout the trade.

J. Brass Containing 5-15% Zinc

Brasses containing up to 15% zinc have good corrosion resistance. Like copper, they are ductile and suitable for extreme cold working, but are difficult to machine. Gilding brass, UNS C21000 (5 Zn), and commercial bronze, UNS C22000 (10 Zn), are used for bullet

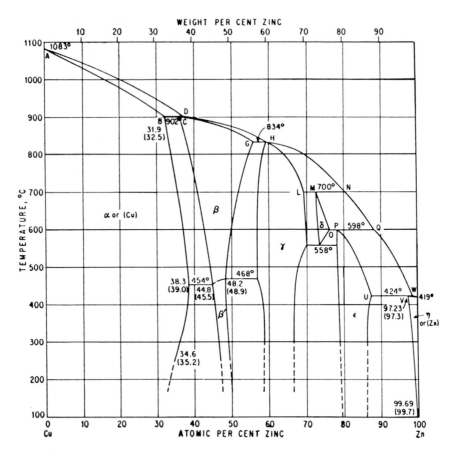

Fig. 9.12 Copper-zinc alloys.

jackets and articles exposed to the weather. Color is often the basis of selection between them. Gilding brass is nearly copper colored; commercial bronze has a true "bronze" color. Red brass, UNS C23000 (15 Zn), also called "rich low brass," is used, for imitation gold jewelry. It has the best corrosion resistance of all the brasses and is superior to copper for handling hard water, and is consequently used for plumbing hardware, pipe, radiator cases, and condenser tubing.

K. Brasses Containing 20-36% Zinc

Brasses containing between 20 and 36% zinc are very readily cold worked. Their advantages over brasses containing less zinc lie in low cost, since zinc is cheaper than copper, and in superior machinability. Brasses containing 20-36% zinc can also be hot worked, but only if their constituents are very pure. This is a new development, since high-purity zinc must be used and it has not been available commercially in the past.

The medium-zinc brasses are subject to corrosion, which works in either of two ways. When exposed to acids or salt solutions (such as hard water), zinc may be removed electrolytically. This dezincification leaves spongy layers of copper, often called plugs, on the surface. Under the influence of corrosive media and stress, brasses with more than

15% zinc fail by season cracking, sometimes called corrosion cracking. The failure occurs in areas where stress is maximum; to minimize the danger of failure, stresses remaining after cold working should be relieved by annealing. In service, however, load stresses can scarcely by avoided. According to recent tests the load stress that eventually produces season cracking is very much less than the yield stress.

Brass with 20% zinc, called low brass, has fair corrosion resistance, being subject to dezincification and season cracking under only the most severe conditions. Both low brass, UNS C24000, and brazing brass, UNS C25000 (25 Zn) are deep drawn, spun, and cold formed with ease. They are used for musical instruments, drawn eyelets, and for ornamental architectural work. Brass containing 30% zinc is very tough and ductile, since it is high in zinc. It is particularly suited to deep drawing or spinning and is usually called *cartridge brass*, UNS C26000. In the past, cartridge cases were manufactured entirely by cold working. Today, however, a high-purity grade of zinc is used, as mentioned above, and hot working in the initial forming of the part is employed, thus avoiding repeated cold working and annealing cycles.

The substitution of 1% tin in cartridge brass produces a grade known as Admiralty metal, UNS C44300. The usual composition is 70 copper, 29 zinc, 1 tin, with as much as 0.05% arsenic, which tends to inhibit dezincification. Other compositions of Admiralty metals with antimony and phosphorized versions are UNS C44400 and UNS C44500, respectively. The tin gives the alloy fair resistance to corrosion by sea water and has led to its use in marine condenser tubing.

The 34% zinc brass, known as yellow brass, UNS C26800, is the cheapest of all the brasses that may be cold worked. Large quantities of screws, rivets, and tubing are made of yellow brass, although it is subject to corrosion. Parts that are cold formed should be annealed after the final forming to inhibit season cracking.

L. Brasses Containing 36% or More Zinc

Brasses containing more than 36% zinc contain the brittle β phase and are difficult to cold work. They may be hot worked, however, Muntz metal, UNS C28000 (40 Zn), is worked at a red heat and is not malleable when it is cold. It is used only in mildly corrosive media, and then principally because of its low cost.

The addition of a small amount of tin (40 Zn, 0.75 Sn, balance Cu) to Muntz metal produces the composition called Naval brass, UNS C46200. As with Admiralty metal, the tin is added to give resistance to corrosion by seawater.

Extrusion is used not only for tubing, but for many irregular shapes. Brasses that are easily hot worked are often extruded when hot, whereas others are extruded cold. The hot-extruding alloys are most often used, however, because the power used in extrusion is a minimum. An alloy that has excellent plasticity at high temperature contains 37% zinc and 4% lead.

Up to 3% lead is often added to brass to improve machining properties, because it causes the chips to break free. Where it is necessary to do severe cold forming and machining on the same part, 1% lead may be used, since a 3% lead brass would be easily broken in cold forming. An alloy that is easily hot worked and machined is forging rod, UNS C37700, (38 Zn, 2 Pb, 60 Cu). Lead in these percentages does not affect the corrosion resistance or conductivity of brass.

Brasses containing over 40% zinc are not widely used. One application is as a brazing spelter of 50:50 composition (50 Zn, 50 Cu).

Practically no straight brasses, that is, plain copper and zinc alloys, are used for castings. Some of the complex brasses, primarily those with the addition of manganese and other elements, are used in quantity for castings. Scrap brass is often cast, however, if accurate control of the properties is unnecessary.

M. Bronzes Not Hardenable by Heat Treatment

Alloys of copper with materials other than zinc are usually called bronzes, although application of the term is loose, as indicated by the brass called "commercial bronze." Like the copper-zinc alloys, the bronzes may be hardened by cold working. Some of the bronze alloys can also be hardened by precipitation heat treatment.

1. Silicon Bronzes

Silicon bronze has an average composition of 96% copper, 3% silicon, and 1% manganese or zinc. It has the good general corrosion resistance of copper, combined with higher strength, and in addition can be cast, rolled, spun, stamped, forged, and pressed either hot or cold and can be welded by all the usual methods. Common uses are parts for boilers, tanks, stoves, or wherever high strength and good corrosion resistance are required. It is exceptionally easy to weld by any common method, but the silicon content tends to make it difficult to solder and braze.

2. Phosphor Bronzes

Copper-tin alloys deoxidized with phosphorus are called either tin bronze or phosphor bronze. The range of tin is from less than 1 to 11% and the phosphorus ranges from an almost negligible amount up to 0.5%. Strength and hardness increase with tin content, but so does the cost. The tin bronzes have high strength, resilience, and resistance to wear and fatigue. These properties recommend them particularly for use as a spring material. Up to 4% lead may be added to improve machinability. Shafts and bushings with moderately good wear resistance are made of the free-machining (leaded) phosphor bronzes.

3. Cupronickels

Copper and nickel form a complete series of solid solutions, and show no sudden change in properties with change in composition. Alloys with 20-30% nickel are used under severely corrosive conditions. Large quantities are used for tubing in oil field applications and for marine condensers. Although nickel is an expensive alloying agent, its use is often justified because of the superior resistance of its alloys to corrosion.

4. Nickel Silver

This contains zinc, in addition to copper and nickel. Although there are many variations in composition, an alloy containing 64% copper, 18% nickel, and 18% zinc, UNS C75200, is by far the most widely used. It has a silvery color, good corrosion resistance, and fairly high electrical resistivity. Nickel silver is used as a base for plated jewelry and tableware, ornamental metal work, food-handling equipment, marine fittings, and electrical resistance wire and springs.

5. Spinodal Alloys (C72600-C72900)

Data on these alloys are given in the Appendix at the end of the chapter.

N. Bronzes Hardenable by Heat Treatment

A few of the alloys of copper can be hardened upon quenching as steel is hardened. Most are hardened by solution heat treatment and precipitation aging.

1. Aluminum Bronzes

Aluminum bronze is a versatile alloy since its properties may be controlled over a wide range, not only by variation of alloying agents, but also by heat treatment.

   With up to 7.5% aluminum, the copper-aluminum alloys are extremely ductile. When a metal must be severely cold worked, aluminum bronze is often substituted for brass because of its superior strength and resistance to corrosion. An alloy of 7.5-9.5% aluminum and up to 1.5% iron with copper is used as cast, without hardening, for good corrosion and shock resistance. Increase of aluminum to 12% and iron to 4% provides an alloy that, in the as-cast condition, has a very high endurance limit and resistance to shock. It is used for heavily loaded gears in machine tools, steel mill drives, and construction machinery. Cams, rollers, and slides utilize its excellent resistance to wear. Aluminum bronze high in aluminum (9.5-12%) and iron (1.5-4%) may be hardened to a degree comparable with the alloy steels, yet may have superior resistance to abrasion. Though brittle, it is valuable in such parts as dies and sliding machine members. Hardening of the 10% aluminum alloy, UNS C63000, occurs upon quenching from 900°C. Greater toughness can be restored by annealing to 550-700°C and requenching.

2. Beryllium Bronzes

Beryllium bronze (beryllium copper) is a precipitation-hardening alloy usually composed of about 2% beryllium and the remainder copper, UNS C17200. It is usually solution heat treated and cold worked to some extent by the supplier before fabrication by forming and machining. It may be hardened by aging in a furnace for several hours at 260-320°C. When hard, it has better wear resistance than phosphor bronze. Heavily loaded bushings, seats, or springs subject to vibration or shock are usually made of beryllium bronze, even though it is relatively expensive. The springs have a remarkable freedom from hysteresis and elastic drift and retain their springiness to a greater degree than springs of any other corrosion-resistant material.

O. Data Source

Property data on copper and copper alloys is readily available from the Copper Development Association (405 Lexington Avenue, New York, New York 10174). Each supplier contributes to CDA as the single most reliable source of information on copper and copper alloys.

VI. GERMANIUM*

Germanium is produced as a by-product of cadmium and zinc refining and is extracted by hydrogen reduction of the purified oxide. The pure metal is a semiconductor that has a nonmetallic crystalline structure similar to the diamond. Its largest application is for

---

*Contributed by ROBERT S. SHANE.

crystal detectors and rectifiers for microwave electronic applications. A potential field of application lies in its silver and gold alloys, which expand upon solidification and have low melting points. Such properties are useful in dental and jewelry parts and in solders. Other potential applications may employ its ability to alloy with platinum, iron, copper, and the light metals, the very high index of refraction of its glass, and its very high electrical resistance.

A fuller discussion of germanium preparation and fabrication for use in semiconductor applications is given in Chap. 6.

VII. HAFNIUM*

Hafnium is a dense, hard, ductile metal of silvery appearance. Its physical properties are summarized in Table 9.15.

The principal use of metallic hafnium is as a neutron absorber in some types of nuclear power reactor control rods. It is also used as a strengthening agent in alloys of niobium, tantalum, molybdenum, and tungsten.

Hafnium has two allotropic forms with a transformation temperature of 1777°C.

Chemically, hafnium resembles zirconium so closely that after the discovery of zirconium it was many years before it was realized that hafnium is invariably present with zirconium in nature, in the ratio of about 1:50. Accordingly, hafnium shares many of zirconium's peculiarities, such as being pyrophoric and being a strong getter for all except the noble gases.

As indicated in Sec. XVI, hafnium is a by-product of the purification of zirconium for nuclear applications. Processes for producing metallic hafnium are described in Ref. 1.

Selected properties of hafnium are listed in Table 9.15.

VIII. LEAD AND ITS ALLOYS†

Lead is one of the best known metals; it has a long history in use both as a metal and in alloys. Table 9.16 gives the properties of the metal. An especially valuable book for

---

*Contributed by MILES C. LEVERETT.
†Contributed by ROBERT S. SHANE, based on material supplied by the Lead Industries Association, Inc., New York, New York.

Table 9.15 Selected Properties of Hafnium

Atomic weight	178.6
Melting point	2230°C
Coefficient of expansion	5.9×10^{-6} °C^{-1}
Density	13.36 g/cm^3
Specific heat @ 25°C	0.034 cal/g-°C (0.34 BTU/lb-°F)
Thermal conductivity @ 50°C	0.223 wall/c-°C (12.9 BTU/ft-hr-°F)
Yield strength @ 25°C	154 + MPa (22.3 ksi)
Ultimate strength @ 25°C	409 MPa (59.3 ksi)

Source: From Ref. 2.

Table 9.16 Properties of Lead

general properties

Color . . . . .Bluish gray
Patina . . . . On atmospheric exposure lead takes on a silvery gray
patina except in industrial atmosphere where it changes
to dark gray to black
Atomic number 82
Atomic arrangement . . . . .Cubic face-centered
Length of lattice edge . . . . .4.9398 A
Atomic weight 207.21

weight and density

weight
Cast lead, 20°C., calculated . . . . .0.4092 lb. per cu. in.
  equivalent to (density 11.34 grams per c.c.) . . . .707 lb. per cu. ft.
Rolled, 20°C. (density 11.37) calculated . . . . .0.4103 lb. per cu. in.
  equivalent to . . . . .709 lb. per cu. ft.
Liquid, 327.4°C., calculated . . . . .0.3854 lb. per cu. in.
  equivalent to . . . . .666 lb. per cu. ft.
Sheet lead, 1 ft. square by 1/64 in. thick. . . . . .approximately 1 lb.
Volume of 1 lb. cast lead, 20°C., calculated . . . . .2.44 cu. in.

density
Cast lead, 20°C . . . . .11.34 g. per c.c.
Rolled, 20°C . . . . .11.35 to 11.37 g. per c.c.
Just solid, 327.4°C . . . . .11.005 g. per c.c.
Just liquid, 327.4°C . . . . .10.686 g. per c.c.
Density, vapor (Hydrogen 1), calculated 103.6

thermal properties

Melting point, common lead . . . . .327.4°C. (621°F.)
Melting point, chemical lead .325.6°C. (618°F.)
Elevation of melting point for each
  150 atm. increase in pressure . . . . .1.2°C. (2.16°F.)
Casting temperature . . . . .790 to 830°F.

boiling point at different pressures

Pressure, in atmospheres	0.14	0.35	1.0	6.3	11.7
Boiling point, °C.	1325	1410	1525	1873	2100
Boiling point, °F.	2417	2570	2777	3403	3812

vapor pressure

Temperature,° C.	808	1000	1200	1365	1525	1870	2100
Pressure, mm. Hg.	0.08	1.77	23.29	166.	760.	6.3 atm.	11.7 atm.

specific heat (cal. per g.)

Temp °C	specific heat
−150	0.02805
−100	0.02880
−50	0.0298
0	0.0303
50	0.0309
100	0.0315
200	0.0325
300	0.0338
327.4 (solid)	0.0340
327.4 (liquid)	0.0333
378	0.0338
418	0.0335
459	0.0335

thermal conductivity (cal./cm²/cm/°C./sec.)

Temp °C	specific heat
−247.1	0.117
−160	0.092
0	0.083
100	0.081
200	0.077
300	0.074
400	0.038
500	0.037
600	0.036

Latent heat of fusion . . . . . . 6.26 cal. per g. or 9.9 B.T.U. per lb.
To melt 1 lb. of lead heating from 20°C requires 7100 g. cal. or 27.9 B.T.U.
Latent heat of vaporization .202 cal. per g.
Relative thermal conductivity (silver 100) 8.2

tensile strength and elongation

(laboratory rolled specimens, room temperature, pulling speed ¼ in.
per min. per in. of test section)

grade of lead	tensile strength psi	elongation percent
corroding (99.99 + 0.006 Bi)	1904	37.7
common (99.85 + 0.13 Bi)	1931	34.4
common (0.002 Cu)	2093	43.0
chemical (99.92 + 0.06 Cu)	2961	42.2
chemical (commercial sheet)	2454	47.0
chemical (extruded)	2200	48.0

resistance to bending

(extruded strips under 200 lbs. per sq. in. stress subjected to alternate
90° reverse bends over 5-in. rolls, 11 cycles per min)

grade of lead	cycles to failure	elongation percent
corroding	54	35
common	72	49
chemical	103	52

coefficient of expansion

Linear (−190 to 19°C. mean) .0.0000265 per °C.
Linear (17 to 100°C. mean) . . . . .0.0000293 per °C.
  or 0.0000163 per °F.
Cubical (liquid at melting point to 357°C.) . . . . .0.000129 per °C.
Increase in volume from 20°C. to liquid at melting point 6.1%
Decrease in volume from liquid at melting point to 20°C. calculated
  5.8%
Increase in volume on melting 4.01%
Decrease in volume on solidification 3.85%
Shrinkage on casting taken in practice as 7/64 to 5/16 in. per ft.

low temperature properties

temperature °F.	tensile strength psi	elongation percent	Brinell	impact
cast lead				
room	3000	33	4.3	2.3
−300	6200	40	9.0	3.8
rolled lead				
59	3600	52	. .	. .
−4	7200	40	. .	. .
−40	13300	31	. .	. .
−103	15200	24	. .	. .

electrical properties

Specific electrical conductance
  at 0°C. . . . . .5.05 x 10⁴ cm.⁻¹ ohms⁻¹
  18°C. 4.83
  melting point 1.06
Atomic electrical conductance, calculated . . . . .1.139 x 16⁶
Relative electrical conductance (copper 100) 7.82
Relative electrical resistance (copper 100) 1280
Magnetic susceptibility per g.
  18−330°C. . . . .−0.12 x 10⁶
  300−600°C. . . . .−0.08 x 10⁶
Electrolytic solution pressure
  ions of Pb + + ...6.3 x 10⁻⁵ atm.
  ions of Pb + + + + ...3.0 x 10⁻¹⁴ atm.

mechanical properties

Hardness, Moh's scale 1.5
Brinell no., 1 cm. ball, 30 sec., 100 kg. load
  Common lead . . . . .3.2 to 4.5
  Chemical lead . . . . .4.5 to 6.0
Influence of temperature on Brinell hardness (chemical lead)

Temperature °C.	25	100	150
Hardness	5.3	3.6	2.6

Ultimate tensile strength
  Common lead . . . . .1400 to 1700 lb. per sq. in.
  Chemical lead . . . . .2300 to 2800 lb. per sq. in.

Effect of temperature on tensile properties (lead annealed at 100°C.)

Temperature		Tensile strength psi	Elongation percent	Reduction in area percent
°C	°F			
20	68	1920	31	100
82	180	1140	24	100
150	302	710	33	100
195	383	570	20	100
265	509	280	20	100

creep (room temperature)

stress lb. per sq. in.	creep, percent per hour	
	common lead	chemical or copper lead
200	5 x 10⁻⁵	0.4 x 10⁻⁵
300	3.5 x 10⁻⁴	1.5 x 10⁻⁵
400	11 x 10⁻⁴	3 x 10⁻⁵

fatigue

Fatigue limit (50,000,000 cycles extruded. . . . . .215 lb. per sq. in.)

metal	endurance limit at 5 x 10⁷ cycles, lb. per sq. in.	endurance limit at 10⁷ cycles lb. per sq. in.
lead	215−400	407
lead + 0.026% calcium		1038
lead + 0.038% calcium	685−840	
lead + 0.04% calcium	820−1500	
lead + 0.041% calcium		1180
lead + 0.06% calcium	1120	
lead + 1% antimony	300−1000	
lead + 2% tin	800	
lead + 0.06% copper	600	725
lead + 0.045% tellurium		1000

Source: Courtesy of Lead Industries Association, Inc., New York.

engineers considering the use of lead in design is *Lead for Corrosion Resistant Applications—A Guide* (published by the Lead Industries Association, Inc.).

Alloys of lead used in solders are discussed in Sec. XII. The outstanding properties of lead and its applications in each class may be summarized as follows.

Corrosion resistance: Lead is used as the metal in sheet form as a corrosion barrier and as lead chromate for corrosion resistant paints.

Radiation resistance: Lead is a superior material for attenuating γ rays because of its density, high atomic number, high level of stability, and its ease and low cost of fabrication.

Vibrational control: Lead and imbedded fiber, such as asbestos or synthetic rubber plus lead, make antivibration pads for use under buildings and machinery. Loads up to and above 1000 psi can be accepted.

Sound control: Dense, limp material like lead, in the form of wool, sheet, shot, leaded plastics, or leaded fabrics, dampens noise from sources as different as typewriters and jackhammers.

Lubricity: About 0.15-0.35% of lead is added to steel ingots during pouring. The lead remains in the solidified steel as a fine dispersion and acts as a built-in lubricant and as a chip breaker during machining. The advantages are to the tools that have a longer life span, requiring fewer dressings, and to the work where dimensional tolerances are easier to maintain; frequently, too, grinding and deburring operations can be eliminated while time for heavy cuts can be reduced.

Solid-state properties: Aside from its use in batteries (the most important single use), lead finds increasing use in such advanced components as Josephson junctions, where a lead-indium-gold alloy was shown by IBM to be a way to achieve outstanding memory devices. Lead telluride is used in a thermo-electric power generator. Lead zirconate titanate is highly efficient as a piezoelectric material in both ultrasoncis and electrical energy generation.

Densest of the common metals: Beyond its use as a radiation barrier, lead's density leads to its use as a material for counterweights, ballast, and other design applications where high density is desired.

Low melting point: This property is advantageous for casting shapes and the bene-fit carries over to alloys.

Easily alloyed: Fusible alloys of lead with tin, antimony, arsenic, copper, zinc, silver, alkali metal, bismuth, cadmium, calcium, and other metals are well known. Fusible alloys have many uses stemming from the wide range of melting points from 47 to 310°C. Among these are bearings, type metal, roofing, battery plates, fusible safety plugs for sprinkler systems and boilers, low-temperatures baths, fillers for bending thin-walled tubing, molds, and ammunition. (See Table 9.17.)

IX. MAGNESIUM AND ITS ALLOYS*

A. Production

Magnesium was traditionally produced from seawater by electrolysis. A second method is by thermal reduction of magnesium oxide by ferrosilicon.

Table 9.17 Properties of Typical Lead Alloys

Designation	Form	Tensile strength (psi)	Yield strength, 0.5% offset (psi)	Elongation (%)	Hardness, Rockwell, ½ in Ball (30 kg)	Density (lb/in.^{3})
Commercial pure lead	Rolled sheet	2,500	1300	27 (8 in.)	75	0.41
	Extruded sheet	2,575		58 (2 in.)	78	0.41
Chemical lead	Sand cast	1,800	800	30	4 BHN	0.409
Antimonial lead (1% Sb)	Extruded, aged 1 month	3,000		50	7 BHN	0.406
Antimonial lead (8% Sb)	Rolled 95% 235 C.Q., aged 1 day	4,650 12,350		31.3 4.7	9.5 BHN 26.3 BHN	0.388 0.388

1. Electrolytic Method

Seawater contains about 0.13% magnesium. This seawater is mixed with dolime (calcined dolomitic rock) and placed in large settling tanks or ponds where insoluble magnesium hydroxide settles to the bottom. The magnesium hydroxide slurry is pumped to filters; the cake is then dissolved in hydrochloric acid to produce a magnesium chloride solution that is evaporated in driers to produce granular magnesium chloride. The granules are decomposed by an electrical current in an electrolytic cell into molten magnesium and chlorine gas. The magnesium metal is drawn off and cast into primary ingots, or it can be alloyed with such metals as aluminum and zinc and cast into magnesium alloy ingots. The chlorine gas is collected and converted to hydrochloric acid for recycling in the process.

2. Ferrosilicon Process (Magnetherm, Pidgeon Process)

This method uses a mixture of powdered ferrosilicon and dolime. The mixture is heated in a chamber under vacuum. The silicon and magnesium oxide react, allowing magnesium to vaporize at the elevated temperature. Upon cooling the magnesium vapor condenses into crystals, which are collected, melted, refined, and cast into ingots.

---

*Contributed by ROBERT S. SHANE, based on material from the Second Edition and information supplied by Dow Chemical USA, Midland, Michigan.

B. Characteristics

Magnesium is the lightest of the commercial metals; its density is about two-thirds that of aluminum and one-fourth that of steel. Pure magnesium is seldom used as a structural material because of its low strength, but its alloys are stronger and many of them can be hardened by precipitation heat treatment. The uses for magnesium in alloys for portable or high-speed machinery have multiplied, and its price has dropped correspondingly. The price is close enough to aluminum that magnesium competes with aluminum in many applications.

The strength-weight ratio of the precipitation-hardened magnesium alloys is comparable to that of the strong alloys of aluminum or to the alloy steels. Magnesium alloys, however, have a lower density and stand greater column loading per unit weight. They are also used when great strength is not necessary, but where a thick, light form is desired. Examples are complicated castings, such as housings or cases for aircraft, and parts for rapidly rotating or reciprocating machines. The strength of magnesium alloys is reduced at somewhat elevated temperatures; temperatures as low as 200°F produce considerable reduction in the yield strength.

Despite the active nature of the metal, magnesium and its alloys have good resistance to corrosion by most atmospheres. In industrial or humid areas, however, corrosion is more rapid and will, over a period of years, cause powdering of the surface. The rate of corrosion is slow compared with the rusting of mild steel in the same atmosphere. Immersion in saltwater is very dangerous, but a great improvement in resistance to saltwater corrosion has been achieved, especially for wrought materials, by reducing some impurities, particularly nickel and copper, to very low proportions. Corrosion troubles may be expected even with protective treatment in poorly designed assemblies where moist air is trapped or where rain is allowed to collect. Where such designs are avoided, unpainted magnesium alloy parts that are oily or greasy will operate indefinitely with no sign of corrosion. Magnesium alloy parts are usually painted, however, except where conditions of exposure are mild. To assure proper adherence of the paint, the part is cleaned by grinding, buffing, or blasting and then given an acid dichromate dip.

C. Fabrication

1. Hot and Cold Working

Magnesium alloys harden rapidly with any type of cold work, and therefore cannot be extensively cold formed without repeated annealing. Sharp bending, spinning, or drawing must be done at about 500-600°F, although gentle bending around large radii can be done cold. Slow forming gives better results than rapid shaping. Press forging is preferred to hammer forging, because the press allows greater time for metal flow. The plastic forging range is 500-800°F, and if the metal is worked outside this range it will be easily broken.

2. Casting

The magnesium alloys, especially those that may be precipitation hardened, are used in casting. Sand, permanent mold, and die-casting methods may be used, but plaster of Paris casting has not yet been perfected.

Sand casting in green-sand molds requires a special technique, because the magnesium will react with moisture in the sand, forming magnesium oxide and liberating hydrogen. The oxide forms blackened areas called burns on the surface of the casting, and the liberated hydrogen may cause porosity. Inhibitors, such as sulfur, boric acid, ethyl glycol, or ammonium fluoride, are mixed with the damp sand to prevent the reaction. All gravity-fed molds require an extrahigh column of molten metal so the pressure will be great enough to force gas bubbles out of the casting and cause the metal to take the detail of the mold cavity. The thickness of the casting wall should be at least 5/32 in. under most conditions. Extralarge fillets must be provided at all reentrant corners, since stress concentrations in magnesium castings are particularly dangerous.

Permanent mold castings are made from the same alloys and have about the same physical properties as sand castings. Since the solidification shrinkage of magnesium is about the same as that of aluminum, aluminum molds can often be adapted to make magnesium alloy castings (although it may be necessary to change the gating).

Pressure cold-chamber castings are used for quantity production of small parts. The rapid solidification caused by contact of the fluid metal with the cold die produces a casting of dense structure with excellent physical properties. The finish and dimensional accuracy are very good, and machining is necessary only where extreme accuracy is required. Usually these castings are not heat treated.

3. Welding, Soldering, and Riveting

Many of the standard magnesium alloys are easily welded by gas or resistance-welding equipment, but they cannot be cut with the oxygen torch. Magnesium alloys are not welded to other metals, because brittle intermetallic compounds may form, or because the combination of metals may promote corrosion. Where two or more parts are welded together, their compositions should be the same.

Soldering of magnesium alloys is feasible only for plugging surface defects in parts. The solders are even more corrosive than with aluminum, and should never be required to withstand stress.

Riveted joints in magnesium alloy structures usually employ aluminum or aluminum-magnesium alloy rivets. Magnesium rivets are not often used because they must be driven when hot. Where aluminum alloy rivets are used under corrosive conditions, the rivet should be dipped in a bitumastic paint to insulate it from the magnesium. The rivet holes should be drilled, expecially in heavy sheet and extruded sections, since punching tends to give a flaky structure at the edge of the hole and to cause stress concentrations.

4. Machining

Magnesium and its alloys have the best machining characteristics of the common metals. The power required in cutting them is small, and extremely high speeds (5000 ft/min in some cases) may be used. The best cutting tools have special shapes, but the tools for machining other metals can be used, although somewhat lower efficiency results.

When magnesium is cut at high speed, the tools should be sharp and should be cutting at all times. Dull, dragging tools operating at high speed may generate enough heat to ignite fine chips. Since chips and dust from grinding can therefore be a fire hazard, grinding should be done with a coolant, or with a device to concentrate the dust under water. The magnesium grinder should not be also used for ferrous metals, since a spark might ignite the accumulated dust. If a magnesium fire should start, it can be smothered with cast iron turnings or dry sand, or with other materials prepared especially

Table 9.18 Composition of Representative Magnesium Alloys

UNS number	Alloy		Al	Mn	Th	Zn	Zr	Mg	Usage
M11311		B							Sheet, plate
	AZ31	or	3.0			1.0		Bal.	Extrusion
M11312		C							
M13310	HK31A				3.0		0.7	Bal.	Sheet, plate
M13210	HM21A			0.8	2.0			Bal.	Sheet, plate
M11610	AZ61A		6.5			1.0		Bal.	Extrusion
M11800	AZ80		8.5			0.5		Bal.	Extrusion
M13312	HM31A		1.2	3.0				Bal.	Extrusion
M16210	ZK21A					2.3	0.6	Bal.	Extrusion
M16600	ZK60A					5.5	0.5	Bal.	Extrusion

for the purpose. Water or liquid fire extinguishers should never be used, because they tend to scatter the fire. Actually, it is much more difficult to ignite magnesium chips and dust than is usually supposed, and for that reason they do not present great machining difficulties.

The special techniques that must be used in fabricating magnesium (working, casting, and joining) add considerably to the manufacturing cost. In selecting between aluminum and magnesium for a given part, the base cost of the metal may not give much advantage to either, but usually the manufacturing operations will make magnesium the more expensive.

D. Alloys of Magnesium

Composition and properties of representative magnesium alloys are given in Tables 9.18 and 9.19. The importance of selecting an alloy in accordance with the required service temperature is shown by Table 9.20.

X. NICKEL AND ITS ALLOYS*

A. Production

Nickel is mined in at least 20 countries. It is refined, in addition, solely from imported raw materials in the United States, Norway, the United Kingdom, France, and Japan. World production is about 1.7×10^9 lb/year. Nickel is principally found as sulfide ores (60%) and oxide ores (laterite; 40%). The concentration of nickel in these ores is so low that profitable recovery usually depends on the ability concurrently to recover valuable by-products (iron, copper, gold, silver, platinum group metals, cobalt, and chromium).

---

*Contributed by ROBERT S. SHANE, based on material from the Second Edition and information supplied by International Nickel Co., New York, New York, *The Metals Handbook,* American Society for Metals, Metals Park, Ohio, and High Technology Materials Division, Cabot Corp., Kokomo, Indiana.

Table 9.19 Physical and Mechanical Properties of Magnesium Alloys (Typical)

Alloy and temper[a]	Density (lb/ft)	Melting point (°F)	Room temperature thermal conductivity (cgs units)	Electrical resistivity (μohm-cm)	Tensile strength (ksi)	Tensile yield strength (ksi)	Compressive yield strength (ksi)	Elongation (%)
AZ31B-H10	0.0639	1160	0.18	9.2	35	20	15	17
AZ31B-F					38	29	14	14
AZ31B-O					37	22	16	21
HK31A-H24	0.0647	1202	0.27	6.1	38	29	23	12
HK31A-0			0.25	6.6	33	18	13	20
HM21A-T8	0.0640	1202	0.33	5.0	35	26	21	12
AZ61A-F	0.0647	1140	0.14	12.5	45	33	19	16
AZ80A-F	0.0649	1130	0.12	14.5	45	36	17	11
AZ80A-T5					55	39	32	8
HM31A-T5	0.0651	1202	0.25	6.6	44	38	27	8
ZK21A-F	0.0645	1202	0.30	5.4	42	33	25	10
ZK60A-F	0.659	1175	0.28	6.0	49	37	27	14
ZK60A-T5			0.29	5.7	52	43	31	12

[a]Temper = F, as fabricated; T5 artificially aged; T8, solution heat treated, cold worked, artificially aged; H-24, strain hardened then partially annealed; O, fully annealed; H-10, slightly strain hardened.

Table 9.20 Typical Maximum Service Temperatures for Three Alloys

Alloy	Service temperature	
AZ31B	100°C	210°F
HK31A	315°C	600°F
HM21A	345°C	650°F

When undersea mining becomes practical, the deep sea nodules that contain nickel, cobalt, copper, and manganese may become important sources. Nickel is concentrated by pyrolytic and sedimentary processes. The final refining is done by electrolysis to recover a 99.95% pure product (which may contain 0.3-0.5% cobalt).

B. Characteristics

Nickel is a tough silver-colored metal with about the same density as copper. It is distinguished for its oxidation and corrosion resistance, especially at elevated temperatures, and for its ability to form many alloys. About 55% of the total nickel output goes into ferrous alloys, mostly for austenitic chromium-nickel stainless steels. About 20% is used in nonferrous alloys, like those with copper, cobalt, chromium, and molybdenum and combinations of these. Electroplating uses about 15% of the nickel production. The balance is used in a wide variety of cast products and minor applications, such as chemicals, salts, catalysts, ceramics, coinage, and permanent magnets.

Nickel-copper alloys have high strength, good weldability, and excellent corrosion resistance over a wide range of temperatures and conditions (under 500°C). The nickel-chromium alloys have high strength and excellent corrosion resistance at high temperatures (above 540°C). The nickel-iron-chromium alloys have good high-temperature strength for somewhat less critical usage than the nickel-chromium alloys. In every case where hydrochloric or hydrofluoric acid exposure is contemplated the specific data applicable to a given alloy should be consulted.

Special alloys of nickel are used as electric heater materials because of their resistance to oxidation and retention of strength even at red heat temperatures. Among these are Nichrome (60% nickel, 15% chromium, 25% iron) and Nichrome IV (80% nickel, 20% chromium). Important to the use of any heater material is provision for unhindered expansion and contraction to avoid the concentration of stresses during use.

Nitinol is an approximately 50:50 alloy of nickel and titanium. Discovered by W. J. Buehler and associates at the Naval Surface Weapons Center (White Oaks, Maryland), it has "shape memory." Flat strips can be bent readily at room temperature. Upon heating to a higher temperature they become stronger and regain their flat shape or any other preselected shape that has been annealed into them.

Over 30 different alloys are used to make thermostat metals. Most of these are nickel-iron or nickel-chromium-iron alloys. Some are distinguished by their extremely small coefficient of thermal expansion. An interesting use of such an alloy is in international length standards.

Electrical resistors are constructed from copper-nickel (e.g., 57% Cu, 43% Ni) and copper-manganese-nickel (e.g., 83% Cu, 13% Mn, 4% Ni) alloys. These are characterized by a very low thermal coefficient of resistance. Some superalloys have effective service temperatures in excess of 2000°F.

XI. SILICON*

This is the most abundant metal in the earth's crust. As a metal, silicon is too brittle and too difficult to form in structural shapes. The use of silicon in electronic applications is discussed in Chap. 6. Silicon possesses good resistance to acids (except HF) and atmospheric corrosion.

Ferrosilicon is made by adding iron oxide to the furnace charge when reducing silica by carbon in an electric furnace to produce the metal. Ferrosilicon has been added to steels to form electrical core sheets and as a deoxidizer for steel. Silicon is added to brasses, bronzes, aluminum, magnesium, and other alloys.

XII. TIN, SOLDER, AND BEARING MATERIALS†

A. Mining Tin

Tin has the unusual distinction of being mined in districts that are geographically distant from the areas where it is primarily consumed. Seven countries, namely, Malaysia, Bolivia, Indonesia, Thailand, Australia, Zaire, and Nigeria, account for over 85% of the world's production of tin (3,4). The most important tin mineral is cassiterite, a naturally occurring oxide of tin (SnO_2). Primary occurrences of tin are chiefly in narrow hydrothermal ore bodies in faults and veins of country rock and usually in association with granites or in overlying sedimentary rocks of these intrusive complexes.

Erosion of granites, which are known to be a genetic source of tin mineralization, results in the formation of tin-bearing alluviums. These alluvial deposits are the primary source from which tin is won.

Dredging and gravel pump mining account for most of the output from this type of deposit. Dredges are in operation that can dig to depths of 150 ft (45 m), and these units can handle over 750,000 cubic yards (574,000 m^3) of alluvium per month. Ore grades as low as 0.015% tin have been processed profitably by modern tin-dredging equipment.

Gravel pump mining accounts for about 40% of the world's tin ore production. During this operation, water jets are used to break up the stanniferous ground, and this material is washed into a sump. The tin-bearing gravels are then pumped from the sump to an elevated and riffled launder where the heavy ore is concentrated by gravity. Concentrates may contain 70-75% tin.

Lode deposits in Bolivia and Cornwell are mined by either sinking shafts or driving adits to the ore zones. Ore is broken from the working face of the mine by drilling and blasting. The tin-bearing rock is then crushed or ground to produce a finely divided ore capable of being upgraded by various beneficiation techniques.

B. Smelting Tin Ores (5,6)

Tin ore concentrates are usually sold directly to custom smelters, most of whom employ pyrometallurgical methods to reduce the ore to metal. Much of the smelting is accom-

---

*Contributed by ROBERT S. SHANE.
†Contributed by JOSEPH B. LONG.

plished in fixed-head reverberatory furnaces. A furnace charge consists of tin concentrates with anthracite coal as the reducing agent and limestone as the flux.

C. Refining Tin

After smelting, the crude tin is refined to pure metal, which is heated in cast iron pots to about 600°C. Steam or air is introduced into the melt to oxidize impurities, and these are volatilized or drawn off in the dross and slag. Metal purity of 99.85% of higher is obtained, and pure ingots are cast from the molten metal recovered at this stage. Alternatively, the metal is cast into tin anodes for electrolytic refining of the metal (7). The electrolytic process involves the use of low voltage and low current densities in an acid electrolyte composed of sulfuric acid, phenolsulfonic acid, and cresylic acid. Tin of very high purity is plated onto cathodes, and after a period of electrodeposition, the cathode tin is removed from the electrolyte and the metal melted off the cathode plates and cast into high-purity ingots.

   Commercial tin is usually sold in ingots of 28 or 56 lb (13 or 25 kg) or pigs weighing 100 lb (45 kg). The London Metal Exchange recognizes "standard" tin or 99.75% purity; in the United States commercial tin is designated Grade A with a minimum purity of 99.85%. Some brands of tin consistently provide for a minimum tin content of 99.9%.

   Tin is a silvery white metal with an atomic number of 50 and an atomic weight of 118.70. It is between germanium and lead in group IVB of the periodic table. It has 10 naturally occurring isotopes with ^{120}tin the most abundant.

   Tin has two allotropic forms: white tin, β (beta), and gray tin, α (alpha). White tin crystallizes in the body-centered cubic crystal structure; gray tin has a diamond cubic crystal structure. Transformation of β to α tin may occur with high-purity tin at the equilibrium temperature of 55°F (30°C), but usually considerable undercooling or inoculation by α tin is required to initiate the transformation. Small amounts of impurities in commercial tin are usually sufficient to prevent or retard the allotropic change at temperatures well below zero.

   Uses for tin are probably more widespread than for any other metal. A characteristic of tin is that it is rarely used alone, but more often in conjunction with other metals and often only in small amounts. Tin owes its position in technology to certain basic properties. Physically speaking, tin has a low melting point (450°F; 232°C) and it readily forms alloys. The metal has high resistance to corrosion and good appearance. Tin is usefully amphoteric and can form chemical bonds with carbon.

   Five classes of applications exemplify the unusual properties of tin, namely, as unalloyed metal, as powder, as coatings, as alloys and as chemical compounds.

1. Tin Used as an Unalloyed Metal

In certain applications, the corrosion resistance of pure tin is used advantageously. Sheet tin is employed as a lining for storage tanks for distilled water (8). Even the internal surfaces of valves and piping of distillation systems are coated with thick cast linings of tin to prevent contamination of the purified water (9). Collapsible tubes are extruded from slugs of pure tin and are used for packaging food pastes, pharmaceutical products, and artists' colors (10-12).

   Flat plate glass is produced on the surface of molten tin (13). Glass surfaces are so smooth that grinding and polishing are unnecessary. Molten tin is the ideal metal for this application since tin has a very low vapor pressure at 1050°C, the temperature of the glass as it emerges from the glass-melting furnace.

Great amounts of pig tin are melted and cast into anodes or billets for extrusion into anode forms for electrolyzing in plating baths. Minor amounts of pure tin as well as tin alloys are rolled to foil thicknesses for packaging foods, enrobing candy, or making electrical condensers.

2. Tin Used as Powder

Tin is easily atomized to form spherical powders used in the manufacture of sintered bronze or iron powder metallurgy parts (14). Tin is also alloyed with lead, antimony, or silver, and the atomized powders are mixed with various flux compositions to provide useful paste or cream solders for electronic or plumbing applications. Some tin powders are flame sprayed onto surfaces of machinery used to prepare various foods (15) or to repair surfaces of damaged bearings (16).

3. Tin Used as a Coating

Electrotinned steel (tinplate) is the world's largest use for tin (17). Almost 5 million tons of tinplate are produced annually in the United States to provide containers and packaging materials for foods and beverages, as well as a variety of nonfood items.

Electrotinplate is manufactured on continuous lines from either acid or alkaline electrolytes. It can be produced either with an equal amount of tin on the steel surfaces or with a differential coating on each side of the steel. Nominal tin coating thickness for equally coated tinplate can range between 3.8×10^{-4} and 15.4×10^{-4} mm, depending on the container specifications. For differentially coated tinplate, the maximum coating thickness on one surface is 20.7×10^{-4} mm. Tin mill products are supplied in either cut sheets or rolls.

Less than 1% of all tinplate manufactured in the United States is produced by the older hot-dip method. This method involves the dipping of chemically cleaned and fluxed sheets of low-carbon steel into a molten tin bath to produce thickly coated tinplate. A variety of parts and assemblies are also hot tinned to provide an attractive bright coating corrosion resistant to many media (18).

Pure tin coatings are nontoxic and are used on food-processing equipment (19) and also provide solderable and corrosion-resistant surfaces for wire (20-22). Tin coatings are unattacked by organic vapors from sleeving or other insulation and are the preferred coatings in electrical applications where these conditions are expected.

4. Tin Alloy Coatings

Tin alloy coatings have been developed to provide for harder and more corrosion resistant finishes than pure tin. Bronze alloys of 10-45% tin, balance copper, are plated on furniture hardwares, trophies, and other ornamental pieces (23,24). Bronze coatings also serve as a stop-off finish during nitriding operations.

Tin-lead coatings are applied by either hot dipping or electrodeposition (18,25,26). Coatings may range from 5 to 70% tin depending on their end use. Lower tin (5-40%) alloys are applied to copper wire, brass strip for radiator manufacture, and electronic parts. Low-carbon steel is coated with lead-tin alloys (5-25% tin) to provide a product known as terneplate, which has coating weights in the range of 0.3-1.4 oz./ft.^{2} (0.2-0.4

kg/m^2) (27). Terneplate is used in the manufacture of gasoline tanks, air filters, roofing, downspouts, flashing, cabinets, and consoles for instruments.

Higher tin content (40-70%) solder coatings are used by the electronics industry for printed circuit boards, electrical connectors, lead wires, and capacitor and condenser cases (28). Coatings in thicknesses of 0.0001-0.0003 in. (0.003-0.008 mm) have shown excellent solderability and shelf life.

Tin-nickel coatings are plated from acid electrolytes to provide a hard and corrosion-resistant finish (29). The alloy coating was developed originally as a decorative finish, but its high resistance to ferric chloride and ammonium persulfate have made it useful as a selective etchant resist in printed circuit manufacture. Tin-nickel electrodeposits are inherently brittle, and therefore it is not practicable to perform much fabrication after plating. The excellent corrosion resistance and low coefficient of friction properties of tin-nickel deposits have led to a variety of industrial applications.

Tin-zinc coatings are plated from alkaline cyanide solutions (30). It has been established that a zinc content of 20-25% of the deposit affords the greatest protection to steel. The tin-zinc plate is superior to zinc in environments of long-sustained humidity. In addition, the alloy plate affords galvanic protection for steel in contact with aluminum (31). It may function as a bearing surface, especially if lubricated. Solderability of tin-zinc finishes is not as good as tin but better than cadmium and is far superior to zinc.

Tin-cadmium coatings are normally deposited from a fluoborate bath (32). Alloy deposits of 75% cadmium, 25% tin plated on steel have been reported to withstand over 3700 hr exposure to salt spray before the onset of corrosion of the base metal. The alloy exhibits good corrosion resistance to jet fuels containing mercaptans and high-temperature synthetic oils.

Tin-cadmium alloys are also deposited on steel parts by a mechanical plating process (33). The process consists of tumbling of the parts to be coated in a plating barrel containing a water slurry of fine tin and cadmium powders, glass spheres, and a proprietary chemical compound, which is a mild pickling agent. In operation, the glass beads impact the soft powders against the parts, and in doing so, cold welding occurs between the two clean surfaces. Powder particles weld first to the parts, and increased thickness of the deposit is provided by additional powder particles welding to the coating.

Fully hardened steels are mechanically plated with 50:50 tin-cadmium coatings without danger of hydrogen embrittlement. Various steel automotive powder metallurgy parts, fasteners, springs, clips, sockets, and brackets are mechanically plated in coating thicknesses of 0.0001-0.0003 in. (0.003-0.013 mm), depending on the service conditions required by the part.

Tin-cobalt coatings are electrodeposited from a chloride-fluoride bath (34). Tin content may be as high as 85% in the highly corrosion resistant coating.

A ternary alloy of 2 copper, 8 tin, and 90 lead is used as an overplate for copper-lead bearings; another group of ternary alloys, copper-tin-zinc alloys plated from an alkaline cyanide electrolyte, provide nonmagnetic solderable coatings with low and stable contact resistance.

Tin immersion coatings are applied without electrical current by placing parts in an aqueous solution containing tin salts (35). Deposition takes place by chemical displacement where the tin in solution displaces the base metal, usually copper. Plating thickness of the coating is normally about 0.000015 in. before the process stops. However, thicknesses up to 0.002 in. have been provided by placing the base metal in contact with a dissimilar metal, such as zinc, thereby creating a current and promoting additional plating.

Tin immersion coatings are popular for decorative finishing of small parts and improvement of the solderability of solder-coated printed circuit boards. Thin immersion coatings of tin are also applied to tubing to prevent discoloration of water supplies by copper salts. Aluminum pistons are often immersion tinned to provide metallic lubricity during engine break-in periods.

Aluminum-tin coatings applied by plasma arc spray techniques are useful as oxidation-resistant coatings for refractory metals in service at temperatures above 1090°C (36). These intermetallic coatings are usually applied in the form of a semimolten slurry.

Tin and tin alloys are often metallurgically bonded to a variety of base metals. These clad metals are usually bonded by a solid-phase process that results in a continuous bond with excellent strength. Typical applications include solderable surfaces for alloys, such as Kovar and tin-clad nickel for germanium or silicon cradle supports. Most of the tin-clad materials are supplied only in rolled tempers since the melting temperature of tin is well below the annealing temperatures of the base metals.

5. Tin in Solders (37-39)

Soldering is a group of joining processes that produces coalescence of materials by heating them to a suitable temperature and using a filler metal having a liquidus not exceeding 450°C and below the solidus of the base materials being joined. The filler metal is distributed between closely fitting surfaces by capillary attraction. Typical solders and their uses are given in Table 9.21.

Table 9.21 Solders and Their Uses

Type	Composition	Tensile strength ton/in.^{2}	Uses	MPa
Common, solders				
Tin-lead	60% Tin, lead	3.4	a, b	47
	20-50% Tin, lead	2.7-2.9	a, b	37-40
Antimonial tin-lead	30-50% Tin, lead, antimony	3.1-3.6	b	43-50
Special-purpose solders				
Lead-tin	2% Tin, lead	1.8	c	18
	10% Tin, lead	2.4	d	33
Lead-tin-silver	5% Tin, lead, 1.5% silver	2.5	d, e	34
	1% Tin, lead, 1.5% silver	1.8	d, e	18
	60% Tin, lead, 1.5% silver	4.5	d, e	69
Tin	Tin	0.9	e	12
Tin-antimony	Tin, 5% antimony	2.8	e	39
Tin-silver	Tin, 2% silver	1.7	e	23
	Tin, 5% silver	3.8		52
Tin-lead-cadmium	Tin, 30% lead, 18% cadmium	2.8	f	39

Key to uses: a, electronics; b, general engineering: Avoid high or very low temperature; c, tinplate can seams; d, cryogenics; e, creep resistance; f, low-melting-point solder.

Soldering is an ancient craft that was performed in Roman times. However, today soldering is an advanced technology and many assemblies depend upon the reliability of soldered junctions in assemblies, namely,

1. To provide a mechanical fixing of parts
2. To provide electrical continuity
3. To conduct heat from one position to another
4. To provide integrity and strength to an assembly of components
5. To provide a liquid or gas seal

Soldering offers some distinct advantages over other joining processes. Joints may be made relatively easily. The range of alloys that melt at different temperatures and the various heating methods provide versatility to the process. Overall costs associated with soldering are generally low. Temperatures involved in soldering will not ordinarily alter the properties of the base metals being joined. Repairs of defective solder joints can be made easily and inexpensively.

6. Solder Alloys

Most of the solders used are mixtures of tin and lead. However, elemental additions of antimony, silver, copper, and metals with lower melting points are often combined with the binary alloys to provide solders with different melting ranges and mechanical and physical properties (Table 9.22).

Most tin-rich solders will show an increase in strength with decreasing temperature to about −100°C, where they incur a drop in both strength and ductility. Lead-rich solders do not exhibit this embrittlement at low temperatures. Impact strength of tin solders also drop at about −70°C, and therefore, solders used in cryogenic applications should probably not include more than 20% tin (40).

Alloys used for soldering have relatively low strength but good ductility compared with the base metals they are to join. Room temperature corresponds to about 65% of the solidus temperature of eutectic base tin-lead alloys, so that the tensile and shear strength of these alloys will decrease with increasing temperatures above ambient.

Volume 2 discusses the soldering process and the elements of what is required to provide sound joints and testing for solderability. Basis metal composition, joint design, the choice of fluxes, solder alloys, and heat sources are essential considerations in providing the required reliability and integrity levels in soldered assemblies.

7. Tin in Bearing Alloys (41,42)

One of the important uses of tin is in plain bearings. Tin is an important constituent in bearing construction since the metal has the ability to retain a film of lubricant and prevent metal-metal contact, even if a temporary failure of the lubricant supply occurs. In addition, tin-based white metals for bearings provide excellent conformability and embeddability, which allow particles of foreign material to be tolerated by burial in the bearing surface, thus avoiding damage to shafts of low hardness. Finally, tin improves the resistance of plain bearings to corrosion by the lubricant or contamination that may be present in the oil.

Table 9.22 Tensile Properties of Bulk Solders at 20°C and 100°C (Tested at 0.05 mm/min)

Nominal composition (%)				Tensile strength MPa (lb/in.²) at		Elongation (%) at		Loss in tensile strength	
Sn	Pb	Sb	Ag	20°C	100°C	20°C	100°C	20°	100°C
60	40	–	–	19 (2700)	4 (580)	135	100	–	79%
40	60	–	–	17 (2420)	–	130	–	–	–
10	90	–	–	20 (2850)	8 (1160)	56	100	–	60%
62	36	–	2	43 (6120)	19 (2700)	7	–	–	70%
40	58	2	–	23 (3270)	7 (960)	78	100	–	56%
95	–	5	–	31 (4410)	29 (2900)	25	31	–	35%
96.5	–	–	3.5	37 (5260)	–	31	–	–	–
5	93.5	–	1.5	30 (4270)	19 (2750)	20	25	–	27%
1	97.5	–	1.5	35 (4980)	–	28	–	–	–
–	95	–	5	32 (4550)	–	25	–	–	–

Source: From Ref. 37.

D. Tin in Babbitt Alloys

The traditional white metal alloys used for bearings are divided according to the metal used as their basis, that is, tin based, lead based, and also a number of alloys that are combinations of tin and lead. These alloys are commonly named babbitts, which include compositions ranging from 80% or more tin with no lead to 80% lead and less than 5% tin. Antimony and copper are usually present in babbitts, and cadmium, nickel, arsenic, or tellurium are sometimes added for increased strength properties.

Tin-based babbitts normally limit the lead content to 0.35-0.5% to avoid the formation of a low-melting tin-lead eutectic, which will reduce the bearing strength. Addition of about 1% cadmium in a tin-rich bearing alloy will raise the compressive strength, bulk fatigue strength, and bearing fatigue rating by about 40%. However, addition of more than 0.2% cadmium to a lead-based alloy will create disastrous effects on mechanical properties because of the formation of a ternary tin-lead-cadmium eutectic at 150°C.

Tin- and lead-based babbitt alloys are used at high sliding velocities under light loads. They are usually bonded to steel or bronze backings to improve their load-carrying capacities. Babbitts can be gravity or centrifugally cast onto bushing shells. Many sleeve bearings are manufactured by casting white metals onto tinned steel strip, and then the material is rolled, blanked, formed, and broached to form half-bearings. Babbitt alloys are also used to impregnate sintered alloys used for small bearings.

Lead-based babbitts offer an economic advantage over tin-rich alloys, but this must be balanced against their lower strength and hardness at elevated service temperatures. In addition, lead-based bearing metals are prone to drossing and segregation of the intermetallic phases of the alloys during solidification, and this condition may be exaggerated during centrifugal casting. Cerium, arsenic, or nickel are sometimes added to these alloys to control the amount of segregation during casting.

The group of intermediate alloys that contain significant amounts of both tin and lead include a wide range of bearing materials containing 20-75% tin and 10-65% lead, as well as antimony content in the region of 12-15%. Copper content is usually up to 5%. These alloys offer mechanical properties that, at room temperature, are similar to tin- or lead-based bearing metals. At higher service temperatures these properties decrease rapidly because their structure contains the eutectic composition. Therefore, these alloys are used for less exacting applications and where sufficient bearing area can be used to reduce surface loading and temperature.

In large engines and equipment, the difficulty of obtaining good alignment of shafts and small clearances demands the use of tin- and lead-rich babbitts. These bearings are adequate for applications where unidirectional loads are distributed over a sufficiently large surface area. However, where greater loads and speeds and higher operating temperatures are encountered, increased fatigue strength is required in the bearing metal. Aluminum-tin alloys were developed to meet this need.

1. Aluminum-Tin Bearings (43,44)

Aluminum alloy bearings are run against hardened shafts, and they provide high load-carrying capacity, fatigue strength, and thermal conductivity. The alloys are sensitive to dirt and other foreign particles in lubricating oils. Their successful applications as connecting rod and main bearings in internal combustion engines, in roll neck bearings in steel mills, and in reciprocating compressors and aircraft equipment have only resulted from adherence to strict dimensional tolerances and scrupulous cleanliness to avoid oil contamination. The limited embeddability of these alloys is improved by using an electroplated lead-tin overlay.

The SAE 770 aluminum-based casting alloy has a structure consisting of essentially pure tin around primary grains of aluminum. Nominal compositions allow up to 6.5% tin and small amounts (i.e., 1%) of copper and nickel, which increase the hardness, strength, and scuff resistance. SAE 780 alloy has a similar composition, except 1.5% silicon is added. Often this alloy is used to form bimetal bearings that have a layer of the 780 alloy roll bonded to nickel-plated steel.

An alloy of 1-3% cadmium and varying amounts of silicon, nickel, and copper provides improved bearing properties. This alloy is designated SAE 781 and is bonded to steel backings. A 0.0005 in. (0.013 mm) overlay of lead-10% tin provides improved surface properties for these composite bearings during the engine break-in periods.

Aluminum-tin alloys have been developed that have 20-40% tin and 1% copper (43). Rolling and heat treatment of these high-tin alloys provides a reticulated structure of discrete particles of tin in an aluminum matrix. Fatigue resistance and bearing performance of the 20% tin alloy bonded to steel backings is adequate for virtually all automotive applications. Aluminum alloys with higher tin contents have lower fatigue strength than the low-tin alloys, but they have higher seizure resistance and increased embeddability and conformability.

2. Tin in Bronze Bearings (45)

Bronze alloys normally range in composition from 5 to 20% tin. However, the tin content is of the order of 10-12% in the customary bearing bronzes. Cast copper-tin alloys containing more than about 5% tin show the hard constituent (δ) of tin-copper in a softer (α) matrix. Increasing tin content up to 12% improves structural strength and resistance to deformation by pounding. With more than about 12% tin, the amount of the δ constituent results in a very hard and brittle alloy. Therefore it is only under exceptional circumstances that it is advisable to use copper alloys containing more than 12% tin. Exceptions are in alloys containing 15% tin, which are used in such special applications as locomotive slide valves, and bronzes containing 20% tin. The latter alloy is used in the construction of movable bridges and railroad turntables.

It is usual for copper-tin bronzes to contain varying percentages of other elements, of which phosphorus, lead, zinc, and nickel are the most important. Phosphorus is added in small amounts to deoxidize the metal and improve casting properties. A residual phosphorus content of about 0.05 or 0.1% will assure deoxidation of the bronze, resulting in improved crystal structure and wearing properties.

Lead additions to copper-tin bronzes improve the plasticity and machinability of the alloy but may reduce the toughness and shock resistance of a bearing. Of the group of leaded bronzes, one of the best known alloys has the composition of 80% copper and 10% each tin and lead.

Zinc, like phosphorus, performs as a deoxidizer. The copper-tin alloys containing up to 6% zinc are known as gunmetals. A typical alloy is 88 copper, 10 tin, 2 zinc, which has been used for marine castings. Generally, zinc is not added as a special constituent to alloys for bearing applications since it is considered that the phosphor bronzes are superior.

Nickel additions improve the strength and toughness properties of copper-tin bronzes. Optimum improvement is obtained with nickel additions of 1 or 2%. Further increases in the nickel content provides alloys that have been employed as bearing materials in elevated temperature environments, or in the presence of corrosive media.

3. Tin in Powder Metal Bearings (46)

Sintered metal sleeve and flanged bushings in sizes ranging from 1/32 to 6 in. (0.8-150 mm) diameter and with length from 1/32 to 4 in. (0.8-100 mm) are popular for small motors in home appliances and machine tools, as well as in aircraft and automotive accessories, business machines, and farm and construction equipment. In addition, sintered bearings are useful in inaccessible places where relubrication would be difficult. By the nature of their manufacture, these bearings are porous and these interconnecting pores may take up to 10-35% of the total volume of the part. Lubricating oil is stored in these voids, and in operation, the oil feeds through the interconnecting pores to the bearing surface. Any of the oil forced from the loaded zone of the bearing is reabsorbed by capillary action.

The 90% copper-10% tin bronze is the most commonly used porous bearing material (46). Many variations of composition are possible to meet specific bearing requirements. Fractional horsepower motors under high-speed, light-load conditions demand high porosity in the bearings with a maximum amount of lubrication. Graphite at levels of 1-3.5% is frequently added as an impregnant in sintered bronze bearings to enhance lubrication properties. A low-oil content in a low-porosity material with a high-graphite content has proven satisfactory in applications where oscillating or reciprocating motions are encountered and where it is difficult to provide an oil film.

Powdered iron bearings, which include additions of copper and either tin or solder (47) powders, are suitable in many automotive applications, toys, farm equipment, and machine tools. The load capacity of porous iron bearings is generally higher than that for sintered bronze, but limiting surface speeds are about one-third lower. A common composition for sintered aluminum bearings is 5% copper and 3.5% both tin and lead. These bearings provide for cooler operation, greater tolerance for misalignment, and lower weight than either porous bronze or iron.

Elastomeric materials, carbon-graphite, wood, cermets, ceramics, and plastics are used as bearings in special applications. These nonmetallics are discussed in Chap. 16, Sec. XII.

4. Tin Bronzes (48,49)

Alloys of tin and copper have been used for centuries, and they may possibly be the oldest alloys known. Although the bronzes originally referred to binary alloys of tin and copper, the term has come to be used rather loosely and in recent times has been applied to alloys that do not contain tin. The aluminum bronze alloys are typical, and thus it is becoming customary in accurate presentations to refer to the true bronzes as tin bronzes.

The long-term durability of tin-copper alloys is evidenced by the excellent condition of bronze coins and statuary dating back almost 7000 years. However, modern technology demands more than durability, and now higher quality, greater strength, and uniformity of metal are primary concerns of users of bronze alloys.

Phosphor bronze is a true tin-bronze alloy that has been subjected to a special treatment by the introduction of small amounts (i.e., 0.01-0.45%) of phosphorus. This results in an alloy with greater resiliency, fatigue endurance, and hardness. Slight changes in the tin content and variation in reduction and annealing processes provide for a range of alloys with improved properties, such as higher tensile strengths and increased elasticity, corrosion resistance, and resistance to shock, wear, abrasion, and season cracking. Some properties of the common phosphor bronzes are given in Table 9.23.

Additions of lead to low-tin phosphor bronzes improves the machinability of alloys for screw machine operations. Zinc may also be added to phosphor bronze compositions, but these alloys are only used in special applications.

5. Tin Brasses

The second group of copper alloys containing tin are classified as tin brasses, which are essentially copper and zinc alloys with minor amounts of tin (up to 3%). Naval brass, having a nominal composition of 60% copper, 39.25% zinc, and 0.7% tin, is widely used in marine applications, such as tube support sheets for heat exchangers and steam condensers. Other important applications are for welding rods, valves, stems, propeller shafting, and marine hardware. By substitution of 1.75% lead for an equal amount of zinc, an alloy is produced that has been named leaded Naval brass. This material has more than twice the machinability of the unleaded alloy.

Alloys of the nominal composition 71 copper, 28 zinc, 1 tin are known as Admiralty metal. Often either arsenic, antimony, or phosphorus in amounts of about 0.05% are added to Admiralty metal to prevent dezincification. This alloy is commonly used in evaporator and heat exchangers, and condenser tubes when the feed water velocity is low. (See also Sec. V.K in this chapter.)

Despite its name, manganese bronze is a tin brass containing 58.5% copper, 39.2% zinc, 1% iron, 1% tin, and 0.3% manganese. This alloy has excellent hot-working properties, although it is only suited for moderate cold working. Its high tensile strength and good resistance to corrosion make it useful as a material for propellers, pump rods, shafting, valve rods and stems, clutch disks, and screening used in paper manufacture and coal mining.

Single-phase copper-silicon alloys containing from 0.8 to 3.5% silicon are known as silicon bronzes. Tin as well as other alloying elements, such as manganese, zinc, iron, lead, and nickel, are usually present in this class of alloys. The small amounts of tin do not have an adverse effect on the hot-working properties of the alloy. Low-silicon (1.5%) bronzes containing 1.6% tin exhibit excellent malleability, even when hard drawn, and this makes for relatively easy cold-heading or roll-threading operations.

A copper-nickel-tin alloy has been developed for electronic applications where solderability is a requirement (50). This alloy has the nominal composition of 9% nickel, 2% tin, balance copper. Its reverse bending fatigue life at 10^8 cycles makes it comparable to other copper-based spring materials at equivalent tempers. Transverse and longitudinal properties are similar, which affords freedom in design and economy in its use.

A copper-tin-magnesium alloy containing about 5% tin and 1% magnesium has been developed, which on heat treatment provides a hardness of about 220 HV, a tensile strength of nearly 710 MPa, and an electrical conductivity of 35-40% IACS (51). Hardness levels and tensile strength can be further improved by cold working in the solution-treated condition and then aging.

Table 9.23 Properties of Common Phosphor Bronzes

Alloy Number UNS	Nominal composition			Cold-working capacity	Hot-working capacity	Forms	Tensile strength range	
	Copper	Tin	Phosphorus				ksi	MPa
C50500	98.75	1.25	0.35 max.	Excellent	Good	Flat	40-75	275-515
						Wire	72-79	495-545
C51000	95	5	0.35 max.	Excellent	Poor	Flat	47-107	325-735
						Rod	70-75	480-515
						Wire	50-140	345-965
C52100	92	8	0.35 max.	Good	Poor	Flat	55-120	380-825
						Rod	80	550
						Wire	60-140	415-965
C52400	90	10	0.25 max.	Good	Poor	Flat	66-128	455-885
						Wire	66-147	455-1020

6. Pewter (52)

Modern pewter alloys for both functional and decorative articles consist primarily of tin with small amounts of copper, antimony, and occasionally bismuth. In the past pewter alloys were adulterated with varying amounts of lead, but now lead content in pewter is strictly limited, principally on health grounds.

Pewter alloys are soft compared with such metals as copper or brass. The high-tin alloys may be readily worked by hammering or pressing. Pewter is quite fluid at casting temperatures and may be cast easily by gravity and centrifugal or pressure die casting techniques, as well as the lost wax process (53).

Many pewter hollowwares are manufactured by spinning flat disks cut from pewter sheet. During this process, the metal disk is formed over a mandrel held in the spinning chuck of a lathe. The sheets from which these disks are cut are produced by rolling chill-cast ingots or bars of the metal. It has been established that tearing of pewter sheet can be reduced to very low levels if, at an intermediate thickness in the rolling schedule, the sheet is either turned through 90° (cross rolled) or heat treated in the temperature range of 150-200°C. Rolling is then continued down to final thickness (54).

7. Fusible Alloys (55,56)

Tin combines readily with lead, cadmium, bismuth, indium, and other metals to form binary, ternary, quaternary, and quinary alloys that find a number of industrial applications. The alloys in this classification have melting temperatures below that of tin (232°C). Selection of fusible alloys for various uses is influenced by their physical properties, particularly their relatively low melting temperatures and their varying degrees of hardness and tensile strengths. Depending on composition, hardness values for a range of fusible alloy composition may be from 9 to 22 HB, tensile strengths from 21 to 90 MPa (3000-13,000 psi), and elongations from 0 to more than 200%. Fusible alloys are subject to creep at room temperature under relatively light continuous loads.

Many of the fusible alloys contain bismuth, which has the unusual property of expanding on solidifcation. By selection of the appropriate bismuth content, fusible alloys can be made to contract, expand, or even remain dimensionally stable on solidification. Alloys of 35-45% bismuth will exhibit a slight shrinkage during solidification. These alloys expand in the solid phase, which is sufficient to compensate for initial solidification shrinkage. Lead-free alloys containing 50% or more bismuth expand during solidification with only slight shrinkage while cooling to room temperature.

8. Type Metals (57)

Type metals are lead-based alloys containing 10-25% antimony and 3-13% tin. Although other metals are occasionally present in type metal alloys, they are introduced only to slightly modify the alloy properties and rarely exceed 1% of the alloy composition. Type metals can be corrected for composition, cleaned, and remelted with simple equipment and at a low cost and virtually no loss in composition. Hardness can be controlled by varying the percentages of tin and antimony added to the lead. Tin also improves the fluidity of the alloys and permits lower casting temperatures and sharp definitive castings.

9. Tin-Alloyed Gray Irons (58-62)

Tin is added to flake or nodular graphite irons to suppress the formation of free ferrite in castings. Fully pearlitic microstructures can be achieved in flake gray irons with the addition of small amounts of tin, normally 0.03-0.10%, depending on the composition of the iron. Amounts of tin in malleable irons are typically 0.007-0.02%. High levels of tin retard first and second stages of graphitization. In nodular spherical graphite irons, tin is added in amounts of less than 0.10%, which has a powerful effect on the formation of pearlite and promotes graphite nodule formation. More than 0.10% can result in non-spheroidal graphite in cellular boundaries.

Tin alloyed irons have good resistance to scaling and oxidation (growth) at elevated temperatures. Sections are more uniform in hardness, which increases machinability. Strength and wearing properties of iron castings are enhanced by alloying with tin. Virtually all the tin is retained in the casting.

10. Tin in Battery Alloys (63-65)

The lead-calcium-tin alloy system has provided for a new genre of lead-acid storage batteries that have been named "maintenance-free." The ternary alloy has some important advantages over the more familiar antimonial lead alloys used for grids. Batteries using lead-calcium-tin plates have shown improved cranking performance over an extended period due to higher discharge voltages, prolonged shelf life, and large reductions in battery failure due to positive grid oxidation and reduction. Equally important is the effective operation of the Pb-Ca-Sn battery grids throughout a reasonable working life without the need for water additions.

11. Other Alloys

Titanium- and zirconium-based alloys often include small amounts of tin to improve weldability, strength, and corrosion properties (66,67). Silver dental amalgam alloys contain 25-27% tin and consist largely of the intermetallic Ag_3Sn. The material must have good strength, fatigue resistance, and lack of chemical reactivity with oral fluids, foods, or tooth material (68). Brazing alloys that include tin are often used as a replacement for filler metals containing cadmium (69). Although these tin alloys may be of higher cost than their cadmium-containing equivalents and may be slightly more difficult to use because of their higher melting points and wider melting range, they do not present health hazards associated with the evolution of fumes from brazing with the cadmium alloys.

12. Tin in Chemical Compounds (70)

The use of tin in the form of its chemical compounds has shown rapid growth over the past 25 years. Increasing demands of inorganic and organic tin chemicals has transformed the tin chemical industry from one mainly based on recovered secondary tin to one largely dependent on primary ingot tin.

Industrial applications of inorganic tin chemicals can be classified into direct and indirect uses. The latter represents the use of inorganic tin chemicals in electrolyte solutions for the electrodeposition of tin and tin alloys and as intermediates in the manufacture of organotin compounds. Direct uses of tin chemicals are where tin in the manufactured product is as an inorganic compound, as distinct from metallic or organometallic form.

13. Inorganic Tin Compounds (71)

Stannic oxide finds widespread use as an opacifier for glazes, and in combination with other metal oxide systems provides for a variety of pigments for ceramic wares (72).

After ceramic and glass applications, catalysts represent the next most important direct use of inorganic tin chemicals. Tin halides, particularly stannous and stannic chloride, have been employed extensively as catalysts for homogeneous reactions, namely, curing and other polymerizations, esterifications, halogenations, and liquid-phase hydrogenations (particularly the stannous chloride-platinum complexes). Stannic chloride has long been recognized as an effective Friedel-Crafts catalyst for homogeneous acylation, alkalation, and cyclization reactions (71).

Other inorganic tin compounds find use as oxidative catalysts of aromatic compounds for the production of organic acids and acid anhydrides. Other tin-containing catalysts, such as tin-platinum or tin-rhenium systems, have a markedly beneficial effect in the petrochemical industries in dehydrogenation, dehydrocyclization, cracking, isomerization, and the hydrogenation of hydrocarbon.

Novel catalysts based on coprecipitated tin-copper oxide gels have been developed. These catalysts exhibit high activity for the low-temperature (about 100°C) oxidation of carbon monoxide and the reduction of nitrogen oxides by carbon monoxide at 150°C (73,74).

Tin oxide electrodes are used for direct melting or for electrical boosting in conventional gas- or oil-fired glass-melting furnaces (75).

In situ pyrolysis of stannic chloride or dimethyltin chloride produces a stannic oxide film on glass containers that improves their impact strength and resistance to abrasion (76,77). Thicker films of SnO_2 glass find a variety of uses where electrical conductivity combined with optical transparency is required (78-80).

Inorganic tin chemicals are used in a number of tin and tin alloy plating solutions. These are discussed in Sec. XII,D.4.

Stannous fluoride has been proven to be an effective anticaries agent in various dentrifice formulations. It is also used as a topical application directly to teeth in the prevention of dental decay (81).

Plastics are sensitized by solutions of stannous and palladous chlorides prior to their electroless metallization with copper or nickel (82). Other well-established applications of stannous salts include the manufacture of dyes, stabilization of perfumes in toilet soaps, and as anthelminthics for treatment of parasitic worms in animals and birds.

14. Tin Salts of Organic Acids

Stannous salts of organic acids, such as the acetate, oxalate, oleate, stearate, and the 2-ethylhexoate, are homogeneous tin catalysts used for curing silicone elastomers and the production of polyurethane foams (83).

15. Organotin Compounds (84-86).

Organic compounds of tin are very versatile and are used in such diverse applications as wood preservatives, marine antifoulants, food-packaging materials, fungicides, and stabilizers for plastic materials.

Triorganotin compounds have biocidal properties that are put to good use in suitable paints to give long-term protection against many marine fouling species (87). Paints have been developed based on a nonsaponifiable resin system that will provide a supporting framework in the film after leaching out of the organotin toxicant. Other slow-release systems comprise nitrile rubber that has been impregnated with organotins.

Certain triorganotin compounds are effective against a wide range of fungal plant diseases (88). These compounds ultimately break down to give nontoxic inorganic forms of tin, so that there is no long-term environmental hazard in their use. Other agricultural uses of these compounds include their use as pesticides, disinfectants, slime control during paper making, and in antimildew preparations for paints and adhesives.

Another important use for triorganotin compounds is in the preservation of timber. These preservatives are often applied by controlled immersion and also by vacuum or pressure impregnation of wood (89).

Diorganotins are particularly useful in the plastics industry, both as effective stabilizers and as catalysts in the production of polyurethane foams and coatings or in the curing of certain silicone elastomers.

Basically, the tin stabilizers fall into two groups: sulfur-containing mercaptide compounds that impart excellent heat stability to polyvinyl chloride (PVC) and non-sulfur-containing compounds used when resistance to light and weathering are the main considerations.

Applications of tin-stabilized polyvinyl chlorides include a range of bottles for various commodities, pipe and fittings, furniture coverings, flooring, clad siding, and fencing and also for sheet for roofing applications.

Monoalkyl organotin compounds are used as nontoxic stabilizers in certain grades of PVC films; a tetraalkyl compound is used as a corrosion-inhibiting additive to lubricating oils that may be normally corrosive to bearing metal alloys.

XIII. TITANIUM*

Titanium combines low density (4.5 g/cm^3) with excellent corrosion resistance and wide occurrence in mineral deposits. These deposits are based on either rutile or ilmenite. Despite the abundance of the element in the earth's crust, production of the metal is comparatively expensive because of the complex reduction process, which involves producing titanium tetrachloride from the ore, purifying it, and reducing the purified tetrachloride by magnesium or sodium in an inert atmosphere. The reduction product is titanium sponge, which is mixed with scrap and pressed into billets that are melted in underground vacuum electric arc furnaces to form ingots. These ingots are then fabricated into general mill products from which finished shapes are made.

Titanium's low density (between aluminum and iron), high strength (comparable to heat-treated steels), and thermal conductivity and corrosion resistance (comparable to nickel-chromium stainless steel) make it a very desirable metal for structural design where its cost can be justified. Currently titanium sales are 35% for commercial aircraft, 28% for industrial (corrosion) applications, and 37% for military aircraft and missiles. Marine and

*Contributed by ROBERT S. SHANE, based on material from the Second Edition, and supplied by High Technology Division, Cabot Corp., Kokomo, Indiana, RMI Co., Howmet (Pechiney, Ugine, Kuhlman), *Barron's Magazine,* and *The Metals Handbook,* ninth edition, American Society for Metals, Metals Park, Ohio.

chemical applications are expected to increase, and the use of titanium and its alloys is expected to increase far into the future.

A. Properties

Titanium (Fig. 9.13 and Tables 9.24 and 9.25) undergoes a transformation at about 885°C from a close-packed hexagonal crystal (α) to a body-centered cubic crystal (β). This transformation is strongly influenced by the interstitial elements oxygen, nitrogen,

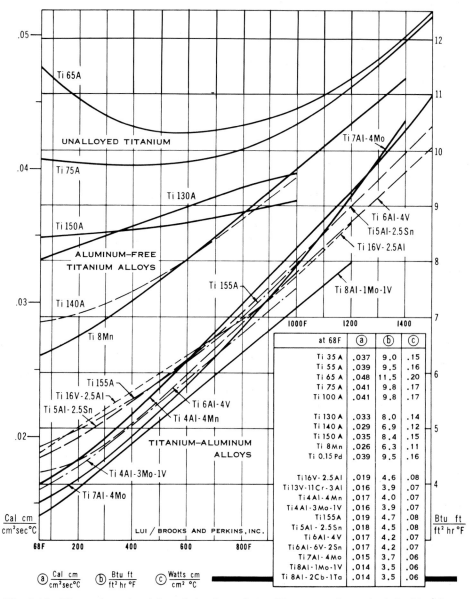

at 68F	(a)	(b)	(c)
Ti 35 A	.037	9.0	.15
Ti 55 A	.039	9.5	.16
Ti 65 A	.048	11.5	.20
Ti 75 A	.041	9.8	.17
Ti 100 A	.041	9.8	.17
Ti 130 A	.033	8.0	.14
Ti 140 A	.029	6.9	.12
Ti 150 A	.035	8.4	.15
Ti 8Mn	.026	6.3	.11
Ti 0.15 Pd	.039	9.5	.16
Ti16V- 2.5Al	.019	4.6	.08
Ti 13V-11Cr-3Al	.016	3.9	.07
Ti4Al-4Mn	.017	4.0	.07
Ti4Al-3Mo-1V	.016	3.9	.07
Ti155A	.019	4.7	.08
Ti 5Al - 2.5Sn	.018	4.5	.08
Ti 6Al-4V	.017	4.2	.07
Ti6Al- 6V-2Sn	.017	4.2	.07
Ti 7Al- 4Mo	.015	3.7	.06
Ti8Al-1Mo-1V	.014	3.5	.06
Ti 8Al-2Cb-1Ta	.014	3.5	.06

(a) $\dfrac{Cal\ cm}{cm^2 sec\,°C}$ (b) $\dfrac{Btu\ ft}{ft^2\ hr\ °F}$ (c) $\dfrac{Watts\ cm}{cm^2\ °C}$

Fig. 9.13 Thermal conductivity of titanium alloys. (Courtesy of Brooks & Perkins, Inc., Detroit, Michigan.)

Table 9.24 Typical Properties of Titanium and Alloys

Alloy name	Normal composition	Tensile strength, ksi (MPa)	0.2% yield, ksi (MPa)	Elongation (%) 2 in.
Unalloyed Ti, grade 1 UNS R 50250	Titanium-99.175 min N-0.03 max O-0.18 max C-0.10 max Fe-0.20 max H-0.015 max Others-0.05 max each, 0.30 max total	35 (241)	25 (172)	24
Unalloyed Ti, grade 2 UNS R 50400	Titanium-98.885 min N-0.03 max H-0.015 max C-0.10 max O-0.25 max Fe-0.30 max Others 0.05 max each 0.30 max total	50 (345)	40 (276)	20
Unalloyed Ti, grade 3 UNS R 50550	Titanium-98.885 min N-0.03 max Fe-0.30 max C-0.10 max H-0.015 max O-0.35 max Others-0.30 max total 0.05 each max	65 (450)	55 (380)	18

Alloy	Composition			
Unalloyed Ti, grade 4 UNS R 50700	Titanium-98.635 N-0.05 max Fe-fo max C-0.10 max H-0.015 max C-0.040 max Others-0.30 max total 0.05 each max	80 (550)	70 (480)	15
Ti-5Al-2.5 Sn UNS R 54520	N-0.05 max Fe-0.50 max C-0.10 max Al-4.0-6.0 H-0.02 max Sn-2.0-3.0 O-0.20 max Others-0.30 max total 0.05 each max bal.-Ti	115 (780)	104 (717)	15
Ti-6Al-4V UNS R-56400	N-0.05 max Fe-0.40 max C-0.10 max Al-5.5-6.75 H-0.015 max V-3.5-4.5 O-0.20 max Others-0.05 max total 0.1 each max bal.-Ti	130 (895)	120 (825)	6
Ti-6Al-2Sn-4Zr-2Mo UNS 54620	N-0.05 max Al-6.0 C-0.05 max Sn-2.0 H-0.0125 max Zr-4.0 O-0.12 max Mo-2.0 Fe-0.25 max Others 0.30 max total 0.10 max each bal.-Ti	130 (895)	120 (825)	10

Table 9.25 Properties, Specifications, and Applications for Wrought Titanium Alloys

Nominal Composition, %	Crucible Steel	Harvey Aluminum	Reactive Metals	Republic Steel	Titanium Metals Corp.	AMS No.	Forms Available (a)	Form	Condition (b)	RT Elastic Modulus, 10⁶ Psi (c)	RT Tensile Strength, Psi	RT Yield Strength, Psi	RT Elongation, %	Test Temp, F	Ext. Tensile Strength, Psi	Ext. Yield Strength, Psi	Ext. Elongation, %	Applications and Characteristics
99.5Ti	A-30	HA-1930	MST-30	RS-25	Ti-35A	—	B,b,P,S,s,T,W,E	S	Annealed	14.9/12.1	38,000	27,000	30	600	20,000	10,000	50	Airframe, chemical, marine, high formability
99.2Ti	A-40	HA-1940	MST-40	RS-40	Ti-55A	4902, 4941, 4951	B,b,P,S,s,t,W,E	S	Annealed	14.9/12.3	60,000	45,000	28	600, -423	28,000, 175,000	13,000, —	45	Airframe skins, parts for chemical industry; good formability, moderate strength. (AMS 4941, welded tube; AMS 4951, weld wire)
99.1Ti	A-55	HA-1950	MST-55	RS-55	Ti-65A	4900A	B,b,P,S,s,T,W,E	S	Annealed	15.0/12.5	75,000	60,000	25	600	33,000	19,000	33	Chemical and marine. Good formability, weldability and corrosion resistance
99.0Ti	A-70	HA-1970	MST-70	RS-70	Ti-75A	4900B, 4921	B,b,P,S,s,T,W,E	S	Annealed	15.1/12.5	90,000	75,000	20	600	43,000	27,000	28	Engine forgings
98.9Ti				—	Ti-100A		B,b,W,E	S	Annealed	15.5/12.6	100,000	85,000	17	600	47,000	30,000	25	
0.15 to 0.20 Pd	A-40 Pd	HA-1940 Pd	MST-Ti-0.2Pd	—	Ti-0.15Pd		B,b,P,S,s,T,W,E	S	Annealed	14.9/12.3	62,000	46,000	27	600	28,000	13,000	30	Special corrosion resistant alloy for moderately reducing environments
Alpha Alloys																		
5Al, 2.5Sn	A-110AT	HA-5137	MST-5Al-2.5Sn	RS-110C	Ti-5Al-2.5Sn	4910, 4926, 4953, 4966	B,b,P,S,s,W,E	S / b	Annealed / Annealed	16.0/13.4 / 16.0	125,000 / 115,000	117,000 / 110,000	18 / 20	600 / 1000	82,000 / 75,000	65,000 / 56,000	19 / 18	Weldable with good oxidation resistance; high temperature strength (600-1100 F) and stability
5Al, 2.5Sn (low O)	A-95AT	HA-5137 ELI	MST-5Al-2.5Sn ELI	RS-110C-L	Ti-5Al-2.5Sn ELI		B,b,P,S,s,W,E	S	Annealed	16.0/13.4	110,000	95,000	20	-423, 1000	229,000, 84,000	206,000, 75,000	15, 21	Special grade for cryogenic service down to -423 F
5Al, 5Sn, 5Zr			MST-5Al-5Sn-5Zr		Ti-5Al-5Sn-5Zr		B,b,P,S	S	Annealed (e)	16.0/14.2	125,000	120,000	18	1000	78,000	60,000	20	Turbine engine and airframe parts requiring high creep strength
7Al, 12Zr			MST-7Al-12Zr		Ti-7Al-12Zr		B,b,P,S	b	Annealed (f)	16.0/14.3	135,000	130,000	15		93,000	67,000	23	
7Al, 2Cb, 1Ta (d)			MST-721		Ti-7Al-2Cb-1Ta		B,b,P,S,W,E	S	Annealed (g)	17.7/15.1	165,000 / 126,000	159,000 / 120,000	14 / 17	600	130,000	119,000	18	Parts requiring high impact strength
8Al, 1Mo, 1V		HA-8116	MST-811	RS-811	Ti-8Al-1Mo-1V		B,b,P,S,s,W,E	b	Annealed (j) / Annealed (k)	18.5 / 18.0	160,000 / 145,000 / 150,000 / 141,000	150,000 / 138,000 / 142,000 / 130,000	17 / 18 / 15 / 13	1000	100,000 / 88,000	81,000 / 71,000	25 / 20	Airframe and turbine engine parts for service up to 650 F. Good weldability, creep resistance
Alpha-Beta Alloys																		
8Mn	C-110M		MST-8Mn	RS-110A	Ti-8Mn	4908A	P,S	b	Annealed	16.4/14.4	137,000	125,000	15	800	80,000	59,000	15	Good formability, moderate strength
2Fe, 2Cr, 2Mo					Ti-140A	4923	B,b,S,s	b	Annealed / SHT	16.7/14.7	137,000 / 179,000	125,000 / 171,000	13 / 16	800	75,000 / 136,000	55,000 / 112,000	30 / 16	Good formability, moderate strength
2.5Al, 16V			MST-16V-2.5Al				B,b,P,S,s,W	S	Aged		105,000	45,000	16					Excellent formability (as solution treated); heat treatable to high strength
3Al, 2.5V	C-130AM		MST-3Al-2.5V	RS-130		4925A	S,s,T	S	Annealed	15.0/13.5	180,000	165,000	10	600	140,000	125,000	10	Tubing alloy, formable and weldable
4Al, 4Mn		HA-4145	MST-4Al-4Mn				B,b,r,W	b	Annealed	15.5/13.0	100,000 / 148,000	85,000 / 135,000	20 / 15	800	70,000 / 110,000	50,000 / 90,000	25 / 17	Heavy section aircraft components including large fasteners. Jet engine compressor parts
4Al, 3Mo, 1V			MST-4Al-3Mo-1V	RS-115	Ti-4Al-3Mo-1V	4912, 4913	P,S,s	b / S / S	Aged / Annealed / Aged	16.5/14.0	162,000 / 140,000 / 195,000	143,000 / 120,000 / 167,000	10 / 10 / 6	600	125,000 / 152,000 / 145,000	100,000 / 120,000 / 115,000	21 / 7 / 8	Structural aerospace parts. Good formability; heat treatable to high strength; stable
5Al, 1.25Fe, 2.75Cr				RS-140	Ti-5Al-4Fe6Cr		B,b,P,S	b	Annealed	16.8/15.5	155,000	145,000	15	800	122,000	102,000	20	Airframe and related components
5Al, 1.5Fe, 1.4Cr, 1.2Mo					Ti-155A	4929	B,b,P	b	Aged	17.6/16.2	190,000	175,000	6	800	144,000	117,000	10	Ordnance, turbine engine and airframe components
6Al, 4V	C-120AV	HA-6510	MST-6Al-4V	RS-120A	Ti-6Al-4V	4969, 4911, 4928A, 4935	B,b,P,S,s,W,E	b / S,b	Annealed / Aged	17.0/14.6 / 16.5/13.5	154,000 / 195,000 / 138,000	145,000 / 184,000 / 128,000	20 / 6 / 2	800	115,000 / 150,000 / 90,000	100,000 / 125,000 / 90,000	16 / 14 / 8	Rocket motor cases; turbine engine disks, rings, blades; airframe forgings; pressure vessels; ordnance equipment
6Al, 4V (low O)	C-125AVT	HA-6510 ELI	MST-6Al-6V-2Sn	RS-120A-L	Ti-6Al-4V ELI	4967	B,b,P,S,s,T,W,E	S	Annealed / Annealed	16.5/13.5 / 15.0/13.4	135,000 / 165,000	127,000 / 150,000	15 / 15	600	105,000 / 132,000	95,000 / 117,000	12 / 20	Cryogenic uses (pressure bottles and tanks)
6Al, 6V, 2Sn, 1(Fe, Cu)		HA-5158			Ti-6Al-6V-2Sn		B,b,P,W,E	b	Aged	16.5/14.5	190,000	180,000	16	600	150,000	132,000	15	Ordnance, aircraft and missile forgings; rocket motor cases
7Al, 4Mo	C-135AMo	HA-7146	MST-7Al-4Mo	RS-125	Ti-7Al-4Mo		B,b,P,W,E	b	Annealed / Aged	16.2/14.2 / 16.9/15.0	160,000 / 185,000	150,000 / 175,000	10	600	127,000 / 150,000	108,000 / 123,000	12	Turbine engine and airframe parts to 800 F; missile forgings and ordnance equipment
Beta Alloys																		
1Al, 8V, 5Fe			MST-1Al-8V-5Fe				B,b,P	b	Annealed / Aged	16.5/14.7 / 16.5/14.5	177,000 / 221,000	170,000 / 215,000	8 / 8	600	128,000 / 140,000	115,000 / 123,000	19 / 18	High strength fasteners
3Al, 13V, 11Cr	B-120VCA		MST-13V-11Cr-3Al	RS-120B	Ti-13V-11Cr-3Al	4917	B,b,P,S,s,W	S / b	Annealed / CR+aged	14.2/13.2 / 14.8/13.8	135,000 / 260,000	130,000 / 245,000	16 / 4	800	115,000 / 160,000	100,000 / 120,000	12	Aerospace components; welded pressure vessels; honeycomb panels; fasteners. Excellent formability; heat treats to high strength

(a) B = billet, b = bar, P = plate, S = sheet, s = strip, T = tubing, W = wire, E = extrusions.
(b) SHT = solution heat treated; CR = cold rolled. (c) Room temperature 600 F. (d) Formerly 8Al,2Cb,1Ta; all data given are for the 8-2-1 composition. (e) 1600 to 1650 F, 1/2 to 4 hr, air cool. (f) 1300 F, 1 hr, air cool. (g) 1650 F, 1 hr, air cool. (h) 1450 F, 8 hr, furnace cool. (i) 1450 F, 8 hr, furnace cool + 1/4 hr, air cool (duplex). (j) 1450 F, 8 hr, furnace cool + 1/4 hr, air cool (triplex). (k) 1850 F, 1 hr, air cool + 1375 F, 1/4 hr, air cool (triplex). (l) 1850 F, 1 hr, air cool + 1100 F, 8 hr, air cool. Data supplied by Defense Metals Information Center (DMIC), Battelle Memorial Institute, Columbus, Ohio, and the five producers listed above. Part II of this Data Sheet, next month, will present processing data and fabricating conditions.

Source: American Society for Metals, Metals progress data sheet, *Metal Progress*, March 1964.

Table 9.26 Corrosive Media Resisted by Titanium and Its Alloys

Seawater
Wet chlorine
Hypochlorite solutions
Nitric acid
Acetic acid
Sulfuric acid at low concentration and temperature
Chlorinated organic compounds
Inorganic chloride solutions (but see Table 9.27)
Sulfur dioxide
Stearic acid
Marine environments

and carbon, which are α stabilizers and raise the transformation temperature, and by metallic alloying elements, which may raise or lower the transformation temperature. Obviously the service temperature must be set below the transformation temperature (also called *transus* temperature).

B. Corrosion Resistance

Titanium resistance to corrosion is markedly enhanced by addition of 0.2% palladium. These alloys are UNS R52400 and UNS R52250. Under hydrogenating conditions, hydrogen embrittlement will occur and violent reactions will occur between titanium and liquid oxygen or red fuming nitric acid. In general, titanium and its alloys (due to a thin stable tenacious film that forms rapidly in air or other oxidizing media, particularly when even small amounts of water are present) have outstanding resistance to a wide variety of corrosive media (Tables 9.26 and 9.27).

C. Fabrication of Titanium

Titanium can be cast to equal the strength of forged parts at an enormous saving of material. The metal is commercially available as sheet, plate, strip, billet, seamless tubing, welded tubing, and pipe. Hence it can be readily formed, hot or cold, welded, machined, extruded, die forged, adhesively bonded, and processes by powder metallurgy. It is important to keep in mind that all processing will have a serious effect on mechanical and other properties. Expert advice should be sought when specific processing is planned. (See Table 9.28.)

Table 9.27 Corrosive Media Contraindicated for Titanium

Dry chlorine
Sulfuric acid
Hydrofluoric acid
Aqueous concentrated ammonium chloride
 (above 150°F)
Anhydrous ammonia (above 300°F)

Table 9.28 Heat Treatment and Fabrication of Wrought Titanium Alloys

Nominal Composition, %	Forging Start	Forging Finish	Beta Transus Temp, F ±25 F	Stress Relief	Annealing Treatment	Solution Treatment (or Aging)	Formability, Minimum Bend Radius, T(a)	Weldability	Form	Condition	Test Temp, F	Type of Test	Stress, 1000 Psi (10⁷ Cycles)	K_t factor and Stress Range(b)
99.5Ti	1500-1700	1200-1500	1630	1000-1100 F, 1/2 hr, AC	1250-1300 F, 2 hr, AC	Not heat treatable	1-2	All unalloyed grades are weldable	—	—	—	—	—	—
99.2Ti	1500-1700	1200-1500	1675	1000-1100 F, 1/2 hr, AC	1250-1300 F, 2 hr, AC	Not heat treatable	1-3		Bar	Annealed	600	Rotating beam	22	1
99.1Ti	1600-1700	1200-1600	1690	1000-1100 F, 1/2 hr, AC	1250-1300 F, 2 hr, AC	Not heat treatable	1-3		Bar	Annealed	70	Rotating beam	63	1
99.0Ti	1700-1750	1300-1600	1740	1000-1100 F, 1/2 hr, AC	1250-1300 F, 2 hr, AC	Not heat treatable	2-3		—	—	—	—	—	—
98.9Ti	1750	1550	1760	1000-1100 F, 1/2 hr, AC	1250-1300 F, 2 hr, AC	Not heat treatable		Weldable	—	—	—	—	—	—
0.15 to 0.20 Pd	1700	1550	1675	1000-1100 F, 1/2 hr, AC	1250-1300 F, 2 hr, AC	Not heat treatable	1-3	Weldable	—	—	—	—	—	—
Alpha Alloys														
5Al, 2.5Sn	1600-1950	1400-1750	1900	1000-1200 F, 1/4-2 hr, AC	1325-1550 F, 10 min-4 hr, AC	Not heat treatable	3-5	Weldable	Sheet	Annealed	70	Direct axial	93	1(A-0.9)
									Bar	Annealed	70	Rotating beam	27	3.2
5Al, 2.5Sn (low O)	1600-1900	1400-1750	1910	1000-1200 F, 1/4-2 hr, AC	1325-1550 F, 10 min-4 hr, AC	Not heat treatable	3-5	Weldable	—	—	—	—	—	—
5Al, 5Sn, 5Zr	1800-1900	1800	1815	1100 F, 1/2 hr, AC	1650 F, 4 hr, AC	Not heat treatable	3-5	Weldable	—	—	—	—	—	—
7Al, 12Zr	1825-1925	1800	1825	1000 F, 1/2 hr, AC	1600-1650 F, 1/2-4 hr, AC	Not heat treatable	3 1/2-5	Weldable	—	—	—	—	—	—
7Al, 2Cb, 1Ta(c)	1950	1850	1900	1100-1200 F, 1/2 hr, AC	1300 F, 1 hr, AC	Not heat treatable (Consult producers for other treatments)	4-6	Weldable	Bar	Annealed	70	Rotating beam	81	1
8Al, 1Mo, 1V	1950	1850	1900	1100-1200 F, 1 hr, AC	1650 F, 1 hr, AC (Consult producers for other treatments); 1450 F, 8 hr, FC + 1450 F, 1/4 hr, AC (duplex); 1450 F, 8 hr, AC + 1375 F, 1/4 hr, AC(d); 1450 F, 8 hr, FC + 1850 F, 5 min, AC + 1375 F, 1/4 hr, AC(d)	Not heat treatable	3-5	Weldable	—	—	—	—	—	—
Alpha-Beta Alloys														
8Mn	Forging not recommended; 1700	1300	1475	900-1100 F, 1/2-2 hr, AC; 900-1000 F, 1/2-1 hr, AC	1250-1300 F, 1 hr, FC to 1000 F; 1200 F, 1/2 hr, AC	SHT not recommended	2 1/4-4; 3-5	Not weldable	Sheet	Annealed	70	Direct axial	90	1(A-∞)
								Not weldable	Sheet	Annealed	600	Reverse bend	44	4.6
2Fe, 2Cr, 2Mo	1400 (max)	1350 (min)	—	—	1200 F, 1/2 hr, AC	1400-1480 F, 1 hr, WQ or AC (900-950 F, 2-8 hr, AC)	3-5	Not weldable	—	—	—	—	—	—
2.5Al, 16V	Forging not recommended; 1600-1750	1300-1600	—	Solution treating is recommended for both annealing and stress relieving	—	1360-1400 F, 10-30 min, WQ (960-990 F, 4 hr, AC)	2	Not weldable	Sheet	Annealed	70	Direct axial	32	1(A-1)
3Al, 2.5V	1750	1650	1700	1300 F, 2 hr, FC	1300 F, 1 hr, AC; 1300 F, 2/4 hr, FC	SHT not recommended	2-3		—	—	—	—	—	—
4Al, 4Mn	Forging not recommended; 1400-1750	1400-1500	1755	1000-1100 F, 1 hr, AC	1225 F, 4 hr, SC to 1050 F, AC	1400-1500 F, 1/2-2 hr, WQ (800-1000 F, 8-24 hr, AC)		Not weldable	Bar	Annealed	70	Rotating beam	90	1
									Bar	Aged	70	Rotating beam	106	1
4Al, 3Mo, 1V	1650-1750	1650	1725	1100 F, 1 hr, AC	1450 F, 1 hr, SC to 1050 F, AC	1625-1650 F, 1/4 hr, WQ (925 F, 8-12 hr, AC)	2 1/4-4	Weldable under special conditions	Sheet	Aged	70	Direct axial	124	1(A-0.6)
5Al, 1.25Fe, 2.75Cr	1750-1900	1650	1755	1200 F, 2 hr, AC	1200 F, 4-24 hr, AC	1350-1500 F, 2 hr, WQ (900-950 F, 5-6 hr, AC)	3 1/2-5	Weldable under special conditions	Bar	Annealed	70	Rotating beam	88	1
									Bar	Aged	70	Rotating beam	105	1
5Al, 1.5Fe, 1.4Cr, 1.2Mo	1750-1900	1400-1750	1820	900-1200 F, 1.4 hr, AC (Usual: 1 hr, 1100 F, AC)	1300-1550 F, 1-8 hr, SC to 1050 F, AC	1600-1625 F, 1 hr, WQ (1000 F, 24 hr, AC)		Not weldable	Bar	Annealed	70	Rotating beam	100	1
									Bar	Aged	70	Rotating beam	110	1
6Al, 4V	1750-1900	1650	1820	900-1200 F, 1.4 hr, AC (Usual: 1 hr, 1100 F, AC)	1300-1550 F, 1-8 hr, SC to 1050 F, AC	1550-1750 F, 5 min-1 hr, WQ (900-1000 F, 4-8 hr, AC)	3-5	Weldable	Bar	Annealed	70	Rotating beam	75	1
									Bar	Aged	70	Rotating beam	92	1
6Al, 4V (low O)	1725		1735	1100 F, 2 hr, AC	1300-1400 F, 1-2 hr, AC	SHT not recommended		Weldable	—	—	—	—	—	—
6Al, 6V, 2Sn, 1(Fe, Cu)	1800-1950	1550-1800	1840	900-1300 F, 1-8 hr, AC	1450 F, 1-8 hr, SC to 1050 F, AC	1650-1750 F, 1/2-1 1/2 hr, WQ (900-1200 F, 4-16 hr, AC)	3-5	Weldable	Bar	Annealed	70	Rotating beam	100	1
7Al, 4Mo	1725	1550			1250 F, 1 hr, FC to 900 F, AC	SHT not recommended		Weldable under special conditions	—	—	—	—	—	—
Beta Alloys														
1Al, 8V, 5Fe	1500	1450	1525	1000-1100 F, 1 hr, AC		1375-1425 F, 1 hr, WQ (925-1000 F, 2 hr, AC)		Not weldable; Weldable under special conditions	Bar	Aged	70	Direct axial	60	Threaded bolts 3.9
3Al, 13V, 11Cr	1800-2150	1400-1800	1325	SHT is synonymous with annealing for this alloy. SHT is also recommended for stress relieving.		1400-1500 F, 1/4-1 hr, WQ or AC (900 F, 2-96 hr, AC); 1450 F, 1/3 hr, AC + CR + 800 F, 24 hr, AC	2-4	Weldable	Bar	Aged	70	Rotating beam	34	—

AC – air cool; SC – slow cool; WQ – water quench; FC – furnace cool; CR – cold rolled; SHT – solution heat treat. (a) T – bend radius (in.) divided by thickness of bend specimen (in.) (b) A – alternating stress divided by mean stress.

(c) Formerly 8AI, 2Cb, 1Ta; all data given are for 8-2-1 composition. (d) Triplex treatment. The three treatments listed above apply to sheet and plate. For forgings, use the following cycle: 1850 F, 1 hr, AC + 1100 F, 8 hr, AC. Data

supplied by Defense Metals Information Center (DMIC), Battelle Memorial Institute, Columbus, Ohio, and the five producers listed last month in Part I of this Data Sheet, which covered properties, specifications and applications.

Source: American Society for Metals, Metal progress data sheet, *Metal Progress,* April 1964.

Table 9.29 Typical Properties of Machinable Tungsten Alloys[a]

Property	Grade W-10[b] (nominal)	Grade W-2[c] (nominal)
Density	17.0 g/cm³ 0.61 lb/in.³	18.5 g/cm³ 0.67 lb/in.³
Hardness (Rockwell C)	26-30	27-32
Tensile strength	118,000 psi	110,000 psi
Tensile yield strength (0.2% offset)	90,000 psi	80,000 psi
Elongation (½ in. gage length)	7%	4-8%
Compressive yield strength (1% offset)	125,000 psi	120,000 psi
Modulus of rupture (transverse rupture strength)	250,000 psi	225,000 psi
Compressibility	50% of length No cracks at 500,000 psi	40% of length No cracks at 400,000 psi
Fatigue strength (rotating beam, 10^8 cycles)	48,000 psi	40,000 psi
Ultimate shear strength	80,000 psi	70,000 psi
Young's modulus of elasticity	45,000,000 psi	50,000-54,000,000 psi
Charpy impact (unnotched specimen)	50 ft-lb	8 ft-lb
Chemical composition	90% W; 10% Ni, Cu, Fe	97.4% W; 2.6% Ni, Cu, Fe
Thermal Expansion (in./in./°F) at		
400°F	3.08×10^{-6}	2.35×10^{-6}
750°F	3.52×10^{-6}	2.59×10^{-6}
1200°F	3.75×10^{-6}	2.71×10^{-6}
Poisson's ratio	0.31	0.27
Electrical conductivity	17% Cu	18% Cu
Specific heat	0.037 cal/g/°C	0.033 cal/g/°C
Thermal conductivity	0.180 cal/cm²/cm/sec./°C at 300°C	0.230 cal/cm²/cm/sec./°C at 300°C

[a]Properties shown are characteristic of Kennertium sections of up to 1½ in. thickness. Some properties will vary according to the size and shape of the piece. On request, Kennertium is available with low magnetic permeability.

[b]Exceeds requirements of Mil Spec T-21014 Type II, Class 1 and AMS 7725.

[c]Exceeds requirements of Mil Spec T-21014 Type II, Class 4.

Source: Courtesy of Kennametal Inc., Latrobe, Pennsylvania.

XIV. TUNGSTEN*

The principal commercial sources of tungsten are the minerals in which it is combined with calcium, iron, or manganese, from which the pure metal is obtained by reduction of the oxide with hydrogen. It is compacted, sintered in a hydrogen atmosphere, and worked at successively lower temperatures until it becomes ductile. It has unique properties (Table 9.29). The melting point is the highest of all elements (6115°F), the density is equal to gold, its vapor pressure is very low, its modulus of elasticity very high (53 $\times$ 10^6 psi), and its conductivity is relatively good. These properties give the metal its unique position for lamp filaments.

Other uses include alloy steels, high-speed tool steels, hot-working tool steels, and steels for strength at high temperatures (see Chap. 8). It is also used in nonferrous alloys, such as Haynes Alloys No. 25 and No. 188 and Haynes Stellite (High Technology Division, Cabot Corp.). It is used for electrical contacts, hard surfacing welding electrodes, spark plugs, radar grids, counterweights for aerospace applications, weights for self-winding watches, control components for guided missiles, radiation shielding (it is equal to 1½ times its thickness in lead), gyroscope rotors, torsional vibration dampers, impact-type vibration dampers, vibration exciters, governors, centrifugal clutches, and inertial guidance systems. As the carbide, tungsten is used in cemented carbide cutting tools.

Tungsten, with the addition of small amounts of nickel, copper, and iron, can be easily machined, like a high-nickel alloy steel. The manufacturer's recommendations should be closely followed.

XV. ZINC†

A. Characteristics and Refining

Zinc is a fairly heavy, bluish white metal used principally because of its low cost, corrosion resistance, and alloy properties. Its density is slightly less than that of copper, but it costs only about half as much.

The principal zinc ore is the sulfide, zincblende. The ore is concentrated by flotation, after which it is roasted in the presence of air to yield zinc oxide. The oxide may be reduced either in a furnace or electrolytically. In furnace reduction, a mixture of zinc oxide and coal is fired, the zinc is vaporized, reduced at high temperature, and then caught and liquefied in a condenser. The purity of the zinc produced varies with the type of ore but is usually not less than 98%. Zinc of more than 99.99% (special high grade) purity may be produced in large quantities by fractional distillation of the less pure zinc.

---

*Compiled by ROBERT S. SHANE from the Second Edition and information from *The Metals Handbook,* ninth edition, American Society for Metals, Metals Park, Ohio, and furnished by Kennametal, Inc., Latrobe, Pennsylvania, and the High Technology Division, Cabot Corp., Kokomo, Indiana.
†Compiled by ROBERT S. SHANE from the Second Edition and material from the Zinc Institute, Inc., New York, New York, and Texasgulf, Inc., Stamford, Connecticut, and *The Metals Handbook,* ninth edition, American Society for Metals, Metals Park, Ohio.

The electrolytic reduction of zinc is begun by dissolving the oxide in sulfuric acid. The resulting zinc produced depends upon the purity of the electrolyte; if the electrolyte has been chemically purified, zinc of more than 99.99% purity can be produced by this method also.

The strength of zinc is not great, and static loads well below the ultimate strength will cause creep even at room temperature. The temperature range through which zinc and its alloys retain their strength is relatively narrow. The upper limit is the recrystallization temperature, 100-115°C, above which they have very little strength. The lower limit for many parts is a freezing temperature, below which zinc alloys are brittle. If a part must stand shock at low temperatures, it should not be made of zinc.

In the less pure grades of zinc, the most common impurities are lead, tin, and cadmium. Their general effect is to harden the zinc and reduce its ductility and erosion resistance. The quantity and kind of impurities present determine the exact properties and cost. There are characteristic uses for each of the several grades of zinc, which range from 98 to 99.99% purity.

A eutectoid composition of 78% zinc and 22% aluminum is called "superplastic zinc." This can be formed, like plastics, by vacuum forming, blow molding, and compression forming. The room temperature properties are similar to those of lead. The mechanical properties are improved by annealing at 340°C followed by slow cooling.

B. Coatings

The protection of iron and steel from corrosion is done more often with zinc than with any other metal coating. The various forms are galvanizing, electrogalvanizing, metallizing, and application of zinc-rich paints, particularly the newly developed water-based inorganic zinc-rich paint KZ-531 (Inorganic Coatings, Inc., W. Chester, Pa.). Table 9.30 details the applications.

C. Die Casting

The major use of zinc as a structural material is in alloys for pressure die casting (Tables 9.31 and 9.32). The special high-grade zinc (>99.99% zinc) plus addition of particular alloying constituents and control of impurities within close limits has contributed to the success of this material.

D. Brass Making

All the grades of zinc may be used in brass making, but the purity is of considerable importance in determining the physical properties (see also Secs. V.I, V.J, V.K, and V.L). Brasses made with 10-36% zinc were formerly hot short and very difficult to hot work. Today, however, special high-purity zinc makes it feasible to hot work these brasses, but zinc of lower purity is still frequently used in brass making.

E. Wrought Forms

The wrought zinc of industry varies considerably in purity, and usually contains up to 1% added copper. It is ductile and easily formed at room temperatures. Many articles, such as dry battery cans, eyelets, and vanity cases are drawn from rolled zinc sheet or strip. Although cold working hardens the metal, the effect of cold work is lost at only moderately elevated temperatures. During the time zinc retains the effects of cold work it

Table 9.30   Zinc Corrosion Protection Coatings

Application	Zinc coating thickness (in.)

Galvanizing
  Prepared steel sheet, strip or wire passes continuously through a bath of molten zinc
  Items to be galvanized are dipped in a bath of molten zinc
  Used very widely
	0.0035

Electrogalvanizing
  Electrodeposition of zinc allows buildup of thicker coats but is usually used in thin coats
  Excellent paint base
	0.00015-0.001

Metallizing After shot blast cleaning, molten zinc is sprayed on the surface; good for field use; excellent paint base	0.003-0.016

Zinc-rich paints
  (80-95% metallic zinc)
  Organic vehicles
  Inorganic vehicles provide same sacrificial protection as a galvanized layer plus the advantages of an exceptionally corrosion-resistant primer
	0.002

may have as much as double the strength along the direction of the distorted grains as across them.

F.   Machining and Finishing

Most zinc parts are not machined, since die castings or wrought parts are formed directly in their final shape. When machining is necessary, it is easily done, especially if the tools are correctly ground. If it is necessary to machine zinc containing coarse grains, the operation should be done at 70-100°C (160-210°F).

Wrought zinc can usually be soldered with soft solder. The die casting alloys, however, are difficult to solder because of the coat of aluminum oxide. If die castings require soldering, it should be done to an insert of copper or brass in the casting, or the part should be nickel plated.

Table 9.31   Nominal Composition of Representative Zinc Die Casting Alloys

Alloy name	UNS No.	SAE No.	ASTM No.	Cu	Al	Mg	Zn
AG 40A	Z 33520	903	B240, B86	0.10 max	3.9-4.3	0.025-0.05	Bal.
AC 41A	Z 35530	925	B240, B86	0.25 max	3.5-4.3	0.020-0.05	Bal.
Zn-11 Al				0.05-1.0	10.5-11.5	0.01-0.02	Bal.
Zn-27 Al				2.0-2.5	25-28	0.01-0.02	Bal.

Table 9.32 Nominal Properties of Representative Zinc Die Castings[a]

Property	AG 40A	AC 41A	Zn-11Al	Zn-27Al
Tensile strength	41 ksi (285 MPa)	47.6 ksi (330 MPa)	40-45 ksi (276-310)	58-64 ksi (400-401)
Yield strength	—[b]	—[b]	30 ksi (207 MPa)	53 ksi (365 MPa)
% Elongation	10	7	1-3	3-6
Hardness, BHN	82	91	105-125	90-120
Density lb/in^3 (mg/m^3)	0.238 (6.6)	0.242 (6.7)	0.218 (6.03)	0.181 (5.01)
Melting range, °F (°C)	717-728 (381-387)	716-727 (380-386)	716-810 (379-432)	714-919 (378-490)

[a]Note that aging has a marked effect on the mechanical properties through an increase of ductility.
[b]No recognized elastic modulus or yield strength.

Zinc alloys are plated or painted for decoration as well as protection. The surface preparation is of the greatest importance in obtaining an adherent coat. Some finishes are baked on, but temperatures above 250°F (recrystallization temperature) are seldom used.

XVI. ZIRCONIUM*

A. Uses

Zirconium is used both as its chemical compounds and as the metal itself. Uses of the chemical compounds of zirconium account for about 95% of the zirconium used in the United States. The main use of this type is as zircon sand in foundries, where its high melting point (2950-3000°C) and other good properties make it a superior refractory material. This discussion will, however, focus on metallic zirconium materials, since these are likely to be of most interest to engineers.

Metallic zirconium and its alloys are used as

Fuel element cladding and structures in nuclear power reactors
High-pressure process tubing in certain nuclear power reactors
Crucibles (mainly laboratory type)
Getters, for removing traces of such gases as hydrogen, oxygen, and nitrogen
    from evacuated or inert gas-filled spaces
Explosive primers, where low ignition temperature and high heat of oxidation
    are advantageous
Photographic flash light powder or foil
Heat exchanger tubes, as an alloy with titanium
Pyrotechnics
A grain refiner in aluminum and magnesium castings
A superconductor, with niobium
In a high-strength copper alloy

Of the uses for metallic zirconium, its use in nuclear reactors, including naval reactors, amounted to about 7 million pounds (3 million kg) in 1978. Other uses of the metal amount to a relatively small fraction of this. In 1973, the latest year for which such figures are available, nonnuclear uses of metallic zirconium added up to about 0.7 million pounds (90).

In 1978, nuclear-grade zirconium prices for "tube shells," that is, hollow cylindrical billets used as input to commercial fuel element cladding manufacture, were in the general range of $15-20 per pound, depending on quantities, specifications, delivery, and other factors. Zirconium for nonnuclear use is less expensive, partly because nuclear use requires close control of the oxygen content of the metal, and partly because nuclear-grade zirconium must be essentially free of hafnium.

Hafnium always occurs with zirconium in nature, typically in the ratio 1 part hafnium to 50 parts zirconium. Hafnium is a strong absorber of thermal neutrons (absorption cross section 105 barns as compared with 0.18 barn for zirconium), and thus acts as a nuclear reactor poison. The requirement for separation of hafnium from zirconium adds to the cost of nuclear-grade zirconium.

---

*Contributed by MILES C. LEVERETT.

B. Occurrence and Production of Zirconium

Zirconium, with its accompanying hafnium, occurs in nature mostly as zircon, a zirconium silicate, which is found in association with rutile and ilmenite, both ores of titanium. The principal U.S. domestic source of zirconium is the titanium-mining operations in Florida; extensive deposits are known in other states also. Australia is the most important foreign source of zirconium ore. Zircon is separated from the titanium ore in the process of concentrating the latter (90,91).

Commercial-grade zirconium may be produced by blending zircon with coke. The mixture is then reacted with atmospheric nitrogen in an electric arc to form zirconium carbonitride. In an older process, the carbonitride is then chlorinated to crude zirconium tetrachloride, which is reduced to zirconium metal with magnesium in an inert atmosphere (the Kroll process); the resulting mixture of zirconium metal and magnesium chloride is then vacuum distilled to remove the magnesium chloride, leaving *zirconium sponge.* More recently, direct chlorination of zircon is used to produce zirconium tetrachloride, which is then reduced by the Kroll process. The zirconium sponge is crushed, compacted into consumable electrodes, and vacuum melted into ingots. These ingots may undergo a second vacuum melting to further reduce impurities before being machined, extruded, forged, and so on.

If nuclear-grade zirconium is desired, zirconium tetrachloride, made as above, is dissolved in a hydrochloric acid-ammonium thiocyanate mixture, which is then subjected to countercurrent solvent extraction with methyl ethyl ketone. Hafnium thiocyanate dissolves preferentially in the solvent, from which the hafnium can be recovered, after several additional operations, as the oxide (see Sec. VII). Zirconium, essentially free of hafnium, remains in the aqueous phase and can be recovered by putting it through the Kroll process.

Very pure, but relatively expensive, zirconium is made by reacting hafnium-free zirconium with iodine vapor. The resulting zirconium iodide, which sublimes at 431°C, is then brought into contact with a filament heated to about 1300°C. At this temperature, the iodide decomposes, leaving very pure zirconium crystals ("iodide zirconium") on the filament. Zirconium made in this way is often called "crystal bar zirconium."

The most important supplier of nuclear-grade zirconium has been Teledyne Wah Chang of Albany, Oregon, which until recently supplied about 80% of the total free world requirements. Other suppliers are Ugine Aciers (a subsidiary of Pechiney Ugine Kuhlman of France) and, recently organized, Western Zirconium, Inc., a subsidiary of Westinghouse Electric Manufacturing Co.

C. Zirconium Fabrication

Zirconium can be machined, formed by bending, rolled, swaged, drilled, tapped, drawn, or stamped in relatively conventional ways. Zirconium welding must be performed in an inert gas atmosphere because of its strong affinity for oxygen. Zirconium extrusion is usually done in a copper jacket to provide protection against the atmosphere and for lubrication. An interesting extension of this technique is the production of zirconium-clad metallic uranium fuel elements for the Hanford nuclear reactor. Hollow cylindrical uranium billets are machined to exact size, then enclosed in a close-fitting Zircaloy* can, which is in turn enclosed in a close-fitting copper can. The assembly is evacuated, welded closed, and extruded through a die over a mandrel. The result is a uranium-Zircaloy-copper tube about 35 ft long. The tube is cut into lengths of about 2 ft, and the protec-

---

*See Section F.

tive copper layer is removed by dissolving it in nitric acid. Zircaloy end caps are brazed over the exposed uranium at the ends of the fuel element. Supporting Zircaloy hardware is attached by spot welding.

Zirconium or Zircaloy chips, turnings, or powder are pyrophoric and must be handled carefully (92). Air, or even water, can support rapid burning of zirconium under the right conditions. Serious explosions have occurred as a result of storing chips or turnings in a moist condition in closed steel drums. The hydrogen that results from the reaction of water with zirconium may have been a factor in these accidents. Serious fires have occurred in zirconium scrap bins. Zirconium powder can ignite in atmospheres of N_2 or CO_2. Good practice dictates that the accumulation of large quantities of metallic zirconium chips or turnings should be avoided if possible. In some plants, a special zirconium incinerator is provided in which zirconium scrap is oxidized under controlled conditions in small batches. Zirconium is, however, not unique among metals in respect to its combustability, and with proper precautions, large quantities of it are handled quite safely and routinely.

Zirconium has no toxic properties of concern in handling it industrially.

D. Physical Properties

Pure zirconium is a ductile metal of silvery appearance. It has two allotropic forms: α, which exists at temperatures up to $862°C$ and has a hexagonal close-packed crystal structure, and β, which exists at higher temperatures and has a body-centered cubic crystal structure (Ref. 93, p. 710). Some other properties are summarized in Table 9.33 (Ref. 93, pp. 710-711). Detailed scientific and technical information on the metallurgy of zirconium and its alloys, their corrosion characteristics, their hydriding behavior, and their thermomechanical properties may be found in Refs. 94 and 95. Less current but more condensed and, hence, more convenient information on the physical and engineering properties of zirconium and its alloys is found in Ref. 93.

The mechanical and corrosion-resistant properties of zirconium are much improved by the addition of alloying agents, such as tin, iron, chromium, and nickel. As a result, one or the other of two alloys, Zircaloy-2 and Zircaloy-4, are usually used instead of the pure metal wherever mechanical properties or corrosion resistance are important. The chemical compositions of Zircaloy-2, Zircaloy-4, and a fifth alloy (designated as ASTM grades R60802, R60804, and R60901, respectively are given in Table 9.34. The me-

Table 9.33   Some Physical Properties of Pure Zirconium

Atomic weight	91.22
Melting point	$1852°C$
Boiling point	$3580-3700°C$
Isotropic coefficient of expansion	$5.92 \times 10^{-6} °C^{-1}$
Density, Zr @ $20°C$	6.51 g/cm^3
Density, Zr-2 @ $10°C$	6.55 g;cm^3
Specific heat @ $25°C$	$0.0678 \text{ cal/g-}°C \ (0.0678 \text{ BTU/lb-}°F)$
Thermal conductivity @ $25°C$	$0.211 \text{ watt/cm-}°C \ (12.2 \text{ BTU/ft-hr-}°F)$
Young's modulus	$14.1 \times 10^6 \text{ psi} \ (9.7 \times 10^4 \text{ MPa})$

Source: From Ref. 93.

Table 9.34 Chemical Requirements of Zirconium for Nuclear Application (Bars, Rod, Wire, Sheet, Strip, and Plate)

Element	Composition (weight %)		
	Grade R60802	Grade R60804	Grade R60901
Tin	1.20-1.70	1.20-1.70	
Iron	0.07-0.20	0.18-0.24	
Chromium	0.05-0.15	0.07-0.13	
Nickel	0.03-0.08		
Niobium			2.40-2.80
Iron + chromium + Nickel	0.18-0.38		
Iron + chromium		0.28-0.37	
Oxygen			0.09-0.13
Carbon	0.027[a]	0.027[a]	0.027[a]
Nitrogen	0.0080[a]	0.0080[a]	0.0080[a]
Hafnium	0.010[a]	0.010[a]	0.010[a]

[a]Permissible upper limit.
Source: From ASTM Specifications B351 and B352, Ref. 96.

chanical properties of these alloys are summarized in Tables 9.35 and 9.36. Similar information for zirconium for nonnuclear applications appears in ASTM standard B551-77.

The approximate mechanical properties of pure zirconium are given in Table 9.37. It is evident that the pure metal is weaker and more ductile than the alloys listed in Tables 9.34 to 9.36. Zirconium alloys respond to cold working, after hot rolling, by increases in yield strength and ultimate strength and by a decrease in reduction of area, as would be expected. Young's modulus and Poisson's ratio are essentially the same for the zirconium alloys as for the pure metal.

Zirconium and its alloys take up hydrogen under some conditions. For example, hydrogen produced by the corrosion process can be taken up by the metal itself rather than being released. A relatively small hydrogen content can embrittle the alloy significantly; hence, conditions favorable to hydrogen production and uptake are to be avoided.

E. Corrosion of Zirconium

Despite its high chemical reactivity, zirconium, particularly its alloys, have excellent corrosion resistance to a wide range of chemical reagents (91), to the atmosphere, and to water or steam. The corrosion resistance of zirconium and its alloys is due to the formation on its surface of a tightly adherent, relatively impermeable oxide film. The most protective films of this type are those formed in a high-temperature steam autoclave; such films are black. Gray, brown, or white films are less protective. If the surface becomes contaminated with chlorides or other casual materials from human hands or other agents, the film that develops may be gray, brown, or white and lack protective character. Hence, once the material has been cleared and rinsed free of cleaning agent, it is handled with gloves and is otherwise carefully protected against contamination if corrosion resistance is important in its destined application.

Table 9.35 Mechanical Properties of Zirconium Alloy Bars, Rod, and Wire, Tested in the Longitudinal Direction (for Nuclear Application)

Grade	Condition	Temperature	Tensile strength, min. ksi (MPa)	Yield strength (0.2% offset), min. ksi (MPa)	Elongation in 2 in. (51 mm), min. %
R60001	Annealed	RTa	42 (296)	20 (138)	18
R60802	Annealed	RT	60 (413)	35 (241)	14
R60802	Annealed	600°F (316°C)	31 (214)	15 (103)	24
R60804	Annealed	RT	60 (413)	35 (241)	14
R60804	Annealed	600°F (316°C)	31 (214)	15 (103)	24
R60901	Cold worked	RT	74 (510)	50 (344)	15
R60901	Annealed	RT	65 (448)	45 (310)	20

aRT: room temperatures.
Source: From Ref. 96, ASTM Standard B351.

Table 9.36 Mechanical Properties of Zirconium Alloy Sheet, Strip, and Plate (for Nuclear Applications)

Grade	Condition	Direction of test	Temperature	Tensile strength, min. ksi (MPa)	Yield strength (0.2% offset), min. ksi (MPa)	Elongation in 2 in. (51 mm.) min. %
R60001	Annealed	Long.	RTa	42 (296)	20 (138)	18
R60001	Annealed	Trans.	RT	42 (296)	30 (207)	18
R60802	Annealed	Long.	RT	60 (413)	35 (241)	14
or						
R60804	Annealed	Trans.	RT	57 (392)	44 (303)	15
R60802	Annealed	Long.	600°F (316°C)	31 (214)	15 (103)	24
or						
R60804	Annealed	Trans.	600°F (316°C)	29 (200)	18.4 (127)	30
R60901	Cold worked	Long.	RT	74 (510)	50 (344)	15
R60901	Cold worked	Trans.	RT	74 (510)	56 (385)	15
R60901	Annealed	Long.	RT	65 (448)	45 (310)	20
R60901	Annealed	Trans.	RT	65 (448)	50 (344)	20

aRT: room temperature.
Source: From Ref. 96, ASTM Standard B351.

Table 9.37 Approximate Mechanical Properties of Pure Zirconium

Property	Pure zirconium[a]
0.2% Yield strength, ksi	
Room temperature	8
250°C	4.6
500°C	5.1
Ultimate tensile strength, ksi	
Room temperature	25
250°C	12
500°C	10
Young's modulus, 10^6 psi	
Room temperature	14
250°C	11.8
500°C	9.6
Poisson's ratio	0.34
% Reduction of area	
Room temperature	43
250°C	64
500°C	62

[a]"Crystal bar" zirconium.
Source: From Ref. 93.

F. Zirconium Alloys for Nuclear Application

The Zircaloys were developed for nuclear reactor applications. Zircaloy-1 was superseded by Zircaloy-2 before it could be applied in the nuclear-powered submarine project. Zircaloy-2 was developed to provide a zirconium material of better corrosion and mechanical properties than those of the pure metal (97). In nuclear reactors, fuel element cladding must be corrosion resistant at temperatures above 600°F for many years. Zircaloy-3 was superseded by Zircaloy-4 before it could be applied. Zircaloy-4 was developed when it was found that the conditions in pressurized water nuclear reactors led to excessive hydrogen absorption by and early failure of Zircaloy-2. Conditions in boiling water reactors are less severe, and Zircaloy-2 is therefore satisfactory in boiling water reactors. Additionally, Zircaloy-2 is less subject to corrosion in a steam atmosphere than is Zircaloy-4. The fifth alloy is stronger and less subject to creep than the others. It was developed primarily for use in nuclear reactor process tubes, which must operate at high pressure and temperature [e.g., 1400 psi (9.65 MPa) and 300°C] for many years and where it is necessary to minimize creep as well as to achieve high strength and corrosion resistance. The more recent Canadian Candu reactors use an alloy of this type.

In nuclear power reactors, corrosion control is achieved by maintaining high water purity (particularly with respect to chlorides, which are usually kept below 50 ppb), except for intentional additives, which are used for nuclear control of for oxygen or pH control in some pressurized water reactors. Such additives are LiOH and H_2.

G. Design Application of Zirconium Alloys

An extensive literature exists on the mechanical behavior of zirconium alloys (95). Creep and crack growth, for example, have been extensively investigated. The most extensive application of zirconium alloys is in nuclear power reactors where zirconium alloys form the fuel element cladding, the fuel bundle channels (in boiling water reactors), and, in graphite and heavy water moderated power reactors, the coolant pressure tubes.

Design rules for zirconium having the authority of those established for some metals by the ASME do not exist. However, equivalently conservative practices are followed by the designers. Table 9.38 gives design stress level data for the pressure tubes in several pressure tube-type nuclear reactors (98,99).

Table 9.39 gives data from Ref. 98 on the strength of Zircaloy-2 and zirconium-2.5% niobium alloy tested at 300°C in various ways and after various treatments with respect to heat treating and cold work.

It will be noted that, in tubular form, and particularly in burst tests, zirconium alloys exhibit higher strength than in wire, plate, bar, or sheet form. This difference is ascribed in part to the different grain structure produced during tube fabrication and in part to the strengthening effect of biaxial loading in closed end tests as opposed to the uniaxial loading under which strips are tested.

It appears from Tables 9.38 and 9.39 that design practice is that maximum design stress levels in zircaloy tubing should not exceed about one-third the bursting strength of the tubes under operating conditions.

In the Canadian nuclear reactor applications, the life-limiting consideration of the zircaloy tubes is neutron-enhanced creep deformation (98), as earlier noted. The Zr-2.5% Nb alloy has superior creep resistance as compared with Zr-2 and Zr-4 and, hence, is

Table 9.38 Design Stress Level Data for Zircaloy Pressure Tubes

Reactor	Pressure tube material	Design temperature (°C)	Design stress, ksi (MPa)	Year completed
N-Reactor, Hanford	UMS R60802	350[a]	10 (69)	1963
NPD, Canada	UMS R60802	270[a]	13.5 (93)	1962
Douglas Point, Canada	UMS R60802	290[a]	16 (110)	1968
Pickering 1 & 2, Canada	UMS R60802	290[a]	16 (110)	1971
Pickering 3 & 4, Canada	UMS R60804	290[b]	21 (145)	1972-1973
Gentilly, Canada	UMS R60804	270[a]	26 (180)	1972
Bruce 1-4, Canada	UMS R60804	300[b]	21 (145)	1976-1979

[a]Cold worked.
[b]Heat treated.
Source: Data from Refs. 98 and 99.

Table 9.39 Strength of Two Zircaloys at 300°C as a Function of Heat Treating, Cold Work, and Irradiation

Material	Test	0.2% yield, ksi (MPa)	Ultimate tensile strength, ksi (MPa)	Burst strength, ksi MPa)
UMS R60802	Long.	45 (310)	54 (372)	
Cold worked	Trans.	50 (345)	53 (359)	
	Burst	60 (414)		63 (435)
UMS R60804	Long	53 (365)	76 (524)	
Cold worked	Trans.	77 (530)	81 (559)	
	Burst	75 (517)		(585)
UMS R60804	Long.	69 (475)	(593)	
Heat treated	Trans.	95 (655)	(675)	
	Burst	92 (635)		(751)
UMS R60802 Cold worked Irradiated[a]	Burst	78 (536)		(541)
UMS R60804 Cold worked Irradiated[b]	Burst	110 (758)		(768)

[a]Irradiation to~10^{21} nuetrons/cm^2 above 1 MeV.
[b]Irradiation to ~6×10^{21} neutrons/cm^2 above 1 MeV.
Source: Data from Ref. 98.

favored for use in pressure tube reactors designed since the early 1960s. See Table 9.34 for composition information.

Zirconium and its alloys do not display a nil-ductility transition temperature such as is found in ferritic materials.

Reference 100 contains a comprehensive discussion of the metallurgical aspects of the use of zircaloys in nuclear reactor fuel elements.

XVII. OTHER NONFERROUS ALLOYS OF COMMERCIAL IMPORTANCE*

A. Antimony

Antimony is used as an alloying element. It hardens lead, pewter, costume jewelry alloys, and bearing alloys. It expands on solidification. Antimony is used with bismuth to make thermoelectric materials. In copper-based alloys antimony is added in small amounts to prevent dezincification. Antimony is added to ductile iron to promote formation of nodular graphite and to gray iron to promote pearlite formation. Antimony is used as an ingredient of III-V semiconductors, for example in indium antimonide and gallium antimonide.

---

*Compiled by ROBERT S. SHANE from the Second Edition and many additional sources.

B. Bismuth

Bismuth is produced mainly as a by-product in the refining of copper, lead, or zinc. It occurs combined as the oxide or, more usually, the sulfide. Refining is accomplished by roasting to obtain the oxide and reducing with carbon. The pure metal is soft, very brittle, and of low thermal conductivity. It is used mainly for solders and fusible alloys. Woods metal (a 4 parts Bi, 2 Pb, 1 Sn, 1 Cd alloy) melts at 60.5°C and is widely used for steam safety plugs and automatic sprinkler systems. The bismuth-lead eutectic does not change volume upon solidification at 254°F and is used for pattern duplication. A bismuth-tin-lead-antimony alloy, known as matrix alloy, pours in the range 480-600°F and is nonshrinking. It is used for short-run dies, as a mounting medium for complex dies, and for die repair. Some high-bismuth alloys are used for seals between glass and copper or brass. Some bismuth compounds are used for catalysts, semiconductors, and in pharmaceuticals.

C. Boron

Hot wire boron is prepared by reduction of boron halides on a hot tungsten wire. Boron is also prepared as a powder by electrolysis of a molten salt and by reduction of boric acid by magnesium. The hot wire boron is used as a metal matrix reinforcement (see Sec. XVIII). Boron powder is used to deoxidize alloys, for example as a component of ferromanganese or ferronickel, thereby decreasing the amount of molybdenum required to appreciably increase the hardenability of medium-alloy steels. Boron in aluminum alloys improves the thermal and electrical conductivity and contributes to grain refining. Borides, such as TiB_2, are used as reinforcing agents. The isotope, ^{10}B, is used as a neutron absorber in atomic pile control rods. Boron carbide and nitride are very hard abrasives and nozzle materials.

D. Cadmium

Cadmium is usually obtained as an impurity in zinc ores and is extracted by distillation. Its principal use is as a galvanic protective plating on small steel articles. It is also widely used in low-melting alloys for solders, silver-brazing alloys, and fusible plugs. It is used to 1% with copper as a hardener for trolley wires, and with silver or nickel in cadmium-based bearings that will withstand higher temperatures than tin- or lead-based bearings. Cadmium is also used for cathode ray phosphors, catalysts, nuclear reactor controls, fire detection devices, and photovoltaics. Cadmium plating liquors and scrap cadmium must be disposed of to avoid entering community sewers and water supplies.

E. Calcium

Production of calcium by electrolysis is difficult because of rapid oxidation and high hygroscopy of the salt. Aluminothermic reduction in a vacuum has been successfully performed. The metal is employed as a deoxidizer for copper, and for reducing oxides of such metals as chromium, titanium, uranium, and zirconium. The silicon and silicon-manganese alloys are excellent decarburizers and desulfurizers. Calcium is used as an alloying or modifying agent for aluminum, beryllium, copper, tin, lead, and magnesium alloys. It is also used as a getter for residual gases in high-vacuum applications, including vacuum tubes.

F. Cerium

Cerium is one of the "rare earth" metals that occur together in certain minerals. It can be separated only by laborious fractional crystallization or precipitation. Electrolysis produces mischmetal, an alloy of several rare-earth metals. Both produce brilliant sparks and are used in lighter flints, tracer bullets, and ignition devices. Pure cerium is also used in high-temperature magnesium alloys and nichrome electrical resistance wire. Cerium oxide is used in the polishing of optical glass.

G. Chromium

The most important chromium ore is the ferrous chromite. Ferrochrome, a 1-8% chromium alloy with iron, is used for adding chromium to steel, and it is obtained by direct reduction of the ore with carbon. The pure metal is obtained from the oxide by reduction with aluminum. The most important use of the pure metal is in electroplating. Thin coatings are used for their lustrous appearance and thick coatings for wear resistance. Alloy applications are equally important, because chromium is a constituent of low-alloy steels, stainless steels, cutting steels, heat-resisting steels, and electrical resistance alloys. It is also used in a number of superalloys for high-temperature service.

H. Columbium*

This element (also called niobium) is a "refractory" metal (melting point 2468°C; 4475°F) that is generally used in alloys. In the steel industry it is a carbide stabilizer in stainless steel (type 348 stainless steel), and it refines the grain and strengthens low-alloy steels. In nonferrous superalloys (see Secs. IV, X, and XIV), columbium improved resistance to thermal shock, hot ductility, and strength. Columbium has good resistance to chemical corrosion. Columbium has a low thermal-neutron cross section. It can be made superconducting at 18.45 K by alloying with titanium or tin. Columbium can be machined (with due appreciation of its tendency to gall), drilled, threaded, ground, blanked, punched, drawn, spun, and welded.

I. Gallium

Minute quantities of gallium are obtained from sulfuric acid plant flue dust, from the residue of zinc distillation, or from rare minerals. The pure metal is noted for its low melting point, 85°F, and its high boiling point, 4170°F. The long liquid range is utilized in high-temperature quartz thermometers. It also has excellent corrosion resistance and forms exceedingly white mirrors on glass. Gallium's chief use is in the electronics industry, where it is combined with elements of group III, IV, or V to form semiconducting materials.

J. Gold

Gold is the most ductile and malleable of all metals and is usually found in the pure state mixed with gravel or as veins in other rocks. The pure metal is separated by panning and sluicing, which depend on the high density (specific gravity, 19.3), or by amalgamation

---

*With important contributions by Kawecki Berylco Industries, Inc. (Cabot Corp.), Reading, Pennsylvania.

with mercury and later extraction by volatization of the latter. Vein deposits are crushed and the gold separated either by amalgamation or a cyanide process. The latter is best suited to small particles. It consists of treatment with a soluble cyanide solution, forming a soluble complex gold cyanide, separation of the solution, and reducing the gold by precipitation with another metal. Silver may be present in appreciable quantities, and it is usually separated by electrolysis.

Besides the use of gold as a money standard and for coinage, it is widely employed for jewelry, dental work, decorative purposes (as gold leaf), and as electroplated coatings. Its value has often obscured its high resistance to all chemicals, except the halides, which has resulted in many applications to chemical processing equipment.

K. Indium*

This metal looks like silver and tin, but is soft and ductile like lead. It is obtained as a by-product of zinc and lead production. The principal use of indium is in lead or cadmium bearings. The indium is electroplated over the lead or cadmium, and the coating is diffused with the surface, where it promotes oil retention and reduces wear. Other uses are as a plating over silver to reduce tarnishing, an 18% alloy with woods metal to obtain a 117°F melting solder, alloying with lead and tin for solder to resist alkaline corrosion, a solder for glass and metal useful over a narrow temperature range, and a neutron indicator in the atomic pile. Surfaces of nonconductors can be made electrically conductive and infrared barriers by applying a highly transparent coat of indium oxide.

L. Iridium†

Iridium is one of the hardest, most brittle, most corrosion resistant, and most difficult to work of all metals. It is used mainly in platinum alloys as a hardener for jewelry and electrical contacts, and in insoluble electroplating anodes. The 10.1% iridium alloy is used for standard lengths and weights because of its low rate of thermal expansion and excellent resistance to atmospheric corrosion. Alloys with osmium are used for fountain pen tips. The very high strength at elevated temperatures and extremely high melting point (2443°C; 4429°F) permit use as a crucible for the melting of nonmetallic substances for crystal growth.

M. Lithium

Lithium is used as a deoxidizer and getter for inert atmospheres. It is used in alloys with aluminum, magnesium, zinc, and lead. It is used in heat transfer applications, tritium breeding, and in battery anodes. Compounds of lithium are used as sources of hydrogen and oxygen, as a reducing agent, as special soaps used in glass, ceramics, and lubricants, as refrigerant dryers, and as catalysts.

---

*With important contributions from Indium Corporation of America, Utica, New York
†With contributions from Mathey Bishop, Inc., Malvern, Pennsylvania, Climax Molybdenum Co., Greenwich, Connecticut, and Schwarzkopf Development Corp., Holliston, Massachusetts.

N. Manganese*

The most important source of manganese is the ore containing its dioxide. The pure form is obtained by the Goldschmidt process (reduction with aluminum at high temperature), but iron alloys for addition to ferrous metals are obtained by reduction with carbon in the blast furnace.

Pure manganese is hard and brittle, and the metal is used almost entirely as an alloying element. It is added to steels either as the ferromanganese alloy having up to 80% Mn or as Spiegeleisen having up to 20% Mn and some carbon. The latter is used as a deoxidizer and to add carbon after the Bessemer process. The manganese scavenges sulfur in steels (also nickel alloys), increases strength of low-alloy steels, and provides abrasion resistance in austenitic steels containing 10-14% Mn. Manganese additions are also made to aluminum and magnesium for imparting hardness and strength without loss in corrosion resistance, and to silicon bronzes and high-strength brasses. A manganese-copper-nickel (10 Mn, 80 Cu, 10 Ni) alloy, called manganin, is used for standard electrical resistances because of its small temperature coefficient of resistance. Another alloy, 72% Mn, 18% Cu, and 10% Ni, has a very large thermal coefficient of expansion and is used for thermostatic bimetals.

O. Mercury*

Although mercury is sometimes found free in nature, the most important source is the sulfide from which it is extracted by heating and distillation (boiling point 356.6°C).

Pure mercury is the only metal that is a liquid at room temperatures. This property and its high density are employed in barometers, thermometers, and other scientific instruments, and as a confining seal for vacuum pumps. The electrical conductivity of the liquid and the gas is employed in small tilt-type electric switches, in mercury vapor lamps, and in mercury arc rectifiers. Properties of the gas are employed in diffusion (vacuum) pumps and in high-temperature "topping" turbines. Mercury vapor is readily absorbed through the respiratory tract, the gastrointestinal system, or unbroken skin. Mercury acts as a heavy metal poison, but its effect is only recognized after prolonged exposure. Precautions are mandatory during handling and disposal.

P. Molybdenum*

The commercial source of this metal is its sulfide, from which it is purified by successive reaction with oxygen and then hydrogen. In industry it is marketed in many forms, such as ore concentrates, alloys, molybdates, and oxides.

The pure metal is ductile and has a high modulus of elasticity and high strength. It oxidizes rapidly at fairly low temperatures, and coating it with ceramic materials, titanium, and alloys through electrodeposition and thermal decomposition is under study for high-temperature uses.

Molybdenum is a very important hardener for steel (see Chap. 8). It raises grain-coarsening temperatures, increases hardenability, minimizes temper brittleness, and imparts high-temperature creep strength. It is equally effective in cast irons. Its alloys

---

*With contributions from Mathey Bishop, Inc., Malvern, Pennsylvania, Climax Molybdenum Co., Greenwich, Connecticut, and Schwarzkopf Development Corp., Holliston, Massachusetts.

with nickel have excellent heat and corrosion resistance for chemical equipment. Typical applications of molybdenum alloys are to jet engines, incandescent-lamp supports, electrical contacts, heating elements, welding electrodes, and x-ray targets.

Molybdenum has been added to stainless steels type 316 (17 Cr, 12 Ni, 2.5 Mo) and type 317 (19 Cr, 13 Ni, 3.5 Mo) for use in marine environments. However, for brackish water and seawater exposure an austenitic stainless steel (18 Cr, 24 Ni, 2.5 Mo) has been developed. For the extremely rigorous use in seawater condensers, a ferritic stainless steel (26 Cr, 3 Mo, 2.5 Ni) has been put in use. It contains very low levels of the interstitial elements (C and N), as a result of argon-oxygen decarburization refining, and is stabilized with titanium to ensure resistance to intergranular corrosion in the welded condition.

Molybdenum alloy (0.5 Ti, 0.07 Zr, 0.01 C, balance Mo) has a tensile strength twice that of the pure metal above 2000°F. Like the pure metal, the alloy becomes brittle above its recrystallization temperature (2500°F), but this is 700°F higher than the pure metal. In fabricating either the alloy or the pure metal the following points should be kept in mind: (1) fabrication *must* be below the brittle-ductile transition temperature; (2) addition of heat during fabrication is helpful; (3) molybdenum is notch sensitive; stress risers should be avoided; (4) molybdenum has directional properties that relate to the rolling direction; (5) the working environment must be compatible with molybdenum; (6) embrittlement by contact with foreign solids and vapors should be avoided; and (7) early consultation with the molybdenum supplier is highly desirable to head off problems.

Q. Osmium*

This is the heaviest of all metals (specific gravity, 22.48), which melts at 4900°F and is harder than glass and quartz. It is not worked successfully either hot or cold, but can be sintered at 3600°F. The principal uses are for hard corrosion-resistant alloys, for pen and phonograph needle tips, and for specially hard electrical contacts. Osmium should not be heated in the presence of oxygen since it forms a very toxic oxide, OsO_4, that boils off at 130°C. A naturally occurring alloy of osmium and iridium is the source of material used for fountain pen tips, phonograph needles, electrical contacts, and instrument pivots (see Table 9.40).

R. Platinum*

Best known and most widely used of the platinum group metals, platinum is very malleable and ductile. It is virtually nonoxidizable except in acids generating free chlorine, such as aqua regia. Certain oxides, peroxides, and caustic alkalis will attack platinum at high temperatures, but it is resistant under normal conditions to most other alkalis, oxides, and salts.

Platinum is widely used, either solid or clad, for chemical equipment. It is also used in chemical catalysis, dentistry, electrical contacts, electrical resistance heaters, jewelry, temperature-sensing devices, and so on (see Table 9.40).

---

*With important contributions from Mathey Bishop, Inc., Malvern, Pennsylvania.

Table 9.40 Platinum Metals—Physical Properties

Property	Platinum	Palladium	Ruthenium	Rhodium	Iridium	Osmium
Atomic weight	195	106	107	103	192	190
Specific gravity	21.45	12.02	12.45	12.41	22.65	22.57
Melting point, °F	3214	2829	4190	3560	4429	5522
Melting point, °C	1768	1554	2310	1960	2443	3050
Thermal conductivity, BTU/ft2/°F/hr	493	522	725	1044	1015	609
Coefficient of expansion, in./in./°F $\times 10^{-6}$	4.94	6.53	5.1	4.6	3.8	2.6
Specific heat @ 32°F, BTU/lb/°F	0.0315	0.0584	0.0551	0.0589	0.0307	0.0309
Specific heat @ 212°F	0.0325	–	–	–	–	0.0314
Resistivity ohms/cir. mil ft						
@ 32°F	58.86	64.0	47	30.1	–	54
@ 68°F	63.60	64.8	84	–	–	57.1
@ 212°F	1.90	–	–	–	–	–
Mean temp. coeff. of resistance per °C, 0-100°C	0.00392	0.0038	0.0042	0.0046	0.0043	0.0042
Tensile strength, annealed, ksi	20	25	–	100	160	–
Tensile strength, 50% RA, ksi	35	47	–	–	–	–

S. Rhenium*

Rhenium has been produced in limited quantities from molybdenum smelter flue dust. The metal looks like silver, has a density exceeding gold, has extremely good corrosion resistance, and a melting point of 3170°C, exceeded only by tungsten. It forms hard acid-resistant alloys with tungsten, molybdenum, chromium, and tantalum. Because of its high melting point it shows promise for high-temperature thermocouples, lamp filaments, electric contacts, and similar uses as its availability increases. Surface films or mirrors of rhenium are laid down by electroplating or vapor deposition. Deposits can be laid down on many common base metals, on precious metals, and on refractory metals. In massive form rhenium remains untarnished for several years; in sponge or powder form rhenium oxidizes readily and is a very valuable selective hydrogenation catalyst ingredient.

T. Rhodium

This metal also looks like silver, and it can be worked at room and elevated temperatures. It is harder than silver and resists attack by chemical and industrial fumes. Platings of the metal accurately reproduce the subsurface. Uses include plating of searchlight reflectors, and alloys for electric contacts, rayon spinnerets, resistance-furnace heaters, and thermocouples (see Table 9.40)

U. Ruthenium

Another of the platinum group, this metal is hard to work but can be forged at 2700°F. The main application is as an alloying element to harden and increase corrosion resistance of platinum and palladium. Fabrication is usually by powdered metallurgy techniques.

V. Selenium

This metal is produced from the residue of electrolytic copper refining. Its metallic form is gray with a specific gravity of 4.81 and a melting point of 428°F. Selenium cells employing its characteristic change in electrical conductivity upon exposure to visible light are its most common application. They are used in counters, scanning and sorting machines, safety devices, and light meters. The metal is also used in electric rectifiers and for decolorizing glass. Its principal alloying application is in copper alloys, where 0.5-0.6% adds to machinability. Similar characteristics are imparted by adding 0.25% to invar and stainless steel. An important new use is as a coating for xerographic drums.

W. Silver

Silver is the most abundant of the precious metals. Aside from its use in tableware and other household objects, silver's high electrical conductivity and excellent corrosion resistance makes it a material of choice for electronic and electrical equipment. Alloys of silver have been used in soldering and brazing because they wet the substrate metals well and have high strength and corrosion resistance. Silver is hardened by alloying with copper, such as in sterling silver (7.5% Cu), coinage silver (10% Cu), silver-copper

---

*With important contributions from Cleveland Refractory Metals Div., Chase Brass and Copper Co. (Kennecott Copper Corp.), Cleveland, Ohio.

eutectic alloy (28% Cu). The latter has the highest combination of strength, hardness, and electrical properties of the silver alloys.

X. Sodium

This metal is made on a large scale by electrolysis of sodium chloride. Most of the production goes into the manufacture of chemicals. Metallurgical uses are relatively minor but include hardening of lead-bearing alloys, removal of arsenic and antimony from lead-tin alloys, deoxidization of metal baths, and descaling of steel. The pure metal is being used as a coolant and heat-transfer medium from 400 to 800°C in aircraft engines. Experiments are under way to utilize its electrical conductivity, which is three times that of copper on a weight basis (see also Sec. XX). Sodium is also being tested in advanced concept batteries, as in the sodium-sulfur cell.

Y. Tantalum*

The chief ore of tantalum is the mixed tantalate and columbate of iron and manganese. Pure tantalum is prepared by fused-salt electrolysis or by reduction of its halides with lithium or sodium. Because of its tendency to form nitrides and carbides, it is prepared by powder-metallurgy methods and sintered in a vacuum. Cold-working methods may be employed for fabrication. Uses of the pure metal include surgical sutures and plates, chemical apparatus, electronic anodes and grids, rectifiers, condensers, and lightning arresters. Its carbide is used for cutting tools, dies, and parts of extraordinary hardness and wear resistance. It also alloys with tungsten for higher mechanical strength and higher annealing temperature.

The powdered metal is consolidated by sintering, vacuum arc, or electron beam melting for further processing into wrought products. The melting point of tantalum is 2996°C (5425°F), but it is a reactive metal and cannot be used in the atmosphere above 260°C (500°F). Protective atmospheres (argon, helium, and vacuum) must be used above that temperature. Metal working is by methods akin to those used with a fully annealed steel. Machining and joining of tantalum and its alloys has been extensively studied. Details are available from major suppliers. Compared with other metals (except gold and platinum) tantalum has excellent corrosion resistance to a wide variety of corrosive substances. Hence it is widely used for acid-resistant heat exchangers, condensers, chemical lines, and other chemical processing equipment, except in the presence of hydrofluoric acid, hot, concentrated sulfuric or phosphoric acid, and some alkalis according to concentration.

Z. Tellurium

This metal, like selenium, is obtained from the residue of electrolytic copper refining. It is practically a nonconductor of electricity and has no light-sensitive properties. Industrial uses are as a minute alloying element to increase the work hardenability and recrystallization temperature of lead and tin, increase the chill of iron, and improve the machining properties of copper alloys and stainless steel. Like selenium, tellurium has semiconductor properties. With lead and other elements tellurium forms important thermoelectric alloys.

---

*With an important contribution by Kawecki Berylco Industries, Inc. (Cabot Corp.), Reading, Pennsylvania.

AA. Thallium

All domestic production of thallium is from cadmium flue dust. The metal is malleable, soft, and has a low melting point. It tarnishes rapidly upon exposure to air, but is added to lead to improve corrosion resistance. A thallium-lead alloy is used in electrical fuses. Many of its uses are nonmetallic in such items as fungicides, insecticides, and rat poison. Its glasses have high refractive index, and its addition to selenium cells improves sensitivity. It is also added to mercury vapor rectifiers and lamps. Thallium is also used in alloying to lower the melting point of mercury, as in arctic thermometers and low-temperature switches. Because of the high toxicity of thallium skin contact and inhalation of the dust are to be avoided.

BB. Thorium

Thorium is a gray-white metal that is of low mechanical strength, ductile, soft, and easily cold worked. It is used in photoelectric cells, x-ray targets, and glow-tube electrodes. Small additions are used to restrict grain growth and prolong high-temperature life of nickel-chrome and platinum resistors. Its oxide is used to restrict grain growth in tungsten filaments and for highest melting point furnace linings. The metal forms a radioactive series, like uranium, but of shorter life than radium. Thorium, as a solid or liquid in elemental, intermetallic, or oxide form, is a fertile material for the generation of fissionable uranium-233. The oxide is dispersed in nickel to strengthen it. The oxide is used in gas mantles.

CC. Uranium

Uranium is most commonly used in nuclear reactors. The naturally occurring element contains 0.72% ^{235}U, which is fissionable; the balance is the stable ^{238}U. Uranium-233 can be produced by neutron irradiation of ^{232}Th; it is also fissionable. Depleted ^{238}U from which the fissionable isotope has been removed is stable and is used as counterweight material for rotating machines because of its high density (19 mg;m^3). Finely divided uranium is pyrophoric and must be handled carefully. Uranium is a heavy metal poison, and care must be taken to avoid ingestion of fumes or dusts.

DD. Vanadium

Pure vanadium is a ductile metal that may be drawn into wire, but metal of the necessary purity is not produced commercially. More than 90% of the production is of ferrovanadium, which is added to high-speed tool steels, and is a carbide-forming element for steels to be used at elevated temperatures. It is also added to gray cast iron to improve strength and hardness, and to alloyed irons to improve machinability. More limited use has been made of vanadium in highly magnetic noncobalt alloys, in some bronzes, and in Hastelloy. About 3% of the metal produced is used as a catalyst, replacing platinum in chemical processes.

EE. Conclusion

The foregoing discussion of each of the nonferrous metals has been intended as a general summary of the facts important in designing to use each of these metals. When a material must be selected, this general information should narrow the field to a group of metals or alloys that meet the requirements. To determine the one metal or alloy that is best, however, may often require a knowledge of facts that have not been presented here.

The physical properties of the hundreds of nonferrous alloys may best be obtained from design handbooks or from manufacturers' specifications. The designer should also determine the current cost of the material, as well as the fabricating cost, and compare these with the costs for other metals.

XVIII. METAL MATRIX COMPOSITES*

A recent development has been the evolution of stiffer, stronger metals by incorporation of certain fibers in the molten metal and solidifying. The fibers may be short staple or long staple, but the effect is dramatic.

Ductile metals, such as aluminum, magnesium, steel, and superalloys, are stiffened and strengthened by the fibers. The result is a capability for using thinner, narrower structural sections in design. The fibers can be oriented in the metal to give greatest reinforcement along the axes of greatest stress.

Fibers that have been shown to be successful are graphite, alumina, and boron-coated tungsten. However, it is possible to incorporate a second phase material in the melt that will result in direction solidification and internal reinforcement. This process is called *eutectic cooling.*

XIX. LIQUID METALS†

A. Uses

Liquid metals are used industrially mainly in applications where a highly effective, high-temperature heat-transfer medium is needed without accompanying high pressure. Typical metals and uses have been as follows.

Sodium, used as a coolant in hollow-headed and hollow-stemmed internal combustion engine valves. The motion of the valve causes the sodium to splash between the upper (hot) end and the lower (cool) end of the valve, thus cooling the upper end.

Lead, used as a bath or jacket material to control the temperature of chemical reactors.

Mercury vapor, used as a thermodynamic working fluid in power generation. High thermodynamic efficiency due to the wide temperature range available was the impetus in this application.

Mercury vapor, used as a heating medium of precisely controllable temperature in the distillation and cracking of petroleum.

Sodium, used as a nuclear reactor coolant in breeder or fast reactors.

Molten tin, used as a smooth, level surface on which to float and solidify molten glass in one process for the making of plate glass.

Sodium, used as a chemical reagent in the manufacture of various chemicals, of which the largest volume use has been the manufacture of lead tetraethyl.

The largest application of liquid metals for heat transfer is that of sodium as the coolant in developmental breeder or fast nuclear reactors. Such reactors have been built and operated in the United States, the United Kingdom, France, Germany, Japan, and the Soviet Union. Sodium melts at 97.8°C, boils at 883°C, and has a density of 0.928

---

*Compiled by ROBERT S. SHANE.
†Contributed by MILES C. LEVERETT.

Table 9.41 Heat Transfer Coefficients of
Three Liquids

Liquid	Film coefficient of heat transfer $(BTU/hr\text{-}ft^2\,°F)$
Sodium	25,000
Water	8,000
Diphenyl	2,000

g/cm^3 at 100°C. In pumping liquid metals, advantage is usually taken of their electrical conductivity, which makes electromagnetic pumps feasible.

The standard reference on liquid metals is the *Liquid Metals Handbook* (101), where extensive property and application information will be found.

B. Technology

Liquid metals have excellent heat-transfer properties. For example (102), at a velocity of 20 fps, and in typical geometry, the liquid film heat-transfer coefficients of three heat-transfer liquids are given in Table 9.41.

Because of the very high film coefficient in flowing sodium heat-transfer systems, the overall heat-transfer coefficient through a metallic wall is likely to be controlled by fouling the surface or by the thermal resistance of the wall material itself.

Sodium, which is currently the liquid metal coolant of most interest, can be contained in steel, stainless steel, and other common alloys so long as it is essentially free of oxygen. For this reason, and because it may react violently with air or water, sodium must be handled in hermetically sealed or inert gas blanketed systems.

XX. CLAD METALS (METALLIC COMPOSITES)*

A clad metal is made by metallurgically bonding two or more distinct metals or alloys. The cladding of metals dates back many centuries to processes for bonding decorative gold or silver surfaces on base metals. Today, cladding processes have been refined to a sophisticated technology used in hundreds of industrial applications. It may be considered a special method for making a composite.

The cladding of metals is based on the technology of solid-state welding without the use of intermediate brazing alloys or adhesives. On the other hand, cladding may be employed to yield a surface to which it is easier and faster to braze or weld. Virtually any combination of ductile metals can be clad, with individual components representing from 2 to 98% of the total composite thickness. However, in this section we will omit consideration of cladding by hot dipping, as in galvanizing, or cladding by electro- or electroless plating, or by sputtering, or by vacuum diffusion.

As a rule of thumb, designers may approximate composite properties, such as tensile strength, electrical conductivity, and thermal conductivity, by using the arithmetic sum of the volume percentage of each component multiplied by its specific property value.

---

*Contributed by ROBERT S. SHANE with major contributions from published literature supplied by Texas Instruments Co., Metallurgical Materials Division, Attleboro, Masschusetts.

A. Manufacture of Clad Metals

Clad metals can be made by several distinct processes, each of which is aimed at pressing two very clean metal surfaces together so that a bond is formed with electrons shared between the mating surfaces. High-pressure rolling mils or explosive processes furnish the necessary bonding pressure. Subsequent thermal treatments are used to induce diffusion across the mating surface, improve bond strength, and provide stress relief for further cold-working operations (see Chap. 21). It should be noted that the ratio of component materials remains constant during all post bonding operations. Subsequent to creation of the clad composite, finishing operations are performed, such as rolling to intermediate and final thickness, annealing to temper, cleaning, buffing (as required), edge trimming, or slitting. Extremely close thickness tolerances (to ±0.0025 mm) can be achieved by use of Sendzimir mills in the rolling operation.

B. Benefits to the Designer of Using Clad Metals

The objective of cladding, as with other composites, is to give the optimum combination of needed functional properties. The desired result is a significant improvement in reliable product performance at minimum cost. Some of these benefits are described in the following paragraphs, which exemplify a wide range of applications of clad metals.

1. Molten Caustic Tank Cars

A large saving was achieved in the lining of steel tank cars with pure nickel. Molten sodium hydroxide is poured into the car (equipped with nickel steam heating coils) and shipped to the user where the contents are reliquified and used directly or piped to the user's storage container. Enormous savings in solidification, powdering, packaging, and loss in transit result. This is in addition to the reduced contamination of the sodium hydroxide and the much longer life of the costly tank cars. It is of interest to note that the cylindrical part of the car is fabricated by rolling; the dished ends are made by explosive bonding.

2. Aluminum-Clad Aluminum Alloys

Pure aluminum has greater corrosion resistance than desirable (for strength and weight reasons) aluminum alloys. By cladding the aluminum alloy with pure aluminum the desirable weight saving and strength enhancement is achieved without undue corrosion penalty.

3. Automobile Trim (Appearance and Corrosion Resistance)

For many years automobile trim was made of stainless steel. Unfortunately the stainless steel was noble to the low-carbon steel of the body, and a galvanic cell was created that produced corrosion in areas adjacent to the trim. The problem was aggravated by the extensive use of saline deicers on roads during northern winters. The problem was solved by creating a composite of aluminum and stainless steel. The aluminum clipped to the low-carbon body and provided cathodic protection; the stainless steel on the outside provided appearance and dent protection.

4. U.S. Minor Coinage (Density, Appearance, and Resource Conservation)

To match the look and feel of silver coins yet retain compatibility with vending machines, a composite was created that was one-eighth cupronickel, three-fourths copper, and one-eighth cupronickel. The short supply of silver was overcome.

5. Cable Shielding (Mechanical Strength and Corrosion Resistance)

Cable shielding that was once made of copper or bronze alloys is now made of a composite whose outer layers are 0.02 mm copper and whose core is 0.09 mm stainless steel. The copper provides electrical conductivity; the stainless steel provides strength. The combination reduces cost by allowing thinner material to be used while still meeting corrosion requirements.

6. Welding Transition Material

To overcome the severe problems of joining aluminum to steel in order to gain strength, save weight, and achieve corrosion resistance, aluminum-clad steel is inserted between the steel and the aluminum. This allows secure joining by spot welding since aluminum bonds to aluminum and steel bonds to steel.

7. Radiotube Anode (Strength, Emissivity, and Heat Dissipation)

Electron tube anode plate material must have good heat dissipation, emissivity, and formability. To meet these diverse requirements a five-layer clad metal composite was created comprising aluminum bonded to steel, which is bonded to copper, which is bonded to another layer of steel, to which a final layer of nickel is bonded. Copper provides improved thermal conductivity; nickel eliminates secondary emission; and steel provides structural strength. After part fabrication, the anode plates are heated to a temperature at which the aluminum layer combines with its steel interface to form aluminum-iron intermetallic compounds that provide a very desirable highly emissive surface for one side of the anode plate. The cross section of the composite is 0.03 mm aluminum; 0.08 mm steel, 0.05 mm copper, 0.08 mm steel, and 0.02 mm nickel. The entire composite is only 0.26 mm thick.

8. Copper-Clad Aluminum Coaxial Cable

The copper center conductor in coaxial cable has been almost totally replaced by copper-clad aluminum wire for reasons of savings in weight and cost and substantial electrical and mechanical advantages. A special design advantage is elimination of connector pullout due to temperature change because the thermal expansion properties match those of the aluminum outer conductor. A typical coaxial center conductor has an outer layer of 0.08 mm thick copper bonded to a 2.74 mm aluminum core to give a final diameter of 2.90 mm.

9. Diode Lead

A composite wire developed for many applications because it has high mechanical strength, high electrical conductivity (88% of IACS copper), and magnetic properties, turns out to be excellent for diode leads. Thus, to a 1.0 mm copper core is bonded a 0.05 mm steel layer, which is then clad with an 0.08 mm layer of copper.

10. Thermally Conductive Lead Frame

Copper alloy lead frames for densely packed integrated circuitry, which require great heat dissipation to ensure chip reliability, are being supplanted by copper-clad stainless steel. The advantages lie in the superior fracture resistance, lower cost, and compatibility with existing tooling due to thermal compatibility. A typical composite is 0.20 mm stainless steel clad on both sides with 0.03 mm copper. The stainless steel is a low expansion rate iron-nickel alloy Invar.

C. Economics of Clad Metals

The many ways of saving money through the design usage of clad metals comprise: (a) straight replacement of expensive metals or alloys, (b)reduction of usage of expensive metals or alloys by using them only at the surface where their special properties are needed, (c) lower cost design by elimination of costly subassemblies and their entire procurement chain, (d) simplified manufacturing operations, (e) extended or lower cost warranty coverage.

D. Design Considerations

The initial step in any design choice of materials is to list the desired functional properties. Next, the metals or alloys that exhibit those properties are tabulated. Finally, an appropriate combination is bonded. Although necessarily incomplete since each design application is unique, the following factors must be considered in developing an appropriate composite.

1. Physical requirements
   a. Electrical
   b. Thermal
   c. Magnetic
2. Mechanical requirements
   a. Tensile strength
   b. Yield strength
   c. Stiffness
   d. Formability
   e. Fatigue properties
3. Service environment requirements
   a. Corrosion
   b. Temperature
   c. Humidity
4. Manufacturing
   a. Forming methods
   b. Tool design
   c. Joining
   d. Heat treatment
   e. Scrap rate, scrap recovery
   f. Appearance
   g. Finishing process
5. Interaction of composite layers
   a. Galvanic corrosion
   b. Thermostatic deflection
   c. Diffusion to produce intemetallic compounds or alloys during heat treatment.

It becomes evident that the ability of draw upon one's own and others' experience is an important part of the design process if the optimum use of clad metals is to be achieved.

APPENDIX: SPINODAL ALLOYS*

Spinodal copper-nickel-tin alloys are being offered as replacement alloys for the traditional high-strength alloys of beryllium coppers, phosphorus bronzes, and alloys of other
metal alloy systems. They are characterized by excellent corrosion resistance, high yield
strength, high stress relaxation characteristics, and high temperature strength retention
(see Fig. 9.14). Parts currently being produced are: ocean cable parts (cast housings, hardware); connectors; relay parts; springs; high-strength, corrosion-resistant fasteners; eye-
glass frames; screwmachine products. Spinodal Cu-Ni-Sn alloys are available in the following forms: castings (up to 10,000 lb), strip, wire, extruded rod, extruded heavy-walled
tube, extrusion shapes (up to 1200 lb), rolled shapes, forgings.

How does a spinodal alloy differ from a precipitation hardening alloy such as beryllium copper? In precipitation hardening, a "hardening-particle" phase precipitates out of
a solid solution, which causes the alloy to increase in strength with time at temperature.
The process is referred to as "aging." With spinodal alloys the heat-treatment process is
similar but results in the formation of two phases. Spinodal decomposition occurs when
an elevated temperature phase separates into two phases of slightly different compositions and atomic lattices due to the fact that the equivalent free energy of formation of
each of the two new phases makes either of them equally likely to form.

Following are the chemical compositions and mechanical properties (see Tables
A-1 and A-2) of spinodal Cu-Ni-Sn alloys registered with the Copper Development Asso-
ciation. These alloys will appear in the next edition of CDA's *Application Data Sheet,
Standard Designations for Copper and Copper Alloys*, possibly in 1986.

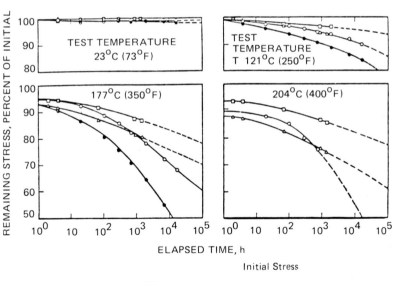

Symbol	Alloy	Initial Stress kal	$\sigma_0/\sigma_{\gamma 01}$
○	Cu-4% Ni-4% Sn	66.9	0.80
△	Cu-9% Ni-6% Sn	118	1.00
□	Cu-15% Ni-8% Sn	143	1.00
●	CA172(Cu-1.9% Be)	129	0.73

---

*From Material Information Services Newsletter, No. 81-7, July 1981. Courtesy of General Electric,
Schenectady, New York.

Table 1 Chemical Compositions, %, of Registered Spinodal Alloys

Alloy	Descriptive Name	Cu	Ni	Sn	Mn	Mg	Cb	Zn max	Fe max	Pb max	P max	Other max	Other Total
C72600	Cu-4Ni-4Sn	Rem.	3.5-4.5	3.5-4.5	0.20 max	--	--	0.50	0.20	0.05	0.05	--	0.50
C72700	Cu-9Ni-6Sn	Rem.	8.5-9.5	5.5-6.5	0.05-0.30	0.15 max	0.10 max	0.50	0.50	0.02	--	--	0.50
C72800	Cu-10Ni-8Sn+Cb	Rem.	9.5-10.5	7.5-8.5	0.05-0.30	0.005-0.15	0.10-0.30	1.00	0.50	0.005	0.005	Al 0.10 B 0.001 Bi 0.001 Si 0.0025 Sb 0.02 Si 0.05 Ti 0.01	0.10
C72900	Cu-15Ni-8Sn	Rem.	14.5-15.5	7.5-8.5	0.05-0.30	0.15 max	0.10 max	0.50	0.50	0.02	--	--	0.50

Table 2 Mechanical Properties of Registered Spinodal Alloys

ALLOY	DESCRIPTIVE NAME	CONDITION	ULTIMATE TENSILE STRENGTH ksi min	ksi max	MPa min	MPa max	TENSILE YIELD STRENGTH ksi min	ksi max	MPa min	MPa max	ELONGATION %
C17200	Beryllium Copper	Hard + Age	190	220	1310	1520	165	--	1140	--	1 min
C51000	Phosphor Bronze	XS	107	--	740	--	80	--	550	--	3
C72600	Cu-4Ni-4Sn	1/2 Hard + Age	90	--	620	--	75	--	520	--	12
C72600	Cu-4Ni-4Sn	SS + Age	102	--	700	--	85	--	590	--	7
C72700	Cu-9Ni-6Sn	78% CW + Age	160	--	1100	--	135	--	930	--	2.5
C72800	Cu-10Ni-8Sn+Cb	1/2 Hard + Age	170	--	1170	--	145	--	1000	--	7 min
C72900	Cu-15Ni-8Sn	78% CW + Age	163	200	1120	1380	155	170	1070	1170	2

REFERENCES

1. S. B. Ampian, Hafnium in *U.S. Bureau of Mines Bulletin* 667, Mineral Facts and Problems, 1975.
2. L. B. Prus, Hafnium. *Reactor Handbook,* 2nd ed., Chapter 36. Interscience, New York, 1960, pp. 783 ff.
3. J. J. Fox, *A Second Technical Conference on Tin,* 3 vols., The International Tin Council, 1970. London.
4. W. T. Dunne, The Occurrence and Production of Tin, *Tin and Its Uses,* 92, 3, 1972.
5. D. V. Belyayev, *The Metallurgy of Tin,* Macmillan, New York, 1963.
6. P. A. Wright, *The Extractive Metallurgy of Tin,* Elsevier, New York, 1966.
7. T. S. Mackey, *The Electrolytic Tin Refining Plant at Texas City, Texas,* paper presented at the AIME Meeting, Washington, D.C., February 20, 1969.
8. R. F. Kark, Tin and Ultrapure Water, *Tin and Its Uses,* 101, 9, 1974.
9. *Tin Lined Tubing Distribution Systems,* Publication of the Barnstead Co., Catalog A 08, 1972.
10. The Use of Tin in Collapsible Tube Production, *Tin,* 2, January, 1956.
11. Tin Foil: Today's Uses, *Tin and Its Uses,* 69, 1, 1965.
12. A. Delwasse, Production of Tin and Tin-Lead Foils, *Tin and Its Uses,* 115, 5, 1978.
13. Float Glass Has a New Aspect, *Tin and Its Uses,* 77, 6, 1968.
14. D. J. Madeley and C. A. MacKay, A Preliminary Study of the Properties of Iron/ Solder Compacts Produced by Warm Pressing, *Powder Met. Inst.,* 7, 4:170, 1975.
15. J. Cauchetier, The Spraying of Tin and Tin Alloys, *Tin and Its Uses,* 66, 6, 1965.
16. S. Gebalski, *Anti-Friction Bearings Prepared by Metal Spraying,* Paper Presented at the Second International Metal Spraying Conference, Birmingham (England), September 29, 1958.
17. W. E. Hoare, E. S. Hedges, and B. T. K. Barry, *The Technology of Tinplate,* Edward Arnold, London, 1965.
18. C. J. Thwaites, *Hot Tinning,* International Tin Research Institute Publication 102, 1965.
19. P. M. Dinsdale, Hot Tinning for the Food Industry, *Tin and Its Uses,* 116, 9, 1978.
20. P. M. Dinsdale, Hot Tinned Wire for the Electronics Industry, *Tin and Its Uses,* 114, 1, 1977.
21. L. N. McKenna and A. T. Quintana, Electrotinning of Copper Conductors for the Wire and Cable Used in Telephone Central Offices, *Tin and Its Uses,* 117, 7, 1978.
22. Continuously Electroplated Wire for the Electronics Industry, *Tin and Its Uses,* 109, 5, 1976.
23. F. A. Lowenheim, Bronze and Other Tin Alloys. *Modern Electroplating,* 3rd ed., John Wiley & Sons, New York, 1974, p. 528.
24. *Working Instructions for Speculum Plating,* International Tin Research Institute Publication 134, 1953
25. *Electrodeposition of Tin-Lead Alloys,* International Tin Research Institute Publication 325, 1968.
26. C. J. Evans, Developments in Tin-Lead Plating, *Tin and Its Uses,* 92, 8, 1972.
27. C. A. MacKay and B. T. K. Barry, *Terneplate, Production, Properties and Applications,* Pb '74, Fifth Int. Conf., Paris, November 1974.
28. C. A. MacKay, Surface Finishes and Their Solderability, *Welding and Metal Fabrication,* Brazing and Soldering Suppl., 55, Jan./Feb., 1979.
29. *Electroplated Tin-Nickel Alloy,* International Tin Research Institute Publication 235, 1968.
30. *Tin-Zinc Plating,* International Tin Research Institute Publication 202, 1963.
31. The Protection of Steel Bolts Used in Aluminum Structures, *Tin and Its Uses,* 25, 3, 1951.

32. I. T. Turner, The Application of Cadmium-Tin Alloy Plating to Threaded Fasteners, *Plating,* 677, July, 1965.
33. Mechanical Plating with Tin-Cadmium Alloy, *Tin and Its Uses,* 112, 4, 1977.
34. J. Hyner, Tin-Based Chromium-Like Finishes, *Plating and Surface Finishing,* 64, 2:2, 1977.
35. F. A. Lowenheim and R. M. MacIntosh, Immersion Tinning, *Modern Electroplating,* 3rd ed., John Wiley & Sons, New York, 1974, p. 412.
36. Refractory Metals: Protection by Aluminum-Tin Coatings, *Tin and Its Uses,* 71, 8, 1966.
37. C. J. Thwaites, *Soft Soldering Handbook,* International Tin Research Institute Publication 533, 1978.
38. H. M. Manko, *Solders and Soldering.* McGraw-Hill, New York, 1964.
39. *Soldering Manual,* American Welding Society Publication, Miami, 1978.
40. R. I. Jaffee, E. J. Minarcik, and B. W. Gonser, Low-Temperature Properties of Lead-Base Solders and Soldered Joints, *Metal Progress,* 843, Dec., 1948.
41. P. G. Foresster, *Babbitt Alloys for Plain Bearings,* International Tin Research Institute Publication 149, 1963.
42. C. J. Thwaites, Developments in Plain Bearing Technology, *Tribologia e Lubrificazione,* 10, 94, September, 1975.
43. *Aluminum-Tin Alloy Bearings,* International Tin Research Publication 463, 1973.
44. J. B. Long, Solid Aluminum-Tin Bearing Alloys, *Tin and Its Uses,* 108, 11, 1976.
45. *Bearing Bronzes,* Copper Development Association Publication, 15, 1062.
46. Porous Bronze Bearings, *Tin and Its Uses,* 57, 1, 1962.
47. S. K. Barua and P. A. Ainsworth, *Further Aspects of Sintering Iron Powder Compacts Containing Tin and Copper,* 3rd European Powder Met. Conf., Part I, 1971, p. 323.
48. C. J. Evans, *Continuous Casting of Bronze,* International Tin Research Institute Publication 546, 1977.
49. *A Guide to Copper and Its Alloys,* Copper and Brass Research Association Publication 9, 1959.
50. *Copper Alloy-725,* International Nickel Corporation Publication, Data Sheet, 1971.
51. P. A. Ainsworth and C. J. Thwaites, *A High-Strength Tin Bronze with Improved Electrical Conductivity,* International Tin Research Institute Publication 411, 1970.
52. *Modern Pewter,* International Tin Research Institute Publication 494, 1977.
53. *Working with Pewter,* International Tin Research Institute Publication 506, 1979.
54. R. Duckett and P. A. Ainsworth, *Some Factors Affecting the Directional Properties of Rolled Pewter Sheet,* International Tin Research Institute Publication 464, 1973.
55. *Fusible Alloys Containing Tin,* International Tin Research Institute Publication 175, 1967.
56. O. J. Seeds, *Modern Uses of Bismuth and Bismuth Alloys,* Met. Soc. AIME, preprint 2A-RF-4, 1966.
57. *Type Metals, Their Characteristics and Their Performance,* Imperial Metal and Chemical Co. Publication, Chicago, 1966.
58. C. J. Thwaites and B. F. Muller, *Tin Alloyed Iron Castings,* International Tin Research Institute Publication 545, 1978.
59. C. J. Thwaites, How Tin Affects Cast Iron, *Metal Progress,* 88, 3:100, 1965.
60. Further Use of Tin-Alloyed Cast Iron in the Motor Industry, *Tin and Its Uses,* 106, 1, 1975.
61. Recent Research on Tin Alloyed Cast Iron, *Tin and Its Uses,* 113, 10, 1977.
62. S. K. Chattergee and R. R. Dean, Recent Developments in Tin-Alloyed Cast Iron, *Tin and Its Uses,* 107, 3, 1976.
63. J. A. Young and J. B. Barclay, *Cast Lead-Calcium-Tin Grids for Maintenance Free Batteries,* paper presented at the 85th Meeting, Battery Council International, 1973.

64. P. M. Dinsdale, Lead-Calcium-Tin Alloys for the Lead-Acid Battery, *Tin and Its Uses*, 11, 12, 1976.

65. P. M. Dinsdale, Commercial Developments, Part II, *Tin and Its Uses*, 111, 13, 1967.

66. C. J. Evans, Advances in Titanium—The Place of Tin, *Sheet Metal Industries*, 51, 7:384, 1974.

67. R. F. Smart, Zirconium-Tin Alloys, *Metal Industry*, 535, December 27, 1957.

68. C. J. Evans, Tin in Dentistry, *Tin and Its Uses*, 91, 3, 1972.

70. J. B. Long, Tin Containing Brazing Alloys, *Tin and Its Uses*, 118, 13, 1978.

71. *Tin Chemicals for Industry*, International Tin Research Institute Publication 447, 1972.

72. M. J. Fuller, Industrial Uses of Inorganic Tin Chemicals, *Tin and Its Uses*, 103, 3, 1975.

73. R. R. Deam and C. J. Evans, Tin in the Ceramics Industry, *Tin and Its Uses*, 114, 9, 1977.

74. M. J. Fuller and M. E. Warwick, The Catalytic Oxidation of Carbon Monoxide on Tin (IV) Oxide, *J. Catalysis*, 29, 441, 1973.

75. M. J. Fuller and M. E. Warwick, SnO_2-CuO Gels: Novel Catalysis for Low Temperature Oxidation of Carbon Monoxide, *J. Chem. Soc.*, 6, 210, 1973.

76. W. B. Hampshire and C. J. Evans, Tin Oxide Electrodes in Lead Glass Manufacture, *Tin and Its Uses*, 118, 3, 1978.

77. (a) Strengthening the Glass Container, *Tin and Its Uses*, 77, 3, 1968. (b) Glassware Strengthened with Tin Chemicals, *Tin and Its Uses*, 100, 3, 1976.

78. Conductive Tin Oxide Films on Glass by Chemical Vapor Deposition, *Tin and Its Uses*, 107, 10, 1976.

79. Surface Conductive Glass, *Tin and Its Uses*, 27, 11, 1952.

80. Conductive Glass Developments, *Tin and Its Uses*, 40, 12, 1959.

81. An Effective Anti-Caries Dentifrice Based on Stannous Fluoride, *Tin and Its Uses*, 54, 3, 1962.

82. C. J. Evans, Plating on Plastics, *Tin and Its Uses*, 98,7, 1973.

83. C. J. Evans, Tin Chemicals as Catalysis for New Silicones, *Tin and Its Uses*, 89, 5, 1971.

84. P. J. Smith, Organotin Compounds and Applications, *Chem. Brit.*, 11, 6:208, 1975.

85. A. K. Sawyer, *Organotin Compounds*. Marcel Dekker, New York, 3 vols., 1971.

86. J. J. Zuckerman, *Organotin Compounds: New Chemistry and Applications*. American Chemical Society, 1976.

87. C. J. Evans and P. J. Smith, Organotin-Based Anti-Fouling Systems, *J. Oil Color Chem. Assoc.*, 58, 160, 1975.

88. G. Huber, Stages in the Development of Plant Protection Agent, *Tin and Its Uses*, 113, 7, 1977.

89. A. J. Crowe, R. Hill and P. J. Smith, *Tributyltin Wood Preservatives*, Tin Research Institute Publication 559, 1978.

90. S. G. Ampain, Zirconium, in *U.S. Bureau of Mines Bulletin* 667 Mineral Facts and Problems, 1975.

91. Kirk-Othmer, *Encyclopedia of Chemical Technology*, 2nd ed., Vol. 22. Interscience, New York, 1970, pp. 614 ff.

92. National Fire Protection Association Bulletin No. 482M-1976, Standard for the Processing, Handling and Storage of Zirconium. National Fire Protection Association, 470 Atlantic Ave., Boston, MA 02210.

93. B. Lustman and J. G. Goodwin, Zirconium and Its Alloys, in *Reactor Handbook*, 2nd ed., Chap. 32. Interscience, New York, 1960, pp. 708 ff.

94. D. L. Douglas, *The Metallurgy of Zirconium*, Supplement 1971. International Atomic Energy Agency, Vienna, 1971. (Obtainable from Unipub, P. O. Box 433, Murray Hill Station, New York, NY 10016.)

95. H. Stehle, E. Steinberg, and E. Tenckhoff, Mechanical Properties, Anisotropy and Microstructure of Zircaloy Canning Tubes, in *Zirconium in the Nuclear Industry*, ASTM STP 633, Lowe and Parry, editors. American Society for Testing and Materials, (1916 Race Street, Philadelphia, PA 19103), 1977, pp. 486-507.

96. ASTM standards on zirconium and its alloys:
    B350 Zirconium and Zirconium Alloy Ingots for Nuclear Application
    B351 Hot Rolled and Cold Finished Zirconium and Zirconium Alloy Bars, Rod and Wire for Nuclear Application
    B352 Zirconium and Zirconium Alloy Sheet, Strip and Plate for Nuclear Application
    B353 Wrought Zirconium and Zirconium Alloy Seamless and Welded Tubes for Nuclear Service
    American Society for Testing and Materials, 1916 Race Street, Philadelphia, PA 19103.

97. S. Kass, The Development of the Zircaloys. Proceedings of USAEC Symposium on Zirconium Alloy Development, Pleasanton, CA, Nov. 12-14, 1962, pp. 101 ff. GEAP-4089. Available from Office of Technical Services, U.S. Department of Commerce, Washington, D.C. 20025.

98. W. Evans, P. A. Ross-Ross, J. E. LeSurf, and H. E. Thexton, Metallurgical Properties of Zirconium Alloy Pressure Tubes and Their Steel End Fittings for Candu Reactors. *Peaceful Uses of Atomic Energy*, Vol. 10. International Atomic Energy Agency, Fourth International Conference, 1971, pp. 513 ff. United Nations Publications, Room LX-2300, New York, NY 10017.

99. N-Reactor Safety Analysis Report, Sect. 5.3, United Nuclear Industries, Inc., P.O. Box 490, Richland, Washington 99352, 1978.

100. S. H. Bush, Zirconium Clad Uranium Dioxide Fuel Element for Light Water Reactors. Internal report, Bettelle Northwest Laboratories, Richland, Washington 00352.

101. *Liquid Metals Handbook*, NAVEXOS P-733 (Rev.) 2nd ed., 1952. U.S. Atomic Energy Commission and Department of the Navy. R. N. Lyon, Editor-in-Chief.

102. A. Amorosi, Selection of Reactors, in *Nuclear Engineering Handbook*. McGraw-Hill, New York, 1958, pp. 12-58. Harold Etherington, Editor.

GENERAL READING

Magnesium

American Society for Metals, *Metals Handbook,* Vol. 2, *Properties and Selection: Nonferrous Alloys and Pure Metals,* Metals Park, OH, 1979. See pages 525-609.

Dow Chemical Co., *Fabricating with Magnesium,* Midland, MI 48640, 1982.

Dow Chemical Co., *Primary Magnesium Ingot and Magnesium Alloy Ingot,* Midland, MI 48640, 1982.

Dow Chemical Co., *Magnesium: Designing Around Corrosion,* Midland, MI 48640, 1982.

Dow Chemical USA, *Digest of Specifications for Magnesium Products,* Midland, MI 48640, 1982.

Dow Chemical USA, *Operations in Magnesium Finishing,* Midland, MI 48640, 1982.

Tin

Anon., *Properties of Tin,* International Tin Research Institute Publication 218, 1965.

P. M. Dinsdale, *Guide to Tin,* International Tin Research Institute Publication 540, 1978.

E. S. Hedges, *Tin and Its Alloys.* Edward Arnold, Ltd., London, 1960.

C. L. Mantel, *Tin, Its Mining, Production, Technology and Applications.* Hafner, 1970.

10

Miscellaneous Inorganic Materials
Principles of Use and Design

I. INTRODUCTION: PROPERTIES AND STRUCTURE*

The most useful properties of any material are its structure-sensitive ones. Early humans discovered this truth when they worked with inorganic materials or ceramics. And although they did not realize it, the shaping of ceramic articles involved moisture-dependent plasticity and thixotropy; the decorative textures were based on vitrification and devitrification, the nucleation of crystalline phases and gases, and discrete variations of viscosity, expansivity, and surface tension; and the colors evolved were due to oxidation, structure imperfection in crystals, and ionic states and excitations.

The use of metals was discovered soon after the first widespread use of fire-hardened clay. Here again, metals and the usefulness of the properties they exhibit owe their utility to property-structure relationships. Early workers with metals found that, although metals are stiff below a certain stress, they become plastic above a certain stress, and that, at various temperatures, metals change to a liquid state. They applied such knowledge in hammering, hot working, annealing, and casting objects of copper and other metals. Later, it was realized that there were three basic ways of modifying the properties of a metal: (1) work hardening, (2) alloying, and (3) heat treatment. All result in property changes based upon structural changes brought about by recrystallization, augmentations to the crystal lattice (e.g., of iron), and others. Metallurgists in recent times have unraveled many such change relationships showing the importance of crystal interfaces and grain boundaries to many physical effects. It has been proved that the mechanical strength of a lattice is affected by the presence of imperfections, which may be of several kinds, for example, a missing atom somewhere in an otherwise regular array, foreign atom replacements, shifts in the alignment, or an extra row of atoms that does not extend through the whole lattice. Most metals would be about a thousand times stronger if their lattices were completely regular and free of dislocation, since the motion of these dislocations through crystal lattices causes slip to occur. By adding small foreign atoms, such as carbon in steel, this motion is prevented. Work hardening, dispersion strengthening, and precipitation hardening by heat treatment are other methods discovered centuries ago.

---

*Contributed by DONALD G. GROVES.

In placing this brief background of past endeavors in the perspective of today's work in materials science and engineering, it is of interest to note that Cyril Stanley Smith has observed (1) that

> Jewelry from the royal graves at Ur in Mesopotamia (around 2600 B.C.) exhibits higher standards, both aesthetic and technical, than most objects made today, and shows that *their makers could reproducibly exploit most of the properties of metals that only now are being scientifically explained.* [Italics added.]

and Robert A. Huggins states (2)

> In order to have more than a transient impact, we need to understand where, and more importantly, why, important materials are currently used. Knowledge of their functions, *in fundamental terms of the structure and properties of the materials employed, in devices and of the various phenomena that control them will be important.* [Italics added.]
>
> These questions are, of course, near the heart of the current paradigm in materials science; if this information were elucidated, sophisticated materials engineering might develop alternatives with the same property-related function as the commodity whose availability is under question.

Also, in two landmark studies (3,4) of materials science carried out by the National Academy of Sciences-National Research Council, the following statements are made, which are essentially as true today as when they were made in the late 1960s:

> The engineering use of ceramics for a variety of needed candidate applications is still limited. Specifically, the uniformity, reproducibility, reliability and mechanical performance of ceramic hardware are presently below intrinsic capability and are strongly influenced by size . . . . It should be evident that the success of any ceramic processing program is dependent upon the ability to correlate processing parameters with selected features of the character of the product. This situation indicates the additional need for the capability of correlating features of the character of the product with its properties and behavior, since the ultimate use of the product is dependent upon the latter. In order to achieve this evaluation, improved means are necessary for determining and expressing the character, properties, and behavior of the product.

and

> There is a widespread misconception about the function of descriptions of preparation and values of measured properties in the characterization of materials. True, every characterization method uses some property of atoms, ions or molecules, to determine the compositions, structures and defects in materials. Nevertheless, it needs to be proved, always, that the property measurement reflects *directly* and unambiguously the relevant compositional or structural features of a material before it can be accepted as a valid characterization method. *Most property measurements do not fall in this category and require, therefore, an independent means of characterizing the structure or composition of the material.* Clearly, much of solid-state research is concerned with the effort to understand properties, but . . . if fails to distinguish between property studies on characterized and uncharacterized materials. In short, the fundamental objective of analytical chemistry and structural analysis have not been heeded. [Italics added.]

The above-mentioned remarks are important since the reality of the situation today is that as yet the level of our understanding, which links properties with structure, is primarily empirical and rarely quantitative. There are several reasons, both real and otherwise, for this. Basically, the task of characterizing those features of the composition and structure (including defects) of a material that are significant for a particular preparation, study of properties, or use and suffice for reproduction of the material) is a rough, tough, and usually costly proposition. Moreover, there are still some inadequacies in the available means for characterization.

As T. D. Schlabach points out (5),

> The goal of structural characterization is to describe precisely the location and arrangement of constituent atoms within a solid, the bonds holding them together, and the nature, location and arrangement of the various point, line, area and volume defects present. It spans the range from the grossest features of the solid such as external morphology, to the finest details of atomic arrangement such as the omission of single atoms or ions from their expected location in a lattice. Starting with nominally grosser features, it is essential to know the nature and number of major phases present and whether they are crystalline or noncrystalline. Next are the geometric features of size, shape, relative orientation and location associated with the grains, second or dispersed phases and pores, voids or microcracks present. Following this, characterization of the various boundaries present must be made including the surface, grain and subgrain boundaries, phase boundaries, twin boundaries and stacking faults. Here it becomes important to know also if, and what, phases or impurities are present at such boundaries. Next, the density, type, location and configuration of dislocations present within the material must be characterized. Finally, the point defects in the form of vacancies, interstitials and foreign atoms and their local clustering must be determined. All of these structural and defect features can and do affect the mechanical as well as the physical properties of solids.

Obviously, and from a pragmatic viewpoint alone, not everyone recognizes the need for involvement in this kind of a characterization of materials effort. Moreover, in many engineering materials, which are invariably complex in nature, *complete* characterization is work enough for the next millenium. Consequently, characterization has come to have different meanings according to the definer. For instance, for the specification engineer and materials supplier the thrust is placed on those minimum requirements needed to define the material in terms of various properties, such as elongation, hardness, and tensile strength, with these being measured in a specified manner according to a standard. Very often, however, such standards do not ensure that the material is uniformly reproducible from lot to lot and heat to heat batches. Unfortunately, this is the situation that exists today in many endeavors, a situation that is obviously not of much comfort and utility to the design engineer, who is usually most interested in such properties as yield point, creep, modulus of elasticity, fracture toughness, oxidation and corrosion resistance, conductivity, and thermal shock. It does not necessarily follow that the few properties measured in conformance to a standard will ensure that the properties coveted by the designer will be reproducible in the material. However, the selection of materials in many designs has often been based on this assumption or chosen on the basis of handbook data, which are frequently inadequate.

In view of this, Schlabach (5) advises

> What we can ask of standards and specifications, that are related to characterization, is that the test methods used to measure property conformance be standardized as much as possible, be capable of good intra- and inter-laboratory repro-

ducibility, and be chosen, wherever possible, to reflect critical and subtle features of the composition and structure of the material in question.

The critical role that materials processing and manufacturing plays in improving the properties of materials cannot be neglected. In such processing it is essential to learn to control and refine the structures of materials. In metals, especially the high-strength, microalloy steels, considerable progress has been made by controlling processing conditions, including rolling. Also in the secondary processing of metals, such as casting and directional solidification, progress is also evident. Additionally, perhaps the most sensational breakthroughs that have emerged in recent years are in the transistor field. Here, with the knowledge revealed by studies of the microstructure and atomic structure of certain materials, it has become possible to take advantage of the structure-properties-performance linkage. Transistors are produced by highly controlled processing techniques that result in the precise control of composition and internal structure of silicon and other semiconductors. However, pragmatic data on performance in a sample component are still used to qualify a material.

In conclusion, although many engineering materials are too complex to characterize completely, and except in a general sense, not all property data can necessarily be predicted from structure alone, any more than performance can be from properties alone, the urgent need exists in the entire field of materials science, development, and engineering to work out through the reciprocal flow of scientific and empirical information the structure-properties-performance relationship of materials.

II. NONMETALLIC MATERIALS*

Until about 1910 chemical science was concerned largely with the composition and structure of naturally occurring materials, and with increasing and perfecting their usefulness. It seemed little likely, at that time, that physical and chemical knowledge would be so much enlarged, and techniques so rapidly developed in the next few decades, that useful man-made materials would challenge those found in nature in number, versatility, and complexity. This is, however, a major task to which chemical science has since been particularly devoted. Historically, the process has had two major goals: first, to understand the principles through which important naturally occurring substances are built; and second, to apply and extend these principles in the laboratory and in the factory to prepare new products. The effects set in motion by early studies have, in truth, carried far beyond the dreams of those who pioneered in these investigations. Nature is versatile as well as bountiful. However, profound structural changes occur in nature only at great intervals. New forms of matter, or modifications of old forms, are not discovered frequently, nor when most wanted. In the test tube or in the reaction kettle additions and alterations in reactants and in conditions can be made almost at will. It is not surprising that the term "tailor-made molecules" has come to be more than a mere catchword or advertisement slogan.

In this chapter, therefore, the major coverage will be devoted to a description of inportant types of synthetic nonmetallic materials and to the general principles governing

---

*Prepared by ROBERT S. SHANE from the Second Edition with contributions by M. M. SPRUNG, General Electric Co., Schenectady, New York, and others as noted.

their structures and properties. Those natural products that are still of great industrial importance, or that serve to illustrate principles of constitution or synthesis, will also be considered in their proper relationship to analogous or cognate synthetic substances.

A. Classification

Any classification of nonmetallic materials must necessarily be somewhat arbitrary. One conventional and useful classification recognizes a division into "inorganic" and "organic" types, with subdivisions into materials of natural and synthetic origin. Within such a loose framework most of the common nonmetallic materials can be disposed as follows.

I. Inorganic nonmetallic materials
    A. Natural
        1. Stone
        2. Minerals and ores
        3. Clays and loamy deposits
        4. Salts
    B. Synthetic
        1. Cement
        2. Concrete
        3. Plaster and gypsum
        4. Glass
        5. Ceramics
            a. Brick
            b. Porcelain
            c. Vitreous enamels
        6. Fused silica and alumina
        7. Graphite
        8. Carbides, nitrides, and mixed ceramic "alloys"
II. Organic nonmetallic materials
    A. Natural
        1. Components of natural gas
        2. Components of petroleum
        3. Components of carbonaceous deposits
        4. Simple carbohydrates
        5. Wood
        6. Cellulosic products other than wood
        7. Shellac
        8. Leather
        9. Cork
      10. Natural rubber
      11. Natural resins
      12. Natural fibers
    B. Synthetic or modified
        1. Paper
        2. Regenerated cellulose
        3. Cellulose derivatives
        4. Artificial leather

     5. Synthetic resins
     6. Synthetic rubbers (elastomers)
     7. Synthetic fibers
     8. Composites

B. Relationship with Metals

The chemistry, physics, and mechanics of metals are highly complex subjects, as is clear from preceding chapters, but most metallic systems can be defined in terms of a number of rather precisely related and measurable properties. For example, most metallic systems have (1) a definite melting point, (2) a limited number of transition temperatures, attributable to definable changes in structure, (3) measurable and reproducible mechanical properties, such as tensile strength, thermal expansion, modulus of elasticity, and yield strength,* and (4) a characteristic crystalline pattern under normal conditions. Frequently, nonmetallic materials also conform to these relationships. In the field of plastics, for example, a specialized technology is devoted to mechanicoelastic (i.e., metal-like) behavior.

     Because of the great diversity and multiformity of nonmetallic materials, however, gross departures from these and similar quantitative relationships are often encountered. This can be illustrated by a rather extreme, but not uncommon, example. A block of pure quartz and a specimen of "bouncing silicone putty" both contain silicon and oxygen atoms bound to one another in intimate, specific relationship. Otherwise, they show little or no similarity. The quartz is representative of an ordered, crystalline solid; the putty is as strikingly a physically disordered, amorphous structure. The quartz exhibits Newtonian behavior; that is, it obeys Hooke's law: strain is proportional to stress. The putty is a non-Newtonian substance.

III. TYPES OF CHEMICAL BONDING†

Matter may be viewed as held together (see also Chap. 2) by three fundamental types of bonds: (1) ionic or heteropolar, (2) covalent or homopolar, and (3) secondary valence or cohesive.‡ Metals occupy an intermediate position, as will be shown presently.

A. Ionic or Heteropolar Bonding Forces

In the ultimate view, any substance is composed of atomic nuclei and electrons held together in such an arrangement that the potential energy of the system is a minimum. Particularly in the case of many inorganic compounds, the atoms, or certain closely associated atomic groupings, are ionized in the solid. This ionization consists in the giving up of one or more electrons by one atom or group and their acceptance by another. These atoms or groups therefore permanently carry either positive or negative charges (Fig. 10.1). In this structural sense there is no a priori reason to expect discontinuities or boundaries in a solid so composed of spatially arranged ions, except through imper-

---

*However, alloys with "memory" that, after distortion, return to a previous shape, are a conspicuous exception (e.g., Nitinol).

†Contributed by ROBERT S. SHANE.

‡This classification, a convenient and enormously useful one, is yet essentially qualitative and arbitrary. In the rigid mathematical expressions of quantum mechanics, no real discontinuity between these types of binding forces is recognized.

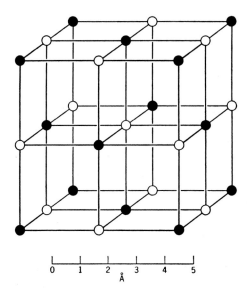

Fig. 10.1 The crystal structure of rock salt (NaCl). One type of circle represents Na^+ ions, the other Cl^- ions. The distance between like atomic centers along any principal axis is 5.63 Å (5.63×10^{-8} cm). (From *Atomic Structure of Minerals*, by W. L. Bragg, Cornell University Press, Ithaca, New York, 2nd printing, 1950.)

fections. A "perfect" crystal of this type would contain a regular and indefinitely extensive arrangement of negatively and positively charged centers. The electrostatic bonds formed between the ions of opposite polarity are understandably called "heteropolar" bonds.

In this view, for example, a crystal of pure rock salt contains only ionic bonds. A regular arrangement of positive sodium ions (Na^+) and negative chlorine ions (Cl^-) exists in the crystal. The molecular weight of such a substance is a purely fictitious concept. No individual atom or ion bears a unique relationship to any second atom or ion. Through the regularity of spatial arrangements, a given ion will in general bear an identical relationship to each of several neighboring ions, and each of these, in turn, will bear an identical relationship to several of its neighbors. Thus, the entire cycle may be viewed as constituting a single, giant molecule of indeterminate molecular size and weight. An inspection of the sodium chloride crystal model (Fig. 10.1) will make this more apparent.

B. Covalent or Homopolar Bonding Forces

In contrast to this situation, a symmetrical gas, ideally in its normal state and at infinite dilution, presents a good example of covalent bonding forces. As with an ionic crystal, bonds are formed by the special disposition of valence electrons, that is, by the few electrons that are particularly available for chemical bonding because of their location (ordinarily) in the outermost of the several concentric groupings of electrons, called "shells," surrounding each elementary nucleus. With covalent compounds, however, complete transfer of electrons from one atom to another does not occur. Instead, a process of "sharing" is visualized. Sharing involves the joint or coordinated action of electrons derived from the valence shells of each of the atoms involved. In the quantum mechanical sense, the resulting electronic orbitals involve contributions (not necessarily equal) from each of the atoms concerned. In another sense, the valence electrons simply act in pairs to form the covalent bond. Since complete electrostatic displacements do not occur,

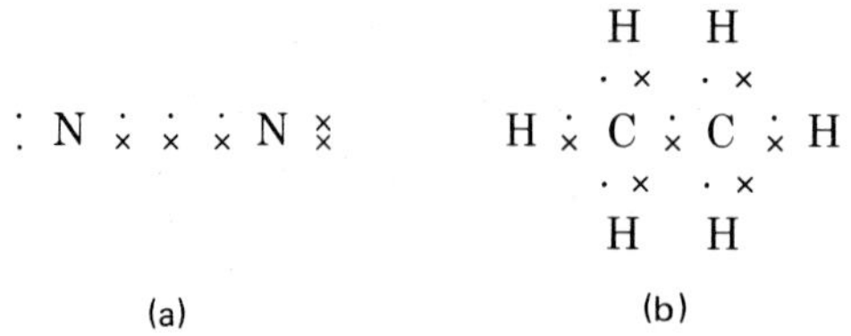

H   H
· × · ×
∴ N ×̣ ×̣ ×̣ N ×̣   H ×̣ C ×̣ C ×̣ H
· × · ×
H   H

Fig. 10.2 (a) Nitrogen gas. N denotes nitrogen atom. (b) Ethane. C denotes carbon atom. H denotes hydrogen atom. Crosses and dots denote valence electrons originally "belonging to" one or the other of the paired atoms.

(a)   (b)

the bond is also said to be homopolar; that is, it does not involve the attractions of centers of opposite polarity.*

Under idealized conditions the valence electrons of the homopolar bond may be considered to be equally shared by the paired atoms forming the bond; for example, in a simple diatomic† gas, such as nitrogen (Fig. 10.2a), or even in a more complex organic vapor, such as the hydrocarbon, ethane (Fig. 10.2b). The crosses and dots represent, schematically, valence electrons originally "belonging to" one or the other of the paired atoms. In the nitrogen gas each of the atoms originally had five electrons in its outermost shell. On combining to form the nitrogen molecule, three valence electrons from each atom are shared. In ethane, eight atoms are involved—two carbon and six hydrogen. Each carbon originally had four valence electrons and each hydrogen had one. In the molecule there are six covalent C–H bonds formed by the sharing of all the valence electrons of hydrogen with carbon, and one covalent C–C bond formed by sharing of the last available valence electrons of the carbon atoms with one another.

The molecular weight, in such cases, is exactly defined as the sum of the atomic weights concerned and is readily determined by measurement of vapor density, to which it bears an exact proportionality.

C. Secondary Valence or Cohesive Bonding Forces

It is necessary to elaborate somewhat on the nature of secondary valence forces before proceeding further. Three sources of attraction between neutral molecules are commonly recognized. They are (1) the permanent dipoles‡ that exist in most nonsymmetrical molecules and owe their origin to separation of electrical charges; (2) the polarization induced in neighboring molecules by these permanent dipoles; and (3) the electric interactions of rapidly shifting charge distributions in the molecule. The last of these are the only important sources of intermolecular attraction in simple symmetrical molecules and are of greatest importance in most other cases. The name of van der Waals, who pioneered in examination of forces of attraction between neutral molecules, is almost universally associated with them.

The distribution between primary and secondary valence forces is a vital one, and must be appreciated clearly before gross chemical and physical properties can be understood. Primary valence forces (which may be either homopolar or heteropolar, as previously described) cause atoms to combine with one another to give stable molecules.

---

*For the sale of simplicity, the modern concepts of resonance and hyperconjugation are ignored in this treatment.

† Diatomic = two atoms per molecule. Except for the rare gases of the atmosphere (e.g., argon and neon), which are monatomic, all elemental gases are diatomic in their normal states (e.g., hydrogen, oxygen, nitrogen, and chlorine).

‡ The nature and consequences of dipole attraction will be discussed at greater length in the section on electrical properties (see V.D).

Complete mutual saturation of force fields, however, never results from such combinations. Otherwise, molecules would exert no attractions upon one another, and would eventually fly apart into ultimate space. It is the secondary valence forces that exert attractions between molecules. In a liquid or in a homopolar solid these are van der Waals forces. In a broad sense they are the forces of attraction associated with residual force fields remaining after atoms combine through primary valence forces into more or less stable molecules.

In a gas, such as methane, the van der Waals forces are weak. They cannot prevent ceaseless motion of gas molecules, motion occasioned by the kinetic energy of the gas. However, in solid methane, the kinetic energy of translation is almost, if not entirely, overcome. The van der Waals forces now are able to hold the molecules in a stable crystalline pattern. Thus, in solid methane, both primary, covalent forces and van der Waals forces are clearly definable.

Crystals, such as those of solid methane, are called molecular crystals. Molecules occupy the lattice points. The intermolecular binding forces in molecular crystals may be van der Waals or they may be dipole orientation forces. In the rare gases, in diatomic elements, in carbon dioxide (CO_2), in carbon monoxide (CO), in the hydrogen halides, HCl, HBr, and HI (but not HF), in methane, and in many other organic compounds the crystal lattice forces are entirely or nearly entirely van der Waals. Dipole orientation forces are more important in water, in hydrogen fluoride (HF), in methanol (CH_3OH) and similar alcohols, in hydrogen peroxide (H_2O_2), and so on. These and many other polar molecules attract one another more strongly than can be accounted for by van der Waals forces alone.

D. Metallic Bonding Forces

As suggested above, the metallic bond may be considered as a special, transitional type, bearing some relationship both to the spatial ionic bond and to the nonspatial covalent bond. The outstanding property of metals lies in the high degree of mobility of certain of the component electrons. Thus, in a typical metal, such as copper, the valence electrons are able to shift rapidly between preferred positions in the lattice. In another sense a metal consists of positive ions that are held together, not by fixed negative ions, but by rapidly shifting electrons. It is by virtue of flow of these electrons under an applied electrical stress (potential difference) that metals conduct electricity.

E. Crystalline and Amorphous Solids

Crystalline solids possess order and symmetry and usually show well-defined boundary conditions. In general, crystalline solids obey Hooke's law; that is, deformation under an applied stress is proportional to the stress and reversible within relatively wide limits. A liquid, in contrast, has no elements of external symmetry and cannot remain at equilibrium under a shearing force. Many solids behave as though intermediate in structure between liquids and truly crystalline solids. Such solids are termed amorphous, a term that unfortunately must be used to cover a very wide range of behavior and of types. The molecules in completely amorphous solids are arranged at random. Definite physical boundaries cannot be recognized. Conchoidal (shell-like) fracture is one common evidence of this condition. On x-ray analysis, completely amorphous solids yield indefinite patterns, similar in some respects to those of liquids, again suggesting a lack of internal

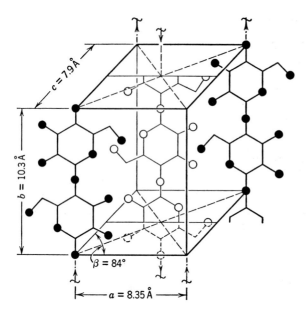

Fig. 10.3 The unit cell of native cellulose, according to Meyer and Misch. The circles represent oxygen atoms; the points where lines running from the oxygens join or change direction represent carbon atoms. Hydrogen atoms are not shown. (From X-ray Examination, by W. A. Sisson, in *Cellulose and Cellulose Derivatives*, E. Ott, ed. Copyright 1943, Interscience Publishers, New York.)

organization. Many materials of great usefulness in industry and art are amorphous. Some important examples are discussed below.

To illustrate the structural complexities with which one must sometimes be concerned, consider a naturally occurring organic substance of very considerable molecular weight, such as native cellulose. Known facts concerning the structure of cellulose may be summarized somewhat as follows. The molecules consist of very long chains of carbon, hydrogen, and oxygen atoms, held together by covalent, primary valence bonds. A complex mixture of chains of varying lengths is always present, but to each individual chain can theoretically be assigned a definite molecular weight related to its chain length and equal to the sum of the weights of all the atoms in the chain. The individual chains are somewhat randomly oriented, but in an approximate sense they are roughly parallel to one another in a direction referred to as "the fiber axis." Between such oriented chains, relatively strong cohesion forces (secondary valence forces) operate, holding the molecules into bundles sometimes termed micelles. X-ray examination has disclosed a definite crystalline pattern for the cellulose fiber that is built up from these imperfectly oriented chains. A schematic diagram of the unit cell* of cellulose is shown in Fig. 10.3. This fiber by no means has the properties of a single crystal, such as those that can be obtained from metals. It is rather a partially crystalline aggregate containing small crystal areas (crystallites) separated by amorphous areas. These relations are shown schematically in Fig. 10.4. This highly complex state of affairs is typical of many organic substances, both natural and synthetic.

---

*A definition of this term is given in the following section.

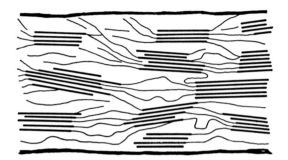

Fig. 10.4 Schematic representation of the gross structure of cellulose. The heavy, approximately parallel lines are crystalline areas (crystallites). The fine, wavy lines represent unoriented chains, which taken together constitute amorphous areas. Individual molecular chains may begin, end, or run through the crystallites. (By permission, from *Colloid Chemistry,* Vol. 5, by Jerome Alexander, Reinhold Publishing Corp., New York.)

IV. STRUCTURE OF INORGANIC MATERIALS*

A. Minerals

The crystalline nature of many minerals was recognized as early as the eighteenth century. Crystallography, although now treated as a separate science, has developed in parallel with scientific mineralogy. Most minerals can be fitted into a formal crystallographic system. For example, quartz (which may be nearly pure silicon dioxide, SiO_2) crystallizes in the rhombohedral system, generally as six-sided prisms, sometimes terminated by six-sided pyramids. Galena, lead sulfide (PbS), occurs in the isometric system, usually cubic in form, but truncations sometimes give octahedral, dodecahedral, or even trisoctahedral forms.

One of the more important problems in the study of a mineral specimen is to establish the crystallographic system to which it belongs. The disposition or arrangement of points in a crystal at which the pattern repeats itself is called a space lattice. Within a space enclosed by these points a complete unit of pattern is found, and this is called the unit cell. In all, there are 14 special kinds of space lattices, which constitute the 14 crystallographic systems. These are shown in Fig. 10.5.

The composition of a mineral is usually determined by ordinary methods of chemical analysis. Igneous rocks, frequently used in construction, consist principally of the oxides of silicon and aluminum, with lesser inclusions of iron, calcium, magnesium, and so on; slates, which are geologically derived from clayey sediments, are very similar to these aluminum silicates in gross chemical composition; sandstones are composed essentially of grains of quartz, cemented together by iron oxides, lime, or clay; limestones are basically calcium carbonate ($CaCO_3$); and granites contain quartz, feldspar (largely potassium aluminum silicate), mica, and sometimes hornblende (a complex of calcium, iron, aluminum, and magnesium silicate). Most minerals are understandably hard, durable, impervious, and stable toward heat, light, and moisture.

The silicate minerals display almost every conceivable variety of crystallographic complexity, but in almost all known silicate rocks, the silicon atoms are surrounded by four oxygen atoms at nearly ideal tetrahedral angles, and the silicon-oxygen bond dis-

---

*Contributed by ROBERT S. SHANE.

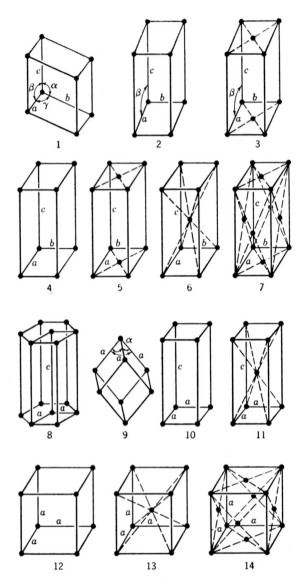

Fig. 10.5 The 14 crystallographic systems; (1) triclinic, this model has the lowest elements of symmetry; (2) simple monoclinic; (3) a-face centered monoclinic; (4) simple orthorhombic; (5) a-face centered orthorhombic; (6) body-centered orthorhombic; (7) face-centered orthorhombic; (8) hexagonal; (9) rhombohedral; (10) simple tetragonal; (11) body-centered tetragonal; (12) simple cubic; (13) body-centered cubic; (14) face-centered cubic. (By permission from *Atomic Structure of Minerals,* by W. L. Bragg, Cornell University Press, Ithaca, New York, 2nd printing, 1950.)

tance is fixed at about 1.6 Å (1.6 × 10^{-8} cm). Often, regular ring structures are found in the crystal. Sometimes there are zigzag chains of alternate Si and O atoms. Three-dimensional silicon-oxygen hetworks occur in quartz and in some related minerals.

Mica is a mineral of special interest, particularly to the electrical industry. Mica is coming into increasing use as a filler and reinforcing agent in organic matrix composites. It serves the double function of increasing stiffness and decreasing permeability to vapors. The platelike structure makes it possible for mica to be used where some flexibility is required. The crude mineral can be split into sheets that may have a thickness as small as 0.0005 in. It has unusually high thermal stability and under some conditions may be used up to 500°C. In continuous segments it has a very high dielectric "strength" and very low dielectric loss factor.* Chemically, micas are complex silicates of magnesium, aluminum, iron, and potassium. *Muscovite* corresponds to $KAl_2(AlSi_3O_{10})$ $(OH)_2$, *phlogopite* to $KMg_3(AlSi_3O_{10})$ $(OH)_2$. Crystallographically, it is possible to detect a succession of thin double sheets in mica, with the potassium atoms located between them. A model of a mica crystal is shown in Fig. 10.6.

Asbestos, another remarkably useful inorganic material, is a hydrated magnesium silicate. One common variety (*chrysotile*) corresponds approximately to $3MgO \cdot 2SiO_2 \cdot 2H_2O$. Common asbestos and a few closely related minerals of the *amphibole* and *serpentine* groups are fibrous in nature. The fibers vary in length from a few tenths of an inch to well over an inch, the long-fiber variety naturally being preferred technically. These fibers are composed of crystals with one axis, along which stretches an indefinite chain or alternate silicon-oxygen-silicon atoms, parallel to the fiber bundle. Crystallization can be observed to be imperfect in one direction, a rare phenomenon in minerals, but common among organic substances (e.g., cellulose). The fibrous structure, combined with nonflammability, makes asbestos peculiarly suited for thermal and electrical insulation and, properly compounded, for many structural purposes. It must be noted that asbestos must be handled to prevent release of fibers in production, fabrication, or use. Asbestos fiber has been shown to be carcinogenic when inhaled.

B. Carbides, Nitrides, Glasses, and Ceramic Materials

1. Carbides and Nitrides

In silicon carbide (SiC) and boron carbide (B_4C) the binding forces are predominantly homopolar, rather than ionic. The hard abrasive quality of these materials is associated with high symmetry and close packing of the constituent atoms. Silicon carbide is produced in an electric furnace at about 2000°C. Two forms exist, the β-cubic form and the α-hexagonal form, which is stable at high temperatures. The β crystal slowly changes to the α crystal during prolonged exposure to elevated temperature with adverse effect on the mechanical properties. Silicon nitride is a material made by reacting nitrogen gas with finely divided silicon at 1400-1450°C. A silicon preform is nitrided to achieve the part that is stable to around 2000°C. Boron nitride is a cubic crystalline material made under high pressure at about 2000°C. Boron carbide produced at 2500°C is also stable at about 2000°C. These "structural ceramics" and others will be discussed extensively in Chap. 14.

---

*For definitions of these terms, see Chap. 13.

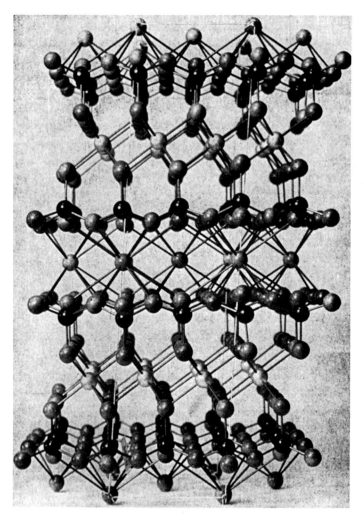

Fig. 10.6 Schematic representation of the structure of mica, according to W. L. Bragg. The unit cell is outlined by the fine lines that form a rectangle from the two topmost atoms to the opposite two at the bottom. (From *Atomic Structure of Minerals,* by W. L. Bragg, Cornell University Press, Ithaca, New York, 2nd printing, 1950.)

2. Glass*

A typical glass can be represented by an arrangement of atoms as in Fig. 10.7. (See also Chap. 14 for an extensive treatment of glass.)

3. Ceramic Materials

The term "ceramics" applies to a variety of inorganic products characterized by the high-temperature treatment universally employed during their manufacture. In a sense cement

---

*Contributed by MARVIN G. BRITTON.

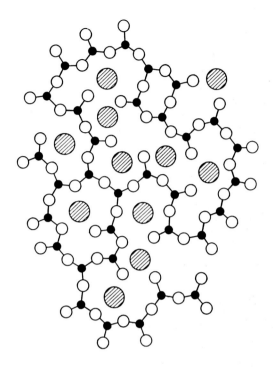

● Si   silicon atoms

○ O   oxygen atoms

◍ Na   metal atoms

Fig. 10.7 Representation of the arrangement of atoms in a typical glass, after Zacharia-sen. (By permission, from *Physical Chemistry*, by Eastman and Rollefson, McGraw-Hill Book Co., New York, 1947.)

and concrete are also ceramics since heat is used to prepare the cement. This subject will be discussed later. The principal raw material concerned is clay; but since clays differ widely in their composition, purity, firmness, plasticity, fusibility, and other properties, the term is wide and frequently misused. The clays alone or mixed with silica, flint, or feldspar, for example, are fired in kilns at temperatures frequently in excess of 2000°F (1050°C), then cooled slowly.

The characteristics of clays that are of primary interest to the ceramics industry are frequently less their exact chemical composition than their colloidal properties, such as particle size, shape, nature of surface, plasticity, and thixotropy. Thixotropy is a property common to many colloidal systems. It is this property that enables some suspensions and gels to be converted (reversibly) by mechanical agitation to fluids of relatively low viscosity. A dilute bentonite gel provides a common example. Ordinary paints also are thixotropic; this allows them to flow readily when agitated by brushing but to increase greatly in viscosity as the brush is lifted. It is reasonable to assume from such behavior that the physical structure of thixotropic clay suspensions is in accord with this conception. With pottery or china, the importance of thixotropy can easily be appreciated, since the clay suspensions used must be very carefully balanced to flow evenly, yet to "set" properly during firing.

Common products of the ceramics industries are china, porcelain, porcelain enamel, ceramic tile, firebrick, high-temperature inorganic insulating materials, electrical conduit, ordinary brick, tile and clay pipe, chemical stoneware, and others. Refractories are other important products of the ceramics industry. Because they must have very high melting points, and since impurities lower the melting point, they are almost always pure substances. Common examples are alumina (Al_2O_3), Aluminum silicate [$Al_2(SiO_3)_3$], zirconia (ZrO_2), magnesia (MgO), iron chromite ($FeCr_2O_4$), silicon carbide (already mentioned), and, of course, silica. Engineering ceramics are discussed at length in Chaps. 14 and 28.

C. Organic Materials

Organic chemistry concerns itself with nearly a million individual, distinguishable compounds. The only certain common denominator is the presence of at least one atom of carbon in each. One function of this branch of chemistry is, of course, to establish intelligible relations within this otherwise vast and unwieldy subject matter.

The homopolar or covalent bond is the "intrinsic mortar" that holds most organic compounds together. Carbon, situated in the fourth group and in the first short series of the periodic table of the elements, is, in a sense, the most "neutral" of elements. It accepts and donates electrons with equal facility. Because this process of "sharing" electrons may be repeated over and over again, it is possible to build up a great variety of molecules, including many very large or "giant" molecules. When these large molecules are made by repetitive combination of smaller molecules, they are called polymers. These will be discussed more fully in Secs. IV.C.7 and IV.C.8 in Chap. 11. Such macromolecules are both the end result of many important vital processes (hemoglobin, proteins, cellulose, keratin, silk, and natural rubber are a few common examples) and of many commercially important synthetic processes (of which synthetic rubber, nylon fibers, alkyd paint finishes, and phenolic plastics are examples).

The structure of organic compounds may be illustrated by a few typical examples listed in Table 10.1.

In the above structures only homopolar bonds are indicated. Molecular weights are the arithmetic sum of the atomic weights involved. Most of these compounds are neutral, but diethylamine and pyridine are basic, butyric acid is acidic, and polyglycine contains both acidic (-COOH) and basic (-NH$_2$) groups. Organic acids and bases form ions in conducting media, ordinarily through electrolytic dissociation. Salts that are obtained on evaporation of such solutions may, in certain simple cases, crystallize in typically ionic systems. However, most simple organic compounds form molecular lattice crystals. With simple organic compounds, as was stated above, the points on the crystal lattice are usually occupied by molecules, or by segments of molecules. The molecules are held in the crystal by lattice forces that are similar to, if not identical with, van der Waals forces. In crystals of aliphatic compounds, structures related to that of diamond are frequently detected.* The tetrahedral valence bonds of carbon are charac-

---

*The crystal structures of diamond and graphite are shown in Fig. 10.9.

teristic of most aliphatic compounds and often provide one important element of crystal symmetry. For aromatic compounds, the basic building block is the six-carbon benzenoid ring, as it is known to exist in graphite.* These six-carbon atom rings are almost invariably planar, and the forces between them are weak, corresponding again to secondary valence or van der Waals forces. The more complex aromatic molecules approach most nearly the actual graphite structure. When polar groups, such as hydroxyl, aldehyde, ketone, or halogen, are present in organic molecules, the molecules pack together more closely in the crystal, owing to the stronger intermolecular forces such groups exert. When even more highly polar groups are present, such as the carboxyl or amino groups, and more especially their salts, genuine electrostatic forces may come into play. Since these forces are of the same order of magnitude as those that operate in ionic crystals, they cannot always formally be distinguished from them. Depending upon the size and shape of the purely organic part of the molecule, relative to these strong polar groups, the total crystal will assume a configuration intermediate between that of a purely ionic and that of a purely molecular crystal. The resulting structure is often highly complex. Since there are only incomplete analytical data for many such crystal systems, significant structural details cannot often be found.

1. Petroleum and Natural Gas

Petroleum deposits of the simplest type ("Pennsylvania crudes") are predominantly complex mixtures of aliphatic saturated hydrocarbons, of which the simplest generic formula is C_nH_{2n+2}. In natural gas, which is frequently associated with petroleum deposits, the lowest members of this series are found. These are methane, CH_4, ethane, C_2H_6, propane, C_3H_8, and butane, C_4H_{10}. These hydrocarbon gases are also sometimes present in crude oil, along with numerous higher members of the series. Individual members up to heptacontane, $C_{70}H_{142}$, have been synthesized. On distillation of petroleum, various "fractions" are obtained, whose use ranges from volatile "petroleum spirits" through gasoline, kerosene, fuel oil, and lubricating oils to semisolid and solid petroleum jelly and paraffin wax fractions. Many oil fields, notably the American midcontinent oil fields, contain a much smaller fraction of straight-chain aliphatic hydrocarbons and a larger proportion of asphaltic components than do Pennsylvania crudes. The asphaltic components contain, as major constituents, cyclic aliphatic hydrocarbons—naphthenes. Examples are dimethylcyclopentane and ethylcyclohexane (see Fig. 10.8). Many petroleum crudes, such as California crudes, also contain a considerable percentage of aromatic hydrocarbons.

2. Coal Tar

When bituminous coal is subjected to *coking* (a process involving "destructive distillation"), gaseous and liquid products are obtained in addition to coke. The crude liquid, called coal tar, when subjected to fractional distillation and purification, yields a variety of useful products—neutral, acidic, and basic. The neutral products include a series of cyclic aromatic hydrocarbons. The unsaturation of aromatic compounds, as indicated previously, is of a special type, since these hydrocarbons are, in general, very stable thermally and inert toward many chemicals. Examples are benzene (C_6H_6), toluene (C_7H_8), xylene (C_8H_{10}), cumene (C_9H_{12}), naphthalene ($C_{10}H_8$), and anthracene ($C_{14}H_{10}$).

---

*The crystal structures of diamond and graphite are shown in Fig. 10.9.

Table 10. 1 Examples of the Structure of Organic Compounds

Compound	Structure	Type	Descriptive notes
Propane	$CH_3CH_2CH_3$	Saturated aliphatic hydrocarbon	Representative of a large number of open-chain compounds that contain only carbon and hydrogen, all having the generic formula C_nH_{2n+2}. Generally they are rather stable, except toward oxidation and thermal decomposition ("cracking").
1-Pentene	$CH_3CH_2CH_2CH{=}CH_2$	Unsaturated aliphatic hydrocarbon	Contain less hydrogen than the above type. The unsaturation increases chemical reactivity.
Cyclohexane		Cycloaliphatic hydrocarbon	Representative of saturated closed-ring hydrocarbons. The reactivity is similar to that of openchain analog.
Toluene		Aromatic hydrocarbon	"Aromatic" signifies the presence of a closed ring of atoms of particular and peculiar reactivity. Although low in hydrogen content, they are not "unsaturated" in the same sense as are similar open-chain compounds, and they are more stable towards oxidation and cracking. The hydrogens are more easily replaced by other atoms or groups.

Name	Structure	Class	Description
Ethanol	CH_3CH_2OH	Alcohol	Representative of hydroxyl (OH) derivaties of saturated aliphatic hydrocarbons. The OH group is a center of reactivity. Relatively easily oxidized. The OH group is replaced by other groups fairly easily
Ortho cresol	(ring structure: HO—C, C—CH₃, CH, CH, CH, CH)	Phenol	Phenols are "aromatic alcohols." They are also weak acids. The hydroxyl group vastly increases the ease of replacement of the ring hydrogens by other groups.
Formaldehyde	H—C=O, H	Aldehyde	Aldehydes are oxidation products of alcohols. They are in turn easily oxidized to acids, easily reduced to alcohols. They react readily with a variety of other molecules, including alcohols, phenols, amines, and amides, to form more complex compounds.
Acetone	CH_3—C(=O)—CH_3	Ketone	Similar to aldehydes, but less reactive.
Butyric acid	$CH_3CH_2CH_2C$(=O)—OH	Acid	Representative of organic acids. Relatively weak compared with mineral acids. Oxidation products of alcohols and of aldehydes. Relatively stable and very difficult to reduce. The hydrogen of the carboxyl group (COOH) is replacable by various groups, giving esters, acid anhydrides, acid chlorides, and so on. Carboxylic acids form salts with alkalis or organic bases.

Table 10.1 (continued)

Compound	Structure	Type	Descriptive notes
Diethyl ether	$CH_3CH_2-O-CH_2CH_3$	Ether	Formally, ethers are "anhydrides" of alcohols (anhydride = minus water). They are relatively unreactive. Some ethers are useful as solvents.
Chlorobenzene		Aromatic halide	Halogen derivative of aromatic hydrocarbon. Relatively stable toward heat, light, and most chemicals.
Diethylamine	$CH_3CH_2\overset{\textstyle H}{N}CH_2CH_3$	Amine	Amines are organic derivatives of ammonia. Three general types: primary, RNH_2; secondary, R_2NH; tertiary, R_3N. [R represents any hydrocarbon group, such as methyl (CH_3), ethyl (C_2H_5), and phenyl (C_6H_5). Amines are roughly of the same basicity as ammonia. They form salts with organic and inorganic acids.

Pyridine

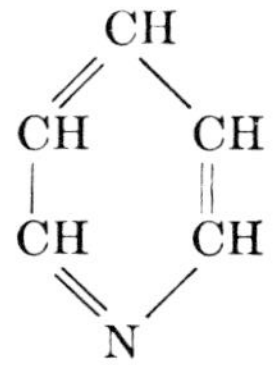

Aromatic heterocyclic amine

"Heterocyclic" indicates a ring of atoms including at least one atom other than carbon. Pyridine is a tertiary amine (see above), since all three valences of nitrogen are occupied with carbon.

Polyisobutylene

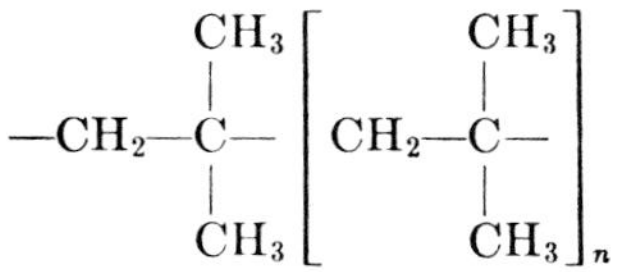

Polymeric hydrocarbon

A polymer contains a larger number of identical repeated units. Here the repeated unit is $-CH_2-C(CH_3)_2-$, related to the "monomer," isobutylene [i.e., $CH_2 = C(CH_3)_2$] from which the polymer is formed. Polymers are generally mixtures. The average molecular weight depends on the average value of n.

Polyglycine

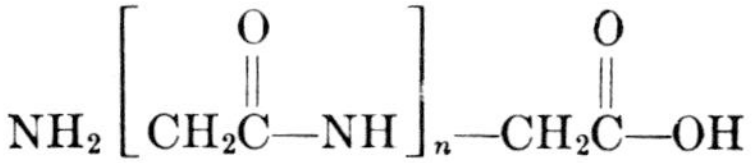

Polypeptide (polyamide)

Polypeptides are related closely to the building blocks of the proteins. They are also polymers. The "monomer" in this case is glycine, NH_2CH_2COOH, both an amine and an acid; that is, an amino acid. The polymer has one amino group and one acid group for every n amide ($-CONH-$) groups.

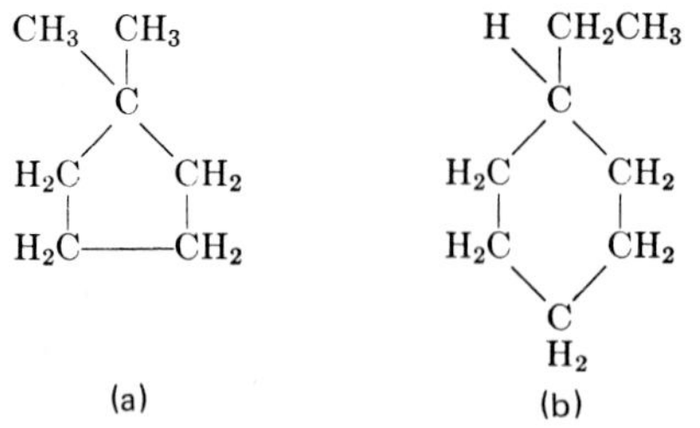

CH₃ CH₃ ... structure image

Fig. 10.8 Schematic structure of (a) 1,1-dimethylcyclopentane, and (b) ethylcyclohexane.

Note the low ratio of hydrogen to carbon compared with the saturated paraffins, C_nH_{2n+2}.

The acidic components of coal tar contain a mixture of phenols, which are hydroxy derivatives of the aromatic hydrocarbons. Examples are phenol (carbolic acid) (C_6H_5OH), cresol (C_7H_7OH), xylenol (C_8H_9OH), and naphthol ($C_{10}H_7OH$).

The basic compounds present in coal tar are principally heterocyclic aromatic amines. The most important members are pyrrole (C_4H_5N), pyridine (C_5H_5N), picoline (C_6H_7N), lutidine (C_7H_9N), quinoline (C_9H_7N), and carbazole ($C_{12}H_9N$).

The predominance of carbon-carbon bonded cyclic structures in coal tar distillates suggests that this linkage is present in the original coal, and this has been partially confirmed experimentally. Carbon has a valence of 4, and if bonded symmetrically, an indefinite structure of close-packed, equispaced carbon atoms at normal, or "tetrahedral-valence," angles to one another can be visualized. Such a structure undoubtedly exists imperfectly in carbonaceous deposits, highly modified by partial hydrogenation, chain rupture, and other changes. In diamond, which is nearly pure, crystalline carbon, this ideal structure is, in fact, realized. Diamond belongs to the isometric crystal system and shows fourfold (i.e., regular, tetrahedral) coordination in its space grouping of atoms. In graphite, which is an allotropic form of diamond, the carbon atoms are arranged in flat planes. Related groupings of six carbon atoms each can be identified. The relatively weak van der Waals forces between parallel planes help account for the platelike fracture and slipperiness of graphite.* Schematic representation of the crystal structures of diamond and graphite are shown in Fig. 10.9.

3. Carbohydrates

The carbohydrates constitute a large group of molecules, widely distributed in nature, that contain only carbon, hydrogen, and oxygen. The simplest carbohydrates are sugars, for example, glucose, which contains a ring of five carbon atoms and one oxygen (see Fig. 10.10. Cellobiose, obtained by careful hydrolysis of cellulose, has two glucose units linked together through an oxygen atom (see Fig. 10.11). The starches and celluloses contain many such simple sugar units linked together by oxygen (Fig. 10.12). The application of x-ray analysis to cellulose and related carbohydrates has been of great help in establishing their intimate structures. A high degree of crystallinity is found in native cellulose specimens. Figure 10.3 shows that the unit crystallographic cell contains

---

*The inorganic, high-melting solid, boron nitride (BN), which has a structure very similar to that of graphite, is also so similar to it physically that it has been given the name "inorganic graphite."

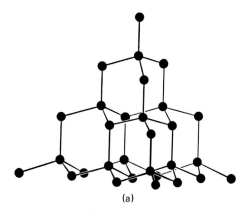

(a)

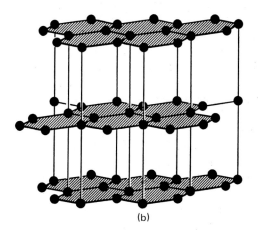

(b)

Fig. 10.9 (a) The crystal structure of diamond. The carbon atoms are virtually in contact, with a distance between carbon centers of 1.54 Å. (b) The crystal structure of graphite. The fine horizontal lines join atoms of successive sheets, which are in the same vertical relationship. The carbon-carbon distance in the planes is 1.42 Å; that between planes is 3.40 Å. (From *Atomic Structure of Minerals,* by W. L. Bragg, Cornell University Press, Ithaca, New York, 2nd printing, 1950.)

Fig. 10.10 Schematic structure of glucose.

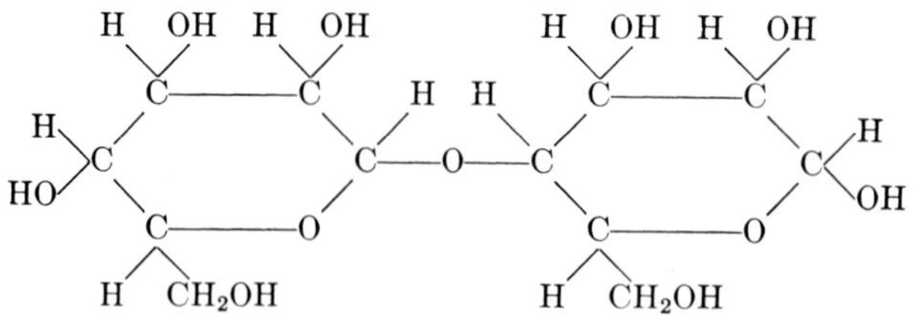

Fig. 10.11   Schematic structure of cellobiose.

four simple glucoselike units. These simple units correspond empirically to $C_6H_{10}O_5$, and cellulose is accordingly $(C_6H_{10}O_5)_x$, where x may be 1000-10,000. It has already been shown that the crystalline regions, or crystallites, in cellulose are composed of bundles of chains (micelles) oriented approximately parallel and separated by amorphous areas. The average diameter of a micellar unit is about 60 Å (60×10^{-8} cm), which corresponds to an area including between 100 and 200 molecular chains. The average length of the micellar units is about 600 Å. The remarkable properties of cellulose are partially a result of the presence of such imperfectly oriented fiber bundles.

4. Shellac

The lac insect, *Tachardia lacca,* produces a viscous exudation from certain sap trees native to India. When refined, this becomes lac resin, called shellac. The lac resin contains a mixture of polyhydroxy acids and esters, components of which are aleuritic acid $[C_{15}H_{28}(OH(_3COOH]$, shellolic acid $[C_{13}H_{16}(OH)_2(COOH)_2]$, and an ester of formula $C_{32}H_{54}O_9$. Shellac is useful as a surface coating because, on standing (drying), the hydroxyl (OH) and carboxyl (COOH) groups take part in esterification reactions; that is, they combine with one another, with loss of water, to form esters:

$$RCOOH + R'OH \rightarrow RCOOR' + H_2O$$

This causes the molecular weight to be built up considerably. The resulting structures (polyesters) are much less sensitive to moisture and solvents than are simple esters, although perhaps still not entirely satisfactory in this respect. Since shellac is a good adhesive, has a low dielectric constant, and resists carbonization, it has been widely used with fillers, such as flake mica and asbestos, and as an impregnant with cotton or glass cloth for electrical insulation (Hercolite cylinders, varnish cambric, commutator slot compound, and the like).

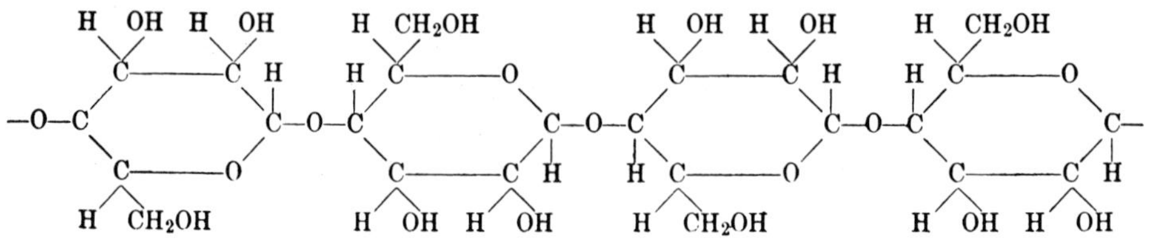

Fig. 10.12   Schematic structure typical of starches and cellulose.

$$-CH_2-\underset{\underset{CH_3}{|}}{C}=CH-CH_2-\left[CH_2-\underset{\underset{CH_3}{|}}{C}=CH-CH_2\right]_n-$$

Fig. 10.13 Schematic structure of multilinked isoprene unit that forms linear chains in natural rubber molecules.

5. Natural Fibers

Among naturally occurring fibers, cotton, flax (linen), hemp, jute, ramie, and wool are cellulosic in structure. The repetitive polysaccharide structure has been described above. Silk and wool are examples of natural protein fibers. Both consist of long chains of polypeptidelike units. These may be represented schematically as

$$-R-\underset{\underset{O}{\|}}{C}-NH-R'-\underset{\underset{O}{\|}}{C}-NH-R''-\underset{\underset{O}{\|}}{C}-NH-R'''-$$

where R, R′, R″, and so on, are groups derived from the various amino acids that are known to occur in nature.

6. Natural Rubber

Natural rubber is obtained in the form of a latex from the sap of *Hevea brasiliensis* and a few other plants, such as guayule. The latex contains spherical colloidal particles in the range of $0.1-2 \times 10^{-6}$ m in diameter. Crude rubber is coagulated by heat or by addition of electrolytes. Pure gum rubber is a hydrocarbon of empirical composition $(C_5H_8)_x$, where x is usually over 1000. The rubber molecule consists of coiled, linear chains of multilinked isoprene (Fig. 10.13). Certain inversions of this regular arrangement sometimes occur. Geometric space relations are also of great importance, since gutta-percha, which has the same chemical structure as rubber but differs in the spatial arrangement of the individual isoprene units, is hard and horny and exhibits low elasticity. The reversible extensibility and the vulcanizability of rubber are associated with the length of the linear chains, the geometric relations, and the presence of the unsaturated linkage in each isoprene segment. In the long search for rubber substitutes, these considerations have been kept in mind, and the natural rubber molecule has served as a model for many of the synthetic materials.

7. Polymers in Materials Science*

A new branch of chemistry has developed rapidly during the last four decades, the science and technology of *high polymers*. In general, these materials consist of the elements C, H, N, and O, and they are conventionally classified as *organic* polymers. However, in some instances other elements, such as B, Si, P, S, F, Br, and Cl, are present in significant proportions and have a more or less important influence on the ultimate properties of the products. Nevertheless, it is customary to refer to this group of compounds as organic polymers or organic macromolecules. Together with the large family of metallic compounds and ceramic systems, these organic polymers are essential engineering mater-

---

*Contributed by HERMAN F. MARK.

ials in the construction of buildings, vehicles, engines, appliances, textiles, packaging and writing sheets, plastics, rubber goods, and household articles of all kinds.

The rapid growth of these relatively new engineering materials in the recent past is due to several factors.

1. The basic raw materials for their production are readily available in large quantities. Natural organic polymers, such as cellulose (paper and textiles), proteins (wool, silk, and leather), starch (food and adhesives), and rubber (tires and many other goods), are mainly available as products of the farm or the forest, whereas the raw materials for the production of synthetic polymers come essentially from coal, natural gas, or oil. The simplest building units are called *monomers,* some of which (ethylene, propylene, isobutylene, butadiene, and styrene) are found in natural gas or are by-products of the manufacture of gasoline and lubricating oils. They are relatively inexpensive and are available in very large quantities. Many other monomers are simple derivatives of ethylene, benzene, formaldehyde, phenol, urea, and other basic organic chemicals. They are, in general, also widely available. Thus, the basic building units for organic polymers, as defined above, represent a large variety of compounds, are readily available, and are of moderate cost (6).

2. During the last forty years, intense research activities in many laboratories have succeeded in elucidating the mechanism of the reactions by which long-chain molecules are formed from the above-mentioned basic units. They are called *polymerization reactions* and represent either chain reactions of a highly exothermic character or step reactions in the course of which chains of systematically repeating units are formed. During the same period systematic engineering efforts developed a number of relatively simple unit-type processes that permit the translation of polymerization and polycondensation reactions into large-scale industrial operations. Today, several types are well-developed standard procedures that allow the conversion of monomers into polymers rapidly, conveniently, and at low cost. In fact, in many cases, the actual conversion cost from monomer to polymer is only a few cents per pound, a condition that has greatly contributed to the rapid expansion of this field (7).

3. Only a few well-developed processes and machines existed 40 years ago to convert organic polymers from the state of a latex, a sheet, or a molding powder into the ultimate commercial product. Today there are many continuous automatic, rapid, and inexpensive methods for spinning, casting, blow molding, injection and compression molding, stamping, and vacuum forming that give each polymer with attractive properties an almost immediate chance of being converted into useful and salable consumer goods. Obviously the existence of manifold applications and the availability of standardized and automated methods of developing a new polymer into many different channels stimulate the synthesis and development of useful new members of the organic polymer family (8).

4. The large number of available monomers and the larger number of polymers and copolymers made from them has provided an almost continuous spectrum of composition and structure of organic macromolecules. On the other hand, the systematic exploration of their mechanical, optical, electrical, and thermal behavior has provided an equally dense spectrum of characterized practical properties. Study has led to a relatively profound and dependable understanding of structure-property relationships. This has the great advantage that new polymers or copolymers with specified properties need not be looked for by empirical, more or less random, synthetic efforts, but can be designed on paper. A successful elimination of many possibilities can be effected before work is

actually started in the laboratory. This approach has been so successful that, in many instances, one can speak of a *molecular* engineering approach in the synthesis and development of new polymer materials (9).

All fundamental and applied efforts of monomer syntheses, polymerization techniques, and manufacturing processes can eventually be condensed in a few guiding principles that represent, so to speak, the essence of our present understanding and know-how. Such principles are, of course, only qualitative generalizations and, in each individual case, have to be supplemented by quantitative considerations and numerical refinements, but they give a convenient and clarifying "helicopter view" of the present state of our knowledge and its practical applicability. As a consequence they are good working hypotheses or guide posts, if they are used with caution and with the realization of their character as approximation and illustration.

In the following paragraphs we shall try to enumerate and discuss the most important principles of this type and to indicate the most prominent applications of them.

a. Molecular Weight

A factor of great importance in the synthesis and application of organic polymers is molecular weight (MW) (10). The words macromolecule, giant molecule, high polymer, and polymer indicate that the molecules of this class of compounds are large and hence consist of *many parts.* In fact, all existing experience indicates that many valuable and interesting properties of natural and synthetic polymers can only be obtained *if their molecular weight is sufficiently high.* Since many important materials consist of chain molecules with repeating units, one can also introduce the concept of the *degree of polymerization* (DP), which represents the *number of basic units* in a given macromolecule. In general, molecular weight and degree of polymerization are used interchangeably in the sense that they are related by the equation

$$MW = DP \times (MW)_u \tag{1}$$

where $(MW)_u$ is the molecular weight of the repeating unit or monomer.

It has also been established by many contributors throughout the last four decades that polymeric materials do *not* consist of strictly *identical molecules,* but always represent a *mixture of many species,* each of which has a different molecular weight or DP. The character of a given material is indicated by a molecular weight *distribution function,* which can conveniently be expressed by a curve in which the frequency or percentage of each species is plotted against the molecular weight or the DP of this particular species. The narrower the distribution curve of a given polymer, the more homogeneous is the material. As a consequence of this polymolecularity of all polymeric compounds, one cannot simply speak of "a" molecular weight or of "a" DP, but must operate with an *average* molecular weight MW or an *average* degree of polymerization DP (10).

Many tests have established that the most important mechanical properties, particularly tensile strength, elongation to break, impact strength, and reversible elasticity of polymers, definitely depend on the average molecular weight or the DP, in the sense that up to a certain, relatively low DP value no strength at all is developed (but note that certain properties, notably toughness, depend on the distribution of MW). Then, there is a steep rise of mechanical performance with DP, until at still larger molecular weights, the curve flattens out and one enters a domain of diminishing return of strength on further DP increments. Figure 10.14 shows the characteristic shape of this curve, which is typical for all polymers and differs for each individual material only in the numerical details.

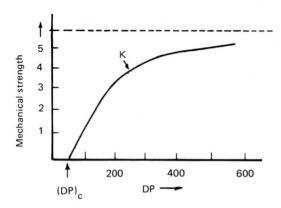

Fig. 10.14 Tensile properties as a function of molecular weight or DP.

Each polymer has, for instance, a critical DP value DP_c, below which it is essentially a friable powder, but the numerical value of DP_c is different for each polymer; polyamides start to develop strength at DP values as low as around 40, whereas cellulose needs values around 60, and many vinyl polymers need still higher values of around 100. The knee K of the curve occurs at a DP of 150 for polyamides, 250 for cellulose, and 400 for many vinyls. However, *all* polymers exhibit *no strength* below DP values of 30 and approach limiting strength of DP values above 600. Even if there are still small gains in the higher molecular weight range of the curve, they are difficult to attain practically because very long chains produce high viscosities in the dissolved and molten state and become increasingly difficult and impractical to process. The characteristic shape of the curve in Fig. 10.15 has the consequence that virtually all practically useful polymers fall in the DP range from 200 to 2000, which, in general corresponds to molecular weights from 20,000 to 200,000.

Evidently the curve in Fig. 10.14 is of great interest and importance for those who want to prepare and launch a new and useful polymer. If they find, at a certain point in research, that the molecular weight of the samples is around 5000, attention and effort will be concentrated on changing the polymerization conditions in order to penetrate into a higher molecular weight range. If, on the other hand, they establish that the samples have molecular weights around 60,000, they will focus interest on other factors,

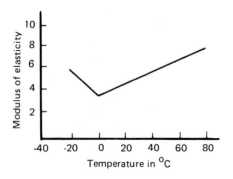

Fig. 10.15 Modulus of rubber as a function of temperature.

such as molecular weight distribution, branching, and influence of reactive groups or stereoregulation, but not on a further increase of molecular weight or of chain length.

The curve of Fig. 10.14 can be represented by an equation of the type

$$MS = (MS)_i - \frac{A}{DP} \tag{2}$$

where

$(MS)_i$ = mechanical strength of infinitely long chains
MS = mechanical strength measured for a given DP
A = constant

The parameters A and $(MS)_i$ are characteristic for any given polymer and are connected with $(DP)_c$, the critical DP, through

$$(DP)_c = \frac{A}{(MS)_i}$$

Hence Eq. (2) may also be written as

$$MS = (MS)_i \left[1 - \frac{(DP)_c}{DP} \right] \tag{3}$$

It should be added that Eqs. (2) and (3) lose their significance when DP becomes smaller than $(DP)_c$. Furthermore, at least at this stage, they are only empirical, but it was shown by Flory and others (11) that they can be rationalized by considering that the bonds along each individual chain are much stronger (chemical bonds) than those between chains (van der Waals bonds). As a result, short chains slip very easily along each other and offer little or no stress transferring action. This explains the existence and significance of $(DP)_c$. On the other hand, if the chains become very long the accumulated resistance of many van der Waals bonds against slippage becomes so great that eventually chemical bonds start to break. From this point on, further lengthening of the chains has little influence on improved mechanical performance. By considering these arguments in the form of equations, it is not only possible to interpret the form of Eq. (3) rationally, but also to arrive at improved expressions that represent in greater detail the conditions for the mechanical failure of chain polymers.

Thus far, we have only discussed the influence of the *length* of the chain molecules; let us now pass to a consideration of other chain properties that are also important for the mechanical and thermal behavior of polymeric systems.

c. Crystallization

One of the first important notions of the basic properties of chain molecules was that they exhibit a tendency to form crystallike bundles or aggregates. If one uses the word crystallite or crystallize in connection with polymers, it must be understood that one wants to convey the idea that certain volume elements of the polymeric system have reached a state of three-dimensional order that, in *certain respects,* resembles the crystals of normal materials, such as sugar, naphthalene, stearic acid, or maleic anhydride. However, crystalline domains in a polymeric material do not have the regular shape of normal crystals; they are much smaller in size, contain many more imperfections, and are connected with the disordered, amorphous areas by through-going polymer chains, so that there are absolutely no sharp boundaries between the laterally ordered (crystalline) and disordered (amorphous) parts of the system. On the other hand, polymeric crystals have relatively sharp melting points, high densities, and high moduli of rigidity; they resist dissolution and swelling; and they are virtually impenetrable to diffusion of small molecules. They give relatively sharp x-ray and electron diffraction patterns and show characteristic absorption peak splitting in the infrared absorption spectrum. In some instances (cellulose, polyvinyl alcohol, and certain proteins), there are reasonably sharp boundaries between the crystalline and amorphous portions of a polymeric material; in other cases (linear polyethylene, polyesters, and polyacetals) it appears more appropriate to consider the entire system as crystalline, but with the understanding that flaws and imperfections (twisted or folded chain segments) are more or less uniformly spread out over its length and width and are responsible for the deviation of its behavior from that of a perfect crystal (12).

But however the incomplete three-dimensional order may be visualized in any special case, there can be no doubt that the tendency to crystallize plays a very important role in the thermal and mechanical behavior of polymers; it will be useful to connect this tendency with the chemical composition and the structural details of the individual macromolecules. There are, of course, many factors contributing to such a complicated process as three-dimensional order; here it is appropriate to discuss only the most important of them. They seem to be the following.

1. Structural regularity of the chains, which can readily lead to the establishment of repetitive identity periods
2. Free vibrational and rotational motions in the chains, so that different conformations can be assumed without penetrating high-energy barriers
3. Presence of specific groups that produce strong lateral intermolecular bonds (van der Waals bonds) and regular, periodic arrangement of such groups
4. Absence of bulky, irregularly spaced substituents that inhibit the chain segments from fitting into a crystal lattice and prevent the laterally bonding groups from getting close enough for substantial interaction

How much does each of them contribute to the crystallizability of a given material?

(1) Structural regularity   The simplest polymer molecules are linear polyethylene and polyformaldehyde; their chains can readily assume a planar zigzag conformation characterized by a sequence of trans bonds and can therefore produce a very short repetitive identity period along the length of the chain. This characteristic favors the establishment of lateral order, particularly if the macromolecules are oriented by stress or shear; in fact, both polymers are easily orientable and can attain very high degrees of crystallinity. The polyethylene chains are nonpolar, and all intermolecular attraction is due to dispersion forces; the rotation around the C–C bond is inhibited by an energy barrier of about 2.7 kg-cal/mol of bonds. This limited flexibility and the dispersion forces between adjacent chains are responsible for the high melting point, high rigidity, and low solubility of this material. In the case of polyformaldehyde the rotation about the C–O bond is less inhibited than that about the C–C bond, but the dipole character of the $\begin{smallmatrix} & O & \\ C & & C \end{smallmatrix}$ group produces polar forces between adjacent chains, which act over a longer range and are stronger than the dispersion forces. As a consequence, polyformaldehyde has a higher rigidity and a higher melting point than polyethylene; it is also less soluble. Table 10.2 presents a few pertinent data on the relationship between crystallization temperature, melting point, and stiffness (13).

Table 10.2 Crystallization Temperature, Melting Point, and Flexural Elastic Modulus of Selected Polymers

Polymer	Crystallization temperature T_g (°C)	Melting point (°C)	Flexural elastic modulus
Polyethylene	−20	141	$0.6\text{-}1.15 \times 10^5$
Polyethylene	−107	95	$0.08\text{-}0.6 \times 10^5$
Polyformaldehyde	−82	181	4×10^5
Nylon 6/6	45	267	4×10^5

If a substituent such as CH_3, Cl, or CN is attached to the chain of linear polyethylene in a 1,3 sequence, there exist two simple ways to establish regular geometric placement of the substituents along the length of the chain: the *isotactic* placement, in which all substituents (or at least a long row of them) have the *same* configurational position (either d or l), and the *syndiotactic* placement, in which there is a *regular alternation* between d and l over the entire molecule, or at least over long stretches of it. Any deviation from these two cases or any mixture of them is called *atactic* and refers to more or less random geometric positions of the substituents along the length of the chain. Natta (14), who first succeeded in preparing pure (or almost pure) representatives of these cases with polypropylene and other α-olefinic polymers, has demonstrated that the *stereoregulated* or *stereospecific* species are rigid, crystallizable, high-melting, and relatively insoluble materials, whereas the *atactic* or *irregular* species are comparatively soft, low-melting, and easily soluble polymers that do not crystallize under any conditions. The spectacular influence of this structural regularity on crystallization and with it on most mechanical and thermal properties has been established by many investigators for numerous vinyl, acrylic, and allylic polymers, and it has become an important principle in the designing and "tailoring" of new polymers, particularly since an appropriate choice of experimental conditions, such as catalyst, solvent, temperature, and additives (15), makes it possible to aim at a definite prevalence of one of the three structures.

An equally important influence of structural regularity on ultimate properties has been discovered in the polymerization of conjugated dienes. If a butadiene molecule becomes the unit of a long chain, the following *three* structures are possible:

1,4-cis CH_2 —CH=CH— CH_2 (I)

1,4-trans CH_2 —CH≡CH— CH_2 (II)

1,2 H —C— C H_2 CH_2 =CH   (III)

Through the use of appropriate initiators, solvents, temperatures, and additives, it has been possible to synthesize each of these structures in almost pure form. Even though the same chemical monomer is used, the three different structural arrangements shown above produce noticeably different behavior (16).

Structure (I) is a soft, easily soluble elastomer with a glass point T_g of about $-60°C$ and a high retractive force; it crystallizes when stretched more than 200%, and the crystalline phase has a melting point of about 20°C (*Hevea* rubber belongs to this species if isoprene is the monomer instead of butadiene).

Structure (II) is a hard, relatively insoluble polymer that crystallizes readily without elongation and has a melting point of about 70°C and a glass temperature T_g of $-20°C$ (*Balata* and gutta-percha belong to this species if isoprene is the monomer instead of butadiene.

Structure (III) exists in an isotactic, syndiotactic, and atactic form, all of which have been prepared by Natta and his collaborators with the aid of Ziegler catalysts. The stereoregular forms are rigid, crystalline, and relatively insoluble materials; the atactic species are soft elastomers with slow and sluggish recovery characteristics. Many other vinyl and acrylic polymers have been obtained in atactic and stereoregulated forms, and in all cases the structural character has a pronounced influence on the ultimate properties, in the sense that regularity favors crystallizability and rigidity, high melting points, and resistance to dissolution (17).

Extensive work on copolymers of all kinds fully confirms the importance of structural regularity on crystallization tendency and, consequently, on properties. Those copolymers built by a regular alternation of the two components A and B that can be represented by

–A B A B A B A B A B A B A B A B A B A B A B A B–

show a distinct tendency to crystallize, whereas the corresponding 1:1 copolymers with random geometric distribution of the two components

–A B B A A A B A B B A A B A B B B A A B A B B A A B–

are intrinsically amorphous and represent nonrigid, soluble, and low-softening resinous materials.

(2) Chain flexibility The word *flexibility* in the context of linear macromolecules refers to the activation energies required to initiate vibrational and rotational motions around single bonds in a macromolecule, as a consequence of which the chain can assume different conformations at moderate temperatures in a relatively short time. The energy barriers that separate different individual conformations in organic molecules have been assiduously studied in many ordinary, low-molecular-weight organic compounds with the aid of determination of specific heats, infrared absorption, and magnetic resonance. The results have led to a knowledge of the stability of different conformational isomers and the rate with which the equilibrium between them is established. This information has been applied, with the necessary caution, to macromolecules and has led to the following general conclusions:

Linear polymers that contain only or mainly single bonds between C and C, C and O, or C and N allow rapid conformational changes; if the molecular structure is regular, and/or if there exist considerable intermolecular forces, then the materials are crystallizable, relatively high melting, rigid, and relatively insoluble, whereas if the molecular structure is irregularly built, the material is amorphous, soft, and rubbery.

Ether and imine bonds or double bonds in the cis form reduce the energy barrier for rotation of the adjacent bonds and "soften" the chain in the sense that the polymers are less rigid, more rubbery, and more easily soluble than are the corresponding chains of consecutive carbon-carbon bonds. This is particularly true if the "plasticizing" bonds are irregularly distributed along the length of the chains so that they inhibit crystallization, rather than favoring it.

Cyclic structures in the backbone of a chain drastically inhibit conformational changes and lead to difficult or slow crystallization. This effect can go so far that, under most practical conditions, the polymers remain amorphous.

(3) Groups that produce intermolecular attraction If flexible chains with structural regularity are allowed to form aggregates or bundles, the stability of these supermolecular entities will depend on a firm cohesion between neighboring chains. *Specific groups* that establish strong intermolecular bonds between such chains will favorably affect crystallization. This effect will be particularly pronounced if these groups are arranged along the macromolecules at regular distances so that they can get into each other's neighborhood without causing any valence strain in the chains themselves. In fact, it has been found that all groups that carry *dipoles, highly polarizable groups,* or that permit the development of interchain hydrogen bonds, favor crystallinity and, concomitantly, all the valuable properties that are a consequence of the presence of crystalline areas (18).

Thus, in polyvinyl chloride and polyvinylidene chloride the C−Cl dipoles increase the lateral cohesive energy density of the system and, with it, the rigidity, softening temperature, and resistance to dissolution and swelling. This effect is enhanced in syndiotactic polyvinyl chloride because of the spatial regularity of the C−Cl bonds. An example of the beneficial influence of polarizable groups can be found in polyethylene terephthalate, where the phenylene rings are polarized by the C=O dipoles, with the result that the lattice structure is firm and rigid. The consequence of interchain hydrogen bonding for crystallinity, rigidity, high melting, and difficult solubility is very pronounced. Polyvinyl alcohol, with hydroxyl groups at every other carbon atom, is a well-known example. Even atactic chains can establish frequent hydrogen bonds if they are first parallelized by stretch or shear, and the polymer is, therefore, a high-melting, rigid fiber former. The syndiotactic species is even more rigid, higher softening, and less soluble in water. Another case of strong lateral hydrogen bonding is provided by cellulose, where its effect is further enhanced by the rigidity of the glucopyranose rings in the β-d junction; as a consequence, cellulose is highly crystalline, infusible, and insoluble (or soluble only with very great difficulty) and has an unusually high modulus of rigidity for an organic polymer. Perhaps the most striking effect of lateral hydrogen bonding between regularly spaced groups is demonstrated by the linear polyamides, where the −CO−NH− groups are responsible for the establishment of these bonds. Since these groups can be spaced at different distances by introducing paraffinic $-CH_2-$ chains of different lengths between them, it is possible to establish the fact that small and regular distances between successive amide groups along the chains lead to rigid, high-melting, and relatively insoluble types, whereas reduction of the lateral hydrogen bonding by large and/or irregular distances between the amide groups produces low-melting and even rubbery types, which are easily soluble in many organic liquids.

A combination of pronounced chain rigidity with lateral hydrogen bonding would lead to polymers of the structure which have, in fact, been prepared, and which represent polyamides that are extremely rigid, high melting, and resistant to heat, dissolution, and swelling.

(4) Bulky substituents   The vibrational and rotational mobility of an intrinsically flexible chain can be inhibited by bulky substituents; the degree of stiffening depends on the size, shape, and mutual interaction of the substituents. Methyl, carboxymethyl, and phenyl groups have, in general, a noticeably inhibiting influence on the segmental mobility of linear macromolecules, as shown by the relatively high glass transition and heat distortion point of polymethyl methacrylate as compared with polymethylacrylate, or of polystyrene as compared with polyethylene. Larger aromatic groups have an even stronger influence, and polyvinyl-naphthalene, -anthracene, and -carbazol are amorphous polymers of remarkably high heat distortion points and unusual rigidity; they are also relatively insoluble. In general, substituents from ethyl to hexyl exhibit a softening influence, because they increase the average distance between the main chains and prevent their dipole groups from approaching close enough to each other for favorable interaction. The substituents in these cases are open chains with from two to six carbon atoms. They have, themselves, a considerable internal mobility, and they act as internal (or chemical) plasticizers rather than as stiffeners. Striking examples of this behavior are the polyacrylic and methacrylic esters from ethyl to hexyl and the polyvinyl esters from propionate to hexoate, where rigidity, softening temperature, and resistance to dissolution and swelling decrease as the number of carbon atoms in the chain of the substituent increases. If these chains become still longer (from 12 to 18 carbon atoms) and remain unbranched (normal paraffinic alcohols or acids), a new phenomenon occurs, namely, the tendency of the side chains to form crystalline domains of their own, in which the side chains of neighboring macromolecules arrange themselves in bundles of laterally ordered units with each other (18).

The result is a firm and resilient system as if two brushes were pushed into each other and their bristles forced to form tightly packed bundles. In all these cases the softening range of the material is close to the melting point of the side-chain crystallites. If the side chains are not homopolar, but contain polar or even hydrogen bonding groups, rather rigid, high-softening, and solvent-resistant polymers are formed (for instance, in the case of polyphenyl methacrylamides or polytrichlorostyrene). Thus it can be seen that the proper choice of substituents in relation to their influence on crystallizability permits the preparation of a large variety of polymers, which in turn can be usefully applied to the synthesis of rubbery and fibrous materials.

d. Molecular Engineering of Elastomers

A particularly important class of polymeric compounds are the soft, tough, stretchable, and resilient *elastomers*. They represent a combination of properties that cannot be matched by an inorganic material, such as metals or ceramics. In fact, it is now recognized that rubberlike elasticity can only be exhibited by systems that consist of long, flexible-chain molecules having weak intermolecular forces and interconnected by primary valence bonds, at suitable intervals, so as to form a three-dimensional network. In addition, it is also desirable that these chain molecules should show the ability to undergo a reversible partial crystallization induced by stretching. In these terms the phenomenon of rubberlike elasticity, that is, the ability to recover quickly and completely from imposed large strains, is ascribed to the uncoiling and recoiling of these long, flexible-chain molecules. As for the phenomenon of partial crystallization on stretching, this has considerable bearing on the stress-strain curve and on the rupture and tear phenomena exhibited by the elastomer.

To treat rubberlike elasticity in a quantitative manner, it is necessary to define the statistical segment as that portion of the chain molecule that lies between cross links. Its mass is often referred to as the molecular weight between network junctions (M_n). It is obvious that the elastic behavior of the material is governed by the length of these segments. For instance, it can readily be seen that the force required to obtain a given extension will be an inverse function of this length; that is, the force will increase with an increase in the number of cross-links per unit volume of the elastomer.

(1) Thermodynamic considerations As in all molecular phenomena rubberlike elasticity can be treated by means of both a thermodynamic and a kinetic approach. In either case, the network segment takes the place of the individual molecule, and an analogy may be made between it and the molecule in a gas or in a solution. The linear extension of such a chain can thus be compared to the volume compression of a gas. The flexibility of an elastic chain is due to the possibility of rotational movements around its single carbon-carbon bonds, which permit the chain to be extended by uncoiling and, conversely, to retract by recoiling to its original conformation. The latter phenomenon, that is, retraction, is caused by the kinetic energy of the chain atoms, whose transverse motions tend to force the extended chain to form a random cell.

The thermodynamic treatment of this phenomenon postulates that the unperturbed, randomly coiled network chain is in a state of highest probability or entropy, and that this entropy is decreased when the chain is extended by imposition of an outside force. In these terms rubberlike elasticity can be considered to be solely an entropy manifestation, unless the intermolecular forces between chains are large enough to introduce additional energy changes during extension and contraction, in close analogy to the behavior of a nonideal gas under the influence of van der Waals forces.

This postulate can be expressed by the equation

$$F = \left(\frac{\partial E}{\partial L}\right)_{T,P} - T\left(\frac{\partial S}{\partial L}\right)_{T,P} \tag{4}$$

where

    F   = external force of extension
    L   = length of the rubber specimen
    E   = internal energy
    S   = entropy
    P,T = pressure and temperature

Equation (4) can also be written as

$$F = \left(\frac{\partial E}{\partial L}\right)_{T,P} + T\left(\frac{\partial S}{\partial T}\right)_{P,L} \tag{5}$$

and is readily seen to be analogous to the equation of state for a gas,

$$P = -\left(\frac{\partial E}{\partial V}\right)_T + T\left(\frac{\partial P}{\partial T}\right)_V \tag{6}$$

In the absence of any intermolecular forces, the first term on the right side of Eqs. (5) and (6) vanishes and they simplify to the ideal expressions

$$P = T\left(\frac{\partial P}{\partial T}\right)_V \tag{7}$$

and

$$P = T\left(\frac{\partial F}{\partial T}\right)_{P,L} \tag{8}$$

A study of the force-temperature relations yields the best information concerning the extent to which changes in the internal energy E and the entropy S contribute to the force of retraction. These relations have been investigated repeatedly, and it was found that, except at very low elongations, the force increases with increasing temperature. In other words, stretched rubber tends to retract on heating, as would be expected from entropy considerations. Figure 10.15 represents the modulus of an elastomer as a function of temperature. The negative slopes at very low elongations are due to the imposition of the normal thermal expansion of the material, which predominates under these conditions. The elongation at which the two effects balance is known as the thermoelastic inversion point; it occurs at about 10% elongation.

Using the extrapolation of the force-temperature plots to zero temperature, it is possible to obtain the appropriate values of $(\partial E/\partial L)_{T,P}$ and to appreciate the influence of the intermolecular forces. Experimentally, it has been found that the major contribution to the elastic force is due to the entropy change, with only a minor effect of the internal energy due to intermolecular forces. Hence the entropy theory of rubberlike elasticity appears to be well justified.

(2) The kinetic approach The relation between the applied force F and the extension α of a chain network can also be treated by means of kinetic considerations, where the extension is related to the restriction in possible spatial distributions of the freely rotating statistical chain segments, and this leads to the following equation for rubberlike elasticity:

$$\frac{F}{A} = NRT\left(\alpha - \frac{1}{\alpha^2}\right) \tag{9}$$

where

      F = external force of extension
      N = number of moles of network segments per unit volume
      A = original cross-sectional area of the sample
      α = ratio of extended length to original length
      T = temperature

Again, it can be seen that Eq. (9) is quite analogous to the equation of state for an ideal gas,

$$P = \frac{NRT}{V}$$

An attempt to obtain experimental verification of Eq. (9) has shown that there exist some deviations from the theory in the domain of low elongations. These deviations have to do with energy contributions, but there are much more serious discrepancies at higher elongation, such as 200% and more (Fig. 10.16).

(3) Influence of crystallization The marked upward deviation from the theoretical force values of the stress-strain curve observed at higher elongations has been interpreted as a consequence of crystallization. Some elastomers (such as natural rubber, high *cis*-polybutadiene, polyisobutylene, and polychloroprene), which possess relatively regular chain structures, exhibit the phenomenon of reversible, partial crystallization induced either by stretching or cooling. Hence, at sufficiently high elongations, some portions of the chains become parallelized and are in a favorable position to form crystallites. However, these crystalline domains are inextensible, and they cause a sharp rise in the stress-strain curve, because rubber, upon extension, has produced its own reinforcing filler (rubber crystallites).

The occurrence of crystallization upon stretching of all sufficiently regular elastomers has been proved experimentally by the use of x-ray and electron diffraction and by infrared absorption. Although only diffuse halo patterns are obtained from stretched noncrystallizing, amorphous elastomers, such as SBR and EPR, distinct diffraction spots are observed in the case of natural rubber or other crystallizing elastomers. It has also been found that the interference pattern is greatly increased in sharpness and intensity at higher degrees of extension; this is because the crystalline areas become larger and more frequent as the elastomer is stretched.

Those elastomers that crystallize on stretching also generally crystallize on cooling in the unstretched state. They freeze and melt at the temperature at which the network segments have lost sufficient kinetic energy to be able to participate in the crystallization process. In view of the long-chain structure of elastomer molecules, this process is imperfect and the material can be only partly crystalline at best, leading to melting or freezing over a temperature range rather than at the sharp melting points of ordinary organic compounds. The phase transformations which a material, such as natural rubber, undergoes with changes in temperature can readily be demonstrated by the volume-temperature relations.

The imperfect crystallization process in elastomers is reflected in the sensitivity of the crystalline melting point to the conditions of crystallization. Thus the crystallites that are formed more rapidly, at lower temperatures, are also more imperfect and exhibit a lower melting point. This was strikingly demonstrated by the work of Roberts and

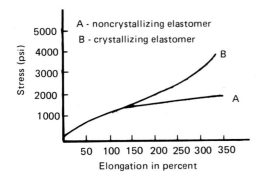

Fig. 10.16 Stress-strain curves of non-crystallizing and crystallizing elastomers.

Mandelkern, as shown in Table 10.3. Thus, by careful cooling at 14°C (and slow heating) the most accurate T_m value for natural rubber is found to be as high as 28-29°C.

Since the crystallites formed on stretching can also help to distribute the stresses to which a number of chains are subjected, they play an important role in rupture phenomena and tend to increase tensile or tear strengths in extension. It is not surprising, therefore, that those elastomers that show the ability to crystallize on stretching also exhibit high tensile strengths in the gum vulcanizates. On the other hand, those elastomers that cannot crystallize show relatively poor tensile strengths and require the use of reinforcing fillers, such as carbon black, in order to attain sufficient tenacities for practical use.

Table 10.4 shows a list of the more important elastomers from the viewpoint of their ability to crystallize on stretching; it demonstrates the importance of regular structure on the strength of soft vulcanizates.

e. Molecular Engineering of Fiber Formers

The manifold industrial uses of fibers require an extremely wide variety of basic materials. In the domain of natural products we use such diverse substances as flax, cotton, silk, wool, rubber, metals, asbestos, and glass as fibers. The range of just one property, for example, *stiffness,* which is exhibited by commercial fibers of these materials, is shown in Fig. 10.17, which presents the initial stress-strain properties of fibrous materials derived from natural sources. One can see that the initial slope of the stress-strain curve traverses the range from rubber, with a modulus of only about 100 psi, to steel, with a modulus of 30×10^6 psi.

Synthetic materials presently in use cover the modulus range from rubber to flax, that is, from 100 psi to about 700,000 psi. The lower limit is represented by the elastomeric spandex-type fibers, which have pronounced rubber properties, and the upper by highly oriented aromatic polyamides, such as Kevlar. Most classic synthetic fibers (particularly rayon, cellulose acetate, the conventional nylons, polyesters, vinyls, acrylics, and polyolefins) fall between these extremes.

The modulus of rigidity is only one of many diverse properties important to the successful end use of a fiber. Technology can effect significant adjustments and changes of many fiber properties through mechanical or thermal treatment during the fiber manufacturing process. However, all these physical processes lead only up to certain limits; to obtain really fundamental changes in fiber properties, it is necessary to go back to chemistry and synthesize a new polymer.

Table 10.3 Effect of Crystallization Temperature on the Melting Point of Natural Rubber

Crystallization temperature (°C)	Melting temperature (°C)
−18	21.0-22.0
0	22.5-23.5
+8	22.5-23.0
+14	28.0-29.0

Table 10.4 List of the Most Important Elastomers and Their Characteristics

Name	Chemical name	Vulcani- zation agent	Crystallization upon stretching	Gum strength
Natural rubber	Cis-1,4-poly- isoprene	Sulfur	High	Excellent
SBR	Poly(butadiene- costyrene)	Sulfur	None	Poor
Butyl rubber	Poly(isobutylene coisoprene)	Sulfur	High	Good
Neoprene	Polychloroprene	MgO or ZnO	Fair	Good
Nitrile rubber	Poly(butadiene coacrylonitrile)	Sulfur	None	Poor
Silicone	Polydimethyl siloxane	Peroxides	Small	Poor
Thiokol	Polyalkylene disulfides	Zinc oxide	Fair	Moderate
Urethane rubber	Polyester urethanes	Di-isocyanates	Medium rate	Good
Polybutadiene rubber	Polybutadiene	Sulfur	Good for high cis structure	Poor to fair
EPR	Poly(ethylene copropylene)	Peroxides and sulfur	None	Poor

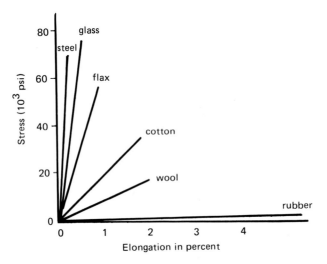

Fig. 10.17   Range of rigidity of fiber formers.

Table 10.5 shows a few particularly important fiber properties, namely, melting point, modulus, recovery power, tensile strength, moisture absorption, and dyeability; they will be discussed briefly in the light of the previous discussion in order to show how the general principles can be put to immediate practical application.

Let us consider the melting point first. All melting points, both those of polymers and of low-molecular-weight materials, depend on the ratio of the heat of fusion ΔH to the entropy of fusion ΔS:

$$T_m = \frac{\Delta H}{\Delta S} \tag{10}$$

Specifically (see Fig. 10.18), the melting point of a polymer is related to the strength and regularity of the sites of intermolecular attraction, which affects the heat of fusion, and it is also related to the stiffness of the polymer chains, which affects the entropy of fusion. As Fig. 10.18 shows, loss of regularity, as in random copolymerization, leads to a lowered melting point.

Figure 10.19 clearly demonstrates the effect of the intermolecular forces provided by hydrogen bonds in polyamides by plotting the melting point versus the concentration of amide groups per 100 Å of extended chain length. The melting point is essentially a linear function of the amide group concentration; increase of heat of fusion

Table 10.5 Key Fiber Properties Determined
by Polymer Composition

1. Melting point
2. Modulus
3. Elasticity and recovery from strain
4. Tensile strength
5. Moisture absorption, dyeability, comfort

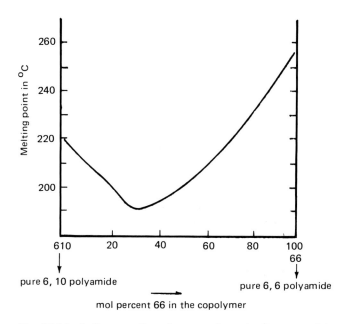

Fig. 10.18 Influence of random copolymerization on melting and softening of linear polyesters.

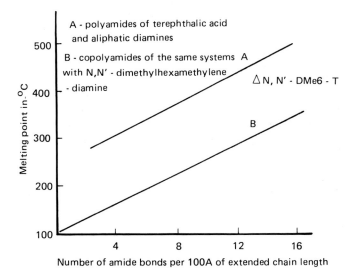

Fig. 10.19 Melting points of polyamides as a function of the distance of the Co-NH groups along the length of the backbone chain.

is directly proportional to the frequency of hydrogen bonds being produced by the increase in amide concentration. One might expect the melting point to extrapolate to that of linear polyethylene at zero amide concentration, because in this case there are only relatively weak van der Waals forces (which determine the heat of fusion of this polymer), but in reality, the extrapolated melting point is somewhat lower, an effect probably caused by the polymolecularity of the samples used to determine the melting points.

The effect of chain stiffness is well known in the case of Terylene or Dacron (these are polyesters using terephthalic acid as a component). This stiffness results from the presence of paraphenylene rings along the main chain. Since macromolecules of this type do not coil very flexibly in their amorphous state, they have a lower entropy of fusion and, consequently, a higher melting point. The influence of chain stiffening on softening, melting, dissolving, and swelling will be discussed in more detail in the next paragraph.

The melting point of a polymer is also affected by the reduction of the regularity with which the monomer groups are spaced along the backbone chain, such as by random copolymerization. This is shown in Fig. 10.18, where the melting points of random copolymers of 6:10-6:6 nylon are plotted against the composition of the material. The lowest melting point (190°C) is reached in the neighborhood of a 70:30 composition. Disturbing the regularity of the amide group spacings lowers the efficiency of hydrogen bonding, reduces ΔH in the basic melting point Eq. (10), and thereby lowers the temperature of fusion. Monomer unit rigidity and regularity also have a pronounced effect on the melting points of polyesters. Compounds based on ethylene glycol and sebacic acid are amorphous and soften at about 70°C, whereas polyesters made of ethylene glycol and terephthalic acid are crystalline and melt around 250°C. Thus the stiff *para*-phenylene ring in the main chain contributes a 180°C increase in melting point.

Figure 10.18 shows very clearly the progressive softening effect of the reduction in stiff segment content and regularity. All factors that affect the melting point also affect the modulus, for much the same reasons, but there are also other contributing structural characteristics, namely, orientation and degree of crystallinity. Thus the modulus of rigidity of a perfect 66 nylon or Dacron crystal has been estimated to be in the range of 20×10^6 psi. However, the density of commercial nylons and polyesters shows them to be only about 50% crystalline. This low crystallinity accounts for the fact that observed moduli are only a small fraction of the theoretical value, since at small deformations it is the amorphous parts of the fiber that yield under the load and cause elongation. Thus, high fiber modulus requires regularly spaced crystallizable groups along the chains and a large fraction of them oriented parallel to the fiber axis (18).

In the discussion of fiber modulus, we have limited ourselves to values referring to room temperature. Let us now consider how the modulus of rigidity of a fiber changes with temperature. Figure 10.20 shows the behavior characteristic of all polymers. Since the initial modulus is chiefly a property of the amorphous regions in the fiber, its value is relatively high at low temperatures and does not show much change with temperature in this first part of the modulus-temperature curve. This low-temperature region finds the amorphous domains of the polymer in the glassy state, and at relatively small deformations, the restoring force results from an increase in the energy of the bonds being deformed.

At the high-temperature end of the curve, although the modulus is considerably reduced, it is again relatively temperature independent. In this low-modulus region, the amorphous portions of the fiber are fluid and quite readily deformed, and the restoring

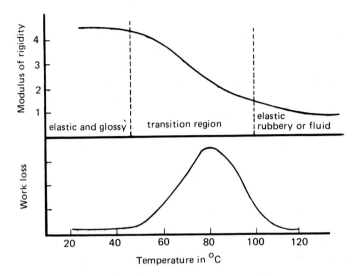

Fig. 10.20 Dynamic modulus and work loss of polymers versus temperature.

force produced by extension results from the decrease in entropy of the uncoiled polymer chains. At this end of the curve, the magnitude of the restoring force and of the modulus is proportional to the concentration of cross-links between polymer chains as well as to the amount of essentially nondeformable crystalline region. The cross-links may be either individual hydrogen bonds or covalent bonds or small crystallites. There is little hysteresis on cyclic loading of the fiber in either of these two regions, and recovery from deformation is good in both.

However, in the intermediate range, which is known as the *glass* or *second-order transition* region, the onset of sluggish responses in the partially melting amorphous regions results in a high hysteresis and in relatively poor recovery. The temperature of maximum work loss is close to the glass transition temperature T_g.

As the degree of crystallinity of a polymer is reduced, the curve tends to move to the left as well as to decrease in modulus level. If T_g is finally well below room temperature, the modulus becomes quite low, and if the cross-linking efficiency through crystallites or through covalent cross-linking is adequate, one has an elastomer with rapid response to deformation and good recovery. Stress-sensitive cross-links, such as hydrogen bonds, do not lead to maximum recovery, because they rupture and eventually re-form in new conformations that will contribute to poor recoveries. Without cross-links of any kind, a viscous liquid with no recovery is observed.

It is not surprising that the same structural characteristics that influence fiber modulus also affect the glass transition temperature, and it follows that, ideally, the glass transition temperature of a good fiber former should not be too close to the normal end use temperature. Table 10.6 illustrates the effects of changing structural polymer features by reducing crystallizability through disrupting chain regularity and reducing hydrogen bonding. A 50% substitution of N-alkylated 610 polyamide in regular 610 polymer and a 60% substitution of ethylene glycol sebacate in ethylene glycol terephthalate produces a marked reduction in fiber modulus, reduces the glass transition temperature

Table 10.6 Reduction of Regularity and Hydrogen Bonding Lowers Modulus and T_g and Increases Extensibility

	Properties at 25°C		
Polymer	Modulus (10^3 psi)	Elongation to break (%)	T_g (°C)
610	280	25	~50
610/n-alkylated 610 (50:50)	6	400	~0
2Gt/2G10 (40:60)	7	300	~-20

to below 20°C, and gives a typically elastomeric fiber for the polyamide and the polyester.

It is possible to improve the elastic properties if, instead of preparing a random copolymer, one blends two homopolymers in the melt (for instance, polyethylene glycol terephthalate and polyethylene glycol sebacate in a 40:60 ratio), allowing a limited extent of ester interchange to occur. The result is a block copolymer (compare Table 10.7), which is also an elastomer, but which has a melting point about 50°C higher than that of the random copolymer (170 versus 120°C).

Further improvements in properties of elastomeric fibers have been achieved in the block polymer framework by partly equilibrating relatively long chains (4000 molecular weight) of polyethylene glycol with polyethylene glycol terephthalate. Here, because of the chemical inertness of the ether links with respect to interchange reactions, one obtains a substantial *weight percent* modification at a relatively small mole percent modification and melting point depression. Thus, with a 60 weight percent polyethylene glycol modification, the melting point-elasticity relationship is even better than for the partially interchanged block copolymer. This effect is shown in Table 10.8 and compared with the properties of two other polyesters discussed above. There is a very substantial improvement in stress decay for the block copolymers as compared with the random copolymers. Long-distance blocks of the terephthalic acid polymer are required in the chain to produce crystalline domains that act like cross-links and minimize creep under sustained stress.

In discussing structural factors that influence modulus and melting point, one is forced to consider factors that also govern elasticity and recovery from deformation.

Table 10.7 Comparison of the Properties of a Random Copolymer with Those of a Block Copolymer of the Same Chemical Composition

Copolymer	Melting point (°C)	Elongation to break (%)	Modulus (psi)
Random[a]	120	300	7000
Block[b]	170	200	1000

[a]Random copolymer made of 40% ethylene glycol terephthalate and 60% ethylene glycol sebate.
[b]Block copolymer made of the same components in the same percentage.

Table 10.8 Melting Point, Extensibility and Stress Decay for Various Copolyesters

Polymer[a]	Melting point (°C)	Elongation to break (%)	Short-term stress decay (%)
A	120	300	20
B	170	200	13
C	200	350	8

[a]A = random 40:60 polyethylene glycol terephthalate-cosebacate. B = block 40:60 same composition. C = block 40:60 polyethylene glycol terephthalate-copolyethylene oxide.

For a typically elastomeric fiber with high elongation and complete recovery from strain, as used in a support or restraining fabric, such as for instance Lycra, one needs certain molecular segments with *low* modulus and a glass transition temperature (T_g) *below* the use temperature. On the other hand, all important natural and synthetic apparel fibers are characterized by a balance of structural design to give much higher moduli and T_g values *above* normal use temperature.

Among the factors that govern fiber tensile strength, molecular weight is of prime importance. Most linear polymers with T_g above room temperature give high tensile strength if their molecular weight is above 10,000 and if the chains are well oriented parallel to the fiber axes. The 66 nylon of molecular weight 18,000 can yield fibers having well above 100,000 psi tensile strength, and there are many other linear polymers that give similar values at normal temperatures. However, if tensile testing is done at a variety of temperatures or at varying rates of loading, other details of the polymer structure begin to show their influence.

Up to now the behavior of homopolymers, random copolymers, and block copolymers has been used to show structure-property relationships. In block and random copolymers the chemical modification to the polymer is *in the chain.* Other types of copolymers can be made by creating free radical sites along the trunk polymer chain and initiating a vinyl polymerization in the presence of a desired monomer. The products of such an operation are *graft copolymers* in which the polymer backbone is unchanged and the chemical modification has the form of branches. High-energy radiation, chemical free radical initiation, and even mechanical action and heat can be utilized to promote grafting processes. Grafts on preformed fibers may be uniformly distributed throughout the cross section of a fiber or may be confined to surface regions by using knowledge of the diffusion and solubility characteristics of various monomers in the substrate.

By proper selection of the modifying monomer in the light of its own polymer proportions, it is possible to increase or decrease fiber modulus and recovery properties or to impart such properties as adhesivity, water absorption, and increased dye receptivity. In grafting a monomer on a preformed fiber, the crystal structure of the backbone polymer remains essentially unaffected, as the acrylic acid graft on preformed nylon 66 fibers demonstrates. Table 10.9 shows how the balance of some key fiber properties is altered by grafting 20% acrylic acid onto nylon 66 fibers and by converting the graft to the sodium or calcium salt of polyacrylic acid. These properties are compared with those of a random copolymer of 66 and 6 nylon. It can readily be seen that

Table 10.9 Property Comparison of Grafted Fibers Versus the Ungrafted Systems and Random Copolymers (66 Nylon Base)

Fiber	Moisture regain (%), 50% RH	Static propensity (log R)	Fiber stick temperature (°C)	Wet crease recovery (%)	Dye rate
66 nylon	2.5	13.3	240	70	Normal
20% acrylic acid graft on 66 nylon (Na salt)	7.5	8.0	380	94	Rapid
20% acrylic acid on 66 nylon (Ca salt)	5.0	13.0	420	70	Normal
66/6 (80:20)	3.5	13.0	200	65	Rapid

1. A significant increase in moisture regain is obtained with the hydrophilic graft.
2. The static propensity as indicated by the log of the electrical resistance of a fabric is markedly reduced with the sodium salt of the graft.
3. Because of the highly swollen nature of the graft in water, the sodium salt of the graft also shows improved wet crease recovery.
4. The rate at which dyes are absorbed is considerably higher with the graft, as with the random copolymer.
5. The divalent calcium ion produces ionic cross-links in the fiber, yielding a marked increase in fiber stick temperature and melting point.

In contrast, the random copolymer contributes improvement only in dye rate, but at a marked sacrifice in melting point.

The data presented and the examples chosen for illustrating characteristic behavior have been taken primarily from condensation polymers and their fibers, but the principles discussed and the conclusions drawn hold for fibers made from all classes of linear polymers. In vinyls and acrylics steric isomerism is important in governing intermolecular forces. Random, block, or graft copolymerization affects them in a manner analogous to condensation polymers; chain stiffening is preferably carried out with the aid of side groups, since it is difficult to insert rings in the backbone chain of addition polymers.

f. Molecular Engineering of Polymers for Building Construction

There are many indications that the conventional wisdom that fields for the application of organic polymers are gradually being amply supplied with various materials that compete with each other and do not offer any good prospect for substantial further expansion is erroneous. Besides textiles, packaging materials, rubbers, plastic coating, moldings, casting, and adhesives, which together represent a very large volume of synthetic polymer consumption (more than 34×10^9 lb plastics and more than 5.5×10^9 lb synthetic rubber for the United States alone in 1979), other large areas of industrial activity exist that would be able to absorb tremendous quantities of organic polymers if the properties and costs could be so adjusted that their systematic and large-scale application would present a technical and economic advantage for these industries. These are transportation (surface vehicles, boats, and aircraft) and the building industry. At present the transportation design usage is obvious; the building usage will be discussed here.

The construction of buildings of all kinds—small and large homes, office buildings, factories, and laboratories—could utilize very large quantities of many types of organic polymers if certain properties or (even better) certain combinations of properties could be conveniently built into these materials.

In order to get a rough estimate of the quantities that are involved, it should be mentioned that the house and home building activities in the United States in 1979 amounted to a total investment of about $100 billion, of which about 20% were spent for construction materials, such as metals, concrete, bricks, glass, and wood. If plastics are to be used in the building trade, they can serve as *structural* (load-bearing) units or as *finishing* materials that service a building with water, air, gas, and electricity and provide for the finishing of floors, ceilings, and walls. Until now, organic polymers have been used mainly in the finishing sector, as pipe, floor tiles, coatings, insulating foams, window and door frames, paneling, and roofing. About 20% of all plastics and coatings are already being used in these capacities, and it is probable that this volume will keep on expanding during the next years. The more important expansion, however, can be seen in the use of organic polymers as structural units, provided that certain improvements in properties can be effected.

Table 10.10 lists the most important demands for a structural polymer, states the present position of the art, and indicates the probable future outlook.

It is obvious that a successful invasion of this field cannot be brought about by the development of *one single* property—for instance, tensile strength—to an extravagantly high value, but that one must aim at a combination of *several properties* that make a polymeric material particularly valuable and attractive for a special use. For a rubber tube one will require softness, flexibility, elasticity, abrasion resistance, impermeability to air and resistance against temperatures up to about 150°C. On the other hand, polyvinyl chloride tubing has successfully entered the cold water and soil pipe fields.

For use as structural units in the building industry one may characterize a favorable compromise of properties as follows.

1. High modulus of rigidity; if possible higher than 700,000 psi.
2. High softening or melting point; if possible higher than 500°C.
3. High tensile strength; if possible higher than 100,000 psi.
4. High elongation to break; if possible above 10%. Favorable values of (3) and (4) give a large requirement of energy to break or tear a piece of the material. This, in turn, manifests itself as a high impact strength and a high abrasion resistance.
5. High resistance to the action of solvents and swelling agents, even at elevated temperatures.
6. High resistance to deterioration through heat, radiation, agressive chemical reagents, and biological agents.

Numerous profitable applications will appear for any material with a low specific gravity and low cost if a favorable combination of the preceding properties can be incorporated into it.

Table 10.10 Key Properties of Polymers for Building Construction

Demanded property	Materials available at present	Outlook
Reasonable price	Improvement needed	Fair
Light weight	Excellent	Fair
Structural strength	In general, satisfactory	Good
Fire retardant	Needs improvement	Good
Long lifetime	Almost satisfactory	Very good
Easily transported	Excellent	Very good
Easily installed	Satisfactory	Very good
Easily repaired	Satisfactory	Very good
Corrosion resistant	Excellent	Very good
Moisture resistant	Satisfactory	Very good
Solvent resistant	Excellent	Very good

It will be useful, then, to recapitulate very briefly the different principles by which well-balanced compromises of these properties have already been achieved. Let us also explore the possibilities for superior combinations. In a general and simplified sense, there are three main principles that have been very useful in the past and that should be good working hypotheses for future efforts:

1. Crystallization
2. Cross-linking
3. The use of inflexible chain molecules

Let us therefore estimate the value of each of these principles for the present purpose and, at the end, consider how combinations of them could be used to arrive at improved products.

The principle of crystallization has already been discussed in some detail; it has long been known to be a very valuable property of linear, flexible macromolecules whenever one wants a good combination of thermal and mechanical properties.

It has long been known that a combination of the favorable properties enumerated in Table 10.10 can also be obtained in an entirely different manner, namely, by the *chemical cross-linking* of long flexible-chain molecules. Rubber is a convenient example. As one reduces the original segmental mobility of the individual chains during vulcanization by the establishment of localized but strong carbon-sulfur and sulfur-sulfur cross-links, the material becomes more rigid, higher softening, and less soluble. If one continues to introduce more and more cross-links, their average distance along the flexible chains decreases, the system is progressively stiffened, and one finally winds up with hard rubber or Ebonite, a material that is very rigid, has an extremely high softening range, and is completely insoluble and unswellable.

It should be added here that an effect similar to that of cross-linking can also be obtained by the incorporation of a *reinforcing filler* into a polymer. The word *reinforcing* refers to the fact that there is a very intimate contact between the filler and the chains of the polymer in molecular dimensions and that there exist strong adsorption forces that fix and immobilize the polymer chains at the surface of the filler particles.

Although, at first glance, the effect of cross-linking is very similar to that of crystallization, there are several important differences.

1. In a crystalline system the rigidity is the result of many, regularly spaced lateral bonds between the oriented chains; each of these bonds is weak, and the ultimate effect comes from their *large number* and their *regularity.* In a cross-linked system the bonds between the long flexible chains are *strictly localized,* and each of them is *strong;* in their entirety they are *randomly arranged* in the system. As a result crystallization is a *reversible* phenomenon, whereas cross-linking is *irreversible* and is the preferred technique for the production of *thermoset resins.*
2. Crystallization is a *physical* effect that takes place at all temperatures and is strongly influenced by physical processes, such as orientation and swelling; cross-linking, on the other hand, is a *chemical* phenomenon that needs the presence of certain special reagents, is strongly accelerated by elevated temperatures, but is not very much influenced by orientation or swelling.

Many important products that are hard, infusible, and insoluble are made with the aid of cross-linking. Although there are obvious and significant differences between the ways in which crystallization and cross-linking act, it is useful to have the two different principles with which to attain a desirable combination of valuable properties. Even more encouraging is the fact that there exists a third independent principle, namely, the use of stiff linear-chain molecules.

Crystallization and cross-linking produce stiffness, high softening temperature, and difficult solubility by the establishment of *firm lateral* connections between intrinsically *flexible chains;* one can, however, also attain the same result by incorporating the stiffness in the *individual chains* and constructing them in such a manner that their segmental motion is intrinsically restricted; the hardening effect of bulky substituents has already been mentioned for polystyrene, which is amorphous, has no cross-links, but is still a hard, relatively high softening (90°C) polymer. The absence of crystallinity results in a complete transparency; the absence of cross-linking results in a reversible moldability and easy flow characteristics. Similar conditions prevail in the case of polymethyl methacrylate, which is linear, amorphous, and has intrinsically flexible backbone chains that, however, are stiffened by the two substituents (CH_3- and $-COOCH_3$) at every other carbon atom and, as a result, is a hard, brilliantly transparent, relatively high softening (95°C) thermoplastic polymer that has found many useful and valuable applications. The only weakness of both materials is their low resistance to fire, swelling, and dissolution. It seems that the bulky and eventually polar substituents are capable of producing favorable mechanical and thermal properties connected with the overall mobility and flexibility of the chain segments but cannot offer sufficient resistance to the penetration of the system by solvents or swelling agents, because this process is a strictly localized phenomenon and depends on the affinity of the substituents for the particular solvent molecules.

Similar effects are produced if the backbone chains themselves are rigid and if their substituents are so arranged that crystallization is prevented. Classic examples of such polymers are cellulose acetate and cellulose nitrate. In fact, both materials are hard, transparent, high-melting, and amorphous *thermoplastic resins* that have been widely and successfully used for many years.

Recently, the principle of using intrinsically rigid chains has been studied in more detail and has found several new and interesting embodiments. Based on rigid monomeric units, such as

Bisphenol

Tetraminodiphenyl

Terephthalic acid

Methylenebis[phenylisocyanate]

Paraphenylene diamine H_2N—⟨ ⟩—NH_2

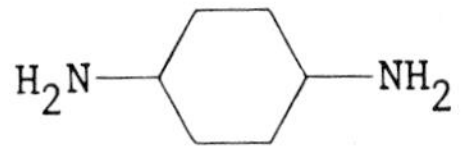

Pyromellitic anhydride

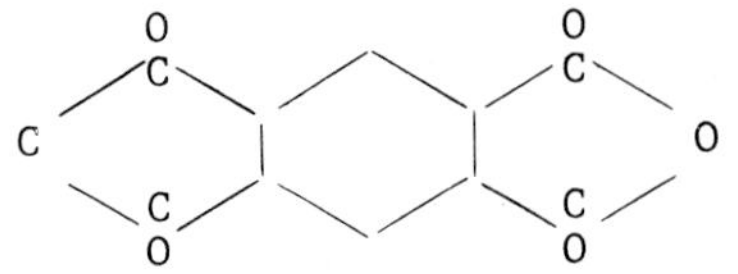

and others, a series of polymers have been synthesized that are substantially amorphous and not cross-linked, but that represent very hard, high-softening, and solvent-resistant materials. Earlier examples of this type are the polycarbonates and the linear epoxy resins, both of which are based on bisphenol; more recent representatives are the polybenzimidazoles, polyimides, and polyphenyl oxazoles, which exhibit unusually high resistance to softening, swelling, and decomposition. In fact, some of these newer materials can withstand temperatures up to 500°C for long periods without softening and deterioration and are completely insoluble in all ordinary organic solvents up to 300°C.

Other rigid molecules now being studied for possible application in the field of high-temperature resistant materials are based on other aromatic chains, such as polyphenylene,

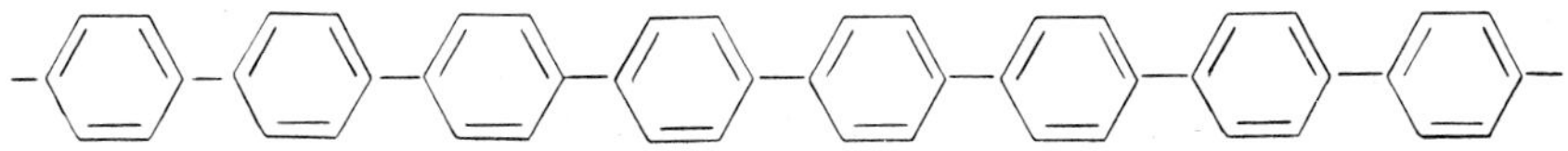

which cannot fold even at rather high temperatures, because rotation about the carbon-carbon single bond between the paralinked phenylene rings can only lead to different angles between the planes of consecutive rings, but not to a kink or bend in the main chain. In fact, representatives of this species are very rigid, very high melting, possess a pronounced tendency to crystallize, and are highly insoluble. This combination of valuable properties has not yet been fully brought to fruition because the presently known polyphenylenes are in a relatively low-molecular-weight range.

Parapolyphenylene oxide and polyphenylene sulfide are other cases that show the chain-stiffening action of a paraphenylene unit.

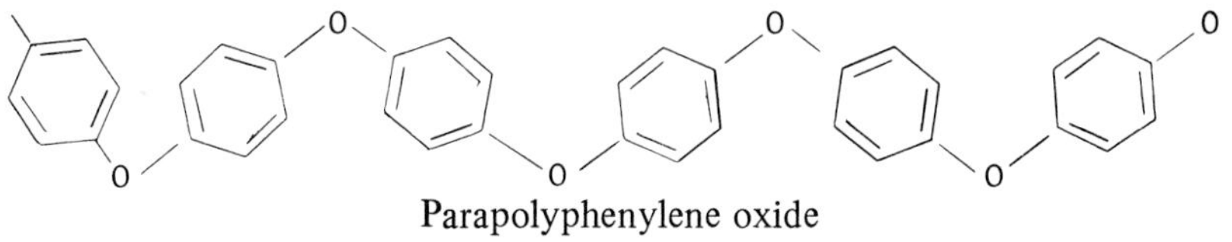

Parapolyphenylene oxide

Rotation about the bonds between an ether oxygen atom and the adjacent carbon atoms of the rings does change the plane of the angle between two units and thus leads to bends and kinks in the chain, but the rotational freedom is noticeably inhibited by the presence of the aromatic rings on each side of the oxygen or sulfur atom. As a consequence, it has been found that chains of this type represent high-melting, rigid, and relatively insoluble materials.

Another interesting way to arrive at chains made up of condensed rings is the synthesis of so-called ladder polymers. The first case of such a structure was prepared by heating polyacrylonitrile to an elevated temperature. This causes the formation of rows of six rings by an electron pair displacement:

which involves stiffening, insolubility, and discoloration. Further heating leads to evolution of H_2 and to aromatization

whereby a black, completely infusible and insoluble material is obtained. This, in its structure, corresponds to a linear graphite in which one carbon atom of every ring has been replaced by nitrogen.

These examples clearly show that there are many possibilities for the formation of long stiff chains and that, in all cases, the properties of the resulting materials confirm expectations.

The existence of *three different* and *independent* ways to establish favorable compromises of valuable properties stimulates the attempt to explore *combinations* of these principles and to see whether such combination might lead to even better results. To conveniently survey these combinations, let us consider a triangle (Fig. 10.21) in which the three principles of crystallization (A), cross-linking (B), and chain stiffening (C) are represented by the three corners A, B, and C.

Corner A is populated by a large number of crystallizable, thermoplastic polymers with flexible chains that have proved to be particularly successful as fiber and film

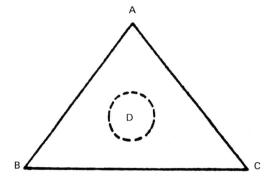

Fig. 10.21   Illustration of three principles that influence polymer properties. A represents the principle of crystallization. B represents the principle of cross-linking. C represents the principle of chain stiffening.

formers. Representative materials are polyethylene, polypropylene, polyoxymethylene, polyvinyl alcohol, polyvinyl chloride, polyvinylidene chloride, and such polyamides as nylon 6 and 66.

In corner B are located the typical thermoset, highly cross-linked systems, such as hard rubbers, urea, melamine, and phenolformaldehyde condensates, highly reticulated polyesters, polyepoxides, and polyurethanes.

Finally, corner C is representative of the amorphous, thermoplastic resins with relatively high rigidity and high softening range, such as polystyrene, polymethyl methacrylate, ABS resins, polystyrene derivatives, and, more recently, polycarbonates, linear polyepoxides, polyethers, and polycondensation products with inflexible chains.

One can now pose the question whether a combination of *two* principles can help to arrive at products with more attractive and valuable properties. Let us therefore explore the *sides* of the triangle in Fig. 10.21.

In fact, the line from A to C accommodates several interesting fiber and film formers. One of them is polyethylene glycol terephthalate (Terylene or Dacron), in which the paraphenylenic units of the acid introduce enough chain stiffening to bring the melting point of this polymer up to about 260°C. This is as high as the melting point of nylon 66, even though the polyester has *no lateral hydrogen bonding* available to stiffen its solid crystalline phase. Thus chain stiffening coacts with crystallization to enhance properties without bringing either of these two principles to an extremely high value. Cellulose is another polymer in which excellent fiber and film-forming properties are developed by a combination of chain stiffening and crystallinity; these two effects together produce a polymer that is extremely rigid, does not melt at all, and is soluble only in a very small number of particularly potent liquid systems. The presence of substantially rigid chains has the favorable consequence that high tensile strength and elevated temperature softening characteristics become apparent even at relatively low degrees of crystallinity; this places cellulose somewhere in the middle of the line connecting A and C. Many of the spectacular improvements of cellulosic filaments were founded on taking advantage of the dual origin of its fiber-forming potential. Another example of a beneficial combination of A and C is cellulose triacetate, in which the capacity to crystallize superimposes several favorable properties on the normal cellulose acetate—particularly insolubility in many organic liquids and heat settability through additional crystallization.

All those rubbers that are slightly or moderately cross-linked and that crystallize on stretching progressively are situated on the side A-B; examples are natural rubber, high *cis*-polybutadiene and polyisoprene, butyl rubber, and neoprene. Depending upon the degree of cross-linking, they are more or less close to B.

Until now we have considered systems of *one component,* that is, a specific polymer or copolymer. However, in technology it is customary to produce stiffening and temperature resistance by the addition of a hard, finely divided *solid filler,* such as carbon black, calcium carbonate, silica, or alumina. Such addition of a crystalline or pseudo-crystalline reinforcing filler is a kind of crystallization of the system, in the sense that the flexible chains of the original polymeric matrix are restricted in their segmental mobility by the presence of the very small and hard particles of the filler, to the surface of which they are attached by strong adsorptive forces. Thus the presence of a reinforcing filler simulates and replaced crystallization and brings the system closer to point A.

The line from C to B has been populated with certain useful polymers in an attempt to increase rigidity, high softening, and insolubility of stiff chain systems by additional cross-linking. There are several well-known cases for the combination of these two principles, for example, the raising of the heat distortion point of acrylic and methacrylic polymers by the incorporation of allyl methacrylate or ethylene glycol dimethacrylate, and the "curing" of epoxy polymers based on such stiff chain elements as bisphenol and cyclic acetals of pentaerythritol. Currently, the properties of the more recent amorphous systems have been improved with intrinsically stiff chains by cautious cross-linking, and these materials have been moved from point C somewhat toward point B. The principal reason for this approach is the improvement, for such systems, of resistance to dissolution and swelling at elevated temperatures.

The advantageous combination of two principles poses the question whether a proper combination of *all three of them* could lead to still further improvements in capabilities. Much exploratory work has been done in this field; in Chap. 11 certain interesting results will be described. One successful application of all three principles is the aftertreatment of cotton with certain cross-linking agents or the spinning of rayon in the presence of such agents. Cellulose has rigid chains that can be brought to a moderate degree of stiffness by the swelling of cotton or the appropriate spinning of rayon; the results are fibers of satisfactory strength, elongation, hand, and dyeing characteristics, but of insufficient recovery power. The introduction of a cautiously controlled system of cross-links with the aid of bifunctional reagents leaves all other desirable properties unchanged and improves substantially the recovery power and wrinkle resistance. Similarly promising results have been obtained with mildly reticulated amorphous stiff-chain systems of the epoxy and urethane type in which crystallization has been replaced by a reinforcing filler.

Returning to Fig. 10.21, it can be seen that combinations of all three principles are situated somewhere in the interior of the triangle around D. A thorough and systematic exploration of this area will lead to many new and interesting polymeric systems with properties superior to those which are at our disposal today.

8. Examples of Synthetic Resins

a. *Vinyl Polymers*

This important group of synthetic high polymers is obtained by a process known as addition polymerization. Vinyl monomers are usually simple unsaturated molecules having the general structure, $CH_2=CHX$, where X may be any of a large number of atoms or groups (see Table 10.11A). The first step in the polymerization process (initiation) consists in activating the vinyl monomer, by addition of energy, or a free radical. or an ionic catalyst such as $AlCl_3$ or BF_3.

One or more of three reactions can occur:

$$CH_2=CHX \xrightarrow{\text{energy}} CH_2=CHX*$$

(a transitory, energy-rich intermediate),

$$\text{or} \quad CH_2=CHX + R \cdot \rightarrow RCH_2CHX \cdot$$

(a free organic radical intermediate)

$$\text{or} \quad CH_2=CHX + AlCl_2 \text{ or } BF_3 \rightarrow MeCl_3CHXCH_2^{+}$$

(an active ionic intermediate).

Table 10.11A Examples of Vinyl Polymers

Polymer	X	Type
Polyethylene	H	Hydrocarbon
Polyvinyl chloride	Cl	Alkyl halide
Polyvinyl acetate	$O-\overset{\displaystyle O}{\overset{\|}{C}}-CH_3$	Ester
Polystyrene	C_6H_5	Hydrocarbon
Polymethyl acrylate	$\overset{\displaystyle O}{\overset{\|}{C}}-OCH_3$	Ester
Polyacrylonitrile	$C\equiv N$	Nitrile
Polyvinyl carbazole		Aromatic amine

Occasionally, more than one hydrogen of the ethylene monomer is substituted by an "X" group, as in the following:

Polyisobutylene	$\left[-CH_2-\underset{\underset{\displaystyle CH_3}{\|}}{\overset{\overset{\displaystyle CH_3}{\|}}{C}}-\right]_n$	Hydrocarbon
Polyvinylidene chloride	$\left[-CH_2-\underset{\underset{\displaystyle Cl}{\|}}{\overset{\overset{\displaystyle Cl}{\|}}{C}}-\right]_n$	Alkyl halide
Polytetrafluoroethylene	$(-CF_2CF_2-)_n$	Alkyl halide
Polychlorotrifluoroethylene	$(-CFClCF_2-)_n$	Alkyl halide

In the second step (*growth*), many units of the monomer combine to propagate a growing chain. This is formulated as follows for the free radical mechanism (similar formulations may be made for the other mechanisms):

$$RCH_2CHX \cdot + n(CH_2=CHX) \rightarrow RCH_2CHX(CH_2CHX)_n \cdot$$

The final step (*termination*) involves elimination of the active center by abstraction of energy, removal of ionic charge, or mutual destruction of free radicals in pairs, by reaction with one another, or through some formally similar mechanism.

$$RHC_2CHX(CH_2CHX)_n \cdot + RCH_2CHX(CH_2CHX)_m \cdot \rightarrow$$

$$RCH_2CHX(CH_2CHX)_n(CHXCH_2)_mCHXCH_2R$$

Important examples of vinyl polymers are listed in Table 10.11A.

b. Uses of the Vinyl Polymers

The vinyl polymers are exceedingly versatile materials and accordingly find numerous uses in industry and in the domestic economy. Some of the principal uses of these important polymers are listed in Table 10.11B.

Table 10.11B Some Important Vinyl Polymers and Their Uses

Polymer	Outstanding Properties	Applications
Polyethylene	Water resistance Toughness Low dipole movement Low electrical loss factor	Electrical insulation Coaxial cable Packaging Moisture-proofing Coating ice-cube trays
Polyisobutylene	Flexibility Moisture resistance Elasticity	"Butyl" rubber Bullet-proof tanks Pour-point depressors Cable coatings
Polyvinyl chloride	Plasticizability Low flammability Flexibility Toughness	Cable jackets Lead-wire insulation Rubber substitute Fabric coating
Polyvinyl ethers	Pressure sensitivity Solubility	Pressure-sensitive tapes
Polyvinylidene chloride	Solvent resistance Acid resistance Toughness Nonflammability	Woven fabrics Seat covers and upholstery Acid-resistant tubing Belts and gaskets
Polystyrene	High resistivity Low dielectric constant Moisture resistance Chemical resistance	Electrical insulation Radar components Lenses Instrument panels Refrigerator-cabinet components
Acrylic polymers	Transparency Clarity Conformance to shape	Lenses Airplane covers and shields Costume jewelry
Polyvinyl carbazole	Acid and alkali resistance Low power factor Thermal resistance	Capacitor material Electronic parts Mica bonding

c. Copolymers

Frequently, two or more monomers are caused to polymerize together. Usually a random interpolymer (copolymer) is formed. The typical structure of a copolymer is as follows:

$$-CH_2CHXCH_2CHYCH_2CHXCH_2CHXCH_2CHY-$$

but in special cases the monomer units may alternate regularly, as in

$$-CH_2CHXCH_2CHYCH_2CHXCH_2CHY-$$

By choosing the proper components, proportions, and conditions of polymerization, optimum properties may be achieved for the systems involved. Thus, with vinyl chloride and vinyl acetate, the copolymers (Vinylite) are more versatile and more useful than either of the homopolymers or physical mixtures of the two.

d. Condensation Polymer

This class of synthetic high polymers is formed by a reaction that may be formally represented as the repetitive elimination of small molecules from two or more reacting species, as follows:

$$n(X-R-X) + n(Y-R'-Y) \rightarrow (-R-R'-)_n + 2n(XY)$$

If at least one of the reacting species contains more than two reactive groups (*functionality* greater than 2), the resulting chain molecule can grow in more than two directions. This results in a three dimensional or cross-linked structure. Glyceryl phthalate is a common example. The reaction proceeds in steps, schematically, as shown in Fig. 10.22.

As the reaction continues, polymer growth occurs, both linearly and by crosslinking. The growing molecules finally reach a size and complexity such that they can effectively react no further. This is the gelation point and is frequently associated with abrupt changes in viscosity as well as with rapid changes in molecular weight and complexity.

Technically it is important to carry out polycondensation reactions so that the gel point is reached at a predetermined and controllable time, for example, when the polymer is about to be ejected from a compression mold, injection mold, or extruder. A number of important condensation polymers are described in Chap. 11.

(2) Cellulose plastics Raw cellulose is practically useless as a plastic, except as a filler in the form of wood flour, cotton flock, or rag stock. Regenerated cellulose, however, which is prepared by solution and controlled reprecipitation of purified cotton, is exceedingly useful as a packaging material (cellophane). Under similar conditions, regenerated cellulose can be drawn into fibers (rayon). Cellulose plastics, in the strict sense, are based upon chemically treated cellulose. The reactions involved are principally esterification and etherification; the products, therefore, are cellulose esters and ethers. Prominent examples are cellulose nitrate (nitrocellulose), cellulose acetate, cellulose propionate, cellulose acetate-butyrate, methyl cellulose (Methocel), and ethyl cellulose. The reagents used attack the cellulose molecules at the free hydroxy groups present on the polysaccharide rings. Cellulose plastics are useful as safety films, surface coatings, hot-melt compositions, dispersing and emulsifying agents, transparent sheet, molding powders, aircraft enclosures, dielectric film and foil, imitation ivory, artificial leather, and so on.

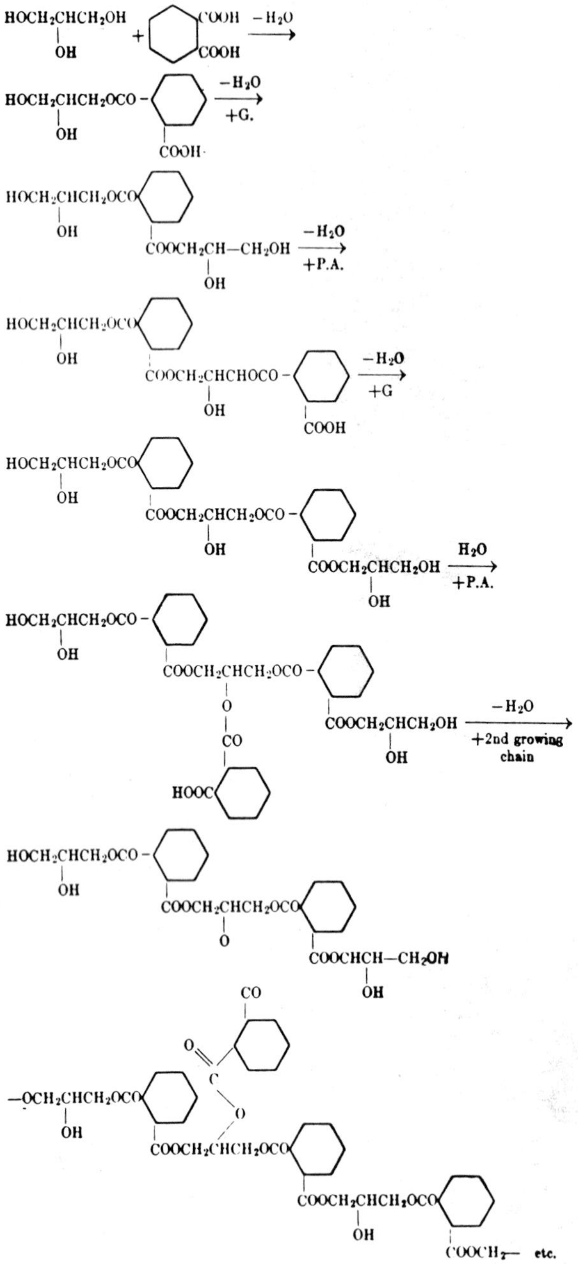

Fig. 10.22 Schematic representation of cross polymerization of glyceryl phthalate. G = glycerol. P.A. = phthalic acid.

(3) Silicones The silicones represent a family of synthetic materials where the skeletal framework consists of alternate silicon and oxygen atoms, rather than carbon atoms, as follows.

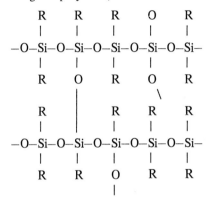

The R groups are organic and are subject to variation almost at will. Thermosetting resins are obtained when a functionality greater than 2 is introduced, just as in the better known organic polymers, as follows.

The silicon-oxygen (Siloxane) skeleton confers the property of unusual heat resistance, especially if the organic R groups are small or in themselves are thermally stable. The silicones find use in the electrical industry as high-temperature insulation, frequently in combination with glass cloth or fibers, mica, or asbestos. The silicones are also available as oils, rubbers, greases, and waxes and as liquids capable of depositing thin, water-repellent films. More detailed discussion of the silicones will be found in Chap. 16.

V. RELATION OF PROPERTIES TO STRUCTURE (SUMMARY)*

The gross properties of solids are determined by the types of bonding that predominate.

Ionic solids have high melting points and low coefficients of expansion. They form conducting ions when melted or dissolved. Mechanically they are hard and durable. Solubility and chemical properties vary over wide limits, depending upon the types of ions present and their arrangements.

Atomic solids, that is, solids that form regular, extensive covalent networks, are also hard and high melting and have low coefficients of expansion and high densities. On melting, or when dissolved, simple molecules are formed rather than ions. With some of these solids, diamonds, for example, neither melting nor solution can be accomplished without some deep-seated chemical change being brought about simultaneously. Atomic solids are usually very resistant chemically.

Molecular lattice solids, where the binding forces are chiefly van der Waals, are comparatively soft and of low mechanical strength, unless highly polymeric. They have

*Contributed by ROBERT S. SHANE.

low melting and boiling points, high coefficients of thermal expansion, and the melt or solution contains discrete molecules.

A. Thermal Properties

From the previous discussion it is apparent that thermal properties are determined by both gross and fine structural characteristics. Solids that form regular, predominantly ionic crystals (for example, ordinary inorganic salts and some silica minerals) have high melting points and low thermal conductivity, compared with metals. Melting is an abrupt physical transition. The latent heat of fusion is generally high, owing to the necessity of overcoming strong interionic forces.

Atomic solids, which form regular, indefinitely extensive, homopolar structures, also are exceptionally stable thermally, for example, diamond, graphite, silicon carbide, boron carbide, and aluminum nitride. These solids also have low thermal conductivity. In solids that can exist as supercooled liquids, the resultant randomness may lead to increased thermal resistance, as in fused silica, fused alumina, and many glasses and ceramics. Colloidal and fibrous varieties of these and similar materials are accordingly sometimes used as thermal insulation.

Organic crystals, where the lattice energy is the result of weak van der Waals forces, are easily melted. Because of the low force fields surrounding individual molecules in the melt, the boiling points are correspondingly low. Among organic compounds of similar structures, the boiling point generally decreases as the symmetry of the molecule increases; the melting point, on the contrary, increases with increasing symmetry.

In most simple compounds, whether ionic, covalent, or metallic, these relations are sharply defined, reversible, and reproducible. This is no longer true for many complex structures. In most linear organic high polymers there is no sharp melting point, but rather a temperature range over which softening and flow occur. This property is, of course, of practical utility in the extrusion and molding of thermoplastic resins. In thermoset resins, however, no appreciable softening of flow occurs below the point of thermal decomposition. Therefore, molding must be accomplished during a critical interval, when the reacting system of concern is still capable of flow and can be converted rapidly and irreversibly into its final infusible condition.

The specific character of these thermal patterns can frequently be varied within relatively wide limits by changes in the intimate structure of the materials concerned. For example, in polyamide resins ("nylon") the melting point is lowered and the softness, flexibility, and solubility in certain solvents are increased as the dibasic acid component of the resin is varied from adipic acid, having 6 carbon atoms, to sebacic acid, which has 10 carbon atoms. The acrylic and methacrylic plastics become softer, lower melting, and more soluble as the length of the carbon chain of the alcohol used for esterification increases. In the silicones, similar phenomena are observed as the length and nature of the organic groups attached to silicon are altered. The longer these organic groups are, and the more internal mobility they are capable of, the more the properties of the silicones depart from those inherent in the inorganic siloxane backbone structure. This backbone, present as a three-dimensional tetracoordinated structure in pure quartz, gives one extreme of properties. More profound alterations occur when one of the four spatial bonds of silicon to oxygen is replaced (hypothetically) by the covalent bond of an organic radical, as in a monoalkyl silicone. These thermosetting resinous compounds are still rather hard, brittle, and high melting compared with most organic high polymers.

The replacement of a second oxygen bridge by an organic radical results in a linear polymer, usually oily or rubbery in nature, and more organic than inorganic in its behavior. A third such hypothetical replacement gives a molecule with only one remaining Si–O bond, and incapable of polymerization; and the final replacement gives a tetraorganosubstituted silane (SiR_4), a completely covalent molecule with properties not strikingly different from those of the analogous tetraorganomethane, CR_4.

B. Mechanical Properties

The following physical and mechanical characteristics are common to most ionic solids and to many symmetrical, close-packed covalent solids: (1) hardness of a relatively high order; (2) noncompressibility; (3) low distortion limits; (4) brittleness; (5) cleavage during fracture along fairly definite planes, related to crystallographic features or to grain boundaries; (6) elastic deformation, but within narrow limits; and (7) high yield values.

1. Deformation and Flow

In plastic materials, whether inorganic (glasses) or organic (high polymers), not only deformation but also flow is encountered when a stress is applied. This is because plastics may be considered, within broadly definable limits, to have the properties both of solids and of liquids.

In an ideal liquid the response to a shearing force is such that the shearing stress is proportional to the rate of shear or the velocity gradient. The yield value is zero. This represents simple or Newtonian flow. An ideal elastic solid, at the other extreme, obeys Hooke's law; that is, the stress is directly proportional to the strain and is completely reversible.

Most plastic solids exhibit elastic deformation if the applied stress is below the yield value. Above the yield value, permanent deformation, that is, flow, occurs. In plastic flow, the deformation is proportional to the time of application of stress and is permanent. The transition from elastic to plastic behavior is usually not sharply defined.

Many liquids, emulsions, suspensions, and colloidal gels also exhibit marked departure from Newtonian flow. In these systems the rate of flow is likewise a function not only of the shearing stress, but also of time and the previous history of the sample. In many systems of this sort (a bentonite-water gel is a good example) pseudo-Newtonian flow is exhibited above the yield point, but even this pattern is usually not entirely reversible. The "yield point" in such colloidal systems is generally associated with the overcoming of some type of physical "structure" (thixotropy). This phenomenon may be viewed as a disentanglement of discrete structures that, below the yield point, literally block one another's paths, and so inhibit flow.

Very dilute solutions of high polymeric organic substances sometimes exhibit true Newtonian flow. More concentrated solutions are always markedly non-Newtonian, and thixotropy is commonly encountered even at moderate concentrations. Presumably, both association of the chain molecules through van der Waals forces and actual physical entanglement of the flexible chains are responsible for these effects.

2. Elasticity

Many thermoplastic high polymers are characterized by a high order of elastic extensibility. This behavior is presumably associated with the ability of such molecules to turn, bend, and uncoil around covalent bond centers. The multiplicity of such bonds in a chain molecule accounts for the remarkable flexibility and extensibility of many such materials.

In natural and synthetic rubbers, the reversible rotation, bending, coiling, and uncoiling of molecular chains around myriad bond centers is realized to an exceptional degree. Elastic extensibility to the extent of 600-800% is encountered. Above the elastic limit, of course, the ordinary laws of mechanics of rigid bodies apply. The stress-strain curve becomes steep, and the break point is usually reached rapidly. X-ray examination indicates that the molecules of rubber and many synthetic elastomers undergo successively greater orientation on stretching. Fully stretched specimens frequently give highly crystalline diffraction patterns. This may be looked upon as due to a "freezing" of the normally loose molecular structure, by physical means, into a regularly oriented, crystalline state. The process is physically analogous to solidification from the melt. Both processes evolve heat—the energy of crystallization of the system—and both lead from a random to an ordered state. Many organic high polymers having low elasticity, on the other hand, are partially crystalline in their normal, unstressed states. Examples are native cellulose, nylon, and polyethylene.

Many plastics, on application of stress, undergo first a small, strictly elastic deformation, followed by a typical "creep" phenomenon, analogous to creep in metals. On removal of the stress, partial, instantaneous recovery takes place, followed by a slower, irregular recovery, again formally analogous to elasticity and anelasticity in creep recovery in metals. This behavior is, of course, evidence that both plastic and elastic effects occur simultaneously in the specimen.

3. Tensile Strength

Approximate values for the tensile strength and for Young's modulus of representative plastics and fibers at room temperature are listed in Table 10.12.

The tensile strength of a high polymer is affected by a number of factors, among which the following are of importance.

1. The polar nature of the molecule. The more polar the chain of molecules, the higher the cohesive forces, which are reflections of van der Waals and dipole interaction forces. Thus, a polar plasticlike cellulose is much stronger than a nonpolar material, such as polyisobutylene.*
2. The average molecular weight. Most physical and mechanical properties of high polymers are additive; that is, they increase regularly with the chain length.
3. The degree of orientation and crystallization. Most polymers are tougher when oriented. Good examples are nylon and polyethylene terephthalate (Dacron or Mylar).
4. The extent to which close packing of chains can occur. This effect is highly pronounced in cellulose derivatives, where variations in strength with the degree of esterification of etherification can be correlated with this phenomenon.

---

*Where hydrogen-containing polar groups, such as OH, NH_2, and COOH, are present on the polymer chain, "hydrogen bonding" plays an important role in determining the net attractive forces. The hydrogen bond is intermediate in strength between normal primary chemical bonds and ordinary secondary valence bonds. The hydrogen bond involves the interaction of the hydrogen on one molecule with a negative atom (usually oxygen) on a second, or, not infrequently, on the same molecule.

5. Plasticization. In general, plasticization decreases the tensile strength markedly, but increases the flexibility. This, of course, is the desired function of the plasticizer. Plasticization is accomplished externally by addition of a compatible liquid or a second, flexible resin. It can also be accomplished internally by attachment of mobile groups at either regular or random positions along the polymer chain. Thus, Butvar is more flexible and less tough than Formvar, and a phenolic resin derived from octyl phenol ($HOC_6H_4C_8H_{17}$) is less brittle than one derived correspondingly from cresol ($HOC_6H_4CH_3$).

4. Cold Flow

Most solids are subject to "cold flow," that is, occurrence of permanent set and relaxation, on long continued subjection to tensile or compression forces. This phenomenon, which again bears considerable resemblance to creep in metals, is particularly pronounced with certain plastics. Cold flow can be observed either by holding a specimen under tension or compression at fixed displacement and observing the decrease in stress with time (relaxation), or by applying a constant load and observing the gradual increase of deformation with time (flow). In general, the creep relations will be found to be similar to those encountered in metals but to be many times greater in actual magnitude.

With most natural and synthetic rubbers, creep and relaxation are considerable even for fairly highly vulcanized varieties. The measurement of compression set of such elastomers is standard in control laboratories. Low "cold flow" may even be important at higher temperatures. Thus, in some applications of silicone rubber, where high compression stresses must be withstood, the extent to which permanent set occurs is important at temperatures far above those to which ordinary rubber can be submitted. "Cold flow" of such rubbers may therefore be measured at temperatures as high as 150-175°C.

"Cut through" is another manifestation of cold flow. In an electrical motor, where the windings may be under tension, cutting stresses may be applied to insulating components of the system. Resistance to such cutting forces and the ability to "relax" or undergo stress relief without rupture of the insulation is obviously important.

C. Chemical Properties

Enough has been said previously to indicate that the chemical properties of a nonmetallic material are intimately related to its chemical structure and can frequently be predicted on that basis. For example, certain water-soluble ionic salts are useful as windows for optical measuring instruments, but such salts will certainly be subject to attack by a variety of chemical reagents. Among inorganic materials that are useful from a structural point of view are those traditional minerals, like marble, whose mechanical properties especially recommend them. However, where acid and alkali resistance is demanded, as in chemical stoneware, inert, refractory oxides, nitrides, or silicates in carefully selected combinations are required (see Chap. 14).

With organic materials, these relationships are more complex and more predictable. In general, the presence of "polar" substituents in an organic substance enhances its chemical reactivity. Thus, the unsubstituted hydrocarbons are more stable chemically than most of their substitution products, such as the alcohols, aldehydes, ethers, halogen compounds, acids, esters, and amines. The presence of such polar groups also enhances water susceptibility. Thus, polyethylene, an unsubstituted hydrocarbon, is highly resistant to moisture. Polyvinyl alcohol, on the contrary, is readily attacked by water, as is

Table 10.12 Approximate Mean Values of Tensile Strength, Flexural Strength, and Young's Modulus of Representative Plastics and Fibers at Room Temperature

Plastic fiber	Tensile strength (psi $\times$ 10^3), ASTM D638	Flexural strength (psi $\times$ 10^3)	Young's modulus (psi $\times$ 10^5)
Polyethylene	1-3	2-6.5	0.21-.27
Polytetrafluoroethylene	2.5-6.5	1.6	
Cellulose acetate butyrate	2.6-6.9	1.8-9.3	
Cellulose acetate	2.3-8.1	2.2-11.5	
Ethyl cellulose	2-8	4-12	
Polyvinyl chloride	6-9	8-15	3.5-4.0
Polyvinylidene chloride	7.5-9	4-7	0.4-0.8
Polyether sulfone	12.2	18.6	
Polybutadiene	.2-1	elastic	
Polystyrene	6-8.1	9-15	4.0-4.7
Polyvinyl chloride-acetate	6-7.5	10-16	3-5
Polyallyldiglycol carbonate	5-6	6-13	2.5-3.3
Polyphenylene oxide	11	15	3.6-4.0

Material			
Polyvinyl butyral	0.5-3	10	3.5-4
Polyacetal	8.8-10	13-14	4.1
Polycarbonate	8-9.5	11-13	3.2-3.5
Polyvinyl formal	10-12		3.5-6
Polyester	6-13	8.5-23	3.0-6.4
Polyepoxy	4-13	8-21	.4-1.5
Polymethyl methacrylate	8.7-11	14-17	3.5-4.5
Phenol formaldehyde	5-9	11-17	4-5
Polyimide	10.5-14	15-17	4.6
Nylon 6/6	9.5-12.4	14.6	1.5-4.1
Hard rubber	4-10		2.0-3.5
Glass-bonded mica	6	70-115	
Viscose rayon	28-47		
Native cotton	42-125		
Silk	45-85	12-18	
Polyester	150-600		
Nylon fiber	300-600		
Glass fiber	200-210	70-90	
Aramid fiber (Kevlar)	410		
Graphite fiber	375-500	9-12	
Sintered α-silicon carbide	54		600

polyvinyl acetate. Polyacrylic acid, a highly polar material, is practically impossible to separate from the water with which it is normally associated. On the other hand, the presence of a halogen in an organic polymer, as in polyvinyl chloride, polyvinylidene chloride, polytetrafluoroethylene, and polychlorotrifluoroethylene does not materially increase the water susceptibility.

The susceptibility of an organic molecule to oxidation (for example, by atmospheric oxygen or by specific chemical reagents) is also frequently predictable on the basis of structure. A good rule of thumb states that compounds that are already highly oxidized will resist further attack by oxidizing agents. Aromatic polynitro compounds are an example. Polymeric ketones and acids are others. On the other hand, the presence of unsaturated centers markedly increases oxidative sensitivity. Natural rubber, which has one double bond per chain unit, is more sensitive to oxidation than butyl rubber, which contains only a small amount of unsaturation, or polyisobutylene, which contains essentially none.

Polysulfide resins, which contain sulfur in a low state of oxidation, are easily attacked by oxidizing agents but have considerable moisture resistance. Polysulfones, on the contrary, are relatively resistant to oxidizing agents, but being more polar, are generally sensitive to hydrolysis.

Molecular weight relations, geometry, orientation, and packing also markedly affect chemical resistance. Specialized textbooks should be consulted for details of these and similar relationships.

D. Electrical Properties*

The possibility of interaction between electrical energy and matter stems from the fact that all matter is fundamentally a collection of electrons and positively charged atomic nuclei normally occupying equilibrium positions with respect to one another. A molecule in its normal state is neutral electrically only in the sense that there is an exact balance in magnitude between positive and negative charges. In a completely symmetrical molecule, the charges are also in geometric or spatial balance; that is, the center of gravity of negative charges coincides with that of positive charges. In a stricter sense, this is exactly true only when the molecule is isolated from any disturbing electrical influence, such as an external field or a nearby second molecule. The monatomic rare gases come closest to electrical neutrality under normal conditions. Other symmetrical gases, at low pressures, also approximate to this condition. In general, however, the centers of gravity of positive and negative charges do not coincide. Electrically unbalanced molecules are said to contain permanent dipoles, and to have permanent dipole moments. The presence of unbalanced charges, whether resulting from permanent dipoles or induced by neighboring molecules, leads to the existence of stray force fields, already described as secondary valence forces.

The magnitude of a molecular dipole moment can be obtained from measurements of the dielectric constant, which can be looked upon as a measure of the ability of a substance to oppose an electric field, which it does either by linear displacement of electrons or by orientation of permanent dipoles.

It has already been suggested that the unique properties of metals are explained by the exceptional mobility of valence electrons. A loose definition of a "dielectric" is

---

*Insulating materials and insulating behavior are discussed in greater detail in Chap. 13.

a substance that does not show metallic conduction.* Most organic substances and many non-saltlike inorganic substances are therefore dielectrics.

When a nonconductor comes under the influence of an external field, the displacements the electrons undergo are defined mathematically by the Clausius-Mosotti equation:

$$P = \frac{\epsilon - 1}{\epsilon + 2} \frac{M}{d}$$

$$= \frac{4}{3} \pi N \alpha_0$$

where

P = polarizability per mole of dielectric
ϵ = dielectric constant
M = molecular weight
d = density
N = Avogadro's number = 6.06×10^{23}
α_0 = ideal or intrinsic polarizability of the molecule

1. Polar and Nonpolar Molecules

Inherent in the Clausius-Mosotti expression is the assumption that a molecule has an electrical moment induced in it by any external field. If the molecule is symmetrical, the moment disappears when the field is removed. Such molecules are nonpolar, for example, the rare gases (helium, argon, neon, and so on), symmetrical diatomic gases, such as hydrogen, nitrogen, and oxygen, and certain atomic lattice solids, such as diamond and silicon. However, most organic molecules and many inorganic molecules contain permanent electrical dipoles.† Such molecules, which possess permanent dipole moments, are called polar molecules. For polar molecules the expression for the polarizability contains a second term, as follows:

$$P = \frac{\epsilon - 1}{\epsilon + 2} \frac{M}{d}$$

$$= \frac{4}{3} \pi N \left(\alpha_0 + \frac{\mu^2}{3kT} \right)$$

where

μ = permanent electric moment
k = gas constant per molecule = 1.372×10^{-16}
T = absolute temperature K
α_0 = molecular polarizability due to an external field
μ^2/kT = molecular polarizability due to the internal charge displacement

The latter type of polarizability is temperature dependent; the former is temperature independent. If the dielectric constant is determined at a series of temperatures, the per-

---

*The consideration of semiconductors is found in Chap. 6.
†It may be helpful to think of the dipoles as internal charge displacements similar to an electrical doublet.

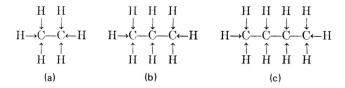

Fig. 10.23 Schematic representation of vector cancellation of dipoles in (a) ethane, (b) propane, and (c) *n*-butane.

manent dipole moment can be calculated, since the temperature-independent part then can be subtracted out.

Polar molecules are too numerous to list or even to classify. A few common examples are the hydrogen halides (HF, HCl, HBr, and HI), water (H_2O), carbon monoxide (CO), and unsymmetrical halogenated methanes, such as $CHCl_3$, CH_3Cl, CH_3Br, CH_3I, and $CHBr_3$.

Since the moment of a homopolar molecule may be considered to be the vector sum of all the individual bond moments,* it is often possible to calculate the dipole moment of a molecule from its molecular structure. The values for the bond moments must of course first be known from the study of other molecules.

A diatomic molecule can be completely nonpolar only when the two atoms are identical; otherwise the centers of opposite charge will not coincide, as in CO, HCl, and ICl, all of which are polar. Triatomic molecules of the general structure B-A-B are non-polar only in the rare case where their atoms lie on a straight line with the two B-A distances equal (e.g., CO_2, which may be written $O \leftarrow C \rightarrow O$). More complex molecules are nonpolar only when special structural effects operate to cancel out all bond moments. In tetramethylmethane, $(CH_3)_4C$, the four methyl groups are located at tetrahedral angles about the central carbon atom, and the C–H moments accordingly all cancel out vectorially. Similarly, carbon tetrachloride, CCl_4, has zero moment because the four tetrahedrally arranged C–Cl moments exactly cancel. Benzene (C_6H_6) is nonpolar because it exists as a plane hexagonal ring with a center of symmetry. The six C–H bond moments cancel one another. On replacing a H atom in benzene with any other atom or group, such as in C_6H_5Cl, C_6H_5OH, $C_6H_5NO_2$, or even in C_6H_5D (where D = deuter-ium, the hydrogen isotope of mass 2), the molecule acquires a permanent **moment** whose magnitude varies with the nature of the group introduced. The normal, straight-chain, paraffin hydrocarbons, $CH_3(CH_2)_nCH_3$, usually have zero moment. The vector cancellation of dipoles is illustrated in Fig. 10.23 for ethane, propane, and *n*-butane. At sufficiently low temperatures, many solids consisting of dipolar molecules show nearly zero moments, because the dipoles are "frozen," that is, they cannot rotate fast enough to be affected by a field of ordinary magnitude.

Ionic crystals, which are clearly polar, show correspondingly high polarizabilities, and when melted frequently become conducting.

It can be shown that forces due to primary chemical bonds, whether ionic or co-valent, act only over very short distances, usually of the order of magnitude of atomic

---

*A bond moment is the electrical moment resulting from the union of two dissimilar atoms or groups and is considered to be a property of the particular bond, more or less independent of other bonds present in the molecule.

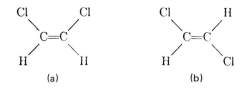

Fig. 10.24 Schematic representation of isomers in dichloroethylene, $C_2H_2Cl_2$. (a) *cis*-Dichloroethylene, which has a moment of 1.9 Debye units; and (b) *trans*-dichloroethylene, which has zero moment.

diameters (1-2 Å). van der Waals forces also are effective over relatively short distances (approximately 4-5 Å). The forces that result from dipole-dipole interaction are of relatively longer range. One result of this is that the total energy of such a system, at a given temperature, may depend upon its shape, as well as its intimate molecular structure.

The magnitudes of ordinary molecular dipole moments are of the order of 10^{-18} cgs units, corresponding to one electrostatic unit of charge displaced by 1 Å of distance $= 10^{-10} \times 10^{-18}$ esu $\times$ cm $= 1$ Debye unit.

2. Dipole Moment and Chemical Structure

The measurement of dipole moment provides a method for determining the structure of many molecules. As an example, it was early realized that the water molecule cannot be linear (H–O–H) but must rather be triangular $\left({}_H{\diagdown}^{O}{\diagup}_H \right)$ because of its high dipole moment. Measurements of dipole moments have also furnished clues as to the space relations of "geometric isomers." For example, in dichloroethylene, $C_2H_2Cl_2$, there are two isomers formulated as illustrated in Fig. 10.24. *cis*-Dichloroethylene has a moment of 1.9 Debye units, whereas the transisomer has zero moment, since the two C–Cl and the two C–H dipoles exactly cancel one another. In disubstituted benzene derivatives, the magnitude of the moment is determined by the location of the two substituents. A good example is furnished by the three isomeric dichlorobenzenes (Fig. 10.25). In the para derivative, the two C–Cl dipoles, being exactly 180° out of phase, precisely cancel one another.

3. Mechanism of Conduction in a Dielectric*

It will now be useful to inquire briefly into the behavior of a neutral molecule subject to an external electrical field. It is well known that a substance that is an insulator toward a dc field may show an appreciable ac conductivity, when a periodic polarization, that is, displacement of electrons in phase with the applied voltage, is possible. Nonpolar molecules, as has been indicated, are not readily polarized under alternating current. Polarization is much more pronounced for molecules that possess relatively large permanent dipoles. Here the molecules are readily oriented or partially oriented parallel to the ac field. If the frequency of the alternating field and the viscosity of the medium are both low, molecular oscillations that occur in phase with the field give rise to an alternating current.

If a dc field is applied to a randomly oriented polar dielectric, the dipoles are oriented by the field and, at moderate field strengths, will maintain this orientation indefinitely. If the field is removed, the kinetic energy of the molecules will soon com-

---

*See Chap. 13 for a detailed discussion of electrical insulation.

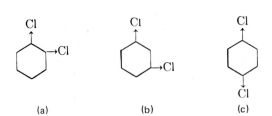

Fig. 10.25 Three isomeric dichloro-benzenes. (a) *ortho*-Dichlorobenzene, $\mu = 2.25$ Debye units. (b) *meta*-Dichlorobenzene, $\mu = 1.48$ Debye units. (c) *para*-Dichlorobenzene, $\mu = 0$ Debye units.

pletely destroy the imposed orientation. The time required for the orientation achieved under steady direct current to fall to 1/e th of this value (where e is the base of natural logarithms) is the "relaxation time," a quantity that varies with the nature of the molecule, with the viscosity of the medium, and with the inverse of the absolute temperature. At low ac frequencies, it is also reasonable to assume that complete relaxation may occur between pulses. At high frequencies, however, orientation cannot keep up with field oscillations, and at very high frequencies, the dipoles are not able to follow the field at all. The dielectric constant measured under these conditions corresponds to the temperature independent part only, that is, to that part owed to electronic polarization alone.

When a dielectric comes under the influence of an ac field, where the current ideally lags by 90° behind the voltage, the actual phase angle is less than 90° by an amount equal to the so-called loss angle. The in-phase component produces heat and consumes power in relation to the loss angle Δ. At ordinary frequencies the expression $\tan \Delta = \epsilon''/\epsilon'$ applies where ϵ' is the dielectric constant and ϵ'' is the dielectric loss factor.

Starting with the above familiar picture, dielectric theory is able to predict, with considerable accuracy, the electrical properties of relatively simple liquids. However, for polar organic high polymers the maximum dielectric loss factor is much lower than simple theory demands and occurs at much lower frequencies than for simple polar liquids. The dielectric constant is also frequently much higher. These effects are, in the first approximation, owed to the interactions of multiple dipoles located along the linear chains of these very large molecules. In most high polymers, the chains are twisted and folded, more or less at random, into complex configuration. The individual dipoles must accordingly relax, when under electrical stress, through rotation and uncoiling of these complicated chain segments. In a sense, then, the polymer behaves as a mixture of different molecules, each with its own relaxation time, each corresponding to one of the many possible configurations that the linear or cross-linked macromolecule can assume. The relatively higher dielectric constant of polar polymers can then be explained by the assumption that only where the appropriate chain segments happen to line up properly through chance molecular motion can the individual dipoles pair off, and thus decrease the net response to the field, as occurs easily with simple liquids.

From the explanation given above, it is apparent that the best electrical insulators are those materials that are the least polar. For example, among the lowest power-factor materials for use in liquid-filled transformers are hydrocarbon oils. In gas-filled transformers, certain stable, symmetrical compounds are found to be the most desirable. Among organic high-polymers useful as insulation, hydrocarbons again show the lowest electrical losses, such as polyethylene, polystyrene, polybutene, and some synthetic rubbers.

Changes in the chemical structure of an inorganic glass or of an organic high polymer can often be correlated with changes in electrical properties, such as polarizability, relaxation time, and dielectric loss factor. Alterations in electrical relaxation time often go hand in hand with changes in mechanical relaxation time and in flexibility. When polyethylene is chlorinated, for example, the flexible highly crystalline hydrocarbon changes gradually to an amorphous, rubbery material, then to a leathery substance, and finally, at about 60% total chlorine, to a brittle solid. Simultaneously, the loss factor, tan Δ, increases, reaches a maximum, and then decreases again. The shape of the tan Δ versus frequency curves indicates an increase in dipole orientation relaxation time, which runs parallel with an increase in the brittle temperature.

4. Dielectric Strength

It is common to classify insulating materials according to their dielectric strength. This characteristic is simply defined as the ac (or in some cases, dc) voltage that can be applied to the material before puncture or dielectric breakdown occurs. The dielectric strength varies with the thickness of the specimen, the size and shape of the electrodes used, and the conditions of the test. As practically determined, this property is not subject to greatly accurate reproducibility, yet it is a very important means of testing the suitability of a material for an intended electrical application.

In Table 10.13, some average electrical properties are listed for a number of high polymers commonly used as electrical insulation. It is to be emphasized that the values given represent, in general, observed ranges and are not to be construed as representing precisely the properties of any particular specimen.

VI. CONCLUSION*

It has been necessary, in the course of this chapter, to touch upon a large number of subjects and to make brief excursions into a number of related fields of chemistry and physics. The field of nonmetallic materials is much too broad to be treated otherwise. However, the coverage of any single aspect of this difficult subject has been necessarily sketchy, and possibly elementary at times. For details on many of the problems encountered it will frequently be desirable to consult the specialized literature. A large number of these problems are not only of great academic interest, but they are also of considerable practical significance to many industrial operations.

Polymers as a part of materials science have had a vastly expanded treatment in this edition in keeping with the increased use of polymers in contemporary design. The principles which have been discussed in this chapter will be exemplified in the next and other chapters on adhesives, elastomers, composites, and design considerations that relate to polymers. There has been some unavoidable duplication as various topics have been presented.

In subsequent chapters, some of the materials and processes discussed theoretically in the present chapter will be treated in more detail from a point of view much closer to that of the operations engineer.

---

*Contributed by ROBERT S. SHANE.

Table 10.13 Electrical Properties of Polymers[a]

Polymer	Volume resistivity, 50% relative humidity	Dielectric strength step by step, 1/8 in.	Dielectric constant			Power factor		
			60 (Hz)	10^3 (Hz)	10^6 (Hz)	60 (Hz)	10^2 (Hz)	10^6 (Hz)
Polystyrene	10^{17}-10^{19}	500	2.4-2.7	2.4-2.7	2.4-2.7	0.0002	0.0002	0.0002
Polytetrafluoroethylene	10^{16}	400	2.0	2.0	2.0	0.0001	0.0001	0.0001
Polyethylene	$>10^{13}$	400	2.3	2.3	2.3	0.0004	0.0004	0.0004
Natural rubber	10^{13}-10^{16}	500	3.0	3.0	3.0	0.006	0.004	0.06
Glass-mica	10^{15}	450	7-8	7-8	7-8	0.004	0.003	0.002
Polyvinyl carbazole	10^{15}-10^{16}	800	3.0	3.0	3.0	0.0015	0.0007	0.0007
Polyvinyl butyral	$>10^{14}$	400	3.6	3.6	3.3	0.007	0.007	0.009
Polyvinyl formal	$>10^{14}$	400	3.6-3.7	3.3	3.0	0.007	0.01	0.02
Butyl rubber	10^{14}-10^{16}	600	2.9	2.8	2.8	0.01	0.015	0.02
Shellac compounds	10^{9}-10^{10}	400	2-4	3-4	4-5	0.01	0.01	0.03
Ethyl cellulose	10^{12}-10^{14}	400	2.5-4	3-4	3-4	0.01	0.01	0.04
Silicone rubber	10^{10}-10^{11}	350	2.5-4	2.5-4	2.5-4	0.015	0.01	0.02
Cellulose acetate	10^{10}-10^{13}	250	5.5-6	5-6	4-5.5	0.02	0.03	0.05
Nylon	10^{13}-10^{14}	400	4-5	4-5	3-4	0.014	0.02	0.04
Polymethyl methacrylate	10^{14}	350	3.5-4.5	3-3.5	2.7-3.2	0.05	0.04	0.03
Phenol formaldehyde, filled	10^{11}-10^{12}	300	5-6	5-6	4.5-5	0.08	0.05	0.02
Polyvinyl chloride, plasticized	10^{11}-10^{13}	350	5-6	4-6	3-4	0.12	0.10	0.10

[a]See Chap. 13 for a detailed discussion

REFERENCES

1. Cyril Stanley Smith: Materials; *Scientific American,* September 1967.
2. Robert A. Huggins: Basic Research in Materials; *Science,* 20 February 1976, Vol. 191, No. 4228.
3. Committee on Ceramic Processing, Publication 1576, National Academy of Sciences, Washington, D.C., 1968.
4. Committee on Characterization of Materials, Publication MAB 229M, National Academy of Sciences, Washington, D.C., 1967.
5. T. D. Schlaback: The Essential Role of Materials Characterization, private communication, January 1970.
6. K. H. Meyer: *Natural and Synthetic High Polymers,* 2nd ed., Interscience Publishers, New York, 1950.
7. F. W. Billmeyer, Jr.: *Textbook of Polymer Chemistry,* Interscience Publishers, New York, 1957.
8. Golding, Brage: *Polymers and Resins,* Van Nostrand, Princeton, New Jersey, 1959.
9. H. F. Mark: *Scientific American* 197, No. 3, 81-89, 1957.
10. H. F. Mark and A. V. Tobolsky: *Physical Chemistry of High Polymeric Systems,* 2nd ed., Interscience Publishers, New York, 1950.
11. P. J. Flory: *Principles of Polymer Chemistry,* Cornell University Press, Ithaca, New York, 1953 (Ch. I).
12. P. J. Flory: *Principles of Polymer Chemistry,* Cornell University Press, Ithaca, New York, 1953 (Ch. III).
13. L. E. Nielsen: *Mechanical Properties of Polymers,* Reinhold Publishing Corp., New York, 1962.
14. G. Natta: *Advances in Catalysis,* Volume XI, 1958.
15. N. G. Gaylord and H. F. Mark: *Linear and Stereoregular Addition Polymers,* Interscience Publishers, New York, 1959 (Ch. VII).
16. G. Natta, *et al., Chim. ind. (Milan) 41,* 398, 1959.
17. N. G. Gaylord and H. F. Mark: *Linear and Stereoregular Addition Polymers,* Interscience Publishers, New York, 1959 (Ch. V).
18. P. J. Flory: *Principles of Polymer Chemistry,* Cornell University Press, Ithaca, New York, 1953.

ADDITIONAL READING

Billmeyer, F. W., Jr., *Textbook of Polymer Sciences,* 2nd ed., Interscience, New York, 1971.
Brit. Plastics, 36, No. 11, 628-633, 1963.
Brown, T. L., and LeMay, Jr., H. E., *Chemistry,* 2nd ed., Prentice-Hall, Englewood Cliffs, New Jersey, 1981.
Burke, J. J., Gorum, A. E., and Katz, R. N. (eds.), *Ceramics for High Performance Applications,* Brook Hill, Chestnut Hill, Massachusetts, 1974.
Collected Papers of Wallace Hume Carothers, H. F. Mark and G. S. Whitby (eds.), Interscience, New York, 1940.
Flory, P. J., *Principles of Polymer Chemistry,* Cornell University Press, Ithaca, New York, 1953.
Groves, D. G., and Hunt, L. M., *The Ocean World Encyclopedia,* McGraw-Hill, New York, 1980.
Harper, C. A., *Handbook of Plastics and Elastomers,* McGraw-Hill, New York, 1975.
Hill, R., *Fibers from Synthetic Polymers,* Elsevier, Amsterdam, 1953.

Murr, G. E., and Stern, C. (ed.), *Frontiers in Materials Science,* Dekker, New York, 1976.
Shalaby, S. W., and Pearce, E. M., *Chemistry of Macromolecules,* American Chemical
    Society, Washington, D.C., 1974.
Stevens, M. P., *Polymer Chemistry,* Addison-Wesley, Reading, Massachusetts, 1975.
Tanford, C., *Physical Chemistry of Macromolecules,* Wiley, New York, 1961.

11

Plastics and Elastomers (Rubber)*

I. INTRODUCTION

Organic polymers are chain molecules consisting of segments that are based mainly on carbon. These polymers can be made by the modification of natural polymers (e.g., cellulose derivatives) or the polymerization of simple organic compounds (e.g., polyethylene or polystyrene). The response of organic polymers to heat is determined by their chemical structure. Polymers made from linear chains may undergo one or more thermal transitions toward the liquid state and are described as thermoplastic. On the other hand, solid, cross-linked polymers cannot be obtained in the liquid state upon heating and are referred to as thermosetting for they may undergo further cross-linking upon heating.

Prior to discussing the interplay of structure, properties, processing, and end use of commercial thermoplastic and thermosetting polymers in the main part of this chapter, chemical notations for chain molecules constituting materials are given in Sec. II. Section IV gives guidelines for materials selections for the production of common forms of shaped articles.

II. CHEMICAL NOTATIONS OF CHAIN MOLECULES

This section provides simple, schematic presentations of the structural, chemical formulas of typical polymeric materials. The given formulas are limited to the main systems, which are treated extensively in the chapter. In addition, the structures of complex monomers, oligomeric intermediates, and/or prepolymers are given in a few cases.

A. Thermoplastic Polymers

1. Aliphatic Polyolefins

$$-(CH_2CH_2)_p-$$ polyethylene (PE)

---

*Contributed by SHALABY W. SHALABY and BARBARA GREENBERG SCHWARTZ. The authors express their appreciation to the American Chemical Society for granting permission to use excerpts from Parts I and II of the Society interaction course entitled Polymer Science and Technology–An Interdisciplinary Approach.

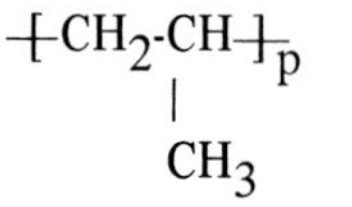

$+CH_2\text{-}CH+_p$ polypropylene (PP)

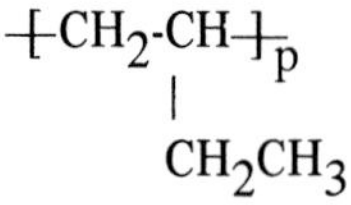

     $|$

     CH_3

$+CH_2\text{-}CH+_p$ polybutylene, (polybutene-1) (PB)

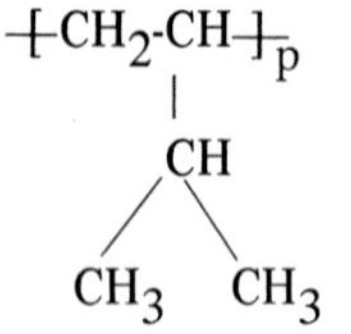

     $|$

     CH_2CH_3

$+CH_2\text{-}CH+_p$ polymethylpentene, poly-(4-methylpentene-1) (PMP)

     $|$

     CH

   CH_3 CH_3

2. Polystyrene

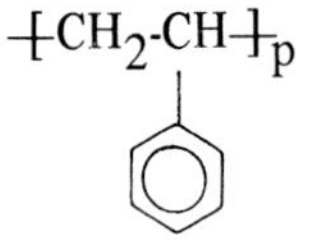

$+CH_2\text{-}CH+_p$ PS

     $|$

     (phenyl ring)

3. Chlorocarbon Polymers

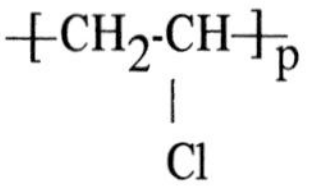

$+CH_2\text{-}CH+_p$ polyvinyl chloride (PVC)

     $|$

     Cl

$+CH_2\text{-}CCl_2+_p$ polyvinylidene chloride (PVdC)

4. Fluorocarbon Polymers

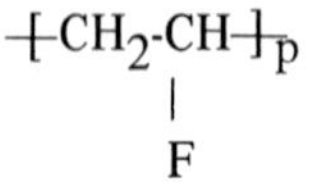

$+CH_2\text{-}CH+_p$ polyvinyl fluoride (PVF)

     $|$

     F

$+CH_2\text{-}CF_2+_p$ polyvinylidene fluoride (PVdF)

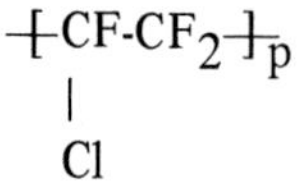

$+CF\text{-}CF_2+_p$ polychlorotrifluoroethylene (PCTFE)

     $|$

     Cl

$+CF_2\text{-}CF_2+_p$ polytetrafluoroethylene (PTFE)

5. Acrylic

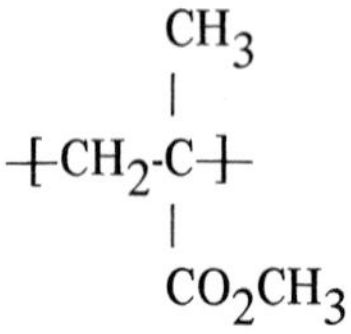

     CH_3

     $|$

$+CH_2\text{-}C+$ poly(methyl methacrylate) (PMMA)

     $|$

     CO_2CH_3

$+CH_2$-$CH+$ polyacrylonitrile (PAN)
  |
  CN

6. Polyacetals

 $+CH_2$-$O+_p$ polyformaldehyde (PF)

7. Polyamides

 $+(CH_2)_5CONH+_p$ nylon 6

 $+(CH_2)_{11}CONH+_p$ nylon 12

 $+NH(CH_2)_6$-NH-CO-$(CH_2)_4$-$CO+_p$ nylon 6,6

 $+NH(CH_2)_6$-NH-CO-$(CH_2)_8$-$CO+_p$ nylon 6,10

8. Polyesters

 $+(CH_2)_5$-CO-$O+_p$ poly-ϵ-caprolactone (PCL)

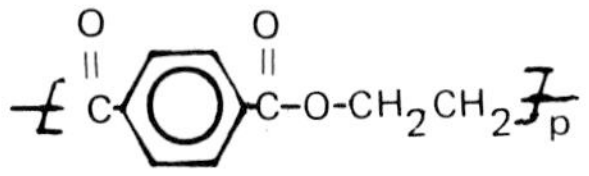

 polyethylene terephthalate (PET)

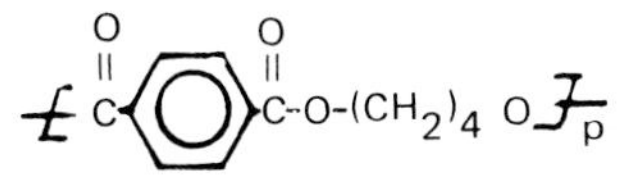

 polybutylene terephthalate (PBT)

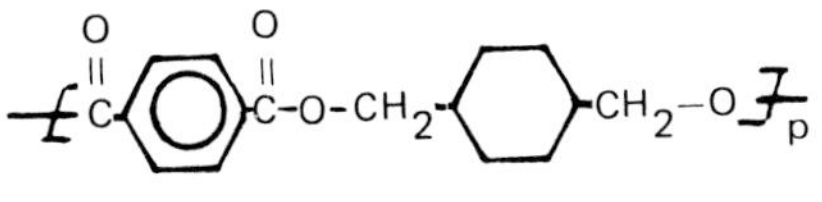

 polycyclohexanedimethylene terephthalate (PCHDM-T)

9. Aromatic Polycarbonate

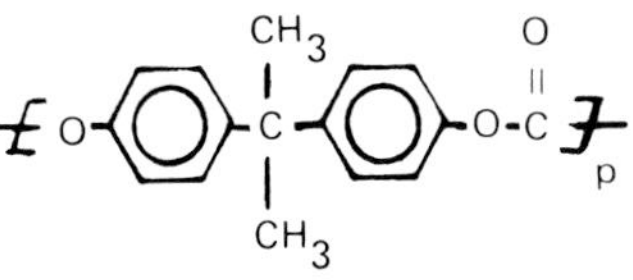

10. Aromatic Polyether

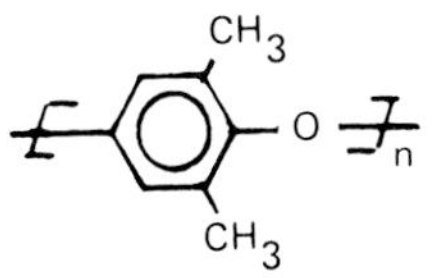

11. Aromatic Sulfone Polymers

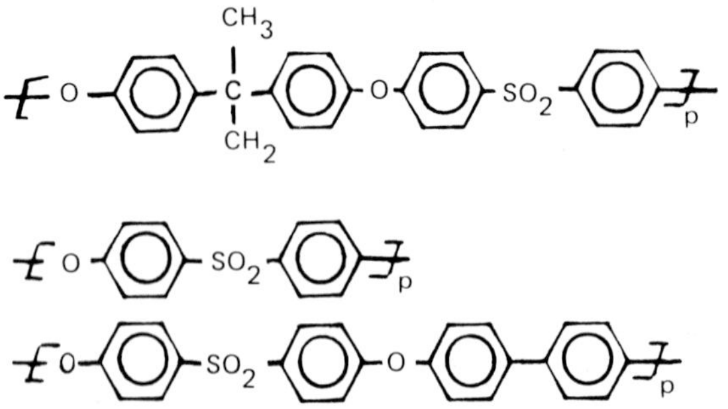

polysulfone (PSul)

polyethersulfone (PE-Sul)

polyphenylsulfone (PP-Sul)

B. Thermosetting Resins

Monomers are first converted to thermoplastic prepolymers, which are then cured to cross-linked infusible materials. The five types of thermosetting resins or oligomeric intermediates and their final network structures are illustrated below.

1. Phenol-Formaldehyde (Phenoplasts)

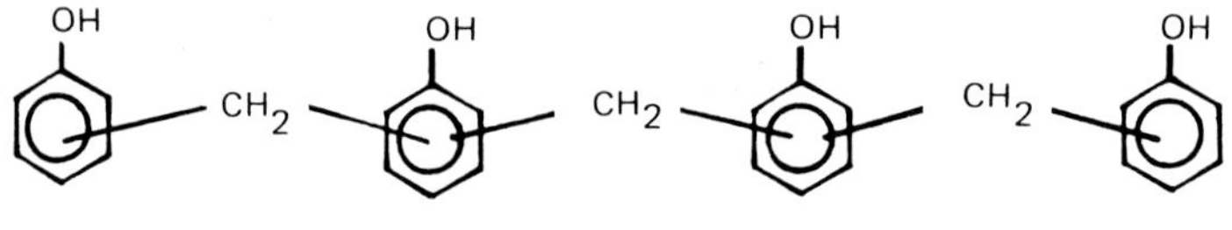

prepolymer

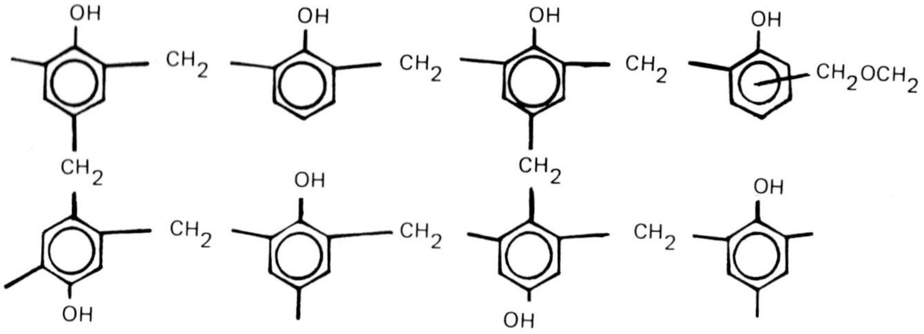

final network

2. Amine-Formaldehyde Resin (Aminoplasts)

The two most common types are those made from melamine and urea and are called melamine-formaldehyde and urea-formaldehyde resins.

a. Urea-Formaldehyde

Urea is reacted with one or two parts of formaldehyde to give monomethylol urea or dimethylol urea, respectively. These oligomeric intermediates can then be cured to a three-dimensional network.

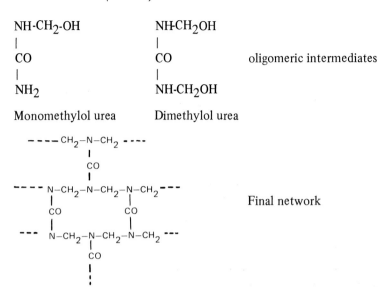

NH-CH$_2$-OH
|
CO
|
NH$_2$

Monomethylol urea

NH-CH$_2$OH
|
CO
|
NH-CH$_2$OH

Dimethylol urea

oligomeric intermediates

Final network

b. Melamine-Formaldehyde

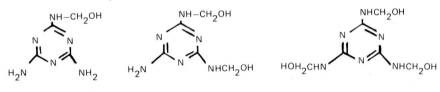

Monomethylol melamine   Dimethylol melamine   Trimethylol melamine

oligomeric intermediates

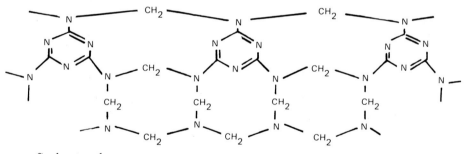

final network

3. Polyesters

a. Alkyd Resins (Saturated Polyesters)

These are the reaction product of a saturated diacid or a suitable derivative with a mixture of diol and a small amount of trifunctional alcohol to allow for cross-link formation. This is typically illustrated below by the reaction product of phthalic anhydride, ethylene glycol, and glycerol.

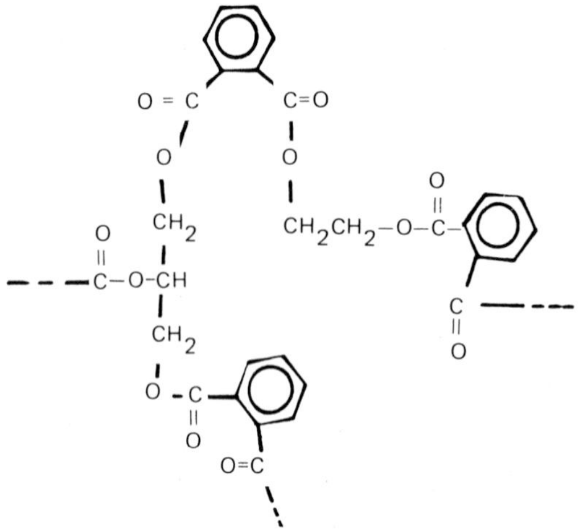

O = C   C=O

(structure: phthalate/glycol ester crosslinked network)

```
        O=C          C=O
          \           |
           O          O
            \         |           O
         O  CH2       CH2CH2-O-C- ⬡
         ||  |                 ||
- - - -C-O-CH                   C  - - - -
            |                   ||
           CH2                  O
             \
            O - C- ⬡
               ||
               O
              O=C
                \
```

b. Unsaturated Polyester Systems

The thermoplastic prepolymer is first made, typically, by reacting a mixture of phthalic anhydride (P) with an unsaturated anhydride, such as maleic anhydride (M), with a glycol (G), such as propylene glycol. The prepolymer is reacted with a vinyl monomer, such as styrene (S), to provide a cross linked network, where polystyrene segments link the polyester chains at the maleic anhydride sequences.

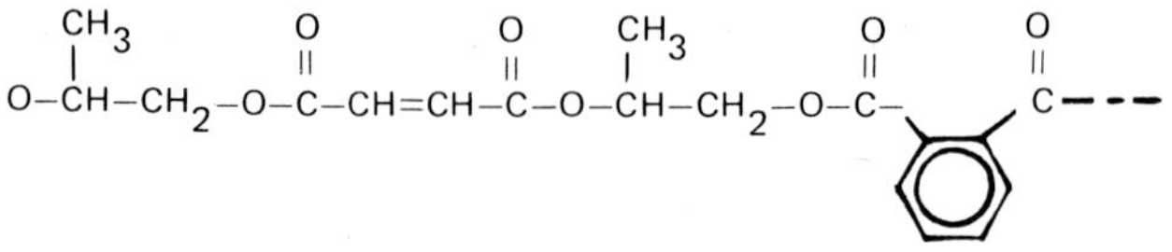

```
  CH3      O         O   CH3      O       O
   |       ||        ||   |       ||      ||
O-CH-CH2-O-C-CH=CH-C-O-CH-CH2-O-C-⬡-C- - -    prepolymer
```

```
 -M-G-P-G-M-G-P-G-M-
   |       |      |
 (PS)    (PS)   (PS)      schematic representation of final network
   |       |      |
 -M-G-P-G-M-G-P-G-M-
```

4. Epoxy Resins

The most common prepolymers are those made by reacting bis-phenol-A with epichlorohydrin to form bisphenol-A diglycidyl ether. A second group of prepolymers is made by the reaction of epichlorohydrin with a low-molecular-weight novolac type of phenolformaldehyde resin. Typical examples of the latter are the glycidyl ethers of m-cresol novolac resin. Both types of glycidyl ether prepolymer are shown below. Curing of either type of prepolymer can be achieved with an anhydride or amine to form a cross-linked network.

Prepolymers

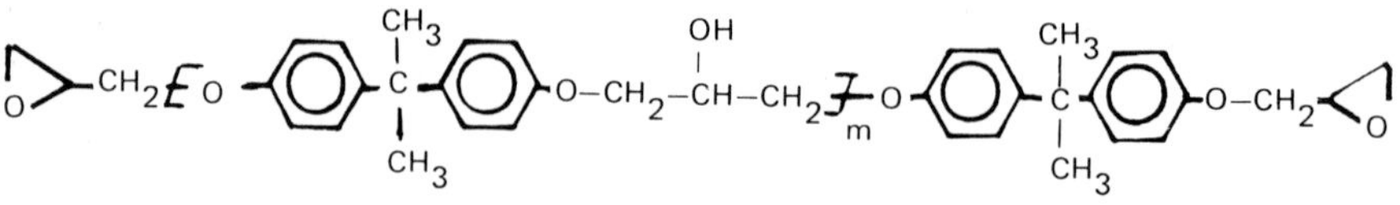

```
         CH3                    OH                  CH3
          |                     |                    |
▷-CH2-O-⬡-C-⬡-O-CH2-CH-CH2-O-⬡-C-⬡-O-CH2-◁
          |                                |
         CH3                  m            CH3
```

Bisphenol-A diglycidyl ether

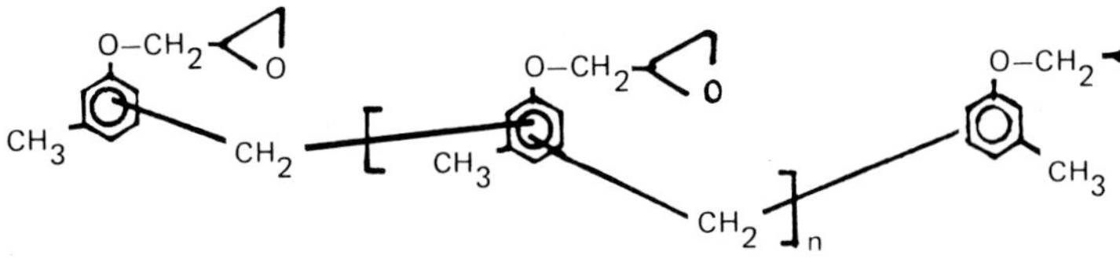

Glycidyl ether of m-cresol novolac

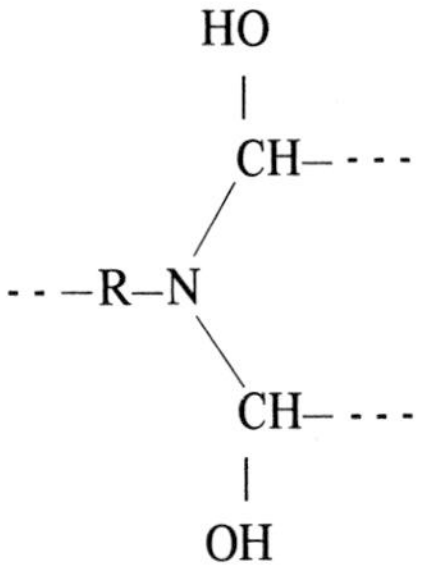

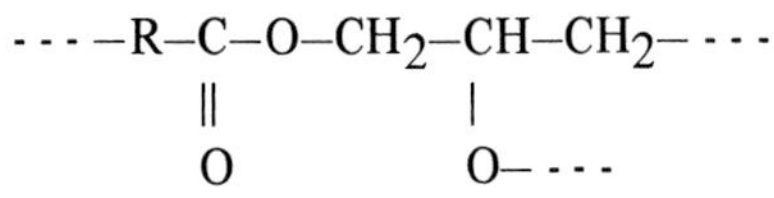

Amine cured   Anhydride cured

5. Furan Resin

An intermediate that has been made available in recent years for the production of these resins is 2,5-bis-hydroxymethylfuran (BHMF). BHMF and a schematic presentation of a typical network structure of the cured resin are given below. The BHMF is capable of self-condensing into a cross-linked network.

BHMF

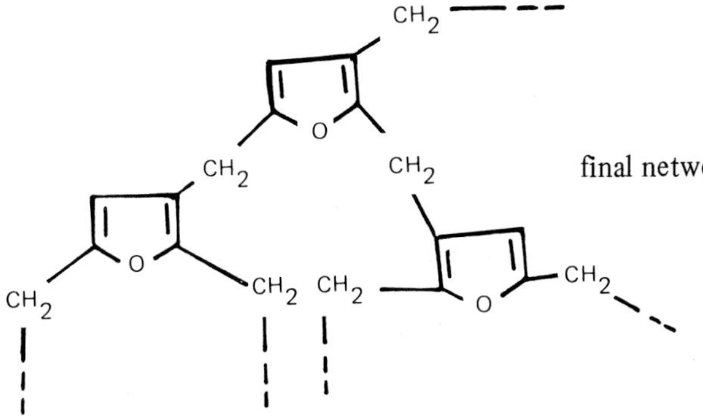

final network

III. INTERPLAY OF STRUCTURE, PROPERTIES, PROCESSING, AND END USE*

Frequently, authors find it easier to treat the subject under two separate headings, namely, the effect of structure on the polymer properties, and then the pertinence of

---

*See also Sec. VIII by Herman F. Mark.

these properties to the processing and uses of polymeric materials. However, in such a style of presentation, the reader may not develop a full appreciation of the inverse consecutive dependence (ICD) of a shaped article performance on polymer processing, physiochemical and mechanical properties of raw or unprocessed polymers, and the inherent chemical and physical properties of the chain molecules constituting these materials.

In an attempt to demonstrate the validity of ICD, a few typical examples of commercial polymers will be treated in this section, starting with simple homochain polymers (e.g., linear polyethylene) and concluding with cross-linked heterochain polymers (e.g., urea-formaldehyde resins).

A. Thermoplastic Materials

Organic thermoplastic materials are generally based on polymers composed of linear, essentially linear, or linearly interlinked chain molecules. The simplest form of thermoplastic is linear high-density polyethylene in which the chains are made of nonpolar methylene groups and can easily align in a crystalline form. In a certain temperature range, the lattice energy is overcome and the chains become mobile. Above this temperature range (melting temperature, T_m), the polymer will be in the liquid state and can be made to flow in the presence of a unidirectional force. Upon cooling the liquid polymer below its T_m, it will revert to the solid state characterized by three-dimensional order. The reversible change of polyethylene from a liquid to solid or vice versa can be easily predicted and controlled. However, the interconversion of these states becomes progressively complicated as one modifies the chain molecule by introducing at regular intervals (1) methyl groups (e.g., polypropylene); (2) ethyl groups (e.g., polybutene-1) or n-propyl (e.g., polypentene-1) groups; (3) an oxygen atom between the methylene groups, e.g., polyoxymethylene and polyoxyethylene (e.g., polyethylene oxide); and (4) an amide group, e.g., nylon 6.

Thermoplastic polymers are usually hard at room temperature and become soft upon heating. This allows their use to fabricate different molded and extruded articles. Upon cooling, these articles harden and maintain stable dimensions at room temperature. It is to be noted that the fabrication or processing cycle can be repeated since the change of thermoplastics from liquid or "liquidlike" soft state to a hard solid state is reversible. Unlike thermosets, no chemical changes take place during the processing cycles of thermoplastics. Thermoplastics are commercially available from material suppliers in pellet, powder, or granular form. Plastic processors produce variously shaped articles or parts by molding, casting, and extrusion (including fiber melt spinning). From extruded or cast sheets, rods, and tubes, other parts can be produced by vacuum forming and forging and machining. In most cases plastic articles are made from formulations that contain plasticizers (to soften rigid polymers), fillers (to impart desired mechanical properties through reinforcement), and coloring matters. Formulations are available for design engineers to meet most chemical, physical, and electrical requirements. Thermoplastics can be converted or shaped into useful articles as small as pins, screws, and connectors and as large as houseware items, aircraft turrets, and cockpit enclosures. Some thermoplastics can be easily extruded or melt spun into multifilament textile yarns and monofilaments.

1. Polyethylene

a. General

Polyethylene is produced in two principal grades, the high-density PE (HDPE) and low density (LDPE). In addition, ultra-high molecular weight PE (UHMWPE) and a few ethylene copolymers are commercially available.

Properties of different types of PE depend not only on their nonpolar hydrocarbon chains and associated chemical properties but also on molecular weight, level of branching, and degree of crystallinity (hence, density). Polyethylenes having a wide range of properties corresponding to a range of attainable density are commercially available. It is to be noted that molecular weight distribution (MWD) can also affect polymer properties. Polyethylenes with narrow MWD are more resistant to low-temperature brittleness and environmental stress cracking than those having a broad MWD.

Polyethylene can be purchased in many grades with a range of density depending on the polymerization technique used. In addition, for any particular density of PE, different molecular weight grades are available. These polymers will vary in processability and mechanical properties. Hence, it is imperative that the converter should have some knowledge of the molecular weight of a particular PE prior to its processing. Measurements of melt viscosity using a capillary rheometer, such as the Melt Indexer, are commonly used as a means of estimating molecular weight.

b. Chemical and Physical Properties of Low- and High-Density PE

Both types of PE are white and translucent. Many of the physical properties of polyethylene are outstanding for use in engineering design. These include their (1) relatively low T_m and, hence, low processing temperature; (2) low density; (3) low-temperature toughness, which allows their use in many low-temperature applications; and (4) low dielectric constant and high electrical resistivity, that are desired properties for electrical applications. In general, polyethylenes are resistant to most chemicals except oxidizing acids, free halogens, and certain ketones. Polyethylene is practically insoluble in common organic ketones. Polyethylene is practically insoluble in common organic solvents. However, the solubility of PE increases sharply with the increase in temperature. Although the solubility of PE at elevated temperatures decreases with increase of density, both types of polyethylenes can be dissolved in some hydrocarbons and chlorinated hydrocarbons.

c. Processing and General Applications of Low- and High-Density Polyethylenes

PE can be easily fabricated by a number of techniques, including extrusion (into pipes, sheets, cable coverings, or films), injection molding, blow molding, extrusion coating, vacuum forming, sintering, and flame spraying. The ease of processing of polyethylene, its toughness, and good chemical and electrical properties led to its extensive use.

(1) Films This is the most important form in which PE is used. PE films are used extensively in the packaging industry. Low-density PE is commonly used for the production of these films. However, for applications where the films are required to have greater thermal resistance, high-density PE films are used. High-clarity, thin PE films are used in packaging clothing. Due to high water permeability, PE films are used to package fresh

vegetables. Heavy gauge (0.006 to 0.010 in.) PE films are used to package agricultural products.

(2) Coatings Practically all PE coatings are made from the low-density grade polymer. Polyethylene-coated, regenerated cellulose film is used in many applications where a vacuum pack is required, particularly when transparency is also desired.

(3) Injection molding Both low- and high-density polyethylenes have been used for the injection molding of household articles. Toughness and chemical inertness of polyethylene make it useful for the production of laboratory and hospital articles. The high toughness of low-density polyethylene makes it quite useful for toys. In the area of packaging products, the low-density grades of PE are used whenever resilience (e.g., snap-on caps and plug fittings) is called for, and high-density grades of PE are used when rigidity (e.g., screw caps) is called for.

(4) Blow molding The high softening temperature and good barrier properties of high-density PE make it quite suitable for the production of blow-molded articles where some rigidity is required, such as in bleach bottles and similar household containers. Apart from packaging applications, where component parts are often molded and then assembled, blow molding makes it possible to convert low- and high-density polyethylene into toys and similar articles at a lower cost than by injection molding.

(5) High-density PE monofilaments Melt extrusion (or spinning) of PE followed by cold drawing to increase orientation and breaking strength leads to strong monofilaments. These can be used, depending on the diameter and extent of orientation, for the production of filter cloth, woven materials for deck chairs, ropes, tennis nets, and fishing nets.

(6) Electrical applications The hydrophobicity, chemical inertness, excellent electrical properties, and low water permeability of polyethylene make it an excellent substitute for natural rubber and paper insulation in many electrical application, such as telephone cables.

(7) Textile finishing Polyethylene can be used to provide a glossy, hydrophobic surface to woven cellulose fabrics. It may also be used for treating carpet backing for improved abrasion resistance.

(8) UHMW Polyethylene High-density PE with an average molecular weight of 2-5 million is commonly known as ultra-high molecular weight polyethylene, or simply UHMWPE. Shaped articles made from UHMWPE exhibit enhanced chemical, physical, mechanical, and biological properties. These include (1) excellent resistance to common chemical reagents and solvents, (2) nonadherent, self-lubricating surface with low coefficient of friction, (3) high abrasion resistance, (4) good fatigue endurance, (5) high impact strength, (6) good noise damping properties, and (7) retention of physical and mechanical properties when used as a prothesis in various biologic environments.

The polymer is produced by a modified Ziegler or Phillips process. It is sold as a powder or shaped forms, such as sheets, plates, bars, rods, and tubes. Standard metalworking equipment is easily used to machine these forms into finished articles.

The extremely high molecular weight and melt viscosity of UHMWPE put some constraints on its processibility. Compression molding is commonly used for producing sheets and plates. Ram extrusion is used for the production of thin sheets, bars, rods, and tubes. Using processing aids, UHMWPE can be converted to pipes and rods in a twin-screw extruder. Shaped articles with complex geometries can be made from UHMWPE by forging (as in the case of metals). The excellent wear properties of UHMWPE make it quite useful in the production of machine components, such as gears, bushings, wheels, and chain guides. The polymer can also be used in the production of noise-damping stock tubes for automobile screw machine equipment. The exceptional surface properties and biologic inertness make UHMWPE quite attractive for certain biomedical applications. These include prosthetic devices for hips, knees, and fingers.

d. Ethylene Copolymers

(1) Ionomers Copolymers of ethylene with sodium methacrylate and similar monomers differ from the homopolymer in that they (1) are less crystalline, (2) are more transparent, and (3) behave as cross-linked systems (due to their ionic components). The ionic components in these systems do not have to originate from acrylic acid-type comonomers. They can be present in copolymeric structures bearing sulfonic (through direct sulfonation of polyethylene) or phosphoric acid moieties (via phosphonylation of polyethylene). Initial applications of ionomers have developed in skin packaging, blister packaging, extrusion-coated foil packaging, adhesive for laminating metals, and blow-molded bottles.

(2) Ethylene copolymers with ethyl acrylate and vinyl acetate Depending on the chain composition of these copolymers, one can obtain materials with a broad range of crystallinity, optical clarity, flexibility, toughness, and adhesive properties. Compared with polyethylene, the low crystallinity of some of these copolymers makes them more transparent, soft, and resilient. This allows the effective use of these copolymers for food and beverage packaging, medical tubing, wire cable and insulation, syringes, and squeeze toys. The polar ester groups in the copolymers impart (1) sufficient dielectric loss for use in high-frequency sealing, and (2) adhesive properties needed for blending with waxes in paper coating and packaging applications.

2. Polypropylene

a. General

A polypropylene chain can assume any of three conformations: isotactic, syndiotactic, and irregular atactic. Molecules of commercial PP are predominantly isotactic but with short lengths of atactic and syndiotactic segments incorporated in them. In addition to PP, copolymers of ethylene and propylene are also available commercially. The ethylene content in these copolymers can be either low (about 7% by weight) or high (about 15% by weight).

Polypropylene is sold in the powder or granular form. Semifabricated forms, such as films, foils, sheets, blocks, rods, and tubes, are produced commercially.

In addition to the extensive use of PP as a molding resin, its use as a textile fiber or monofilament has expanded considerably in recent years.

b. Chemical and Physical Properties

Although the chemical properties of PP are generally similar to those of PE, the two polymers are somewhat different in that (1) PP is more sensitive to oxidation and radiation degradation, (2) PP is usable at higher temperature, (3) PP is more resistant to oils and greases, (4) PP is less sensitive to environmental stress cracking, and (5) permeability of PP to oxygen, carbon dioxide, and water vapor is intermediate between those of low- and high-density PE. Both PE and PP are insoluble at room temperature in organic solvents. At 20°C, concentrated sulfuric acid can cause deterioration in the properties of PP; at 100°C PP is attacked by 30% hydrochloric acid.

The physical properties of "virgin" high-molecular-weight PP are determined mostly by the isotactic content of the chain. In PP the stereoregular (isotactic) fraction of chains determines the polymer softening temperature, the softening hardness, and the tensile strength. On the other hand, an increase in the amorphous (atactic) fraction leads to a reduction of these properties. It is to be noted that the properties of the virgin polymer can be altered markedly in fully processed, shaped articles. Thus, the actual crystallinity of a finished article is quite dependent on the rate of cooling of the molten polymer and degree of orientation of the amorphous and crystalline fractions. Slow-cooled (from the melt) articles will allow the development of large spherulites and maximum crystallinity; this leads to rigid and more temperature resistant systems. Quick-quenched articles develop polymer systems that are less crystalline, tougher, and more transparent than those obtained by slow cooling.

Although molecular weight has a considerable effect on the flow properties of molten PP and impact strength of solid articles, its effect on stiffness and hardness is minimal. Thus, the melt viscosity and impact strength increases noticeably with the increase in molecular weight. In practice, the weight-average molecular weight of PP is estimated in terms of its melt index (MI) at 230°C and 2.16 kg load (ASTM-D1238). An inverse relationship of molecular weight and melt index exists.

Some of the chemical and physical properties of PP are summarized in Table 11.1. A few of the properties of PP deserve special attention, such as the impact strength at low temperature and its modification, deformation, electrical properties, oxidative stability during melt processing, and photo-oxidative stability. The impact strength of PP decreases with decrease in temperature particularly below 10°C. This reduction in impact strength can be minimized by (1) increasing the cooling rate of the molten polymer upon its fabrication and (2) addition of small amounts of butyl rubber or polyisobutylene (as toughening agents). Compared with polyethylene, PP is more resistant to creep. This is particularly noticeable above 100°C and high dead weight conditions. Polypropylene has

Table 11.1 Typical Physical Properties of PP

Properties	Units	Typical values
Density	g/cm^3	0.90-0.91
Crystalline melting temperature	°C	165-175
Softening temperature (Vicat)	°C	150
Thermal conductivity, 20-150°C	cal/cm-s-°C	3.5-5 $\times$ 10^4
Specific heat	cal/g-°C	0.46
Tensile properties of molded specimen		
Tensile strength	psi	4000-5200
Elongation to break	%	400-900
Impact strength		
(Izod) of injection-molded bar	ft-lb/in. notch	2.5-4.0
Hardness (Shore)	D scale	71-75
Dielectric constant	10^6 C/sec	2.0-2.5

excellent electrical properties compared to those of PE. Although the electrical properties of PP are unaffected by molecular weight or stereoregularity, they may be influenced by the presence of catalyst or antioxidant residues. Due to the oxidative degradation of PP at elevated temperatures, antioxidants are used to prevent degradation during processing. For improved, long-term photo-oxidative stability, small amounts of ultraviolet stabilizer and antioxidants are incorporated in many commercial PP formulations.

c. Processing and General Applications

Polypropylene can be molded in all modern injection-molding machines. It can be readily extruded into film, pipe, wire cable and covering, sheet, monofilament, and multifilament yarn (for textile applications). Semifabricated PP tubing can be blown into bottles and other hollow articles. Compression molding is used mostly for the production of semifabricated blocks and sheets. Usual machining techniques for thermoplastics are applicable to PP. On the other hand, due to the chemical inertness of PP, joining techniques, such as welding and cementing, have been applied with limited success unless the surface is given a preliminary oxidative treatment.

Due to their similarity in some chemical and physical properties, HDPE and PP are used in several common applications. On the other hand, PP is considered superior to PE in other applications that call for (1) high Shore-D hardness (PP, 75 versus HDPE, 67-70), (2) high softening temperature (PP, 140-150°C versus HDPE; 122-130°C), and (3) minimum mold shrinkage and high gloss, where PP is preferred.

When tested in thin sections PP displays excellent resistance to repeated flexing. This gives PP the ability to form an integral hinge, which is simply a very thin web of material in which orientation of the chain molecules has produced a great increase of strength in a direction perpendicular to the hinge length. Although hinges can be produced by extrusion, machining, or thermoforming, the chain orientation is initiated during the molding process and completed by subsequent flexing of the hinge (first slowly and then at normal speed). Properly designed and molded PP hinges withstand flexing almost indefinitely and retain high tensile and tear strength.

Polypropylene can accept textured finishes. Thus, moldings with textured finishes can be produced from molds on which the appropriate pattern has been engraved photochemically. Since such molds do not require polishing, a low-cost cast mold can be used.

The high toughness and resilience of PP make it quite suitable for the production of snap-fit assemblies. This and the capability of forming an integral hinge and accepting textured finishes make PP most useful in applications where reduction in the number of components and assembly operations are required. Typical examples of the applications of PP include (1) molded articles (car components, TV parts, tool handles, hospital equipment, valves and pipe fittings, toys, tableware, and hinged containers), (2) pipes and sheets, (tanks, reactors, fans, and filters), (3) wire coatings, (4) oriented films (packaging), (5) fibers (ropes, fabric, and carpets), (6) tape yarn (woven sacks and carpet backing), and (7) monofilaments (brushes and sutures).

d. Polypropylene Fibers

Isotactic (94-98% isotacticity), high-molecular-weight (245,000-2,400,000 daltons) PP can be converted by melt spinning and drawing to a strong monofilament or multifilament yarn. Although the polymer melts at 160-175°C, it is spun at a temperature range of 200-250°C, depending on the polymer molecular weight and the desired diameter

Table 11.2 Typical Properties of Commercial Fibers (Tested as Multifilament Yarn)

Properties	Values
Breaking strength, psi	60-95 $\times$ 10^3
Tenacity, g/d	5-7
Elongation at break, %	15-35
Elastic recovery at 5% elongation, %	88-98
Modulus of elasticity on 10% extension, g/d	70-90
% Shrinkage in boiling water after 20 min	0-3
Moisture region	$<0.3\%$

of the fiber sought. Since the polymer crystallizes rapidly, extruded fibers cannot be generated in an amorphous or low-crystallinity state. Thus, extruded fibers with about 40% crystallinity are usually oriented by drawing at a temperature range of about 60-120°C to develop optimum tensile properties. Fibers drawn at a stretch ratio of 1:5 at an appropriate temperature can develop up to about 70% crystallinity; these are characterized by excellent dimensional stability under typical end-use conditions as textile fibers. Additional modifications of the fiber properties to achieve specific changes in the tensile properties can be realized by heat setting or annealing the drawn fibers. Annealing can be conducted at different temperature, strain, or stress depending on the nature of the desired changes. Properties of commercial PP fibers are given in Table 11.2.

3. Polybutylene

a. General

The homopolymer and a few copolymers of 1-butene are made by the Ziegler-Natta polymerization technique. The homopolymer is crystalline, and its chains are mostly isotactic. It differs from PP by having a pendent ethyl instead of a methyl group. In the solid state, the polymer can be obtained in four (I-IV) crystalline modifications. One of these, the twinned hexagonal form (I), is the most stable and displays a T_m of 125-130°C. Fast or slow cooling of the polymer melt leads to a metastable crystalline form (II) characterized by a T_m range of 115-120°C. Transformation of this form to the more stable form (I) takes place upon standing for several (5-7) days at room temperature, or can be accelerated by pressure or uniaxial orientation.

b. Properties, Processing, and Applications

Although PB is very similar to PP in its chemical properties, permeability to water vapor and gases, and thermo-oxidative stability, it is somewhat different in the following properties: (1) upon tensile failure it exhibits strain-hardening behavior with uniform sample deformation, instead of necking; (2) it is soluble in aromatic and chlorinated solvents at relatively low (about 60°C) temperatures (may be related to its low T_m); and (3) polybutylene accepts high filler loading (up to 85% based on total weight) before significant embrittlement is observed. Upper and lower use temperatures for PB are about 105 and −20°C.

The main applications of commercial PB resins are in the production of films and pipes. As a film, PB has good tear resistance, impact strength, and puncture resistance. The blow-film process is most suitable for PB film production; conventional LDPE equipment is used. Pipes made from PB are characterized by flexibility and resistance to creep, environmental stress cracking, chemicals, and abrasion. These pipes can be processed by conventional single-screw extruders, using vacuum or pressure sizing for dimension control. In addition, PB can be injection molded into smaller articles, such as fittings for use with PB pipes.

4. Poly-(4-Methylpentene)

a. General

Structurally, PMP can be viewed as a derivative of polypropylene where one of the methyl hydrogens of the repeat unit is replaced by an isopropyl group. 4-Methylpentene-1 is polymerized like propylene using a Ziegler-type catalyst to give a crystalline stereoregular polymer. The crystalline PMP is predominantly isotactic and in fabricated articles, the crystallinity is usually about 40%. Annealing of the unfabricated polymer can result in up to 65% crystallinity.

b. Properties, Processing, and Applications

The basic properties of PMP are similar to those of a typical polyolefin. In addition, PMP has some distinctive properties as a thermoplastic, aliphatic polyolefin that make it more useful in certain applications. These properties include (1) an above room temperature glass transition (T_g); (2) a high melting temperature; (3) high transparency; and (4) high water vapor and gas permeabilities. A summary of these and other properties is given in Table 11.3.

Table 11.3 Physical Properties of PMP

Properties	Units	Typical values
T_g	°C	30-50
T_m	°C	245
Density	g/ml	0.83
Coefficient of linear thermal expansion, −50 to 120°C	in./in./°C	1.2×10^{-4}
Thermal conductivity (20-70°C)	cal/cm-s-°C	4×10^{-4}
Specific heat (20-150°C)	cal/g-°C	0.52
Direct transmission factor corrected to 1 mm thickness	%	99
Equilibrium water content (immersed at 20-60°C)	%	<0.05
Water vapor permeability	(g/m^2 24 hr) @ 38°C and 90% RH for 0.001 in thickness	95-110
Gas permeabilities	cm^3, cm/cm^2-sec-mm Hg $\times 10^9$	
Oxygen		2.7
Nitrogen		0.7

With T_g of 30-50°C, the deformation of PMP is very time dependent and quite sensitive to small temperature changes when tested in this temperature range. Above this range and at higher temperatures, due to its high T_m, the deformational properties of PMP are superior to those of HDPE and PP. Due to their structural similarity, PMP and PP are comparable in terms of the (1) dynamic fatigue and environmental effects, (2) impact behavior, (3) electrical properties, and (4) chemical resistance to acids, bases, and organic solvents.

The melt processing of PMP can be achieved using conventional injection molding, extrusion, and blow-molding equipment. It is to be noted that the polymer (1) has a narrow melting range, (2) has a low melt viscosity with thixotropy, and (3) should not be kept for a long period in the molten state to avoid degradation and color development. Compression molding can be used to make blocks and sheets for machining and welding into a prototype. PMP sheets can be press formed but not readily vacuum formed. On the other hand, extruded films and foils can be vacuum formed and oriented at carefully controlled, predetermined temperatures.

Similar to PP, PMP is used in lighting, electrical, medical, and packaging applications but is not recommended for extended outdoor exposure.

5. Polystyrene and Copolymers

This class of polymers includes four major commercial products: (1) styrene homopolymer, (2) butadiene-styrene copolymer, (3) styrene-acrylonitrile copolymer (SAN), and (4) acrylonitrile-butadiene-styrene copolymers (ABS). The homopolymer is available commercially as (1) unmodified, general-purpose PS, (2) toughened PS, and (3) expanded PS. Formation of the styrene polymers is achieved via a free-radical mechanism using bulk, solution, emulsion, suspension, and dispersion polymerization processes. Under these conditions, PS is obtained as an amorphous, hard resin. The bulky polar phenyl groups along the main chain have restricted mobility and are responsible for the stiffness and, hence, high T_g (about 100°C). Polystyrene made by free-radical polymerization is essentially atactic, without any particular regularity of steric placement of phenyl groups about the chains; the molecules do not pack into a crystalline lattice.

a. Polystyrene

Although amorphous PS is one of the most important thermoplastics, its useful applications are limited to those that expose the fabricated polymers to temperatures below the T_g. This led to exploring the formation of isotactic crystalline PS by stereospecific polymerization. It was found that, in spite of the fact that crystallinity of the isotactic PS allows for higher use temperatures than those noted for the atactic resin, the crystalline polymer is far more brittle and, hence, is commercially unattractive.

(1) Typical properties (a) General-purpose PS: It is sold as the unmodified homopolymer and is commercially available in different molecular weights and forms. Typical properties include: (1) high transparency (about 90%); (2) specific gravity of 1.04; (3) a tensile strength of 6000-8000 psi, (4) elongation of 1-2.5%, (5) tensile modulus of 4-5 X 10^5 psi, (6) impact strength (for molded articles) of 0.3-0.5 ft-lb/in., and (7) good resistance to water, acids (except concentrated sulfuric acid), bases, alcohols, and detergents. Polystyrene is susceptible to several organic solvents, and the effect of these solvents can be increased by the presence of an external load. Chlorinated

hydrocarbons mar the surface and, in the presence of an external load, can cause failure of PS articles. Aliphatic and aromatic hydrocarbons are generally capable of dissolving or swelling PS.

(b) High-impact (toughened) PS: The two main approaches to improve the impact strength of PS entail (1) mixing PS with a rubber and (2) formation of PS in the presence of a styrene-soluble rubber (interpolymerization). The mixing of the rubber and PS is achieved, most commonly, by mechanical mixing of dry polymers on a two-roll mill, in an internal mixer, or in a suitable extruder. Interpolymerization of the rubber dissolved in styrene can be conducted by a bulk solution or suspension process. A useful rubber for mixing or interpolymerization is a 75-25 butadiene/styrene copolymer.

Typical properties of impact PS include (1) an impact strength of 1-4 ft-lb/in. of notch; (2) a tensile strength of 2500-5000 psi; (3) a tensile modulus of less than 300,000 psi; (4) an elongation of 10-50%; and (5) a toughness (measured by a falling weight on a flat sheet) of 25 in.-lb for 0.02 in. thickness. Compared with unmodified PS, the impact grades are characterized by high toughness but low tensile strength and modulus. In addition, the impact PS has a few technologically undesirable properties, such as (1) low rigidity, (2) low heat deflection temperature, (3) limited transparency, and (4) poor weatherability (associated with the unsaturated rubber modifier). Impact PS can be used in the production of structural foam in which the foamed product has an integral skin and cellular core.

(c) Expanded PS: This is sold in the form of expandable beads that can be processed into low-density foams for packaging. The two basic methods for the production of expandable beads comprise either polymerizing and gassing the beads in one step, or polymerizing the PS first followed by impregnating the beads at elevated temperature and pressure. Pentane is a common blowing agent.

Polystyrene can be processed by all conventional thermoplastic techniques. *Injection molding,* the most widely used technique, is achieved at a typical processing temperature of 360-500°F. *Extrusion* is used for processing solid articles of general-purpose or impact PS using a typical processing temperature range of 360-450°F. In addition, *extruded foam sheets* can be produced using suitable chemical (e.g., azodicarbonamide) or physical (e.g., nitrogen) blowing agents and a nucleating agent (e.g., talc) to control the cell size. *Biaxially oriented* films are produced by extrusion and then drawing. *Coextrusion* of general-purpose and impact PS is used for products having tough core and glossy skin. Polystyrene articles can be finished or decorated using (1) offset printing, (2) painting, (3) hot stamping, (4) vacuum metallizing, or (5) ultrasonic and solvent welding.

Of the many applications of PS, packaging is the most common. Biaxially oriented PS films are thermoformed into blister packs for use in food packaging. The use of PS in housewares and appliances is extensive. Extruded profiles of solid or foamed impact PS are used in the furniture and construction industries. A fast-growing market for PS is in consumer electronics and medical products.

b. Styrene Copolymers

(1) Butadiene-styrene copolymers These are a family of copolymers in which styrene is the major comonomer. They are noted for their high clarity and good impact

strength. They can be processed like impact PS. The BDS polymers are suitable for many of the applications at present satisfied by PS, PVC, and cellulosics.

(2) Styrene-acrylonitrile copolymers Styrene is the major comonomer used for the formation of these copolymers. The polymers are noted for their strength and good chemical resistance as compared with PS. They are used extensively for cups, typewriter keys, and refrigerator parts.

c. Acrylonitrile-Butadiene Styrene Copolymers

These constitute a family of about 15 engineering thermoplastics formed basically from three monomers: acrylonitrile, butadiene, and styrene. Chemically, ABS is composed of discrete rubber particles grafted with styrene-acrylonitrile copolymeric chains and dispersed in a matrix of a SAN copolymer. The rubber particles are usually made of butadiene-styrene copolymer rich in the butadiene moiety.

Different ABS grades are made by changing the butadiene to styrene, styrene to acrylonitrile, and rubber to SAN ratios in the rubber particles, SAN copolymer, and the final rubber/SAN blend, respectively. In a typical ABS, the rubber component contributes impact strength, toughness, and low-temperature retention of properties; the acrylonitrile-rich copolymeric SAN moieties impart chemical and heat resistance and high strength; and the styrene-rich copolymeric SAN chains are required for materials with high gloss and rigidity.

In addition to controlling the ABS properties through changing the compositional ratios of the comonomers, specific properties can be achieved by varying the system morphology by careful manipulation of the processing and postprocessing conditions, by using a fourth monomer, such as α-methylstyrene (this can be incorporated into SAN), or by alloying with other polymers, such as PVC or polycarbonate.

The ABS family of polymers include general-purpose grades as well as (1) high-performance *electroplating* grades, which meet severe thermocycling demands such as those of the automotive industry; (2) high *heat-resistant* grades with a heat-deflection temperature of about 230°F, used in the automotive and housing industries; (3) *flame-retardant* grades containing halogenated flame retardants for use in electrical applications. (4) *structural foam* grades for use in business machines; and (5) *transparent* grades exhibiting up to 80% light transmission for toys and surgical suction pumps. The largest market for the ABS general-purpose grade is in pipe fittings. They are also used in the production of parts for appliances, sporting goods, and luggage.

Conventional thermoplastic processing techniques are used for the fabrication of different ABS grades. Injection-molding grades are processed on either ram or screw injection-molding machines. Extrusion grades can be converted into pipes, sheets, or blow-molded shapes using single-screw or twin-screw extruders. Structural foam ABS can be converted on a standard screw injection-molding machine using an internally compounded blowing agent or a molding machine using low-pressure nitrogen to achieve the desired cellular structure.

6. Polyvinyl Chloride and Related Vinyl Polymers

These represent a wide range of materials varying considerably in formulation and properties, yet having vinyl chloride, vinylidene chloride, or a vinyl acetal as the main precursor of their polymeric chains. The most important members of this group of polymers are: (1) polyvinyl chloride, both plasticized and unplasticized; (2) polyvinylidene chloride; (3) vinyl chloride-vinyl acetate copolymers (VC-VA); (4) vinyl chloride-vinylidene chloride copolymers (VC-VdC); (5) vinyl formal; and (6) vinyl butyral. These polymers are prepared commercially by (1) suspension polymerization (mostly for PVC); (2) emulsion polymerization, using a redox-type free-radical initiator (for PVC and copolymers); (3) bulk polymerization (mostly for PVC, where the monomer precipitates as it forms); or (4) solution polymerization, where a solvent, such as acetone, is added during the polymerization and a chain transfer agent is normally required.

a. PVC

(1) Chemical and physical properties of PVC   Structurally, PVC is similar to polyethylene with the exception of having one hydrogen atom on alternate carbon atoms replaced by a chlorine atom. As normally made (by free-radical polymerization), the structure is largely atactic. This and the large size of the chlorine atom do not allow the chain packing into a crystalline form. Hence, PVC is virtually an amorphous material. The steric requirements imposed on the polymer chain by the bulky chlorine atoms and the increased intermolecular interaction resulting from the dipolar carbon-chlorine bonds are responsible for the high glass transition temperature of PVC. The glass transition of commercial PVC takes place over a wide temperature range, 75-105°C. The high viscosity of PVC in the liquid state makes it difficult to process at temperature well below 160°C (where the chain undergoes chemical changes that are mostly due to dehydrochalogenation). This led to the development of many vinyl chloride copolymers and plasticized PVC formulations that can be melt processed at a lower temperature with minimum degradation as compared with the unplasticized (or rigid) PVC. The copolymers are commonly made using vinyl acetate, vinylidene chloride, or propylene and may have lower glass transition and softening temperatures than the rigid PVC. The thermal decomposition of rigid PVC during processing is associated with discoloration and loss of strength. Hence, heat stabilizers are used in many of the PVC formulations, and properties of these materials are quite dependent on the type and concentration of the stabilizers. The dipolar nature of PVC limits its use as an insulator in applications involving high frequencies, where its loss factor is very high. On the other hand, compared with polyethylene, PVC can be readily welded by radiofrequency techniques and is self-extinguishing, a rare property among thermoplastics.

The homopolymer and most of its copolymers have excellent resistance to aqueous solutions of acids or bases and all but the most severe oxidizing agents. Sulfuric acid below 90% concentration has no effect on PVC below 60°C, whereas stronger acids cannot be used above 50°C. Although oils, fats, alcohol, and aliphatic hydrocarbons can be used in contact with rigid PVC, the polymer swells in aromatic and chlorinated hydrocarbons, ketones, and esters.

Although rigid PVC is extremely hydrophobic and has excellent barrier properties against water vapor, plasticized PVC exhibits increased permeability. The effect of plasticization on the gas permeability for O_2, N_2, and CO_2 is less pronounced. Typical permeability data of unplasticized and plasticized PVC are given in Table 11.4.

As noted above, PVC is used in formulations containing plasticizers and stabilizers. In addition, colorants, fillers, and several types of additives are used to achieve

Table 11.4 Permeability of Unplasticized and Plasticized PVC

Permeability	Unplasticized	Plasticized
Gas permeability (cm^3-cm/cm^2-sec-cm $H_g \times 10^9$)		
Carbon dioxide	0.06	0.2-2
Oxygen	0.01	0.04-0.4
Nitrogen	0.002	0.01-0.1
Water vapor permeability (g/m^2-24 hr) @ 38°C		
@ 90% RH for 0.001 in. thick film	25	50-180

specific properties. On the other hand, different grades of plasticized PVC (depending on the molecular weight, moleular weight distribution, and method of preparation) can be made to have the typical range of properties as shown in Table 11.5.

The deformation behavior of unplasticized PVC is dependent on temperature, particularly above 40°C. For this reason, unplasticized PVC has the lowest maximum temperature of usefulness as compared with other thermoplastics, although it is one of the most rigid thermoplastics at room temperature. In the absence of stress concentrations, PVC is ductile.

Like other physicomechanical properties, the impact strength is dependent on a number of molecular and processing parameters. The effect of plasticizers or other additives on the impact strength is particularly interesting and important. It is adversely affected by small amounts of additives that are soluble in PVC, such as a plasticizer. A soluble additive concentration of 5% by weight can result in a decrease in the impact strength of PVC that corresponds to a change of 15°C in temperature. Insoluble additives, such as mineral fillers (e.g., titanium dioxide), improve the impact strength of PVC. An addition of 10% mineral filler can lead to a considerable improvement in impact strength. Other useful insoluble impact modifiers include natural and certain synthetic elastomers. Additives and plasticizers also have a significant effect on the dielectric properties of PVC and can lead to some improvement by reducing the power loss at high frequencies. Nevertheless, the electrical properties of PVC are inferior to those of other thermoplastics.

(2) Compounding and processing Since PVC undergoes thermal degradation during melt processing, it is usually compounded with additives and/or plasticizers to aid its processing and end-use performance. *Heat stabilizers* are common to all PVC

Table 11.5 Typical Properties of Unplasticized PVC

Properties	Units	Values
Glass transition temperature	°C	75-105
Density	g/ml	1.30-1.58
Coefficient of linear expansion @ −20 to 60°C	in./in./°C $\times 10^{-5}$	5.0-10.0
Thermal conductivity (20-40°C)	cal/cm-sec-°C $\times 10^{-5}$	3.5-6.0
Specific heat (20-80°C)	cal/g/°C	0.25-0.35
Dielectric strength (1/8 in. thickness)	V/mil	350-500

products; they minimize or prevent thermal degradation during melt processing and may also extend the service life of finished goods. Typical stabilizers include inorganic and organometallic compounds based on tin, lead, calcium, or zinc. *Plasticizers* are used to impart flexibility. The level of plasticizer may vary from 10 to 100 parts per hundred parts of resin, to obtain semirigid to very soft products. Phthalate esters are the most widely used type of plasticizers. Other types include adipate, azelate and phosphate esters, and epoxidized soybean oil. Waxes and metallic salts of fatty acids are used as *lubricants* to facilitate the melt flow of PVC compounds during processing and to prevent adhesion to metal surfaces. For improving the impact strength of PVC, a number of *impact modifiers* have been used. These include chlorinated polyethylene, ABS, methyl methacrylate-butadiene-styrene copolymers, and ethylene-vinyl acetate copolymers. To improve the melt homogeneity of PVC compounds, *processing aids,* such as styrene copolymers (with acrylonitrile or MMA), are used. *Fillers,* such as calcium carbonate, are used to increase the heat-deflection temperature, improve electrical properties, and reduce cost. Organic and inorganic *pigments* are used to impart color, opacity, and weatherability.

Polyvinyl chloride and its compounds can be processed by (1) *extrusion* using single- and multiple-screw extruders to produce solid profiles, cellular profiles, piping, blown film, and flat sheets; (2) *injection molding* using conventional equipment for flexible PVC and high shear-rate screw machines for rigid and semirigid grades; (3) *compression molding* to produce phonograph records—vinyl chloride/vinyl acetate copolymers are most commonly used for this application; (4) *blow molding,* as in the production of PVC bottles; (5) *injection blow molding* to minimize scrap generation; (6) *calendering* to produce flexible and rigid PVC sheeting (flexible products are based on high-molecular-weight resins and rigid ones are made from low-molecular-weight PVC or vinyl acetate copolymers); (7) *powder coating,* where powdered PVC (flexible or rigid) compounds are applied by fluidized bed, electrostatic spray, or electrostatic fluid bed techniques to form solid coatings on metals or other substrates; and (8) *liquid processing* in the form of plastisols (or organosol) solution coating in an organic solvent or latex.

(5) Applications About 55% of the products used are rigid and 45% are flexible. The most important use of rigid PVC is for piping systems (water supply, agriculture, and chemical). Filled and unfilled rigid PVC is also used, increasingly, in the construction industry. Flexible PVC is used for gaskets. Other applications of PVC include (1) floor coverings, (2) bottles, (3) phonograph records, (4) packaging materials, and (5) sporting goods.

b. Vinyl Chloride Copolymers and Related Systems

The three most common types of vinyl chloride copolymers are those made using *vinyl acetate, propylene* or *vinylidene chloride* as comonomers. Depending on the copolymer composition, a broad range of properties can be achieved. The key feature of these copolymers is their ease of melt processing as compared with unplasticized PVC, which undergoes thermal degradation (if not stabilized) above 160°C. *Chlorination* of PVC may be used to obtain materials with properties approaching those of vinyl chloride/vinylidene chloride copolymers. Both systems find use in the production of pipes for hot water due to their increased hydrophobicity and hydrolytic stability at elevated temperature as compared with rigid PVC.

A polymer closely related to PVC is polyvinylidene chloride, which differs from PVC in that it (1) is crystalline (T_m = 210°C); (2) has a high density; (3) has high thermal stability; and (4) has low modulus. Other polymers related to PVC include polyvinyl formal (PVFm) and polyvinyl butyral (PVBt). To illustrate the dependence of the polymer properties on molecular chain composition and the type of additives or fillers used, typical properties of PVC are compared in Table 11.6 with those of modified PVC, a vinyl chloride copolymer, PVBt, and PVFm.

7. Fluorocarbon Polymers

This is a class of polymers with paraffinic chains in which some or all of the hydrogens are replaced by fluorine. Polymers of this class offer unique performance characteristics due to their (1) chemical resistance to concentrated acids and bases; (2) insolubility in common organic solvents; (3) desirable mechanical properties over a wide range of temperatures (−200 to 260°C); (4) outstanding electrical properties (low dielectric constant, very low conductance, and high volume and surface resistivity; (5) low coefficient of friction; and (6) excellent thermal and flame resistance. Commercial fluoropolymers are either thermoplastic or elastomeric materials. Thermoplastic fluoropolymers include (1) polytetrafluoroethylene (PTFE); (2) polychlorotrifluoroethylene (PCTFE); (3) polyethylene-co-tetrafluoroethylene (PETFE); (4) polyvinyl fluoride (PVF); and (5) polyvinylidene fluoride (PVDF). The elastomeric fluoropolymers, which are discussed later in this chapter, include fluorinated ethylene-propylene copolymers (FEP).

a. *Polytetrafluoroethylene*

The PTFE molecule consists of a carbon chain saturated with fluorine atoms with hardly any chain defects, that is, side groups similar to those encountered in polyethylene. The chain symmetry and linearity and the steric requirements dictated by the fluoro group are responsible for the helical conformation acquired by these chains as they pack into a highly crystalline solid form. The molecular weight of commercial PTFE is over a million. The crystallinity of the polymer is 90-95%. However, this extremely high crystallinity is never completely recovered after processing; slow- and quench-cooled PTFE exhibit about 75 and less than 50% crystallinity, respectively.

(1) Polymer manufacturing   Tetrafluoroethylene is polymerized under pressure in the presence of excess water (to help remove the heat generated by the exothermic polymerization), using a free-radical initiator. Depending on the specific polymerization conditions, PTFE can be obtained as (1) a granular powder, (2) a coagulated dispersion powder (commonly known as paste polymer), and (3) aqueous dispersions. Many semi-fabricated forms, including films, sheets, tapes, tubes, and rods, are commercially available.

(2) Chemical and physical properties and processing   Although PTFE is quite resistant to common concentrated acids, bases, and organic reagents, it is attacked by alkali metals, elementary fluorine under certain conditions, and a few fluorinated compounds at high temperature.

Despite the exceptional thermal stability and chemical inertness of PTFE, its high T_m (about 325°C) and high melt viscosity preclude its processing by conventional molding and extrusion techniques. The fabrication of PTFE usually entails cold pressing the polymer particles at about 360-380°C and is commonly known as sintering. Details of

Table 11.6 Typical Properties of PVC and Related Systems

Properties	PVC, VA copolymers as molding resins		30% glass-filled PVC	Propylene-vinyl chloride copolymer	PVdC	PVFm	PVBt	Chlorinated PVC
	Rigid	Flexible[a]						
T_g, °C	75-105	75-105	75-105	70	–	105	49	110
T_m, °C	–	–	–	–	210	–	–	–
Compression molding temperature, °F	285-400	285-350	–	290-360	220-350	300-350	280-320	350-400
Injection molding temperature, °F	300-415	320-385	270-405	350-400	300-400	300-400	250-340	325-440
Density, g/ml	1.30-1.58	1.16-1.35	1.54	1.28-1.40	1.65-1.72	1.2-1.4	1.05	1.49-1.58
Tensile strength, psi $\times 10^3$	6.0-7.5	1.5-3.5	9.5	5.0-8.0	3.0-5.0	10.0-12.0	0.5-3.0	7.5-9.0
Elongation, %	40-80	200-450	2	100-140	Up to 250	5-20	150-450	5-65
Tensile modulus, psi $\times 10^5$	3.5-6.0	–	8.7	2.5-4.5	0.5-0.8	3.5-6.0	–	3.6-4.8
Impact strength, ft-lb/in. of notch	0.4-20.0	–	1.0	0.4-32.0	0.3-1.0	0.8-1.4	–	1.0-5.6
Heat deflection temperature, °C @ 264 psi fiber stress	140-170	–	155	150-170	130-150	150-170	–	202-234
Dielectric strength, V/mil (1/8 in. thick)	–	275-290	335-380	335-380	400-600	455	325	–
Water absorption, % (24 hr, 1/8 in. thick)	0.04-0.4	0.15-0.75	0.008	0.07-0.4	0.1	0.5-3.0	1.0-2.0	0.02-0.15

[a]Plasticized or unplasticized, depending on the VA content.

the processing conditions of PTFE depend on the physical form of the virgin polymer. *Granular* PTFE can be molded into billets, sheets, and rings; this is accomplished through preforming, sintering, and cooling. Rods and thick-walled tubes are made by ram extrusions; the loose polymer is compacted in a long straight tube by a reciprocating ram. The compacted polymer moves along the tube to a heated zone for sintering and fusion to a homogeneous rod or tube. Articles with complex configuration are obtained by machining the resulting rods or tubes. *Coagulated dispersion polymers* are used in producing thin sections (tubes or rods) by extrusion using a processing aid or lubricant (e.g., naphthalene). Thus, a mixture of PTFE and the aid or lubricant are lightly compacted into a cake, transferred to the extrusion chamber of the extruder, and then extruded at a constant ram speed to produce thin-walled tubes or thin rods of unsintered PTFE. The lubricant (or aid) is extracted prior to sintering (and then cooling) the extrudates. The cold working of the polymer during extrusion is associated with chain orientation and results in a fibrous structure that retains the mechanical integrity of the polymer during sintering. This also leads to a final product with excellent fatigue properties. *Aqueous dispersions* of PTFE are used for the (1) impregnation of porous materials, such as asbestos or glass fabric, and (2) coating of metals, ceramics, or glass. The dispersions are applied by dipping, spraying, or flow coating. Depending on the intended application, the applied dispersion may or may not be sintered.

Among the desirable properties noted for PTFE, as an engineering thermoplastic, one may include (1) a broad working temperature range of 250-260°C, (2) excellent resistance to almost all chemical environments, (3) low coefficient of friction, and (4) low dielectric constant (2.05). On the other hand, in some applications PTFE performance is less than desirable for it has poor wear and low deformation resistance, The latter drawback may be corrected by the use of a suitable filler to better the PTFE mechanical properties with a minor compromise of chemical inertness and dielectric properties. Typical properties of PTFE bulk polymer and mechanical properties of shaped articles are outlined in Table 11.7. In addition, it is to be noted that PTFE and other fluorocarbon polymers undergo appreciable degradation when exposed to high-energy radiation (e.g., γ rays).

(3) Application Due to its high cost, high purity, and unusual physical properties, PTFE is used in low volume for highly specialized applications in the chemical, electrical, electronic, aircraft, bakery, and confectionery industries. Due to its high coefficient of thermal expansion, it has been used successfully for the production of high-temperature seals and gaskets. Typical applications of PTFE include (1) valves and O-rings; (2) linings for pipes, flexible hoses, and stopcocks; (3) spacers and connectors in high-frequency cables; (4) dry and self-lubricating bearings; and (5) coatings for cooking utensils.

b. Polychlorotrifluoroethylene

This polymer differs from PTFE in that one of the fluorine atoms in each monomeric unit is replaced by a chlorine atom. This substitution causes (1) a decrease in T_m—PCTFE melts at 218°C and can be easily melt processed; (2) an increase in the steric requirement about the main chain and lower packing density in the solid state—PCTFE has a density of 2.13 g/ml; and (3) an increase in the electronic anisotropy about the main chain—this leads to an increase of the dielectric constant (2.3-2.7) and an increase in the intermolecular cohesion. This latter parameter is responsible for the high strength and stiffness of

Table 11.7 Physical and Mechanical Properties of PTFE

Properties	Units	Typical values
Density	g/cm^3	2.15-2.24
Thermal properties		
$\quad$ T$_m$	°C	327
$\quad$ Thermal conductivity, 20-35°C	cal/cm-sec-°C	6×10^{-4}
$\quad$ Specific heat, above 40°C	cal/g-°C	0.23
$\quad$ Thermal expansion	in./in.-°C	1×10^{-4}
$\quad$ Deflection temperature @ 66 psi		
$\qquad$ fiber stress	°C	250
Flammability		Does not burn
Electrical properties		
$\quad$ Volume resistivity		
$\qquad$ (@ 50% RH and 23°C)	Ohm-cm	$>10^{18}$
$\quad$ Dielectric strength (1/8 in. thick)	V/mil	430
Mechanical properties		
$\quad$ Tensile strength	psi $\times$ 10^3	2-5
$\quad$ Elongation	%	200-400
$\quad$ Flexural strength	psi $\times$ 10^3	1.7
$\quad$ Tensile modulus	psi $\times$ 10^4	5.8
$\quad$ Impact strength	ft-lb/in. of notch	3.0
Water absorption		
$\quad$ (24 hr 1/8 in. thick)	%	0.00
Optical properties		
$\quad$ Direct transmission factor	%	68
$\quad$ Direct transmission factor for		
$\qquad$ 1 mm thickness		~0
$\quad$ Scattering coefficient		80

PCTFE as compared with PTFE. With the exception of some loss in resistance to halogenated solvents, the chemical resistance of PCTFE to most reagents approaches that of PTFE.

At a thickness of $\frac{1}{8}$ in., PCTFE can be made optically clear. The polymer is available in pellet form for molding and extrusion using conventional equipment. During extrusion, PCTFE is known to have high melt viscosity and undergoes slight thermal degradation and loss in properties. The polymer films can be laminated, thermoformed, heat sealed, printed, and vacuum metallized.

Extruded, molded, and machined PCTFE products are used in chemical processing equipment and cryogenic and electrical applications. Films of PCTFE are used in packaging and other applications that require excellent barrier characteristics.

c. *Polyethylene-co-tetrafluoroethylene*

This is predominantly a 1:1 alternating copolymer, and its chain can be regarded as a hybrid of polyethylene and PTFE molecular chains. Taking this into account, one may expect PETFE to have intermediate physical properties between those of its "structural parents," PE and PTFE. Indeed, this copolymer (1) melts at 270°C, (2) has a density of 1.7 g/ml, (3) is a tough material with high impact strength, (4) has a working temperature range from about −260 to 150°C, and (5) has a dielectric constant of 2.6. The chemical and electrical properties and weathering resistance approach those of PTFE.

   The copolymer can be processed by conventional thermoplastic techniques. Wire and cable and injection-molded products are important among the growing applications of PETFE. It requires corrosion-resistant processing equipment.

d. Polyethylene-co-chlorotrifluoroethylene

Similar to the case of PETFE, this copolymer is essentially a 1:1 copolymer of ethylene and chlorotrifluoroethylene and may be regarded as a hybrid of polyethylene and PCTFE. The polymer melts at 245°C, has a density of 1.68 g/ml, and its dielectric constant is 2.5. The copolymer has a working temperature range from about −260 to 180°C. In many other respects, including processing, some physical properties, and applications, PECTFE is similar to PETFE. The copolymer tensile strength, wear resistance, and creep resistance are significantly greater than those of PTFE.

e. Polyvinylidene Fluoride

The polymer chain is made of alternating CH_2 and CF_2 moieties. The polymer (1) melts at 170°C, (2) has a density of 1.78 g/ml, (3) has a dielectric constant of 8-9 and high loss factor relative to other fluoroplastics, and (4) has useful properties over a temperature range of −100 to 150°C. The polymer resistance to chemical reagents approaches those of perfluoroplastics.

   The polymer is available as a powder, pellets, or dispersion (in organic solvents) form. It can be extruded, injection molded, transfer molded, and applied as a coating (by the dry powder or dispersion coating technique). It is used mainly in electrical insulations, chemical processing equipment, and metal finishes.

f. Polyvinyl Fluoride

This is the fluoro analog of PVC but can be easily obtained by free-radical polymerization as a highly crystalline material. It is available commercially in a film form only. It is a tough, flexible film with outstanding weathering resistance. It maintains useful properties over a temperature range of −100 to 150°C. Because of its toughness and excellent abrasion resistance and low moisture permeability, it is laminated to plywood, galvanized steel, and metal foils for improved properties.

8. Acrylic Polymers

Acrylic polymers are materials made strictly or largely from methyl methacrylate (MMA), ethyl acrylate (EA), acrylonitrile (AN), or related monomers. Polymethyl methacrylate and copolymers made mostly from MMA are thermoplastic materials characterized by (1) clarity, (2) exceptional weatherability, and (3) good combinations of stiffness, low density, and moderate toughness. They constitute the most important group of acrylic resins. The second group of acrylic polymers are those made by the copolymerization of ethyl acrylate with one or more unsaturated monomers, including MMA, AN, butadiene, and vinyl ethers. Although the homopolymer of ethyl acrylate (EA) is of limited commercial value, its copolymer can be made in the form of glassy thermoplastics, lattices for surface coating, or thermosetting adhesives. Polyacrylonitrile and copolymers containing a major fraction of acrylonitrile moieties represent the third group of acrylics. The acrylonitrile polymers can be obtained as fiber-forming materials, flexible rubbers, or lattices for surface coating. Depending on the final form of the acrylic polymer, bulk, solution, suspension, dispersion, and emulsion free-radical polymerization can be used. Separate discussions of the three groups of acrylics are given below.

a. Methyl Methacrylate Polymers

That poly(methyl methacrylate) is by far the most important member of this group of acrylics and its properties are not significantly altered by minor inclusions of comonomeric units (e.g., those of styrene) in the polymer main chain justifies limiting the discussion in this section to the MMA homopolymer.

(1) Chemical and physical properties Although PMMA is characterized by good resistance to water, alkalis, aqueous inorganic salt solutions, and most dilute acids, it is attacked by hydrofluoric acid and concentrated sulfuric and nitric acids. The resistance of PMMA to detergent solutions and oils is excellent, but it undergoes crazing in the presence of alcohols, ketones, aromatic hydrocarbons, and chlorinated solvents. The equilibrium water contents of PMMA at 20°C at 60 and 100% relative humidity are 0.9 and 2.1%, respectively. It has excellent barrier properties to carbon dioxide, oxygen, and nitrogen; in all cases, a permeability of less than 1.0×10^{-11} cm^3 (NTP) cm/cm^2-sec-cm Hg can be achieved. In addition to its high clarity (92% light transmission), hardness, rigidity, excellent appearance, and good electrical properties, PMMA is characterized by having other useful physical and mechanical properties. This makes it one of the most versatile thermoplastics. Some of these properties are outlines in Table 11.8.

(2) Processing and applications The PMMA resin is commercially available in the form of pellets, powders, sheets, and films. It is also available as cast rods and tubes. Acrylic sheets are made by casting monomer in plate glass cells or by casting between continously moving, highly polished metal belts. The conversion of the monomer to polymeric glass takes place, usually in the presence of a free-radical initiator. Acrylic films are made by blowing extruded parisons to the desired film thickness and slitting the resulting tube. Acrylic molding and extrusion compounds are produced by three methods; cast blocks made by bulk polymerization are reduced to extrudable particles, melted, and extruded into strands. These are then pelletized to the desired dimensions. If a continuous polymerization is used, the molten polymer is transferred to an extruder directly and pelletized. Bead polymer is obtained by suspension polymerization. Rods and tubes can be produced by casting from the monomer or a syrup. Small rods and tubes can be obtained by extrusion. Cast products are usually made from high-molecular-weight polymers and display a high level of toughness. Acrylic resins in pellet or bead form can be injection molded or extruded in commercial equipment. Predrying is necessary for the production of high-quality molded articles. Cycle times and the temperature of melt and of the mold must be controlled closely to attain maximum part uniformity and good dimensional tolerance. Depending on the composition of the molding compound, configuration of the molded article, and direction of the flow, the mold shrinkage range should be from 0.2 to 0.8%. Two- and three-stage screws used with the vacuum extraction system in conventional extrusion equipment are quite satisfactory

Table 11.8 Some Physical, Mechanical, and Electrical Properties of PMMA

Properties	Units	Typical values
Physical		
Density	g/cm^3	1.18-1.19
Softening temperature	$^{\circ}C$	100-120
Coefficient of linear thermal expansion		
(−30 to 30°C)	$^{\circ}C^{-1}$	6.0×10^{-5}
Thermal conductivity	cal/cm-sec-°C	4.7×10^{-4}
Specific heat, 20-80°C	cal/g-°C	0.35
Mechanical (molded specimen)		
Tensile strength	psi $\times 10^3$	7-11
Tensile modulus	psi $\times 10^5$	3.8-4.5
Elongation at break	%	3-10
Flexural strength	psi $\times 10^3$	13-19
Flexural modulus	psi $\times 10^5$	4.2-4.6
Modulus of elasticity	psi $\times 10^5$	4.5
Impact strength	ft-lb/in. notch	0.3-0.5
Hardness	Rockwell scale M	85-105
Electrical		
Volume resistivity	ohm-cm (@ 50% RH)	10^{14}
Dielectric strength (1/8 in. thickness)	V/mil	400-500

in terms of output and quality of product. Acrylic sheets, tubes, and rods can be machined on standard metal- or wood-working machine tools. Sheets can also be shaped by thermoforming. Acrylic parts can be bonded using a suitable solvent. Parts can be welded by hot-surface, hot-gas, ultrasonic, spin, or vibration welding.

Major end uses of PMMA and similar MMA copolymers include signs, glazing, light fixtures, and automotive lenses for their optical properties. This and the polymer's resistance to many chemicals and detergents, toughness, stiffness, and high deflection temperature make it quite useful for the production of sanitary ware; appliance panels, knobs, and housings; aircraft canopies; and instrument panels.

b. Ethyl Acrylate Polymers

The most useful types of ethyl acrylate polymers are those made by (1) the emulsion copolymerization of 95% ethyl acrylate with 5% α-chloroethyl vinyl ether to produce a rubber that can be easily vulcanized; (2) the bulk copolymerization of ethyl acrylate with methyl acrylate to obtain a highly flexible material that can be converted to conduits for low-temperature application (usable down to −100°C); and (3) emulsion copolymerization of ethyl acrylate with butadiene and other monomers to prepare surface coatings and adhesives.

c. Acrylonitrile Polymers

Common types of acrylonitrile polymers are those denoted commercially as nitrile molding resins and used for the production of acrylic fibers. Separate discussions of these are given below.

(1) Nitrile resins (NR) These are made by copolymerization of 70 parts acrylo-nitrile with 20-30 parts styrene or methyl methacrylate and 0-10 parts butadiene. The nitrile fraction of the chain contributes to the polymer's excellent barrier properties and resistance to chemicals. Films made from NR are used for food packaging (not beverages). The incorporation of butadiene moieties into the copolymer composition increases the material impact strength and toughness. Since polyacrylonitrile is unsuitable for melt processing, its use in a copolymeric form with styrene and MMA is necessary to obtain thermoplastic materials that can be processed with ease using conventional injection-molding and extrusion equipment. In fact, the processing versatility of the nitrile resins has been demonstrated in (1) the extrusion and coextrusion of sheets and films; (2) the extrusion and injection-blow molding of oriented and unoriented containers; (3) thermo-forming; and (4) injection molding of many types of articles.

It is to be noted that processing of NR is typical for polymers that do not have sharp melting temperatures. The NR is viscous at the processing temperature of 390-420°F. This calls for the use of minimum shear rate during melt processing. The melt viscosity of NR below 390°F is too high, and above 420°F the polymer undergoes chemical changes leading to discoloration. Injection-grade NR has a lower molecular weight (high melt index) than those used for extrusion. Some of the properties of different forms of NR articles are given in Table 11.9. The major uses for NR are in the food and nonfood packaging applications.

(2) Acrylic fibers These are based on polymers containing more than 85% acrylo-nitrile units and 15% other monomers, such as vinyl chloride. Copolymers made from 25-85% acrylonitrile are commonly known as modacrylics. Polymers used for the pro-

Table 11.9 Typical Properties of NR

Properties	Form and/or type of NR	Values
Tensile strength @ yield	Molding grade	9-10 $\times$ 10^3 psi
Elongation @ yield	Molding grade	3-5%
Flexural modulus	Molding grade	4.5-5 $\times$ 10^5 psi
Notched impact	Unmodified molding grade	0.2-0.5 ft-lb/in.
	Impact-modified molding grade	1.5-2.0 ft-lb/in.
	Impact-modified extrusion grade	2-4 ft-lb/in.
Heat deflection temperature	MMA copolymer	165-170°F
@ 66 psi	Styrene copolymer	180-200°F $\pm$
Gas transmission	Film	cc/mil/in.^{2}/day
(@ 50% RH, 73°C)		
Oxygen		0.8
Carbon dioxide		1.6
Water vapor transmission		4-5 g/mil/
(@ 90%, 100°F)		in.^{2}/day
Electrical properties	Molding grade	
Volume resistivity		1.9 $\times$ 10^{15}ohm-cm
Dielectric strength		340-400 V/mil

duction of acrylic fibers are made commercially in a continuous process entailing free-radical polymerization of an aqueous dispersion of the monomer.

Although acrylic fibers can be made by melt, wet, or dry spinning, the chemical instability of the molten polymer limits industrial fiber production to the latter two methods, entailing the use of polymer solutions. Due to poor polymer solubility in common organic solvents, special solvents are used to prepare the spinning solutions. Typical useful spinning solvents include N,N-dimethylformamide (DMF) and N,N-dimethylacetamide (DMAC). In wet spinning, the degassed and filtered solution containing 10-30% of polymer is sent to a spinneret and passed through a spinning bath to coagulate the yarn and remove the solvent. This is followed by passing the multifilament yarn through a stretching bath to develop the desired orientation and tensile properties. In dry spinning, a solution of the polymer in DMF is fed to the spinneret (using a special screw or gear pump) at 80-150°C. The filaments coming from the spinneret go through a column in which air circulates at 230-260°C. During this process, the solvent evaporates and the yarn is drawn to develop the desired orientation and tensile properties. Naturally, the most obvious application of acrylic fiber is its use as textile fibers. Typical fiber properties are given in Table 11.10.

9. Acetal Polymers

Commercial acetal polymers consist of a homopolymeric acetal, polyformaldehyde (or polyoxymethylene) (Delrin), and acetal copolymers made from formaldehyde and ethylene oxide (Celcon). Structurally, acetal polymers are similar to polyolefins, the principal difference being that the backbone of the acetal molecule is made in part or fully of alternating carbon and oxygen atoms. The acetal polymers are outstanding among thermoplastics for their high rigidity, which is combined with high tensile yield strength and good fatigue endurance. Copolymeric and stabilized polyoxymethylene have reasonable thermal stability, and their dimensional stability is excellent in a wide range of environments.

Table 11.10   Typical Properties of Acrylic Fibers

Properties	Values
Tensile properties	
Tensile strength, g/denier*	2.5-4.5
Knot strength, g/denier*	2.0-3.5
Elongation, %	27-48
Elastic recovery @ 3% elongation, %	90-95
Apparent Young's modulus, kg/mm^2	260-650
Thermal properties	
Softening temperature, °C	190-240°C
Melting temperature, °C	Not distinct
Water regain @ 20°C and 65% RH	1.2-2.0
Specific gravity	1.14-1.18

*Denier is a unit of fineness of yarn where 9000 meters weigh 1 g.

   Acetal polymers (or resins) are available in grades suitable for injection molding, extrusion, and extrusion blow molding. Glass-filled acetal resins can be used when maximum stiffness is called for.

a. Chemical and Physical Properties

Acetal resins have excellent resistance to alkalis and resist most common organic solvents. On the other hand, this class of polymers undergoes chain degradation and loss of physicomechanical properties in the presence of mineral and some organic acids and even acid-generating materials, such as moist chlorine. Certain organic solvents, such as acetone and ethyl acetate, can cause deterioration of the mechanical properties of the acetals. Due to the oxidative instability of the acetal repeat units, oxidizing acids, such as nitric acid, cause noticeable degradation of acetal resins.

   The high symmetry of the polymer chain in acetal resins is responsible for their observed high crystallinity and inherent stiffness. Despite the chain's high oxygen content, the high crystallinity of acetal resins, and chain packing density, the acetal resin barrier properties to gases and water vapors are good. These and a few important physical and mechanical properties for a typical acetal resins are given in Table 11.11

   As semicrystalline moderate to high-molecular-weight polymers, typical acetal resins exhibit the physical and mechanical properties shown above. In addition, it is to be noted that these resins are suitable for molding and extrusion, and their properly

Table 11.11 Typical Physical and Mechanical Properties of Acetal Resins

Properties	Values
Density, g/cm^3	1.410
Thermal properties: T_m, $^{\circ}$C	163
Thermal expansion ($10^{-5}/^{\circ}$C)	8
Specific heat, cal/g $^{\circ}$C	0.35
Thermal conductivity, cal/cm-sec-$^{\circ}$C	5.5×10^{-5}
Heat distortion temperature (66 lb/in.^{2})-$^{\circ}$F	338
Flammability: burns slowly with drips	
Gas permeability (cm^3-cm/cm^2-sec-cm Hg) $\times 10^3$	
Carbon dioxide	0.004
Oxygen	0.003
Nitrogen	0.001
Water vapor transmission (g/m^2 24 hr) for 1 mil film	42
Equilibrium water absorption	0.8%
Mechanical and electrical properties	
Tensile strength (lb/in.^{2})	10,000
Elongation at break (%)	15-75
Flexural strength (lb/in.^{2})	14,000
Impact strength (ft-lb/in. of notch)	1-2
Dielectric strength (V/mil)	450
Dielectric constant (10^6 C/sec)	0.004

shaped articles or test specimens can be shown to (1) not only have greater rigidity than polyolefins, but the deformational behavior at 20°C is not so strongly influenced by thermal history, (2) have high impact strength that changes only slightly when the molding conditions are varied, (3) have low coefficients of friction under light applied load in the unlubricated state when mated with metals (0.15) or another acetal polymer (0.35), (4) have electrical properties that are not quite as good as those of polyolefins but adequate for many noncritical applications, and (5) tend to be translucent rather than transparent; the direct transmission factor for 0.1 mm thick sheet is essentially zero.

b. *Processing and General Applications*

The normal molding temperature range is between 200 and 300°C, and best performance is achieved on machines with fast ram speeds and high variable pressure. One-piece injection-molded articles do not normally require finishing. Rods, tubes, and sheets can be extruded on conventional equipment. Extruders with high length-diameter ratios (L/D = 20) equipped with a metering screw with a constant-pitch section are recommended. Thin- and thick-walled containers can be formed by blow molding. Molded components can be easily machined, welded, and painted. In their most common form, as injection-molded articles, fabricated acetal resins with their excellent retention of mechanical properties up to about 100°C and in a wide variety of chemical environments appear to have great potential as replacements for metals, such as die-cast zinc and aluminum. Due to their low coefficient of friction, acetal resins are used for the production of bearings and gears that can be run without lubrication for short times under high loading. However, for extended use at high loads, water or nonacidic oils or greases should be used as lubricants. For their good creep properties, acetal resins are used for molding snap fits with good holding power, such as bushings. Other applications of acetal resins include their use in the production of instrument housings, doorknobs, pump and air impellers, and hinges.

10. Polyamides

Commercial thermoplastic polyamides are made mostly of linear aliphatic chain molecules with recurring secondary amide groups. These polymers are formed by the (1) ring opening of lactams (AB-type chains), (2) self-condensation of aminocarboxylic acids (AB-type chains), or (3) the condensation of diamines with dicarboxylic acids (AB-BB-type chains). These three types of polyamides are frequently referred to as nylons, a generic name. For the AB type, the number of carbon atoms of the lactam or amino acid monomer is used to designate the nylon type. Thus, caprolactam (6C) and 11-amino undecanoic acid (11C) can be used to produce nylon 6 and nylon 11, respectively. For naming the AA-BB-type polyamide, combinations of the number of carbon atoms of the diamine and the diacid are used; for example, hexanediamine (6C) and decanedicarboxylic acid (10C) form nylon 6,10.

Properties of the different nylons and hence their application depend primarily on their chemical structure, molecular weight, and molecular weight distribution. The chemical structure and the amide contribution to the total mass determines the polymer melting temperature, tensile strength, modulus, and hydrophilicity. The molecular weight and molecular weight distribution contribute mostly to the physicomechanical properties of the polymers. Most commercial nylons are used for the production of both molded articles and fibers.

a. Chemical and Physical Properties

A basic structural parameter that affects the chemical properties of nylons is the amide group frequency along the main chain. Thus, the type and frequency of the group determine to a great extent (1) the affinity of the polymer to water, (2) the chain's chemical reactivity, and (3) the physicomechanical properties of bulk polymer and shaped articles.

Most nylons are unaffected by water at room temperature or boiling water. However, at higher temperatures, especially above the T_m, hydrolysis and degradation occur. Nylons are usually stable to dilute alkalis; 10% sodium hydroxide solution at 85°C has no effect on nylon 66 over a period of 16 hr. On the other hand, as in simple amides, nylons undergo rapid acidolysis and degradation upon heating with aqueous mineral or monocarboxylic acids at elevated temperatures, particularly above the polymer T_m.

Heating a typical commercial nylon, such as nylon 6, in an autoclave with ethylene oxide results in a hydroxyethylated polyamide. Depending on the degree of amide substitution, the hydroxyethylated nylon can be obtained as an elastic material with high vapor permeability.

Ethylene carbonate can be combined with nylon 6 or 66 to form water-soluble products through reaction with the carboxyl end groups and amide —NH— of the chain molecule.

Elastic methoxymethylated polyamides have been obtained by treating, for instance, nylon 66 with formaldehyde and methanol in the presence of an acidic catalyst. Formaldehyde reacts with polyamides in the solid state or in formic acid solution to form N-methylol derivatives. These are thermosetting and undergo cross linking upon heating.

Nylons can be grafted with several vinyl and acrylic monomers when exposed to high-energy radiations. Using ionizing radiation (electron beam or γ rays), nylon 6 has been grafted with styrene, acrylonitrile, or acrylic acid to produce materials with a broad range of properties. The acrylic acid grafts in nylon 6 fibers are readily interconverted to the corresponding sodium or calcium salts by treatment with sodium carbonate or calcium acetate. The sodium salts of the grafts are highly hydrophilic as measured by moisture regain (20% moisture regain at 100% relative humidity, versus less than 10% for a nylon control).

Although the moisture absorption of a polyamide depends on its chemical composition (as reflected in the hydrocarbon fraction), for most commercial polymers, the equilibrium moisture pickup ranges from about 1% (nylon 12) to about 8% (nylon 66) at room temperature and 100% RH. It is to be noted that the moisture absorption of a particular nylon can vary to some extent with the degree of crystallinity, the temperature, and the relative humidity. At a high temperature, the moisture content is considerably lower; thus, at 265°C, a nylon 66 melt contains 0.16% of moisture at equilibrium.

Commercial nylons are usually soluble in phenols, formic acid, and chloral hydrate at room temperature. At elevated temperatures, alcohol-halogenated hydrocarbon mixtures can be used to dissolve nylons. Nylon 66 is soluble in methanol under pressure.

Nylons undergo not only thermal (usually above 300°C) and thermo-oxidative degeneration (normally at or below 300°C in the presence of oxygen) but also photo and photo-oxidative degradation (in the presence of light and oxygen). Thermal degradation of nylon 66 at 305°C leads to the formation of carbon dioxide, water, ammonia, hexamethyleneamine, and a few primary monoamines. Thermal degradation of nylon 6 leads principally to its depolymerization to caprolactam. Exposure to sunlight causes deterioration of the properties of many nylon-based fibers. The degradation is accompanied by a loss of strength and elasticity of the fibers. Stabilization of nylons toward photooxidation (in the presence of light and air) can be accomplished by use of ultraviolet screeners that are effective at wave lengths up to at least 4000 Å, or by inorganic antioxidants, wuch as Cu^+ and Mn^{2+} salts. Titanium dioxide (TiO_2), as a delustrant in nylons, acts as a photosensitizer and increases the photodegradation at wave lengths above 3000 Å. Exposure of nylon to high-energy radiation, such as γ rays, leads not only to the formation of cross-links but also to some chain scission. Radiation degradation can be more damaging in the presence of oxygen, particularly as oxygen diffuses into the polymer after the free radicals are initially formed by irradiation.

Due to their regular symmetrical structure and chain flexibility, aliphatic polyamides are usually highly crystalline. In the solid state, the hydrogen bonding in nylon results in high rigidity and high heat deflection temperatures. However, in nylon melts, where the effect of hydrogen bonding is virtually nonexistent, the flexibility of the polyamide chain is responsible for low melt viscosity. This is associated with excellent processability and ease in developing strength through orientation of extruded fibers and films. The alternation of paraffinic segments with highly polar amide groups along the nylon chain contributes to some of the desirable properties of polyamides, such as (1) high lubricity and low friction coefficient, (2) high toughness and impact strength, (3) wear resistance, and (4) good dyeability. In the presence of water, most aliphatic nylons exhibit a decrease in glass transition temperature accompanied by changes in many physical and mechanical properties. The amide-hydrocarbon ratio and the number of methylene groups constituting the hydrocarbon residue in polyamides are two of the most important parameters that have a direct bearing on the physical properties and, hence, applications of nylons. An illustration of the effect of this parameter on the T_m of AB-type polyamides is given in Table 11.12.

b.   Processing and General Applications

Nylons can be melt processed using conventional molding and extrusion equipment. In addition, shaped articles can be made from nylon by thermoforming. Because of the

Table 11.12   Melt Temperatures of Homologous AB Polyamides

$$+ (CH_2)_n - \overset{\overset{\textstyle O}{\textstyle \|}}{C} - NH +_p$$

Polyamides from	n	Tm (°C)
Caprolactam	5	223
Enanotholactam	6	233
Capryllactam	7	200
ω-Aminopelargonic acid	8	209
Azacycloundecan-2-one	9	188
ω-Aminoundecanoic acid	10	190
Laurolactam	11	179

broad range of properties offered by the different nylons it is instructive to discuss the applications of polyamides on the basis of the main properties utilized in these applications, that is, chemical, physical, and mechanical, as shown in Table 11.13.

c. Mechanical Applications

Many of the engineering components that have been traditionally metallic are now being made from injection-molded or pressed and sintered nylons. Among small components, this allows for rapid production of parts at relatively low cost, while benefiting from some of the properties associated with nylon, such as strength with lightness, resilience, corrosion resistance, and freedom from noise in service. Nylon gears and bearings are commonly used in small mechanisms where feed lubrication is difficult, or a sealed unit is required, or when it is undesirable to lubricate, as in food processing. These components are generally made in injection molds designed to produce the final article with a minimum of machining. The flexibility, wear resistance, and low friction of nylon led to its use in the production of rollers and wear plates in many forms of mechanical designs. The combination of easy processing by injection molding, lightness, toughness, and corrosion resistance makes nylon useful for the production of rotating components, such as in fans, propellers, and impellers. An expanding area of application for nylon is in housings and casings for industrial tools and domestic appliances. This is particularly so where a combination of high-temperature resistance, impact resistance, and electrical insulation is required. Generally, glass-filled nylon 66 is used when high rigidity and improved dimensional stability are desired in the housing. When toughness and flexibility are considered more important, unfilled nylon 66 or nylon 11 (or 12) is used. Small and medium-sized housings can usually be injection molded; large housings are often centrifugally cast.

Table 11.13 Types of Properties that Determine Application

Chemical
   Chemical stability at a pH range of 4-14
   Resistance to hydrocarbon solvents

Physical
   High T_m and heat deflection temperature
   High crystallinity and good dimensional stability
   Low specific gravity
   High glass transition temperature
   Low melt viscosity
   High specific resistance

Mechanical
   High tensile strength and modulus
   High impact strength and toughness
   Low friction coefficient and good wear resistance

---

d. Electrical Applications

The tendency of a polyamide to absorb moisture, an undesirable feature in electrical applications, favors the use of more hydrophobic polyamides, such as nylon 11 and 12. These polymers offer better mechanical and thermal properties than polyolefins. Typical applications of nylons include their use as insulators in many electrical parts and appliances.

e. Chemical Applications

For compatibility with various types of chemicals, extruded nylons are widely used in the form of tubing. Thin-walled tubings are used for low-pressure fluid lines. For high-pressure hydraulic lines, nylon-reinforced hoses are used.

f. Packaging

Nylon of various types finds application in the area of packaging. Included are (1) extruded nylon strips as substitutes for steel strappings, (2) heat-stabilized grades of blown film for cook-in-the-bag food items, (3) films laminated with polyethylene for vacuum packing of perishable foodstuffs, and (4) nylon-polyethylene laminates for ready-to-use flexible pouches.

g. Applications of Modified Polyamides

In addition to their use as engineering plastics and fibers, various forms of polyamides are used in the coating and adhesive industries. Copolymers of nylon are made to be appreciably soluble in alcohols and can be made into solutions for adhesives and depositing moisture-absorptive and water-vapor-permeable coatings on fabric. Hot-melt adhesives usually consist of a blend of one or more vegetable-oil-based polyamide resins, together with a plasticizer. Vegetable-oil based polyamides can be combined (as curing agents) with epoxy resins to give thermosetting adhesives. Soluble polyamides, particularly those based on vegetable oils, have been used in flexographic ink formulations, required for plastic packaging material such as polyethylene and cellulose acetate.

h. Nylon Fibers

One of the most important applications of nylon is its spinning to multifilament yarns for the production of textiles. Melt-spun nylon monofilament is used where natural threads and metallic wires have been traditionally useful. Most textile yarns are based on nylon 6, 66, or 11 and can be easily produced by melt spinning predried chips. The yarn is then drawn to impart improved orientation and the development of crystallinity. The resulting yarn (monofilament or multifilament) has a high tenacity and, depending on the type of nylon, molecular weight, and processing conditions, tenacities can range from 3.0 g/denier to 10 g/denier. Fiber strength is only slightly affected by wetting, dropping to $2.6 - 9$ g/denier. Elongation of commercial nylon fibers ranges between 16 and 65%; the fibers have excellent recovery properties after up to 5% extension.

11. Polyesters

The three most important commercial thermoplastic polyesters are polyethylene terephthalate (PET), polybutylate terephthalate (PBT), and polycyclohexanedimethylene terephthalate (PCHDMT). These polymers are made by the condensation of terephthalic acid or its dimethyl ester with ethylene glycol, 1,4-butanedial, and 1,4-cyclohexane dimethanol (70:30 isomer mixture) to produce PET, PBT, and PCHDMT, respectively. Both PET and PCHDMT are commonly used in the production of textile fibers to take advantage of their high T_m and ability to develop excellent tensile properties after orienting and annealing their yarns. Although both PET and PBT can be used for injection molding (in presence or absence of fillers), the lower T_m and high crystallizability of PBT make it a more desirable resin than PET as an engineering thermoplastic. A fourth polyester, which is made by the polymerization of ϵ-caprolactone, is finding some use as a molding resin.

a. Chemical Properties

Crystalline polyester articles made by extrusion (fibers or films) or molding (filled or unfilled) are known to (1) be hydrophobic (water absorption at equilibrium is less than 1.0%); (2) resist dilute mineral and organic acids; (3) undergo degradation in aqueous alkaline solutions, especially in hot concentrated ones; (4) react with alcohols and amines (depending on temperature, these reagents may cause excessive degradation of the polyester chain and, hence, deterioration of the polymer mechanical properties); (5) generally, be more resistant to photo and photo-oxidative degradation as compared with other heterochain thermoplastics, such as the nylons. The effects of ehcmical and photo-oxidative degradation on polyesters are illustrated for yarns of PET, PCHDMT, and nylon 66 in Tables 11.14 and 11.15.

b. Physical Properties, Processing, and Applications

The physical properties of the four polyesters, namely, PET, PBT, PCHDMT, and PCL (polycaprolactone), as well as some related systems, are discussed with some reference to their method of production and processing.

Table 11.14 Effect of Acids and Alkalis on Pet Yarn

Chemical	Concentration (%)	Temperature (°F)	Time (hr)	Tenacity[a] (g/d)
Control	—	70	—	6.1
Sulfuric acid	1	70	1000	6.0
Sulfuric acid	1	250	100	4.4
Hydrochloric acid	37	70	100	5.3
Hydrochloric acid	10	160	10	6.6
Sodium hydroxide	40	70	10	6.6
Sodium hydroxide	10	210	1	4.4
Sodium hydroxide	1	250	10	1.3

[a]Tenacity is a measure of the fiber strength and is proportional to the force required to achieve tensile failure.

Table 11.15   Half-Life of Continuous Yarns in a Weather-O-Meter[a]

Yarn composition	Tenacity (g/d)	Elongation (%)	Time (hr) to lose 50% of the initial	
			Tenacity	Elongation
PCHDM-T	2.8	20	122	96
PET	4.7	17	270	60
Nylon 66	5.6	17	145	175

[a]An instrument used to study photo and photo-oxidative degradation.

c. PET Production, Properties, Processing, and Applications

The polymer is made in most commercial processes in two stages starting with dimethyl terephthalate and an excess of ethylene glycol in the presence of a suitable catalyst, usually $Mg(OAc)_2 + Sb_2O_3$. The product of the first stage, the bis-hydroxyethyltere-phthalate, is then heated at 270-285°C under reduced pressure until the desired molecular weight is achieved.

Depending on the percentage of crystallinity and degree of orientation, the glass transition of PET occurs at 80-150°C. Unoriented, unannealed PET is usually a clear glass with a T_g of about 80°C. The crystalline polymer melts at 252-265°C, depending on the percentage of crystallinity and thermal history. The polymer is generally delivered as clear amorphous pellets that must be dried before processing. Rotary vacuum driers, horizontal driers, or hopper driers with mechanical agitation to prevent the tacky, hot (above T_g) pellets from sticking together may be used to reduce the moisture level to the desired 0.005% level. Since drying is conducted above the polymer T_g, crystallization takes place during this process. The dry polymer can then be melt processed at 260-275°C. Conventional extrusion equipment may be used in producing monofilaments (or multifilament textile yarn), strapping, tubing, and films. To stabilize the molded, shaped article dimensions and improve the mechanical properties, annealing above the polymer T_g (80°C) is usually done. On the other hand, films and fibers are oriented at different temperatures to develop optimum physicomechanical properties. Since PET is used mostly for the production of biaxially oriented films and textile fibers, the discussion of PET shaped articles pertains to these forms. Typical illustrations of fiber and film properties are given in Tables 11.16 and 11.17. It may be concluded that biaxially oriented PET films have (1) excellent tensile properties, (2) lower water transmission and gas permeability as compared with films made from other heterochain thermoplastics, and (3) good electrical properties.

Although PET is used extensively for fiber and film production, injection-molded articles can be obtained in either an essentially amorphous or a semicrystalline state. For the production of amorphous articles, extremely pure, high-molecular-weight PET is used; the molded articles are not exposed to a temperature above 50°C (to limit crystallization). Amorphous molded articles are usually transparent and have good impact strength. On the other hand, semicrystalline molded articles can be made from PET with moderate molecular weight, containing a nucleating agent. The crystalline articles are somewhat stronger than the amorphous ones, as shown in Table 11.18.

Table 11.16 Typical Properties of Biaxially Oriented Pet Films

Properties	Values
Specific gravity	1.38-1.41
Tensile strength, psi $\times 10^3$	20-35
Elongation, %	60-165
Tearing strength, g/mil	12-27
Water absorption at 24 hr, %	<0.8
Rate of water transmission at $23°$C-g-mil/100 in.^{2}/24 hr	1.0-1.3
Permeability to gases, ml/100 in.^{2}/mil	
Carbon dioxide	15-25
Nitrogen	0.7-1.0
Oxygen	3.0-4.0
Dielectric strength, V/mil	7500 (1 mil)
Volume, resistivity, ohm/ml	10^{18}
Heat-sealing temperature, $°$C	218-232

Table 11.17 Properties of Fiber-Grade PET and Properties of a Typical Textile Yarn

Properties	Values
Polymer	
Molecular weight, daltons	18,000-25,000
T_g, $°$C (amorphous)	78-80
Crystallization temperature (T_c), $°$C	125-180
T_m, $°$C, by DSC	225
T_m, $°$C, by microscopy	260-265
Fiber[a]	
Tenacity, g/denier	2.5-6.0
Elongation, %	12-25
Elastic recovery at 2% extension, %	90-96
Average stiffness, g/denier	8-25
Specific gravity	1.38
Moisture regain (at 65% RH and $25°$D), %	0.4

[a]The polymer can be spun at about 260-280°C, and the amorphous fibers are drawn at 180-200°C.

Table 11.18 Properties of Amorphous and Semicrystalline Molded Specimens

Properties	Amorphous	Crystalline
Specific gravity	1.30-1.34	1.32-1.38
Tensile strength, psi $\times$ 10^3	8	11
Elongation to break, %	250	250
Impact strength, Izod (ft-lb/in.)	1.0	0.8
Maximum water absorption, %	0.7	0.6

d. PCHDMT Production, Properties, Processing, and Applications

The polymer is made in two stages, a melt polymerization and solid-state or melt post-polymerization. The melt polymerization in conducted at 220-240°C under atmospheric nitrogen, using dimethyl terephthalate and slight excess (5%) of 1,4-cyclohexanedimethanol (a 70:30 trans/cis mixture is most commonly used). The resulting low-molecular-weight polymer is then postpolymerized in the melt at 300-310°C under reduced pressure to attain the desired viscosity. Alternatively, the powdered low-molecular-weight polymer is postpolymerized in the solid state at 260°C under reduced pressure.

Prior to melt processing the polymer must be dried as discussed above for PET. Due to the high (about 290°C) T_m of PCHDMT, its application is essentially limited to the production of textile fibers using conventional extrusion equipment. The extrusion is usually conducted above 300°C. The resulting fibers are drawn at 110-200°C. Additional thermal treatment to increase the polymer crystallinity and dimensional stability can be conducted at 175-220°C. Some of the properties of PCHDMT textile fibers are comparable to those of PET. Of the properties in which the two fibers differ, one may note (1) the relatively high T_g and hence better retention of properties at elevated temperature by PCHDMT, (2) the comparatively high tensile strength and elongation associated with PET, and (3) the noticeably greater hydrolytic stability of PCHDMT; this may be related to its high T_m and crystallinity.

PCHDMT Copolymers (PCHDM-T1) are made from mixtures of dimethyl terephthalate and isophthalate. They have a lower T_m than PCHDMT itself and, thus, can be melt processed under more practical conditions. One of these copolymers is sold under the trade name Kodar. It is extremely clear and tough. It is suggested as a substitute for PVC in packaging applications (*C & E News*, Sept. 22, 1975, p. 6). Compared with PVC, these copolymers are stated to (1) have lower density, (2) require a shorter cycle time and lower heat energy for thermoforming blisters, and (3) display higher toughness and, hence, are more useful for producing thin films. Some of these copolymers can be converted to biaxially oriented films with comparable properties to those of PET. Properties of a film made from 83:17 terephthalate-isophthalate PCHDM copolymers are compared with PET in Table 11.19.

e. PBT Production, Properties, Processing, and Applications

The polymer is obtained by the condensation of dimethyl terephthalate with an excess of butanediol, in presence of a suitable catalyst, such as titanium tetrabutylate. The polymerization is conducted in two stages as described above for PET, but the maximum reaction temperature is kept at 250°C (to avoid thermal degradation). Compared with

Table 11.19 Properties of 0.5 Mil Biaxially Oriented Films of PCHDM-TI and PET

Properties	PCHDM-TI	PET
Density, g/ml	1.23	1.39
Tensile properties		
Yield strength, psi $\times 10^3$	10	14
Breaking strength, psi $\times 10^3$	17	26
Elongation at break, %	45	105
Modulus, psi $\times 10^5$	4.0	6.8
Elmendorf tear strength, g/mil	6	18
Mullen burst strength	28	37
Water absorption (24 hr immersion at 25°C), %	0.30	0.55

PET, PBT is known to have excellent molding properties, for it has relatively low T_g and T_m and a greater tendency to crystallize. Molded PBT specimens with over 40% crystallinity can be easily obtained. The amorphous T_g and crystalline T_m are about 60-70°C and 225-232°C, respectively, depending on the thermal history of the samples.

Although PBT can be converted into strong fibers, it has become more popular as an injection-molding resin. Both conventional screw injection molding and plunger-type machines can be used for the production of molded PBT at about 235-250°C, using a short cycle time. Molded, high-molecular-weight PBT has (1) good strength, (2) high toughness, (3) excellent dimensional stability, (4) low water absorption, (5) low static and dynamic coefficient of friction, (6) good chemical resistance, and (7) significantly lower heat deflection temperature than PET. Reinforced PBT with 10, 20, and 30% glass fiber can be converted to molded articles with improved mechanical properties as compared with the unfilled polymer. Typical properties of filled and unfilled PBT molded articles are given in Table 11.20.

Table 11.20 Physical and Mechanical Properties of Molded PBT

Properties	Unfilled	Filled with 30% glass fiber
Specific gravity	1.31-1.38	1.52
Tensile strength, psi $\times 10^3$	8.2	16.0-19.6
Elongation, %	50-300	2-4
Flexural strength, psi $\times 10^3$	12.0-16.7	23.0-26.5
Tensile modulus, psi $\times 10^5$	2.8	13
Flexural modulus, psi $\times 10^5$	3.3-4.01	11.0-12.0
Impact strength, ft-lb/in. of notch		
Deflection temperature at 66 psi fiber stress, °C	115-190	225
Water absorption, % (0.5 in. thick, 24 hr)	0.08-0.09	0.06-0.08

f. Polycaprolactone (PCL) and Related Polymers

Polycaprolactone is formed by the ring-opening polymerization of ϵ-caprolactone in the presence of a suitable catalyst (e.g., dibutyltin oxide) at temperature below 200°C. The polymer has a T_g and T_m of about −40 and 70°C, respectively. Although it can be easily melt processed to several types of molded and extruded articles, its low melting temperature limits its use in many conventional thermoplastic applications.

Other cyclic esters, such as glycolide and lactide, can be easily converted to high T_m polylactones. Thus, glycolide is polymerized in the presence of a tin catalyst to produce a polymer that melts at about 225-235°C. This can be melt spun, and the extrudates can be drawn to produce strong fibers that undergo extensive degradation in moist environments. Similarly, a 10:90 lactide-glycolide mixture can be converted to a fiber-forming copolymer that melts at about 205°C. Fibers made from these two polymers are not suitable for use as textile fibers but are successfully used for the production of absorbable surgical sutures.

12. Polycarbonates

This is a special class of polyesters based on carbonic acid, as compared with the traditional polyesters made from organic diacids (or their derivatives) discussed in the previous section. Commercial polycarbonates (Lexan, Merlon, and Tuffak) are made by condensation of phosgene or diphenylcarbonate with a bisphenol. General-purpose polycarbonate is derived from bisphenol-A, but for special uses formulations using small amounts of other polyhydric phenols are employed. These include resins characterized by high melt strength for extrusion and blow molding.

a. Chemical and Physical Properties

Polycarbonates are relatively hydrophobic as compared with polyamides for they absorb about 0.35% water at equilibrium and retain their properties in water and dilute solutions of weak alkalis and acids. However, the physical and mechanical properties of polycarbonates can be adversely affected by ketones, aldehydes, esters, aromatic hydrocarbons, ammonia, organic bases, and concentrated alkaline or acidic solutions. In standard grades, polycarbonate exhibits exceptionally high impact strength and good electrical properties. Thus, the high thermal transition temperatures [for poly-(2,2-bis-(4-phenylene)propane carbonate) the T_g and T_m are about 150 and 220-230°C, respectively], high modulus (or rigidity), low water absorption, and good dimensional stability, permit their use over a wide temperature range, −50 to 130°C, at 264 psi. The creep resistance of polycarbonate is good over a broad temperature range, and parts can be molded consistently to tolerances of 0.002 in./in. Due to the high T_g of polycarbonate, crystallization is not easily achieved in normal moldings; this leads to transparency in molded articles. Highly crystalline materials can be obtained only by special techniques, such as annealing at 180°C for more than 24 hr.

The highly aromatic nature of polycarbonate and the absence of secondary and tertiary carbon atoms accounts for the good oxidative stability of this polymer. It is usually stable in air at temperatures up to 150°C over long periods of time. At higher temperatures some oxidation and cross-linking occur. Ultraviolet light is strongly absorbed by polycarbonate and causes crazing and degradation; this effect is restricted to the surface. Thus, films may become brittle or weathered; molded parts are not seriously affected.

b. Processing, Mechanical Properties, and Applications

Polycarbonate resin (from bisphenol-A) is easily processed as a typical thermoplastic. Although it is often injection molded or extruded into flat sheets, it can be processed by (1) profile extrusion, (2) structural foam molding, (3) blow molding, and (4) thermoforming using vacuum, pressure drape, or strip heating methods. Secondary operations and finishing methods include solvent and adhesive bonding, painting, printing, and hot stamping, ultrasonic welding, and heat staking. Although polycarbonate resists water at ambient temperature, it degrades in the presence of moisture under melt-processing conditions. Hence, it is usually necessary to dry the polymer prior to processing. This is achieved by circulating hot air at 125-130°C.

Although unfilled polycarbonate is an excellent engineering thermoplastic, reinforcement with 5-40% glass fiber is commonly used to (1) improve the polymer creep resistance by up to 4000 psi at temperatures as high as 100°C, (2) decrease the mold shrinkage, (3) increase the tensile and flexural modulus, and (4) increase the tensile strength. Some of the physical and mechanical properties of filled and unfilled polycarbonates are given in Table 11.21. It may also be noted that foamed polycarbonate with excellent properties can be obtained by conventional foam-processing techniques. Foamable polycarbonate's properties are a direct extension of those of the base resin. However, in the foamed state, chemical resistance is improved because of low molded-in stresses. A number of foamable grades are available with a flexural modulus as high as 800,000 psi.

In spite of its high melt viscosity, predried polycarbonates can be molded on conventional molding machines at 240-300°C. Extrusion can be also achieved in this temperature range, and good results have been obtained with single, flighted screws with equal pitch and increasing core diameter. Rods, pipes, sheetings, and films can be easily processed by extrusion. Molded parts are easily finished by drilling, turning, milling, planing, and polishing. They can also be metallized and printed. Coatings can be cast

Table 11.21 Some Physical and Mechanical Properties of Filled and Unfilled Molded Polycarbonates

Properties	Unfilled	Filled (10-40% glass fiber)
Crystalline melting T_m, °C	Not detectable	Not detectable
Amorphous transition temperature T_g, °C	150	150
Specific gravity	1.20	1.24-1.52
Tensile strength, psi $\times 10^3$	8.0-9.5	12.0-25.0
Elongation, %	100-130	1-5
Tensile modulus, psi $\times 10^5$	3.0-3.5	5.0-17.0
Flexural modulus, psi $\times 10^5$	3.2-3.5	5.0-14.0
Deflection temperature (at 264 psi fiber stress), °C	130-140	143-149
Dielectric strength, V/mil (1/8 in. thickness)	380	450
Water absorption, % (24 hr, 1/8 in. thickness	0.15-0.18	0.07-0.20

from solutions or applied by fluidized-bed techniques. Polycarbonate can be used for the production of films and molded articles for the electrical, building, lighting, and automobile industries. Examples of specific applications include tail lights for cars, parts for x-ray equipment, and household utensils.

13. Aromatic Ether Polymers

These are phenylene (or arylene) oxide-based resins made by the oxidative coupling of phenolic monomers, such as 2,6-dimethylphenol. The polymerization is achieved in an oxidizing medium, such as a copper salt in pyridine. Aromatic ether polymers are available commercially as (1) polyaryl ether, or (2) modified phenylene oxide resins (Noryl). The latter system is a proprietary composition made of aromatic polyethers and other thermoplastics.

Chemically, the aromatic polyethers have an exceptional resistance to hydrolysis and are unaffected by acidic and basic solutions at room temperature. The hydrolytic stability is attributed to the polymer hydrophobicity and absence of ester or amide group. The stiffness and steric requirements about the main chains of aromatic polyethers are responsible for the noncrystalline nature of these polymers and their high T_g (above 100°C). The high T_g of the aromatic polyethers and their high aromatic content are responsible for their low tendency to absorb water. However, they soften or dissolve in certain chlorinated or aromatic hydrocarbons. Electrical-grade resins exhibit good dielectric properties, which are virtually unaffected by changes in relative humidity. Commercial polyether resins can be converted to shaped articles with good tensile properties and excellent dimensional stability at a wide temperature range. Due to their high T_g, the polyethers are characterized by good creep resistance. Both unfilled and filled resins, when molded, display minimum mold shrinkage for they undergo virtually no crystallization.

In terms of processability, polyaryl ether and Noryl can be melt processed using conventional molding and extrusion equipment. In addition, they are known to have good flow properties and high thermal and thermooxidative stabilities. All types of injection-molding machines have been used successfully. However, screw machines are preferred over plunger-type or combination presses because of their rapid, uniform heating of the cylinder and reduced pressure loss through the cylinder. Physical and mechanical properties of filled and glass-filled aromatic polyether compositions are given for typical commercial specimens in Table 11.22. The data illustrate the similarity between Noryl (modified polyphenylene oxide) and polyaryl ether, with a few exceptions. These include the T_g, heat deflection temperature, and water absorption.

Aromatic ether polymers are used extensively in electrical and electronic applications for their desirable thermal and electrical properties and low water absorption. Platable-grade resins are used in the automotive industry. Extrudable resins find application as protective coatings in wiring devices.

14. Aromatic Sulfone Polymers

These are highly or fully aromatic polymers with sulfonyl groups constituting an intrinsic part of the chain. Commercial sulfone polymers have also an isopropylidene and/or ether group in their main chains. The three common sulfone systems are (1) polysulfone, with an isopropylidene bisphenyl structure; (2) polyether sulfone having sulfonyl and ether groups interlinking p-phenylene groups; and (3) polyphenyl sulfone, which is the same

Table 11.22 Physical and Mechanical Properties of Polyaryl Ether and Filled and Unfilled Phenylene Oxide-Based Resins

Properties	Units	Modified phenylene oxide resins, molding grade			Polyaryl ether
		Extrusion grade	Unfilled	Glass-filled, 20-30%	
Glass transition temperature T_g	°C	105-120	105-120	105-120	160
Heat deflection temperature 264 psi, fiber stress	°C	212-265	212-265	270-300	300
Tensile strength	psi $\times 10^3$	6.5-9.6	7.8-11.5	14.5-17.0	7.5
Elongation	%	50-60	50-60	4.0-6.0	25-90
Tensile modulus	psi $\times 10^5$	3.55-3.80	3.55-3.80	9.25-12.0	3.2
Flexural strength	psi $\times 10^3$	12.8-13.5	12.8-13.5	18.5-20.0	11.0
Flexural modulus	psi $\times 10^5$	3.6-4.0	3.6-4.0	7.5-11.0	3.0
Density	g/ml	1.06-1.10	1.06-1.10	1.21-1.36	1.14
Dielectric strength (1/8 in. thick)	V/mil	400-550	400-550	420-600	430
Thermal expansion	in./in./°C $\times 10^3$	5.2	5.2	2.2	3.6
Water absorption, 24 hr., 1/8 in. thick	%	0.066	0.066	0.06	0.25
Impact strength	ft-lb/in. of notch	5.0	5.0	2.3	8.0

as polyether sulfones with the exception of having the ether interlinking p-phenylene and biphenylene groups. The steric requirements about the chains of the sulfone polymers and their inherent stiffness and high aromatic content are reflected in their (1) high glass transition temperature; (2) glassy, noncrystalline nature; and (3) high oxidative, thermo-oxidative, and hydrolytic stabilities.

The glass transition temperature of the sulfone polymer occurs between about 300 and 430°F. This, coupled with their noncrystalline nature and thermal stability, allows the molding of these polymers in conventional thermoplastic molding equipment at about 640-720°F. Although mold temperatures of 200-300°F are commonly used, higher mold temperatures generally assist mold filling and reduce mold-in stresses. The amorphous nature of these polymers leads to low mold shrinkage; this important feature makes them quite useful in applications requiring close dimensional tolerance.

a. Polysulfone

Polysulfone has high resistance to acids, alkalis, and salt solutions. This polymer is attacked by polar organic solvents, such as ketones, aromatic hydrocarbons, and chlorinated hydrocarbons, but has exceptional hydrolytic stability.

The polymer has a T_g of 375°F, and it exhibits good creep resistance. It has an Izod impact strength of 1.3 ft-lb/in. of notch, and a tensile impact strength of 200 ft-lb/in.[2] The dielectric constant of the polymer is very low because of its hydrophobicity; the electrical properties are independent of changes due to high humidity or immersion in water, even at temperatures as high as 350°F. Polysulfone is used in electrical, electronic, and biomedical applications. It is also used in chemical processing and pollution control equipment, as an unfilled or glass-filled grade.

b. Polyether Sulfone

With the exception of its high T_g (430°F), this polymer has comparable chemical and physical properties to those of polysulfone. The high T_g of polyether sulfone makes it more desirable in applications requiring dimensional stability at high service temperatures. The Izod impact strength of this polymer is 1.6 ft-lb/in. of notch, which is slightly higher than that of polysulfone.

c. Polyphenylsulfone

Although polyphenylsulfone is similar, in many respects, to the above two polymers, it is distinguished by having (1) a high deflection temperature of 400°F, which allows a continuous use temperature of 375°F; (2) a very high impact strength, 12 ft-lb/in. of notch; and (3) a high resistance to environmental stress cracking.

B. Thermosetting Resins

1. Phenolic Resins

The phenolic resins are not only the most versatile of the synthetic thermosetting plastics, but they were the first synthetic resins to be commercially exploited on a large scale. These materials are a class produced by the condensation of a phenol (such as carbolic acid), or mixture of phenols (such as cresylic acid, from cresol and xylenol) with formaldehyde, accompanied by the elimination of water.

As with other thermosetting resins, it is necessary to be able to provide the commercial user with a material that is both tractible and fusible, with the ability to later cross-link. These low-molecular-weight products are referred to as A-stage resins. In the case of phenol-formaldehyde, two types of A-stage resins may be formed: novolacs and resols. Novolacs are prepared by reacting phenol with part of the required formaldehyde under acid conditions to obtain a thermoplastic. Production of a resol is accomplished by reacting a phenol with an excess of formaldehyde under basic conditions. (Unlike novolacs, the resols have the capability to cross-link on their own by an adjustment of the pH, and are known as one-step resins.) In the heated mold both novolacs and resols begin to harden; they are transformed to rubberlike materials, in which form they may be swollen, but not dissolved by a variety of solvents; the transformed solid material is known as a B-stage resin. Upon further reaction during the molding operation, the final cross-linked product, a C-stage resin, is obtained; it is rigid, insoluble, and infusible. If the resin is delivered to the user in novolac form, a hardener (e.g., "hexa," i.e., hexamethylene-tetramine; cystamine), which supplies the balance of the needed formaldehyde, must be added prior to the hot molding.

The rate at which a phenolic resin will reach the desired degree of cross-linking is dependent upon temperature, pH, moisture, and formulation of the system.

1. A low-pH system will result in faster gelation than a high-pH system. Slowest reactivity occurs at a pH of about 4. If one increases the alkaline content of the system, it will level off to a fairly rapid gel time. Increasing the acidity of a system can result in an explosively fast reaction.
2. Increases in temperature dramatically cut the gel time of the resin.
3. Varying the amount of cystamine in a novolac system affects both the rate of increase in the viscosity and the final viscosity at gelation.

Major applications for phenolic resins fall into the following categories: molding compounds, laminates, and bonding resins. When molded under heat and pressure, phenolic resins yield strong, heat-resistant, insoluble products that will neither crack nor split on aging. However, a phenolic resin is rarely molded without the addition of modifiers. Alone, phenolic resins tend to be brittle and to display extensive shrinkage upon polymerization, due to the elimination of water. Final product dimensions are therefore difficult to control, and parts tend to adhere to hot mold surfaces, creating problems in ejection. To overcome these problems, the formulation of the molding compound contains phenolic resin, filler or reinforcing fiber (e.g., metals, minerals, and cellulosic fibers), a hardening agent, such as cystamine, plasticizer, dye or pigment, and a mold release agent.

Due to the large variety of compounds that can be manufactured, the phenolics are classified by the final product properties. General-purpose phenolic resins have both good strength and electrical resistance and are inexpensive. These materials, which can have a high surface gloss, are employed for a majority of molded products, including pot handles and knobs. In general, the filler employed for these compounds is wood flour.

By changing to a fibrous filler, the user can obtain medium- and high-impact phenolic compounds. The use of cotton, nylon, and glass fibers can increase the impact strength of a phenolic-based material up to 20-fold.

Laminated phenolic structures are manufactured employing the one-step resins previously mentioned (resols). When dissolved in alcohol or water, the resin may be used to impregnate such materials as paper and cloth. The amount of resin in the final product,

Table 11.23 Properties and Uses for Reinforced Phenolic Resins

Reinforcement medium	Properties and uses
Paper	Insulation: terminal strips, insulating sleeves, high-voltage applications
Fabric (e.g., cotton)	Noise, shock absorbance: gears, bearings
Asbestos	Heat resistance: high-temperature applications
Glass	Heat resistance, impact strength
Decorative grades	Serving trays

which generally runs from 30 to 70%, is controlled by the concentration of the solution and by some form of squeeze roll in the process. Oven heat is then employed to advance or "body" the resin. Final cure is done in heated platen presses where the finished laminate is made.

The properties of a phenolic laminate will depend not only on the resin itself (including the catalyst used and the final molecular weight of the resin), but on any of the following: the solvent employed, the viscosity and resin content of the resin-solvent (varnish) mixture, the type of reinforcement, and process conditions, such as temperature and pressure. Addressing the latter, processing is generally done at temperatures up to 375°F and pressures of about 2000 psi.

Phenolic laminates generally have good strength, high rigidity, good machinability, and good insulating properties. Table 11.23 indicates that the wide variety of phenolic resins available, combined with varied reinforcements, provides a diversity of products. It is the industrial, rather than the decorative grades, that predominate in the industry. This is due to the limited, dark colors available in phenolics.

The use of phenolics as bonding resins applies to friction materials, abrasives, wood particles, and other inorganic fibers. Resinous adhesives, based on resorcinol, can be made to harden at room temperature when activated (by paraformaldehyde) to a water-proof, heat-resistant glue.

Miscellaneous uses for phenolic resins include coatings, cast products, vulcanizing agents, and antioxidants for some thermoplastics.

2. Aminoplasts

The term "amino resin" refers to materials that have their chemical basis in ammonia compounds. Aminoplasts and amino plastics, however, are more specific in their reference to only the thermosetting class of materials. Although a wide variety of aminoplasts exists, the major commercial resins are based on the reaction of formaldehyde with urea and with melamine; both of these reactions will be dealt with in some detail.

The formation of resins from formaldehyde and an amino material is accomplished by adding the formaldehyde to initiate a condensation reaction. In the presence of an acid catalyst the reaction can proceed through, forming a dimer, a polymer, or an infusible network; a by-product of this process is water. What is necessary to the successful manufacture of these resins is close control over the stoichiometric amounts of reactants, the pH of the reaction mixture, and the reaction time and temperature.

a. Urea-Formaldehyde Resins

Urea, or carbamide, is the most important building block for the amino resins, in that urea-formaldehyde is by far the most widely used of the amino resins, and urea itself is a source for melamine (the second most widely used).

The addition of formaldehyde to urea crystals results in a methylol compound; this reaction is catalyzed by either acids or bases. Since, however, the product of this step is further reactive under acid conditions, it is usually carried out in alkaline media. Then, acid catalysis and the application of heat causes the monomer units to condense, giving off water, and curing.

Urea-formaldehyde resins were originally manufactured to compete with the already existing phenolics. The advantages and disadvantages of each are compared in Table 11.24.

A wide range of uncured urea polymers for varied end uses may be obtained by changing the proportions of the reaction components (including the catalyst) and the reaction temperature. If the uncured polymer is then blended with additives, it can be made suitable for molding compounds.

Molding compounds from urea resins may contain the urea resin and hardener; a filler, which for urea-formaldehyde resins is limited to α-cellulose or wood flour; a stabilizer to increase shelf life; a plasticizer, whose main role is to reduce cure shrinkage and improve flow properties; lubricants (in very low concentrations); and a pigment.

Urea-formaldehyde molding resins have found their way into the manufacture of buttons, trays, and toilet seats, as well as electrical applications.

These resins have also been employed as adhesives, an important application of which is the manufacture of chipboard, where the resin system is dissolved in a solvent, such as formalin, and added to wood chips; the entire mixture is then hot pressed to provide polymerization. The urea resin may also be mixed with water and a latent catalyst (which reacts with the water to harden the system), in which case the user has a "cold-set" adhesive.

Other uses for urea resins are found in the textile industry (to finish fabrics), as foams, and due to its excellent oil and grease resistance, as containers for cosmetics.

Table 11.24   Use of Urea and Phenolic Plastics

Ureas	Phenolics
Not recommended for use above 77°C	Can be used at temperatures in the range 205-260°C
Absorbs moisture, swells, can lead to crazing	Better moisture resistance
Should not impart taste or odor to product	Can impart taste and odor to product
Availability of a wide range of colors	Only limited, dark colors available
Better electrical insulation in cases where sparking can occur	Better electrical application where higher strength is required

b. Melamine-Formaldehyde Resins

A number of fairly complex methods for the manufacture of melamine exist, some of these include urea or cyanamide as a source material. In any case, the result is a white, crystalline, nonhygroscopic solid, which may then be reacted with formaldehyde to produce a methylol compound. Resinification at this point is dependent on the pH of the system, which is normally about 10-10.5. (An increase in the speed of the reaction can be achieved by lowering the pH of the reaction mixture.)

A comparison of the properties of the melamine-formaldehyde resins with those previously stated for urea-formaldehyde resins is given in Table 11.25.

Aside from the popular use of melamines in dinner ware and kitchen appliances, the melamine-formaldehyde resins, when combined with various fillers, may be employed in automobile ignition systems, heavy-duty circuit breakers, and switch components.

The excellent physical characteristics of the melamine resins make them good candidates for lamination. Melamine laminates may be employed in decorative panels, countertops, and wall coverings. In order to provide wider use of the resin and cut costs, the melamines are often modified or used in conjunction with a cheaper resin, such as a phenolic.

Other applications for melamines include

1. Adhesives
2. Textile treatments
3. Surface coatings

3. Epoxy Resins

The characteristic molecular structure of an epoxy resin is the three-membered ring compound, referred to as 1,2-epoxide or oxirane:

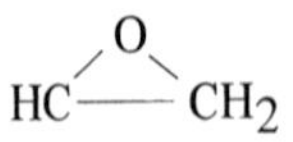

$$HC \overset{\displaystyle O}{\underset{\textstyle \diagup \quad \diagdown}{\text{------}}} CH_2$$

This ring may be opened during an addition reaction by hydrogen-bearing molecules, leading to polymerization. The most important commercial epoxy resins are derived from the reaction between bisphenol-A and epichlorohydrin. Other epoxy systems include those where bisphenol-A is replaced by phenol, resorcinol, or aliphatic glycols.

Table 11.25 Comparison of Melamine- and Urea-Formaldehyde Polymers

Melamine-formaldehydes	Urea-formaldehydes
Wide variety of colors	Wide variety of colors
Wide variety of fillers, including cotton fabric (for strength), asbestos (for same order of heat resistance as phenolics), and minerals (for electrical resistance)	Limited to the use of only two fillers
Excellent water resistance	Absorbs water
Hardest surface of any synthetic resin	

Epoxies, like polyesters, achieve curing by addition reactions; heat is evolved during the reaction, but no gaseous by-products are formed. Whether liquid or solid, the resin cures through the addition of curing agents or hardeners. These additives fall either in the class of catalytic systems or coreactive systems. Hardeners will incorporate themselves into the final product.

Catalytic curing agents cause the resin to homopolymerize, and although the amount of catalyst will affect the speed of the reaction (and final product quality), the specific amount added is not as critical as in polyesters. The homopolymerization is a direct result of the presence of either a Lewis acid or base in the curing process. The most frequently employed acids are complexes of boron trifluoride with amines or ethers; important Lewis bases include tertiary amines.

On the other hand, since coreactive or hardener systems become linked into the epoxide chain, specific attention must be paid to making additions stoichiometrically. Table 11.26 includes some typical hardeners employed in epoxy systems.

As can be seen from Table 11.26, the number of commercially available hardening agents is large, and the choice of one over another will not only depend on the desired final properties of the material, but on ease of handling, pot life, and cure rates. It should be stressed that the time and temperature of cure will play an important role in the properties of the final material, since rapid highly exothermic cures tend to leave a high level of stress in the cured resin.

Epoxy materials are desirable for their excellent properties, which are better understood from a consideration of their molecular structure.

Table 11.26 Some Typical Hardeners Used in Epoxy Systems

Class	Properties	Examples
Primary aliphatic amines	Short pot time, fast cure, high reaction exotherm, skin irritant	Diethylene triamine, triethylene tetramine, triethylamine, dimethylaminopropylamine, diethylaminopropylamine, piperidine
Anhydrides	Less toxic than primary aliphatic amines, low reaction exotherm, requires heat cure, can yield final product with high heat distortion temperature	Phthalic anhydride, pyromellitic dianhydride, hexahydrophthalic anhydride
Secondary cyclic amines	Used with primary aliphatic amines to modify pot life, exotherm, cure rate	Tridimethylaminoethylphenol, dimethylaminophenol

1. Toughness and durability are derived from the bisphenol-A moiety.
2. Epoxies are resistant to chemicals, including solvents, acids, and alkalis, due to the presence of the ether linkages.
3. The resins can withstand temperatures in excess of 205°C with good retention of mechanical properties, because of the bisphenol-A.
4. Their high peel strengths and good adhesive properties stem from the presence of the hydroxyl and epoxy groups.

Unfortunately, although they have many advantages for use in commercial products, the epoxy resins as a class tend to be brittle and rigid and thus not yield a product with the desired impact strength. Furthermore, the liquid epoxy resins may be extremely viscous, making their handling difficult. One can, however, modify the epoxide, or rigid, portion of the polymer chain to provide increased toughness and flexibility, as well as easier handling. Table 11.27 summarizes the various means by which the epoxy resins may be modified.

The major commercial applications for epoxies include coatings, adhesives, laminates, molding, potting, encapsulation, and as a matrix for structural composites.

4. Polyesters

An ester is the result of the reaction of an organic acid with an alcohol (see also Sec. II.A.11). This reaction yields water as a by-product. If the organic acid contains more than one $-COOH$ group (polybasic) and the alcohol contains more than one $-OH$ group (polyhydric), then the product will be a polyester. A controlling factor in the final yield of these reactions will be the extent to which the water is removed; the presence of the water can cause a reverse reaction to occur.

In order for the resin to be capable of polymerizing to a three-dimensional thermosetting structure, the polyester must be unsaturated; that is, either the acid or the alcohol or both must contain unsaturated or doubly bonded carbon atoms ($C=C$ or $C=O$). The polyester can then cross-link by one of two means: (1) reaction with an adjacent polyester of the same structure, or (2) more often, reaction with a monomeric substance also containing an unsaturated double bond. The addition of a monomer tends to "thin"

Table 11.27   Modifiers of Epoxy Resins

Modifier	Example	Result
Diluents	Phenyl glycidyl ether	Reduce resin viscosity, simplify handling, retards cure
Fillers	Polyamide liquids	Reduce cure shrinkage, lower coefficient of expansion, reduce exotherms, increase thermal conductivity
Plasticizers	Polysulfide rubbers	Increase flexibility, toughness
Reinforcing fibers and fabrics	Graphite cloth	Increase strength, rigidity

the product and provide a medium of low enough viscosity for subsequent processing. Such monomers include styrene (a low-viscosity liquid) and diallyl phthalate. The further addition of a catalyst, whether it be in the form of a chemical additive, heat, or light, permits the respective unsaturated bonds of the monomer and the ester to react, forming a hard, infusible resin.

The final chemical and physical properties of the polyester product will depend mainly on (1) the composition and amount of the constituent acid, alcohol, and cross-linking monomer employed, (2) the molecular weight of the resulting resin, and (3) the inhibitor (a material added to the resin by the manufacturer to prevent premature cross-linking).

From the engineer's point of view, polyesters provide a variety of useful properties.

1. Low cost and ease of production
2. A wide range of both formulations and resin forms (liquid, paste, and so on), adaptable to many processing requirements
3. Excellent mechanical, thermal, and electrical properties, as well as dimensional and chemical stability
4. Ease of modification for a variety of molds and colors
5. Ability to be employed in specialized fields, such as radar applications

As seen in Sec. II.A.11, polyesters can be produced in thermoplastic form, such as fibers (Dacron) and films, (Mylar), which are both high-molecular-weight and oriented, and plasticized completely saturated polyesters. Polyester foams are also available; these materials contain a high concentration of hydroxyl groups cross-linked with isocyanates.

a. Effect of Molecular Structure on Physical Properties

As previously stated, the final material properties are mainly influenced by the chemical composition of the constituents. In the choice of an acid for a particular polyester for-mulation, it should be remembered that the molecular weight of the acid will determine the frequency of ester linkages in the polyester chain. Ester bonds are particularly polar, and therefore will display a strong affinity for adjacent polyester chains. This attraction enhances the physical strength of the polymer by adding rigidity to the material.

The choice of the acid constituent will not only affect the resulting strength of the final product, but it will also influence such other properties as water and heat sensi-tivity, flexibility, and the degree of crystallinity. An increase in the chain length of the alcohol will lead to increased spacing between the ester bonds from the acid components. These bonds are the most probable sites at which cross-linking will occur; a greater dis-tance between them will increase the flexibility of the final polymer.

The cross-linking monomer selected will play an integral part in the copolymeriza-tion process. Since the reaction occurs at the time of product formation, it is imperative that the compatibility of this component with the chosen acid and alcohol be known.

Table 11.28 shows some of the possible choices for acid, alcohol, and cross-linking monomer components, and the influence of each component on final polymer properties.

As in the case of most other polymer systems, an increase in the molecular weight is associated with increased physical strength and hardness, as well as resistance to heat and chemicals. A higher molecular weight resin implies a high incidence of cross-linking (tightly knit network).

Table 11.28   Acid, Alcohol, and Cross-Linking Monomer Components

Substance	Features
Acids or anhydrides	
Maleic anhydride	Low cost
Fumaric acid	Low cost, toughness, chemical resistance
Adipic acid	Flexibility
Sebacic acid	Resilience
Phthalic anhydride	Low cost; good, average properties
Isophthalic acid	Toughness; heat and chemical resistance
Chlorendic acid	Self-extinguishing
Tetrahydrophthalic anhydride	Tack-free surface
Diols (from petroleum)	
Propylene glycol	Low cost, hardness[a]
Ethylene glycol	Low cost, usually with other diols[a]
Dipropylene glycol	Lower water absorption, increased heat resistance[a]
Diethylene glycol	Higher impact strength, increased heat resistance[a]
Tripropylene glycol	Increased flexibility
Triethylene glycol	Increased toughness
Butylene glycol	Chemical resistance
Glycerol pentaerythritol	Introduces chain branching, chemical resistance, slightly improved strength
Cross-linking monomers	
Styrene	Low cost, compatibility, strength
Vinyl toluene	Heat-deflection resistance, low volatility
Acrylic or methacrylic acid or esters	Light stability
Diallyl phthalate	Heat and chemical resistance, low volatility
Diallyl isophthalate	Higher heat resistance
Dichlorostyrene	Fire and heat resistance

[a]Less tendency to crystallize.

Inhibitors, which will slow or prevent the chemical reactivity that initiates cross-linking, are generally added to the resin by the manufacturer in order to promote the shelf life of the material or to reduce the rate of the final reaction exotherm. These inhibitors can "protect" the resin from heat, light, or contaminating factors.

Alternatively, when activation of the resin is required, a catalyst system or initiator is added (which will overcome the effects of the inhibitor).

Table 11.29 provides a list of some of the more common catalyst systems.

b. General Properties and Applications

The application of polyesters to a specific process or product is directly related to one of the following classes of properties: physical, electrical, chemical, and weathering. In

Table 11.29 Catalysts

Room temperature[a]
  Benzoyl peroxide
  Cumene hydroperoxide
  Cyclohexanone peroxide
  Methyl ethyl ketone peroxide (MEKP)

Moderate temperature (80-120°C)
  Benzoyl peroxide
  Cumene hydroperoxide
  MEKP
  Lauroyl peroxide

High-temperature (175°C)
  Dicumyl peroxide
  Tertiary butyl perbenzoate

[a]Can be used with promoters to speed reaction.

general, polyesters are known for their high impact resistance, hardness, good electrical properties, and heat and abrasion resistance.

It should be noted that many of the plastics employed for products are "reinforced" to enhance physical properties. By reinforcing synthetic resins, the scope of the use of these products has widened into areas including engineering and structural applications. The reinforcing material is generally a fiber, such as glass, carbon, alumina, or graphite. As an example of the benefits of reinforcement, glass is known to have high tensile strength and modulus of elasticity, and these, as well as other properties are imparted to the polyester upon reinforcement. A few of the various applications for reinforced polyester resins are fishing rods, table tops, luggage, washing machine tubs, refrigerator parts, chemical piping, and polyester-glass cloth, mats, and other laminaters.

c. Alkyd Resins

The preceding section on polyesters covers a variety of combinations of acids and alcohols. Alkyd resins are a specialized class of polyesters, and have gained enough importance in industry to deserve being placed in a class of their own.

The term "alkyd" generally applies to an oil-modified polyester, that is, the product of a reaction between a polyhydric alcohol (commonly, ethylene glycol), a polybasic acid (such as phthalic acid), and the fatty acids of various animal- and vegetable-derived oils (tung, linseed, soya, castor, coconut, and others). The oils employed are all triglycerides, which contain unsaturated groupings.

The monobasic fatty acid controls the extent of polymer growth and contributes to the final physical and chemical nature of the product by virtue of its own inherent properties. For instance, the degree of unsaturation in the fatty acid group will determine the susceptibility of the alkyd resin to oxygen attack, allowing it to dry to a film.

Aside from their low cost, the value of alkyd resins lies in the ease with which they can be modified.

d. Modification of Alkyd Resins

In order to achieve a variety of properties, alkyds may be modified with resin, phenolic resin, epoxy resin, silicones, and monomers, such as styrene. Such modification allows a range of applications, including coatings for primers, refrigerators, and farm equipment

Alkyd molding compounds are further subject to modification by using fillers. These mineral and fibrous additives include clay, asbestos, glass fibers, and cellulose.

Alkyd resins are characterized by high rate of cure, good adhesion, durability, flexibility, and lack of volatiles.

5. Furans

Furans are derived from furfural or furfuryl alcohol, both of which resinify rapidly under strongly acidic conditions. Furfuryl alcohol is currently the more important commercially. It is rapidly resinified in the presence of acid catalysts.

Since the alcohol is highly reactive, with a large exotherm, the resinification is done in two stages. In the first stage the resinification is stopped while the system is still liquid by cooling and neutralization to a pH of 5-8. Water from the intermolecular dehydration is removed. At the time of use, the system is recatalyzed by addition of acid (the commonly used acids are phosphoric acid, p-toluene sulfonic acid, p-toluene sulfonyl chloride, and isopropyl sulfuric acid). To avoid eruptive blowing, the resin is cooled during the final cure. Design should provide for shallow pathways to the surface in order to allow venting of moisture without development of blowholes.

The cured resin is dark colored, very thermally stable (service temperature can be as high as 130°C), and extremely resistant to alkalis, acids, and many common solvents (chloroform, chlorobenzene, toluene, acetone, ether, and esters). The furans, on the other hand, are susceptible and readily degraded by strong oxidizing agents (bromine, iodine, hydrogen peroxide, nitric acid, sulfuric acid stronger than 60% concentration, sodium hypochlorite solutions, and concentrated chromic acid). Cured furan resins are very resistant to ultraviolet radiation and outdoor environments, including marine exposure. Furan resins are somewhat brittle. This is overcome by dispersing fibrous reinforcement material (e.g., glass or asbestos fiber) in the resin.

The principal uses of furan resins are in the chemical industry as a chemically inert coating and for corrosive environments.

IV. MATERIAL SELECTION FOR SPECIFIC APPLICATIONS

Regardless of the particular process being employed, and whether the material in question is a thermoplastic or a thermoset, it is imperative to understand the fundamental material properties that will influence both process and final product quality. (See Figs. 11.1-11.10.) Some of these properties, together with the questions that should be asked when selecting a particular material for a specific application, follow.

1. How does the *viscosity of the material* change with changes in temperature and pressure? Are there shearing forces involved in the application that will degrade the material? How will these shearing forces affect the viscosity? Finally, what kind of control is maintained over material viscosity such that the minimum requirements for final product quality are met?

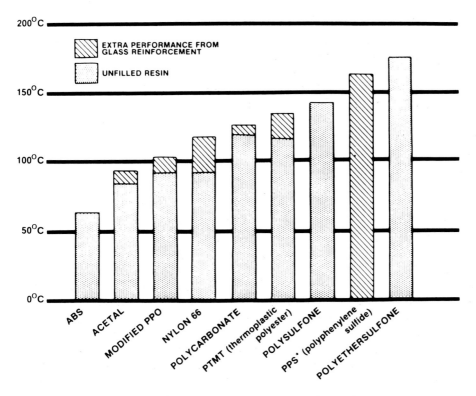

'MINERAL FILLED GRADES CAPABLE OF HIGHER CONTINUOUS USE TEMPERATURES

Fig. 11.1 Maximum continuous-use temperature of some engineering thermoplastics based on Underwriters Laboratories ratings.

2. How does the *melting point* change as a function of pressure? How high is the melting point relative to the process and application temperatures? What is the thermal history of the material under consideration? Has the material been degraded?

3. Where is the *glass transition temperature* relative to both the application and processing temperatures? What is the thermal history of the material? Is the material semicrystalline, a blend, or a copolymer?

4. How does the *specific volume* vary with the temperature, pressure, the rate of change of temperature, and the rate of change of pressure? How will this relate to material shrinkage during processing?

5. How will the ability of the material to *transfer heat* influence heating and cooling rates, reaction times, and mold design? If curing is involved, and an extremely exothermic reaction is part of the material cure process, will heat removal take place quickly enough so as to avoid a runaway reaction (versus furan resins)?

6. Will the energy expended to raise the material to operating temperatures (*specific heat*) be used efficiently?

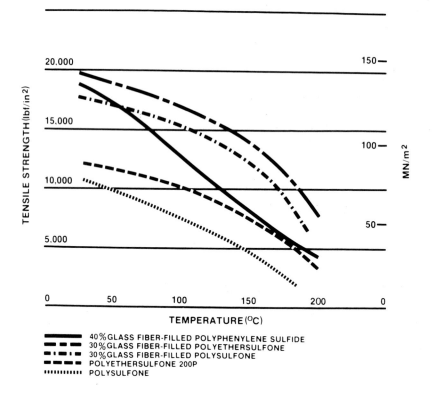

Fig. 11.2   Effect of temperature on tensile strength of some high-temperature thermo-plastics.

7. How will the inherent *elasticity* of a material influence final product properties? Is the elasticity dependent on the time scale of the specific process; that is, is the product viscoelastic?

8. Is the total cure time efficient for the process? How are other material properties, such as viscosity, related to the *reaction time* of the material?

Further consideration must be given to the particular operating conditions of the process and the compatibility of those conditions with material properties. Of course, a major factor in all process design problems is *cost*; the most desirable material may not be economical for a specific application, and it may become necessary to settle on a material of lesser quality, but lower cost (as long as product standards are met).

The next section will focus on addressing the foregoing questions in the design of a total process. We do not endeavor here to delve deeply into each process; that will be done in Chap. 27. Rather, an attempt is made to acquaint the reader with problems and considerations associated with selecting materials for different processes.

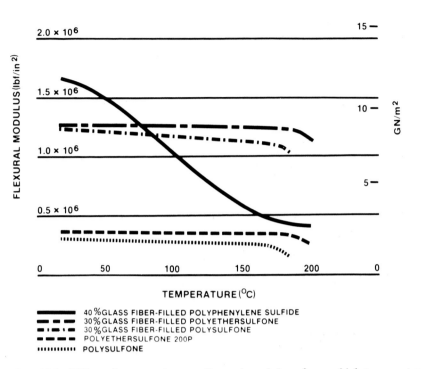

Fig. 11.3 Effect of temperature on flexural modulus of some high-temperature thermoplastics.

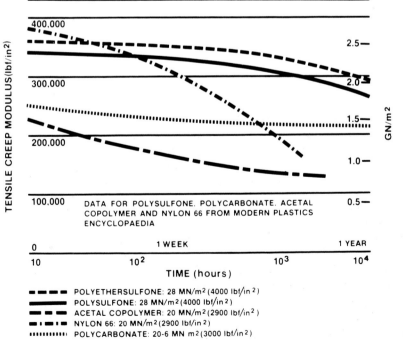

Fig. 11.4 Time dependence of tensile creep modulus of some thermoplastics at 20°C.

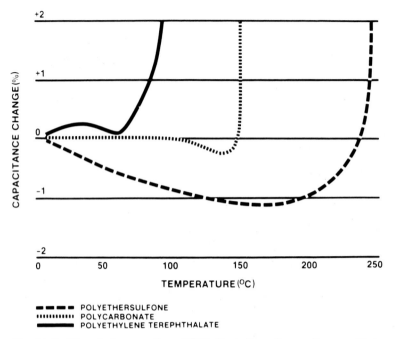

Fig. 11.5 Electrical properties of PES. Variation of capacitance with temperature of some thermoplastics.

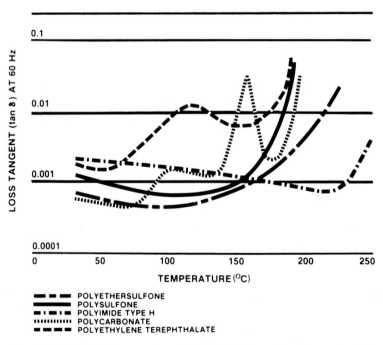

Fig. 11.6 Electrical properties of PES. Variation of loss tangent with temperature of some thermoplastics.

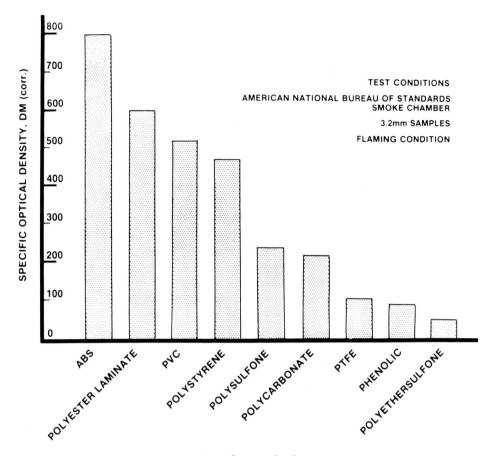

Fig. 11.7   Smoke emission on burning of some plastics.

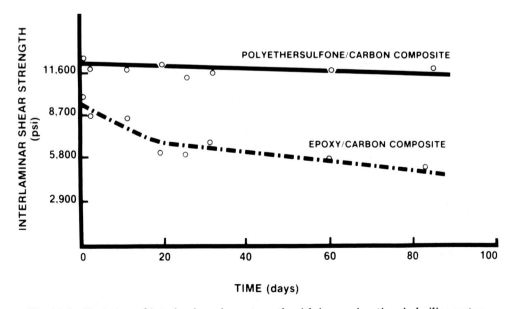

Fig. 11.8   Variation of interlaminar shear strength with immersion time in boiling water.

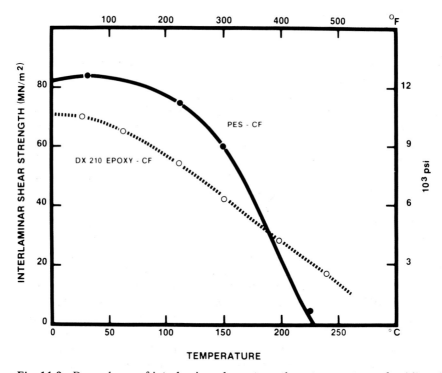

Fig. 11.9 Dependence of interlaminar shear strength on temperature of unidirectional carbon fiber composites.

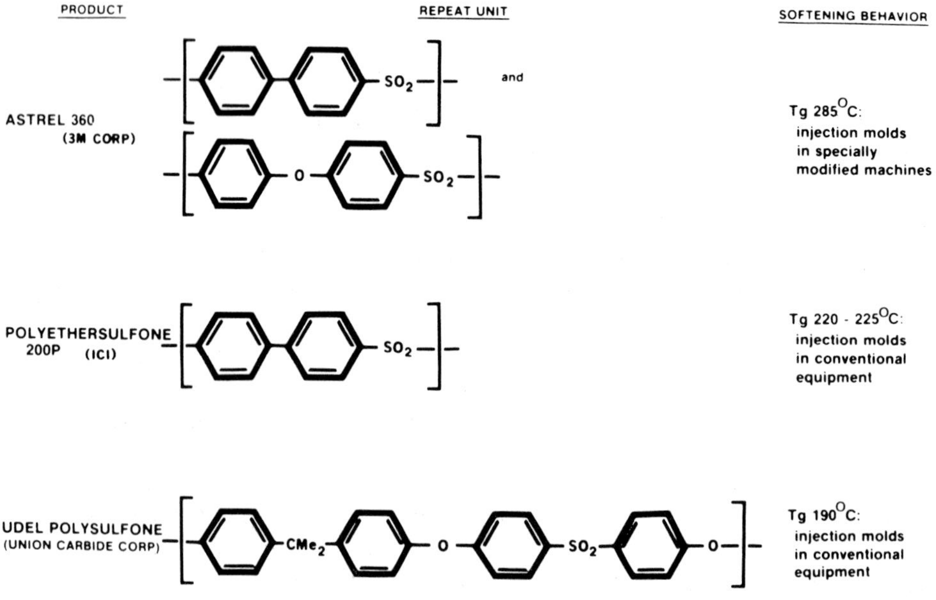

| PRODUCT | REPEAT UNIT | SOFTENING BEHAVIOR |

Fig. 11.10 Available polyether sulfones.

V. THERMOSET PROCESSING

A. Casting

Most raw materials for the thermosetting polymers employed in casting are in the form of low-molecular-weight liquids, which are poured into an open mold and left to harden, sometimes at elevated temperature. One of the main things to consider in casting thermosets is the shrinkage that occurs on hardening. All plastics will shrink to some degree during processing, and in thermosets the greatest shrinkage will develop during the initial stage of cure. On the beneficial side, this shrinkage can aid in the removal of a cast item from its mold. For this reason, epoxies, which have a small, predictable shrinkage, are excellent candidates for casting materials. On the other hand, polyesters, which can experience as much as 12% shrinkage (by volume), far exceed the amount of material contraction desirable. Although a factor in mold design, extensive shrinkage can result in nonuniformity from product to product, and in the extreme, warpage and cracking.

During a cure reaction, certain materials, such as acrylics and polyesters, can generate enough heat to achieve a temperature of 260°C. The very center of the casting will attempt to rapidly expand due to heat buildup; at the same time curing causes molecules to contract as they bond. If left uncontrolled, the simultaneous expansion and contraction can lead to cracking and brittleness in the final product. In the worst case, this instability may lead the casting to either foam or ignite. The means of controlling these reactions include the use of heat-conductive molds (or fillers) and reduction of the amount of promoter or catalyst added to the material. Alternatively, one can select a material, such as a phenolic, that is inexpensive, responsive to machining, and most importantly, has a low reaction exotherm (cures at room temperature).

In casting, flow into the mold is of prime interest. The use of a material with a low viscosity, such as methyl methacrylate (about 0.6 cP at room temperature) can lead to leakage from the casting cell. This problem with methacrylates is overcome by partially reacting the material and heating it to a prepolymer syrup.

In heating the material to a prepolymer, another problem in employing methacrylates is solved; any dissolved oxygen in the material is released; bubbles are not left in the final product. Furthermore, removal of the dissolved gas early in the process causes some material contraction; thereby the final product shrinkage is minimal.

B. Hot Compression Molding

Hot compression molding is probably the oldest method of producing thermoset formed products. Materials for this process are available in varied forms, such as powder, putty, pellets, and briquettes.

A simple means of reducing the cycle time for this process is through preheating the mold. Most of the plastics employed in hot compression molding are poor heat conductors; however, if careful control of the mold temperature is not maintained, premature curing of the resin may result.

Extreme overcure during molding is a problem readily discernible in urea, melamine, and epoxy resins through changes in color and the occurrence of crazing, brittleness, and blistered surfaces. Materials such as these must therefore be processed at comparatively low temperatures (for example, embrittlement is observed in urea resin when held at temperatures exceeding 77°C for long periods of time). Conversely, a material such as a filled phenolic is more heat stable and can withstand high temperatures for short periods of time.

Some thermosets have an intrinsic tendency to absorb moisture. Moisture absorption and retention, such as in ureas, can lead to premature curing, as well as difficulty of flow of the material in the mold. On the other hand, moisture delays curing in phenolics and can be responsible for a considerable increase in cycle time. In either case, the finished product resulting from a moisture-laden material is subject to porosity, blisters, and other surface blemishes.

C. Transfer Molding

Transfer molding, which is a modified version of compression molding, has the distinct advantage of short cycle times due to rapid final curing (accomplished by the frictional heat associated with the high pressures employed) after an initial softening by heat and pressure in a transfer chamber. The transfer molding process is therefore subject to the same considerations given to the hot compression-molding process.

D. Lamination

Central to the selection of a material for use in lamination is an understanding of the pressure required to cure that material. Although curing temperatures are always of prime importance, the particular temperature employed is a function of the hardener or catalyst that has been chosen for the material.

Epoxies, which cure by addition, may be laminated at very low, or *contact* pressures, ranging from 0 to 25 psi.

Polyesters, which cure at pressures below 400 psi, are referred to as *low-pressure laminates.* Often, however, parts fabricated in this way from polyester must be reinforced with a material, such as glass fiber, to lend superior strength to the product.

Last are those materials that cure between 1200 and 2000 psi. These *high-pressure laminates* include the ureas, melamines, and phenolics, although castings of these resins are made at atmospheric pressure.

In employing the phenolics, it must be remembered that the solid resin must be used in conjunction with a solvent, such as alcohol, in order for it to be applied to the resinforcement. If strict control and monitoring of solvent evaporation is not maintained, undercure and a defective product can result. Other potential problems associated with phenolics include delamination, porosity, and blistering.

Another important consideration in lamination concerns the resin ability to maintain uniform gel times and uniform exotherm characteristics; controlled curing conditions are critical to final product quality. In order to reduce the possibility of runaway cures (and blistering) it is therefore desirable to select a resin with a moderate exotherm.

Furthermore, the resin should experience a moderate amount of shrinkage to control both warpage and incidence of a wavy pattern developing in the laminate (brought about by the texture of the reinforcing web).

Other factors to consider include the incidence of trapped air, expanding vapor, undercure, inhibition, and a highly reactive resin.

E. Reaction Injection Molding

As with the previously discussed processes, there exist some materials that are more suited for the reaction injection molding process than others. High-viscosity material can lead to short shots (material solidifying before the mold is completely full) and voids. A high-volatile content can cause blistering and embrittlement. A slow cure time may yield parts that are warped.

All the above problems may be somewhat alleviated by processing changes, and in particular, factors related to the temperature, pressure, and time of the process. For instance, a material with a tendency to blister may have its problem rooted in entrapped gases; slowing the rate of injection might give the gases more time to escape.

VI. PROCESSING THERMOPLASTIC MATERIALS

A. Injection Molding

Since many of the thermoplastic processes are similar in the problems that most often arise and factors that must be considered, only one of the processes, injection molding, will be discussed in detail. Most commonly used materials for the other processes, along with major considerations are laid out schematically in Sec. VI.B.

The injection molding of thermoplastics involves melting a plastic and injecting that melt into a closed mold. The mold itself may either be air or water cooled or refrigerated to ready the product for ejection. Cycle times in this process can be as low as 4 sec.

In selecting a material to be injection molded, there is a fine balance of polymer and process properties that must be maintained; the overriding consideration in this process is time. One asks: how much time will it take to melt the polymer and bring it to the minimum required viscosity such that short shots do not occur? Can the material be held for that length of time at increased temperatures without degrading? Is the pressure used to inject the material into the mold enough so that the mold will fill properly in the allotted amount of time; Is an adequate pressure drop effected to allow the polymer to shrink in the mold; Finally, is the polymer material temperature lowered to allow hardening to the final shape in the allotted time?

In consideration of the above, both polyethylenes and polystyrenes make excellent materials for injection molding; their flow viscosity is such that it allows proper filling of molds that have very complex shapes and designs. On the other hand, molten nylons are generally so fluid that molds must be especially designed to prevent the possibility of material leakage prior to cooling.

As with other processes, undue shrinkage can lead to part thinning and product nonuniformity. Methacrylate polymers do not shrink extensively, solving this problem. Other process problems include

1. Surface blisters, usually caused by the generation of volatiles from entrapped solvents, water, or air
2. Bubbles and voids, caused by shrinkage upon cooling, and polymer decomposition

B. Other Thermoplastic Processes

   1. Blow molding

Use	Consider
Polyethylene	Variety of colors and designs available
Nylon	Keep moisture content below 0.2%
Acetals	The viscosity of acetals is not highly depen-dent on temperature
Acrylics	The high melt viscosity of acrylics provides low shrinkage

   2. Extrusion. Use: polyethylene, polypropylene, polystyrene, PVC, polycarbo-nates, and nylon (see Blow molding).
   3. Thermoforming. Use: rigid grades of polyethylene, PVC, and acrylics.
   4. Calendering. Use: ABS polymers to allow for surface embossing and PVC.
   5. Casting. PMM especially good
   6. Materials for special consideration
      a. Ultra-high molecular weight polyethylene. The most common process em-ployed is compression molding; most extrusion operations require special equipment. The product will be tough, hard, and have good fatigue life.
      b. Tetrafluoroethylene. This material is difficult to melt, plasticize, and dis-solve. It is highly crystalline, and has a high melt viscosity for processing.

VII. RUBBER*

In the beginning, there was only natural rubber. Until the advent of the automobile industry with its need for large quantities of tires and other products, there was little incentive to search for a substitute for natural rubber. Even then, the increased demand was met by plentiful raw materials from the rubber plantations.

   Although the idea of producing a truly synthetic rubber with roughly the same properties of the natural product had long intrigued researchers, it was not until World War I that any significant progress was made. German scientists, their country cut off from natural-rubber sources, developed a rubberlike material called "methyl rubber." It was a start, but not a successful one; methyl rubber was significantly inferior to natural rubber and was so expensive to make that production was discontinued with the renewed availability of natural rubber after the war.

   Prior to World War II, three significant events marked the evolution of synthetic rubber materials. Interestingly, two of these—the development of Thiokol (polysulfide) in the late 1920s and neoprene (chloroprene) in the early 1930s—did not result from overt attempts to make a rubber. Nevertheless, these materials are used today in many products because of their superior oil and solvent resistance. The third event was the development of Buna rubbers (copolymer of butadiene and styrene, Buna S, and copoly-mer of butadiene and acrylonitrile, Buna N) in Germany. Although the quality of these

---

*Contributed by PETER J. LARSEN.

materials was poor compared with that of natural rubber, the technology involved, much improved and modified, formed the basis for major synthetic rubber production in the United States in the early 1940s.

When World War II made much of the world again dependent on substitutes for many materials, including natural rubber, SBR (known then as GR-S: Government-Rubber-Styrene), an improved version of the original German Buna S emerged as the principal substitute material. Other contenders at the time were GR-N (neoprene), GR-I (butyl, or polyisobutylene), GR-A (nitrile, or polyacrylonitrile), and GR-P (polysulfide) (see Table 11.30 for nomenclature).

Today, SBR still dominates the field, accounting for approximately one-half of all rubber, natural and synthetic, used in the United States. The demand for SBR has been responsible for the building of a massive production capability for this material. This, in turn, has caused the sounding of frequent predictions from some industry observers that the use of natural rubber (NR) will fall off sharply.

Proponents of this thinking became particularly vocal around 1960 with the development of ethylene-propylene rubber (EPM) and ethylene-propylenediene rubber (EPDM), low-cost synthetic materials. EPDM is more nearly chemically inert than NR, giving it excellent resistance to ozone and to atmospheric aging in general. But EPDM does not have the tensile strength, the flexibility, or the desirable handling characteristics (green strength and high tack) of natural rubber, so it never materialized as a one-for-one replacement for NR.

The predicted fall-off in demand for natural rubber has not happened. In fact, current demand exceeds production; every pound of natural rubber produced is sold. Consumption of SBR is high because more than half of its production goes into passenger-car tires in the United States. Natural rubber is used almost exclusively in the more demanding areas, such as truck, bus, aircraft, and off-highway tires.

A. Natural Rubber

Among some 2000 species of plants known to contain rubber, only a few have ever produced it in substantial quantities for commercial use. Two of these, the "rubber tree" (*Hevea brasiliensis*), growing principally in Africa and Southeast Asia, and the guayule shrub (*Parthenium argentatus* Gray), have been the only sources of commercial rubber, the former much more so than the latter. Oddly enough, although the products of both plants are identical, *Hevea* grows in the rain forest regions of the world and guayule, an inconspicuous shrub less than 3 ft high, grows in the semiarid regions of Mexico and southwestern United States.

In 1910, guayule provided about 10% of the world's natural rubber and continued to be a minor source for almost 40 years. Although the guayule program got a big lift from the needs of World War II, the project was abandoned around 1946. There appeared to be little need for another rubber source. *Hevea* was in good supply and under no political threat. Furthermore, it was believed by some that synthetic rubber would eventually make natural rubber obsolete. This has not happened. Actually, the increasing demand for natural rubber now has some forecasters predicting a worldwide shortage by the late 1980s.

Consequently, a great deal of activity is now taking place in reevaluating the potential of the guayule bush. Techniques for extracting the rubber are much improved over those used in the 1940s, and a better quality rubber can be produced more economically

Table 11.30 Selection and Service Guide for Rubbers

Selection and service guide for rubbers

Common or trade name	Natural rubber	Synthetic natural	GRS or Buna S	Polybutadiene	Butyl	Chlorobutyl	EP	EPDM	Neoprene	Hypalon	Nitrile or Buna N	Epichloro-hydrin	Acrylate	Thiokol	Silicone	Fluoro-silicone	Fluoro-carbon	Urethane
Chemical type	Natural polyisoprene	Isoprene	Styrene butadiene	Butadiene	Isobutene isoprene	Chlorinated isobutene isoprene	Ethylene propylene	Ethylene propylene diene monomer	Chloroprene	Chloro-sulfonated polyethylene	Nitrile butadiene	Epichloro-hydrin	Polyacrylate	Polysulfide	Poly-siloxane	Fluoro vinyl methyl siloxane	Fluorinated hydrocarbon	Polyester or polyether urethane
ASTM D1418 Designation	NR	IR	SBR	BR	IIR	CIIR	EPM	EPDM	CR	CSM	NBR	CO, ECO	ACM, ANM	PTR	MQ PMQ VMQ PVMQ FC FE GE	FVMQ	FKM	BG
ASTM D2000/SAE J200 type, class	AA	AA	AA-BA	AA	AA, BA	AA, BA	AA BA CA	AA BA CA	BC BE	CE	BF BG BK CH	CH	DF DH	AK	FC FE GE	FK	HK	AU EU
PHYSICAL																		
[1]Density (g/cm³)	0.93	0.93	0.94	0.94	0.92	0.92	0.86	0.86	1.24	1.10	1.00	1.36-1.27	1.10	1.34	1.1-1.6	1.4	1.4-1.95	1.05-1.30
Hardness range (Shore A)	30-100	40-80	40-100	45-80	30-100	30-100	30-90	30-90	40-95	50-95	20-90	40-90	40-90	20-80	25-80	40-80	60-90	35-100
Permeability to gases	C	C	A	C	A	A	A	A	B	B	B-A	A	B	C	D	D	B	B
Electrical resistivity	A	A	A	A	A	A	A	A	C	B	D-C	C-B	B	B	A	B	B	B
Odor	B-A	B	B	C	B	C-B	B	B	C-B	C-B	C-B	C-B	C-B	D	B	B	C-B	B
Taste	C-B	C-B	C-B	C-B	C-B	C-B	B	B	C-B	A	C-B	C-B	B	D-C	A	B	C-B	B
Nonstaining	A	A	D-B	A	C	A	B	B	A	A	B-A	B	C-B	C-B	B-A	B-A	C-B	C-B
Bondability	A	A	A	A	C	A	C	B	A	A	B-A	C-A	B	C-B	B-A	B-A	C-B	C-B
MECHANICAL																		
[2]Tensile strength (max psi)	4,500	4,000	3,500	3,000	3,000	3,000	3,000	3,000	4,000	4,000	3,500	2,500	2,500	500-1,500	600-1,500	1,200	2,500	5,000-8,000
[3]Abrasion resistance	A	A	B	A	B	B	B	B	B-A	B	A	C-B	C-B	D	C-B	D	B	A
[4]Flex resistance	A	A	B	C	A	A	B	B	B	B	B	B	B	D	B	D	B	A
[5]Tear resistance	A	B	C	C	B	B	D	C	B	C-B	C	C-A	D	D	D-C	D	B	B-A
[6]Impact resistance	A	A	A	B	B	B	B	B	A	B	B	C-A	D	B	D-C	C	C	B-A
[7]Deformation capacity	A	A	B	A	B	A	C-B	B-A	B	C-B	B	C	C-B	B	B-A	C-B	B-A	B-A
[8]Elasticity	A	A	B	A	C	C-B	C	C	B	C	B	C-B	C-B	C	D-A	C-B	C	C-A
[9]Resilience	A	A	B	A	C	C-B	C-B	B-A	B	C	B	C-B	C-B	C	A	B	C	C-A
[10]Creep, stress relaxation	A	B	A	A	C	C-B	C-B	C-B	A	C	B-A	B	C	C-B	C-A	B-A	B	C-A
THERMAL																		
Recommended max temp (°C)	70	70	100	70	100	100	125	125	100	125	100-125	125	150	70	200-225	200	250	100
[11]Low-temp stiffening	B	B	C	A	C	C	B	B	C	B	B	C	D	B	C-B	B	D	C
Heat-aging resistance	B-C	B-C	D	A	B-A	B-A	A	A	B-A	B-A	B	B-A	A	C-B	A	A	A	B-A
Flame resistance	D	D	D	D	D	D	D	D	B-A	B-A	D	B-D	D	C-B	A	A	A	D
RESISTANCE TO																		
Weather	C-B	C	C-B	C	A	A	A	A	B	A	C-B	B	A	B	A	A	A	A
Oxygen	C	B	C	B	A	A	A	A	B	A	B	B	A	B	A	A	A	A
Ozone	C-D	C-D	C-D	C-D	A	A	A	A	B	A	C-D	A	A	A	A	A	A	B
Radiation	B	B	B	B	B	B	B	B	B	B	A	C-D	B	C	C-B	C-B	C-B	B
Water	A	A	B-A	B	B-A	B-A	A	A	B	B	A	B	B-C	D	A	A	B	C-B
Steam	D	B	C	B	B-A	B-A	A	A	B	B-A	C-B	B	D	D	C-B	C-B	B	D
Alkali dil/conc	A/C-B	C-B/C-B	C-B/C-B	C-B/C-B	A/A	A/A	A/A	A/A	A/A	A/A	B/B	B/D	B-C/C-C	B/B	A/A	A/B	B/C	B/D
Acid dil/conc	A/C-B	C-B/C-B	C-B/C-B	C-B/C-B	A/A	A/A	A/A	A/A	A/A	A/A	B/B	B/C	C/C	C/NR	A/B	A/B	B/C	C/D
[13]Oil, gasoline, kerosene	NR	NR	NR	NR	NR	NR	NR	NR	C	C-D	A	A	A	A	D-C	A	A	A
Benzene, toluol	NR	NR	NR	NR	NR	NR	NR	NR	C-D	B-C	B	B-A	C-B	B-A	NR	B-A	B-C	A
Animal, vegetable oils	D-B	D-B	NR	B	B-A	A	NR	B-A	B	B	A	A	C-B	A	A	A	A	A
Oxygenated solvents	B	B	NR	NR	NR	B-A	B-A	B-A	B	B	D	C	D	B-A	B-C	B-A	NR	D
Halogenated solvents	NR	NR	NR	NR	NR	NR	NR	NR	D	D	C-B	C-B	B	C-A	B-C	A	A	C-B
Alcohol	B-A	B	B	B	A	A	B-A	B-A	A	D	C-B	C	B	A	NR	B-C	A	C-B
Synthetic lubes (diester)	NR	NR	NR	NR	NR	NR	NR	NR	D	D	C-B	C-B	D-C	C-B	NR	C-A	A	C-B
Hydraulic fluids																		
Silicates	B-A	B-A	B-A	B-A	B-A	B-A	B-A	B-A	B	B	B	B	B	D-C	D	B	B	C
Phosphates	B	B	B	B	A	D	C	C	C	C	D	NR	D-C	D-C	B	B	A	C

A = Excellent   B = Good   C = Fair   D = Poor   NR = Not Recommended

1. The higher the density, the more rubber is required to make a given part. For example, compare neoprene and natural rubber. Even at the same price per pound, neoprene would be more expensive to use.

2. While tensile strength per se is not necessarily important, retention of strength at elevated temperatures suggests retention of other mechanical properties as well.

3. Abrasion-resistance ratings apply to a wide range of temperatures as well as type of abrasion (such as rubbing and impingement).

4. A high resistance to crack growth indicates good general durability—necessary where physical abuse is expected.

5. Tear resistance, along with crack-growth resistance, is desirable where physical abuse is expected.

6. Rubbers that strain-crystallize at extreme deformations are much more durable in impact than those that don't. Low-temperature flexibility also helps improve impact performance.

7. A high deformation capacity usually indicates a high fatigue resistance to flexing.

8. The lower the permanent set, the better the structural integrity and the better the retention of initial dimensions.

9. The higher the resilience, the less the degradable heat buildup in a flexing or dynamic situation.

10. The better the resistance to creep, the longer the life of the part, particularly where clearances are to be maintained.

11. Resistance to stress relaxation is essential in seals and other components under steady stress in service.

12. Good low-temperature flexibility is a must for most shock absorbers. The first jolt is critical, regardless of subsequent softness.

13. Resistance to oils and greases is essentially a surface effect: parts with poor resistance to these substances but that have appreciable bulk will not be degraded by such exposure.

Data courtesy Hughson Chemicals, Lord Corp., Erie, Pa.

today. Actually, guayule rubber compares quite favorably with *Hevea,* even when compounding normally associated with *Hevea* is used. There is no question that properties can be made identical.

Many factors favor the production of rubber from guayule. The material is "in-house," it grows in an area of little use for anything else and one where local economic conditions could stand improvement, and its production is not subject to the same labor sensitivity as is the *Hevea.* Commercialization of guayule could have a stimulating effect on natural rubber use; it would remove the reluctance to switch from a synthetic to a readily available natural rubber.

Another factor favoring both guayule and *Hevea* natural rubber is that both of these materials, derived from replenishable sources, can be valuable as fuels when their useful life is over. Coal averages 13,000 Btu/lb; rubber averages about 15,000 Btu/lb; and oil, whose fuel use days may soon end, is only slightly better.

	Hevea-based	Guayule-based
Modulus at 300%	1770	1050
Tensile strength (psi)	4050	3645
Elongation at break (%)	490	635
Set at break (%)	13	14
Rebound, Bashore (%)	48	40
Hardness, Shore A	60	54
Tear strength (lb/in.)	436	178

B. "Rubber" Versus "Elastomer"

The rubber industry differentiates between the terms "elastomer" and "rubber," on the basis of how long a deformed material sample requires to return to its approximate original size after the deforming force is removed and of its extent of recovery.

ASTM D1566 defines an elastomer as, "a macromolecular material which, at room temperature, is capable of recovering substantially in shape and size after removal of a deforming force."

The same standard is more specific and quantitative in defining rubber as, "a material that is capable of recovering from large deformations quickly and forcibly . . . [and which], in its modified state, free of diluents, retracts within one minute to less than 1.5 times its original length after being stretched at room temperature to twice its length and held for one minute before release."

Thus, by these definitions, all rubbers are elastomers, but all elastomers are not rubbers, since no return time or deformation hold time is specified in the elastomer definition. Also, some plastics qualify as elastomers, according to the rather loose definition of that category.

Rubbers are a class of materials that serve an enormous number of engineering needs in fields dealing with shock, noise and vibration control, sealing, corrosion protection, abrasion protection, friction production, electrical and thermal insulation, waterproofing, confining other materials, and load bearing. Although the term "rubber" originally meant the material obtained from the rubber tree, *Hevea brasiliensis,* today it means any material capable of extreme deformability, with more or less complete recovery upon removal of the deforming force. Today, synthetic materials, such as neoprene, nitrile, styrene butadiene, and butadiene rubber, are grouped together with natural rubber. Chemists have even developed a "synthetic" natural rubber (polyisoprene) that essentially duplicates the chemistry and properties of the product of nature.

C. Structure of Rubbers

In contrast to the ordered and rigid crystalline arrangement of atoms and molecules in metals, the atoms of rubbers are arranged in long, chainlike configurations, which are in constant, thermally induced motion. The result is a tangled mass of kinked, twisted, and intertwined elements similar to a snarled mass of fishing line. Along the chain, the atoms remain substantially the same distance apart, but the spatial distance from one point on a chain to another is always less than that measured along the chain's length.

Statistically, at a specific temperature, there is one most probable spatial distance between any two points on a given chain. When an applied force changes this distance, the thermal movement of the system sets up a force to restore the distance to what it was originally. This action accounts for the elasticity, or recovery, of a piece of deformed rubber. It also explains why the modulus of an elastomer, when heated, increases in the direction of strain.

Within the elastic solid, the tangled chain segments are relatively free to move with respect to one another, except to the extent that they encounter mechanical entanglement or, upon vulcanization, are "hooked" together at chemically reactive sites on the chains and attain structural integrity.

D. Controlling Deformation

The response pattern of rubber to a deforming force is a function of the degree of ease with which the chainlike segments of molecules can move relative to one another. This motion can be hindered, for example, by any filler substance put into the mass of tangled, twisting, disordered chains of molecules; the result is a stiffer rubber compound. Conversely, anything that is put in to lubricate the system makes the compound softer because a lubricant increases the ease of chain movement. The structure of the molecule itself also affects stiffness. The smaller and fewer the chemical attachments along the chains, the less hindrance to relative movement, and the greater the resiliency and elasticity.

Compared with most other engineering materials, rubbers have a relatively high degree of elasticity and deformability. They are further unique in that these properties are not special characteristics of a specific type of chemical substance; rather, they result from a particular type of molecular structure. It does not seem to matter, in this context, what is put together; the arrangement is what is important.

E. Use of Rubber for Design

1. Selection

In selecting a rubber for an application, a number of things must be considered, and almost any choice becomes a compromise. As in any material selection problem, the considerations include mechanical or physical service requirements, operating environment, a reasonable life cycle, manufacturability of the part, and cost.

Further, within the framework of each family of rubbers exists a wide range of available properties. These are made possible by compounding; that is, incorporating additives that improve a weak property, make the compound easier to process, or reduce cost without significantly affecting properties. In addition to the varieties of rubbers available, almost any physical or chemical property can be altered to some extent. Thus, selecting the best material for a given application can involve considerable investigation.

2. Industry Standards

Among the available standards on various aspects of rubber, the following are recommended as aids to identification and selection:

ASTM D1418: Rubber and Rubber Lattices—Nomenclature. This standard decribes all available rubbers in terms of their chemical compositions.

ASTM D1566: Standard Definitions of Terms Relating to Rubber. This reference helps to ensure unambiguous communication among producers, molders, and designers of rubber parts.

ASTM D2000: Standard Classification System for Rubber Parts in Automotive Applications. This standard, despite its title, is not limited to automotive parts, and is probably the most important document of all.

The D2000 classification system, which is also available as SAE J200, is based on the premise that properties of all rubber compounds can be arranged into characteristic material designations. These designations are determined by "type," based on resistance to heat aging, and "class," based on resistance to swelling in oil. Basic levels are thus established that, together with values describing additional requirements, permit complete characterization of all rubber materials.

Specification D2000 can also be used to describe a material not yet available, but desirable, in these same terms. The standard allows the rubber technologist and the design engineer to discuss materials in a mutually understood fashion without the rubber technologist's divulging the chemical makeup of the material, which would be of little value to the engineer anyway.

The material descriptions that follow are designed to help make a quick preliminary selection of a base material. Studying Table 11.30 can further aid the selection. ASTM Standard D2000 can then be used to "name" or specify an available material (or one that should be developed) for the application. These brief descriptions are grouped into two categories: those listed as having no requirement for oil resistance, and those that do. The no-requirement group are all in the class A designation. The headings include the common name, ASTM D1418 designation (chemical composition), and material designation type and class, according to ASTM D2000.

3. Specifying Rubber Materials by Performance

Listed in Table 11.31 are the type and class designations assigned to rubbers in accordance with ASTM D2000 and SAE J200. The type designation is determined by a thermal test that established a maximum service temperature; letters A through J indicate the range from 70 to 275°C. Class designations, based on maximum volume swell with immersion in a prescribed (ASTM #3) oil test, are also letters; letters A through K represent the 10 classes set up in these specifications.

Table 11.31 Basic Requirements of Elastomers for Establishing Type and Class

Type	Test temperature (°C)	Class	Volume swell (maximum %)
A	70	A	—a
B	100	B	140
C	125	C	120
D	150	D	100
E	175	E	80
F	200	F	60
G	225	G	40
H	250	H	30
J	275	J	20
		K	10

aNo requirement.

Type and class designations are written together. For example, AK defines a requirement for a rubber that can be used at 70°C continuously and that will not swell more than 10% when immersed in an ASTM reference.

Table 11.32 lists the rubber materials that are most often used in meeting typical

Table 11.32 ASTM 2000, SAE J200 Designation

Type/Class	Typical rubber
AA	Natural rubber, styrene butadiene, butyl, ethylene propylene, polybutadiene, polyisoprene
AK	Polysulfide
BA	Ethylene propylene, styrene butadiene (high temperature), butyl
BC	Chloroprene
BE	Chloroprene
BF	Nitrile
BG	Nitrile, urethane
BK	Polysulfide, nitrile
CA	Ethylene propylene
CE	Chlorosulfonated polyethylene
CH	Nitrile, epichlorohydrin ethylene/acrylic
DA	Ethylene propylene
DF	Polyacrylate (butyl acrylate type)
DH	Polyacrylate
FC	Silicone (high strength)
FE	Silicone
FK	Fluorinated silicone
GE	Silicone
HK	Fluorinated rubbers

requirements as spelled out by ASTM D2000 and SAE J200. This list is not limiting; other polymers may meet the same specification.

F. Natural Rubber (D1418:NR;D2000:AA)

The commercial base for natural rubber is latex, a milklike serum, generated by the tropical tree, *Hevea brasiliensis*. The latex is collected in much the same fashion as maple sap. However, latex should not be confused with the sap of the tree. Latex is secreted in the inner bark of the tree, and a tree can be severely harmed if a tapping cut is deep enough to draw sap as well as latex. Naturally occurring latex is a dispersion of rubber in an aqueous serum containing various inorganic and organic substances. The rubber precipitated out of this solution can be characterized as a coherent elastic solid.

It is against natural rubber that all other rubbers should be measured. For centuries it was the only rubber available, and it was used extensively even before the discovery of vulcanization in 1839.

Synthetic rubbers have been developed either by accident or as the result of pressures of political upheaval and consequent unavailability of the natural product. However, no synthetic material has yet equalled the overall engineering characteristics and consequent wide latitude of application available with NR.

As with other rubbers, many grades and types of NR are available; these are produced by varying impurity levels, collection methods, and processing techniques. Natural rubber is generally considered to be the best of the general-purpose rubbers—those that embody properties and characteristics suitable for broad engineering applications. Superior compounds can be evolved over a wider stiffness range with natural rubber than with any other material and at a competitive price that will, in all probability, become even more significant as the energy situation worsens.

Natural rubber is likely the best choice for most applications, except those where an extreme performance or exposure requirement dictates the use of a special-purpose rubber, often at some sacrifice of other, less critical characteristics.

Natural rubber has a large deformability capacity. This, coupled with its ability to strain crystallize, gives it added strength while deformed. Its high inherent resilience, which is responsible for a very low heat buildup in flexing, makes NR a prime candidate for shock and severe dynamic loads. Properties negatively affected by heat, such as flex, abrasion resistance, cut resistance, and endurance, generally can thus be much improved. NR has low compression set and stress relaxation; these characteristics favor its application in sealing devices where maintenance of sealing forces and the surface conformability of high-quality soft stocks are important. Further advantages are excellent green (uncured) strength, building tack, and general processing characteristics.

Natural rubber does have some shortcomings, as do the other general-purpose rubbers. There are rubbers, for example, that maintain initial properties at greater extremes of temperatures. The useful service temperature of natural rubber is generally considered to range from −65 to (in special cases) 250°F, which covers most applications. Other drawbacks of NR, such as poor oil, oxidation, and ozone resistance, can be minimized or virtually eliminated, either by proper design attention and/or by compounding. These degradative forces are essentially surface effects that can be tolerated by using thicker cross sections, shielding, or antioxidants and antiozonants.

Natural rubber often can be the first choice for many high-performance applications if it can be made to live in the service environment. It remains the best choice for tires, shock mounts and other energy absorbers, seals, isolators, couplings, bearings and other motion accommodation devices, and springs and other dynamic applications.

G. Synthetic Rubber

1. Synthetic Natural Rubber (IR; AA)

The synthetic rubber that comes closest of all to duplicating the chemical composition of natural rubber is synthetic polyisoprene. It shares with natural rubber the properties of good uncured tack, high unreinforced strength, good abrasion resistance, and those general characteristics that make for good performance in dynamic applications. However, because of some of the inherent impurities in the natural product that affect vulcanization characteristics in a positive fashion, natural rubber scores somewhat better on overall ratings. A significant disadvantage of IR is its lack of green strength. IR can be used interchangeably for natural rubber in all but the most demanding applications. Although IR is currently a little more economical, the cost picture may change with increasing crude oil costs. Specific product areas are about the same as for natural rubber.

2. Styrene Butadiene (SBR: AA,BA)

This material emerged as a high-volume substitute for NR during World War II because of its suitability for use in tires. Despite the fact that the basic feedstock for SBR is crude oil, it has remained quite competitive in cost because of the extensive production capacity for SBR in the United States.

SBR continues to be used in many of the applications where it earlier replaced NR, even though it does not have the overall versatility of natural rubber and the other general-purpose materials. For most applications, SBR must be reinforced (hence, stocks are stiffer) in order to have acceptable tensile strength, tear resistance, and general durability. SBR is significantly less resilient than NR, so it has higher heat buildup on flexing. Further, it does not have the processing and fabricating qualities of NR, lacking both green strength and building tack.

An important reason for the past high-volume use of SBR is that it has done a creditable job in passenger car tires; it has good abrasion resistance and general durability. Recently that picture has changed because of the need for the greater green strength and building tack of natural rubber in radial tires and for the better low-temperature flexibility of natural rubber for snow tires. High-performance tires, such as for trucks and aircraft, have always been made from natural rubber providing it was available.

Specific product use of SBR rubbers is somewhat the same as for natural rubber except for products needing high-quality soft stocks and for other applications that are more demanding.

3. Polybutadiene (BR; AA)

This general-purpose, crude-oil-based rubber is even more resilient than natural rubber. It is the material that made the solid golf ball possible. It is also superior to natural rubber in low-temperature flexibility and exhibits less dynamic heat buildup. However, it lacks the toughness, durability, and cut-growth resistance of NR. It can be blended with natural rubber or SBR to improve low-temperature flexibility. It should be noted that silicones have superior low-temperature flexibility, but this is achieved at a much higher price and at a sacrifice in other properties, such as tensile strength, tear resistance, and general durability.

Large-volume polybutadiene use is in blends with other polymers to enhance their resilience and reduce heat buildup. It is also used in products requiring high resiliency over a broad temperature range, such as industrial tires and vibration mounts.

4. Butyl (IIR, CIIR; AA, BA)

The two types of rubber in this category are both based on crude oil. The first is poly-isobutylene, with an occasional isoprene unit inserted in the polymer chain to enhance vulcanization characteristics. The second is the same, except that chlorine is added (approximately 1.2% by weight), resulting in greater vulcanization flexibility and enhanced cure compatibility with general-purpose rubbers.

Butyl rubbers have outstanding impermeability to gases and excellent oxidation and ozone resistance. The chemical inertness is further reflected in lack of molecular weight breakdown during processing, thus permitting the use of hot-mixing techniques for better interaction of polymer and filler.

Flex, tear, and abrasion resistance of butyl rubber approach the values of natural rubber; moderate strength (2000 psi) unreinforced compounds can be made at a competitive cost. Butyl rubber lacks the toughness and durability of some of the general-purpose rubbers.

Its excellent impermeability to air is responsible for the high-volume use of butyl rubber in automotive inner tubes and tubeless tire inner liners. Here, they are dominant. The butyls are also used in belting, steam hose, curing bladders, O-rings, shock and vibration products, structural caulks and sealants, water barrier applications, roof coatings, and gas metering diaphragms.

5. Ethylene Propylene (EPR, EPDM; AA, BA, CA)

Like the butyls, the EP rubbers are of two types. One is a fully saturated (chemically inert) copolymer of ethylene and propylene (EPR); the other (EPDM) is the same as this plus a third polymer building block (diene monomer) attached to the side of the chain. EPDM is chemically reactive and is capable of sulfur vulcanization. EPR must be cured with peroxide.

These materials were originally expected to be the economic ultimate in rubber materials—one that would put natural rubber out of business. Economic superiority was never achieved, however, mainly because of the increasing yield per acre from natural rubber plantations.

The physical properties of EPR and EPDM are not as good as those obtainable with NR. However, property retention is better than NR on exposure to heat, oxidation, or ozone. Bonding is somewhat more difficult, especially with EPR. These materials have broad resistance to chemicals but not to oils and other hydrocarbon fluids. Electrical properties are good.

Typical applications of EPR and EPDM comprise automotive hose, body mounts and pads, O-rings, conveyor belting, wire and cable insulation and jacketing, window channeling, and other products requiring excellent weathering resistance.

H. Oil-Resistant Rubbers

1. Neoprene (CR; BC, BE)

Except for polybutadiene and polyisoprene, neoprene is perhaps the most rubberlike of all, particularly with regard to dynamic response. Neoprenes comprise a large family of rubbers that have a property profile approaching that of natural rubber and with better resistance to oils, ozone, oxidation, and flame. They age better and do not soften on heat exposure, although the tensile strength at high temperature may be lower than that of NR. These materials, like NR, can be used to make soft, high-strength compounds. A significant difference is that, in addition to neoprene being more costly than NR by the pound, the density of neoprene is about 25% higher than that of natural rubber. Neoprenes do not have the low-temperature flexibility of natural rubber, which detracts from their use in shock or impact applications.

General-purpose neoprenes are used in hose, belting, wire and cable, footwear, coated fabrics, tires, mountings, bearing pads, pump impellers, adhesives, and seals for windows and curtain-wall panels.

2. Chlorosulfonated Polyethylene (CSM; CE)

This material, more commonly known as Hypalon (du Pont) can be compounded to have an excellent combination of properties, including virtually total resistance to ozone and excellent resistance to abrasion, weather, heat, flame, and crack growth. In addition, CSM has low moisture absorption and good dielectric properties and can be made in a wide range of colors because it does not require carbon black for reinforcement. The resistance of CSM to chemicals and oil approaches that of neoprene. Although low-temperature flexibility is not good, the brittle point is near $-80°F$.

Hypalon is a moderately expensive, special-purpose material, not particularly recommended for dynamic applications. It is used generally where its outstanding environmental resistance is needed. Typical applications include coated fabrics, maintenance coatings, tank liners, and protective boots for spark plugs and electrical connectors.

3. Nitrile (NBR; BF, BG, BK, CH)

The nitriles are copolymers of butadiene and acrylonitrile, used primarily for applications requiring resistance to petroleum oils and gasoline. The resistance of NBR to aromatic hydrocarbons is better than that of neoprene but not as good as that of polysulfide. NBR has excellent resistance to mineral and vegetable oils, but relatively poor resistance to the swelling action of oxygenated solvents, such as acetone, methyl ethyl ketone, and other ketones. It has good resistance to acids and bases with the exception of those having strong oxidizing effects. Resistance to heat aging is good, often a key advantage over NR.

With higher acrylonitrile content, the solvent resistance of an NBR compound is increased but low-temperature flexibility is decreased. The low-temperature resistance of NBR is inferior to that of natural rubber; although NBR can be compounded to give improved performance in this area, the gain is usually at the expense of oil and solvent resistance. As with SBR, this material does not crystallize on stretching, and reinforcing materials are required to obtain high strength. With compounding, nitrile rubbers can provide a good balance of low creep, good resilience, low permanent set, and good abrasion resistance.

The tear resistance of NBR is inferior to that of natural rubber; electrical insulation is lower. NBR is used instead of natural rubber where increased resistance to petroleum oils, gasoline, or aromatic hydrocarbons is required. The properties of this rubber make it useful for carburetor and fuel-pump diaphragms, aircraft hoses, and gaskets. In many of these applications, the nitriles compete with polysulfides and neoprenes.

4. Epichlorohydrin (CO, ECO; CH)

Epichlorohydrin rubber is available as a homopolymer (CO) and a copolymer (ECO) of epichlorohydrin with ethylene oxide. Reinforced, these rubbers have moderate tensile strength and elongation properties, plus an unusual combination of other characteristics. One of these is low heat buildup, which makes them suitable for applications involving cyclic shock or vibration.

The homopolymer has outstanding resistance to ozone, good resistance to swelling by oils, intermediate heat resistance, extremely low permeability to gases, and excellent weathering properties. This rubber also has low resilience characteristics and low-temperature flexibility only to 5°F. These two characteristics may be undesirable for some applications.

The copolymer is more resilient and has low-temperature flexibility to −40°F, but it has poorer permeability characteristics. The oil resistance of both compounds is about the same.

Typical applications for these rubbers include bladders, diaphragms, vibration-control equipment, mounts, vibration dampers, seals, gaskets, fuel hose, rollers, and belting.

5. Ethylene/Acrylic

This new family of rubbers basically a copolymer of ethylene and methyl acrylate, is currently sold only in masterbatch form by du Pont, under the trade name of Vamac. They provide, at a moderate price, heat resistance surpassed by only the more expensive, specialty polymers, such as fluorocarbons and fluorosilicones. The material has very good resistance to hot oils, hydrocarbon-based or glycol-based proprietary lubricants, transmission fluids, and power steering fluids. It is not recommended for use with esters, ketones, highly aromatic fluids or high-pressure steam. Low-temperature flexibility limits are −29°C for unplasticized compounds and −46°C when compounded with ester plasticizers. A special feature of Vamac is its nearly constant damping characteristic over broad ranges of temperature, frequency, and amplitude.

The polymer is recommended for applications requiring a tough, set-resistant rubber with good low-temperature properties and resistance to the combined deteriorating influences of heat, oil, and weather. It is used in various automotive components and for marine motor lead wire insulation.

6. Fluorocarbon (FKM; HK)

Generally produced as a copolymer of vinylidene fluoride and hexafluoropropylene, the fluorocarbons are high-performance, high-cost rubbers known generally as Viton (du Pont) and Fluorel (3M). These rubbers have outstanding resistance to heat and to many chemicals, oils, and solvents compared with other commercial rubbers. In air, fluorocarbon rubber parts retain at least half their original properties after 16 hr exposure at 600°F. These same compounds offer low-temperature stability to −40°F.

In the reinforced state, FKM rubbers offer moderate tensile strength but relatively low elongation properties. They resist oxidation and ozone, and they do not support combustion. Several versions are available, and conventional compounding produces formulations within a hardness range of 65-95 Shore A. Fluorocarbon rubbers are severely attacked by highly polar fluids, such as ketones, hydrazine, anhydrous ammonia, and Skydrol (phosphate ester) hydraulic fluids. Postcuring is required to develop optimum properties.

Typical applications are seals, gaskets, diaphragms, pump impellers, tubing, and high-vacuum and radiation equipment.

7. Perfluoroelastomer (FFKM)

Chemical resistance of perfluoroelastomer parts is similar to that of PTFE, and mechanical properties are similar to those of the fluorocarbon rubbers. This rubber, produced by du Pont as Kalrez, is essentially unaffected by all fluids, including aliphatic and aromatic hydrocarbons, esters, ethers, ketones, oils, lubricants, and most acids. However, some fully halogenated fluids and strong oxidizing acids may cause swelling. The parts are suitable for continuous service up to 260-290°C and intermittent service to 316°C. Resistance to ozone, weather, and flame is exceptional. Radiation resistance is good, and high-vacuum performance excellent. Kalrez is sold only in finished part form. It commands a very high price and is only used where no other material can do the job.

Kalrez parts are used primarily in fluid-sealing applications. Usage is in demanding applications in the chemical processing, oil production, aerospace, and aircraft industries.

8. Acrylate (ACM, ANM; DF, DH)

These are specialty rubbers based on polymers of methyl, ethyl, or other alkyl acrylates. They are highly resistant to oxygen and ozone, and their heat resistance is superior to that of all other commercial rubbers except the silicones and the fluorine-containing rubbers. Water resistance is relatively poor; the acrylates are not recommended for use with steam- or water-soluble materials, such as methanol or ethylene glycol. Flexibility life is excellent, as is permeability resistance. Oil swell and deterioration characteristics are also excellent at high temperatures.

Low-temperature flexibility is not good. These rubbers decompose in alkaline solutions and are swelled by acids. Low-temperature flexibility and water resistance can be improved, but only with a marked decrease in overall heat and oil resistance. These materials are used extensively for bearing seals in transmissions and for O-rings and gaskets.

9. Urethane (AU, EU; BG)

These rubbers, combinations of polyesters or polyethers and diisocyanates, are unusual in that physical properties do not depend on compounding materials. Urethanes crosslink and undergo chain extension to produce a wide variety of compounds. They are available as castable or liquid materials and as solids or millable gums.

Urethane polymers have excellent tensile strength, load-bearing capacity, and elongation potential, accompanied by high hardness and outstanding abrasion resistance. Other properties include high tear strength, either high or low coefficient of friction, and good elasticity and resilience, even in very hard stocks.

Typical applications include seals, bumpers, metal-forming dies, valve seats, liners, coupling elements, rollers, wheels, conveyor belts, and especially whereever abrasive conditions are present.

10. Polysulfide (PTR; AK, BK)

These polymers have outstanding resistance to oils, greases, and solvents, but they have an unpleasant odor, resilience is poor, and heat resistance is only fair. Abrasion resistance is about half that of natural rubber, and tensile strength ranges from 1200 to 1400 psi. However, these values are retained after extended immersion in oil.

The basic properties of polysulfide polymers are determined by the type of chain structure and the number of sulfur atoms in the polysulfide groups. Increased sulfur concentration improves solvent and oil resistance and also reduces the permeability to gases. These materials are used in gasoline hose, printing rolls, and newspaper blankets. Other uses include caulking materials, adhesives, and binders.

11. Silicone (MQ, PMQ, VMPQ, PVMQ; FC, FE, GE)

Silicone rubber comprises a versatile family of semiorganic synthetics that look and feel like organic rubber yet have a completely different type of structure from that of other rubbers. The backbone of the rubber is not a chain of carbon atoms but an arrangement of silicone and oxygen atoms. This structure gives a very flexible chain with weak interchain forces, which results in a remarkably small change in dynamic characteristics over a wide range of temperature. Silicone rubbers have no molecular orientation or crystallization on stretching and must be strengthened by reinforcing materials.

Silicone rubbers are at the high end of the cost range for rubbers, but they can be made to withstand temperatures as high as 600°F without deterioration. Nevertheless, silicones retain useful flexibility at −150°F. No plasticizers are needed that might cause property sacrifices.

Although the tensile strength of silicone rubbers is lower than that of other rubbers, these materials have outstanding fatigue and flex resistance. They do not require high tensile and tear strength to serve in dynamic applications. Falloff in tensile properties with extended exposure to high temperatures is less than for other rubbers. Resistance to chemical deterioration, oils, oxygen, and ozone is also retained under these conditions. Chemical inertness makes these materials of special interest for surgical and food-processing equipment, where frequent sterilization must be endured.

12. Fluorosilicone (FVMQ; FK)

This type of silicone provides most of the useful qualities of the regular silicones plus improved resistance to many fluids. Exceptions are ketones and phosphate esters; however, FVMQ rubbers can be blended with conventional dimethylsilicones, which have good resistance to these fluids at temperatures to 300°F. The FVMQ rubbers are most useful where the best in low-temperature flexibility is required in addition to fluid resistance, although resistance to organic fluids (especially those containing aromatics) is poorer than that of the FKM-type fluorocarbon rubbers. Although improvements have been made in the general strength of the fluorosilicone rubbers, more durability under both static and dynamic stress would greatly improve their application potential.

Fluorosilicone rubbers have good dielectric properties, low compression set, and excellent resistance to ozone and weathering. They are expensive and definitely special purpose. Typical applications include seals, tank linings, diaphragms, O-rings, and protective boots in electrical equipment.

VIII.   RELATIONSHIP OF POLYMER STRUCTURE TO PERFORMANCE*

Systematic investigation of the structure of natural organic polymers began in the early decades of this century as soon as experimental methods became available to measure very large molecular weights and to establish the arrangement of the giant molecules in the solid state as they exist in such important materials as cotton, flax, silk, wool, leather, starch, and rubber (1).

Several general principles emerged from these studies.

1. The molecules in all these materials have high molecular weights, usually in the range of 10^4-10^6.
2. Most of them have the shape of chains that frequently are linear but also can be branched and, in certain cases, even reticulated.
3. These chains form small areas of high lateral molecular order, usually referred to as *crystallites* or *crystalline domains,* which are embedded in a matrix made up by chain segments of inferior intermolecular organization.
4. In special cases, such as lignin, horn, or nutshells, a considerable degree of intermolecular cross-linking prevails that renders these materials hard, infusible, and insoluble.

A.   Macromolecularity and Polymolecularity

All evidence points to the fact that a certain large molecular weight, that is, a certain considerable chain length, is necessary to endow an organic polymer with interesting properties for the formation of fibers, films, plastics, rubbers, coatings, or adhesives.

Early approaches set a certain molecular weight about 10,000 (2), as the lower limit for useful mechanical behavior, but more systematic work established empirically the character of a generalized strength-molecular weight relationship in the form of a curve, as shown in Fig. 11.11. It actually represents a family of curves, all of which have the same shape but each of which is valid for a certain type of linear polymer. On the ordinate of this figure is plotted the tensile strength of a film made from a particular polymer against its molecular weight or chain length on the abscissas. If one wants to compare many different chainlike molecules with each other it is convenient not to use the molecular weight but, instead, a quantity proportional to it, namely, the degree of polymerization (DP), the average number of monomeric units in the chain molecules under consideration. This quantity is plotted on the abscissa of Fig. 11.11, which shows that, as a general rule valid for all polymers, a certain minimum number of chain elements is necessary to produce any significant mechanical strength at all. Below this minimum number, which is between 20 and 50, the material represents a friable powder that cannot be spun into a filament, cast into a film, or molded into an object having useful

---

*Contributed by HERMAN F. MARK.

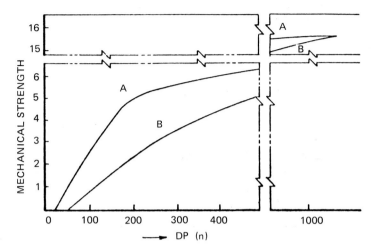

Fig. 11.11 Mechanical strength of polymeric materials as a function of DP.

mechanical strength. This threshold differs for the different classes of polymers; it is lowest (around DP 20) for linear polyamides (curve A) that possess strong intermolecular bonding forces in the form of hydrogen bridges and is largest (around DP 50) for poly-hydrocarbons (curve B), such as polyethylene, the chains of which are entirely nonpolar and develop the weakest type of intermolecular cohesion.

Above this critical lower DP limit for the establishment of mechanical strength the curve rises relatively steeply, indicating that any chain length increment in this range produces a noticeable improvement of strength, until in the DP range between 200 and 500 the curve forms a knee and a third domain is reached in which an additional lengthening of the chain molecules produces only relatively minor increments in the mechanical strength of the material. The position of the knee and the slope of the curve above it differs for different materials; chains having strong intermolecular bonding capacity (polyamides and polyesters) reach their approximate strength ceiling at a lower DP (around 200), whereas polyhydrocarbons need higher DP values (above 500) to es-tablish satisfactory mechanical characteristics. It appears, however, that at very high DP all linear polymeric materials reach about the same strength level. The general shape of the curves in Fig. 11.11 is characteristic for a hyperbolic relationship between strength and DP and can be empirically expressed by the equation

$$T_n = T_\infty - \frac{T}{n} \tag{1}$$

where T_n is the strength pertaining to the DP of n, T_∞ is the strength developed by very high DP (virtually infinite chain length), and A is a constant characteristic for each in-dividual polymer—it expresses the loss of mechanical performance as one shortens the chains, that is, as one proceeds to lower values of n. The lower critical limit T = 0 is reached for a DP of N_0, which is given by the equation

$$T_n = 0$$
$$= T_\infty - \frac{A}{n_0}$$

$$n_0 = \frac{A}{T_\infty} \qquad (2)$$

Since T_∞ has about the same value for all polymers, the constant A characterizes in principle how mechanical strength is developed to higher and higher values as the length of the chains is gradually increased. Qualitatively this can be understood in terms of the average overlapping between neighboring chains in the highly oriented state of a sample. If the chains are very short, DP below n_0, then the overlapping is so insignificant that virtually no intermolecular cohesion is produced and the material is a friable powder that cannot be brought into any useful form by spinning, casting, or molding. As the chains are lengthened, more and more overlapping is produced and the strength increases roughly proportional to the DP, as actually shown by the rising part of the curves in Fig. 11.11. When, later, the intermolecular cohesion between the oriented chains reaches higher and higher values because of increased overlapping, a state is finally approached in which, under stress, there is no more chain slipping, but instead, those individual chains exposed to the highest stress concentration break and the tensile strength has no more to do with intermolecular forces between the chains but is now controlled by the force necessary to sever the covalent bonds in the backbone of the chains themselves. This is the reason that, in this domain, all polymers, regardless of their detailed structure, have the same strength since the backbones of all of them are formed by C–C, C–N, and C–O bonds, which have approximately the same dissociation energy and the same force constant.

This characteristic influence of DP on strength can also be considered in terms of the concentration of chain ends in a given volume element. If a chain end is situated in one of the crystalline areas of a semicrystalline polymer it creates a lattice disturbance of the type shown in Fig. 11.12 and produces in its immediate neighborhood tensile and compressive stress accumulations.

In a given cross section, if there are too many such accumulations, the stress to produce a microcrack at this point will become substantially reduced, a crack will form and there will be a break at this cross section as a result of crack propagation (3).

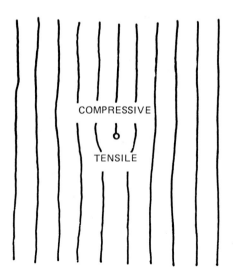

Fig. 11.12 Stress field around a chain end.

This weakening influence of the chain ends is also responsible for another important fact. It is known that all polymers, natural and synthetic, as they are used in the production of fibers, films, and other commercial articles, do not consist of macromolecules of identical DP but are mixtures of many species, each of which has a different DP. Hence they cannot be described by a single DP or molecular weight but rather by an *average* figure that takes account of the *polymolecular character* of the system and replaces the actually existing molecular weight distribution function by one of its averages. In this sense the term n in Eq. (1) does not describe the actual number of monomeric units but the average of this number taken over all molecules of the investigated sample.

It has now been found by extensive tests with many different polymers that the mechanical strength is very sensitive to the presence of low-DP constituents. It is therefore necessary to remove these low-molecular-weight species where they exist (as in such materials as cellulose, polyethylene, or polystyrene). In light of the detrimental influence of chain ends, this experience becomes understandable, because even a small weight percentage of low-DP species introduces a disproportionately large number of chain ends and causes widespread lattice distortions that gravely reduce the mechanical strength.

Summarizing, it can be stated that the molecular weight (DP, chain length) and the molecular weight distribution of a polymeric material influence the mechanical properties in a decisive manner.

B. Principles for the Design of Useful Polymers

For most practical applications of polymers in industry—fibers, films, plastic materials, coatings, and adhesives—it is necessary to endow the material not with a single outstanding property but rather with a *balance* of different properties, each of which must exceed a certain minimum value. In general terms these properties are

> Modulus of rigidity (resistance to shearing stress)
> Tensile strength (resistance to pulling stress)
> Impact strength (toughness, resistance to crack propagation under load)
> Melting point or softening range
> Resistance against the action of solvents
> Resistance against chemically deteriorating agents

Considerable progress has been made in the past to arrive at property combinations that have been satisfactory for many applications, but the available organic polymers are only now finding application in such large industrial uses as the construction of buildings and vehicles.

It might, therefore, be useful to review briefly those known principles with the aid of which one has been able to achieve reasonably well balanced compromises between valuable thermal and mechanical properties and to explore whether and how one possibly could arrive at still better combinations. In a general and simplified sense one can say that the two most important principles that have been useful in the past and should be good working hypotheses for future efforts are *crystallization* and *cross-linking*.

Let us, therefore, review their importance for the present purpose and, at the end, consider how far they lead us toward the achievement of high melting point and satisfactory rigidity.

1. Principle of Crystallization

The capacity to crystallize has long been known to be a very valuable property of linear, flexible macromolecules whenever one wants a good combination of thermal and mechanical properties. Today it is clear that the necessary requirement for crystallization is regularity of chain configuration and conformation (4). Even linear polyethylene, a completely nonpolar material with weak interchain bonding, is rigid, relatively high melting (130°C), strong, tough, abrasion resistant, and insoluble in anything at room temperature only because it possesses a *strong tendency to crystallize,* which means in this context, to form lamellae, bundles, or domains of high internal lateral order. The same is true for *isotactic* polypropylene, with a melting point of 175°C, and for *isotactic* polystyrene, which melts around 230°C. If the macromolecules contain polar groups and are of regular chemical architecture, even better combinations of mechanical and thermal properties have been realized, as in polyvinyl alcohol, polyvinylidene chloride, polyoxymethylene, and many aliphatic polyesters and polyamides, like polycaprolactam or 66 and 610 nylon.

In all these cases we have intrinsically flexible chain molecules of regular architecture that have a distinct tendency to form crystalline domains inside of which a systematic accumulation of the interchain forces rigidifies and reinforces the entire structure to such an extent that the system becomes hard, high softening, and soluble with difficulty.

The most important property of a crystalline polymer is its *melting point.* It is well known that all melting points, those of polymers and of low-molecular-weight materials alike, depend on the ratio of the heat of fusion ΔH to the entropy of fusion ΔS:

$$T_m = \frac{\Delta H}{\Delta S}$$

As a consequence, the melting points of polymers are related to the strength and regularity of the sites of intermolecular attraction, which affects the *heat* of fusion, and to the stiffness of the backbone chains, which affects the *entropy* of fusion. The effect of intermolecular forces provided by hydrogen bonds in polyamides can clearly be seen by plotting the melting points versus the number of amide groups per 100 Å of extended chain length. It has been found by such plots that the melting points are essentially a linear function of the amide group frequency because the heat of fusion increases proportionally with the frequency of hydrogen bonds produced by the increase in amide group concentration. One might expect the melting points to extrapolate to that of linear polyethylene at zero amide concentration because in this case only relatively weak van der Waals forces determine the heat of fusion of this polymer. In reality the extrapolated melting point is somewhat lower, which probably is caused by the polymolecularity and other irregularities of the samples used to determine the melting characteristics.

The effect of chain *stiffness* is well known in the case of Terylene or Dacron, which are polyesters using terephthalic acid as one component, which results in the presence of paraphenylene rings along the main chain. Since macromolecules of this type do not coil as flexibly in their amorphous state, they have a lower entropy of fusion and, consequently, a higher melting point, even though the intermolecular forces (ΔH) are not very strong.

The melting points of polymers are also noticeably affected by the reduction of the regularity with which the monomer groups are arranged along the backbone chain, such as occurs by random copolymerization. This is a reduction of tacticity (i.e., development of atacticity). It has been found, for instance, that the melting points of random copolymers of 610 and 66 nylon have a distinct minimum of 190°C at a 70:30 composition. Disturbing the regularity of the amide group spacings lowers the efficiency of the intermolecular hydrogen bonding, that is, it reduces the H (heat of fusion) in the basic melting point equation and, thereby, lowers the temperature of fusion.

Atactic polypropylene is a soft, heptane-soluble, amorphous resin, whereas the isotactic (ordered) species is a crystalline polymer with a T_m of 175°C. Between these there exists a series of semicrystalline materials with melting points roughly proportional to the degree of stereoregularity.

Another important thermal characteristic of a polymer is its *glass point* T_g; this refers to the amorphous noncrystalline part of a sample and is that temperature or temperature range at which, on cooling, the segments of the individual polymer chains lose so much of their mobility that a stress accumulation, such as an impact, can no longer be relaxed by segment diffusion but leads to a *break,* whereas above T_g it would have produced a *deformation.* Therefore, T_g is also occasionally called the *brittle point.* If all bonds of a chain have low conformational energy barriers, T_g is low. Such cases are silicones and poly-cis-dienes with T_g values below −60°C. On the other hand, large substituents and stiff chain segments—polyimides, aramides, and arylates—lead to T_g values above 200°C (5). Monomer unit rigidity and regularity have a pronounced effect on the melting points of polyesters; compounds based on ethylene glycol and sebacic acid are amorphous and soften at about 70°C, whereas polyesters made of ethylene glycol and terephthalic acid (PET) are crystalline and melt around 250°C. Thus, a 180°C increase in melting point is contributed by the stiff paraphenylene ring in the main chain. The melting point-composition diagram for random copolymers made of these two materials shows very clearly the progressive softening effect of the reduction in stiff segment content and regularity.

All factors that affect the melting point also affect the modulus of rigidity for much the same reasons, but in addition, there are now other contributing structural characteristics, namely, orientation and degree of extended chain crystallinity. More than 50 years ago the modulus and tensile strength of a very high molecular weight polyethylene, which is completely oriented and crystalline, were estimated on the basis of the known force constants of the C–C bond (6). This was recently confirmed.

Recently, Porter et al. (7) succeeded in preparing samples of super extended polyolefines that had moduli in the range of $1\text{-}2 \times 10^7$ psi ($7\text{-}14 \times 10^{10}$ Pa). Similarly, Morgan et al. (8) arrived at moduli and at tensile strengths for aramides in the same ranges.

Recently it was found that spatial regularity in polymeric systems is not only important for mechanical and thermal properties but also for other characteristics. One of these is permeability and semipermeability for molecules or ions in the gaseous or dissolved state. The polyester bottles that are now rapidly replacing the familiar glass bottles and metal cans for the packaging of carbonated beverages owe their low oxygen and CO_2 permeability to a high degree of biaxial orientation and crystallinity (9) in their tubular part. Eminently important is the high semipermeability of very thin membranes for the separation of gases and liquids and for the purification and desalination of water. It is also the result of high crystalline layers with perfect molecular regularity; this has been produced in several polymers, such as aramides, arylates, polyaromatic sulfones, and other rigid chain macromolecules (10).

Regularity in configuration and conformation has also opened up entirely new applications for organic polymers in the field of electro- and photoresponsive materials.

One extremely interesting effect was discovered, when one brings polyvinylidene fluoride $+CH_2-CF_2+_n$ into a planar trans-trans configuration (11). If a film of this material is brought into an electrical field above its T_g and then slowly cooled down into the glassy region, a permanently dipolarized sheet is obtained (see Fig. 11.13) that can be used as an extremely sensitive and effective *transducer* to convert electrical pulses into sound tracks, and vice versa. Because of the size, flexibility, and adjustability of such sheets and films, they have found wide usage in underwater sound detection, loudspeakers, hearing aids, and other new areas of sonar technology. These materials are also *pyroelectric*; they have served to make temperature sensors and as elements of design that produce ac current when put under vibrational pressure.

It has long been known that *polyconjugated* linear-chain molecules exhibit electronic conductivity (12), but it was only recently discovered that highly ordered and crystallized polyacetylenes exhibit a very wide range of electrical conductivity (13) if they are doped with adequate electron donors or electron acceptors. Preferred electron donors are alkali and alkaline earth metals; preferred acceptors are complex halide components, such as AsF_5 and SbF_5. Depending on the type of doping, n and p semiconducting sheets or filaments may be made that offer very interesting possibilities for the designing of new transistor components of microsized dimensions (see Table 11.33).

Photoelectric polymers also capitalize on the phenomenon of regularity and crystallinity (14) and have reached already very interesting practical application through the use of polyvinylcarbazole complexed with aromatic nitro-compounds or tetrathiofulvalenes. It has also been found that polyvinyls that contain aromatoaminic substituents can be converted into excellent photoconductors.

All this makes it very evident that the principle of crystalline order and regularity covers a wide field of phenomena and opens for organic polymers extremely interesting applications in the domain of mechanical, thermal, electrical, optical, and acoustical effects.

2. Principle of Cross-Linking

A random mass of long flexible chains can also be rigidified by the establishment of an irregular network of strong chemical bonds between the individual chain molecules. The denser this network is, the higher will be the softening range of the material, the greater will be its hardness, and the less will it be soluble. It is obvious that the recovery from larger strains and the avoidance of a permanent set will be favored by a system of fixed points that tie the chains into a supermolecular three-dimensional network and prevent them from slipping over each other in an irreversible fashion.

In order to produce such a system of firm and permanent cross-links in an effective and convenient manner, it is necessary to have certain groups available along the

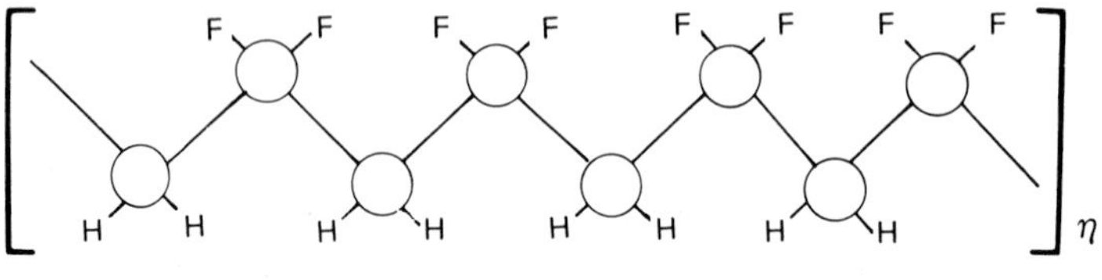

Fig. 11.13 Structure of polyvinyl fluoride.

Table 11.33 Electrical Conductivity (ohm-cm)[a]

Ag, Cu	10^6
Al	10^5
Bi	10^4
Ge	10^{-2}
Graphite	10^{-1}
Si	10^{-5}
Nylon	10^{-11}
Polyethylene	10^{-16}
Quartz	10^{-17}

[a]Doped polyacetylenes range from 10^{-19} to 10^3.

length of the individual chains that can serve as tie points for the network. The classic case is the sulfur vulcanization of *Hevea* rubber, where some of the double bonds in the chains are reacted with elemental sulfur or with sulfur compounds and produce covalent $-C-S-C-$ and $-C-S-S-C-$ bridges between individual macromolecules. These tie points are strong enough to prevent at temperatures up to about $120°C$ the irreversible slippage of entire chains along each other. Thereby, the network is able to return to its original random conformation after being extended.

Saturated olefinic polymers and copolymers have no double bonds left in the backbone chain; it is therefore necessary to introduce them with the aid of a specially chosen monomer if one wants to have the opportunity of cross-linking through the standard sulfur vulcanization process. In the case of butyl rubber, one introduces a small amount (1-3%) of butadiene or isoprene into long chains of polyisobutylene, and in the case of ethylene-propylene rubbers one adds reactivity with sulfur through copolymerization of the original ethylene-propylene composition with a diene, such as 1.4 hexadiene, dicyclopentadiene, or others.

Cross-linking—vulcanizing—with sulfur is a widely known and generally practiced art of rubber compounding and remains, at least for the time being, the preferred method of curing. There exist, however, numerous other ways to transform individual, independent chain molecules into a three-dimensional network with the aid of primary chemical valences. Simple olefine polymers, including polyethylene, polypropylene, polybutylene, and their copolymers, can be cross-linked with peroxides in the absence or presence of other ingredients, such as sulfur or sulfur-containing organic compounds. In these cases the covalent cross-linking is produced by a free-radical initiated chain reaction. The individual steps of the reaction—propagation, chain transfer, and termination—provide convenient mechanisms for the formation of covalent cross-links between any kind of polyhydrocarbon chain.

If the linear macromolecules of the initial system of independent chains carry reactive groups, there exist, evidently, many more possibilities for chemical cross-linking. Chains containing hydroxyl, amino, or epoxy groups can be cross-linked with dicarboxylic acids, anhydrides, dialdehydes, diisocyanates, or diepoxides; chains that carry COOH or SO_3H groups can be reticulated with diamines, diepoxides, or inorganic bases, and chains that contain halogens can be tied into three-dimensional networks through reaction with organic or inorganic bases.

As long as one wants to maintain soft rubberiness, the concentration of cross-links must be kept at a low level—1 in 200-300 units of the main chains—but even higher degrees of cross-linking still produce elastic networks. Their modulus is, of course, larger, and their extensibility is correspondingly lower. Since the mechanism of most cross-linking processes is known, there exist, in general, no serious difficulties in controlling the *degree* of network formation in a given polymer-curing agent system. Much less control exists of the spatial arrangement of the cross-links, the so-called *topology* of the network. As a first, qualitative approach it is obvious that the distribution of cross-links should be uniformly random, postulating that there should be no areas or domains of high and/or low cross-linking density. Under-cross-linked parts of the network will be too soft and will favor swelling and chemical or thermal degradation. Over-cross-linked portions will be too hard, will give rise to stress accumulation with resulting brittleness, and will have relatively slow postextension recovery rates with unfavorable consequences for the hysteresis characteristics of the elastomer.

Since many cross-linking processes are based on a chain reaction mechanism, it must be expected that the formation of the individual tie points is not statistically independent, but on the contrary, a formed cross-link acts as a nucleating agent for the formation of more junctions in its neighborhood.

The chemical character of fix-points can be very different: in certain cases they are normal covalent bonds, such as C–C, COO–C, C–N–C, or C–S–C. In other cases they

H

are heteropolar bonds between ionic groups. Even the strong physical adsorption of chain segments at the surface of dispersed filler particles, such as carbon black, silica, or alumina, can help tie the chains together into a three-dimensional network, and in certain cases, a system of hydrogen-bonding groups acting in a nonpolar, hydrocarbon-type environment can prevent chain slippage to a considerable extent over a certain range of temperatures and strains.

Much progress was made in the technology of cross-linking by the use of *segmented* linear and branched polymers where the two components possess adequately different thermal and mechanical properties. If we have, two segments, for example, one with a T_g of 100°C the other with a T_g of −50°C, then the material will be a readily extrudable and moldable *plastic* above 100°C; it will be elastic and resilient rubber between 100°C and −50°C, and it will behave as a hard resin below −50°C. If the molecular weight of the individual chains is relatively low (say below 60,000), it will be brittle; if the molecular weight is high (say above 200,000), it will be hard and tough.

Altogether, the establishment of an ideal fix-point system is a delicate and most interesting problem whose solution lies in the concept that best overall results are obtained if one operates with *several types* of cross-links having different degrees of firmness and bond strength.

1. It appears desirable to have a thin and dilute system of very strong and completely irreversible junctions that hold under all conditions and prevent the development of any permanent set. This network should be so thin that there exists no danger of over-cross-linked areas. Strong chemical bonds—covalent or heterovalent—are such fix-points.

2. These cross-links alone would give a very soft elastomer with a low 300% modulus and must, therefore, be supplemented by a system of fix-points that hold up to a certain strain at a given temperature but are eventually severed and re-form after the chains have assumed another, more relaxed conformation. This arrangement of fix-points can be much denser than that of point 1 because the dynamic opening and reclosing of these junctions provides an effective mechanism for stress equalization even if there are over-cross-linked domains in the original sample. Reversible cross-links of this type can be provided by sulfide or polysulfide bonds, by hydrogen bonds, and by highly adsorptive fillers.

3. Both of the former mechanisms are operating with strictly localized fix-points. It seems advantageous to establish a third, more diffuse mechanism for progressive reinforcement by designing the chain segments in such a manner that they can crystallize reversibly under the influence of larger strains. This contributes another independent and uniformly spread out capacity for self-reinforcement of the elastomer; it is controlled by the strain itself and can be made automatically reversible over a desired temperature range.

Present experience indicates that a carefully adjusted interaction of these three types of cross-links results in the most favorable combination of macromolecular events during stretching and relaxation. It leads to an optimum compromise of such important properties as high tensile and tear strength, good resistance against flex cracking, abrasion, and swelling in ambient organic liquids.

Although at first glance the effect of cross-linking appears to be very similar to that of crystallization, there are several important differences.

1. In a crystalline system the rigidity is the result of many, regularly spaced lateral bonds between the oriented chains; each of these bonds is weak and the ultimate effect comes from their large number and their regularity. In a cross-linked system the bonds between the long flexible chains are strictly localized and each of them is strong; in their entirety they are randomly arranged in the system. As a result, crystallization is a *reversible* phenomenon, whereas cross-linking is *irreversible* and occurs during the production of thermosetting resins.

2. Crystallization is a *physical* effect that takes place at all temperatures and is strongly influenced by mechanical processes, such as orientation and relaxation; cross-linking, on the other hand, is a *chemical* phenomenon that needs the presence of certain special reagents and is strongly accelerated by elevated temperatures but is not very much influenced by orientation or relaxation.

Many important products that are hard, infusible, and insoluble are made with the aid of cross-linking, such as all hard rubbers, urea-, melamine-, and phenol-formaldehyde condensation products, polyesters that use glycerol, trimethylol propane, or pentaerythritol as components, and resins that are hardened by the grafting of styrene on a polyester backbone that contains aliphatic double bonds. Although there are obvious and significant differences in the way in which crystallization and cross-linking act, it was very welcome to have *two different principles* to attain a desirable combination of valuable properties. However, there exists a third independent principle to arrive at even better results, namely, the use of *stiff linear chain molecules.*

3. Synthesis and Application of Linear Rigid Polymers

Crystallization and cross-linking produce stiffness, high softening, and difficult solubility by the establishment of *firm lateral* connections between intrinsically *flexible* chains; one can, however, also obtain the same result by incorporating the stiffness in the individual chains and constructing them in such a manner that their segmental motion is intrinsically restricted. One step in this direction is the introduction of bulky substituents into chains with flexible backbones. Polystyrene, for instance, is amorphous, has no cross-links, but still is a hard, relatively high softening (90°C) polymer. The absence of crystallinity results in complete transparency, the absence of cross-linking in reversible moldability, and easy flow characteristics. Similar conditions prevail in the case of polymethyl methacrylate, which is linear, amorphous, and has intrinsically flexible backbone chains, which, however, are stiffened by the two substituents (CH_3 and $COOCH_3$) at every other carbon atom and, as a result, is a hard, brilliantly transparent, relatively high softening (95°C) thermoplastic polymer that has found many useful and valuable applications. The only weakness of both materials is their low resistance against swelling and dissolution; it seems that the bulky and eventually polar substituents are capable of producing favorable mechanical and thermal effects that connected with the overall mobility and flexibility of the chain segments but that cannot offer sufficient resistance to the penetration of the system by solvents or swelling agents, because this process is a strictly localized phenomenon and depends on the affinity of the substituents for the particular solvent molecules.

Much more drastic effects are produced if the backbone chains themselves are *rigid* and if their substituents are so arranged that crystallization is prevented. Classic examples of such polymers are cellulose acetate and cellulose nitrate, where the glucosidic backbone chains possess a considerable intrinsic stiffness and the irregularly arranged acetyl and nitrate groups of incompletely substituted specimens do not permit the formation of a crystalline order. In fact, both materials are hard, transparent, high melting, and amorphous thermoplastic resins that have been widely and successfully applied for many years; cellulose acetate is still rather sensitive to the action of solvents and swelling agents, just like polystyrene and polymethylmethacrylate, whereas cellulose nitrate only dissolves in a few selected systems.

Recently the principle of using intrinsically rigid chains has been studied in more detail and has found several new and interesting embodiments. Based on rigid monomeric units, such as those shown in Fig. 11.14 and others, a series of polymers have been synthesized that are either amorphous or semicrystalline and not cross-linked but represent very hard, high-softening, and solvent-resistant materials. Earlier examples of this type are the polycarbonates and the linear epoxy resins, which are both based on bisphenol; more recent representatives are polyphenylene oxides, polyphenylene sulfides, polyxylylenes, polybenzimidazoles, polyimides, polyphenyloxazoles, and others, all of which exhibit unusually high resistance against softening, swelling, and decomposition. In fact, some of these newer materials can stand temperatures up to 500°C for long periods without softening or deterioration and are completely insoluble in all organic solvents up to 300°C.

Other rigid molecules now being studied for possible application in the field of high-temperature-resistant materials are based on aromatic chains, such as polyphenylene

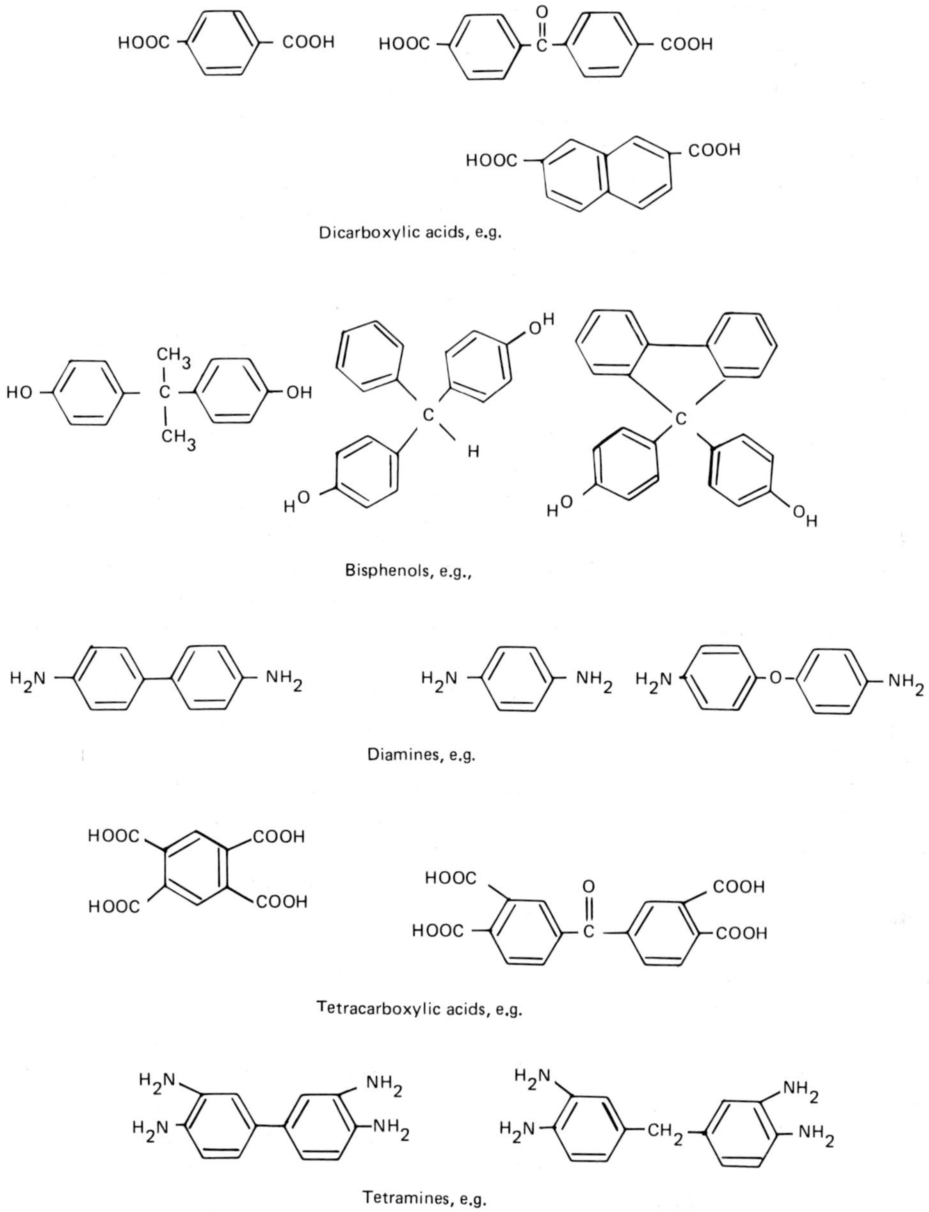

Fig. 11.14 Examples of commonly used rigid monomeric units.

which cannot fold even at rather high temperatures because rotation about the carbon-carbon single bond between the para-combined phenylene rings can only lead to different angles between the planes of consecutive rings but not to a kink or bend in the main chain. In fact, representatives of this species are rigid, high melting, possess a pronounced tendency to crystallize, and are highly insoluble.

All these stiff-chain polymers excel by their thermal and mechanical properties. but for the same reason they are very difficult to process. It is therefore necessary to use somewhat unusual processing methods, such as forging, pultruding, or sintering, to bring these materials into useful shapes, such as rods, tubes, tires, or plates (15). Another way to overcome processing difficulties is to carry out the form-giving steps with a precursor and arrive at the desired structure of the molecule in its final shape. This is done, for instance, with fibers, films, and shaped objects of polyimides and certain ladder molecules. Still another method to circumvent complicated blending and form-giving processes is the in situ polymerization of vinyls, dienes, epoxies, and diisocyanates in the presence of large amounts of reinforcing globular or fibrous fillers. A particularly interesting case of this technique is the BR 200 series of highly filled ultra high molecular weight polyethylenes of du Pont (16). Rotation about the bonds between an ether oxygen atom and the adjacent carbon atoms of the rings does change the plane of the angle between two units and thus leads to bends and kinks in the chain, but the rotational freedom is noticeably inhibited by the presence of the aromatic rings on each side of the oxygen or sulfur atom. As a consequence, it has been found that polymers of this type are high melting, rigid, and relatively insoluble materials.

Another interesting way to arrive at chains made up of condensed rings is the synthesis of so-called ladder polymers. The first such structure was prepared by exposing polyacrylonitrile fibers to elevated temperatures in the presence of air, which causes the formation of rows of fused six-membered rings by an electron pair displacement:

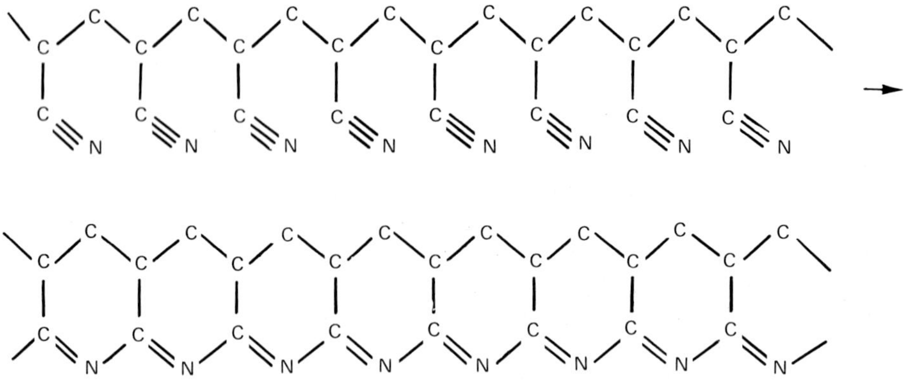

which results in insolubility, discoloration, and mechanical stiffening. Further heating in nitrogen up to 1000°C leads to evolution of H_2 and to aromatization.

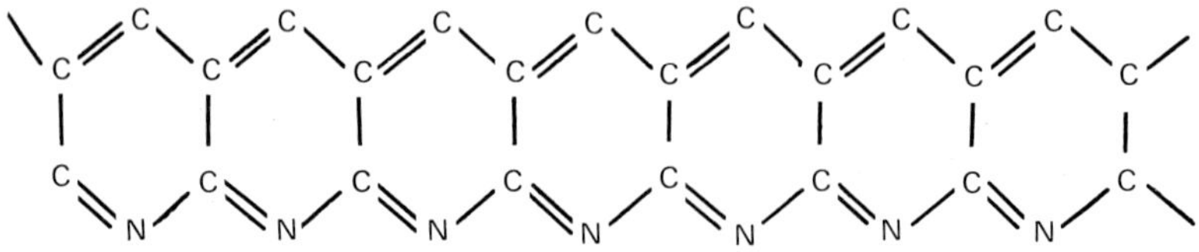

whereby a black, completely infusible and insoluble material is obtained that, in its structure, corresponds to a linear graphite in which one carbon atom of every ring has been replaced by nitrogen.

These examples show that there are many possibilities for the formation of long stiff chains and that, in all cases, the properties of the resulting materials confirm the expectations.

The existence of three different and independent ways to establish favorable compromises of valuable properties stimulates the attempt to explore combinations of these principles and to see whether they might lead to even better results. To get a convenient survey of such combinations, let us reconsider Fig. 10.21 in which the three principles of crystallization (A), cross-linking (B), and chain stiffening (C) are represented by the three corners A, B, and C.

The corner A is populated by a large number of crystallizable, thermoplastic polymers with flexible chains that have proved to be particularly successful as fiber and film formers. Representative materials are polyethylene, polypropylene, polyoxymethylene, polyvinyl alcohol, polyvinyl chloride, polyvinylidene chloride, and many linear polyamides and polyesters.

In corner B are located the typical thermoset, highly cross-linked systems, such as hard rubbers, urea-, melamine, and phenol-formaldehyde condensates, highly reticulated polyesters, polyepoxides, and polyurethanes.

Finally, corner C contains the amorphous, thermoplastic resins with relatively high rigidity and high softening range, such as polystyrene, polymethyl methacrylate, polyvinyl carbazole, polyimides, polybenzimidazoles, and the different ladder polymers.

One can now pose the question whether a combination of two principles can help to arrive at products with more attractive and valuable properties, and one is thereby led to the exploration of the *sides* of the triangle.

In fact, the line from A to C accommodates several interesting fiber and film formers. One of them is polyethylene glycol terephthalate (Terylene or Dacron), in which the para-phenylenic units of the acid introduces enough chain stiffening to bring the melting point of this polymer up to about 260°C, which is as high as the melting point of 66 nylon although a polyester has *no lateral hydrogen bonding* available for the reinforcement of its solid crystalline phase. Thus, chain stiffening cooperates with crystallization to produce attractive properties without bringing either of these two principles to an extremely high value. Cellulose is another case in which excellent fiber- and filmforming properties are built up by a combination of chain stiffening and crystallinity; these factors together produce a polymer that is rigid, does not melt at all, and is soluble only in a small number of particularly potent liquid systems. The presence of substantially rigid chains has the favorable consequence that high tensile strength and high softening characteristics become apparent at relatively low degrees of crystallinity; this places cellulose somewhere in the middle of the line connecting A and C. Many of the recent spectacular improvements of cellulosic filaments are founded on a capitalization of the dual origin of its fiber-forming potential. Another example for a beneficial combination of A and C is given by cellulose triacetate, in which the capacity to crystallize superimposes several favorable properties on the normal cellulose acetate, particularly insolubility in many organic liquids and thermoset capability through additional crystallization.

On the side AB are situated all those rubbers that are slightly or moderately crosslinked and crystallize on stretching progressively; examples are natural rubber, high cis-

cis-polybutadiene and polyisoprene, butyl rubber, and neoprene. Depending upon the degree of cross-linking, they are more or less close to B.

Until now we have only considered systems of one component, namely, a specific polymer or copolymer; in the technology of elastomers it is, however, customary to produce stiffening and temperature resistance by the addition of a hard, finely divided *solid filler,* such as carbon black, silica, or alumina. This addition of a crystalline or pseudocrystalline reinforcing filler is a kind of crystallization of the system in the sense that the flexible chains of the original polymeric matrix are restricted in their segmental mobility by the presence of the very small and hard particles of the filler to the surface of which they are attached by strong adsorptive forces. Thus the presence of a solid, hard, and high-melting fiber is another way to stiffen a polymeric system to a desired degree. *Composite systems* of this kind have been studied and used for a long time. They have been recently brought into the focus of scientific and practical interest because they promise to deliver a large system of engineering materials that approach metals in such important properties as hardness, thermoresistance, and toughness with only a fraction of their specific gravity. Additionally, they offer substantial advantages in processing and corrosion resistance (17). The aircraft, automobile, and rail car industries are looking at them as the major new building materials of the 1980s. *Sheet-molding compounds* and *high-performance composites* dominate research and development activities on a large scale and add to our triangle another independent dimension. Most efforts are, understandably, concentrated on the basic large-volume resins, such as polyolefines, ABS, PVC, polyamides, polyesters, polyurethanes, and polyepoxies. To enhance their properties we have at our disposal a large number of useful fillers. The globular systems comprise all kinds of carbon black, together with finely powdered silica, alumina, calcium carbonate, and aluminum silicates. Particularly interesting are hollow microspheres of glass, which combine very low specific gravity with high rigidity and excellent uniformity.

Fibrous fillers, used in the form of prepregs or as short chopped extrudable additives, include glass fibers, cellulosics, aramides, carbon fibers, and recently the FR fiber of du Pont, which consists of alumina. This latter has such a high melting range (around 2300°C) that it has been successfully used to reinforce Mg, Al, Zn, and Pb. Thus are provided new materials for the construction of planes, vehicles, and chemical equipment that offer hitherto inaccessible combinations of thermal, chemical, and mechanical properties.

The line from C to B has been populated with certain useful polymers in the attempt to increase rigidity, high softening, and insolubility of stiff chain systems by additional cross-linking. Well-known cases for the combination of these two principles are the raising of the heat distortion point of acrylic and methacrylic polymers by the incorporation of allylmethacrylate or ethylene glycol dimethacrylate and the "curing" of epoxy polymers based on such stiff-chain elements as bisphenol and cyclic acetals of pentaerythritol. Additional attempts are now under way to improve the properties of the newer amorphous systems with intrinsically stiff chains by cautious cross-linking. These materials are moved thereby from point C somewhat in the direction of point B. The principal reason for this approach is, for such systems, to increase the resistance to dissolution and swelling at elevated temperatures.

The advantageous combination of all three of them could lead to still further improvements in properties. This is a field in which, at present, much exploratory work is done and in which certain interesting results have already been obtained. Successful applications of all three principles are the aftertreatment of cotton with certain cross-linking agents or the spinning of rayon in the presence of such ingredients. Cellulose has rigid chains that can be brought to a moderate degree of crystallinity by the swelling of cotton or the appropriate spinning of rayon. This results in fibers of satisfactory strength, elongation, hand, and dyeing characteristics but of insufficient recovery power. The introduction of a carefully controlled system of cross-links with the aid of bifunctional reagents leaves all other desirable properties unchanged and improves substantially the recovery power and wrinkle resistance of the fibers. Similarly promising results have been obtained with mildly reticulated amorphous stiff chain systems of the epoxy and ure-thane type, in which crystallization has been replaced by a reinforcing filler.

More recently the cross-linking of partially crystalline poly-bis-maleiimides and the high-temperature cross-linking of cyclized segments of highly oriented and semicrystalline polyacrylonitrile fibers are other cases where a combination of all three concepts has led to useful new results.

REFERENCES

1. K. H. Meyer and H. Mark, *Der Aufbau der Hochmolekularen Organischen Substanzen,* Acad. Verl., Leipzig, 1930. H. Staudinger, *Der Aufbau der Hochmolekularen Organischen Substanzen,* Springer Verlag, Berlin, 1932.
2. W. H. Carothers, *Collected Papers,* Interscience, New York, 1940. P. J. Flory, *Principles of Polymer Chemistry,* Cornell University Press, Ithaca, New York, 1953.
3. L. E. Nielsen, *Mechanical Properties of Polymers,* Academic, New York, 1960.
4. A. V. Tobolsky and H. Mark, *Polymer Science and Materials,* John Wiley and Sons, New York, 1971.
5. P. J. Flory, *Statistical Mechanics of Chain Polymers,* John Wiley and Sons, New York, 1968.
6. H. Mark, *Physics and Chemistry of Cellulose,* Berlin, 1932.
7. R. S. Porter et al., in *Ultra High Modulus Polymers,* A. Ciferri and I. M. Ward (eds.), Applied Science Publishers, London, 1979, p. 77.
8. P. W. Morgan, S. L. Kwolek, J. R. Schaefgen, and L. W. Gulrich, *Macromol.* 10, 1381, 1390 (1977).
9. H. von Hassell, Polyester bottles, *Plastics Techn.* January 1950, p. 69.
10. H. K. Lonsdale and H. E. Podell, *Reverse Osmosis Membrane Research,* Plenum Press, New York, 1972.
11. J. B. Lando, H. G. Olf, and A. Peterlin, *JPS* A1, 941 (1966); and later R. G. Kepler, *Bull. Am. Phys. Soc.* 21, 373 (1976).
12. J. Economy, New and specialty fibers, *JAPS,* Symp. Vol. 29, 71 (1976).
13. A. G. McDiarmid, C. K. Chiang et al. *Phys. Rev. Lett.* 39, 1098 (1977).
14. H. Mark, Electro- and photoresponsive polymers, *JPS,* Symp. Vol. 62 (1978).
15. D. M. Gale, *JAPS* 22, 1955, 1971 (1978).
16. J. V. Milewski, *Plastics Eng.* Nov. 1978, p. 23. Also RP-200 Series, du Pont; Central Res. Dept. Wilmington, Del. 19898.
17. E. M. Pearce and H. F. Mark; *Int. J. Pol. Mat.* 5, 5 (1976).
18. A. K. Dhingra, A. R. Champion, W. H. Krueger, and H. S. Hartman, Advanced Materials, du Pont, Textile Fiber Dept. Pioneering Research; Exp. Station, Wilmington, Del. 19898, 1979. Also, *Modern Plastics* October 1979, p. 14.

BIBLIOGRAPHY

Books

Allcock, H. R., and F. W. Lampe. *Contemporary Polymer Chemistry*. Englewood Cliffs, New Jersey, Prentice-Hall, 1981.

Billmeyer, F. W., Jr. *Textbook of Polymer Science*. New York, Wiley, 1962.

Brydson, J. A. *Plastics Materials*. London, D. Van Nostrand, 1966.

Carter, M. E. *Essential Fiber Chemistry*. New York, Dekker, 1971.

Crosby, E. G., and S. N. Kochis. *Practical Guide to Plastics Appplications*. Boston, Cahners Books, 1972.

Deanin, R. D. *Polymer Structure Properties and Applications*. Boston, Cahners Books, 1972.

Golding, B. *Polymers and Resins— Their Chemistry and Chemical Engineering*. London, Van Nostrand, 1959.

Harper, C. A. (ed.). *Handbook of Plastics and Elastomers*. New York, McGraw-Hill, 1975.

Lawrence, J. R. *Polyester Resins*. New York, Reinhold Publishing, 1960.

Lenz, R. W. *Organic Chemistry of Synthetic High Polymers*. New York, Interscience, 1967.

Milby R. *Plastics Technology*. New York, McGraw-Hill, 1973.

Miles, D. C., and J. H. Briston. *Polymer Technology*. New York, Chemical Publishing, 1965.

Ogorkiewicz, R. M. (ed). *Engineering Properties of Thermoplastics*. New York, Wiley Interscience, 1970.

Shalaby, S. W. (ed). *Thermal Methods in Polymer Analysis*. Philadelphia, Franklin Institute Press, 1978.

Shalaby, S. W. *Plastics Products Design Handbook,* Part A, ed. By E. Miller, New York, Dekker, 1981a.

Shalaby, S. W. *Thermal Characterization of Polymeric Materials,* ed. by E. A. Turi, Chap. 3. New York, Academic Press, 1981b.

Shalaby, S. W. *Technological Aspects of Polymeric Materials,* Part III, in *Polymer Science and Technology: An Interdisciplinary Approach*. Washington, D.C., American Chemical Society, submitted.

Shalaby, S. W., and H. E. Bair. *Thermal Characterization of Polymeric Materials,* ed. by E. A. Turi, Chap. 4. New York, Academic Press, 1981.

Shalaby, S. W., and E. M. Pearce. *Chemistry of Macromolecules,* Part I, in *Polymer Science and Technology: An Interdisciplinary Approach*. Washington, D.C., American Chemical Society, 1974.

Stevens, M. P. *Polymer Chemistry*. Reading, Massachusetts, Addison-Wesley/W. A. Benjamin, Inc., 1975.

Encyclopedias

Encyclopedia of Chemical Technology, Second Edition,

Encyclopedia of Polymer Science and Technology. New York, Wiley Interscience, 1971.

12
Adhesives*

An adhesive is a substance capable of holding materials together by surface attachment, according to the ASTM definition. The solid materials held together are called adherends. Extensive lists of definitions in the adhesive field can be found in appropriate handbooks (1).

Adhesive bonding offers a number of advantages over other methods for fastening materials together. Adhesives offer the possibility of bonding a variety of dissimilar materials that may be difficult to fasten together by other means. Thin sheets that easily distort and brittle materials, such as glass, are often difficult or impossible to join together by other means. Economical and rapid assembly is characteristic of this type of bonding; finished products characteristically show smooth external surfaces with elimination of gaps and protrusions, such as those that come from the use of rivets or bolts. Weight reduction can be achieved by use of adhesives instead of mechanical fasteners, with the added advantage of a uniform distribution of stresses over the entire bonded area. Stress concentration is minimized, and joint continuity allows full use of the strength of the components. Adhesives can often be used to join heat-sensitive materials that cannot be fastened by methods requiring application of high temperature. Galvanic corrosion between dissimilar substances can often be prevented.

It must also be recognized that there are a number of disadvantages inherent in the use of adhesive bonding. The achievement of a good bond usually requires clamping for a sufficient time for the adhesive to set. Surface preparation must be carefully carried out, and cleanliness in the operation is usually a requirement. Processing conditions must be carefully controlled, and it is often difficult to provide for rapid, nondestructive inspection of the resultant bonds. Hazards, such as fire and toxicity, are present with many solvent-based adhesives.

Since most adhesives are organic materials, care must be taken that the bonds are not exposed to temperatures high enough to cause degradation and subsequent failure of the bond. This point reinforces the need for careful selection of the type of adhesive to meet the intended use requirements. In this connection it is worth remarking that a good adhesive must show good cohesive strength (i.e., it must cohere to itself) as well as good adhesive properties (i.e., it must adhere to the materials it is bonding) under

---

*Contributed by RAYMOND F. WEGMAN and DAVID W. LEVI.

conditions of use. For example, below 0°C ice is a reasonably good adhesive for wood. It adheres well and has sufficient cohesive strength. However, when the ice melts the resultant water has little cohesion and is useless as an adhesive.

I. NATURE OF ADHESION

A. Bonding Forces

The process of adhesion ultimately involves the establishment of some type of attraction between the atoms and molecules comprising the adhesive and the adherend. These types of attraction may involve primary bonding forces, which tend to be quite strong, weaker secondary (van der Waals) forces, and hydrogen bonding as an intermediate case.

Primary bonding may be covalent in nature. In this case electron pairs are shared between adjacent atoms. Ionic (electrostatic) forces are the type of primary bonds found in ionic crystals. Metallic bonds are similar to the covalent type except that they are characterized by great mobility of the conduction electrons, which contribute a part of the bonding energy. (See Chap. 2.)

Secondary bonding involves dipole-dipole interaction, induced dipole interaction, and dispersion forces. This type of bonding is of great importance in adhesion, since often nonpolar or chemically inert surfaces are involved. Hydrogen bonds are often described as a special case of dipole interaction since they are due to the sharing of a proton by two electronegative atoms (usually oxygen). Their action extends farther than ordinary van der Waals forces, and bond strength can be of the same order as a weak primary bond. The types of bonding have been reviewed by Salomon (2).

The above forces are considered to be universal in character, and if sufficiently good contact is obtained on the surface between the adhesive and the adherend, dispersion-type forces alone should be adequate to give a satisfactory adhesive bond (3). In dealing with liquid adhesives (as is usual), obtaining the required molecular contact with the solid adherend requires wetting of the solid surface by the liquid.

B. Surface Considerations

To understand the conditions required for adequate wetting, we first consider the surface energies of liquids alone. Free surface energy arises at a liquid boundary because of the imbalance of intermolecular attractive forces. The molecules in the body of the liquid are equally attracted in all directions. However, on a free surface the attractive forces only operate downward into the body of the liquid. To extend the surface area, molecules must be removed from the interior of the liquid to the surface, so that work must be done in overcoming the inwardly directed forces. These molecules moving to the new surface are raised to a greater potential energy than those in the bulk liquid. This extra energy corresponds to the usual surface tension, that is, the increase in free surface energy per unit increase in surface. Then, for liquids, free surface energy and surface tension are equivalent and experimentally accessible. Actually, the free surface energy does not represent the total energy expended on forming each unit of new surface since the molecules in the surface layer are less densely packed and organized than in the bulk liquid. Hence there will be an increase in entropy so that the free surface energy will be less than the total surface energy. This difference will not be further considered here since we are primarily concerned with behavior at a solid-liquid interface, and the free surface energy is sufficient for this purpose.

In the case of solids, free surface energy cannot be measured simply and directly as was the case with liquids. The methods used are beyond the scope of this discussion. Our interest lies in the wetting of the solid by the liquid adhesive. A diagrammatic representation of the situation is shown in Fig. 12.1, where the free surface energies (γ) are treated as vector quantities, and where SV, LV refer to the solid and liquid phases in equilibrium with the liquid vapor and SL to the solid-liquid interface. When the contact angle $\theta = 0$, the liquid spreads freely and completely wets the solid, so that good contact is made and good adhesive bonding usually is accomplished. When $\theta > 0$ the liquid does not completely wet the surface and poor bonding can be anticipated (see Fig. 12.1).

The free surface energies of most liquid adhesives are below 100 dyne/cm, as is also the case with organic solids, such as polymers. On the other hand, hard solids, such as metals and metal oxides, typically show free surface energies in the 500-5000 dyne/cm range (3). Thus, the latter high-energy solids would attract the liquid molecules more strongly than the liquid molecules attract each other and good wetting would generally be expected. However, since the free surface energies of liquids are of the same order of magnitude as those of many low-energy solids (e.g., many polymers), poor wetting is more commonly encountered in these cases.

Even if good wetting and contact between the adhesive and adherend is obtained, difficulty may be encountered in obtaining a good, durable bond if there is a weak boundary layer on the adherend. For example, in bonding iron or steel, if there is a layer of red rust (hydrated oxide) on the surface, the adhesive will form a bond. However, the rust boundary layer will readily lift off the metal surface so that the bond is of no practical use. Such an example emphasizes the need for careful surface preparation of the adherend to assure that such a weak boundary layer will not make the bond practically useless.

As another example, it is well known that polyethylene can only be bonded satisfactorily after a surface pretreatment, such as plasma exposure, flaming, or chemical attack. It was believed that these treatments gave an easily wetted surface by superficial oxidation. However, it turns out that this alone is not sufficient to account for the enhanced bondability. In addition, it is now believed that cross-linking during the treatment

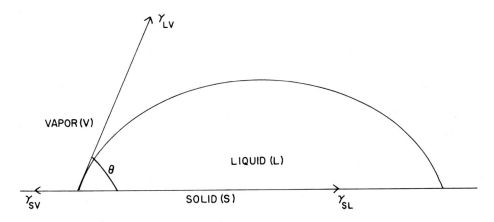

Fig. 12.1 Vector diagram for a drop of liquid on a solid surface.

strengthens this surface layer so that satisfactory bonds are formed (4). Apparently, both wetting and formation of a strong boundary layer are required in this case.

II. TYPES OF ADHESIVES

During the past century, adhesives have advanced rapidly. For thousands of years only the animal and vegetable glues were of any importance. These are still widely used to bond porous materials. During World War I casein glues were used to bond wood in aircraft. The natural adhesives have limited resistance to moisture and mold growth; in the 1930s there was a large movement toward development of new and improved adhesives based upon synthetic resins. In the 1950s, the epoxy resin-based adhesives made their debut and introduced a new era in adhesive bonding. Today the epoxy-based adhesives are still the mainstay of adhesive bonding, with many new special-application adhesives being introduced. Table 12.1 describes some of the major classifications of adhesives.

III. PREPARATION OF SURFACES

Adhesion is a surface phenomenon, and as such, one must be very concerned about the type of surface being bonded. The surface must be clean and free of weak boundary layers and have a high surface free energy.

Recently, there has been an increase in the number of investigations into the state and condition of the surface of the adherend as it is rendered by any given surface treatment.

Hamilton (5) describes various techniques by which one could evaluate surface wettability. These include surface morphology, surface film thickness, surface composition, and surface crystallography. From then on there have been many advances in the characterization of adherend surfaces. Weber and Johnston (6), Russell and Garnis (7), Bijlmer and Schliekelman (8), and Venables et al. (9) describe techniques used to characterize aluminum surfaces. Surface characterization studies have been mostly conducted on metal surfaces, but there is a need for this type of work to be done on plastics and other surfaces to help understand the method of adhesion and of failure to these surfaces.

The preparation of the adherend surface is one of the most critical operations required for adhesive bonding. Surface preparation involves having a working knowledge of the manufacturing processes used to fabricate the part. This may be the method by which the sheet was milled or the method by which a part was formed, that is, cast, machined, forged, molded, or laid up. Surface preparation also involves having a knowledge of properties of the adherends, the end use of the bonded item, and limitations that might restrict the type of surface treatment used.

In recent years, many investigators have conducted various studies in attempts to find the best surface preparation for various adherends. To become involved with the details of surface preparation for each particular adherend would involve a handbook of its own and is beyond the scope of this chapter. To help the reader find sources of information as to how to treat various adherends, Tables 12.2 and 12.3 list some of the more common adherends for which published surface treatment are available and give some of the references.

Table 12.1 Major Classifications of Adhesives

Class	Basic type	Specific type
Animal	Animals	Animal glue, albumen, casein, shellac
Vegetable	Natural resins	Gum arabic, Canada balsam, oils, proteins, and carbohydrates
Mineral	Inorganic	Silicates, litharge sulfur bitumen, mineral waxes, resins, phosphates
Rubbers, elastomers	Natural	
	Synthetic	Butyl, polyisobutylene, polybutadiene blends; polyisoprenes and polychloroprenes; polyurethane, fluorocarbon, silicones, polysulfide, and polyolefins
	Reclaimed	
Thermoplastic	Cellulosic	Acetate, acetate-butyrate, caprate nitrate, and cellulose derivatives
	Vinyls	Polyvinyl acetate, alcohol, acetate chloride; polyvinylidene chloride and polyvinyl alkyl ethers
	Polyesters	Saturated (polystyrene and polyamides)
	Polyacrylates	Methacrylate and acrylate, cyanoacrylates and acrylamides
	Polyethers	Polyhydroxy, polyphenolic esters
	Polysulfones	
Thermosetting	Amino	Urea-, melamine-formaldehydes, epoxy-polyamide, epoxy-polysulfide, epoxy-nylon, nitrile epoxy
	Phenolic	Phenol and resorcinol formaldehydes, phenolic-nitrile, phenolic-neoprene, phenolic-epoxy
	Polyesters	Unsaturated
	Polyaromatics	Polyimides, polybenzimidazole, polybenzothiazole, polyphenylene
	Furanes	Phenol furfural

Table 12.2 References for Surface Preparation of Metals

Metal	References
Aluminum	13-30
Beryllium	31-34
Cadmium	1, 21, 34
Chromium	1
Copper, copper alloys	1, 13, 23, 27, 35
Gold	36, 37
Lead	38
Magnesium, magnesium alloys	13, 27, 39-41
Nickel, nickel alloys	34
Platinum	34
Silver	34, 37
Steel, mild	13, 20, 22, 23, 42-46
Steel, stainless	13, 19, 20, 21, 23, 42, 43, 45, 47-51
Tin	34
Titanium	11, 13, 52-58
Tungsten	43
Uranium	59
Zinc	43

Surface preparation is also the key to durable and reliable structures. Wegman and Bodnar (12) demonstrate the effect of surface preparation on the durability of bonded structure. Other work on durability is discussed in Sec. V.

IV. APPLICATION OF ADHESIVES

The application of an adhesive to the adherend is dependent upon the type of adhesive, the size and shape of the adherends, and the volume of production. The low-viscosity liquid adhesives are generally applied by brushing, spraying, or roller coating. As the viscosity of the adhesive increases, the method of application generally tends toward extrusion and troweling. Automated measuring, mixing, and dispensing equipment is very useful when the volume of production becomes large enough to amortize the cost of equipment.

The solvent-based adhesive must be allowed sufficient open time to permit the evaporation of the solvent from the adhesive prior to assembly. Often the adhesive is allowed to dry and then is either solvent reactivated just prior to assembly or put together dry and then heat reactivated. For other applications, the solvent will be allowed to evaporate and then the two mating surfaces will be put together while the surface is tacky. With porous surfaces, such as paper, fabrics, and wood, the adhesive is coated onto the part and the assembly is made while wet.

The aerospace industry and others using honeycomb structures often will use film adhesives. These generally are made from a mixed adhesive, which requires an elevated temperature to effect the cure. These films are cut to size, placed between the mating surfaces, and then subjected to heat and pressure by placing the assembly in an autoclave, hydroclave, or press.

Table 12.3 References for Surface Preparation of Plastics and Rubbers

Plastics, rubbers	References
Acetal copolymer	20
Acetal homopolymer	1, 60, 61
Acrylonitrile-butadiene Styrene (ABS)	1, 62
Cellulosics	1
Ethylene vinyl acetate	1
Nylon	61, 63, 64
Phenylene oxide	65, 66
Polyaryl ether	67
Polyaryl sulfone	68
Polycarbonate	20, 60, 61
Polychlorotrifluorethylene	1, 22, 61, 64
Polyethylene	60, 69, 70-73
Polymethyl methacrylate	61, 64
Polypropylene	60, 67
Polystyrene	64
Polysulfone	1, 64
Polyvinyl chloride	1, 61, 64, 74
Polyvinyl fluoride	64, 75
Polyvinylidene fluoride	21
Polydiallyl phthalate	1, 20, 21, 59-62, 64
Epoxies	1, 21, 37, 43, 61, 64
Melamines	1, 21, 43, 61, 64
Phenolics	1, 21, 37, 43, 61, 72
Polyesters	1, 21, 43, 46, 61, 64
Polyimides	1, 19, 77, 78
Polyurethane	1, 21, 43, 61, 64
Silicone	21
Urea-formaldehyde	1, 21, 61, 64
Reinforced thermoset	21, 64, 79, 80
Glass-reinforced thermoplastic	81, 82
Plastic foams	21
Rubbers	1, 19, 21, 43, 49, 83-86

In a recent improvement, an adhesive is applied to the surface of one of the adherends and allowed to B-stage. B-stage is defined (87) as an intermediate stage in the reaction of certain thermosetting resins in which the material softens when heated. This method is sometimes used to apply the adhesive to the cell end of honeycomb cores, and in this case, the adhesive is referred to as a *reticulated adhesive*.

V. DURABILITY AND RELIABILITY

Durability refers to the proper functioning of the bond under use conditions for as long as possible. Reliability refers to the mathematical probability that the bond will function according to the design specification for the life of the structure. An excellent review of this topic has recently appeared (88).

Before selecting an adhesive material for a particular application it is necessary to be sure that the bond will be durable under the conditions of thermal and mechanical stress as well as under the environmental (chemical or corrosive) conditions that will be encountered in use.

The general durability of a system should be evaluated by accelerated testing before it is used in an item. Such tests generally involve placing the bond at a temperature somewhat above the value it will encounter in actual use. In addition, the bond is usually concurrently exposed to further adverse conditions, such as an applied stress or immersion in hot water. Strength reduction with time is observed, preferably at several temperature levels.

Various theories have been proposed as a basis for predictive tests in using accelerated testing as a basis for estimating actual service life. Among these theories, Kaelble (88) has included

1. Chemical reaction rate theory
2. Classical cumulative damage theory
3. Statistical cumulative damage theory
4. Physical chemistry theory of cumulative damage

Detailed discussion of these theories is beyond the scope of this chapter. The reader is referred to Ref. 88 for details.

VI. DESIGNING FOR ADHESIVE BONDING

When designing an item that will be adhesively bonded, there are a few major questions that the design engineer must consider.

Stress. What type of stress will the joint experience—shear, peel, cleavage, or tensile? What will be the magnitude of each component stress, and what will be the duration of the stress?

Strength. What is the minimum bond strength acceptable at the most extreme service environment?

Material. From what materials will the item be fabricated, and are these materials bondable?

Surface condition. Can the bonding surfaces be treated to make them bondable, durable, and reliable?

Thermal coefficients of expansion. Are the materials close in thermal coefficient of expansion, to allow them to expand or contract together, or will one change more than the other and cause uncontrollable stresses and premature failure.

Tolerances. Are the tolerances between mating surface large enough to allow a proper bond line thickness for the type of adhesive being used.

Designing for adhesive bonding encompasses many things. Keimel (89) discussed the design of equipment cabinets; Pinckney and Griffith (90) discuss a design concept for a helicopter fuselage. Both these paper references demonstrate the need for complete and thorough planning before and during the design phase. As a designer, do not attempt to do it alone. Consult materials and process engineers—they can help. Include on your design team the adhesives engineer, the process engineer, and the quality assurance engineer. The adhesive engineer can help select adherends, the proper surface treatments, and the proper adhesives for the end-item service. The process engineer will make sure the design can be fabricated without undue expense. The quality assurance engineer will make sure the parts are inspectable and of the quality needed to produce a reliable and durable bonded structure.

VII. TESTING OF ADHESIVE PROPERTIES

Many tests are used to test the various properties of adhesives. Among these may be listed the following ASTM tests (91) and recommended practices (the latest revision should be employed).

1. D897, Test for Tensile Properties of Adhesive Bonds
2. D903, Test for Peel or Stripping Strength of Adhesive Bonds
3. D905, Test for Strength Properties of Adhesive Bonds in Shear by Compression Loading
4. D906, Test for Strength Properties of Adhesives in Plywood-Type Construction in Shear by Tension Loading
5. D950, Test for Impact Strength of Adhesive Bonds
6. D1002, Test for Strength Properties of Adhesives in Shear by Tension Loading
7. D1062, Test for Cleavage Strength of Metal-to-Metal Adhesive Bonds
8. D1101, Test for Integrity of Glue Joints in Structural Laminated Wood Products for Exterior Use
9. D1183, Tests for Resistance of Adhesives to Cyclic Laboratory Aging Conditions
10. D1184, Tests for Flexural Strength of Adhesive-Bonded Laminated Assemblies
11. D1205, Testing of Adhesives for Brake Lining and Other Friction Materials
12. D1304, Testing Adhesives Relative to Their Use as Electrical Insulators
13. D1338, Test for Working Life of Liquid or Paste Adhesives by Consistency and Bond Strength
14. D1344, Testing Cross-Lap Specimens for Tensile Properties of Adhesives
15. D1780, Recommended Practice for Conducting Creep Tests of Metal-to-Metal Adhesives
16. D1781, Climbing Drum Peel Test for Adhesives
17. D1828, Recommended Practice for Atmospheric Exposure of Adhesive-Bonded Joints and Structures
18. D1876, Test for Peel Resistance of Adhesives
19. D2095, Test for Tensile Strength of Adhesives by Means of Bar and Rod Specimens
20. D2182, Test for Strength Properties of Metal-to-Metal Adhesives by Compression Loading

21. D2293, Test for Creep Properties of Adhesives in Shear by Compression Loading
22. D2294, Test for Creep Properties of Adhesives in Shear by Tension Loading
23. D2295, Test for Strength Properties of Adhesives in Shear by Tension Loading at Elevated Temperatures
24. D2339, Test for Strength Properties of Adhesives in Two Ply Wood Construction in Shear by Tension Loading
25. D2557, Test for Strength Properties of Adhesives in Shear by Tension Loading in the Temperature Range from 267.8 to –55C
26. D2918, Recommended Practice for Determining Durability of Adhesive Joints Stressed in Peel
27. D2919, Recommended Practice for Determining Durability of Adhesive Joints Stressed in Shear by Tension Loading
28. D3163, Recommended Practice for Determining the Strength of Adhesively Bonded Rigid Plastics Lap Shear Joints in Shear by Tension Loading
29. D3164, Recommended Practice for Determining the Strength of Adhesively Bonded Plastic Lap-Shear Joints in Shear by Tension Loading
30. D3165, Test for Strength Properties of Adhesives in Shear by Tension Loading of Laminated Assemblies
31. D3166, Test for Fatigue Properties of Adhesives in Shear by Tension Loading
32. D3167, Test for Floating Roller Peel Resistance of Adhesives
33. D2433, Test for Fracture Strength in Cleavage of Adhesives in Bonded Joints
34. D3528, Test for Strength Properties of Double Lap Shear Adhesive Joints by Tension Loading
35. D3632, Recommended Practice for Accelerated Aging of Adhesive Joints by the Oxygen-Pressure Method
36. E229, Test for Shear Strength and Shear Modulus of Structural Adhesives

Other tests that are used for adhesives properties include the wedge test (92-94), hot water aging test for durability (95,96), cyclic creep test (97), and the more advanced use of computer-assisted tests to characterize the mechanical properties (98). The materials engineer must be careful in selecting the test to be used to evaluate an adhesive to obtain those properties most important to the particular end-item usage. Further, it should be remembered that almost all the tests will help select the best adhesive among a group being considered for an application but that end-item testing under service conditions must be the final test.

REFERENCES

1. Shields, J. *Adhesives Handbook*, CRC Press, Cleveland, Ohio, 1970, p. 343 ff.
2. Salomon, G. Chapter 1, *Adhesion and Adhesives*, Edited by R. Houwink and G. Salomon, Elsevier, New York, 1965.
3. Alner, E. J. An introduction to surface energy, in *Aspects of Adhesion 5*, Edited by E. J. Alner, University of London Press, 1969, p. 171 ff.
4. Allen, K. W. Theories of adhesion surveyed, in *Aspects of Adhesion 5*, Edited by E. J. Alner, University of London Press, 1969, p. 15.
5. Hamilton, W. C. *Some Useful Techniques for Characterization of Adherend Surfaces.* Processing for Adhesives Bonded Structures, Applied Polymer Symposia Vol. 19, pp. 105-124, Interscience, New York, 1972.

6. Weber, K. E., and Johnston, G. R. *Characterization of Surfaces for Adhesive Bonding.* SAMPE Quarterly, Vol. 6, No. 1, pp. 16-21, Oct. 1974, SAMPE, Azusa, California.

7. Russell, W. J., and Garnis, E. A. *A Study of the FPL Etching Process Used for Preparing Aluminum Surfaces for Adhesive Bonding.* SAMPE Quarterly Vol. 7, No. 3, pp. 5-12, April 1976, SAMPE Azusa, California.

8. Bijlmer, P., and Schliekelman, R. J. *The Relation of Surface Condition after Pretreatment to Bondability of Aluminum Alloys,* SAMPE Quarterly, Vol. 5, No. 1, pp. 13-27, Oct. 1973, SAMPE, Azusa, California.

9. Venables, J. D., McNamara, D. K., Chen, J. M., Sun, T. S., and Hopping, R., *Oxide Morphologies on Aluminum Prepared for Adhesive Bonded Aircraft Structures,* 10th National SAMPE Technical Conference, Kiamesha Lake, New York, Vol. 10, pp. 362-276, 1978, SAMPE, Azusa, California.

10. Wegman, R. F., Hamilton, W. C., and Bodnar, M. J. *A Study of Environmental Degradation of Adhesive Bonded Titanium Structures in Army Helicopters,* 4th National SAMPE Technical Conference and Exhibition, Palo Alto, California, Vol. 4 pp. 425-432, 1972, SAMPE, Azusa, California.

11. Mahoon, A., and Cotter, J. *A New Highly Durable Titanium Surface Pretreatment for Adhesive Bonding,* 10th National SAMPE Technical Conference, Kiamesha Lake, New York, Vol. 10, pp. 425-440, 1978, SAMPE, Azusa, California.

12. Wegman, R. F., and Bodnar, J. J. *Structural Adhesive Bonding of Titanium Superior Surface Preparation Techniques,* SAMPE Quarterly, Vol. 5, No. 1, pp. 28-36, Oct. 1973, SAMPE, Azusa, California.

13. American Society for Testing and Materials, Phildelphia, ASTM D2651-67 (1973), *Standard Recommended Practice for Preparation of Metal Surfaces for Adhesive Bonding,* Part 22, 1974, Annual Book of ASTM Standards.

14. Snogren, R. C. Hughes Aircraft Company, *Surface Treatment of Joints for Structural Adhesive Bonding,* paper presented at Design Engineering Conference and Show, American Society of Mechanical Engineers, held at Chicago, Illinois, May 9-12, 1966. Paper 66-MD-39.

15. Ross, M. C., Wegman, R. F., Bodnar, M. J., and Tanner, W. C. *Effect of Surface Exposure Time on Bonding of 2024-T3 Aluminum Alloys,* SAMPE Journal, Vol. 10, No. 1, Jan./Feb. 1974.

16. Ross, M. C., Wegman, R. F., Bodnar, M. J., and Tanner, W. C. *Effect of Surface Exposure Time on Bonding of 6061-T4 Aluminum Alloys,* SAMPE Journal, March/April 1974.

17. Ross, M. C., Wegman, R. F., Bodnar, M. J., and Tanner, W. C. *Effect of Surface Exposure Time on Bonding of 5052/H34 Aluminum Alloy,* SAMPE Journal, Vol. 10, No. 3, May/June 1974.

18. Ross, M. C., Wegman, R. F., Bodnar, M. J., and Tanner, W. C. *Effects of Surface Exposure Time on Bonding of 7075-T6 Aluminum Alloy,* SAMPE Journal, Nov./Dec. 1974.

19. Snogren, R. C. TRW Systems Group, *Selecting Surface Preparation Processes for Adhesive Bonding, Sealing and Coating,* paper presented at Design Engineering Conference and Show, American Society of Mechanical Engineers, held at Chicago, Illinois, April 22-25, 1968. Paper 68-DE-45. Also published as *Selection of Surface Preparation Processes,* Parts 1 and 2, Adhesives Age, 12(7&8), July and August 1969.

20. Smith, D. R. Avco Corp., *How to Prepare the Surface of Metals and Non-Metals for Adhesives Bonding,* Adhesives Age, 10(3):25-31, March 1967.

21. Snogren, R. C. *Handbook of Surface Preparation,* Palmerton Publishing, New York, 1974.

22. Martin, J. T. Chap. 12, Surface treatment of adherends, in *Adhesion and Adhesives,* Vol. 2, *Applications,* 2nd Edition, edited by R. Houwink and G. Salomon, Elsevier, 1967.

23. Rogers, N. L. Surface preparation of metals, Journal of Applied Polymer Science (Applied Polymer Symposia), No. 3, 327-340, 1966.

25. Minford, J. D. Effect of surface preparation on adhesive bonding of aluminum, Adhesives Age, 17(7):24-29, July 1974.

26. Minford, J. D. ALCOA Research Laboratories, Chap. 2, Durability of adhesive bonded aluminum joints, in *Treatise on Adhesion and Adhesives,* Vol. 3, edited by Robert L. Patrick, Marcel Dekker, New York, 1973.

27. Schliekelmann, R. J. Chap. 15, Adhesive bonded metal structures, in *Adhesion and Adhesives,* Vol. 2, *Applications,* 2nd Edition, edited by R. Jouwink and G. Salomon, Elsevier, New York, 1967.

28. Carrillo, G. *An Automated System for Phosphoric Acid Anodizing of Aluminum Alloys,* 10th National SAMPE Technical Conference, Vol. 10, pp. 377-396, Kiamesha Lake, New York, 1978, SAMPE, Azusa, California.

29. Locke, M. C., Scardino, W., and Croop, H. *Non-Tank Phosphoric Acid Anodize Method of Surface Preparation of Aluminum for Repair Bonding,* 7th National SAMPE Technical Conference, Vol. 7, pp. 488-504, Albuquerque, New Mexico, 1975, SAMPE, Azusa, California.

30. Russell, W. J. *Chromate Free Process for Preparing Aluminum for Adhesive Bonding,* Durability of Adhesive Bonded Structures, Applied Polymer Symposia, Vol. 32, pp. 105, 117, Interscience, New York, 1977.

31. Dastin, S. J. Bonded beryllium structures, Adhesives Age, 9(5):24-27, May 1966. Dastin is now with Grumman Aircraft Corporation.

32. St. Cyr, M. *Adhesive Bonding of Beryllium,* presented at 15th National SAMPE Symposium and Exhibition, 29 April-1 May 1969, Los Angeles, California, Proceedings, Vol. 15, pp. 719-731 (PLASTEC 12598-35).

33. Stevens, J. H., and Layman, W. E. Jet Propulsion Laboratory, California Institute of Technology, Pasadena, California, *Development of an Adhesively Bonded Beryllium Propulsion Structure for the Mariner Mars 1971 Spacecraft,* JPL Technical Memorandum 33-517, NASA CR-124742, January 1, 1972 (PLASTEC M-21155) (N72-14871).

34. Cagle, C. V. Chap. 21, Surface preparation for bonding beryllium and other adherends, in *Handbook of Adhesive Bonding,* edited by Charles V. Cagle, McGraw-Hill, New York, 1973.

35. Vazirani, H. N. Bell Laboratories, Murray Hill, New Jersey, Surface preparation of copper and its alloys for adhesive bonding and organic coatings, Journal of Adhesion, July 1, 1969, pp. 208-221.

36. Wegman, R. F., and Bodnar, M. J. Bonding rare metals, Machine Design, 31:139-140, October 1, 1959.

37. Peterson, C. H. How to prepare substrates for better bonding, Adhesives Age, 8(7): 22-24, July 1965.

38. Fader, B. Adhesive bonding of lead, Chap. 13 in *Handbook of Adhesive Bonding,* edited by Charles V. Cagle, McGraw-Hill, New York, 1973.

39. Jackson, L. C. Chap. 15, Principles of magnesium adhesive-bonding technology, in *Handbook of Adhesive Bonding,* edited by Charles V. Cagle, McGraw-Hill, New York, 1973.

40. Department of Defense, Washington, D.C., MIL-HDBK-693A, April 14, 1972, *Military Standardization Handbook—Magnesium and Magnesium Alloys.*

41. Jackson, L. C. Effects of surface preparations on bond strength of magnesium, Journal of Applied Polymer Science (Applied Polymer Symposia), No. 3, 341-351, 1966.

42. DeLollis, N. J. *Adhesives for Metals–Theory and Technology,* Industrial Press, New York, 1970.
43. Guttmann, W. H. *Concise Guide to Structural Adhesives,* Reinhold, New York, 1961.
44. Bruno, E. J., Editor. *Adhesives in Modern Manufacturing,* Society of Manufacturing Engineers, 1970.
45. Vazirani, H. N. Bell Laboratories, Murray Hill, New Jersey, Surface preparation of steel for adhesive bonding and organic coating, Journal of Adhesion, 1 (July 1969), pp. 222-232.
46. Kempf, J. N. Preclean metals and plastics for adhesive bonding, Product Engineering, 3(33):44-46, August 14, 1961.
47. Krieger, R. B., Jr., and Politi, R. E. American Cyanamid Co., Bloomingdale Dept., *High Temperature Structural Adhesives,* Proceedings, 9th National SAMPE Symposium, Nov. 15-17, 1965, Dayton, Ohio. Paper V-4 (PLASTEC 8871).
48. Vaughan, R. W., and Sheppard, C. H. *Cryogenic/High Temperature Structural Adhesive,* NASA CR 134465, prepared by TRW Systems Group for NASA-Lewis Research Center. January 1974. Available from NTIS.
49. Sharpe, L. S. Chap. 26, Adhesive bonding, Fastening and Joining Reference Issue, Machine Design, 41(21):119-128 (Sept. 11, 1969).
50. Levine, H. R. Whittaker Corp., Narmo Research and Development Div., *Heteromatic Structural Adhesives,* Proceedings, 9th National SAMPE Symposium, Nov. 15-17, 1965, Dayton, Ohio. Paper V-1 (PLASTEC 8871).
51. Keith, R. E., Randall, M. D., and Martin, D. C. *Adhesive Bonding of Stainless Steels–Including Precipitation Hardening Stainless Steels,* NASA-George C. Marshall Space Flight Center, NASA-SP-5085. NASA-TMX-53574, April 1968. Prepared by Battelle Memorial Institute, Columbus, Ohio, under the supervision of Redstone Scientific Information Center, Redstone Arsenal, Alabama, RSIC-599 (PLASTEC 11101). This report is available from NTIS as AD 653 526.
52. Lively, F. W., and Hohman, A. E. Vought Systems Div., LTV Aerospace Corp., Dallas, Texas, *Development of a Mechanical-Chemical Surface Treatment for Titanium Alloys for Adhesive Bonding,* Proceedings, 5th National SAMPE Technical Conference, Oct. 9-11, 1973, Kiamesha Lake, New York, pp. 145-155.
53. Walter, R. E., Voss, D. L., and Hochberg, M. S. McDonnell Aircraft Co., *Structural Bonding of Titanium for Advanced Aircraft,* National SAMPE Technical Conference, Vol. 2, *Aerospace Adhesives and Elastomers,* Dallas, Texas, Oct. 6-8, 1970, pp. 321-330 (PLASTEC 15020).
54. Newell, E. D., and Carrillo, G. Materials and Process Engineering, Aerospace Group, Rohr Industries, Riverside, California, *A New Surface Preparation Process for Titanium,* Proceedings, 5th National SAMPE Technical Conference, Oct. 9-11, 1973, Kiamesha Lake, New York, pp. 131-135.
55. Bulletin, D., Aircraft Products Co., 1808 National, Anaheim, California 92801. *DAPCOTREAT 4023/4000 Cleaning Process for Adhesive Bonding of Titanium,* April 1, 1974.
56. Paul, R. D., and McGivern, Jr., J. Hamilton Standard Div., United Aircraft Corp., Windsor Locks, Connecticut, *Electrochemical Characterization and Control of Titanium Surfaces for Adhesive Bonding,* Proceedings, 5th National SAMPE Technical Conference, Oct. 9-11, 1973, Kiamesha Lake, New York, pp. 581-588.
57. Stifel, P. M. McDonnell Aircraft Co., St. Louis, Missouri, *Durability Testing of Adhesive Bonded Joints,* Proceedings, 19th National SAMPE Symposium and Exhibition, Vol. 10, Buena Park, California, April 23-25, 1974, 75-81.
58. Keith, R. E. Chap. 12, Adhesive bonding of titanium and its alloys, in *Handbook of Adhesive Bonding,* edited by Charles V. Cagle, McGraw-Hill, New York, 1973.

59. Wegman, R. F. and Bodnar, M. J. Bonding uranium 238, Adhesive Age, June 1960; and Tanner, W. C. Picatinny Arsenal, Adhesives in ordnance applications, Journal of Applied Polymer Science, 6(20):204-209, Mar./Apr. 1962.

60. Bodnar, M. J., and Powers, W. J. Adhesive bonding of the newer plastics, Plastics Technology, 4(8):721-725, August 1958.

61. American Society for Testing and Materials, Philadelphia, ASTM D2093-69, *Standard Recommended Practice for Preparation of Surfaces of Plastics Prior to Adhesive Bonding,* Part 22, 1974, Annual Book of ASTM Standards.

62. Scales, G. M., and Stefanelli, T. M. *A Guide to Surface Preparation and Pretreatments for Adhesive Bonding,* available from Hardman, Inc., Belleville, New Jersey, 07109 ($2).

63. Cagle, C. V. *Adhesive Bonding: Techniques and Applications,* McGraw-Hill, New York, 1968.

64. Cagle, C. V. Chap. 19, Bonding plastic materials, in *Handbook of Adhesive Bonding,* edited by Charles V. Cagle, McGraw-Hill, New York, 1973.

65. Abolins, V., and Eickert, J. General Electric Co., Adhesive bonding and solvent cementing of polyphenylene oxide, Adhesives Age, 10(7):22-26, July 1967.

66. Tanner, W. C. Manufacturing processes with adhesive bonding, Journal of Applied Polymer Science (Applied Polymer Symposia), No. 19, 1972, pp. 1-21 (PLASTEC 17462).

67. Byra, P. M. Polyaryl Ether, *1974-1975 Modern Plastics Encyclopedia,* McGraw-Hill, New York, pp. 62-64.

68. Penney, W. H. 3M Company, Polyaryl sulfone, *1974-1975 Modern Plastics Encyclopedia,* McGraw-Hill, New York, pp. 64-69.

69. Anonymous, 3 Prime factors in adhesive bonding of plastics, Plastics Design and Processing, 8(6):10-22, June 1968.

70. Schrader, W. H., and Bodnar, M. J. Adhesive bonding of polyethylene, Plastics Technology, 3(12):988-996, December 1957.

71. Anderson, M. D., and Bodnar, M. J. Surface preparation of plastics for adhesive bonding, Adhesives Age, 7(11):26-32, November 1964.

72. Devine, A. T., and Bodnar, M. J. Effects of various surface treatments on adhesive bonding of polyethylene, Adhesives Age, 12(5):35-37, May 1969.

73. Ayres, R. L., and Shofner, D. L. Phillips Petroleum Co., Bartlesville, Oklahoma, Preparing polyolefin surfaces for ink and adhesives, SPE Journal, 28(12):51-55, December 1972.

74. Dannenberg, H., and May, C. A. Shell Oil Co., Emeryville, California, Chap. 2, Epoxide adhesives, in *Treatise on Adhesion and Adhesives,* Vol. 2, *Materials,* edited by R. L. Patrick, Marcel Dekker, New York, 1969.

75. Hall, J. R., Westerdahl, C. A. L., Bodnar, M. J., and Levi, D. W. Picatinny Arsenal, Effect of activated gas plasma treatment time in adhesive bondability of polymers, Journal of Applied Polymer Science, 16(6):1465-1477, June 1972.

76. Schonhorn, H., and Hansen, R. H. Surface treatment of polymers for adhesive bonding, Journal of Applied Polymer Science, 11(8):1461-1474, August 1967.

77. George, D. E. du Pont Company, Polyimide, *1974-1975 Modern Plastics Encyclopedia,* McGraw-Hill, New York, p. 90.

78. Anonymous (staff written). Polyimide, in Engineering resins primer, Plastics Engineering, 30(12):27, Dec. 1974.

79. Devine, A. T., Wegman, R. F., Bodnar, M. J., and Tanner, W. C. Picatinny Arsenal, Effect of surface exposure time on bonding of glass reinforced plastic, SAMPE Journal 7(3):16-19, April/May 1971.

80. Jackson, L. C. The Bendix Co. Preparing plastics surfaces for adhesive bonding, Adhesives Age, 4(2):30-32, February 1961.

81. Theberge, J., Arkles, B., and Cloud, P. Liquid Nitrogen Processing Corp., Malvern, Pennsylvania, *Low-Cost Glass-Fortified Thermoplastic Resins for Automotive Applications,* Proceedings, 27th Annual SPI Conference, Reinforced Plastics/Composites Institute, held at Washington, D.C., Feb. 8-10, 1972. Paper 14-C, 12 pages (PLASTEC 16728).

82. Lachowecki, W. Fiberfil Div., Dart Industries, Inc., Fiber glass reinforced thermoplastics, Part 2. Processing and fabricating, Plastics Design & Processing, 9(7):28-31, July 1969.

83. Fust, G. W., and Kays, A. O. Thiokol Chemical Corp., Huntsville Div., Huntsville, Alabama, Adhesive for bonding of insulation in solid propellant rocket motors, Journal of Applied Polymer Science (Applied Polymer Symposia) No 3:219-231, 1966 (PLASTEC 7527-15).

84. Gaughan, J. E. Chapter 16, Bonding Elastomeric compounds, in *Handbook of Adhesive Bonding,* edited by Charles V. Cagle, McGraw-Hill, New York, 1973.

85. Vaccari, J. A. Relative performance of rubbers and elastomers, Materials Engineering, 80(1):26-28, July 1974.

86. DeLollis, N. J., and Montoya, O. Sandia Laboratory, Bondability of RTV silicone rubber, Journal of Adhesion, 3(1):57-67, Sept. 1971.

87. American Society for Testing and Materials, Philadelphia, ASTM D907, *1974 Standard Definitions of Terms Relating to Adhesives,* Part 22, *1974 Annual Book of Standards.*

88. Kaelble, D. In *Treatise on Adhesion and Adhesives, Structural Adhesives,* R. L. Patrick, Editor, Vol. 4, Chap. 9, Performance and reliability, Marcel Dekker, New York, 1976, pp. 183-209.

89. Keimel, F. A. Design—The keystone of structural bonded equipment enclosures, Applied Polymer Symposia Vol. 3, pp. 27-41, 1966, Interscience, New York.

90. Pinckney, R. W., and Griffith, T. W. *A Helicopter Fuselage Design Concept,* 10th National SAMPE Technical Conference, Vol. 10, pp. 889-897, Kiamesha Lake, New York, 1978, SAMPE, Azusa, California.

91. *Annual Book of Standards,* American Society for Testing and Materials, 1916 Race St., Philadelphia 19103.

92. Bethune, A. W. Durability of bonded aluminum structure, SAMPE Journal, Vol. 11 No. 3, July/August/September 1975.

93. Marceau, J. A., Moji, Y., and McMillan, J. C. A wedge test for evaluating adhesive bonded surface durability, Adhesives Age, Vol. 20, No. 10, October 1977.

94. Russell, W. J. *The Rapid Evaluation of Adhesive Bond Durability Using a Modified Wedge Test,* 10th National SAMPE Technical Conference, Vol. 10, pp. 124-137, Kiamesha Lake, New York, 1978, SAMPE, Azusa, California.

95. Wegman, R. F. Durability of some newer structural adhesives, in *Durability of Adhesive Bonded Structures,* Applied Polymer Symposium, Vol. 32, pp. 1-10.

96. Levi, D. W., Wegman, R. F., Ross, M. C., and Garnis, E. A., Use of hot water aging for estimating lifetimes of adhesive bonds to aluminum, SAMPE Quarterly, Vol. 7, No. 3, pp. 1-4, 1976, SAMPE, Azusa, California.

97. Seago, R. *Cyclic Creep Studies of Adhesives,* 2nd National SAMPE Technical Conference, Vol. 2, pp. 135-144, Dallas, Texas, SAMPE, Azusa, California.

98. Frazier, T. B. *A Computer Assisted Thick Adherend Test to Characterize the Mechanical Properties of Adhesives,* 2nd National SAMPE Technical Conference, Vol. 2, pp. 71-88, Dallas, Texas, SAMPE, Azusa, California.

13
Electrical Insulation*

Electrical insulation is a major component of all electrical equipment. It serves to keep the current along its proper path in the conductors.

The word *insulator* usually refers to a product that supplies primarily an insulating function, such as a suspension insulator used to support high-voltage transmission lines from the towers. However, the word insulator can also be used to mean an insulating material. In their more scientific aspects, insulating materials are often called dielectrics. This term is sometimes extended to include piezoelectric and other electronic materials, which are not considered here.

A distinction can be made, also, between insulating materials by themselves and insulation systems in which several materials are combined in a form characteristic of their use in electrical equipment.

In the early days of the electrical industry, natural materials, such as paper, wood, cotton, silk, sulfur, beeswax, and glass, were used for electrical insulation. Perhaps the first advance was made when paper and wood were dried and immersed in oil. At about the same time, asphalt or natural resins, such as copal, were mixed or "cooked" with linseed oil and dissolved in a solvent to make a varnish that could coat paper, textile yarns, and cloth to protect them from the ingress of water and dirt. Such materials were also coated directly on copper to make the first "enameled" wire.

The engineering requirements placed on insulating materials increased markedly as electrical equipment increased in size, voltage, and complexity. Fortunately, tremendous development in the chemical industry took place at the same time, providing synthetic materials and polymers of a wide variety with capabilities that could not have been imagined earlier.

With this development has come the need to blend the different disciplines, so that, for example, the electrical engineer can utilize effectively the materials available and the chemist can understand the specialized needs in electrical applications.

This chapter on electrical insulating materials is divided into three parts: functional properties and processing characteristics, a description of typical materials and their applications, and factors in design and application.

---

*Contributed by KENNETH N. MATHES.

I. FUNCTIONAL PROPERTIES AND PROCESSING CHARACTERISTICS

To provide the functional requirements in electrical equipment, electrical insulation systems must first keep the conductors in place and properly spaced from each other as well as the other components. In fact, it is mechanical failure that very often precedes and causes the ultimate electrical failure.

A. Mechanical and Physical Properties

Because mechanical properties, for example, for plastics, are described extensively in other sections of this book, some of the important mechanical and physical properties of solid insulating materials are only listed here.

> Strength and elongation (modulus)
> Flexibility
> Compressive strength and resistance to cut-through
> Resistance to shock (impact strength)
> Bond strength
> Flammability and fire resistance
> Thermal conductivity and specific heat

The mechanical properties adequate for operational requirements must be maintained over long periods of time—sometimes up to 20 or 30 years or more—over temperatures ranging from -269°C for cryogenic applications to 220°C or even higher in some special situations. Ambient factors, such as moisture, chemical contaminants, biological growth such as fungus, and radiation, must be withstood. In some cases, solids are immersed in oil or other liquids and must not be adversely affected by them.

Temperature cycling may introduce special problems, such as differential thermal expansion between insulating materials and associated conductor, magnetic, and structural materials. In large turbine generators the mechanical stresses from differential thermal expansion between the insulating materials, copper, and iron are combined with very high compressive (centrifugal) force from high-speed rotation. Combined mechanical stresses are quite common and pose special problems for electrical insulating materials.

Compressive and tensile creep, flexural fatigue, and other time-dependent mechanical properties are also very important. In some cases, the effect of these combined mechanical stresses (with electrical stress as well) is best evaluated with the electrical device itself or with models that functionally represent the device.

The physical properties of gaseous and liquid insulating materials, especially flammability, viscosity (liquids), vapor pressure, and those properties that contribute to thermal transfer, are all very important as well.

Although the mechanical and physical properties of insulating materials are very important, the emphasis here will be placed upon electrical characteristics.

B. Electrical Properties

A distinction may be made between those characteristics that have direct, functional importance to the performance of electrical equipment and those that may indirectly provide information about the functional properties. For example, voltage breakdown may be functionally important but insulation resistance may indirectly indicate the influence of moisture on breakdown. In many cases, it is necessary to consider not just single property values but, more importantly, the way a property like voltage breakdown may change with other factors, such as temperature. This approach has been taken in Ref. 1, which gives a great deal of information about the electrical properties of plastics as a function of many parameters, such as time, temperature, and moisture exposure.

A deatiled description of the electrical properties of insulating materials may be found in many published standards, including Ref. 2. The principal properties are described briefly in the following paragraphs.

1. Insulation Resistance

Insulating materials serve to separate conductors at different potentials. Often it is only necessary that the insulation prevent a large flow of current between the conductors (see dielectric strength). However, in some types of equipment in which relatively high-impedance circuits are used (electronic, control, and so on), it may be necessary, for proper operation, to maintain high values of resistance over and through the insulation between conductors.

The value of the dc resistance between the conductors is known as the *insulation resistance* for the particular geometry of insulation involved. Since both geometry and the properties of the material determine insulation resistance, it is usual when considering insulating materials to determine surface and volume resistivity. *Surface resistivity* can be described as the resistance between two opposite edges of any square portion of the insulation surface, and *volume resistivity* is the resistance between any two opposite faces of a unit cube of the insulating material. Since insulating materials are often not homogeneous, the surface and volume resistance may have different values, depending upon the orientation of the voltage stress. The insulation resistance between the terminals mounted on a laminated insulating material, as shown in Fig. 13.1, depends on a combination of the surface resistivity, volume resistivity in the direction of the laminations, and the volume resistivity. Figure 13.2 shows the decrease in resistivity of a typical

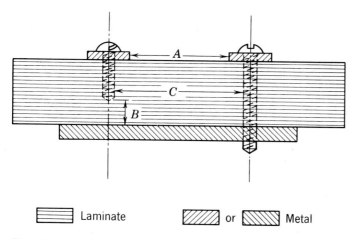

Fig. 13.1 Schematic sketch of laminated insulation illustrating components of insulation resistance. (A) Surface resistance. (B) Volume resistance, perpendicular to lamination. (C) Volume resistance, parallel to lamination.

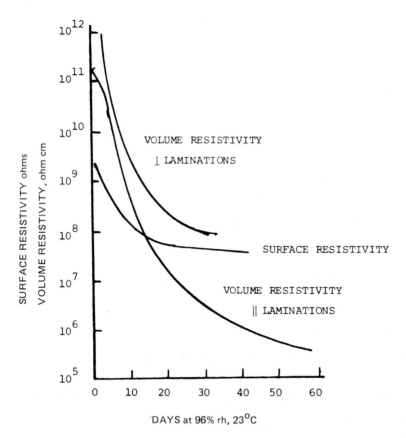

SURFACE RESISTIVITY ohms
VOLUME RESISTIVITY, ohm cm

VOLUME RESISTIVITY
⊥ LAMINATIONS

SURFACE RESISTIVITY

VOLUME RESISTIVITY
∥ LAMINATIONS

DAYS at 96% rh, 23°C

Fig. 13.2 Volume (perpendicular and parallel to the laminations) and surface resistivity versus exposure at 23°C and 96% RH, paper phenolic laminate.

laminated insulating material during exposure at high humidity. The lowest resistivity is most important in determining insulation resistance. If an insulating configuration, such as is shown in Fig. 13.1, is exposed to high humidity, surface resistance decreases rapidly; untimately, however, volume resistance attains a lower value. Consequently, first the surface and then later the volume resistance is the controlling factor determining insulation resistance.

Because insulation resistance is so much affected by moisture and contaminants, it is often measured in an attempt to determine the condition of insulating materials in service. Such measurements provide some indirect indication of the likelihood of actual dielectric failure, but they need careful interpretation. They are most useful when determined repeatedly at periodic intervals.

2. Dissipation Factor and Permittivity (Dielectric Constant)

When alternating voltages are involved, an insulating material can be represented as a combination of resistance and capacitance in parallel. The vector diagram for this equivalent parallel circuit is shown in Fig. 13.3. The energy loss in the dielectric is given by the following equation:

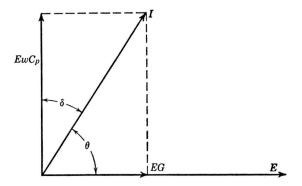

Fig. 13.3 Vector diagram—equivalent parallel dielectric circuit.

$$\text{Loss (watts)} = E^2 f\, C_p \tan \delta\, \epsilon'$$

where

E = voltage

f = frequency-cycles per second

$\tan\delta = \dfrac{1}{2 f \pi C_p}\, R_p$ = tangent "loss angle"–dissipation factor

$\epsilon' = \dfrac{C_p}{C_a}$ = permittivity (dielectric constant, specific inductive capacitance, SIC)

C_p = equivalent parallel capacitance, farads

C_a = a capacitance similar in geometry to that for C_p, in which the dielectric is replaced by a vacuum, farads.

R_p = equivalent parallel resistance, ohms

$\sin\delta = \cos\theta$ = power factor

$\epsilon' \tan\delta$ = loss index

Although the dissipation factor is more useful, the power factor is often used instead, probably because of its use in somewhat different fashion in conventional power circuits. Below values of 0.1, the dissipation factor and the power factor are nearly equivalent.

At high voltages and high frequencies, dielectric loss may be functionally important. The dissipation factor may increase with temperature so that the dielectric loss (heating) becomes progressively greater and thermal "runaway" occurs, leading to ultimate failure. Absorbed moisture in some insulating materials also increases the dielectric loss. Again the losses cause the temperature to increase. In this case, thermal runaway will take place unless the loss of moisture (drying) and the heat flow overcome the runaway. The influence of such parameters is illustrated in Fig. 13.4.

The dielectric loss may also change with frequency, as illustrated in Fig. 13.5 for a "polar" materials. If the operating frequency is close to that for the "absorption peak," the resulting dielectric loss may be functionally important. Both temperature and moisture absorption may affect markedly both the frequency at which the absorption peak occurs and its magnitude. (See Ref. 1 for more details.)

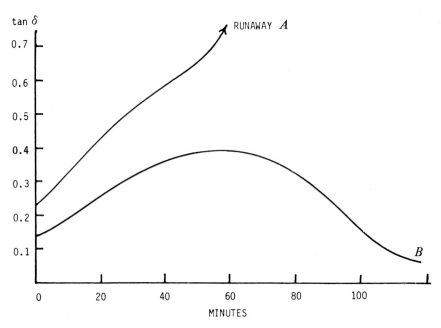

Fig. 13.4 Typical changes in the dissipation factor tan δ with time of voltage application for a "wet" insulating material. (A) Drying is too slow to prevent thermal runaway. (B) Drying progresses fast enough to prevent thermal runaway.

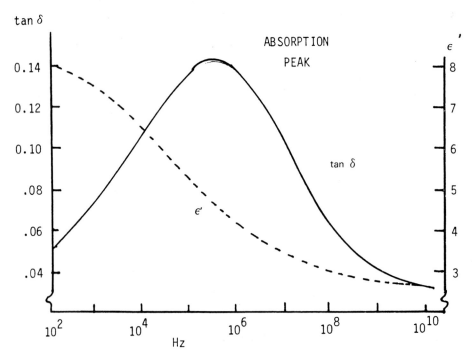

Fig. 13.5 Dissipation factor tan δ and permittivity (dielectric constant) ϵ' versus frequency. Plasticized polyvinyl chloride at 23°C.

692

The measurement of the dissipation factor and permittivity as a function of applied voltage, temperature, moisture exposure, or other factors can provide a convenient and nondestructive means of measuring the effect of such parameters on the performance of an insulating material. Even when the value of tan δ is not of direct functional importance, it may often provide an indication of the likelihood of failure in other ways, such as electric breakdown.

3. Dielectric Breakdown

If the voltage across an insulating material is increased slowly, the way in which the current increases depends upon the nature and condition of the material, as illustrated schematically in Fig. 13.6. For material A, the current increases very slowly and approximately linearly with voltage until a large, sharp increase results in what can be described as "disruptive" dielectric breakdown. In contrast, for material B, the current increases more and more rapidly until current runaway occurs. It is customary, therefore, to define the rate at which the voltage is increased, so that a more definite, though arbitrary, value of dielectric breakdown may be obtained. It is also observed that, with a relatively slow increase in voltage, material B may show a very marked increase in temperature, and the failure is then termed "thermal" breakdown.

Dielectric strength is defined as the potential gradient at which breakdown occurs. It is easily calculated for uniform fields by dividing the breakdown voltage by the insulation thickness. Nonuniform fields are common, however. For example, the voltage stress across the insulation in a shielded cable is greater at the conductor than at the shield, especially if the ratio of the two diameters is large. Difference in permittivity between different insulating materials in a composite or between the material and the air or liquid that surrounds it can result in nonuniform distribution of voltage stress. The stress in a void or at a sharp metal point is a very important, practical example of situations in which the resultant high-voltage stress may cause voltage breakdown.

The arbitrary value of the breakdown strength, as determined by conventional, short-time, breakdown tests, such as ASTM D149, provides a basis for comparison but in itself is generally not very useful for the purpose of design or the consideration of service

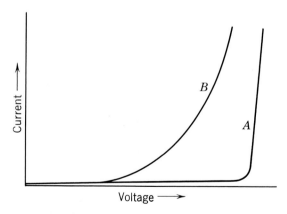

Fig. 13.6 Current flow preceding insulation breakdown. (A) Disruptive breakdown. (B) Thermal breakdown.

Table 13.1   Comparison of Short-Time Breakdown Voltage with Design Stress

Equipment	Material	Short-time breakdown (60 Hz, volts/mil)	Design stress
Small motors	Film-coated magnet wire	1500-3500	10-25
Large motors	Polyester film	1500-5000	40-80
Power capacitors	Polypropylene film	4000-6000	1000-1500
Large transformers	Oil-impregnated paper and board	800-1500	80-150
Power cable	Liquid-impregnated paper	1200-2000	300-500

performance. The failure of an insulation system in actual equipment may involve many parameters, including exposure to combined stresses* that produce deterioration slowly over long periods of time, as will be described in the next section. To illustrate this point, usual short-time voltage breakdown values for materials are compared in Table 13.1 with typical design values for operating voltage stress in several types of equipment.

Moreover, actual equipment may be subject to switching surges and lightning (impulse) voltages, which are much higher than the 60 Hz operating voltage and may cause or start the breakdown failure. Electronic equipment may be exposed to repeated high-frequency, ramp or pulse voltages with many sorts of wave shapes. Performance with these sorts of voltages cannot be predicted from 60 Hz, short-time voltage tests.

The most widely published electrical property—short-time voltage breakdown stress—is perhaps the least useful property in the functional sense. Voltage breakdown *is* a useful property when used in a comparative fashion as illustrated by its use to monitor thermal endurance tests, to be described in the next section.

4. Thermal Endurance

In electrical equipment the rating and associated maximum temperature is limited not only by the properties at high temperature but also by the slow deterioration that takes place over long periods of time at elevated temperatures. Many chemical mechanisms can contribute to such degradation, but it is usually possible to describe thermal degradation of insulating materials and insulation systems in terms of simple, chemical reaction-rate theory. In Fig. 13.7 the log of time to failure for two insulating materials is plotted as a linear function of the reciprocal of the absolute test exposure temperature, $1/K$ (the linear function assumes conformity with reaction-rate theory). The principles for conducting and evaluating thermal aging tests are given in a number of IEEE and international IEC publications (3). The lines through the experimental points for the two typical materials in Fig. 13.7 can be extrapolated (dashed lines). The temperature at which the extrapolated line intercepts an accepted arbitrary test time (often 20,000 hr) is called the thermal index (TI).

---

*Sometimes called multistress, which includes voltage, heat, environmental (moisture), and mechanical (vibration, and so forth).

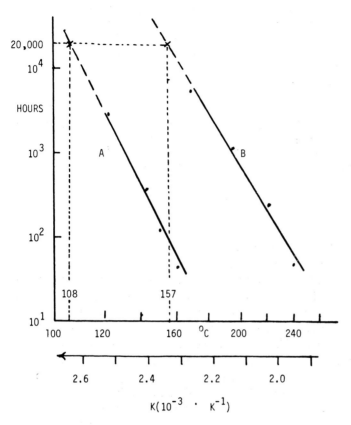

Fig. 13.7 Thermal life: log time to failure versus the reciprocal of the absolute test temperature 1/K. For material A the thermal index $TI_{(20,000)}$ is 108. For material B the thermal index, $TI_{(20,000)}$ is 157.

Since the slope of the lines in Fig. 13.7 as well as the intercept is needed to define more completely the thermal characteristic, a thermal endurance profile (TEP), or the halving interval in °C (HIC), is sometimes determined (see IEC 216-1). The procedure for determining the failure point—the end-point criterion—influences the results obtained. Electrical, mechanical, and other end-point criteria are used that define best the functional requirements in the intended application. Voltage breakdown, tensile strength, and elongation to failure are the most common. An arbitrary percentage of the initial property—50% of the tensile strength or voltage breakdown—or an absolute value, such as 2% elongation, is used. Thus it is desirable to describe the thermal index as $TI_{(electrical)}$, $TI_{(mechanical)}$, or even $TI_{(tensile)}$, since the values using these different end points may be quite different.

It is customary to group the thermal indices of insulating materials as shown in the Table 13.2.

From the foregoing, it is implied that the thermal endurance capabilities of insulating materials and insulation systems must be determined by test. This approach is expensive and not universally accepted. Extensive service experience is also useful and

Table 13.2 Thermal Indices of Insulating Materials

TI range	Preferred number	Old class designation[a]
105	105	A
130	130	B
155-180	150	F
180-200	180	H
200-220	200	—
220 and above	220	C

[a]A class designation for insulating materials is no longer accepted as standard in the United States, but is accepted in international standards. Because of long usage, the letter classes are still referenced rather commonly for insulating materials and are used to designate the insulation systems for rotating machinery and some other types of apparatus. Transformers use another type of designation.

provides perhaps the only true means for relating the thermal indices, which provide only comparative data, with appropriate design temperatures for specific apparatus.

However, it is no longer generally acceptable to define temperature classes for insulation systems by a broad description, such as organic materials, class A, and silicones, class H. Some organic materials can be used at temperatures even above 220°C. It must be recognized that a functionally acceptable design temperature for specific apparatus also depends upon many factors, such as

> Required service life and duty cycle
> Other design parameters, such as operating voltage
> Environmental restrictions, such as moisture, vibration, mechanical shock, and
> > radiation (sunlight, atomic, and so on)

Obviously, the question of thermal endurance is technically complex. Practical means for obtaining "relative" thermal indices and "estimated" service temperatures are currently under development. For some limited and special purposes it may be necessary to publish lists of estimated service temperatures (4).

5. Voltage Endurance

The performance of insulating materials and insulation systems under sustained voltage stress is of great practical importance. Unfortunately, no generally acceptable theoretical approach, similar to the chemical reaction-rate theory for thermal endurance, has so far evolved. It is accepted that the presence of partial discharges (corona*) at the surface or in voids within an insulating material will produce slow degradation that may lead to ultimate failure in service. However, materials differ greatly in this respect. Some materials, like mica, can withstand considerable partial discharge without failure over many years of service, but polyester film and other similar materials may fail very quickly. Although partial discharge is the primary mechanism involved in voltage endurance, others, such as electrolytic factors, have been recognized.

---

*The word *corona* is no longer officially accepted in most standards but is often used.

Voltage aging tests can be made with either insulating materials or insulation systems. ASTM D2275 (2) describes one procedure for materials. Some of the fundamentals related to aging in insulation systems are given in IEC 505 and 510 but, so far, well-accepted specific procedures have not been generally determined. Usually voltage aging is considered to be a negative power function of the voltage.

$$\text{Voltage life} = \frac{A}{V^n}$$

A typical log-log plot of test voltage versus time to failure (life) is shown for two insulation systems in Fig. 13.8. The mica composite is much more resistant to discharges than the plastic film even though the initial breakdown voltage is much lower. For both materials the results are not linear at lower voltages, probably because the partial discharges tend to disappear. Two values have been identified—partial discharge inception voltage pdiV as the voltage is increased and partial discharge extinction voltage pdeV as the voltage is decreased. To take account of these factors the relationship below has been proposed:

$$\text{Voltage life} = \frac{A}{(V - \text{pdeV})^n}$$

Unfortunately, pdiV or pdeV, the discharge inception or extinction voltages, may change with time during the aging test so the relationship is seldom of practical use. The measurement and interpretation of corona are described in Ref. 5.

Discharges or arcs on the surface of insulating materials may lead to slow deterioration or "tracking" and ultimate failure. Arcs from circuit interruption may produce "dry" tracking. Conducting surface contaminants may also cause surface failure. Often such contaminants become conducting when moist, and the resultant degradation is

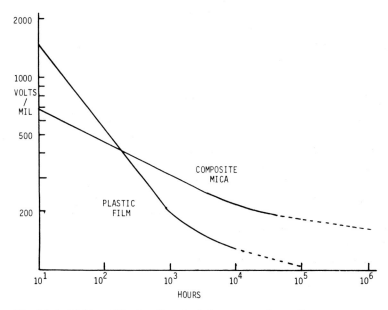

Fig. 13.8 Voltage life: log time to failure versus log voltage.

called "wet" tracking. Wet tracking and associated erosion have limited the application of many insulating materials especially for high-voltage service outdoors. Porcelain and glass have been used traditionally in outdoor electrical application, but specially formulated track- and erosion-resistant resins are now available and useful for many applications where moisture and contamination are involved.

6. Mechanical and Environmental Life

Many factors other than temperature and voltage may influence the life of insulation systems. Mechanical factors, which may be functional or part of the environment, include vibration, mechanical shock, tension or compression (cut-through), and differential thermal expansion. Some of the purely environmental factors include

> Moisture, dirt, and chemical contaminants
> Radiation: light, x-ray, and atomic
> Biological: bacteria, fungus, insects, and rodents

In Fig. 13.2 it is apparent that degradation under high-humidity exposure may continue to progress for more than 2 months. Radiation damage may continue over many years. It is possible here to only mention the possibilities.

The concept of degradation from combined, long-term stresses should be recognized. For example, heat and moisture *together* may produce far more damage to some materials susceptible to "hydrolytic degradation" than either factor alone. Hydrolytic degradation from the combined effects of heat and moisture usually causes embrittlement, but for some materials softening or even liquefaction may take place. AC and especially dc voltage can accelerate hydrolytic degradation. This process is sometimes called "electrochemical degradation."

A possible scenario for insulation failure can be described as follows.

> Thermal aging embrittles the insulation.
> Vibration opens up cracks in the brittle material.
> Moisture and dirt enter the cracks, decreasing the breakdown strength.
> Overvoltage from a switching surge initiates discharges.
> Discharges increase with time so that breakdown finally occurs at the operating 60 Hz voltage.

It is little wonder that failure in insulation systems is fairly common and so difficult to predict or control.

II. INSULATING MATERIALS

Insulating materials can be grouped conveniently as (1) gases, (2) liquids, and (3) solids. Porous solid materials, such as paper, are always filled or impregnated with gas, liquid, or a combination thereof. Vacuum is a special case of a gaseous dielectric in which relatively few gas molecules are present. The interfacial characteristic between gases and liquids and the surface of solids must be considered as well as the volume properties.

A. Gaseous Dielectrics

In addition to air and nitrogen (N_2), synthetic gases, such as sulfur hexafluoride (SF_6), are commonly used. The dielectric loss in gases is very low. The voltage breakdown de-

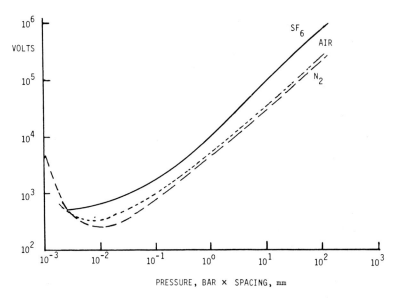

Fig. 13.9 Paschen curves: breakdown voltage of gases in a uniform field versus the product of pressure (bar) times the electrode spacing (mm).

pends upon electrode spacing and pressure. The breakdown voltage is often plotted versus the product of pressure and spacing, as shown in Fig. 13.9.* High-voltage gradients, from sharp points for example, decrease the breakdown voltage markedly, and efforts are made to keep the voltage stress as uniform as possible with the use of large metal shields, voltage grading, or other approaches. Electrical discharges can cause decomposition in many gases, which may adversely affect performance.

The electronegative gas SF_6 has a relatively high voltage breakdown as well as a wide useful temperature range from its relatively low boiling point (depends on pressure) to at least 150°C, where chemical decomposition may become important. Other synthetic gases may be used with higher boiling points to take advantage of the latent heat of vaporization in cooling the equipment. Hydrogen, a relatively poor dielectric, is used in turbine generators to keep "windage" losses low. Helium, argon, and some other gases are used as dielectrics in some special applications. Like hydrogen, the noble gases and especially helium are relatively poor dielectrics.

Nitrogen, SF_6, the fluorocarbon gases, and most of the other synthetic gases may permit higher operating temperatures with many insulation systems if oxygen (air) is rigidly excluded. The suitable operating temperature of a dry-type, power transformer in N_2 or SF_6 may be 20°C or more higher than in air, but functional tests are needed to arrive at a suitable "hot-spot" temperature for each system. Totally enclosed transformers containing N_2 or SF_6 are in widespread commercial use.

The thermal transfer provided by gaseous dielectrics is generally much less efficient and effective than that of liquids. In large turbine generators, both the stator and rotor may be cooled by circulating liquid, such as water, through hollow conductors;

---

*These so-called Paschen curves with uniform electric fields have a Paschen "minimum" below which breakdown cannot take place.

similar techniques have been proposed for transformers. For some transformers, fluoro-carbon and chlorofluorocarbon gases, which boil somewhat below the normal operating service temperature of the windings, are coming into use. The liquid in contact with the windings boils and cools them very efficiently. In this case liquid, gas, and the combination of liquid and gas serve as the dielectric.

In addition to its excellent dielectric properties, SF_6 has arc-quenching characteristics that make it useful in high-voltage circuit breakers. Pressurized SF_6 is in use also for spacer-type, high-voltage cable and totally enclosed switchgear. In such applications, loose particles may greatly decrease the breakdown voltage. So far as possible, particles are kept out of such systems, and electrostatic particle "traps" are used to inactivate residual particles that cannot be completely removed in other ways. The use of gaseous dielectrics will increase as environmental, esthetic, fire-resistance, and other requirements dictate the use of totally enclosed equipment, especially in densely populated areas.

In electronic vacuum tubes and vacuum circuit breakers, the vacuum is used to meet operational requirements. Moreover, a very good (hard) vacuum can withstand an operating voltage several times that of any other dielectric. With a vacuum dielectric, oxidative degradation of associated solid dielectrics is eliminated, but it is necessary to avoid off-gassing of the solids at the maximum operating temperature.

B. Liquid Dielectrics

For a great many years mineral oil-impregnated paper has been used extensively as an inexpensive and very technically suitable dielectric in large, high-voltage transformers, high-voltage cable, and to some extent in capacitors as well as some other specialty applications. For years mineral oils, meeting quite rigid requirements for specific applications, have been specified and were readily available. The scarcity of certain petroleum crudes and even more demanding requirements have changed the situation so that synthetic liquids are becoming increasingly important.

For a long time askarels, chlorinated biphenyl liquids (PCB), of several compositions were used in capacitors where high dielectric constants were needed and in some transformers where the flame resistance of the askarel was important. Environmental considerations have now completely eliminated the use of PCB and they are no longer manufactured in the United States and many other countries of the world. It has become necessary to replace the askarels with a variety of other liquids.

1. Mineral Oil

In large oil-filled transformers, the composition and physical properties of the oil are selected to meet both insulating and cooling requirements. As the previously used naphthenic oils are replaced by the more available paraffinic oils, synthetic liquids, such as dodecylbenzene, may sometimes be blended with oil to improve its electrical characteristics.

2. Cable Liquids

Mineral oil for cables has a high naphthenic content and often is more viscous than transformer oil. Blends of synthetic materials, such as polybutenes and aromatic liquids, are now used to take advantage of the decreased dielectric loss and better resistance to corona.

3. Capacitor Liquids

To achieve the maximum capacitance in the smallest volume, it is desirable to have either a liquid with a high dielectric constant or one that, in combination with the solid insulation (usually a special thin paper, plastic film such as polypropylene, or a combination of the two), can withstand very high voltage stresses. With the use of polypropylene film, the solvency and swelling characteristics of the liquid are also very important. For years the askarels, when combined with suitable stabilizing materials, provided the needed balance of properties. Now that the askarels no longer may be used, a variety of liquid blends are in use, including esters, such as dioctylphthalate (relatively high dielectric constant), and aromatic compounds, such as monoisopropylbiphenyl (MIPB) and phenylxylethane (PXE) with dielectric constants ϵ' of about 2.5. With the low ϵ' liquids it is necessary to increase the voltage stress to be competitive in overall size with capacitors using higher ϵ' liquids. In such cases polypropylene film is usually used. Each manufacturer claims an advantage for its own approach. Castor oil, which has long been used in discharge capacitors, is now also being replaced with synthetic liquids. In such cases polypropylene film (without paper) is usually used.

4. Fire-Resistant Liquids

For many years askarels have been used in railway and some other types of transformers where fire resistance is needed in case of internal arcing or exposure to external flame, especially if a leak developed. Silicones and hydrocarbon liquids, with as few low-boiling constituents as possible, are in use. Despite some degree of flame resistance, both liquids can be made to burn, and the hydrocarbon liquid is especially suspect. The silicone liquid is relatively expensive. To achieve flame resistance, a relatively viscous silicone is used; this causes problems in cooling because the viscosity does not decrease with increase in temperature nearly as much as askarels or conventional hydrocarbon oils. At this time (1984) the problem of fire-resistant liquids suitable for transformer application has not yet been completely solved.

C. Solid Dielectrics

A tremendous variety of solid materials have been used for electrical insulation. Natural materials, such as gutta-percha, copal, shellac, and asphalt, as well as natural fibers, such as cellulose (paper), cotton, silk, and asbestos, have been used for at least 100 years. Inorganic materials, such as sulfur, portland cement, mica, glass, and porcelain, have also been utilized. Combinations of natural drying oils—linseed, tung, and others—combined with natural resins, like copal and asphalt, as well as the early synthetic resins—phenolics and alkyds—were made first as varnishes and paints but were quickly adapted and improved to meet exacting dielectric requirements.

However, it was the explosive development in synthetic resins (see Chap. 11) that has had the greatest impact on development in electrical insulation. Synthetic inorganic materials, such as new types of glass, porcelain, and ceramics, as well as glass and ceramic fibers, have become very important. It is interesting that many synthetic materials found their first applications as electrical insulation, for example, polyethylene in cable and polyester film (Mylar, du Pont) as motor or other types of apparatus insulation. Now, the use of these materials is so extensive that the larger applications have reduced costs to the advantage of the manufacturers and users of electrical equipment.

1. Yarn, Paper, and Fabric

Cotton, silk, and asbestos provide a positive spacing and may in time slowly thermally degrade (asbestos only at very high temperatures) without melting. Polyamide (nylon), polyester (Dacron, du Pont), glass, and ceramic fibers also provide a positive spacing up to their softening points, all above 200°C. Moisture has an especially adverse effect on the electrical properties of asbestos and nylon. Glass and ceramic fibers are relatively easily damaged by mechanical abrasion. Paper used for electrical insulation is made from a wide variety of fibers or even flakes of glass or mica. Different fibers are combined, a variety of fillers or sizings may be included, and the paper may be specially treated or washed to reduce the ionic content. Papers may be calendered and are produced with different structures. A very dense, heat-resistant paper made of aromatic amide fibers, Nomex (du Pont) is widely used for electrical insulation.

   All of the fibrous structures can entrain dirt or other contaminant and will pick up water by capillarity. Consequently, fibrous insulation is almost always impregnated with liquids or resinous materials to prevent or inhibit entrainment of contaminant and to hold the fibers together mechanically.

   Asbestos fiber is subject to environmental restriction if it can be inhaled. Impregnation with resins overcomes this potential problem. Asbestos provides a combination of mechanical and thermal capability unmatched by any other material for many applications. For many years cotton, glass, and a combination of Dacron plus glass, as well as nylon and silk fabrics, have been used as a base or carrier to produce resin or elastomeric (usually silicone rubber) coated sheet or laminate.

2. Tape, Sheet, and Laminate

Cotton, glass, and other woven tapes are used to provide external support for insulated components. Precoated cotton cloth at present is seldom made or used in the United States. A very small amount may be imported. Many types of resin, including polytetra-fluoroethylene, as well as elastomers, have been used to coat glass cloth or glass-Dacron blends. Except for silicone rubber-coated glass, most of these materials have been replaced to a considerable extent by flexible synthetic sheet, such as Mylar and polyimide film (Kapton, du Pont). Polypropylene and polycarbonate film are used in capacitors alone or in combination with paper and in many other applications.

   Large mica flakes are used as a capacitor dielectric and in some other special applications. Smaller mica flakes are pasted together with resin. Reconstructed mica paper (Samica, Samica, United States, or ISOLA, Europe, and Micamat, General Electric) is made from tiny mica flakelets and mechanically reinforced with resinous or inorganic materials. These materials are used in a wide variety of electrical applications. In many cases, the flake mica sheet or mica paper is combined with other thin, sheet materials, such as paper, Mylar, or glass cloth, which provide the physical strength needed in application.

   Multiple layers of thin sheets may be combined or laminated. Traditional laminates are made of paper, cotton, or glass cloth impregnated with phenolic, melamine, polyester, epoxy, or silicone resins and partially cured to what is called the B-stage. These precoated sheets are then placed in a press and laminated under heat and pressure. In the press, the resin flows and then cures so that it is no longer thermoplastic. Low-pressure laminates are made with polyester or epoxy binders that flow readily and in some cases cure at room temperature. Such materials can be made in complex and large shapes with many electrical applications. "Pultrusions," in which glass thread or roving is impregnated and pulled through a hot die, provides parts with special profiles for many applications.

Special laminates for printed circuit board are made with copper sheet on either or both surfaces. Other laminates are made from plastic film coated with an adhesive or a thermoplastic resin, such as perfluorethylene-propylene (FEP) coated on Kapton. Circuit components can be included to make laminated, flexible printed circuit boards or with parallel thin strip conductors to make flat flexible cable.

The uses of sheet insulation are too many and too complex to be described here. However, Fig. 13.10 shows cross sections of two types of motor or generator windings that illustrate the use of sheet insulations. The requirements in these two types of windings are quite different. In form-wound coils the insulated conductors are bonded together and the structure is taped or wrapped with flexible ground insulation followed by an armor tape on the outside. The completed coil is then resin impregnated and often molded or pressed to its final shape before insertion in the machine. In a random or mush-wound coil, a relatively stiff liner is placed in the slot and the wire wound directly in place. The wedge is inserted to hold the winding in the slot and is subsequently varnish or resin treated by processes to be described in a following section.

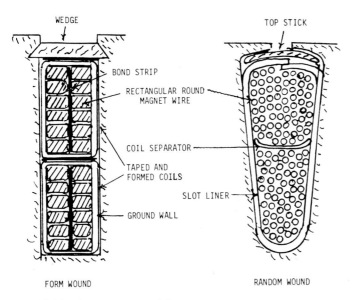

WEDGE

TOP STICK

BOND STRIP

RECTANGULAR ROUND
MAGNET WIRE

COIL SEPARATOR

TAPED AND
FORMED COILS

SLOT LINER

GROUND WALL

FORM WOUND

RANDOM WOUND

Fig. 13.10 Cross sections of form-wound coil and random-wound coil. Typical materials used in form-wound coil are, wedge: wood, cotton, or glass fiber laminate; magnet wire: film coated, glass served, or combinations thereof; ground wall insulation: Mylar or mica wrapper with glass tape armor overall plus varnish treat; fill strip: cotton or glass fiber laminate; bond strip: adhesive-coated glass cloth. Typical materials used in random-wound coil are wedge or top stick: 0.10 in. Mylar, pultruded glass fiber; magnet wire: film coated, of many kinds; slot liner: Mylar, Nomex, Kapton, and/or combinations thereof; coil separator: similar to slot liner.

3. Magnet Wire Insulation

In the United States, NEMA (6) lists 45 specifications for different types of magnet wire (more appropriately called "winding wire" in the United Kingdom). All the fibrous materials, described previously, are applied to magnet wire as yarn but sometimes in other ways as well. Felted asbestos is no longer used but a considerable amount of asbestos insulated wire is still in place in apparatus. Once asbestos-insulated wire (or other asbestos products) are treated to immobolize the fibers, no hazard exists.

Glass fiber has largely replaced asbestos, but it is not as resistant to abrasion even when varnish treated. Paper, fabrics, plastic film, mica paper, and even flake mica tape are all applied to wire.

For many years, oleoresinous enamels have been used to coat magnet wire, but such coatings, without fibrous overcoats, can be used only in simple coil-winding operations where the mechanical, electrical, and thermal requirements are minimal. Starting in about 1936, a combination of polyvinyl formal (Formvar) and phenolic resins, when coated on wire, provided sufficient mechanical toughness so it could be used in many applications without fibrous overcoating. With such relatively thin coatings, it was possible to reduce markedly the size of electrical equipment. Many types of synthetic resins—polyamide, polyurethane, polyester, polyimide, ester-imide, amide-imide, and even polytetrafluoroethylene—are used now in various combinations as wire coatings. Each material has special capabilities. For example, polyurethanes need not be mechanically removed before soldering. The long-term temperature capability of some of the imides may be above 200°C. To achieve the best combination of properties, wire may be first coated with one material and overcoated with another. Resistance to crazing, thermal shock, and varnish solvents, along with mechanical toughness, are all important properties to be achieved in this fashion. New means of coating magnet wire are under development to reduce the emission of polluting solvents during manufacture.

The choice of the most suitable magnet wire insulation for a specific application involves too many factors to be dealt with here. Compatibility with other materials in the insulation system is of great importance. Only experience and functional life-test on the complete insulation system can provide an adequate guide.

4. Coating and Impregnating Resins

For a long time, only resins dissolved in a variety of organic solvents were available for coating and bonding insulating materials or for impregnating windings, such as those illustrated in Fig. 13.10. Such solvent-containing varnishes were divided into two principal classes—air drying or baking. Some air-drying varnishes "cured" primarily by oxidation at room or low temperature. They were made from "drying" oils, such as linseed, combined with asphalt or synthetic resins, such as alkyds. Baking varnishes usually contained resins, as well as drying oils, that would polymerize with heat to give better mechanical properties and resistance to contaminants such as oil.

The advent of solventless, liquid resins, such as the polyesters and epoxies, permitted the development of impregnating varnishes containing little or no solvents. With such materials, especially when combined with vacuum-pressure impregnation (VPA), it became possible to produce reasonably void-free, mechanically strong and moisture-resistant windings far superior to those possible with solvent-containing materials. The solventless resins can also be formulated to provide protective, thick "conformal" coatings over electronic circuit boards. "Encapsulants" can both impregnate and coat electrical components, such as small coils. With the use of a container (which sometimes is later removed) "potting" or "casting" compounds can completely fill the desired space.

By using different fillers, the solventless resins can be given a wide variety of physical properties. Thixotropic fillers prevent resin "runoff" so that conformal coatings and encapsulants can be made to give a heavy surface film. Other fillers decrease shrinkage during cure and improve thermal conductivity. Fibrous fillers add strength to casting resins. In some applications, such as line insulators, a glass fiber-reinforced rod is used as a tensile member around which a resin especially formulated for such outdoor service is cast (Fig. 13.11).

5. Molded and Extruded Insulation

Almost all, if not all the plastics described in Chap. 11 are used as insulating materials in various electrical applications. Many of the thermoplastic materials, such as polyethylene, may be extruded or injection molded and then sometimes cross-linked by radiation or chemically by such additives as dicumylperoxide. Cross-linking enables operation at the somewhat higher temperatures often needed in electrical use. The "nonpolar" (1) polymers have very low dissipation factors over a wide frequency range (Fig. 13.12) and also at temperatures up to their crystalline melting points (somewhat below their physical softening temperatures). For this reason, such nonpolar plastics are especially useful for high-voltage or electronic applications.

Some of the newer thermoplastic materials, such as the polyphenylene-sulfide and polyether-imide resins, have very high softening temperatures and resistance to thermal aging in air. They can be used without cross-linking in electrical applications even above 200°C. However, these more polar thermoplastics do not have as low a dielectric loss as the nonpolar materials.

A wide variety of thermosetting resins and elastomers are fabricated for electrical applications by compression, transfer, and injection molding. The oldest of these, phenolic resins, polymerize by a "condensation" reaction in which water is released, sometimes to the detriment of the electrical characteristics. The advent of epoxy and unsaturated polyesters, which cure without releasing moisture, avoided the moisture evolution problem and also made possible low-pressure molding of very large electrical insulating parts. Pultrusion, in which epoxy or polyester resin-impregnated glass fiber is pulled through and cured in a heated die, has been used to make many structural insulating parts, such as the wedges (Fig. 13.10) used to hold windings in place. In similar ways, resin-impregnated glass roving is wound into tubes or rings or even directly around the outside of rotating motor armatures and commutators. When cured, these products are sufficiently strong and creep resistant to replace insulated steel parts or the steel wire wraps previously used. Sheet-molding compounds (SMC) are also used where a combination of good electrical properties with structural capabilities is needed. Combined strength, dimensional stability, heat resistance, and good electrical properties have made possible the use of such fibrous reinforced plastics (FRP) in many critical electrical applications.

6. Glass and Ceramics

Many types of glass and ceramics are used as electrical insulation. (See also Chap. 14.) Line insulators are a commonly observed example. In such applications, the design imposes primarily compressive stress in which glass and ceramics are strong. These inorganic

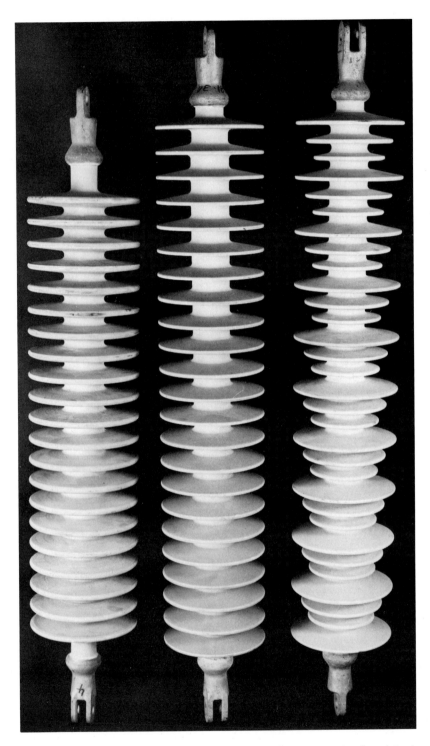

Fig. 13.11 Developments 115 kV plastic line insulators: cast, track-resistant epoxy resin surrounds a structural glass fiber rod.

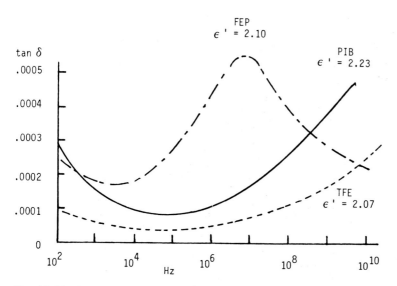

Fig. 13.12 Dissipation factor tan δ versus frequency for nonpolar polymers (*Note*: Permittivity (dielectric constant) ε changes very little, with frequency.) TFE, polytetrafluoroethylene; FEP, polyfluoroethyenepropylene; PIB, polyisobutylene.

materials are also resistant to weathering and degradation from voltage stress on wet and contaminated surfaces. In electronic applications, special glasses and ceramics with extremely low loss are used. Silica glass and high-alumina ceramics are examples. Special types of both glass and ceramic are available that can be machined. Alternatively, the ceramic can be shaped in the "green" form, which is then fired. Glass-bonded mica— tiny mica flakes bonded in a glass matrix—is made in several varieties, some of which can be machined. A fluorine-substituted mica is sometimes used to permit the use of a higher melting point glass and thereby a higher operating temperature. Metal inserts can be molded directly into glass-bonded mica parts. Aluminum, for example, can be cast around the outside of such parts as well.

In general, it is difficult to hold close dimensional tolerances in glass or ceramic parts without expensive grinding operations, and the parts have poor tensile strength and impact resistance. However, glass-bonded mica can be molded to close tolerance and has better impact resistance.

III. FACTORS IN DESIGN AND APPLICATION

Electrical insulating materials must meet a very wide variety of functional requirements when used in different types of electrical apparatus. The needs for a tiny aircraft relay differ markedly from those for a 1000 MW generator with a large rotor turning at 3600 rpm. Huge power transformers, which deliver power to transmission lines at up to a million volts, also pose special insulation problems. In such cases great reliability is needed but to achieve it may require a different approach for each type of equipment. It is possible to test many small, inexpensive relays to destruction in various possible failure modes, but it is obviously impossible to test 1000 MW generators to destruction in statistical quantities.

Unfortunately, small-scale laboratory tests on insulating materials by themselves often are primarily useful just for product identification and quality control or assurance. However, if factors of functional significance for the device and its intended service are considered, then such material tests may be more generally useful. Moreover, tests can be made with simple combinations, component parts, and models characteristic of the design. These, too, can be tested in functionally significant fashion. Figure 13.13 shows what may be called an "evaluation ladder." Results from each step on this ladder can be compared to provide overall a more comprehensive and significant understanding of how insulating materials may perform in service. Many inexpensive functional tests can be made on materials so that fewer expensive tests on models are needed.

Throughout, it is necessary to consider those factors of functional importance in the intended application. For example, voltage breakdown is seldom if ever of importance in solid-state, electronic circuits operating at a maximum of 8 V, but electrolytic corrosion and insulation resistance at high humidity may be. Failures in service, when properly investigated, may provide the best source for selecting and understanding the most significant properties.

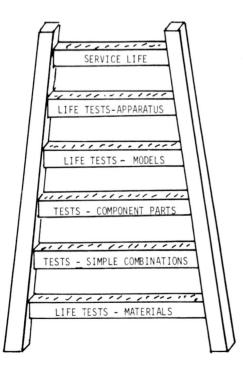

Fig. 13.13 The "ladder" of evaluation. Steps in endurance testing.

REFERENCES

1. Mathes, K. N. (1966). Electrical properties, in *Encyclopedia of Polymer Science,* Vol. 5, John Wiley, New York, pp. 528-628. Also, Mathes, K. N. (1964). Electrical properties, in *Engineering Design for Plastics,* Ed. E. Baer, Reinhold, New York, pp. 437-588.
2. *Electrical Insulation–Test Methods: Solids and Solidifying Fluids,* Part 39, Annual Book of ASTM Standards, Philadelphia.
3. IEEE 1, 98, 99, and 101 from IEEE, 33W 47th Street, New York. IEC 85, 216, 505, 610, and 611 from IEC, 1 rue de Varembé, Geneva. *Note:* These publications are periodically revised and additional ones issued.
4. *Recognized Component Directory,* Underwriters' Laboratories, Chicago. *Note:* This publication is revised periodically. See also UL 746B–*Polymeric Materials, Long Term Property Evaluation,* and UL 1446, *Systems of Insulating Materials.*
5. *Engineering Dielectrics, Corona Measurements and Interpretation,* ASTM STP 669, Philadelphia.
6. National Electrical Manufacturers Association (NEMA), *Standard MW-1000,* Washington, D.C.

ADDITIONAL READING

Clark, F. M. (1962). *Insulating Materials for Design and Engineering Practice,* John Wiley and Sons, New York.

Hill, N. E., et al. (1969). *Dielectric Properties and Molecular Behavior,* Van Nostrand Reinhold, New York.

von Hippel, A. R. (1954a). *Dielectric Materials and Applications,* John Wiley and Sons, New York.

von Hippel, A. R. (1954b). *Dielectrics and Waves,* John Wiley and Sons, New York.

Ozorkiewicz, R. M. (1970). *Engineering Properties of Thermoplastics,* Wiley Interscience, New York.

Raether, H. (1964). *Electron Avalanches and Breakdown in Gases,* Buttersworth, London.

Saums, M. L., and Pendelton, W. W. (1973). *Materials for Electrical Insulating and Dielectric Functions,* Hayden, New York.

Sillars, R. W. (1971). *Electrical Insulating Materials and Their Applications,* Peter Pereginnis, London.

Wada, Y., et al. (1979). *Charge Storage, Charge Transport and Electrostatics with Their Applications,* Elsevier, New York.

14
Ceramics

I. NONVITREOUS CERAMICS*

In modern usage any inorganic, nonmetallic material is considered to be a ceramic. This class of materials offers the design engineer unique combinations of properties that are unavailable in other materials. As a consequence the past several decades have seen the ever-widening use of ceramic materials in critical applications where they can play a unique role. In particular we are beginning to see ceramic materials applied to load-bearing engineering applications. The substitution of "high-performance" ceramic materials, such as silicon nitride, silicon carbide, various glass-ceramics, and oxide ceramics for more conventional engineering materials will come about because they are abundant, low-cost, high-strength and/or low thermal expansion, high-temperature materials. The major impediment to their past utilization as tensile load-bearing components has been their inherent brittleness. Designers are learning to design around the brittleness problem and thus make ceramic materials available for a wider spectrum of engineering applications (1,2). Section I deals with nonvitreous ceramics; Sec. II deals with glasses.

Ceramic materials have always been associated with high temperatures. Today the high-temperature capabilities of the "new" engineering ceramics, such as silicon nitride and silicon carbide, are being focused on the critical areas of energy and materials shortages. To increase efficiency of heat engines requires higher working temperatures. The temperatures desired frequently exceed the service temperature capabilities of current or projected metal alloy systems. Further, the availability of many elements required in high-temperature alloys is at least as problematical as the availability of petroleum. Therefore, there is considerable interest in exploiting abundant ceramic materials in gas turbines and diesel engines as well as in other high-temperature applications. "Older" ceramics, such as fiberglass and mullite, are aiding energy conservation in the form of insulation to increase the thermal efficiency of high-temperature industrial processes, or to substitute for asbestos.

Although energy conversion is, perhaps, the largest single new area of opportunity for ceramic application, there are many other areas where ceramics are now being util-

---

*Contributed by R. NATHAN KATZ, WINSTON DUCKWORTH, and DAVID W. RICHERSON.

ized or considered, where they can yield significant savings of scarce resources or entirely new capabilities. The following are typical examples.

The use of fiber optic cables for data transmission will save copper and can reduce cable weight by a factor of 21.

The use of sodium vapor lamps, which require alumina (or other high-temperature transparent ceramic) envelopes, to replace conventional filament lamps, reduces tungsten wire usage (via 15-20 times longer lamp life) and delivers significantly more lumens per kilowatt.

The use of silicon carbide ignitors to replace continuous pilot lights on gas appliances can save as much as 25% of the gas consumed.

The use of zirconium oxide sensors to monitor and control (via a microprocessor) automotive exhaust emissions and combustion efficiency.

The use of improved ceramic grinding wheels and cutting tools to increase the efficiency of metal removal operations.

The use of synthetic polycrystalline diamond rock drilling bits can cut oil well drilling times in half and reduce costs by more than a third. In some oil well drilling situations this translates to a saving of 9 days and $1 million per well.

An alumina-lined, glass-reinforced plastic-wrapped pipeline can handle highly erosive coal slurries with significantly reduced maintenance costs.

High-purity, high dielectric strength oxide ceramic insulators, such as alumina (either polycrystalline or as single-crystal sapphire) or beryllia, are used as electronic substrates and as such are literally the base for much of modern microelectronic circuitry.

This brief list makes the point that ceramics are used because of their unique optical, electronic, wear-resistant, or high-temperature properties.

The various families of ceramic materials, with typical examples shown in Fig. 14.1, include carbons and salts. One can even include ice, and in fact much ice research has been performed by ceramists (3). The applications cited in Fig. 14.1 tend to range from the most ancient to the most futuristic as one goes from top to bottom. Many of the more ancient applications represent a continuous record of technological growth and response to new demands. For example, the ancient use of ceramics as containers now includes the use of ceramic technology to contain nuclear waste materials, either by fusing them in a glass or hot pressing them in large blocks (4). It is remarkable to reflect that, although a mass production ceramic industry has existed for at least 9000 years (5), ceramic materials are still at the technological forefront in addressing society's most critical needs.

The foregoing discussion provides a perspective from which the relation of ceramic technology to modern engineering practice can be understood, and it is apparent why it is important for engineers to have a reasonable "feel" for the properties of ceramics and be introduced to some of the key issues involved with the successful application of ceramic materials in systems design. These two topics constitute the balance of this section. Other important topics relevant to the successful exploitation of ceramics include an understanding of ceramic processing, nondestructive testing and quality assurance, and single-crystal growth. These topics will be discussed in Chaps. 6, 28, and 30. The thrust of this chapter is toward advanced uses for ceramic materials. For those wishing to learn more about the more traditional applications of ceramics, that is, white ware, sanitary ware, high tension line insulators, and so on, there is an excellent discussion of these

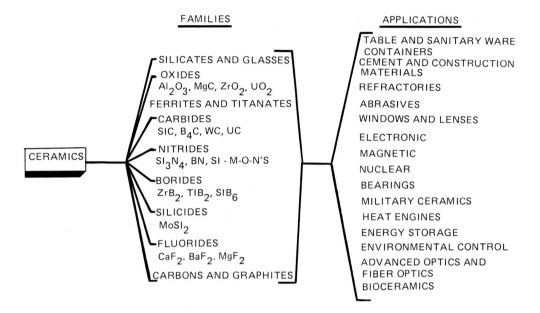

Fig. 14.1 Families and applications of ceramics.

areas in the previous edition of this book (6). We turn now to how the properties of ceramics originate in their crystal and microstructures.

A. Structural Origin of Properties

The properties of ceramics result from a combination of the effects of atomic bonding and microstructure. The effects of bonding are primarily reflected in the intrinsic properties that comprise physical, thermal, electrical, magnetic, and optical properties. Microstructure can also affect some of the intrinsic properties, but has its major effect on mechanical properties.

In Sec. I.A.1 we will review the primary types of bonding and their general effect on properties, emphasizing in particular the difference between metals and ceramics. Later we will discuss how variations in the microstructure can further modify or limit the properties.

1. Atomic Bonding

Atomic bonding involves the outermost electrons and results from the drive of the atoms to situate in the most stable structure. In a pure metal, where all the atoms are the same size and have identical electron structures, the atoms are often stacked in a closepacked array and the valence electrons move relatively freely among the atoms. Due to this sharing of electrons, at any moment each atom in the metal has a more stable electron structure that an isolated metal atom would have. This mutual sharing of electrons provides the bonding force that holds the metal atoms together into a metal crystal. It also provides the basis for most the properties we associate with a metal: high ductility, good electrical conductivity and thermal conductivity, and high thermal expansion.

Atoms in a ceramic material are primarily held together by covalent or ionic bonding, or a combination of the two. Covalent bonding is similar to metallic bonding from the standpoint of sharing of electrons. However, in covalent bonding electrons are only shared by two adjacent atoms. This results in directionality of the bond and dramatically different properties for a covalently bonded ceramic compared with a metal. The differences will be discussed as we progress. Examples of ceramic materials with a highly covalent bond include diamond (C), titanium carbide (TiC), silicon carbide (SiC), and silicon nitride (Si_3N_4).

Ionic bonding involves the actual transfer of one or more electrons between two atoms producing oppositely charged ions bonded by coulombic attraction. Examples of ceramic materials with a highly ionic bond include sodium chloride (NaCl), calcium fluoride (CaF_2), and magnesium oxide (MgO). Note that these compositions are made up of elements from opposite extremities of the periodic table, elements of groups I and II bonded to elements of groups VI and VII. As we move toward the interior of the periodic table, where elements have more valence electrons available for bonding, the ionic nature of the bonds decreases and the covalent nature increases. Aluminum oxide (Al_2O_3), silicon dioxide (SiO_2), titanium dioxide (TiO_2), and zirconium oxide (ZrO_2) exhibit a combination of ionic and covalent characteristics.

2. Bond Effects of Properties

This section describes the effects of metallic, covalent, and ionic bonding on some of the important engineering properties of materials. Particular emphasis will be on the comparison of ceramics with metals to provide the background that will be useful in understanding later discussion on designing with ceramics.

a. Melting Temperature

Melting temperature is a function of the strength of atomic bonding. Weakly bonded alkali metals and monovalent ionic ceramics have low melting temperatures, as shown for sodium metal and NaCl in Table 14.1. More strongly bonded transition metals (e.g., Fe, Ni, and Cr) and multivalent ionic ceramics (BeO, Al_2O_3, and ZrO_2) have much higher melting temperatures. Very strongly bonded metals (W) and covalent ceramics (TiC and HfC) have the highest melting temperatures.

The melting temperature of a ceramic is strongly affected by the presence of impurities in the composition. In most cases, the impurities cause a reduction in the melting temperature. An estimate of the type and degree of effects for a given material and impurity can be obtained by referring to the relevant phase diagrams (7).

b. Thermal Conductivity

Thermal conductivity k is a measure of the rate of heat flow through a material. The amount of heat transfer is controlled by the amount of heat energy present, the nature of the heat carrier in the material, and the amount of dissipation. The heat energy present is a function of thermal energy applied and the volumetric heat capacity of the material. The primary heat carriers are electrons or phonons (quantized lattice vibrations). The amount of dissipation is a function of scattering effects.

The carriers in metals are electrons. Due to the nature of the metallic bond, these electrons are relatively free to move throughout the structure; hence, metals have high

Table 14.1 Melting Temperatures of Metals and Ceramics

Material	Melting temperature or range ($^\circ$C)	Material	Melting temperature or range ($^\circ$C)
Na metal	98	$MgAl_2O_4$ (spinel)	2135
NaCl	801	BeO	2570
Fe metal	1535	MgO	2620
Ni-based superalloy (IN-100)	1263-1335	ZrO_2	2700
$BaTiO_3$ (tetragonal)	1625	TiC	3100
Si_3N_4	1900[a]	UO_2	2800
$Al_6Si_2O_{13}$ (Mullite)	1850	W	3300
Al_2O_3	2045	HfC	3500[a]
SiC	2500[a]		

[a]Decomposes or sublimes.

thermal conductivity. This is especially true for a pure metal, where the atom size and packing are uniform and nothing is present to interfere with the free motion of the electrons. Alloying disrupts the uniformity of the structure and reduces the thermal conductivity by increasing scattering. Electron mobility and heat capacity increase as the temperature increases; therefore, thermal conductivity of metals increases as temperature increases.

The primary carriers of thermal energy in ceramics are phonons and radiation. The highest conductivities are achieved in the simplest structures. These are such structures as diamond and graphite, which are made up of only a single element, such structures as BeO, SiC, and B_4C, which are made up of elements of similar atomic weight, and structures with no extraneous atoms in solid solution. Lattice vibrations can move relatively easily through these structures because the lattice scattering is small. Such ceramics as UO_2 and ThO_2 have a larger difference in the size and atomic weight of the atoms and have lower thermal conductivity due to the increased attenuation of lattice vibration. The thermal conductivities of UO_2 and ThO_2 are less than one-tenth those of BeO and SiC.

Temperature has a strong effect on thermal conductivity of ceramics, as shown in Fig. 14.2. For crystalline ceramics, in which lattice vibrations are the primary mode of heat conduction, increasing temperature progressively increases lattice scattering and decreases thermal conductivity. Glasses, which do not have a repeating lattice, have poor phonon conduction to start with and are thus not affected in the same way by increased temperature. For glasses, heat capacity effects dominate (under adiabatic conditions). The heat capacity increases with temperature, resulting in a slight increase in thermal conductivity, as shown for fused silica glass in Fig. 14.2.

Radiation to the ambient can significantly increase thermal conductivity as a function of temperature for glass, transparent crystalline ceramics, and porous ceramics. Radiation is proportional to a power function of the absolute temperature T and is usually in the range $T^{3.5}$-T^5 for ceramic materials. The increase in thermal conductivity with temperature for insulating firebrick in Fig. 14.2 is due to radiation across the pores.

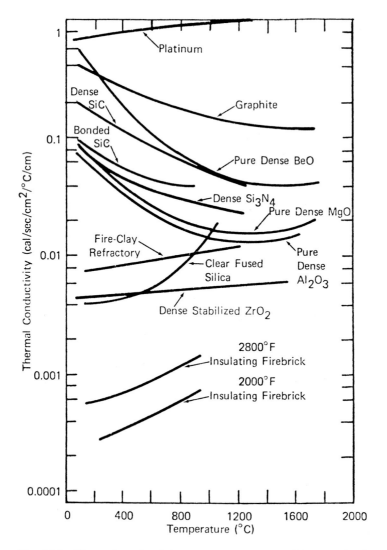

Fig. 14.2 Thermal conductivity versus temperature. (From Ref. 8.)

c. Thermal Expansion

The amplitude of atomic vibration within a structure increases as the temperature increases, resulting in an increase in volume as a function of temperature for almost all solid materials. This increase is normally referred to as thermal expansion and is most often reported in terms of the linear thermal expansion coefficient α:

$$\alpha = \frac{\Delta l}{l_0 \Delta l} \tag{1}$$

where l_0 is the length at room temperature and Δl is the change in length for Δ temperature increase.

The relative thermal expansion of metals is primarily dependent upon bond strength. The greater is the bond strength, the lower the expansion. For instance, Na metal has a higher thermal expansion than Fe, which in turn has a higher expansion than W. Greater thermal energy is required to increase the amplitude of thermal vibration in strongly bonded W than in less strongly bonded Fe or in weakly bonded Na.

The relative thermal expansion of metals and ceramics is dependent upon both bond strength and the configuration of the structure resulting from the bond type. Metals and ionic ceramics have densely packed atomic structures. When any atom within the structure expands due to thermal vibration, it pushes against surrounding atoms. Thus, the total expansion of the structure is the sum of the expansion of all the individual atoms. On the other hand, covalent bonding is directional and produces structures with large open spaces. When a covalent ceramic is heated, a portion of the expansion of the atoms can be absorbed by the open space within the structure or by bond angle shifts. This results in much lower thermal expansion for covalent ceramics than for ionic ceramics or metals. Table 14.2 illustrates differences in the thermal expansion of metals and ceramics.

The thermal expansion of isometric (cubic) single crystals and polycrystalline ceramics is uniform in all directions. However, anisometric crystals have different thermal expansion along different crystallographic axes, depending on the bond strength in these directions and the nature of the structure. An extreme case is graphite. The carbon atoms in graphite are strongly bonded covalently into a sheet or layer structure. The coefficient of thermal expansion parallel to the sheets or layers is very low (1×10^{-6} °C^{-1}) due to the strong bonding and the open spaces in the covalent structure. The carbon-carbon bonds in graphite are all satisfied in forming the sheets. The sheets are held together by very weak van der Waals forces. The thermal expansion in the direction parallel

Table 14.2 Thermal Expansion Characteristics of Various Metals and Polycrystalline Ceramics

Material	Approximate linear thermal expansion coefficient in temperature range 25-1000°C (cm/cm °C $\times 10^6$)	Material	Approximate linear thermal expansion coefficient in temperature range 25-1000°C (cm/cm °C $\times 10^6$)
Pb metal	29	$MgAl_2O_4$ (spinel)	9
Al metal	25	ZrO_2	10
Fe metal	12	$Al_6Si_2O_{13}$ (mullite)	5
W metal	4.2	Fused SiO_2	0.9
Typical superalloy	16-17	$LiAlSi_2O_6$	0.2-0.5
NaCl	40	Si_3N_4	3.2
MgO	9	SiC	4.5
BeO	10	TiC	7.2
Al_2O_3	8		

Table 14.3   Thermal Expansion Coefficients for Anisometric Ceramics

Material	Along C axis ($^\circ C^{-1}$)	Perpendicular to C axis ($^\circ C^{-1}$)
Al_2TiO_5	11.5×10^{-6}	-2.6×10^{-6}
$CaCO_3$	25×10^{-6}	-6×10^{-6}
$ZrSiO_4$	6.2×10^{-6}	3.7×10^{-6}
SiO_2 (quartz)	9×10^{-6}	14×10^{-6}

to these weak bonds (perpendicular to the sheets) is very high (27×10^{-6} $^\circ C^{-1}$). Table 14.3 lists thermal expansion coefficients for some other anisometric crystals.

When polycrystalline components are fabricated from these anisometric materials, internal stresses and microstructural features that affect the mechanical properties may be present. (This will be discussed later in the section entitled Microstructure Effects on Properties.)

Low thermal expansion is an important property for ceramics in applications where rapid temperature changes may occur. Most of us have experienced the cracking or breaking of a ceramic article by too rapid heating or cooling. Some anisometric ceramic materials have been fabricated into polycrystalline articles having extremely low net expansion. Lithium aluminum silicate ($LiAlSi_2O_6$; spodumene, also frequently called LAS), for instance, shows almost no expansion over a wide temperature range and is very resistant to thermal shock damage. Its unique thermal expansion properties have resulted in its use in such diverse applications as heat-resistant cookware and stove tops in the home and $1800^\circ F$ heat exchangers and catalyst supports in the automotive industry.

d. Modulus of Elasticity

The atomic spacing in a material is governed by equilibrium between forces of repulsion and forces of attraction. If an external force or load is applied, a slight change results in the atomic spacing. Up to a certain limit for each material, the atoms return to their original position after the load is removed. This is referred to as elastic deformation, and the load (stress) and the deformation (strain) are related by a simple proportionality constant E. For a given tensile stress σ and strain ϵ,

$$\sigma = E\epsilon \tag{2}$$

The proportionality constant E is called the modulus of elasticity or Young's modulus.

The stronger the atomic bonding, the larger is the stress required for deformation and thus the higher is the modulus of elasticity. Table 14.4 lists the Young's modulus of a variety of ceramics and metals. Comparison of this table with Table 14.1 points out a close parallel between elastic modulus and melting temperature. Hardness of ceramic materials follows the same pattern. All are largely controlled by the strength of atomic bonding.

Elastic modulus decreases slightly as temperature increases. This results from the increase in interatomic spacing due to thermal expansion. As interatomic spacing increases, less force is necessary for further separation.

Table 14.4 Elastic Modulus Comparison for Ceramic, Metallic, and Organic Materials

Material	Elastic modulus[a]	Material	Elastic modulus[a]
Rubber	5×10^2-5×10^5	ZrO_2	20
Nylon	0.4	Al_6O_{13} (Mullite)	21
Polystyrene	0.4	UO_2	25
Al metal	9	Al_2O_3	55
Fe metal	28.5	BeO	45
Ni-based superalloy (IN-100)	30.4	Si_3N_4	44
NaCl	6.4	SiC	60
Fused SiO_2	10	TiC	67
Typical glass	10	Diamond	150
MgO	30		

[a]Elastic modulus $\times 10^6$ except as noted otherwise.

As was discussed under thermal expansion, bond strength varies in different crystallographic directions for noncubic structures. Therefore, elastic modulus also varies. Most commercial ceramics are polycrystalline with random crystal orientation so that reported elastic modulus values represent an average for the various crystal directions. However, even though the ceramics appear to have a single elastic modulus, the engineer must be aware that the individual crystals in the microstructure are anisotropic and that internal stresses may result under loading that can adversely affect the application of the material.

e. Ductility

When the elastic limit of a material is exceeded such that nonreversible deformation occurs, either fracture or plastic deformation occurs. Ductility is the ability of a material to elongate or deform under a load rather than fracturing. This deformation in a material is accommodated by dislocation movement along planes of atoms in the crystal structure. For movement to occur, a dislocation must (1) be present or easily initiated; (2) have an activation energy below the fracture initiation energy for the material; and (3) have an unobstructed path for movement. These conditions are satisfied ideally in a pure metal with a close-packed structure. First, atoms in the close-packed array have many planes available for slip. Second, the atoms have a uniform electron environment due to the mutual sharing of electrons throughout the structure, such that movement of one atom in relation to another does not disturb electrical balance or equilibrium of the structure. Therefore, relatively low activation energy is required to initiate dislocation movement. Third, the atoms are all the same size and type and provide an unobstructed path for movement. Thus, pure metals with close-packed structures have very good ductility and undergo much elongation prior to fracture. For instance, pure copper, aluminum, and iron have elongations at room temperature in the range of 40-60%.

Not all metals have high ductility. Addition of secondary atoms as alloying agents or a secondary dispersion blocks the movement of dislocations and increases the energy required for slip. The highly alloyed superalloys only have elongations at room temperature in the range 5-20%. Oxide dispersion-strengthened alloys (ODS alloys) have similar elongations at room temperature, and these values decrease as temperature increases. Cast iron has effectively no ductility and fractures in a brittle mode.

Ionic-bonded ceramics have close-packed structures similar to the pure metals and thus have many potential slip planes. However, due to the opposite electrical charge of adjacent ions, each ion is stable only in a certain equilibrium position and coordination (number of nearest neighbors), and a higher activation energy than was required for metals is necessary to move oppositely charged ions in relation to each other and cause slip. In most cases, this activation energy is higher than the energy required to initiate fracture through stress concentration at a surface or internal material flaw.

The situation is very similar for covalent ceramics. The directionality of bonding places atoms in equilibrium positions that require high activation energy for slip. In addition, the directionality results in some structures that are not close packed and do not have as well defined planes available for dislocation movement.

Von Mises (9) and Taylor (10) determined that five independent slip systems are necessary for ductility to occur in a polycrystalline material. Highly symmetrical cubic structures, such as NaCl and MgO, have many planes available for slip. At low temperature slip occurs along (110) planes in the [110] direction resulting in only two independent slip systems. At high temperatures slip also occurs along (001) and (111) planes in the [110] direction, resulting in five independent slip systems at high temperatures. However, the temperatures at which these slip systems become operative and the stress required for slip vary depending upon the bond strength of the material. Slip occurs at low temperature and low stress in weakly bonded alkali halides, such as NaCl, but requires much higher temperature and stress for more strongly bonded ceramics, such as Al_2O_3, diamond, and SiC. Table 14.5 shows data reported for stresses and temperatures that produced deformation in various ceramic materials. These data are included only for illustrative purposes and should not be used for engineering calculations without first referring to the original references to review the specific sources of the ceramics test materials and the experimental conditions used for obtaining the data (see Chap. 1).

The degree of ductility or plastic deformation in a ceramic is not only dependent upon the bond strength, the stress, and the temperature, it is also affected by the strain

Table 14.5 Examples of Stresses and Temperatures That Produced Deformation in Ceramic Materials

Material	Strength and type of bond	Stress producing yield (psi)	Temperature (°C)	Reference
KCl	Weak, ionic	200	25	11
CaF_2	Weak, ionic	2,000	160	12
MgO	Moderate, ionic	1,500	1300	13
Al_2O_3	Strong, covalent and ionic	18,000	1270	14
Al_2O_3	Strong, covalent and ionic	4,000	1570	14

rate and by the microstructure (microstructure effects will be discussed in a later section on creep). Typically, decreasing the strain rate decreases the temperature or stress required to initiate deformation. For instance, Kronberg (14) showed that a 10-fold decrease in the strain rate over the temperature range 1500-1700°C for single-crystal Al_2O_3 resulted in a twofold decrease in the yield stress.

It is important to stress that, although ductility has been observed in scientific investigations of certain single-crystal ceramics, from an engineering viewpoint the strains are so small that for practical purposes all polycrystalline ceramics must be considered to be brittle materials.

f. Optical Properties

The optical properties of a material are determined by the level of interaction between the incident electromagnetic radiation and the electrons within the material. If electromagnetic radiation at a given wavelength stimulates electrons to move from their initial energy level to a different energy level, the radiation is absorbed and the material is opaque at that particular wavelength of radiation. Metals have many open energy levels for electron movement and thus are opaque to electromagnetic radiation at most wavelengths. Ionic ceramics have filled outer electron shells comparable to inert gas electron configurations and do not have energy levels available for electron movement. These ionic ceramics are transparent to most electromagnetic wavelengths, so long as they do not have many inclusions, pores, or other internal flaws that reduce transmission due to scattering. Covalent ceramics vary in their level of optical transmission. Those that are good insulators and have a large band gap transmit; those that are semiconductors and have a small band gap can transmit under some conditions, but they become opaque as soon as enough energy is present for electrons to enter the conduction band.

Optical transparency is important in many applications. Glass and a variety of ionic ceramics are transparent in the visible range of the spectrum (0.4-0.7 μm), making possible many applications for windows, lenses, prisms, and filters (see Sec. II). MgO, Al_2O_3, and fused SiO_2 are also transparent in the ultraviolet (0.2-0.4 μm), a portion of the infrared (0.7-3.0 μm), and the radar (10^3-10^5 μm) ranges and are important electro-optical and electromagnetic window materials for tactical and strategic missiles, aircraft, remotely piloted vehicles, spacecraft, battlefield optics, and high-energy lasers. MgF_2, ZnS, ZnSe, and CdTe are also important for these applications.

Color is another optical property that leads to many ceramic applications. Color results from the absorption of a relatively narrow wavelength of radiation with the visible region of the spectrum (0.4-0.7 μm). For this type of absorption to occur, transition of electrons must occur. This occurs primarily where transition elements having an incomplete d shell (V, Cr, Mn, Fe, Co, Ni, and Cu) or f shell (rare earth elements) are present. It can also result from nonstoichiometry.

The oxidation state and bond field are also important in color formation. For instance, neither S^{2-} or Cd^{2+} cause visible absorption, but CdS produces strong yellow. Likewise, Fe^{3+} and S^{2-} are responsible for the color of amber glass.

Ceramic colorants are widely used as pigments in paints and other materials produced and used at low temperatures. They are especially important where processing is done at elevated temperature where other types of pigments are destroyed. For instance, porcelain enamels fired in the range 750-850°C require ceramic colorants. Ceramic colorants having the spinel structure AB_2O_4 (such as blue, $CoAl_2O_4$) are often used in this temperature range. Doped ZrO_2 and $ZrSiO_4$ are used at higher temperatures (1000-1250°C) because of their increased stability to attack by the glass in which they are dispersed. Dopants include vanadium (blue), praseodymium (yellow), and iron (pink) (15).

Phosphorescence is another important optical property displayed by some ceramic compositions. Phosphorescence is the emission of light resulting from the excitation of the material by an appropriate energy source. Ceramic phosphors are used in fluorescent lights, oscilloscope screens, and TV screens.

The fluorescent light consists of a sealed glass tube coated on the inside with a halogen phosphate, such as $Cu_5(PO_4)_3(Cl,F)$ or $Sr_5(PO_4)_3(Cl,F)$ doped with Sb and Mn and filled with mercury vapor and argon. A capacitor provides an electrical discharge that stimulates radiation of the mercury vapor at a wavelength of 2537 Å. This ultraviolet radiation excites a broad band of radiation in the visible range from the phosphor, producing the light source.

For oscilloscopes and TV, the phosphor is excited by an electron beam that sweeps across the phosphor-coated screen. Decay time of the light emission of the phosphor is important. For color TV, decay occurs in approximately $1/10$-$1/100$ of a second. A phosphor having slower decay is necessary for a radar screen. More than one phosphor is required for a color TV, each being selected to emit a narrow wavelength of radiation corresponding to one of the primary colors. The most difficult color to achieve was red. $Zn_3(PO_4)_2$, doped with Mn, and YVO_4, doped with Eu, produce shades of red.

Two of the most important future application areas of optical ceramics are laser and fiber optics. Both are likely to play important roles in communications and many other fields. Important laser ceramics include Cr-doped Al_2O_3 (ruby), Nd-doped $Y_3Al_5O_{12}$, and a variety of doped glass compositions. Optical fibers are made of extremely high quality glass. The objective is to achieve maximum light transmission with minimum loss due to scattering or absorption. To achieve this a fiber configuration and glass refractive index are selected that provide total internal reflection within the fiber. One way of doing this is to clad the optical fiber with a layer of glass of lower refractive index. Another approach is to vary the refractive index from the surface to the core by the use of ion-exchange procedures. This focuses the light rays along the interior of the fiber.

g. Electrical Properties

The electrical conductivity of a material is somewhat analogous to thermal conductivity. The amount of conduction is a function of the amount of energy present (in this case, the size of the electric field), the number of carriers, and the amount of dissipation.

The carriers in metals are electrons. As we have noted previously, these electrons are relatively free to move throughout the structure and result in high electrical conductivity. This is especially true for pure metals, where atom size and packing are uniform and nothing is present to impede the free motion of the electrons. Alloying disrupts the uniformity of the structure and reduces the electrical conductivity. Increase in temperature also disrupts the structure (due to lattice vibration) and results in a decrease in electrical conductivity.

Ceramic materials show a broad range of electrical conductivity behavior. Table 14.6 compares the electrical resistivity (reciprocal of electrical conductivity) of some metals and ceramics at room temperature.

Table 14.6 Electrical Resistivity of Some Metals and Ceramics at Room Temperature

Material	Resistivity (ohm-cm)	Material	Resistivity (ohm-cm)
Metallic conduction		Insulators	
Copper	1.7×10^{-6}	SiO_2	10^{14}
Iron	10×10^{-6}	Steatite porcelain	10^{14}
Tungsten	5.5×10^{-6}	Fireclay brick	10^8
ReO_3	2×10^{-6}	Low-voltage porcelain	10^{12} to -10^{14}
CrO_2	3 ∓ 10^{-5}	Al_2O_3	10^{14}
		Si_3N_4	10^{14}
		MgO	10^{14}
Semiconductors			
SiC	10		
B_4C	0.5		
Ge	40		
Fe_3O_4	10^{-2}		

Source: After Kingery et al., Ref. 8, p. 851.

Some transition metal oxides, such as ReO_3, CrO_2, VO, TiO, and ReO_2, have conduction by electrons much like that which occurs in metals. This results from an overlap of electron orbitals, such that wide unfilled d or f bands are present. Under the influence of an electrical field, these electrons are relatively free to move and carry charge through the material. The presence of impurities can reduce the conductivity slightly by scattering. Increases in temperature have the same effect.

Other metal oxides, such as CoO, NiO, Cu_2O, and Fe_2O_3, have an energy gap between the filled and empty electron bands such that conduction will only occur when external energy is supplied to bridge the energy gap. These materials are semiconductors. An increase in temperature can provide the energy. For instance, NiO, Fe_2O_3, and CoO are insulators at low temperature with conductivities of less than 10^{-16} $(ohm\text{-}cm)^{-1}$. In the range 250-1000 K, the conductivity increases nearly linearly to values in the range $10^{-4}\text{-}10^{-2}$ $(ohm\text{-}cm)^{-1}$.

Semiconducting properties can be achieved in many ceramics by doping or by forming lattice vacancies through nonstoichiometry. Examples of ceramics responsive to this approach include TiO_2, ZnO, CdS, $BaTiO_3$, Cr_2O_3, Al_2O_3, and SiC.

In some ceramic materials, especially oxides and halides, ions are the primary carriers in electrical conduction. The degree of conductivity is largely dependent upon the energy barrier that must be overcome for the ion to move from one lattice position to the next. At low temperatures, conductivity is low. However, when the temperature is high enough to overcome the barrier to lattice diffusion, the conductivity increases. Presence of defects in the structure aids conduction. Controlled impurities can be added to increase the concentration of defects and thus the electrical conductivity.

Many important ceramic applications have resulted from the electrical conductivity properties of ceramics or a unique combination of electrical properties plus other properties, such as temperature capability. Porcelains and glasses are used for both high- and low-voltage insulation. Al_2O_3 is used extensively as a substrate in electrical devices.

Semiconducting SiC is used in heating elements. Cu_2O is used as a rectifier, a device that only allows electrons to flow in one direction. Semiconductor spinel materials, such as Fe_3O_4 and $MgCr_2O_4$, diluted in controlled solid solution with nonconducting spinel materials, such as $MgAl_2O_4$, $MgCr_2O_4$, and Zn_2TiO_4, have been used as thermistors, devices that closely control electrical resistance as a function of temperature. Other important applications include transistors, photocells, detectors, and modulators.

Ceramic insulator materials are often useful due to their dielectric properties. Dielectric refers to the polarization that occurs when the material is placed in an electrical field. The negative charge shifts toward the positive electrode and the positive charge toward the negative electrode. The total polarizability resulting from a combination of electronic, ionic, and dipole orientation effects is referred to as the dielectric constant K'. Dielectric constants for some ceramic materials are listed in Table 14.7.

One important application of dielectric materials is in the fabrication of capacitors. The charge that can be stored in a capacitor is dependent upon the dielectric constant of the material between the capacitor plates. If the material has $K' = 10$, the capacitor can store 10 times as much charge as it could if only vacuum were present between the plates. Materials with a high dielectric constant, such as $BaTiO_3$, make possible substantial miniaturization of capacitors.

$BaTiO_3$ is highly anisotropic and thus has significantly different dielectric properties in different crystallographic directions. When an electrical voltage is applied a mechanical distortion results. This piezoelectric property has led to widespread use of $BaTiO_3$ as a transducer in ultrasonic devices, microphones, phonograph pickups, accelerometers, strain gages, and sonar devices. $PbTiO_3$ is also useful in these applications.

With the above background on the properties of ceramics, we can return again to consider the major problem that the engineer faces in using ceramics in a structure, namely, brittleness. Even when the primary selection criterion in a design is a nonmechanical property, there is generally also some mechanical integrity required. For example, dielectric properties are the primary selection criteria for radomes, but the thermal expansion (for thermal shock resistance), the thermal conductivity (for heat transfer), and the strength (for launch and maneuvering integrity) are also key properties that must be considered. Ophthalmic lenses whose function is purely optical still must possess a governmentally specified impact resistance to protect the wearer's eyes from injury in the event of accidents. Virtually every application of ceramic materials entails some mechanical or thermal loading that can cause overstressing and thus brittle failure, unless proper design procedures are followed. In the following discussion we will briefly outline such procedures.

Table 14.7 Dielectric Constants for Ceramic Materials

Material	K'	Material	K'
NaCl	5.9	Porcelain	5.0
MgO	9.6	Fused SiO_2	3.8
Al_2O_3	8.6-10.6	High-lead glass	19.0
TiO_2	96	$BaTiO_3$	1600

B. Generalized Brittle Materials Design Process

Design with brittle materials requires a very precise definition of the state of stress at every point in the component. Brittle materials can, in fact, be quite strong, but they generally manifest a wide scatter in strength and do not have the ability to redistribute a high local stress concentration by yielding as is the case in metals. Ceramics are, in fact, extremely sensitive to high localized stress gradients due to notches, thermal gradients, or other sources. Therefore, successful design with ceramic materials in highly stressed applications begins with the careful application of modern two- and three-dimensional computerized finite-element thermal and stress analysis techniques (16-18). These computer techniques are the heart of the materials/design/engineering tradeoff process shown schematically in Fig. 14.3.

What makes the design trade-off situation different for ceramic materials utilization is that brittleness manifests itself in a wide mechanical strength scatter. Hence, a probabilistic rather than a deterministic philosophy and design formalism must be applied. A detailed description of such processes in beyond the scope of this text but may be found in Refs. 16-18. To a first approximation the main additional complexity that designing with a brittle material, on a probabilisitic basis, brings to the design situation is that, for each element in the finite element grid, a distribution of properties rather than a fixed value is used for materials strength, and the design is developed until an acceptably low probability of failure for the component is attained. In practice a fine finite-element mesh (the level of precision in definition of local stresses) is required when designing with ceramics as compared with metals. It is worthwhile to point out that the deterministic design commonly used in metallic design practice is really an approximation to probabilistic design, where a failure probability based on (typically) a 3σ standard deviation of property scatter is assumed. That metallic structures rarely fail is because, over centuries of design experience, codes of practice have been developed that have generated a high level of confidence in choosing appropriate approximations.

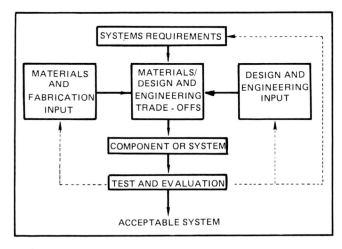

Fig. 14.3 Brittle materials design systems trade-off logic.

The next section of this chapter focuses on the origin of the high scatter of strength in ceramics and the statistical methods for dealing with the data so that they can be factored into the probabilistic stress analysis/design process briefly described above.

1. Origin of Scatter: Fracture Mechanics

The problem can be considered as having three principal elements:

> Griffith's failure criterion
> Subcritical crack growth
> Strength dispersions

Griffith's failure criterion gives the conditions that will trigger brittle fracture in an individual specimen. Subcritical crack growth causes fracture stress to be time dependent, and an intrinsic dispersion of fracture stress values is used in characterizing a ceramic's strength via statistical relations that account for a size dependence. Each of these elements of the problem is discussed in the paragraphs that follow.

a. Griffith's Failure Criterion

When a material is subjected to stress, any discontinuity or flaw in the material will intensify the stress locally. One can characterize the local stress by a stress intensity factor, which is the product of the applied stress and a term s that accounts for the severity of the flaw. In the case of an applied tensile stress that tends to open a crack σ_t, the stress-intensity factor is designated K_I.

The Griffith criterion (19) states that a crack will become unstable when $K_I = \sigma_t s$ reaches a critical value K_{IC} at any site in the material. When this condition obtains, a crack propagates rapidly and spontaneously through the material; that is, the brittle fracture event is triggered. It follows that a safe ceramic component will be one so designed and used that K_I, the product of tensile stress and flaw severity, is maintained below a critical value K_{IC} at every point in the component.

Because brittle fracture is the consequence of a local condition, structural design with ceramic materials requires determination of the precise tensile stresses that will be experienced at all sites in the component. Unlike designing with ductile metals, local stresses, such as those that occur at sites of load transfer or changes in section, cannot be neglected in the stress analysis.

If the flaw in the ceramic that is responsible for fracture is not itself a crack, a crack develops prior to the event. Under uniaxial tension, the severity of the crack that becomes unstable when Griffith's criterion is met is given by the following approximate relation (20):

$$s \simeq \frac{Y(A)^{1/2}}{1.7} \tag{3}$$

where A is the area of the crack in a plane normal to the applied tension and $Y \simeq 2.0$ if the crack extends from a surface, or $Y \simeq 1.77$ if the crack is beneath the surface (21). It can be seen from this relation that stress intensification by a crack depends mainly on its area. Equation (1) introduces a maximum error of less than 5% in determining crack severity. A more exact (and more common) relation is

$$s = Y\sqrt{\frac{a}{Z}} \tag{4}$$

where a is the depth of a surface crack, or the radius of half the minor axis, respectively, of a circular or elliptical subsurface crack, Z is a parameter that varies with crack shape, having a value of 1.0 for a long shallow crack and increasing with increasing crack depth-width ratio (20). For a circular crack $Z = \pi/2$.

Cracks that become unstable are often called "Griffith" cracks. In high-strength ceramics, they tend to be very small, of the order of few thousandths of a centimeter wide. Their small size tends to thwart attempts to detect them nondestructively prior to fracture. Flaws that give rise to Griffith cracks are sometimes intrinsic features of a ceramic's microstructure, such as a pore, inclusion, or large grain. However, they are more apt to be an extrinsic surface feature, such as a scratch or pit introduced in finishing or from abuse in handling or service. In some cases, two or more neighboring flaws combine to trigger the fracture event.

It is important to note that the size of the microstructural feature at a fracture origin does not necessarily correspond to the size of the crack that becomes unstable. The lack of a clear relation between flaw size and the associated Griffith crack size also hampers efforts to predict the strength of a ceramic from features detected nondestructively.

The critical stress intensity factor in the Griffith relation, K_{IC}, is also known as the material's fracture toughness. Ideally, it is considered to be a material constant, varying only with temperature, and related to two separate material properties, Young's modulus E and fracture surface energy γ_f as follows.

$$K_{IC} = (2E\gamma_f)^{1/2} \tag{5}$$

K_{IC} is a more fundamental property of a ceramic than is fracture stress, the quantity commonly reported for strength. Depending on crack severity, strength can vary widely for a given ceramic without K_{IC} being affected. If the severity of the most severe flaw in a ceramic component can be specified with certainty, knowledge of K_{IC} permits computing the stress the component can withstand.

There are several experimental techniques for evaluating K_{IC}. The results from them sometimes disagree, and at present there is uncertainty about the significance of the values obtained. With this reservation, K_{IC} values for different ceramics are listed in Table 14.8.

Table 14.8 Room Temperature K_{IC} Values for Several Ceramics

Material	K_{IC} (MN-m$^{3/2}$)
Soda-lime glass (4)	0.75 ± 0.03
Glass ceramic (9606) (4)	2.38 ± 0.08
Sintered alumina (Alsimag 614) (4)	3.84 ± 0.05
Hot-pressed alumina (4-6)	4.2-5.8
Hot-pressed Si_3N_4 (NC-132) (7,8)	4.0-6.0
Sintered SiC (9)	4.0

b. Subcritical Crack Growth: Time and Size Dependencies of Ceramic Strengths

Ceramic materials often exhibit so-called static fatigue (28). Fracture is delayed, occurring after a stress has been sustained for some period of time. Coincident with static fatigue, strength increases with strain or stress rate. Also, ceramic strengths are found to decrease when specimen size is increased.

   If the stress that will trigger brittle fracture in a ceramic component is to be closely defined for structural design purposes, both these phenomena, the time and size dependencies of failure stress, must be considered, and both phenomena have been interpreted in a manner consistent with the Griffith criterion for brittle fracture.

   The *time dependence* of fracture stress is a consequence of a slow, stable crack growth that increases the severity of existing cracks. It occurs in a reactive environment or at a temperature where the ceramic exhibits creep. In the case of most glasses and oxide ceramics, moisture causes the environment to be reactive, even at room temperature. Under the combined influence of a reactive environment and tension, a crack will grow slowly until the Griffith criterion is met. It then becomes unstable, and very rapid growth ensues.

   The velocity V of subcritical crack growth can be measured, and in a given ambient the following empirical relation is often found to describe most of the stable crack growth that occurs prior to fracture (29):

$$V = BK_I^n \tag{6}$$

where B and n are empirical material constants for the particular environment. This basic relation permits calculating the effect of time on a ceramic's strength. If subjected to a constant tensile stress σ_t, the time to failure t, is given by (28,29)

$$t = \frac{2Z^2}{Y^2 B(n-2)} \left(\frac{K_{IC}}{\sigma_{IC}}\right)^{2-n} \sigma_t^{-n} \tag{7}$$

where σ_{IC} is fracture stress in an inert environment (i.e., in the absence of slow crack growth). Similarly, one can calculate the effect of stress rate or strain rate on strength (fracture stress). The relation between stress rate $\dot{\sigma}$ and strength σ_f is

$$\sigma_f = \left[\frac{2\dot{\sigma}(n+1)}{B(Y/Z)^2(n-2)} \left(\frac{\sigma_{IC}}{K_{IC}}\right)^{n-2}\right]^{1/(n+1)} \tag{8}$$

where $\sigma_{IC} - \sigma_f$ is the strength degradation due to subcritical crack growth. Since ceramic materials rarely deviate much, if at all, from linear elastic behavior, strain rate $\dot{\epsilon}$ is proportional to stress rate; that is, $\dot{\epsilon} = \dot{\sigma}/E$, where E is Young's modulus.

   These relations have not been subjected to extensive experimental verification. Table 14.9 gives strength values for ceramic specimens tested in water and in dry nitrogen under uniaxial tension. Subcritical crack growth occurred during testing with a constant stress rate of 4 MN-m^2/sec in water, and was absent during testing in dry nitrogen with a stress rate of 100 MN-m^2/sec. Table 14.9 also includes calculated strength in water given by Eq. (8). Values for B and n used in the calculation were determined from independent slow crack experiments in water (30). In a separate study (31), Eq. (7) adequately predicted the time to failure (fatigue life) of the alumina ceramic. Surface flaws were responsible for failure of both ceramic materials in dry nitrogen as well as in water. Had subsurface flaws caused failures in dry nitrogen, we would not have obtained the excellent correlation between calculated and measured strengths in water. Beyond

Table 14.9 Ceramic Strengths in Dry Nitrogen and Water

Ceramic	Type of loading	Average strength (MN-m^2)		
		Dry nitrogen	Measured water	Calculated[a] water
Sintered Al$_2$O$_3$	3-Point bending	408	295	288
Sintered Al$_2$O$_3$	4-Point bending	369	271	263
Glass ceramic	4-Point bending	317	204	210

[a]Calculated from dry-nitrogen strengths with Eq. (6). Values of B and n used in the calculation were determined in slow-crack growth experiments as $10^{-27.6}$ and 42, respectively, for the sintered Al$_2$O$_3$ and $10^{-35.3}$ and 56 for the glass ceramic (12).

this, Eqs. (7) and (8) require that the surface flaws that exhibit subcritical crack growth be responsible for failure in the absence of such growth. Also, the equations require that K_{IC} be a material constant, unaffected by test environment or local conditions at the crack-initiation site, and that Z, the crack-shape parameter, not change much during the subcritical growth.

Clearly, if the stress to cause fracture of a ceramic component is to be closely defined for structural design purposes, one must determine whether subcritical crack growth occurs in the service environment, and if it does, account for the growth in specifying a service life as well as a safe stress for the component. Equation (6) or modifications of it, if found applicable, provide a basis to account for time in defining fracture stress. It should also be noted that the selection of a ceramic for a structural component should not be based on fracture stress alone; subcritical crack growth rate is an equally important factor.

Since ceramics generally fail as predicted by the Griffith equation, Eq. (5), it follows that the strength of a ceramic is a function of both the size and location of flaws in the component. If the ceramic contained a homogeneous population of identical flaws, it should fail at a unique tensile stress. In this case, a value of critical stress could be predetermined from strength tests and used as the failure criterion in structural designing. However, we normally observe that strength values of nominally identical specimens when tested alike are dispersed, and the values for individual specimens are in general inversely related to the size of flaws found microscopically at fracture origins.

Recognition that variability in fracture-initiating flaws precludes assigning a unique strength value to a ceramic is of paramount importance in structural designing. One consequence of the variability in worst flaws is a size dependence of strength. Large specimens tend to fail at lower mean strengths than small ones simply because there is apt to be one or more severe flaws among the larger population of a big-specimen. Hence, the effective size is smaller, and consequently the strength is higher in bending than in direct tension (32). This is because only part of the specimen is subjected to tension in bending.

Two approaches are available to the designer for finding out whether K_I achieves the critical level K_{IC} at any site in a component. The more positive approach is preservice proof testing (33). In this approach, the component is subjected to a loading that imposes the maximum tensile stresses throughout that will be encountered in service, and the components having critical flaws with severity greater than K_{IC}/σ are eliminated. If the component survives the test, it is placed in service. Proof testing requires that the component surface be representative of that which develops in service. Proof testing of a component with a finished surface would be meaningless if surface flaws are responsible for fracture either in proof testing or service and the finish is lost from service-incurred chemical attack or physical abuse. Also, account must be taken for any subcritical crack growth that might occur in service (and in the proof testing itself) when this approach is used. We have discussed treatment of the effect of subcritical crack growth on fracture stress in the preceding section.

Proof testing, appropriately conducted, assures the integrity of a ceramic structural member and should be employed in qualifying the member. However, if used as the primary design tool, its cut-and-try nature obviously is not well suited for analytically predicting performance or for optimizing a design, and if the component is large or has a complex shape, the cost and time involved in fabrication to arrive at a satisfactory design through proof testing could be prohibitive.

The other available approach is analytical and less positive than proof testing. It is the statistical approach, based on laws of probability. Statistical theories of fracture strength, of which Weibull's (34) is most prominent, treat the scatter among individual strength values caused by variations in the fracture-initiating flaws in a way to provide a mathematical basis for predicting from strength data the failure probability of any size component with any stress distribution in it.

c. Strength Dispersions: Application of Weibull Statistics

The statistical theories of strength attach special significance to the dispersion of ceramic strength values. They treat the dispersion as an inherent property of the ceramic, reflecting effects from an assumed identical distribution of numerous flaws in any piece of the material. The theories usually assume that fracture of individual specimens occurs in accordance with the Griffith criterion, that is, when K_I reaches a critical level at some site in the specimen.

Because they describe a ceramic's strength by formulations that contain parameters obtained empirically from the dispersion of strength values, the use of statistical theories as a design tool requires that much care be exercised to ensure applicability of strength data. Precautions to be taken include the following:

Data quality: If the dispersion reflects testing errors in assessing strengths of individual specimens, the dispersion obviously is not an inherent property of the material and should not be used for characterizing strength.

Nonrepresentative data: If the processing of strength specimens differ from that of the component in a way that affects the population density or severity of strength-initiating flaws, the use of the strength data obviously would be misleading. In this connection, flaws introduced at corners during grinding and surface finishing often are responsible for failure in ceramic strength specimens. These edge failures constitute a major cause of nonrepresentative data.

K_{IC} *variability*: Since the strength of an individual specimen depends on K_{IC} as well as flaw severity, K_{IC} must be invariant or vary randomly Otherwise, K_{IC} variations will bias strength dispersions.

Flaw location: There must be assurance that the flaws responsible for failure of both the strength specimens and the component have the same location.

If surface flaws are responsible for failure in one case and subsurface flaws in the other, a description of strength obtained from the specimens will not describe the component's strength. As will be shown later, failures in the ceramic can result both from extrinsic surface flaws associated with the surface finish and from intrinsic subsurface flaws associated with microstructural features. For ceramics exhibiting such bimodal failures, no single statistical formulation can describe the ceramic's strength.

Service effects on extrinsic flaws: If surface flaws associated with finishing are responsible for failure of strength specimens, the strength data can only be used if these flaws remain unchanged during service exposure and continue to be responsible for failure.

(1) Treatment of Ceramic Strength Dispersions   To mathematically describe strength in the statistical approach the probability of failure P as a function of stress σ is determined from the dispersion of values in a set of strength data. Individual values are ordered from weakest to strongest, and each is assigned a probability of failure based on its ranking n, as follows.

$$P = \frac{n}{N + 1} \qquad (9)$$

where N is the total number of data points.

Failure probabilities in any P-σ plot will always be less than one and greater than zero. Equation (9) shows that the probability range is determined solely by the number of data points. If 20 specimens are tested, P will range from 0.0476 to 0.9524, and if 100 specimens are tested, the range is extended from 0.0099 to 0.9901. The probability range covered by the available data is very important in the statistical approach because the design stress is based on an acceptable failure probability. For example, if failure of 1 component in 100 is tolerable, the stress corresponding to an 0.01 probability for the component size and stress distribution is chosen for the design stress. Since no basis exists for extrapolating P-σ curves, an adequate number of data points must be obtained so that the acceptable failure probability is within the range of experimentally determined failure stresses.

Weibull's formulation offers a mathematical description of effects of size and stress distribution on P-σ curves for a material (32,34). In principle, the mathematical description requires only an empirical P-σ curve for a specimen of known size fractured in a test that imposes a defined tensile stress distribution. Weibull's basic expression for the failure probability of a material is as follows.

$$P = 1 - \exp\left[-\int_V \left(\frac{\sigma}{\sigma_0}\right)^m dV\right] \qquad (10)$$

The integral is taken over all volume elements dV subjected to tensile stress; σ is the maximum tensile stress in the stress field; m, the Weibull modulus, is a measure of the variability of failure stress. Large values of m (e.g., greater than about 30) reflect little scatter and a small effect of size on the material's strength. The term σ_0 is a normalizing constant.

The above form of the Weibull function applies to failure from subsurface (volume) flaws. If surface flaws are responsible for failure of the material, the function should be integrated over the surface subjected to tensile stress rather than the volume; that is, dV is replaced with dS. Another variation can include a third material constant σ_u for materials that exhibit a finite stress for zero probability of failure. The two-parameter form of Eq. (10) is for the case of zero probability of failure when no tensile stress is present. The three-parameter form uses the quantity $\sigma - \sigma_u$ in place of σ in Eq. (10).

Equation (10) can be manipulated and rearranged to yield

$$\log \log \left(\frac{1}{1 - P} \right) = m \log \sigma + \log \frac{V \log e}{\sigma_0^m} \qquad (11)$$

This is an equation of a straight line that allows convenient graphic representation of data. Its utility is illustrated by consideration of Fig. 14.4 giving plots of log log $1/(1 - P)$ versus log σ (Weibull plots) for an alumina ceramic (35). The Weibull plots are three parallel straight lines. A straight line is obtained when the strength dispersion can be described by Weibull's two-parameter function; m is the slope of the line. That the three sets of data, representing specimens of different size or subjected to different loading geometries, give parallel straight lines means that a single two-parameter function describes the entire range of observed failure stresses provided that the lines have the proper separations (Fig. 14.5).

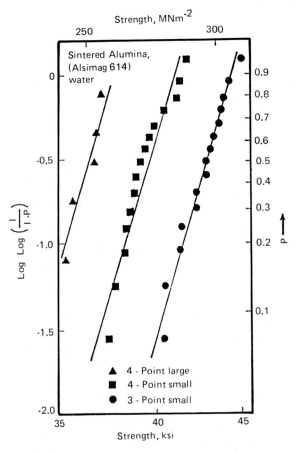

Fig. 14.4 Weibull plots of strength-size data obtained on a alumina ceramic.

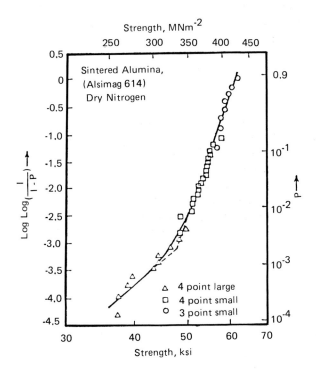

Fig. 14.5 Weibull plot of the data obtained on three different sizes of an alumina ceramic tested in dry nitrogen. Failure probability for the small and large four-point bend data have been normalized to those of the small three-point bend specimen. (Data from Refs. 17 and 18.)

The intercept of any three lines permits evaluation of σ_0. In the case of this particular ceramic, surface flaws were responsible for fracture, so the appropriate form of the Weibull function contains S rather than V, and from Eq. (11) the intercept provides the value of log (S log $e/\sigma_0 m$) from which σ_0 can be extracted. With knowledge of m and σ_0 and assurance of the applicability of Weibull's two-parameter function, Eq. (10) can be used to define the P-σ relation for the σ range covered by the experimental data, regardless of component size, shape, or loading configuration. Using this powerful tool, the designer can calculate the tensile stress that can be permitted in a component with assurance of an acceptable failure probability.

The integration of Eq. (10) has been treated comprehensively by Weibull (34) and others (32), particularly from the standpoint of obtaining the P-σ relation for a component from an experimentally determined P-σ relation when the stress distribution in the component differs from that in the test specimen, for example, use of bend-test data to obtain failure probabilities for a component subjected to uniform tension. This problem is handled mathematically by determining the size of a direct-tension member that has the same P-σ relation as the specimen or component subjected to nonuniform stress. Then, rather than using the actual volume or surface area in Eq. (10), the uniform-tension equivalent or effective size is used (32). The effective size (V_E or S_E) of a simply supported centrally loaded square beam, for example, is

$$V_E = \frac{V}{2(m+1)^2}$$

or

$$S_E = \frac{(m+2)S}{4(m+1)^2}$$

where V and S are the volume and surface area, respectively, of material in the span between supports.

The following general relation is of key importance in structural designing with ceramics whose strengths can be described by a Weibull formulation (36).

$$P_1 = 1 - (1 - P_2)^{V_{E1}/V_{E2}} \tag{12a}$$

Or, in the case of failure from surface flaws,

$$P_1 = 1 - (1 - P_2)^{S_{E1}/S_{E2}} \tag{12b}$$

Equation (12) gives the relation between failure probabilities P_1 and P_2 at a given stress for two members having different effective sizes. By treating effective sizes the equation is independent of stress distribution in either member. To illustrate the use of the equation, suppose a component is to be designed on the basis of data in Fig. 14.4, a 5% probability of failure is acceptable, and the component has an effective surface area one-tenth that of the specimen represented by closed triangles. Substituting in Eq. (12),

$$0.05 = 1 - (1 - P_2)^{1/10}$$

so $P_2 = 0.4$ for the specimen, and the stress corresponding to this failure probability from the plot in Fig. 14.4 is 245 MN-m^2. This then is the maximum allowable tensile stress in the hypothetical component.

Equation (13) is a corollary to Eq. (12). It is the relation between stresses for a given failure probability of two members having different effective sizes.

$$\frac{\sigma_{f1}}{\sigma_{f2}} = \left(\frac{V_{E2}}{V_{E1}}\right)^{1/m} \tag{13a}$$

or

$$\frac{\sigma_{f1}}{\sigma_{f2}} = \left(\frac{S_{E2}}{S_{E1}}\right)^{1/m} \tag{13b}$$

Equation (13) is frequently used to predict the mean strength (0.5 failure probability) of a member from strength data obtained in tests of specimens of a different effective size.

Since Eqs. (12) and (13) are limited in applicability to the range of failure stresses observed in strength tests, they tend to dictate the effective size of specimen(s) that should be strength tested. If the specimen has a significantly smaller effective size than the component, one can expect to test a very large number of specimens to define the P-σ relation covering the failure stress range of interest, but the range can be encompassed with relatively few data points if the effective size of the specimen is nearly equal to that of the component.

(2) Significance of the Weibull Modulus Strength values for ceramic materials are normally reported as the mean failure stress of a series of specimens. The mean value in the statistical approach is the 0.5 failure probability level, a level that will rarely be used as the basis for a design. It signifies that half the components can be expected to fail. If the acceptable failure probability is less than 0.5, the allowable stress must be less than the mean by an amount depending on the material's Weibull modulus m. Further, if the effective size of the component is larger than that of the specimen, this too will reduce the allowable stress, regardless of acceptable failure probability, to an extent also dependent on the material's Weibull modulus.

Table 14.10 shows these reductions in allowable stress for two hypothetical ceramic materials that fail from volume flaws and exhibit the same reported strength of 50,000 psi. One ceramic has a Weibull modulus of 8, and the other has a modulus of 32. It can be seen from the table that, for a failure probability of 0.01, the allowable stress is reduced to 29,400 and 43,800 psi, respectively, for materials with Weibull moduli of 8 and 32 if the component's effective size is the same as that of the specimen. When the component's effective size is 1000 times that of the specimen, these allowable stresses become 12,400 and 35,300 psi.

From this example, it is quite clear that the best ceramic for a given structural application will not necessarily be one with the highest mean failure stress (strength) in a standardized test, but will be one that combines high strength with a high Weibull modulus. Both these properties are subject to upgrading by processing refinements and often by surface-finishing refinements. The prospects for such refinements should not be overlooked in considering a ceramic for structural use.

Such processing-related refinements are discussed in detail in Chap. 28. In this section, we have attempted to give an overview of modern ceramics, delve into the structural origin of their properties, and discuss design with this important class of materials. Understanding the role of processing in determining microstructure and flaw populations, and which may set limitations on component shape, is critical for any engineer who wishes to successfully apply ceramics, and therefore a thorough reading of Chap. 28 is strongly encouraged.

Table 14.10 Effect of Effective Size and Weibull Modulus on Allowable Stress

Failure probability	Allowable stress (10^3 psi)			
	V_E	$10V_E$	$10^2 V_E$	$10^3 V_E$
		m = 8		
0.5[a]	50.0	37.5	28.1	21.1
0.05	36.4	27.0	20.3	15.2
0.01	29.4	22.0	16.5	12.4
		m = 32		
0.5[a]	50.0	46.5	43.3	40.3
0.05	46.1	42.9	39.9	37.1
0.01	43.8	40.8	37.9	35.3

[a]Probability corresponding to mean failure stress, the normally reported strength.

II. NATURE AND PROPERTIES OF GLASS*

A. Structure

The progress in glass manufacture and the development of glass technology up to this century was, by today's standards, incredibly slow. Someone produced the first glass approximately 4000 years ago. The crude material was an opaque or darkly translucent substance and was used for containers and decorative objects. Transparent glass did not come along until about 2000 years ago. The blowing process was invented at about this same time, making possible a greater variety of shapes.

Until 1676 all glass was of the soda-lime type, such as we use today for windows and containers. Then George Ravenscroft, a British chemist, invented lead-alkali glass. This achievement was the first example of intentionally changing the properties of glass by altering its composition. Ravenscroft sought to produce a glass that would be superior for artware—more readily worked and cut than soda-lime glass and more brilliant in appearance.

The next new type of glass was introduced in 1912, 236 years after Ravenscroft. Today inventions come along more frequently (Table 14.11). It is an unusual year without at least one significant new product, and it is rarely more than a few years between announcements of radically new materials and processes. Scientific research is responsible for the improvement in our understanding of the fundamental nature of glass that permits us to recognize and exploit the versatility of the material.

If we view glass from the inside out—in terms of how the substance is assembled—we find that glass is an unusual material. It is mechanically rigid and, in this sense, behaves like a solid. However, the atoms within the glass are arranged in a random, or disordered, fashion. Such disordered structure is characteristic of a liquid. In contrast, each atom in a crystalline solid is held in a definite position within a structural pattern or lattice.

In glass the atoms, though arranged at random, are frozen in position. Thus, glass combines some of the aspects of a crystalline solid and some of the aspects of a liquid. Its state may be described as vitreous or glassy.

The raw materials from which glass is made are crystalline (Fig. 14.6), but when those materials are melted, atoms become detached and move about in a random manner (Fig. 14.7). As the molten material cools, the atoms try to re-establish the ordered pattern they once had. If the melt cools rapidly it thickens and sets, or is fixed in a

---

*Contributed by MARVIN G. BRITTON, P. BRUCE ADAMS, and J. R. LONERGAN.

Table 14.11 Significant Advances in Glass

2000 BC Beads	1940 Foam glass
200 BC Transparent glass blowing	1944 Continuous optical melting
1676 Lead glass	1946 Photosensitive glass
1879 Light bulb	1952 Fused silica
1912 Borosilicate	1957 Glass ceramics
1938 Fibrous glass	1962 Chemcor
1939 96% Silica glass	1963 Photochromic

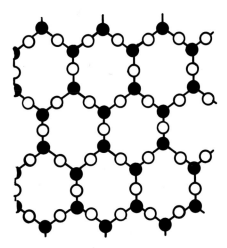

Fig. 14.6 Atoms of crystalline silica are held in place in a regular pattern or lattice. ●—Silicon, ○—Oxygen.

random structure before regular rearrangement can take place (Fig. 14.8). However, if the rate of cooling is slow enough, some crystals will form. The crystalline state is the natural arrangement to which the atoms try to return. But, under the rates of cooling normally encountered in glass manufacture, the atoms are not quick enough and they are trapped in a disordered structure. Growth of crystals in glass is called devitrification. Associated with devitrification is a temperature called the liquidus temperature. Crystals will not form above this temperature, and any that may exist will return to solution. As temperature goes below the liquidus, the tendency to devitrify increases. However, the glass thickens as it cools and this slows devitrification. The resultant of the opposing tendencies of temperature and thickening place the most favorable temperature for devitrification slightly below the liquidus.

Each glass composition has its own characteristic liquidus temperature. In developing a glass composition, it is important that the liquidus be determined so that the glass will not linger near that temperature in any of the temperature cycles employed

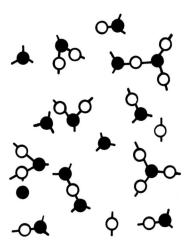

Fig. 14.7 When silica is molten, groups of atoms become detached and are free to move relative to one another.

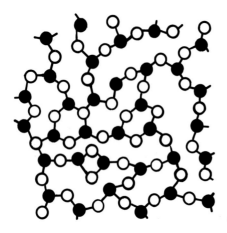

Fig. 14.8 Atoms of silica in the glassy state are frozen in a random, or disordered, manner.

during the manufacturing processes. Commercial glasses are engineered with a high resistance to devitrification.

It may seem strange that glass is a noncrystalline material and fine art glass is called crystal. The designation "crystal" was created many years ago when someone looked at a piece of exceptionally brilliant glass and described it as being "clear as crystal." The term "crystal" refers to external appearance, and the term "crystalline" refers to internal structure. Chemically, glasses are mixtures of oxides. Their compositions can be represented by listing the percentages of the components but not, in most cases, by a chemical formula. Single-oxide glasses are the exception; they can be described by a chemical formula.

The oxides used in glass composition can be divided into three groups. Depending on their function, oxides may be classed as glass formers, intermediates, or modifiers. Silica, or silicon dioxide (SiO_2), is the most frequently used glass former, but the oxides of boron (B_2O_3), germanium (GeO_2), phosphorus (P_2O_5), vanadium (V_2O_5), and arsenic (As_2O_3) also form glasses. The glass formers are those elements whose oxides will form a glassy structure. Aluminum oxide (Al_2O_3), antimony oxide (Sb_2O_3), lead oxide (PbO), and zinc oxide (ZnO) are intermediates. Intermediates may act as formers in some compositions and as fluxes in others.

The modifiers are those oxides that, when added to a glass former, either singly or in combination, will form a mixture having a lower melting temperature than the glass former alone. Modifiers include calcium oxide (CaO), sodium oxide (Na_2O), potassium oxide (K_2O), barium oxide (BaO), and lithium oxide (Li_2O). Because the commonly used alkali oxides tend to produce glasses of low chemical stability or devitrification resistance or both, alkaline earth oxides are usually required as stabilizers. For instance, the combination of silica sand as a glass former and sodium oxide as a flux will produce a glass, but this composition is a rather disappointing glass since it dissolves readily in water. Adding calcium oxide as a stabilizer improves chemical durability and produces a type of glass known as soda-lime glass, the kind used for windows and bottles. Other ingredients may be added to improve manufacturing efficiency or to produce a particular property. Some of the more complicated compositions contain as many as a dozen ingredients (Fig. 14.9).

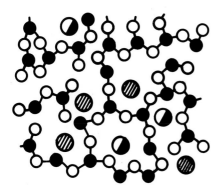

Fig. 14.9 Typical glass contains fluxes and stabilizers to modify properties of glass-forming oxide.

The most frequently used definition of glass is that adopted by the American Society for Testing and Materials:

Glass is an inorganic product of fusion which has cooled to a rigid condition without crystallizing.

B. Types of Glass

1. Soda-Lime Glass

Nearly all commercial glasses fall into one of six basic categories or types. These categories are based on chemical composition. Within each type, except for fused silica, there are several distinct compositions. Soda-lime glass is the most common. This is the glass from which bottles and windows are made. In composition it is similar to the earliest glass, a mixture of the oxides of silicon, calcium, and sodium. Approximately 90% of all glass melted today is soda-lime (or simply lime as it is commonly called). This type is the lowest in cost of all glasses and is readily fabricated into a wide variety of shapes. Resistance to high temperatures and sudden changes of temperature are not good, and resistance to attack by corrosive chemicals is only fair.

2. Lead-Alkali Glass

Lead-alkali glass is somewhat more expensive than soda-lime glass and is favored for electrical applications because of its excellent electrical insulating properties. Thermometer tubing and art glass are also made from lead-alkali glass, which is commonly called lead glass. Lead-alkali glass will not withstand high temperatures or sudden changes in temperature.

3. Borosilicate Glass

Borosilicate glass is the oldest type of glass to have appreciable resistance to thermal shock. It will withstand higher operating temperatures than either soda-lime or lead-alkali glasses and shows markedly superior resistance to chemical attack. Not quite as convenient to fabricate as either lime or lead glass and not as low in cost as lime, borosilicate's cost is moderate when measured against its broad utility. Piping, lamp bulbs, photochromic glasses, sealed beam lamps, laboratory ware, cooking dishes, and dinnerware are examples of borosilicate products (Fig. 14.10).

Fig. 14.10 Excellent resistance to chemical attack and heat shock make borosilicate glass the logical choice for laboratory ware.

4. Aluminosilicate Glass

Aluminosilicate glass, which can be specially heat treated, is another type of heat shock-resisting glass similar to borosilicate in behavior but with the ability to withstand higher operating temperatures than the borosilicates. Compared with the borosilicates, alumino-silicates are appreciably more difficult to fabricate. When coated with an electrically conductive film, aluminosilicate glass is used as resistors for electronic circuitry.

5. Silica Glass

96% Silica Glass is the designation given to a type of glass made by a proprietary process. Resistance to extreme heat shock and to temperatures up to 900°C make this glass the choice for industrial items, such as furnace sight glasses and drying trays. It is also used for outer windows on space vehicles where the glass must withstand the heat of reentry into earth's atmosphere.

6. Fused Silica

Fused silica is the only one of these six categories that consists of a single oxide. This glass consists simply of silica (silicon dioxide) in the noncrystalline or amorphous state. Adding anything to it puts it in another category. Fused silica is the most expensive of all glasses and shows the maximum resistance to heat shock as well as the highest permissable operating temperature (900°C for extended periods; 1200°C for short periods) (Fig. 14.11). Fused silica is clearly superior in a number of respects and is restricted to applications where uncompromising requirements dictate its use, such as mirror blanks for astronomical telescopes, optical waveguides, and crucibles for growing crystals. Fabrication of fused silica is difficult, and the number of available shapes is therefore sharply limited.

The six types of glass can be arranged in three pairs. Soda-lime and lead-alkali are termed "soft" glasses because they soften or fuse at relatively low temperatures.

Borosilicate and aluminosilicate are ordinarily called "hard" glasses because they soften or fuse at relatively high temperatures. 96% Silica Glass and fused silica are the most refractory glasses of all.

The terms "soft" and "hard" as ordinarily applied to glass denote a thermal property and do not have any connection with mechanical hardness. When mechanical hardness is important, the characteristic may be measured by forcing a diamond indenter into the surface of a sample (penetration hardness) or measuring the amount of glass removed by grinding under specified conditions (abrasion hardness).

Fig. 14.11 Vycor brand 96% Silica Glass evaporating dish withstands the shock of ice on the outside and molten bronze at 2000°F on the inside.

The hard glasses are highest in silica content and lowest in thermal expansion rate. The more silica a glass contains, the lower is its thermal expansion and the higher the resistance to heat shock (Fig. 14.12). Soda-lime originated approximately 4000 years ago, lead-alkali came along in 1676, borosilicate in 1912, aluminosilicate in 1936, and 96% Silica Glass in 1939.

7. Special Glasses

In addition to the six basic categories, there are a number of special designations.

Colored glass is produced by adding coloring agents to soda-lime, lead, borosilicate, or 96% Silica Glass. The range of colors is greatest in soda-lime and lead compositions and least in 96% Silica Glass.

Opal glass is white and translucent and may be one of several composition types, some of which are not included in the six basic categories. Opal glass is used extensively for cooking and tableware, because such utensils are less easily soiled, more curable, and more pleasing in appearance, especially when decorated. The translucence of opal glass makes the material well suited for lighting globes and panels where a soft diffused light is required.

Multiform glass, as its name implies, comes in a variety of shapes. It can be of any composition. The distinctive feature about multiform is the process by which it is made. The process consists of melting the glass and then grinding it to a fine powder. The powder is mixed with a liquid to form a thick paste and is then pressed or cast in the desired shape. Firing melts the particles, they flow together, and the end product is a dense glass with properties nearly the same as that of the original material. The multiform process can produce articles with holes or with square corners. Such shapes are not possible by forming directly from fluid glass (Fig. 14.13).

Optical glass must meet exceptionally high standards of purity. It must be free of bubbles, ripples, or other defects that would interfere with the accurate passage or bending of light. Any of the six basic types of glass can be manufactured to optical quality standards. In addition a number of glasses made especially for optical applications differ in composition from all of the six basic categories (Fig. 14.14). Optical waveguides

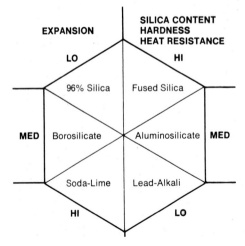

Fig. 14.12 Hard glasses are lower in thermal expansion and higher in both heat resistance and silica content than soft glasses.

Fig. 14.13 The Multiform process lends itself to complex shapes.

Fig. 14.14 Optical glass is manufactured to high standards of purity and quality.

Fig. 14.15 Photochromic glass darkening when exposed to sunlight.

Fig. 14.15 (Continued)

represent a unique combination of extraordinarily pure glass with advanced technology in fiber manufacture; the result is a growing communications industry based on light transmission over long distance.

Photochromic glass darkens when exposed to ultraviolet light and fades when the ultraviolet is removed or when the glass is heated. Some photochromic compositions will remain dark for a week or longer. Others will fade within a few minutes after ultraviolet is removed. This glass was introduced in 1963 (Fig. 14.15).

Photosensitive glass also responds to light but in a different manner from photochromic. When exposed to ultraviolet and then heated, photosensitive glass will change from clear to opal. When exposure is made through a photographic negative, the pattern of the negative is reproduced in the glass. The image thus produced is permanent and will not fade as does photochromic. The exposed glass is much more readily soluble in hydrofluoric acid than the unexposed glass. Immersion in hydrofluoric acid can produce shapes in photosensitive glass by etching away a previously exposed and developed area (Fig. 14.16).

Fibrous glass can be drawn from several different compositions. A glass fiber is nothing more than an extremely thin thread of glass and has much the same properties as the same glass composition has when formed into larger objects. The principal difference in properties is strength; fibers are appreciably stronger than the same glass in more massive forms, partly because fibers are less vulnerable to impact and abrasion. When combined with a plastic resin, the result is a composite material that is tough and strong and finds many uses in furniture, boats, fishing rods, and skis. Mats of glass fibers make

Fig. 14.16 Intricate patterns can be produced in photosensitive glass by exposing to ultraviolet light through a negative followed by developing by heat treatment.

Fig. 14.17 Comparison of a copper telephone cable with the same message-carrying capacity as a single optical wave guide made of high-purity glass.

ideal thermal and acoustical insulation for buildings. Very high purity optical fibers are now being used to transmit light for long distances in optical communication systems (Fig. 14.17).

Foam (or cellular) glass contains many enclosed bubbles and finds many applications as a thermal insulator. Foam glass is made by combining powdered glass with a foaming agent and heating the mixture until the glass melts and the foaming agent gives off gas that forms the bubbles. The resulting material is nearly as light as cork and can be worked with ordinary wood-working tools (Fig. 14.18).

Tempered glass is made stronger than ordinary glass by heat treatment. Soda-lime, lead-alkali, borosilicate, and aluminosilicate compositions can be tempered thermally. Industrial piping, sight glasses, and table ware are some examples of tempered products.

Fig. 14.18 Cellular glass can be shaped to fit closely around pipes and other equipment.

Chemically strengthened glass, introduced in 1962, is at least 10 times as strong as ordinary glass. Any of several chemical treatments may be used. Most chemically strengthened glasses are a soda-aluminosilicate composition. They are used for aircraft windows, pipets, centrifuge tubes, and computer tape reels (Fig. 14.19).

Laminated glass is another method for producing strong products. The general principles were first used by the ancient Chinese. Modern refinements of this technology permit the manufacture of such products as lightweight dinnerware.

C. Glass Properties

The properties of glass can be varied and regulated over an extensive range by modifying the composition, production techniques, or both. We will consider the mechanical, electrical, chemical, optical, and thermal properties separately. Of course, in any glass these properties cannot occur separately. Instead, any glass represents a combination of properties. And, in selecting an individual glass for a particular product, it is this combination that is important. Usually one property cannot be changed without causing a change in other properties. It is the art of the glass scientist to produce the most favorable combination possible, always keeping in mind that the product must be producible in commercial quantities at reasonable cost.

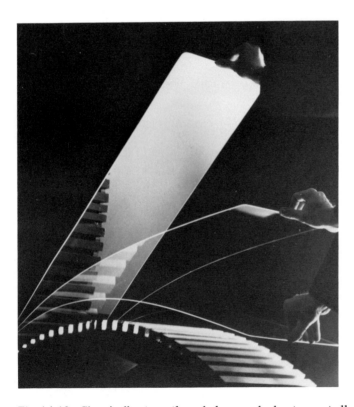

Fig. 14.19 Chemically strengthened glass can be bent repeatedly without damage.

Then it becomes the job of the manufacturing engineer to produce these commercial quantities. The designer has the responsibility of prescribing the shape into which the glass will be fashioned so that utility, cost, and esthetics are each optimized. Thus, the final product represents a combination of efforts and skills as well as of physical and chemical properties.

1. Mechanical Properties

a. Elasticity

Push, pull, or twist a piece of glass hard enough, and it will bend or stretch. Not very much, admittedly, but some bending or stretching is possible. Push gingerly on the middle of a large window pane and feel it bend, or watch the reflections in a large window when the wind is blowing vigorously and you can observe the way the window bends from the force of the wind. Glass is an unusual material in this respect, not because it bends or stretches—most materials do—but because it returns exactly to its original shape when the bending or stretching force is removed. This characteristic of glass classifies it as a perfectly elastic material. If you apply an increasing force, the glass breaks when the force reaches the ultimate strength of the glass. But, at any point short of breakage, the glass will not deform permanently (Fig. 14.20).

To be precise, glass should be classified as a nearly perfect elastic material because under some conditions permanent deformation, or plastic flow, can be produced. For instance, in the indentation hardness determination, the high stress under the point of the indenter produces plastic flow. It is possible in some other cases to observe plastic flow under laboratory conditions. Plastic flow can only be produced by stress almost equal to the ultimate strength of the glass. In a well-designed glass product, stress will be controlled to a level substantially below the ultimate strength and the glass will not deform permanently. Therefore, glass can, for all practical purposes, be considered perfectly elastic.

Mechanical properties deal with the action of forces on a material and the effects that these forces produce within the material. There are three types of force to be considered. A tensile force exerts a pull on the material; a mild tensile force stretches the material; a severe one pulls the material apart. A compressive force acts to squeeze the material. A shear force acts on the material in a manner similar to a pair of shears to slide one part of the material in one direction and another in the opposite direction.

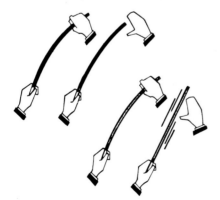

Fig. 14.20 Metal rod bent beyond its elastic limit does not recover. Glass rod shows perfect recovery.

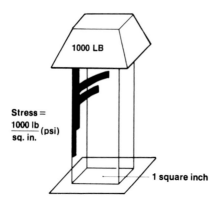

Fig. 14.21   Stress is the amount of force per unit of area.

b.   Stress and Strain

Tensile forces are the most important in glass because they give rise to tensile stresses within the glass, and glass only breaks from tensile stress. Force is measured in pounds or kilograms. Stress is expressed in pounds per square inch or kilograms per square centimeter (other units may, of course, be used as long as the fundamental nature of stress is expressed: force per unit of area). A force of 1000 lb acting on a rod 1 in.^{2} in cross section will produce a stress of 1000 lb/in.^{2} (Fig. 14.21).

The terms "stress" and "strain" are sometimes confused. In fact they are used interchangeably at times as if both represent the same phenomenon. Stress produces strain, and strain cannot exist without stress. However, the two terms represent different physical quantities (Fig. 14.22).

Consider again a force of 1000 lb acting on a rod 1 in.^{2} in cross section. This force will deform the rod—stretch, shorten, or twist the rod, depending on how the force is applied. If this 1000 lb force acts to compress a rod of length L, the rod will be shortened by a small amount, called X. The strain produced is equal to X divided by L, the length of the bar before the force was applied. Strain is a pure number, a ratio not expressed in any units. A glass bar 1 in.^{2} in cross section and 10 in. long compressed by a force of 100 lb will be shortened approximately 0.001 in. The strain, therefore, is 0.0001.

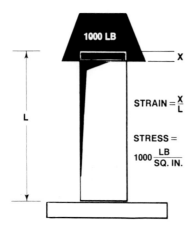

Fig. 14.22   Strain is the proportional deformation produced by the application of stress.

c. Strength

Strength of glass varies widely depending on its method of manufacture and subsequent treatment. Theoretical tensile strength has been estimated to be as high as 4,000,000 lb/in.[2], but commercially produced glass does not even approach this figure. Glass fibers have been measured at 1,000,000 lb/in.[2]

Strength is only slightly affected by composition but is highly dependent on surface condition (Fig. 14.23). Commercially produced glassware can acquire small nicks and scratches in the course of manufacture and later in use. Any applied stress will be concentrated at these points of damage, with the result that the stress at these points will be increased above the amount of the applied stress. The apparent strength of commercial glass rarely exceeds 20,000 lb/in.[2]

When new glass products are being tested for strength, they are usually tested with all surfaces thoroughly abraded so test results will reflect the strength of the product after it has experienced the worst treatment it is likely to encounter in use. Also, when checking out a new product, the test level has to be adjusted to take into account the time-load effect in glass. The strength of glass varies depending on how long a time stress is applied. If a glass item must withstand a load for 1000 hr or more, this load can be only approximately one-third the load that the same item could withstand for 1 sec. After 1000 hr the strength of the glass does not decrease further (Fig. 14.24).

Strength is measured in the laboratory by applying a bending load to a bar. This stretches the lower surface of the bar so it is in tension and squeezes the top surface of the bar so it is in tension and squeezes the top surface so it is in compression. The load is increased until the bar breaks (Fig. 14.25).

The break originates in the lower surface since glass always fails from tension. Note that when there is a serious flaw inside a glass object, it is possible for a break to originate in the interior of the glass. However, in any commercially acceptable glass product, such serious internal flaws would be eliminated during inspection. Therefore, to produce breakage in glass, tension at a surface is necessary (Fig. 14.26).

Glass never disintegrates or explodes; instead, a crack is started at a specific point and progresses to failure. After separating, distinctive contours on the fracture surface record the point of origin, the direction of crack propagation, and other factors present

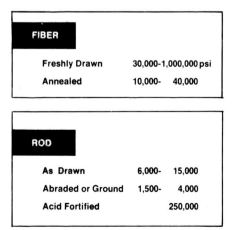

FIBER	
Freshly Drawn	30,000-1,000,000 psi
Annealed	10,000- 40,000

ROD		
As Drawn	6,000-	15,000
Abraded or Ground	1,500-	4,000
Acid Fortified		250,000

Fig. 14.23 Virgin surface is strongest (freshly drawn, fire polished or acid polished). Mold finished surface is somewhat weaker. Abraded or ground surface is weakest.

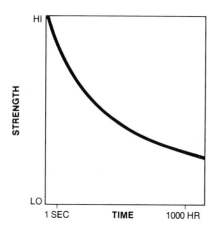

Fig. 14.24 The long-time strength of glass (1000 hr or more) is approximately one-third as much as its short-time strength (one second or less).

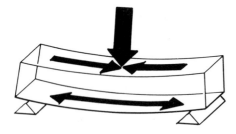

Fig. 14.25 Bending load produces tension in the lower surface of test bar and compression in the upper surface.

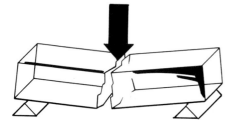

Fig. 14.26 The bar breaks when the tension on the lower surface exceeds the ultimate strength of the glass.

during crack initiation and fracture. The reconstruction of the failure events from these features is known as fractography.

d. Strengthening Glass

What happens when we prestress glass, that is, put all its surfaces under compression? When prestress is applied there will be an equal amount of tension somewhere to maintain the balance. This tension is buried in the middle of the glass, where it is safe from damage, protected by the outer skin of compressed glass (Fig. 14.27). Now, when a bending load is applied to this prestressed bar, the load must first overcome the built-in compression before it can put the lower surface in tension. The result is that the strength of the lower surface is approximately equal to the sum of its strength when stress free plus the amount of built-in compression (Fig. 14.28).

Prestressing is often done by *physical tempering.* In this heat-treating process the glass is heated to a temperature where it is nearly ready to sag out of shape under its own weight. Then the glass is chilled suddenly.

The surfaces harden and start to cool while the interior is still fluid. Then the interior hardens. But, at this point, the surfaces are cooler than the interior. As cooling occurs, the glass shrinks. Since the interior is hotter, it must shrink more to reach room temperature. The surfaces are pulled together by the extra shrinkage of the interior and put into compression (Fig. 14.29).

Glass can also be strengthened by *chemical means.* There are several such methods. One of the most commonly used requires an exchange of ions in the glass surface. The glass is immersed in a molten salt bath, and large ions migrate into the surface, replacing smaller ions. This action crowds the surface and produces compression (Fig. 14.30).

Laminate strengthening is another process. In this method the product is made of a "sandwich" of glasses. The inner glass shrinks more on cooling, causing the outer layer to be put into compression.

Thermal tempering (sometimes called physical tempering or chill tempering to distinguish it from chemical tempering or strengthening) can produce surface compression up to 30,000 lb/in.^{2} under ideal conditions. Chemical strengthening can produce stress well over 100,000 lb/in.^{2} Tempering provides increased mechanical strength in gage glasses, piping, and pressure- and explosion-resistant tubes and globes.

e. Measuring Hardness

Mechanical hardness can be measured by three methods: scratch, penetration, and abrasion. Mechanical hardness of glass is not a fundamental property but a complex phenomenon that is not completely understood. When a glass product is required to withstand wear or abrasion it is best to evaluate it under actual operating conditions rather than to rely on laboratory hardness tests.

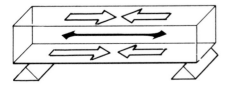

Fig. 14.27 Prestressed bar shows compression in all surfaces. Reactive tension is buried within the bar.

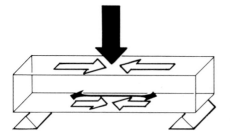

Fig. 14.28 Bending load applied to pre-stressed bar must first overcome built-in compression before surface can be put in tension.

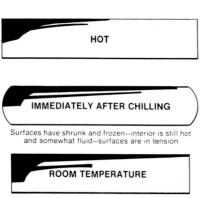

Fig. 14.29 Prestressing by physical tempering consists of heating glass until it begins to soften. Then an abrupt chill shrinks and freezes surfaces. When interior of glass cools and shrinks, surfaces are compressed.

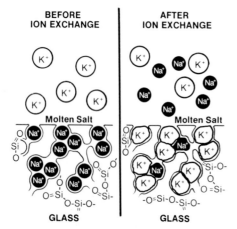

Fig. 14.30 Chemical strengthening by ion exchange causes relatively large potassium ions (K+) to replace smaller sodium ions (Na+), thus crowding surface and producing compression.

2. Electrical Properties

a. Resistivity

Electronic tubes, x-ray tubes, light bulbs, and many other electrical products are made from glass. One reason glass is chosen for these products is because of its excellent *electrical insulating ability*. Glass, like other good insulating materials, provides a high resistance to the passage of electricity. This property is called *volume electrical resistivity* when it measures the resistance to flow of electricity throughout the body of the glass, and *surface electrical resistivity* when it measures the resistance of flow along the surface.

The volume resistivity of glass is approximately 10^{18} times that of copper. When an electrical voltage is applied to a copper bar and a glass bar, 10^{18} as much electricity or electrical current flows through the copper bar as through the glass bar. Volume resistivity is measured in units of ohm-centimeters. Conductivity, or the ability to conduct electricity, is the reciprocal of resistivity (Fig. 14.31).

Surface resistivity is measured by placing two electrodes in intimate contact with the surface of a glass sample. These electrodes are arranged so that the distance between them is equal to the length of each electrode. In this position the electrodes form two opposite sides of a square. Units for surface resistivity are therefore called ohms per square. You may occasionally hear this measurement stated as ohms per square inch or per square centimeter, but the correct unit is simply ohms per square, the square being the one outlined by the electrodes. It can be a square inch or a square mile; the reading will be the same (Fig. 14.32).

b. Dielectric Constant

Dielectric constant is a measure of the ability of a glass to store electricity. If this constant is high the glass will store more than if the constant is low. Dielectric constant is frequently called K; glass having a high dielectric constant is called a high-K glass (Fig. 14.33).

Dielectric constant is a critical property when glass is used for such applications as capacitors. Capacitors are storehouses for electricity, and a high-K glass allows more electricity to be stored in a smaller space.

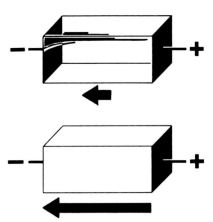

Fig. 14.31 Glass bar shows resistance to passage of electrical current approximately 10^{18} times as great as resistance of copper bar.

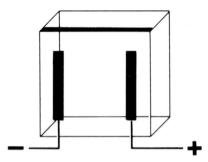

Fig. 14.32 Flow of electrical current along surface of glass is measured between two equal length electrodes that form two opposite sides of a square.

c. Dielectric Loss

Dielectric loss may be thought of as the degree of opacity to an electrical wave. A high-loss glass is relatively opaque; it will absorb a sizable portion of an incident electrical wave. A low-loss glass is relatively transparent.

The units in which dielectric loss is expressed can quickly become confusing. Loss can be labeled dissipation factor, power factor, tangent of loss angle, loss tangent, or sine of loss angle. When loss is low, these quantities turn out to be nearly the same value and can be interchanged. Most glasses are sufficiently low in loss to permit this interchange.

d. Dielectric Strength

Multiply the loss tangent by the dielectric constant and the product is called the loss factor. Dielectric strength is an indication of the amount of voltage a material will stand before it breaks down or punctures. To measure dielectric strength, two electrodes are placed on opposite sides of a sample and aligned with each other. Then, voltage is applied to the electrodes and increased until the sample breaks down.

Glass has such a high dielectric strength that it is difficult to measure this property. Various techniques have been tried, including testing under oil and using special sample shapes designed to prevent flashover around the sides of the sample. Values of dielectric strength are expressed in volts per centimeter or volts per inch.

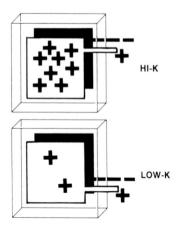

HI-K

LOW-K

Fig. 14.33 Dielectric constant is a measure of the amount of electricity that a glass can store.

Of all the electrical properties, dielectric strength is the most difficult to measure and the results of measurements can be interpreted with the least certainty. When this property is important, it is best to measure it under conditions that approach actual service conditions as closely as possible.

3. Optical Properties

When a beam of light falls on a piece of glass, some of the light is reflected from the glass surface, some of the light passes through the glass, and some is absorbed in the glass. The measure of the proportion of light reflected from the surface is the *reflectance*. The measure of the proportion absorbed is the *absorptance*. The measure of the proportion transmitted is the *transmittance*. Each quantity is ordinarily expressed as a fraction of the total quantity of light in the beam. If the intensity of the beam is represented by the numeral 1, reflectance by R, absorptance by A, and transmittance by T, intensity may be expressed as

$$R + A + T = 1$$

a. Transmission

Optical properties are concerned with the behavior of glass toward light, the visible spectrum that extends like a rainbow from blue on one end to red on the other. However, as the term is usually employed, optical refers also to behavior toward the infrared and ultraviolet regions of the spectrum. The infrared region lies next to the red end of the visible spectrum, and the ultraviolet is on the opposite end of the visible region next to the blue. The chief physical difference between these bands of the energy spectrum is in the wavelength. Ultraviolet waves are shorter than visible waves, and visible waves are shorter than infrared. All of them are so short that we employ extremely small units in measuring them. Wavelengths are expressed in micrometers (microns), nanometers, and angstroms (0.1 nanometer) (Fig. 14.34).

Most glass is transparent, or, to be more accurate, partially transparent. Complete transparency would imply no reflection and no absorption. No glass achieves this uncompromised state, but most glass transmits most of the light that strikes it. For this reason it is easy and convenient to classify glass loosely as a transparent material.

A number of glasses are *selectively transparent*. They transmit light of one wavelength or color more efficiently than any other. A green traffic light is an example. The lamp behind the green lens supplies white light, or light that contains some light of all colors. The green lens absorbs all colors except green, and green is all we see coming through the lens. This selectivity carries over into the ultraviolet and infrared regions. A number of special-purpose glass compositions have been devised to transmit either ultraviolet or infrared while absorbing visible light. These glasses are black in appearance. Also, some glasses are designed to absorb infrared and transmit visible—the heat-absorbing filters such as those found in every film projector. The purpose of these filters is to get as much light as possible on the screen while keeping the slide or film as cool as possible.

b. Refraction

The bending of light when it passes through glass (or refraction) is the phenomenon that makes lenses possible. In a lens all the rays that pass through the glass are refracted

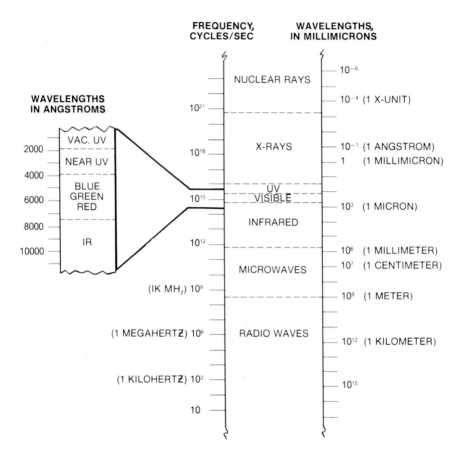

Fig. 14.34 The electromagnetic spectrum. Visible spectrum is bounded on one side by infrared region and on the other by ultraviolet region.

by the lens and brought to focus at a single point. A measure of the amount of bending is the *refractive index*. The higher is the refractive index, the greater the bending.

Not all colors of light are refracted the same. Blue light bends more sharply than red light in the same glass, and the intermediate shades (green and yellow) take a middle course. This difference results in some *dispersion* and prevents all rays that go through a lens from focusing at exactly the same point. The measure of this difference in refractive index is the *dispersion coefficient*.

c. Optical Glass Compositions

Approximately 130 glass compositions are produced for optical uses alone. These glasses comprise a range of refractive indices, dispersion coefficients, and combinations of these two properties to provide the lens designer with a wide array of materials. A simple hand magnifier or a pair of eyeglasses may employ only single-element lenses, but it is not unusual for a sophisticated camera lens to employ as many as seven elements. Each element may be made of a different glass. Reflectance from a glass surface can be regulated by coatings applied to the surface. A metallic coating will produce the maximum reflectance, such as a front-surfaced mirror. Other coatings show selective reflectance, such as the heat-shielding glass that reflects a high proportion of infrared but transmits a high proportion of visible light. Still other coatings eliminate reflectance almost completely; such nonreflective coatings are commonly used on lenses.

4. Thermal Properties

a. Specific Heat

Specific heat is a measure of the amount of heat required to raise a unit mass of a material one degree in temperature compared to that required for an equal mass of water. Specific heats of glasses range from 0.15 to 0.33; values increase with rising temperature. If the value is stated as mean specific heat, the measurement is the average taken over some interval, such as 0-500°C, or the average between the stated temperature and 0°C. If the measurement is stated as true specific heat, the value applies only to the stated temperature (Fig. 14.35).

b. Conductivity

Thermal conductivity is a measure of the ability to conduct heat through the body of that material. In an opaque material thermal conductivity can be measured straightforwardly. In a transparent material, such as glass, any measurement of conductivity becomes complicated by the fact that glass also conducts heat internally by radiation. Values for conductivity rise with rising temperature, and above 400°C the contribution from radiative transfer is appreciable. Radiative transfer depends on the thickness of the sample. Because of this complication, published figures for glass thermal conductivity show considerable disagreement, and because of this uncertainty, thermal conductivity in glass is not used extensively as a design parameter. Other factors are usually more important.

Thermal conductivity is stated in the English system in BTU (British Thermal Units) times inches of thickness per hour per square foot of area per degree Fahrenheit. Room temperature values for glass lie between 4.23 and 10.10, well below the corre-

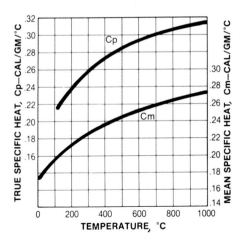

Fig. 14.35 Specific heat of glass increases with rising temperature.

sponding values for metals. However, in some applications glass equipment will transfer heat more effectively than metal equipment in spite of the lower thermal conductivity of the glass. Consider a heat exchanger tube with liquid flowing on either side of the tube wall. The resistance to heat transfer from one liquid to another is the sum of the resistance of the stagnant films on each side of the tube wall, plus the resistance of any scaling or fouling on the wall, plus the resistance of the wall itself. The resistance of the wall is usually smaller than the film resistance and the scale resistance. Therefore a metal tube may be no more efficient than a glass tube, especially if the glass remains free of scale and fouling.

c. Diffusivity

Thermal diffusivity is a measure of the rate at which heat energy diffuses or spreads throughout a material and is particularly important in situations involving transient heat flow. Thermal diffusivity is related to conductivity. Where conductivity measures the propagation of heat energy within a material under steady-state or equilibrium conditions, diffusivity measures propagation under unbalanced or transient conditions. Units for diffusivity are square centimeters per second.

When one end of a rod is heated, the diffusivity of the material is a measure of how rapidly the other end of the rod will heat up. With continued uniform application of heat to one end, the two ends will reach an equilibrium state. In this equilibrium state the conductivity is a measure of how much heat energy flows through the rod, whereas the diffusivity is a measure of how fast the thermal energy from the heated end reaches the unheated end.

d. Radiation

Heat may be transferred by radiation as well as by conduction. Radiant transfer requires no intervening material between the hot material that is radiating the energy and the colder material that is receiving it. The quantity that measures the ability of a material to radiate heat energy is its *emissivity,* which is expressed as a pure number between zero and one. A typical value for glass might be 0.75.

The emissivity of any material is its ability to radiate energy compared to that of a black body. A black body is a theoretical body that absorbs all energy incident on it and reflects none. No material reaches this theoretical state of perfection, but some approach it closely. In an opaque material the relationship between emissivity E and reflectivity R can be expressed as $E + R = 1$. But, in glass, internal reflections enter into the phenomenon, so that the relationship becomes $E + T' + R' = 1$, where E is emissivity, T' is transmittance as modified by internal reflections, and R' is reflectance as modified by internal reflections. The R' is always greater than the reflectivity R, and T' is always less than the transmittance T.

e. Expansion

The rate at which glass expands when it is heated and contracts when cooled is in many applications the most important of the thermal properties. Thermal stress that results from sudden heating or cooling is directly proportional to this rate of thermal expansion.

When two glasses or a glass and a metal are sealed together, the expansion rates must be closely matched to avoid excessive stress. The rate of expansion may be stated as the linear coefficient of thermal expansion or the volume coefficient. The linear coefficient is one-third the volume coefficient and is used almost exclusively. The linear coefficient defines the proportional change in length of a piece of glass per degree change in temperature. Units are inches per inch per degree Celsius, or centimeters per centimeter per degree Celsius. Since the change in length as a proportion of the total length is being measured, the value is the same in either case, but note that *the temperature scale must be specified* since the linear coefficient of expansion for a degree Celsius is $^9/_5$ times that for a degree Fahrenheit. Thus, the most frequently used borosilicate glass has a linear expansion coefficient of 32.5×10^{-7} cm/cm/°C.

Expansion coefficients are ordinarily stated without units or factor of 10. Thus, this glass is commonly referred to as a "32.5 expansion glass," a still more convenient notation (Fig. 14.36). When comparing expansion coefficients of glass and other materials it is well to be sure that the coefficients are stated to the same power of 10. Expansion coefficients for glass are usually stated over the temperature range 0-300°C (Fig. 14.37). However, if one is interested in finding a glass to match a metal to which to seal the glass, the value from room temperature to the setting point of the glass is more important. Up to 300°C glass expands linearly and with no changes in rate, but at higher temperatures the rate of expansion increases. When sealing to a metal the average expansion from room temperature to the setting point influences the stress remaining in the seal at room temperature.

Fused silica has the lowest expansion coefficient of all glasses; soda-lime and lead-alkali glasses have the highest. Higher expansions than those shown here are possible. One glass formulated to seal to copper has an expansion coefficient of 158×10^{-7} °C^{-1}.

5. Chemical Properties

Glass is much more resistant to corrosion than most materials, so much so that it is easy to think of it as corrosion proof. Glass windows after several years' exposure to the elements remain clear and apparently unaffected. Glass bottles hold a wide range of liquids that would dissolve other materials. In the laboratory, reactions are carried out in glass beakers and flasks without damage to the beakers or contamination of the reacting solutions. But, in spite of these indications that glass is indestructible by chemical attack, under certain conditions it will corrode, even dissolve. In these cases, it is important to choose the right type of glass, since some are more corrosion resistant than others.

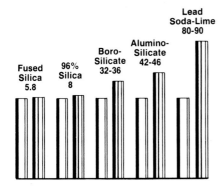

Fig. 14.36 Right-hand bar of each pair indicates relative thermal expansion rates of different glass types.

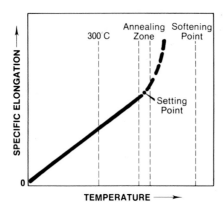

Fig. 14.37 Expansion coefficient is the slope of the curve obtained by plotting specific elongation against temperature.

Only a few chemicals attack glass—hydrofluoric acid, concentrated phosphoric acid when hot or when it contains fluorides, hot concentrated alkali solutions, and super-heated water. Hydrofluoric acid is the most powerful of this group; it attacks any type of silicate glass. Other acids attack glass only slightly; the degree of attack can be measured in laboratory tests, but such corrosion is rarely significant in service for acids other than hydrofluoric and phosphoric.

Acids and alkali solutions attack glass in different ways. Alkalis attack the silica directly; acids attack the alkali in the glass composition. When an alkali solution attacks a glass surface, the surface simply dissolves. This process continuously exposes a fresh surface, which in turn is dissolved. As long as the supply of alkali is sufficient this type of corrosion proceeds at the same rate (Fig. 14.38).

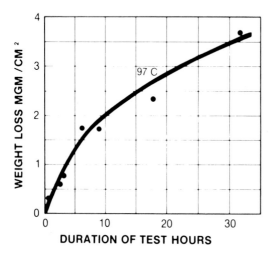

Fig. 14.38 Solubility of a glass of low durability in 5% hydrochloric acid versus time. Acid corrosion rate slows when surface is penetrated.

Acid corrosion behaves quite differently. By dissolving the alkali in the glass composition, a porous surface is left that consists of the silica network with holes where the alkali has been removed by the acid. This porous surface slows the rate of attack since the acid must penetrate this surface layer to find alkali to dissolve (Fig. 14.39).

Corrosion by water is similar to acid corrosion in that alkali is removed from the glass surface. However, water corrosion occurs at a much slower rate. At high temperatures, however, water corrosion can become significant. Gage glasses for steam boilers are a case in point. These products must be protected from the superheated water by a sheet of mica or replaced on a schedule that ensures that they will not be seriously weakened.

Many laboratory tests have been devised for testing corrosion resistance. Some of them aim at accelerating the rate of corrosion by employing high temperatures or by grinding the glass to a specified grain size to expose more surface area to the corroding solution. After treatment for the specified time and at the specified temperature, the weight lost by the glass can be measured or the amount of alkali extracted can be determined.

Many factors influence the rate of corrosion, and no laboratory test to date is capable of predicting service behavior under all conditions. Concentration and rate of agitation of the corroding solution are important factors. As corrosion progresses, the test solution becomes contaminated with components extracted from the glass; this contamination may speed or slow the corrosion rate. Some glass products have a silica-rich skin so the surface will show a different corrosion rate from the interior. A powder test on the glass from such a product will miss this surface effect completely. Conclusions about glasses from accelerated test data will sometimes be reversed by normal service temperature data.

Comparative values of resistance to acids, alkalis, and water may be found in the literature, but these should be employed as guides only. Results must be checked under actual service conditions.

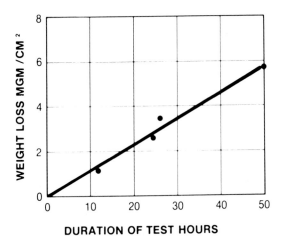

DURATION OF TEST HOURS

Fig. 14.39 Solubility of a soda-lime glass in 5% sodium hydroxide at 95°C versus time. Alkali corrosion proceeds at a constant rate.

Table 14.12 Typical Values for the Properties of Technical Glasses

Glass Code	Type	Color	Principal Use	Forms Usually Available	Corrosion Resistance Class	Weathering	Water	Acid	Thermal Expansion 0-300°C	25°C to Setting Point	Annealed Normal Service °C	Annealed Extreme Service °C	Tempered Normal Service °C	Tempered Extreme Service °C
0010	Potash Soda Lead	Clear	Lamp Tubing	T	I	2	2	2	93.5	101	110	380	—	—
0080	Soda Lime	Clear	Lamp Bulbs	BMT	I	3	2	2	93.5	105	110	460	220	250
0120	Potash Soda Lead	Clear	Lamp Tubing	TM	I	2	2	2	89.5	97	110	380	—	—
0330	Glass-Ceramic	Gray	Bench Tops	RS	I	—	1	3	9.7	—	538		—	—
1720[4]	Aluminosilicate	Clear	Ignition Tube	BT	I	1	1	3	42	52	200	650	400	450
1723	Aluminosilicate	Clear	Electron Tube	BT	I	1	1	3	46	54	200	650	400	450
1990	Potash Soda Lead	Clear	Iron Sealing	—	II	3	3	4	124	136	100	310	—	—
2405	Borosilicate	Red	General	BPU	I	—	—	—	43	53	200	480	—	—
2473	Soda Zinc	Red	Lamp Bulbs	B	I	2	2	2	91	—	110	460	—	—
3320	Borosilicate	Canary	Tungsten Sealing	—	I	[3]1	[3]1	[3]2	40	43	200	480	—	—
6720	Soda Zinc	Opal	General	P	I	2–	1	2	78.5	90	110	480	220	275
6750	Soda Barium	Opal	Lighting Ware	BPR	I	2–	2	2	88	—	110	420	220	220
7040	Borosilicate	Clear	Kovar Sealing	BT	II	[3]3	[3]3	[3]4	47.5	54	200	430	—	—
7050	Borosilicate	Clear	Series Sealing	T	II	[3]3	[3]3	[3]4	46	51	200	440	235	235
7052	Borosilicate	Clear	Kovar Sealing	BMPT	II	[3]2	[3]2	[3]4	46	53	200	420	210	210
7056	Borosilicate	Clear	Kovar Sealing	BTP	II	2	2	4	51.5	56	200	460	—	—
7070	Borosilicate	Clear	Low Loss Electrical	BMPT	I	[3]2	[3]2	[3]2	32	39	230	430	230	230
7251	Borosilicate	Clear	Sealed Beam Lamps	P	I	[3]1	[3]2	[3]2	36.7	38.1	230	460	260	260
7570	High Lead	Clear	Solder Sealing	—	II	1	1	4	84	92	100	300	—	—
7720	Borosilicate	Clear	Tungsten Sealing	BPT	I	[3]2	[3]2	[3]2	36	43	230	460	260	260
7740	Borosilicate	Clear	General	BPSTU	I	[3]1	[3]1	[3]1	32.5	35	230	490	260	290
7760[5]	Borosilicate	Clear	General	BP	I	2	2	2	34	37	230	450	250	250
7800	Soda Barium Borosilicate	Clear	Pharmaceutical	T		1	1	1	50	53	200	460	—	—
7900[1]	96% Silica	Clear	High Temp.	BPTUM	I	1	1	1	8	5 *	800	1100	—	—
7913[1]	96% Silica	Clear	High Temp.	BPRST	I	1	1	1	7.5	5.5*	900	1200	—	—
7940	Fused Silica	Clear	Optical	U	I	1	1	1	5.5	3.5*	900	1100	—	—
7971	Titanium Silicate	Clear	Optical	U	–	1	1	1	0.5	−2	800	1100	—	—
8160	Potash Soda Lead	Clear	Electron Tubes	PT	II	2	2	3	91	100	100	380	—	—
8161	Potash Lead	Clear	Electron Tubes	PT	I	2	1	4	90	99	100	390	—	—
9606	Glass-Ceramic	White	Missile Nose Cones	C	II	—	1	4	57	—	700	—	—	—
9608	Glass-Ceramic	White	Cooking Ware	BP	I	—	1	2	4-20	—	700	800	—	—
9741	Borosilicate	Clear	u v Transmission	BUT	II	[3]3	[3]3	[3]4	39.5	50	200	390	—	—

COLUMN 1
[1]Glasses 7905, 7910, 7911, 7912, 7913 and 7917 for special ultraviolet and infrared applications.
[4]Glass 1720 is available with improved ultraviolet transmittance (designated glass 9730).
[5]Glass 7760 also available with special transmission suitable for sun lamps.

COLUMN 5
B—Blown Ware P—Pressed Ware S—Plate Glass
M—Multiform R—Rolled Sheet T—Tubing and Rod
U—Panels C—Castings

COLUMN 6
[3]Since weathering is determined primarily by clouding

All data subject to normal manufacturing variations

which changes transmission, a rating for the opal glasses is omitted.
[7]These borosilicate glasses may rate differently if subjected to excessive heat treatment.

COLUMN 7
*Extrapolated values.
Code 9608 may be produced in a range of expansion values depending upon intended application.

COLUMN 8
Normal Service: No breakage from excessive thermal shock is assumed.
Extreme Limits: Glass will be very vulnerable to thermal shock. Recommendations in this range are based on

Source: Courtesy of Corning Glass Works.

6. High-Energy-Radiation Effects

High energy radiations include electromagnetic (x-ray and γ) and particulate (α particle, proton, neutron, and electron) radiation. All these will produce discoloration in glass, and some will cause physical damage. Some glasses will fluoresce under ultraviolet light after being exposed to high-energy radiation. Cathode rays (focused electron beam) can cause electrical breakdown. This effect is most noticeable in glasses having high electrical resistivity. Also, spalling may result from cathode radiation. Neutron radiation shows a more pronounced effect on glass structure than other types. Both density and refractive index

	9			10		11			12	13	14		15			16			17	18
Thermal Shock Resistance Plates 15 x 15 cm Annealed			Thermal Stress Resistance °C	Viscosity Data				Knoop Hardness KHN$_{100}$	Density g/cm³	Young's Modulus Multiply By 10³ Kg/mm²	Poisson's Ratio	Log$_{10}$ of Volume Resistivity ohm-cm			Dielectric Properties at 1 MHz, 20°C			Refractive Index	Glass Code
3.2 mm Thick °C	6.4 mm Thick °C	12.7 mm Thick °C		Strain Point °C	Annealing Point °C	Softening Point °C	Working Point °C					25°C	250°C	350°C	Power Factor %	Dielectric Constant	Loss Factor %		
65	50	35	19	392	432	626	983	363	2.86	6.3	.21	17.+	8.9	7.0	.16	6.7	1.	1.539	0010
65	50	35	16	473	514	696	1005	465	2.47	7.1	.22	12.4	6.4	5.1	.9	7.2	6.5	1.512	0080
65	50	35	20	395	435	630	985	382	3.05	6.0	.22	17.+	10.1	8.0	.12	6.7	.8	1.560	0120
—	—	—	178	—	—	—	—	522	2.54	8.8	.26	—	—	—	—	—	—	—	0330
135	115	75	28	667	712	915	1202	513	2.52	8.9	.24	17.+	11.4	9.5	.38	7.2	2.7	1.530	1720
125	100	70	26	665	710	908	1168	514	2.64	8.8	.24	17.+	13.5	11.3	.16	6.3	1.0	1.547	1723
45	35	25	14	340	370	500	756	—	3.50	5.9	.25	17.+	10.1	7.7	.04	8.3	.33	—	1990
135	115	75	37	501	537	765	1083	—	2.48	6.9	.21	—	—	—	—	—	—	1.507	2405
65	50	35	19	466	509	697	—	—	2.65	6.7	.22	—	—	—	—	—	—	1.52	2473
145	110	80	43	493	540	780	1171	—	2.27	6.6	.19	—	8.6	7.1	.30	4.9	1.5	1.481	3320
70	60	40	20	505	540	780	1023	—	2.58	7.1	.21	—	—	—	—	—	—	1.507	6720
65	50	35	18	447	485	676	1040	—	2.59	—	—	—	—	—	—	—	—	1.513	6750
—	—	—	37	449	490	702	1080	—	2.24	6.0	.23	—	9.6	7.8	.20	4.8	1.0	1.480	7040
125	100	70	39	461	501	703	1027	—	2.24	6.1	.22	16.	8.8	7.2	.33	4.9	1.6	1.479	7050
125	100	70	41	436	480	712	1128	375	2.27	5.8	.22	17.	9.2	7.4	.26	4.9	1.3	1.484	7052
—	—	—	33	472	512	718	1058	—	2.29	6.5	.21	—	10.2	8.3	.27	5.7	1.5	1.487	7056
180	150	100	66	456	496	—	1068	—	2.13	5.2	.22	17.+	11.2	9.1	.06	4.1	.25	1.469	7070
160	130	90	48	500	544	780	1167	—	2.25	6.5	.19	18.	8.1	6.6	.45	4.85	2.18	1.476	7251
—	—	—	21	342	363	440	558	—	5.42	5.6	.28	17.+	10.6	8.7	.22	15.	3.3	1.86	7570
160	130	90	49	484	523	755	1146	—	2.35	6.4	.20	16.	8.8	7.2	.27	4.7	1.3	1.487	7720
160	130	90	54	510	560	821	1252	418	2.23	6.4	.20	15.	8.1	6.6	.50	4.6	2.6	1.474	7740
160	130	90	52	478	523	780	1198	442	2.24	6.3	.20	17.	9.4	7.7	.18	4.5	.79	1.473	7760
—	—	—	33	533	576	795	1189	—	2.36	—	—	—	7.0	5.7				1.491	7800
—	—	—	207	820	910	1500	—	463	2.18	6.9	.19	17.	9.7	8.1	.05	3.8	.19	1.458	7900
—	—	—	220	890	1020	1530	—	487	2.18	6.9	.19	17.+	9.7	8.1	.04	3.8	.15	1.458	7913
—	—	—	286	956	1084	1580	—	489	2.20	7.4	.16	17.+	11.8	10.2	.001	3.8	.0038	1.459	7940
—	—	—	3370	—	1000	1500	—	—	2.21	6.9	.17	20.3	12.2	10.1	⁶<.002	⁶4.0	⁶<.008	1.484	7971
65	50	35	18	397	438	632	973	—	2.98	—	—	17.+	10.6	8.4	.09	7.0	.63	1.553	8160
—	—	—	22	400	435	600	862	—	3.99	5.5	.24	17.+	12.0	9.9	.06	8.3	.50	1.659	8161
200	170	130	16	—	—	—	—	657	2.6	12	.24	16.7	10.0	8.7	.30	5.6	1.7	—	9606
—	—	—	—	—	—	—	—	593	2.5	8.8	.25	13.4	8.1	6.8	.34	6.9	2.3	—	9608
150	120	80	54	408	450	705	1161	—	2.16	5.0	.23	17.+	9.4	7.6	.32	4.7	1.5	1.468	9741

mechanical stability considerations only. Tests should be made before adopting final designs. These data approximate only.

COLUMN 9
These data approximate only
Based on plunging sample into cold water after oven heating. Resistance of 100°C (212°F) means no breakage if heated to 110°C (230°F) and plunged into water at 10°C (50°F). Tempered samples have over twice the resistance of annealed glass.

COLUMN 10
Resistance in °C (°F) is the temperature differential between the two surfaces of a tube or a constrained plate that will cause a tensile stress of 0.7 kg/mm² (1000 psi) on the cooler surface.

COLUMN 11
These data subject to normal manufacturing variations.

COLUMN 12
Determined by revised ASTM standard: number of standard not yet assigned.

COLUMN 15
⁶at 10 kHz

COLUMN 17
Refractive index may be at either the sodium yellow line (589.3 nm) or the helium yellow line (587.6 nm). Values at these wavelengths do not vary in the first three places beyond the decimal point.

will change under neutron bombardment. Table 14.12 shows the properties of some of the nearly 700 different glasses produced by the Corning Glass Works.

D. Manufacturing Processes

1. Mixing and Melting

a. Raw Materials

The raw materials for glass are many; nearly every known element has been used at one time or another in the manufacture of glass. Silica sand is the principal ingredient and may be combined with one, two, or many other ingredients. The properties desired in the finished glass dictate the ingredients and the proportion of each.

Since glass is a mixture of oxides, raw materials are chosen that will provide the necessary oxides. In some cases these raw materials are not oxides initially but they react during the melting process to yield oxides. The majority of commercial glasses are composed of the oxides of silicon, boron, aluminum, sodium, potassium, calcium, magnesium, lead, barium, and lithium.

Minerals, frequently used as raw materials, require careful chemical analysis to ensure that no harmful impurities are present. Minerals seldom contain only one constituent (see Table 14.13). For instance, feldspar is used as a source of aluminum oxide, but feldspar also contains silica, alkali, and alkaline earth oxides. An accurate analysis of each source, and sometimes each shipment, of feldspar is essential to correctly formulate the glass batch.

b. Mixing

All materials are granulated to nearly uniform particle size. Previously melted glass, called cullet, is added to every batch. This cullet is of the same composition as the glass being formulated. The whole batch, raw materials plus cullet, is then mixed thoroughly to ensure that each constituent is evenly distributed.

c. Melting

The mixed batch is loaded into a melting unit. In most cases this melting unit is a horizontal continuous tank. Here the batch feeds continually into the back of the tank. Temperatures up to $1600°C$ melt the batch and convert it into glass. The newly melted

Table 14.13 Some of the Raw Materials Used in Glass Manufacture

Glass-forming oxide of	Raw material
Aluminum (Al_2O_3)	Feldspar ($K_2O \cdot Al_2O_3 \cdot 5SiO_2$), calcined alumina ($Al_2O_3$)
Barium (BaO)	Barium carbonate ($BaCO_3$)
Boron (B_2O_3)	Anhydrous borax ($Na_2B_4O_7$), Boric acid ($B_2O_3 \cdot 3H_2O$)
Calcium (CaO)	Dolomite ($CaCO_3 \cdot MgCO_3$)
Lead (PbO)	Litharge (PbO), red lead (Pb_3O_4), lead silicate ($PbO \cdot SiO_2$)
Lithium (Li_2O)	Lithium carbonate (Li_2CO_3), lepidotite ($LiKFAl_2Si_3O_{10}$), petalite ($LiAl(Si_2O_5)_2$)
Magnesium (MgO)	Calcined magnesite (MgO)
Potassium (K_2O)	Calcined potash (K_2CO_3), feldspar ($K_2O \cdot Al_2O_3 \cdot 6SiO_2$)
Silicon (SlO_2)	Silica sand (SiO_2)
Sodium (Na_2O)	Soda ash (Na_2CO_3), sodium nitrate ($NaNO_3$)

glass flows slowly toward the front, or working end, of the tank. The midsection of the tank is a refining section where the glass is held at a high temperature to improve its quality by removing streaks and bubbles. Vertical continuous tanks are sometimes used. In this instance, the batch is loaded on top and melts as it moves downward. At the working end the glass is cooled to a temperature at which its viscosity is suitable for forming. A continuous tank may produce as much as 700,000 lb glass in 24 hr (Fig. 14.40).

2. Forming

When glass moves from the tank to the forming machine, the glass has the appearance of red-hot syrup. As it cools it thickens and finally becomes rigid. All forming processes are adjusted to mold the glass quickly to the desired shape. Blowing, pressing, drawing, and rolling are the traditional forming methods. Centrifugal casting and sagging are employed for a few products.

a. Blowing

Blowing is employed for hollow thin-walled items, such as bottles and flasks. A bubble of molten glass is placed inside a mold. Air pressure within this bubble forces the glass against the sides of the mold. When the glass is rigid, the mold is opened and the item removed. Dimensional control in blown products is moderately good. High-quality surface finish is more easily obtained than with pressed products (Fig. 14.41).

Fig. 14.40 Typical continuous glass-melting furnace (or tank).

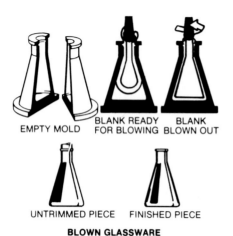

EMPTY MOLD   BLANK READY   BLANK
              FOR BLOWING   BLOWN OUT

UNTRIMMED PIECE   FINISHED PIECE

BLOWN GLASSWARE

Fig. 14.41 Blowing glass forces a bubble of molten glass to take the shape of a mold.

b. Pressing

Pressing provides tighter dimensional control than blowing, and thicker cross sections can be accomplished. A sealed beam lens is a good example of the capabilities of pressing, by which a complicated shape is formed to close dimensional tolerances (Fig. 14.42).

A gob of molten glass is loaded in a mold, and a plunger, which is lowered into the mold, forces the glass to spread and fill the cavity between the mold and the plunger. Any pressed item takes the shape of the mold on the lower, or outside, surface and the shape of the plunger on the upper, or inside surface.

Pressing cannot produce narrow-mouthed shapes, such as bottles. A pressed item must be somewhat larger at the top than the bottom; otherwise the plunger could not be retracted. Items containing flutes or bands can be pressed by using a split mold that swings open after the plunger is retracted (Fig. 14.43).

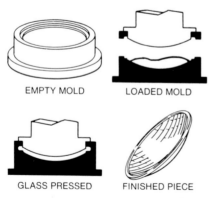

EMPTY MOLD   LOADED MOLD

GLASS PRESSED   FINISHED PIECE

PRESSED GLASSWARE (BLOCK MOLD)

Fig. 14.42 Pressing glass forces a gob of molten glass to spread and fill the space between mold and plunger.

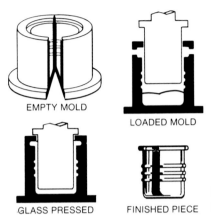

EMPTY MOLD

LOADED MOLD

GLASS PRESSED   FINISHED PIECE

PRESSED GLASSWARE (SPLIT MOLD)

Fig. 14.43 When pressed items contain bulges, the mold is hinged so it can open to allow removal of the ware.

c. Drawing

Drawing produces tubing, rod, and sheet. To form tubing, molten glass is drawn over a mandrel, a hollow cylinder or a cone. Air blown through the cylinder keeps the tubing from collapsing until it becomes rigid. Tubing is drawn horizontally in the Danner machine, on a downward slant in the Vello machine, and straight up in the updraw machine. To make rod instead of tubing, the air pressure inside the mandrel is simply turned off (Fig. 14.44).

d. Flat and Plate Glass

Some flat glass is rolled. A rolling machine resembles a clothes wringer except that the rolls are made of an alloy capable of withstanding the temperature of molten glass. Rolling is a more rapid process than drawing, but the glass produced has a rougher surface. If the product is to be made into plate glass, both surfaces must be ground so they are

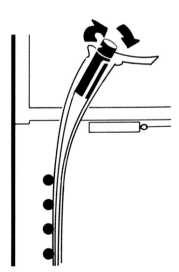

Fig. 14.44 Molten glass is wrapped around a mandrel then drawn out to form tubing.

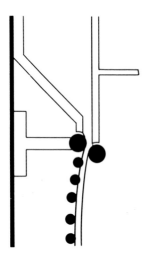

Fig. 14.45 In a rolling machine, rollers squeeze molten glass and form a sheet.

parallel and then polished. Sometimes the rollers are embossed and produce a pattern in the glass (Fig. 14.45).

In the float process, sheet glass is formed by a special process and floated on a bath of molten tin held at a high temperature for a long enough period to produce a smooth fire-polished surface. Float glass does not require mechanical grinding and polishing but is close to ground and polished plate glass in surface quality (Fig. 14.46).

e. Casting

Centrifugal casting is employed for some special shapes. Molten glass flows into the mold while the mold rotates. Centrifugal force pushes the glass up the sides of the mold, where it hardens. Nose cones for missiles and the conical sections of television picture tubes are manufactured in this manner. This method is sometimes called spinning. However, centrifugal casting should not be confused with spinning as used in metal forming, which is an entirely different process (Fig. 14.47).

3. Annealing

A newly formed piece of glass must be cooled gradually by a process called annealing to ensure that the glass will not contain excessive amounts of strain. If a piece of glass

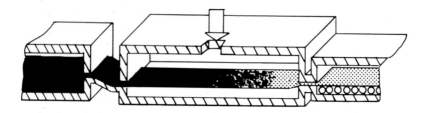

Fig. 14.46 Diagram of the float process for forming sheet glass using an inert atmosphere over a bath of molten tin.

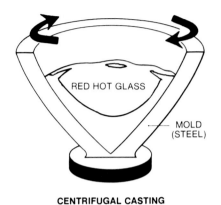

CENTRIFUGAL CASTING

RED HOT GLASS

MOLD (STEEL)

Fig. 14.47 Centrifugal casting employs centrifugal force to distribute molten glass evenly along the sides of a mold.

is not annealed, parts of it have cooled and contracted at different rates and the glass is strained when it has cooled to room temperature.

Annealing is usually performed in a continuous oven, or lehr. The ware moves through the lehr on a belt, and temperatures inside the lehr are regulated so that the ware is first raised to a high enough temperature to relieve any strain that may be present. Then the ware is brought down in temperature, slowly at first and then more rapidly. An understanding of glass viscosity is helpful in understanding the annealing process.

Glasses differ from crystalline solids in that glasses do not have distinct melting points. This difference is explained by the fact that the chemical bonds holding the atoms together in a regular crystalline structure are identical. When the crystalline solid is heated, all the bonds break at exactly the same temperature. Below this temperature, called the melting point, the material is solid; above the melting point the material is a liquid.

In contrast, the bonds in a glass show a range of strengths. When a glass is heated, these bonds break over a range of temperatures. As a result a glass softens gradually as it heats and a distinct melting point cannot be defined. Because of this lack of a specific melting point it is convenient to describe the behavior of glass with temperature in terms of viscosity. Viscosity is the resistance of a liquid to flow. The greater the viscosity, the greater the resistance to flow. Molasses, for instance, is a high-viscosity liquid, water has a low viscosity. The unit of viscosity is the poise. Water at room temperature has a viscosity of 0.01 poise; that of SAE 30 motor oil is 1.0 poise.

Plotting the viscosity of glass against temperature produces a smooth curve extending from an extremely high viscosity at room temperature to much lower viscosity at the temperature where glass is normally worked. In fact, the viscosity at room temperature is so high that it cannot be measured. A number of points on the viscosity-temperature curve have been defined as a convenience in describing the behavior of different glasses. Four of these are shown in Fig. 14.48.

The *working point* is defined as the temperature at which the viscosity is 104 poises, which represents the viscosity at which glass is suitable for working or forming. The *softening point* is that temperature at which a glass fiber less than 1 mm in diameter will stretch under its own weight at a rate of 1 mm/min when suspended vertically. This occurs at a viscosity of 107.6 poises. The *annealing point* is that temperature at

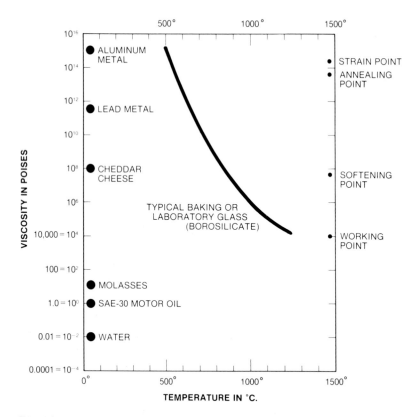

Fig. 14.48 As glass is heated it becomes gradually less viscous.

which strain in thin sections will be removed in 15 min and at which viscosity is 1013 poises. The *strain point* is that temperature below which permanent strain cannot be introduced.

During the first part of the annealing cycle the glass must be raised to the annealing point or slightly above to remove any strain already present. Then the rate of cooling to the strain point must be slow enough to avoid introducing more strain. Below the strain point cooling can be more rapid. Permanent strain cannot be introduced below the strain point, but it is possible to break the glass from temporary strain if cooling is hurried. Thick ware requires a longer annealing schedule than thin ware. Also, high-expansion glass requires a longer schedule than low-expansion glass. A thick piece of soda-lime glass would require the longest schedule, a thin piece of fused silica would require the shortest. A borosilicate beaker with a wall thickness under 1/8 in. will anneal in approximately 20 min, but the 24 in. thick borosilicate mirror disk made for the Mount Palomar telescope required 10 months (Fig. 14.49).

4. Finishing

Some products require secondary, or finishing, operations following primary forming and annealing. Some of these operations involve reheating the glass and shaping by bend-

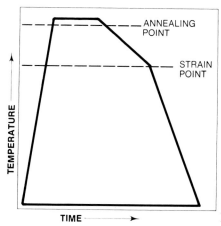

Fig. 14.49 Typical annealing schedule.

ing, sagging, pressing, or blowing. Such hot operations must be followed by an annealing cycle.

Operations that are carried out at room temperature all involve some form of cutting or stock removal. Glass cannot be permanently bent or shaped at room temperature but will spring back to its original shape when the bending force is removed.

5. Cutting

Cutting of glass can be done either mechanically or thermally; both methods are used in production. Tubing, sheet, and rod are produced continuously and must be cut to manageable size after annealing.

Mechanical cutting consists of scoring the glass with a glass cutter wheel then applying force across the score to produce a break. Thermal cutting employs a sharp flame to heat a narrow band of the glass. Then a jet of water is directed at this heated band, causing the glass to break.

Another type of thermal cut employs a ring of sharp flames that heat the glass until it softens and can be pulled apart. This flame cutoff is employed to remove the excess glass, or moil, from the top of blown ware. The open end of a drinking tumbler is a good example of the smooth edge that a flame cutoff leaves.

6. Drilling

Drilling may be done by one of several methods, depending on the number of pieces to be produced and the accuracy required. A tube of mild steel or some other relatively soft metal will serve as a drill. Such a tube is rotated while a slurry of loose abrasive grains and water is fed under the end of the tube. This type of drill produces relatively loose dimensional tolerances and the holes are noticeably tapered.

Tungsten-carbide drill bits work more rapidly, and dimensional control is better than with a loose-abrasive drill. Diamond core drills are capable of the most rapid drilling rates and finest dimensional control. For large-quantity work, they are also the most economical. Initial cost is high, but drill life is long and production rate high.

Ultrasonic drilling is slow and not used for round holes. However, for irregularly shaped holes it is the only method that works at all. It is a loose-abrasive process wherein a tool that is cut to the shape of the hole to be drilled is vibrated against the glass at an ultrasonic frequency while loose abrasive is fed under the tool.

7. Grinding

Grinding is employed to produce dimensional tolerances that are closer than hot forming can produce. Any of the machines used for grinding metals can be adapted for glass grinding: centerless, cylindrical, or surface grinders. Either silicon carbide or diamond wheels can be used. Diamond wheels are more expensive initially, but because of their longer life, are more economical when large production quantities are involved. Loose abrasive grinding is done on a flat, rotating table or grinding mill. A slurry of abrasive grains and water is fed on the surface of the mill, and the ware to be ground is held against the mill. Open ends of cylinders and annular gasket seats on sight glasses are ground by this method. Some stopcocks are ground against their mating shells by rotating the two members against one another while they are covered with a slurry of fine abrasive (usually silicon carbide).

8. Polishing

Polishing produces the smoothest surface possible. A ground surface that is to be *mechanically polished* is first buffed with pumice. The polishing agent, ferric oxide or cerium oxide, is carried on a buff made of felt, leather, or a soft metal.

Fire polishing consists of heating the ware in a flame until the surfaces fuse and flow. Smoothness is as good, or sometimes better, than with mechanical polishing, but dimensional control is not as precise.

Acid polishing is useful to remove minute defects in a glass surface without appreciably changing dimensions. The piece is dipped into a mixture of hydrofluoric and sulfuric acids. The mixture of acids dissolves a thin layer of the glass surface and rounds out or removes damaged spots and blemishes that could act as stress concentration points. Acid polishing tends to increase the strength of a glass product.

8. Lampworking

Lampworking is so named because it consists of forming glass by heating an object in the open flame of a gas-oxygen blast lamp and then blowing, bending, or tooling it to shape.

Complex assemblies may be made by this technique by fusion sealing tubing to pressed or blown parts. Many lampworking operations are done freehand without the aid of molds, and dimensional control is entirely dependent upon the manual skill of the lampworker. In some cases dies or molds of metal, carbon, or graphite are employed (Fig. 14.50).

9. Re-pressing

Re-pressing uses as its starting point a glass rod, gob, or sheet. A section of this rod or sheet, or the whole gob, is heated until soft and then pressed in a mold or die. Low-quality lenses, headers for vacuum tubes, and instrument bearings are made this way.

Fig. 14.50 Complex laboratory apparatus made by hand operations called lampworking.

10. Shrinking

Shrinking of tubing produces extremely close dimensional control of the inside diameter. A mandrel is placed inside a length of tubing, and the tubing is heated until soft. The air between the tubing and the mandrel is pumped out, and the tubing collapses against the mandrel. Tube diameters accurate to within 0.001 in. can be produced in diameters of 1 in. or less.

11. Tempering

Tempering is a prestressing technique that increases the strength of glass. The glass to be tempered is heated to slightly below the softening point and then suddenly chilled with a blast of air. Soda-lime, lead-alkali, borosilicate, and aluminosilicate glasses can be tempered, but the very low expansion glasses (96% Silica Glass and fused silica) cannot.

12. Sagging

Sagging makes trough-shaped pieces, such as dishes or lighting panels, from sheet glass. The sheet to be sagged is placed over a metal or refractory mold; then the glass is heated until it sags and conforms to the shape of the mold. Sagging is most often used for parts of relatively large area and relatively thin cross section. However, thick shapes can also be sagged. Mirror blanks for astronomical telescopes are formed by sagging chunks of glass over a mold.

13. Surface Treatments

There are a number of ways to treat the surface of glass to provide a path for electricity, to indicate volume of contents, to provide a spot for identification or simply for decoration (Table 14.14).

14. Glass-to-Metal Seals

Glass can be sealed directly to metal if proper care is exercised. To make such a seal both the glass and the metal are heated to the temperature at which the glass softens. Then the parts are brought together and cooled. Both the glass and the metal must be chosen so their thermal expansion rates match closely. If the mismatch in expansion rates is large, the parts will shrink on cooling at different rates and the glass may crack (Fig. 14.51). Sometimes, when it is necessary to join a glass and a metal that differ appreciably in expansion rate, a graded seal is used. The graded seal consists of several bands or rings of glass, each of which differs slightly from its neighbors in expansion rate. Expansion of the individual rings progresses from low to high gradually along the length of the cylinder.

a. Design Considerations

In certain seal configurations it is possible to effect a seal when expansions of the glass and metal parts differ widely. If the seal is designed so that the mismatch stress results only in glass compression, this stress can be allowed to go to rather high values since glass does not fail from compression. For example, a glass tube can be sealed inside an aluminum or copper tube, either of which is much higher in expansion than glass.

When the tube and rod cool, the greater contraction of the metal tube simply applies compressive stress to the rod (Fig. 14.52). If the metal part is thin and weak and the glass part thick and strong, there will be but little stress in the glass. The metal will deform from the mismatch and relieve the stress in the glass (Fig. 14.53). The glass dielectric capacitor is an example in which thin aluminum foil of expansion approximately 250 in./in.°C^{-1} is sealed to glass of expansion 102 in./in.°C^{-1}.

Not all metals will make good seals to glass. To seal well a metal must form a tight oxide layer to which the glass can bond. If the oxide layer is flaky or loose, a tight seal is not possible. Sometimes a controlled atmosphere is necessary to develop the required surface on the metal.

The *contraction differential* is the difference in contraction between the two materials being sealed from the setting point of the glass to room temperature. This differential is stated in parts per million. A contraction differential of 100 means that the higher expansion member of the pair will shrink in cooling 100 millionths of an inch for each inch of length more than the lower expansion member. The setting point is taken as 5 degrees Celsius above the strain point. A differential of less than 100 indicates very good sealing conditions. Good seals can be made with differentials between 100 and 500. Above 500 sealing becomes difficult; the shape and size of the seal have to be carefully worked out.

E. Glass Ceramics

Glass ceramics are so named because they begin life as glasses and then are converted into crystalline ceramics. Mixing of raw materials, melting, and forming of a glass ceramic are all done the same as for a glass. Before heat treating, a freshly formed glass-ceramic article is a true glass. It is only after a carefully controlled heat treatment that the difference between a glass and a glass ceramic is apparent. Some glass-ceramic products are known more familiarly by their trademark, such as Pyroceram, a registered trademark of the Corning Glass Works.

Heat treatment converts a glass ceramic from a glass to a dense, fine-grained crystalline ceramic. Conversion takes place in two stages. In the first, the glass ceramic is seeded with nuclei, or centers, around which crystals grow. In the second, crystals grow around these nuclei. The finished article is approximately 90% crystalline and decidedly different from a glass in nearly all its properties.

The difference between a glass that is intended to remain a glass and a glass ceramic in the glassy state lies in the nucleating agent included in the glass-ceramic batch and the proper ingredients for forming crystals. The nucleating agent is a substance that is barely soluble in the glass. At high temperatures it will remain in solution, but at low temperatures it will precipitate and furnish the nuclei necessary as centers for crystal growth. When the glass is raised to the correct temperature and held there for a long enough time, a crystal will grow around each nucleus.

To start the conversion procedure, the formed glass-ceramic article may be lowered to the nucleating temperature after forming or it may be lowered to room temperature, then reheated to the nucleating temperature. Following nucleation the temperature is then raised to the crystal-growth range.

Two conditions are necessary for crystal formation in glass: (1) centers, or nuclei, around which crystals can grow, and (2) sufficient time in a temperature range favorable for crystallization. In a glass not intended for conversion to the crystalline state, crystals

Table 14.14 Surface Treatments for Glass

Treatment	Compatible glass	Application method	Typical applications	Remarks
Vitreous Enamel	Soda-lime and lead types	Spray, silk screen, decal, brush	Decoration on table ware and containers, scales on graduated ware	Chemical and abrasion resistance good. Many colors available. On low-expansion glasses treatment must be limited to small areas and thin coatings
Stain	All	Spray, brush, dip, silk screen, decal	Background for scales on graduated ware, decoration	Good chemical durability and abrasion resistance; transparent; colors limited to range from red, dish brown to yellow
Luster	All	Spray, brush, silk screen, decal	Decoration	Abrasion resistance not as good as vitreous enamels
Metal	All	Spray, brush, vacuum evaporation, flame spray	Lamp reflectors, electrical contacts, decoration	Abrasion resistance fair for fired coatings, poor for vacuum-deposited; fired coatings limited to gold, silver, platinum, palladium

Metal oxide	Best suited for low to medium expansion glasses; can be applied to any glass	Spray	Resistors, electric heaters, static shields, heat reflectors	High chemical and abrasion resistance; good electrical and optical properties available
Lacquer	All	Spray, dip, brush, decal, silk screen	Decoration	Not a fired coating; abrasion resistance and adhesion lowest of all coatings listed
Etch	All	Dip in hydrofluoric acid, brush on etching paste	Scales on graduates ware, decoration	Etched lines can be filled with enamel for improved legibility
Abrade	All	Blast with abrasive grains, such as sand	Diffusing surface, identification spots on parts	
Engrave	All	Rotating soft metal wheel covered with slurry of loose abrasive	Decoration	
Glue chipping	All	Apply hot glue, dry in oven until glue pops off removing glass chips	Decoration	

Source: Courtesy of Machine Design.

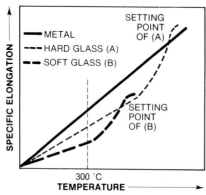

Fig. 14.51 Hard glass (A) will seal to either metal or soft glass (B). Soft glass (B) will not seal to the metal.

Fig. 14.52 Coaxial seal of glass rod inside higher expansion metal tube puts glass in compression.

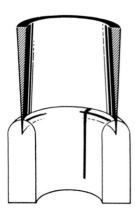

Fig. 14.53 Thin walls of metal tube deform from mismatch stress and leave glass relatively stress free.

may form by accident. These will nearly always be on a surface where chance contamination provides the necessary nuclei. It is not correct to think of a glass ceramic as a glass that can be crystallized and an ordinary glass as a material that will not crystallize. Crystals may grow in any glass if conditions are favorable. The glassy state of random atomic arrangement is only comparatively stable (i.e., metastable), and the atoms will arrange themselves into a regular, crystalline pattern if given the opportunity.

Some glasses when cooled sufficiently will nucleate spontaneously within the temperature range where crystal formation can take place. These glasses are not usually successful commercially because of the difficulty of avoiding crystallization.

Glass ceramics are generally four to five times as strong as glass. They are also mechanically harder than glass. Using the diamond-penetration hardness test, a typical glass ceramic shows a value within the range for hardened tool steel.

Some properties can be manipulated over a wide range. The glass-ceramic composition used for missile nose cones is engineered for low dielectric loss and moderately high resistance to heat shock. The composition used for cooking ware shows extremely high resistance to heat shock but does not have as low a dielectric loss as the missile cone material. Both can be used steadily at temperatures up to 700°C.

Glass ceramics can be ground and polished to the same fine dimensional tolerances as glass. They are free from pores, a rather unusual characteristic in crystalline ceramic materials. Since they are fabricated by the glass-making techniques of blowing, pressing, drawing, or centrifugal casting, products can be rapidly formed in shapes that would be more slowly made by the conventional ceramic processes.

Yet another type of glass ceramic can be shaped by conventional machining methods so that extremely close dimensional tolerances can be achieved. However, such properties as strength and corrosion resistance may be somewhat less in these materials.

Glass ceramics combine some of the properties of glasses and some of the properties of ceramic materials made by conventional methods. The strength of some glass ceramics is approximately equal to the strongest tempered glass. Special treatments can be employed under some conditions that increase the strength even more.

Thermal expansion rate can be varied from twice that of steel to zero. Resistance to prolonged heating at elevated temperature is equal to that of aluminosilicate glasses. Resistance to heat shock can be better than most glasses or conventional ceramics.

Corrosion resistance can be better than most metals but is usually not equal to the most highly resistant glasses. Dimensional stability is the equal of glass.

Acknowledgment

Permission of the Corning Glass Works to use much of the information from "All About Glass" is gratefully acknowledged.

REFERENCES

1a. J. J. Burke, E. M. Lenoe, and R. N. Katz, Eds., *Ceramics for High Performance Applications, II,* Brook Hill, Chestnut Hill, Massachusetts, 1978.

1b. J. J. Burke, A. E. Gorum, and R. N. Katz, Eds., *Ceramics for High Performance Applications,* Brook Hill, Chestnut Hill, Massachusetts, 1974.

2. J. I. Mueller, A. S. Kobayashi, and W. D. Scott, *Design with Brittle Materials,* University of Washington, Seattle, Washington, 1979.

3. W. D. Kingery, *Ice and Snow: Properties Processes and Applications,* MIT Press, Cambridge, Massachusetts, 1963.

4. G. J. McCarthy and M. T. Davidson, Ceramic Bulletin, 55, (1976), 190-194.

5. W. D. Kingery, in *Energy and Ceramics, Materials Science Monographs, 6,* Ed. P. Vincenzini, Elsevier Scientific, New York, 1980, pp. 2ff.

6. J. F. Young, *Materials and Processes,* Chap. 13, John Wiley and Sons, New York, 1954, pp. 548-590.

7. E. M. Levin, H. F. McMurdie, and F. P. Hall, *Phase Diagrams for Ceramics,* The American Ceramic Society, Cleveland, Ohio, 1956.

8. W. D. Kingery, H. K. Bowen, and D. R. Uhlman, *Introduction to Ceramics,* 2nd Ed., John Wiley and Sons, New York, 1976, p. 643.

9. R. Von Mises, Z. Agnew, Math. Msch. 8 (1978), 161.

10. G. I. Taylor, J. Inst. Met. 62 (1938), 307.

11. N. S. Stolof et al., J. Appl. Phys., 34 (1963), 3315.

12. R. N. Katz and R. L. Coble, cited in Ref. 2, p. 729.

13. S. M. Copley and J. A. Pask, in *Materials Science Research,* Vol. 13, Ed. W. Kriegel and H. Palmour III, Plenum Press, New York, 1966, pp. 89-224.

14. M. L. Kronberg, J. Am. Cer. Soc. 45 (1962), 274-279.

15. W. D. Kingery, H. K. Bowen, and D. R. Uhlman, *Introduction to Ceramics,* 2nd Ed., John Wiley and Sons, New York, 1976, pp. 678-689.

16. J. J. Burke, A. E. Gorum, and R. N. Katz, Eds., *Ceramics for High Performance Applications,* Brook Hill, Chestnut Hill, Massachusetts, 1974, Chaps. 3 and 7.

17. J. J. Burke, E. M. Lenoe, and R. N. Katz, Eds., *Ceramics for High Performance Applications II,* Brook Hill, Chestnut Hill, Massachusetts, 1978, Chaps. 1-5, 34, and 39.

18. J. I. Mueller, A. S. Kobayashi, and W. D. Scott, *Design with Brittle Materials,* Universiy of Washington, Seattle, Washington, 1979, Chaps. 6-11.

19. A. A. Griffith, The phenomena of rupture and flow in solids, Phil. Trans. Roy. Soc. (London), A221, 163-198 (1921).

20. G. K. Bansal, Effect of flaw shape on strength of ceramics, J. Am. Ceram. Soc. 59 (1-2), 87-88 (1976).

21. W. F. Brown, Jr., and J. E. Srawley, Plane strain crack toughness testing of high-strength metallic materials, Am. Soc. Test. Mater., Spec. Tech. Publ. 410, 1-15 (1967).

22. G. K. Bansal, On fracture-mirror formation in glass and polycrystalline ceramics, Philos. Mag., 35 (4), 935-44 (1977).

23. H. Hubner and W. Jillek, Subcritical crack extension and crack resistance in polycrystalline alumina, J. Mater. Sci., 12, 117 (1977).

24. G. K. Bansal and W. H. Duckworth, Comments on Ref. 5, J. Mater. Sci., 13, 215-216 (1978).

25. S. W. Freiman, A. Williams, J. J. Mecholsky, and R. W. Rice, in *Ceramic Microstructures '76,* Eds., R. M. Fulrath and J. A. Pask, Westview Press, Boulder, Colorado, 1977, pp. 824-834.

26. G. K. Bansal, in *Ceramic Microstructures '76,* Eds., R. M. Fulrath and J. A. Pask, Westview Press, Boulder, Colorado, 1977, pp. 860-871.

27. A. G. Evans and F. F. Lange, Crack propagation and fracture in silicon carbide, J. Mater. Sci., 10, 1659-1665 (1975).

28. S. M. Wiederhorn, *Fracture Mechanics of Ceramics,* Eds., Bradt, Hasselman, and Lange, Plenum Press, New York, 1973, pp. 613-646.

29. A. G. Evans and H. Johnson, Fracture Stress and its dependence on slow crack growth, J. Mater. Sci., 10, 214-222 (1975).

30. G. K. Bansal and W. H. Duckworth, Effects of moisture-assisted slow crack growth on ceramic strengths, J. Mater. Sci., 13, 239-242 (1978).

31. S. M. Wiederhorn and J. E. Ritter, Application of fracture mechanics concepts to structural ceramics. in *Fracture Mechanics Applied to Brittle Materials*, Ed. S. Freiman, ASTM Special Technical Publication 678, Philadelphia, 1979, pp. 202-215.

32. D. G. S. Davies, Statistical approach to engineering design in ceramics, Proc. Br. Ceram. Soc., No. 22, 429-452 (1973).

33. A. G. Evans and S. M. Widerhorn, Proof testing of ceramic materials—an analytical basis for failure prediction, Int. J. Fract., 10 (3), 379-392 (1974).

34. W. Weibull, Statistical distribution function of wide applicability, Appl. Mech. 18 (3), 293-297 (1951).

35. G. K. Bansal, W. H. Duckworth, and D. E. Niesz, Strength-size relations in ceramic materials: Investigation of an alumina ceramic, J. Am. Ceram. Soc. 59 (11-12), 472-478 (1976).

36. C. A. Johnson and S. Prochagka, *Investigation of Ceramics for High-Temperature Components*, Quarterly Progress Report 3, prepared for Naval Air Development Center under Contract No. N62269-76C-0243.

RECOMMENDED READINGS

Holloway, D. G. *The Physical Properties of Glass*, Wykeham Publications, London, 1973.

Morey, G. W. *The Properties of Glass*, 2nd Edition, Reinhold Publishing, New York, 1954.

Scholes, S. R., and Greene, C. H. *Modern Glass Practice*, 7th Edition, Cahners Books, Boston, 1975.

Shand, E. B. *Glass Engineering Handbook*, McGraw-Hill, New York, 1958.

Volf, M. B. *Technical Glasses*, Pitman & Sons, London, 1973.

15

Advanced Composite Materials*

I. INTRODUCTION

The advanced composites industry is about 20 years old and will in the next decade be fully established and universally accepted. Maturity occurred rapidly due to extensive research and development and sizable expenditures because forecasts for structural improvement were dramatic.

Typical weight savings for military aircraft components are 20%; for commercial aircraft, 25%; for automotive applications, 30-60%; for spacecraft structures, 40%; and for sports equipment, 50%. Material and process developments have lowered the initial high costs for composites; today acquisition costs for composite aircraft and spacecraft components are equal to, or less than, their metallic counterparts. Composite parts for commercial aircraft are being developed and are expected to be 10-15% less costly in production than equivalent metal parts. High production speed and complete automation of composites will be required to provide cost-effective components for the automotive industry. Initial automation and low-cost composites have been achieved for sports equipment.

A. Military Aircraft Composites

Since 1965, many advanced composite structures have been built and successfully laboratory tested; over 30 aerospace structures have passed flight test. Grumman Aerospace was the first to introduce a production flight safe structure, the F-14A horizontal stabilizer (Fig. 15.1), in 1969. Several years later, McDonnell introduced advanced composites for the empennage of the F-15, and later General Dynamics utilized composites for the empennage of the F-16. These programs verified the performance improvements and production suitability of composites.

During the past several years, the composite industry in the United States concentrated on the inhibitors for full-scale implementation: cost, confidence, and durability. The cost issue will always be a concern, but progress has been made. Forecasts for future extremely large volume graphite and Kevlar fibers have overcome escalation,

---

*Contributed by SAMUEL J. DASTIN.

Fig. 15.1   F-14A horizontal stabilizer.

and future projections are attractive for high-technology composite components. Sizable effort has been, and is, in development for low-cost tooling and organic matrix composite manufacture. Automation of the process has shown the process costs for composites can be below those of many equivalent metallic structures. Grumman has developed and commissioned an "Integrated Laminating Center" (Fig. 15.2), which has reduced the production labor to one-half for the composite F-14A horizontal stabilizer.

The confidence issue is slowly but surely being overcome. A major step was the production approval of Grumman's all-composite horizontal stabilizer for the B-1 bomber. After a 5 year in-depth development program, the composite structure satisfied all detailed aircraft specifications, and both Rockwell International, weapons system manager, and the U.S. Air Force accepted the component for production. The composite stabilizer (Fig. 15.3) was 15% lighter and 18% less costly than its metal counterpart. Continued confidence is shown by the use of composite wing skins on the F-18. The advanced Harrier, the AV-8B, is taking maximum advantage of composites, utilizing the material for approximately 29% of the structure's weight.

Future U.S. aircraft will double the percentage of composite structures. Grumman recently completed a study for future tactical aircraft and found that significant size and cost reductions are possible by utilization of composites in up to 75% of structure weight. The advanced design composite aircraft (ADCA) is shown in Fig. 15.4 compared to a metallic aircraft and a composite-substituted metallic aircraft.

Fig. 15.2 Integrated laminating center.

The durability of advanced composites has been under study for several years, with emphasis on moisture resistance. Extensive testing along with in-depth analysis at Grumman have shown that graphite-epoxy laminates can absorb up to 1.5% moisture. In the moisture-saturated condition, the laminate is stable and only degrades if service temperatures are above 260°F. Current material specifications include mechanical tests of moisture-saturated laminates to provide assurance of durability. Evaluation of service data on the F-14A horizontal stabilizer has been undertaken. In service since 1970 with lead aircraft logging over 5000 hr and more than 300 aircraft (over 150,000 flight hr) deployed, durability requirements have been proven. Further, maintenance actions have been few, and life-cycle costs have been low. Similar data have been noted for the composite empennage of the F-15.

Fig. 15.3   B-1 horizontal stabilizer.

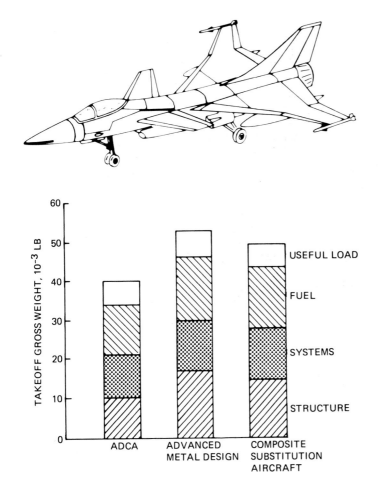

Fig. 15.4 ADCA weight comparison.

Fig. 15.5   Environmental laboratory setup.

Because such data are limited and only qualify the specific composite material and design, Grumman has established a laboratory service life testing facility. The setup (Fig. 15.5) simultaneously simulates loading, humidity, thermal spikes, and temperatures in both accelerated and quasi-real time to provide the lead-the-fleet data for critical composite components. Currently, composites are being evaluated for damage tolerance, effects of defects, and impact and crash worthiness.

B.   Commercial Aircraft Composites

Advanced composite structures of U.S. commercial aircraft have been under study by NASA for the past several years. Initial studies were for secondary structures, such as wing-to-body fairings and control surfaces and for fatigue life extension of metallic components. Service life actual testing showed that nonsandwich components were structurally efficient, durable, and readily maintained and repaired. Corrosion, moisture ingress, and debonding were the major limitations of sandwich structures using an aluminum honeycomb core. Overall, these early studies showed composites will provide significant advantages, and thus current development is for primary structures.

During the past several years, NASA has funded Boeing, McDonnell, and Lockheed to design, test, and implement empennage structures for current commercial aircraft. The prime driver for these programs is energy conservation, and the weight reductions furnished by composites can provide significant strides in that direction. Suitability studies of composite fail-safe designs for wing and body are also underway, and it appears that the commercial aircraft of the 2000s will contain composites in significant quantity. Composite structures successfully flown are given in Table 15.1.

Table 15.1 Composite
Structures Flown on
Commercial Aircraft

Boeing
  737 Spoiler
McDonnell-Douglas
  DC-10 Upper rudder
  DC-10 Aft pylon skin
Lockheed
  L1011 Wing to body fairing
  L1011 Center engine fairing
  L1011 Wing to body fillet

Other areas being evaluated for composites are floor beams, decks, nacelles, and engine components. The recently completed quiet engine study program has highlighted the need for composite components. It appears to be a certainty that future engines will utilize composites, especially for static components. Engine parts already designed and tested using composites are vanes, splitters, and perforated duct skins. Under study at present are frames, augmentor ducts, and nozzle flaps. Previous study of composites for engine blades have shown a foreign object damage (FOD) limitation. The industry is still developing composites for this application utilizing hybrids to provide the resistance to bird ingestion impact.

C. Automotive Applications for Composites

During the past several years, fiberglass-reinforced composites and nonmetallics in general have shown considerable growth in U.S. passenger cars. Current uses of these lightweight materials are for nonstructural applications, but with future requirements for energy conservation, composites will be used structurally. Requirements for fleet average fuel consumption, such as 27.5 miles per gallon (mpg) by 1985 and 31.5 mpg or greater by 1990, will generate an explosive growth for composites.

Several independent studies have forecasted a composites industry of 75 million pounds by 1995, primarily for the automotive industry. Advanced composites using graphite, Kevlar, and hybrids of graphite-fiberglass, Kevlar-fiberglass, and graphite-Kevlar will be initially introduced in trucks, where weight savings are valued greater to permit somewhat more expensive materials.

As volume increases and material costs drop to $5/lb, composites will be a major material for the passenger car industry. Composites are at present being evaluated for several components (Table 15.2). Ford Corporation has completed the development of an all-composite car. The experimental car was the same size as the production 1978 Ford Granada, but weighed 1250 lb less than the conventional car (approximately one-third less) due to the use of graphite epoxy. Table 15.3 lists some of the components made for this experimental car.

Major process developments will be needed to implement composites in the automotive industry. Processes will be needed that provide completed components in minutes rather than hours to satisfy the huge volume required.. Composite sheet-molding

Table 15.2 Composite Automotive
Components Currently Under Evaluation

Driveshafts for trucks and cars
Side rails for trucks
Doors for trucks and station wagons
Door beams for cars
Cross members for car and truck chassis
Suspension arms for cars
Brackets for cars
Leaf springs for trucks

---

compounds and automated compression molding, reaction injection molding, and automated vacuum forming are being evaluated as the candidate processes.

D. Spacecraft Composites

Composites for use in space and on spacecraft have been developed by both NASA and DOD. A recent example is the payload doors of the U.S. space shuttle orbiter. These components are the largest assembled composite, 12 ft wide by 60 ft long. Weight savings are at a premium for space applications, and thus composite growth in this field is rapid. Other special features of composites for spacecraft are the controllable coefficient of thermal expansion, cryogenic stability, load tailorability, and high specific stiffness. Advanced composites for space applications are available in ultrathin thicknesses, that is, 0.001 in. per ply, and thus optimum structures for large-area solar energy collectors will be a future reality.

Near zero coefficient composite tubular structures have been produced by Grumman for the Large Space Telescope (Fig. 15.6). Composites have been successfully used for precision mounts, optical benches, and electromagnetic antennas. Structures for

Table 15.3 Components and Weight Savings for Ford's Experimental Composite Car

Component	Steel (lb)	Graphite (lb)	Weight saving (lb)
Hood	40.0	15.0	25.0
Door, R. H. rear	30.25	12.65	17.60
Hinge, Upper L. H. front	2.25	0.47	1.78
Hinge, Lower L. H. front	2.67	0.77	1.90
Door guard beam	3.85	2.40	1.45
Suspension arm, front upper	3.85	1.68	2.17
Suspension arm, front lower	2.90	1.27	1.63
Transmission support	2.35	0.55	1.80
Driveshaft	17.40	12.00	5.40
Air conditioning, lateral brace	9.50	3.25	6.25
Air conditioning, compressor bracket	5.63	1.35	4.28

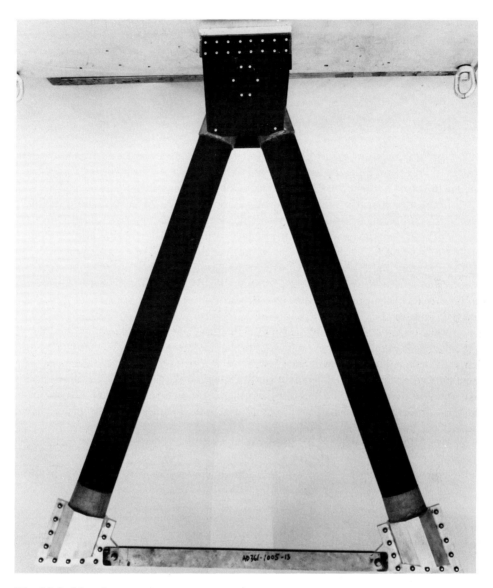

Fig. 15.6 Metering truss composite test component design.

future space applications will be primarily advanced composites; therefore the industry is at present developing the required data base.

E. Composite Sports Equipment

A major outlet for advanced composites is high-performance sports equipment. Graphite, boron, and Kevlar filaments in an epoxy matrix have found acceptance in golf shafts, fishing rods, bows and arrows, and tennis rackets. Current components fabricated with advanced composites are bicycle frames, skateboards, and skis. Hybrid composites are expanding the product lines by reducing component costs without sacrificing performance. Mixtures of Kevlar and graphite epoxy are effective materials for crossbows, ski poles, and surfboards. At present, the sports equipment industry has accepted advanced composites to both maintain and expand the industry, thereby assuring continued utilization.

II. ADVANCED COMPOSITE MATERIAL SYSTEMS

Composite materials are broadly divided into two major classes: organic matrix and metal materials. Polymeric laminates are considered to be a design feature involving processing and are discussed in Chap. 27. Clad metal laminates are discussed in Chap. 9.

In the organic matrix materials, such as boron-epoxy, Kevlar-epoxy, and graphite-epoxy, the reinforcing fibers or filaments are the primary load-carrying element. The matrix of such a composite stabilizes the slender filaments in compression, loads and unloads the filaments by shear, and in general provides the "body" of the composite. Service temperature limitations, environmental suitability, and general manufacturing procedures and producibility are more related to the matrix than to the filament. However, both elements of the composite must be considered when selecting a material for a specific component.

Metal matrix materials, such as boron-aluminum and graphite-aluminum, are unlike the organic matrix composite in that both the reinforcement and the matrix are structural materials in their own right. Transverse properties of metal matrix composites are significantly higher both in strength and stiffness and thus can overcome some of the limitations of organic matrix composites. In general, the metal matrix composites have behavior similar to organic matrix composites in that the filaments carry the major loads and provide the major source of the stiffness. The matrix stabilizes the filaments in compression, loads and unloads the filaments by shear, and provides both short and long transverse properties.

Currently utilized advanced composite material designations are

Boron-epoxy, B/Ep
Graphite-epoxy, Gr/Ep
Fiberglass-epoxy, Gl/Ep
Boron-aluminum, B/Al
Kevlar 49-epoxy, K/Ep
Graphite-aluminum, Gr/Al

Mixtures of more than one type of reinforcement in organic matrices are now being developed. These hybrid composites are mixtures of various types of graphite, graphite and boron, boron and fiberglass, graphite and fiberglass, and Kevlar-fiberglass-graphite. The various types of available graphite-epoxies fall into three basic categories:

LHS: low cost, high strength. Fiber F_{tu}, 400 ksi; E, 34×10^6 psi.
HM: high modulus. Fiber F_{tu}, 350 ksi; E, 50×10^6 psi.
UHM: ultra high modulus. Fiber F_{tu}, 300 ksi; E, 70×10^6 psi.

Hybrid designations for currently considered materials are

Boron-graphite-epoxy, B/Gr/Ep
Graphite-fiberglass-epoxy, Gr/Gl/Ep

Mixed graphite-epoxy, $Gr_{LHS}/Gr_{UHM}/Ep$
Graphite-Kevlar 49-epoxy, $Gr/K/Ep$

Boron reinforcements are commercially available in two basic diameters of 4.0 and 5.6 mils. A third diameter, 8.0 mils, has been manufactured and could be made available in production quantities. Basically the two types of boron utilized are primarily the 4.0 mil for boron-epoxy and the 5.6 mil for boron-aluminum.

III. TYPICAL ADVANCED COMPOSITE MATERIAL PROPERTIES

The typical unitary advanced composite properties for currently considered structural materials are given in Table 15.4. These mechanical properties are for unidirectional (all reinforcement layup at $0°$ and loaded uniaxially in the $0°$ direction) laminates at a 50 volume percent fiber for boron composites and a 60 volume percent fiber for graphite composites. The values given are minimum average requirements of typical material specifications.

The organic matrix composites developed and utilized in the industry have been designed for primary structural applications over the temperature range of cryogenic to 260°F. The metal matrix composites under development are structurally suitable to 600°F. The subject upper limit temperatures are for structures of typical fighter aircraft with a maximum of 200 hr at temperature. The subject composites have been evaluated for permanence properties, aging, electromagnetic and lightning stability, stack gas resistance, and suitability to most aircraft solvent and fluids. Generally, stability is excellent, although various protective finishes are required to satisfy detailed aircraft specifications.

IV. ADVANCED COMPOSITE MATERIAL DESIGN CONSIDERATIONS

Advanced composites are generally designed as filament-controlled laminates. For organic matrix composites, the $0°/90°/±45°$ family of laminates are considered filament controlled. For metal matrix composites, the general family of laminates oriented $0°/±45°$ are utilized. The filament-controlled laminate philosophy can be modified for either selectively reinforced primary metal structures or for secondary or nonstructural all-composite members.

Design allowables for advanced composite structural materials have been utilized on the MIL-HDBK-5 B basis. For boron-epoxy, the Air Force has developed both B and A basis design allowables. These values are given in *Advanced Composites Design Guide* (Third Edition, January 1973). Strength values with and without holes, fatigue strengths, elastic constants, creep strengths, thermal properties, and joint strengths are statistically presented, along with combined loading considerations. Composite strength properties are affected by elevated temperatures and absorbed moisture. Most organic matrix composites are stable when moisture saturated up to service temperatures to 250°F as shown in Fig. 15.7. Although many computer programs are available to select laminate orientations, preliminary designs are usually created using allowables obtained from design graphs, such as those shown in Figs. 15.8 and 15.9 for tensile strength and tensile-compressive moduli for graphite-epoxies. Current structural design considerations require a broad spectrum of specialty. Composites have been thoroughly evaluated under industry

Table 15.4 General Properties Chart of Advanced Composite Materials, Unidirectional Laminates

Mechanical property (minimum average value)	Boron-epoxy		LHS[a] graphite-epoxy		UHM[b] graphite-epoxy		Unidirectional fiberglass (7743)-epoxy		Boron-aluminum		Kevlar 49-epoxy	
	RT	350°F	RT	350°F	RT	350°F	RT	350°F	RT	350°F	RT	350°F
Longitudinal tensile strength, ksi	200	150	190	178	85	80	108	80	215	203	200	150
Longitudinal tensile modulus, 10^6 psi	30.0	26.0	18.5	18.0	41.0	41.0	6.1	5.1	33.2	31.2	11.0	8.0
Longitudinal flexural modulus, ksi	245	180	210	157	105	96	150	69	300	250	90	60
Longitudinal flexural modulus, 10^6 psi	27.0	20.0	17.0	16.0	35.0	31.0	6.2	3.7	30.0	29.0	10.0	8.0
Longitudinal compression strength, ksi	375	140	126	116	65	60	79	62.5	623	539	40	30
Longitudinal compression modulus, 10^6 psi	33.0	31.0	18.0	18.0	40.0	40.0	4.9	4.9	33.2	31.2	8.0	6.0
Longitudinal shear strength, horizontal, ksi	13.0	5.0	7.0	5.0	5.4	4.5	4.3	2.0	22.0	20.0	7.0	5.0
Transverse tensile strength, ksi	7.0	3.0	8.0	4.6	3.0	3.0	8.6	8.0	22.0	18.0	4.0	2.0
Transverse tensile modulus, 10^6 psi	3.0	1.0	2.0	1.0	.9	.9	2.2	0.75	20.1	19.8	1.0	0.75
Transverse flexural strength, ksi	11.0	8.0	9.0	4.6	3.0	1.5	—	—	—	—		
Transverse flexural modulus, 10^6 psi	2.0	0.8	2.0	1.0	1.0	—	—	—	—	—		
Density, lbs/in.^{3}	.075	—	.055		.061		.065		.095		.050	
Layer thickness, in.	0.00510		0.00525		0.00525		0.0092		0.0070		0.0050	

[a]LHS, low cost, high strength.
[b]UHM, ultra-high modulus.

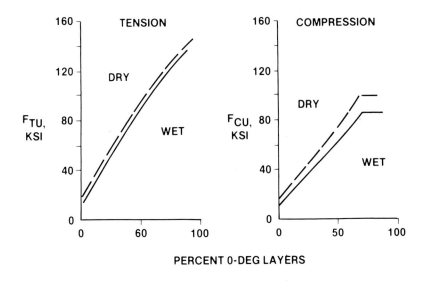

PERCENT 0-DEG LAYERS

Fig. 15.7 Effect of humidity on graphite-epoxy allowables at 250°F, 0/90/±45 laminates with 10% 90° layers.

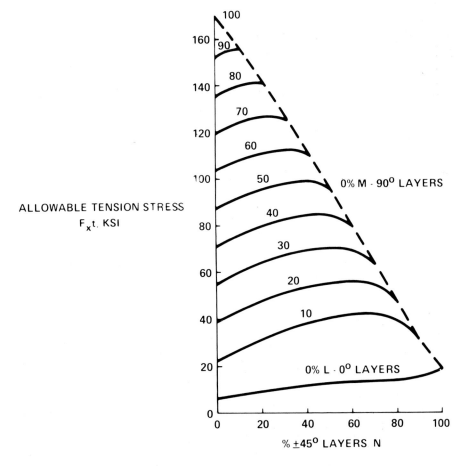

Fig. 15.8 Allowable tension strength of graphite-epoxy laminates at room temperature.

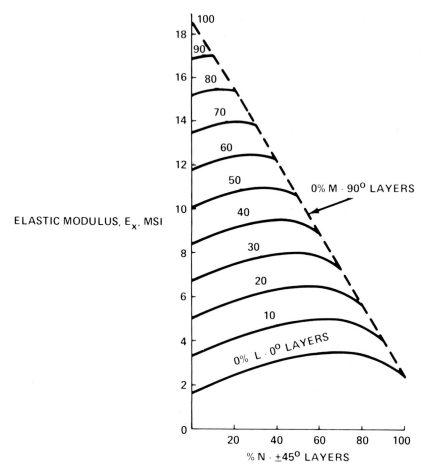

Fig. 15.9 Elastic tension and compression moduli of graphite-epoxy laminates at room temperature.

and government sponsorship and the data base is sizable and rapidly expanding. The data available at present are shown in Table 15.5 and can be obtained from the literature, and from the many composite material suppliers; a partial listing of sources is given in the references of this chapter. An evaluation of the many successful composites structures over the past 15 years has shown that designs based on a conservative approach are necessary for this relatively new material. Typical general design practice used is given in Table 15.6. Composite structures are effectively joined to themselves or to metal components by either bonding or mechanical fasteners. Generally, the large primary structures are assembled by bolting. General design practice for composite joints are given in Table 15.7.

Composite structures provide an extra degree of design freedom compared with metal structures. The composite designer has the opportunity to design the material as well as the structure. Designing the material requires experience, since many composite materials and hybrid materials are available. To provide a rating method for the many

Table 15.5 Composite Structure, Available Test Data

Environmental

Lightning strike, EMI, EMP, lasers	Thermal protection
Acoustic fatigue, mechanical fatigue	Aircraft gluids

Structural

Material design allowables	Design verification
	Flutter
Effects of moisture, temperature on materials	Bolted, bonded joints
	Cutouts

Damage tolerance

Ballistic	Crack protection, propagation
Impact (Fod)	Crashworthiness

Table 15.6 General Design Practice

Practice	Reason
Filamentary controlled laminates: minimum of three layer orientations	To prevent matrix and stiffness degradation
0/90/±45 Laminate with a minimum of one layer in each direction	$0°$ Layers for longitudinal load, $90°$ layers for transverse load, ±45° layers for shear load
A +45° and −45° ply are in contact with each other	To minimize interlaminar shear
45° Layers are added in pairs (±)	In-plane shear is carried by tension and compression in the 45° layers
When adding plies try to maintain symmetry	To minimize warping, interlaminar shear
Minimize stress concentrations	Composites are essentially elastic to failure
±45° Plies at least one pair on extremes of laminate; however, for specific design requirements (applied moments) 0° or 90° plies may be more advantageous in direction of moments	Increases buckling for thin laminates; better damage tolerance
Maintain a homogeneous stacking sequence, banding several plies of the same orientation together	Increased strength

Table 15.7   General Design Practice for Joints

Practice	Reason
3D edge distance and 6D pitch for bolted joints	Bearing strength
Design bonded step joints rather than scarf joints	More consistent results, design flexibility
Bonded joints: no 90° plies in contact with metals	Reduction in lap shear strength
Bonded joints: ±45$^\pi$ plies on last step	To reduce peak loading (E ± 45°, B/Ep = 3.5 × 10^6; E_{Ti} = 17 × 10^6)
When adding plies use a 0.3 in. overlap in major load direction using a wedge-type pattern	Requires approximately 0.3 in. to develop strength

available choices, computer programs and graphic methods are used to measure the structural efficiency of various materials. The major structural requirements are tensile, compressive, and in-plane shear, and thus the critical loading of the structure to be designed is compared among the various composite materials available. Figures 15.10 to 15.12 show composite material efficiency in the most common laminate orientations. The modulus-ultimate strength ratio (E/F) range given in the figures are generally acceptable for efficient structures. After selection of the composite material by this method a preliminary design of the desired structure is created; next the design is reviewed for cost and producibility. Modifications are added, and a detailed design is created along with

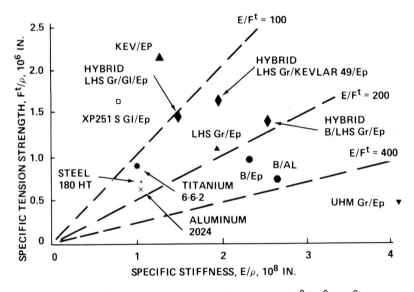

Fig. 15.10   Specific tension properties of composite [0°/90°/±45°] (50%/10%/40%) (with holes).

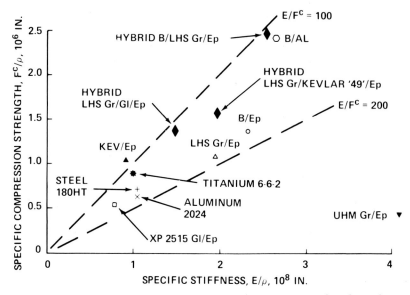

Fig. 15.11 Specific compression properties of composite $[0°/90°/±45°]$ (50%/10%/40%) (with holes).

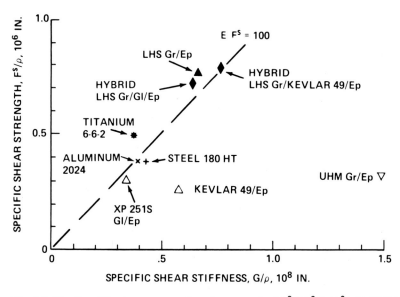

Fig. 15.12 Specific shear properties of composite $[0°/90°/±45°]$ (10%/10%/80%).

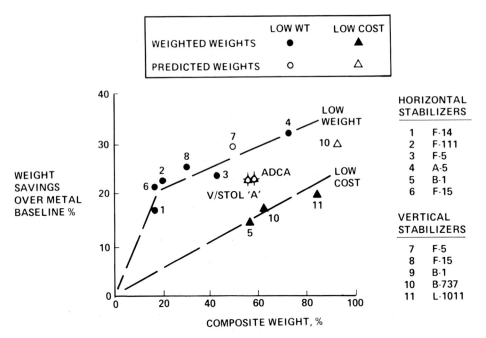

Fig. 15.13   Advanced composite empennage weight savings.

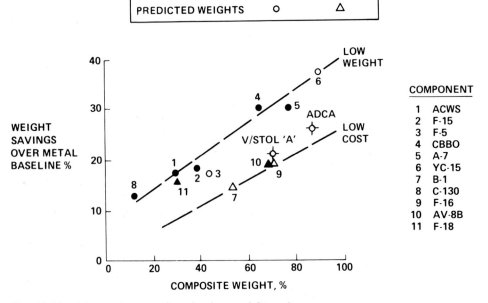

Fig. 15.14   Advanced composite wing box weight savings.

critical design test elements to validate the design. The final design can be one of two types: design for minimum weight or design for minimum costs. Early composite designs stressed low weight; current composite designs maximize low costs. Composite aircraft structures for empennages and wings developed by the industry are shown in Figs. 15.13 and 15.14.

V. ADVANCED COMPOSITE MATERIAL DEFINITIONS

Advanced composites are materials that combine high-strength, high-modulus fibers in an essentially homogeneous matrix. The combined or tailored material is anisotropic, that is, not isotropic, having mechanical and physical properties that vary with direction relative to natural reference axes inherent in the material. Advanced composites components are heterogeneous, that is, a material consisting of dissimilar constituents separately identifiable. Further, advanced composites are anelastic; that is, strain is a function of both stress and time, such that, although no permanent deformations are involved, a finite time is required to establish equilibrium between stress and strain in both the loading and unloading directions.

Other commonly used terms and their definitions are as follows.

Angle ply. Same as Cross ply. Any filamentary laminate that is not uniaxial.

Autoclave. A closed vessel that generates heat and pressure and is used to create composite structures.

B-stage. An intermediate stage in the reaction of a thermoset resin in which the material softens when heated or treated with solvents to permit handling prior to curing. Subsequent processing converts the resin to the C-stage for structural utilization, and subsequent heating or solvents will no longer soften the material.

Balanced laminate. A composite laminate in which all laminae at angles other than 0° and 90° occur only in ± pairs.

Buckling (composite). In advanced composites, buckling may take the form not only of conventional general instability and local instability but also a micro-instability of individual fibers.

Cocuring. Curing a composite laminate and simultaneously bonding it to some other surface during the same cure cycle.

Crazing. Development of a multitude of very fine cracks in the matrix material.

Delamination. The separation of the layers of material in a laminate.

Disbond. A lack of proper adhesion in a bonded joint.

Hybrid. A composite laminate comprised of laminae of two or more composite material systems.

Interlaminar. Descriptive term pertaining to some object, event, or potential existing or occurring between two or more adjacent laminae.

Layup. A process of fabrication involving the placement of successive layers of materials.

Matrix. The essentially homogeneous material in which the fibers of filaments of a composite are embedded.

Polymer. An organic material composed of long molecular chains consisting of repeating chemical units; resin matrices used in composites are polymers.

Prepreg, preimpregnated. A combination of reinforcements (filaments) and a matrix treated to the B-stage. Prepregs are the starting raw material for composites.

Scrim. A reinforcing glass fabric woven into an open-mesh construction used in the processing of prepregs to facilitate handling. Scrim materials are generally left in the completed composite and assure uniform dispersion of layers of filaments.

Selective reinforcement. The reinforcement of selected areas of a basically metallic structure by the addition of advanced composite material for local augmentation of strength or stiffness.

Symmetrical laminate. A composite laminate in which the ply orientation is symmetrical about the laminate midplane.

Tow. A loose untwisted bundle of filaments.

Wet strength. The strength of a composite measured after exposure to humidity, water, or other specified natural environment.

Whisker. A single filamentary crystal of 25 μm diameter with aspect ratios between 100 and 15,000.

VI. COMPOSITE MATERIAL AND LABOR COSTS

In industry a major emphasis is being placed on design to cost. This philosophy is also being utilized for advanced structures. This problem is complex with advanced composites, since material costs are extremely sensitive to both volume and time. Further, labor costs are being reduced by automation and structural design simplification.

Generalized current costs for the first unit are given in Table 15.8. Production costs are lower and tend to follow a 75% learning rate. The B/Ep cost information is taken from production records; other costs are estimated from overall industry data. These costs are presented primarily as a guide, and all specific designs and components must be separately estimated by the cognizant operating departments and estimating groups.

Experience on both production and development programs have shown that advanced composite structures can be of lower total costs than their current (non-cost-

Table 15.8 Typical Unit Primary Costs

Material system	Material cost per pound ($)	Manufacturing labor hours per pound
B/Ep	300	11.0
Gr/Ep (LHS)	40	9.0
Gr/Ep (UHM)	250	10.0
Gl/Ep	3	5.0
B/Al	260	15.0
B-Gr/Ep	75	10.0
Gr-Gl/Ep	25	8.0

Note: Material costs are for production quantities and manufacturing time are for horizontal stabilizers and wing-type structures of large area, 100 ft^2 and above.

optimized) metal counterparts. Low-cost structurally suitable composites are possible using the following.

Mix various filaments (hybrids) within a given structure.
Minimize use of honeycomb sculpturing.
Maximize cocuring of various details (i.e., bond and cure in one operation).
Minimize total numbers of joints and detail parts.
Simplify transitions and variations in skin thickness.
Provide realistic tolerances and inspection requirements.

VII. COMPOSITE TEST METHODOLOGY

The problem of obtaining meaningful data in a cost-effective manner can be solved by the combined efforts of the design and materials disciplines. For example, design engineers have had serious doubts about the structural validity of both flexural and short-beam horizontal shear tests (Fig. 15.15), even though these tests are accepted material engineering standards for development and qualification of composites. The specific objection is the use of unidirectional small-size test specimens, since they have little or no relationship to size, thickness, laminate orientation, loading, and failure mechanisms actually experienced by the complete structure. Conversely, materials engineers object to the greater complexity and lack of standardization of the types of specimens specified by the design engineers (Fig. 15.16). Resolution of this problem can be achieved only by developing standardized tests that are meaningful to both disciplines. One test that might be generally accepted, an off-axis tension coupon test on a multidirectional laminate, would provide insight into behavior under a combined loading situation, and

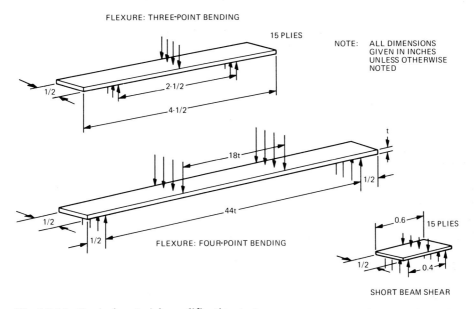

Fig. 15.15 Typical materials qualification tests.

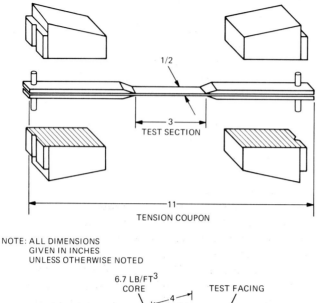

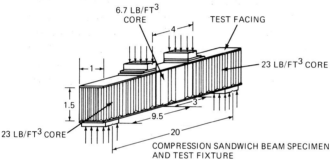

Fig. 15.16 Coupon and sandwich beam specimens.

into the resin-fiber bond; these coupons could be tested both with and without holes
to provide a complete evaluation of the material systems.

Compressive and in-plane shear properties remain the most difficult, controversial,
and least standardized of the mechanical property tests. The compression strength of
filamentary composites is a function of the strength and stiffness of the fiber matrix,
the interface shear and tension strength, and layer instabilities. The interaction of these
characteristics direct that the laminate fabrication procedures for the test item be the
same as that used for the end item.

Compression tests are generally characterized by the method of load application
to the test area and the method of stabilization used. Today, four principal loading techniques are used to obtain compression strength allowables: direct end loading, bending
of a honeycomb sandwich beam, and using a tab-ended plain panel specimen wherein
the compressive load is introduced by shear through a gripping arrangement similar to
that used for introducing load to plate tension specimens, and combinations of these.
Unidirectional composites are generally tested for compression using the sandwich beam;
multidirectional specimens are evaluated for compression properties using the lower
cost shear-loaded plain panel.

The methods of Rosen and Chamis for determining in-plane shear properties have seen greater use in the last couple of years. Both methods analytically determine shear properties from simple tension tests, Rosen by measuring the longitudinal extension and lateral contraction of a ±45° laminate loaded in the 0° directions for shear modulus determinations, Chamis by determining the properties on specimens oriented 10° off axis. Though these two tests are relatively simple and inexpensive to perform, the largest amount of in-plane shear data available are from carefully performed rail shear tests (Fig. 15.17).

Composite parts currently being produced are intended to have extended production and service lives to 20 years and more under adverse conditions. To achieve this, the material systems and processing operations used must provide adequate structural integrity, reproducibility, and durability. These characteristics are dependent upon the properties of the constituents of the material systems. Therefore, dependency studies have been undertaken to establish the chemical composition or "signatures" of resins for quality assurance, and investigation of neat resin mechanical properties to a level never before reported to enhance processing and to establish failure criteria. Variations in matrix systems are known to have affected the response to standard laminating procedures and have thus adversely affected the producibility of composite detail parts. The variations under investigation include type, purity, and concentration of chemical constituents, uniformity of mixing, and resin processing. Evaluation approaches use a variety of currently available analytical methods to assure reproducibility of matrix systems between carefully established limits based on mechanical test data and processing experience under production conditions. Improved material reliability and durability

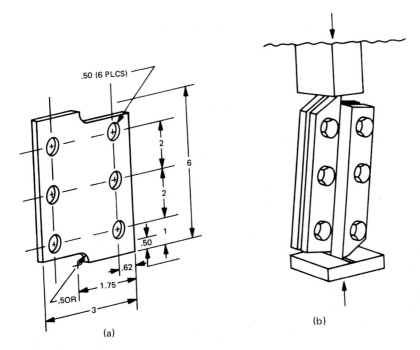

Fig. 15.17 In-plane shear test method.

are to be realized by characterizing approved resin systems and subsequently expanding existing specifications to include chemical composition and processing requirements. Batches of prepregs having undesirable matrix variations can thus be rejected before they are accepted for production using the same practices currently used for materials that do not meet physical and mechanical property requirements. Production samples of the resin and prepreg materials are chemically analyzed using liquid chromatography (LC), gel permeation chromatography (GPC), reverse-phase liquid chromatography (R-PLC), infrared spectroscopy (IR), and differential scanning calorimetry (DSC) to determine differences in chemical composition and processing. Also, test laminates are autoclave molded from each accepted batch of prepreg using controlled processing. Environmental static tests include $90°$ and $±45°$ tension coupons and quasi-isotropic compression beams of $(±45/0_2/90)_s$ fiber orientations after humidity exposure at $150°F$ $(71°C)$ and 82% relative humidity. The specimens are exposed in the form of tabbed plates or fully fabricated beams. Moisture pickup is monitored by traveler specimens of the same fiber orientations and thicknesses until constant weight is attained. In addition, tension coupons in a resin-dominated orientation $(90°/±45°)_s$ are subjected to constant amplitude fatigue at ambient conditions. Diffusion parameters are determined to identify any change in the diffusion kinetics of the matrix system as a result of compositional variations.

A special useful test technique for composites is acoustic emission—nondestructive testing. Historically, the first practical application of acoustic emission monitoring of a composite structure occurred on fiberglass-reinforced plastic rocket motor cases undergoing hydrostatic proof testing using the audible popping noises the motor cases emitted under stress to predict failure pressure. This was accomplished by plotting one average amplitude of the detected noise against the burst pressure. The inverse relationship between the two quantities was explained by assuming that the noise was caused by damage. Thus, if damage was present it would cause early failure and create a high average amplitude of noise compared with the case where no damage was present. A later investigation into the failure of fiberglass-reinforced pressure vessels substantiated this hypothesis.

Using an amplitude sorter while monitoring tension tests of carbon fiber-reinforced plastic specimens, it was discovered that different outputs were obtained for strong and weak specimens. Since energy is proportional to frequency, the data indicate that a weak sample emitted mostly low-frequency emissions and a strong sample emitted fairly broad-band emissions. Furthermore, it was noted that, as failure was approached, the emissions in both samples shifted toward the low-amplitude (low-frequency) end of the spectrum.

Most engineering structures are subjected to cyclic loading, and the possibility of fatigue damage must be considered. Acoustic emission provides an excellent means of detecting and monitoring fatigue damage growth. One of the most important uses to which acoustic emission testing can be put is detecting when a structure is about to fail. The rapid increase in counts immediately prior to fracture gives an experimenter adequate warning of impending failure.

VIII. TYPICAL DESIGN OF COMPOSITE STRUCTURES

The major composite structures are at present in the aerospace field. Although composites are used in construction, automotive, marine, and other industries, most of the

Table 15.9 Composites Usage

Component	State-of-the-Art				A-V/STOL	
	F-14	F-15	F-18	AV-8B	Baseline	Goal
Horizontal stabilizer	√	√	√		√	√
Vertical stabilizer, rudder		√	√		√	√
L.G. doors, speed brakes		√	√	√	√	√
Wing covers			√	√	√	√
Wing substructure				√	√	√
L.E., flaps			√	√	√	√
Fuselage components			√	√	√	√
Nacelles					√	√
Fuselage assemblies						√
LG components						√
% Structure utilized	1	2	10	22	29	56
% Aircraft weight reduction	1	2	5	9	10	16

advancements are in military aircraft structures. Typical of current aircraft and type of structures in use are shown in Table 15.9. The future vertical takeoff aircraft (V/STOL) will expand the use of composite structure since weight savings are critical to vehicle performance. Preliminary studies of one type of V/STOL airframe has shown that maximum weight savings result when at least half the aircraft is designed in composites. Figure 15.18 shows a typical material distribution for a subsonic V/STOL aircraft.

Initial studies at Grumman have shown that the composite wing of this type of aircraft is optimum in a multirib design. The wing skins are plain graphite-epoxy panels

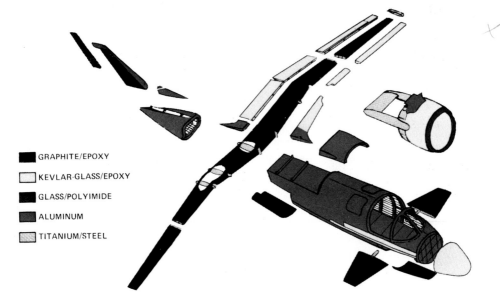

GRAPHITE/EPOXY
KEVLAR-GLASS/EPOXY
GLASS/POLYIMIDE
ALUMINUM
TITANIUM/STEEL

Fig. 15.18 Baseline structure.

including softening strips at the attachment points of the upper cover and an integral rib angle lower cover. The softening strips are low-modulus fibers to permit local high strains to minimize stress concentrations at fastener sites. The integral lower cover permits attachment without cover penetration to minimize potential fuel leakage of the wet wing and to locate fasteners in a low-stress region.

The front and rear spars are channel sections designed to permit web buckling to minimize weight. The ribs are I sections with sine-wave webs to increase stiffness without the use of a honeycomb sandwich. This type of design provides a wing that utilizes 75% composite material and is 22% lighter than an aluminum wing at a comparable production cost.

The design of the V/STOL center fuselage is an integral hat stiffened graphite-spoxy skin with integral graphite-Kevlar hybrid minor frames. The skins are attached to the bulkheads with mechanical fasteners. The bulkheads are graphite-epoxy hat and bead stiffened panels. The fuselage side skins are postbuckle designs, including fiberglass-epoxy damage-tolerant strips as crack stoppers. This type of structure utilizes 80% composite material and is 25% lighter than a comparable aluminum structure.

The design of the V/STOL empennage is an integral multispar hybrid skin structure containing approximately 60% composites and is 24% lighter than a comparable aluminum structure. The design impetus for this structure was low cost of materials operation, and maintenance. The hybrid skins were designed to provide both low-energy impact without repair and ballistic damage with ready repair.

The engine nacelles are integrally stiffened Kevlar-epoxy structures that contain 75% composites and are 30% lighter than an equal aluminum structure; the structures also significantly reduce the radar cross-section of the aircraft. The leading edge of the nacelles are plain panel fiberglass-reinforced polyimide composite, which can operate at 450°F structurally and still provide the deicing function.

The composite design of the future V/STOL aircraft includes advanced aerodynamics, such as supercritical wing and low-drag fuselage. These advances are readily incorporated with composite structures, and thus the overall aircraft weight savings can be as large as 16%.

IX. CONCLUSION

Composite materials and structures are a relatively new technology that is rapidly maturing since components made from composites are significantly more efficient than equivalent metal counterparts. Composite structures are low weight, durable, and suitable for many applications in the fields of transportation, construction, and recreation. Many composite materials are currently available and well characterized, and with time many new and improved materials will be developed. The methodology for design with these anisotropic materials has been established and is in general practice for the design of aircraft structures.

Composite structures have shown at least 20% weight savings over metal structures and lower operating and maintenance costs. Structures in service for longer than 10 years have shown that composites are durable, resist fatigue loading, and are readily maintained and repaired. The early structures were replacement components, and currently composites are designed from the outset. The future for composite materials is assured since the material can significantly improve structures to minimize energy drain and meet economic targets.

X. COMPOSITES: CHALLENGE AND OPPORTUNITY*

This report is based on a study of the current and potential applications of composite materials in the U.S. industrial sector and the technical opportunities that might be addressed by the National Bureau of Standards to enhance commercial utilization of these types of materials. The study includes consideration of the major composite classes: organic, metallic, and ceramic matrix in combination with conventional and novel filler (reinforcement) materials.

The special areas of competence of the National Bureau of Standards that have been considered in this study are those that principally lie in the Center of Materials Science, National Measurements Laboratory. These capabilities have been matched against the challenges to identify opportunities. They are metallurgical science, standards and environmental interaction, fracture mechanics, polymer science and engineering, ceramics, glass, solid-state science, and reactor radiation for analytical studies on materials.

Without further discussion, it is recognized that opportunity for the National Bureau of Standards to enhance commercial utilization of materials of any kind lies in development and making available standard reference samples, development of standard test methods, generation of critical property data, including characterization and publication of results, and program support for industry in the national interest.

The ranking of opportunities will not be attempted here except to partially relate suggestions to perceived current national issues that are listed below with no priority assigned except to suggest that all are of high priority in contemporary U.S. thinking.

Productivity
Energy
Health and safety
Environment
Housing
Consumerism and quality of life
Innovative technology and balance of payments

A. Definition and Caveats

A *composite* is an innovative material that combines a high-strength material and a high-toughness material; it draws upon their mutually beneficial properties while lending support to their respective deficiencies. Multiple high-strength elements are given the task of supporting the majority of applied load. The matrix, on the other hand, bears the responsibility of protecting the high-strength elements from externally induced damage, maintaining their alignment, and, most importantly, transferring the load to the strong elements themselves. Many successful composite systems, as we shall see, have been developed making use of long and short fibers and whiskers, powders, and laminates. The object of composite design is to achieve a combination of properties that offers superior resistance to external stress, whether mechanical, chemical, or thermal (Kelly and Hertzberg, 1976).

---

*Contributed by ROBERT S. SHANE. A report to the Center of Materials Science, National Bureau of Standards. Robert S. Shane, Shane Associates, Incorporated, Wynnewood, Pennsylvania. Purchase Order No. 816639, September 6, 1978.

Excluded from consideration in this study are such important materials and products as concrete, tires, laminates, wood products, electrical circuit boards, and metal and polymeric alloys. We also exclude processing studies except as they fit into the Center's composite program.

By and large, the substantial societal uses of composites at the present time lie with the organic matrix components. Principal emphasis will be given to the organics, but the challenges and opportunities in the metal and ceramic matrix composites will be noted in the appropriate place.

B. Conclusions and Recommendations

1. One of the most important if not the most important problem area is that of characterizing the elements of a composite as well as the composite itself. Characterization is defined as a description of "those features of the composition and structure (including defects) of a material that are significant for a particular preparation, study of properties, or use and suffice for the reproduction of the material" (National Materials Advisory Board, 1967). The problems described in the cited NMAB report, as further clarified in the later report on Organic Polymer Characterization (NMAB 1977), are still present for composites (also see Carpenter, 1977, 1978).

It is recommended that

Important characterizing parameters of each of the major composite classes be identified. This applies to the ingredients of the composite as well as to the composite itself.

Methods for evaluating the characterizing parameters be identified or developed and standardized.

Standard reference materials be developed to assist in validating methods and results of characterization studies.

2. Fracture mechanics has addressed the problem of defect propagation through materials, and substantial progress has been made in understanding mechanisms of failure. In composites it is of great importance to understand the interfacial phenomena whereby imposed stress is distributed by the matrix to the reinforcing element.

Recommendation

Commence and prosecute vigorously studies of the interface between the matrix and the reinforcing element so that rational selection of materials and surface treatments (e.g., sodium potassium alloy treatment of graphite fibers to promote aluminum wetting) may be made and the attainable properties of the composite be realized.

3. A major drawback to more rapid adoption of composites in design is lack of user confidence. Apart from inadequate characterization of the elements of the composite and the composite itself, there is no present nondestructive evaluatory technique to satisfactorily assess the performance capability of the composite; nor is there a good theoretical basis for proof testing.

Recommendations

Commence and vigorously prosecute studies of nondestructive evaluation of composites.

Build a body of data on correlation of laboratory sample testing results with full-scale structural behavior.

Collect and publish user data so that reliability of composite structures may be assessed from statistical data on experience.

Institute a central data bank for composite information on characterization, test methods, test data, and field experience whereby potential users of composites may satisfy themselves as to the experience of others and the basic knowledge on which to acquire confidence in the use of composites.

Commence studies that will lead to a comprehensive theory of fracture in composites akin to the Griffiths law for metals.

4. Specifications and standards for all classes of composites are inadequate to support widespread industrial use of composites.

Recommendation

Institute and prosecute vigorously actions leading to promulgation of standard test methods and compositions. Since this will take some years to accomplish and, as will be seen later, the national need is great, the work on standardization should be started at once using the resources of the voluntary standards organizations as well as those within the National Bureau of Standards.

C. Societal Needs for Composites

From a study published by the Smithsonian Science Information Exchange (Payne, 1978) the following partial list of composite materials applications currently under study has been derived (military applications have been omitted).

Aircraft doors, electromagnetic windows, engine parts, fairings, landing gear,
     propellers, skins, spoilers, windows, wings
Airfoils
Airframers
Automobile components (body, engine, drive train)
Balloons
Battery separators
Bone substitutes
Brake linings
Building components
Chemical-handling hose
Containers
Dental restorations
Drive shafts
Electrical ceramics
Electrical insulation
Electrical switches
Fan and compressor blades
Fencing
Flooring
Flywheels
Gears

Gas turbine vanes and other components
Helmets
Helicopter rotor blades
High-conductivity wire
Hydrofoils
I beams
Joists
Laser-resistant materials
Low-temperature elastomers
Machine tools
Magnets
Mine roof bolts
Packaging
Paving material
Pipes
Prosthetic devices
Portable bridges
Protective clothing
Radomes
Rail car wheels
Railroad crossties
Refractories
Rotor shafts
Seals
Ship and boat components
Shock and vibration absorbers
Solar collectors
Solid lubricants
Structural panels
Superconductors
Tankage
Thermal insulation
Tires
Transmission housings for helicopters
Utility poles
Valve components
Wear-resistant coatings
Windmill components

This study gives available data on present and projected usage by categories. It should be obvious from the data presented that the automobile segment of the transportation industry will be the large user of composites. This is because the national energy crisis demands lighter cars and the American public's taste is for large cars. From Table 15.10, which summarizes the categories of uses of organic matrix composites in 1978, 1983, and 1988 (based on projections), it can be seen that such composites will play an important role in American life. In the detailed discussion of each category, the rationale for these figures will be developed.

Table 15.10 Present and Projected Use of Organic Matrix Composites[a]

	Tonnes		
Market	1978	1983	1988
Civilian aerospace	10,005	12,080	14.795
Household appliances	57,500	77,000	99,900
Business machines	58,000	78,000	101,000
Construction	204,000	269,000	344,000
Industrial, farm, misc. machinery	—	12	29
Marine	172,000	234,600	264,280
Motor vehicles	146,300	204,977	321,710
Power generation	7	45	3,113
Sporting goods	42,238	59,409	79,454
Totals	690,050	935,123	1,228,281

[a]The various combinations of matrix and reinforcements and a discussion of their technical problem areas will follow the section on usage. See note on Fig. 15.12 for method of computing projections.

The principal advantages that accrue from the use of organic matrix composites in design are as follows.

1. Lower energy cost for a given volume of fabricated composite and lower energy cost for moving the generally lighter composite part during its lifetime. (See Fig. 15.19 and Table 15.11.) Thus an automobile hood made of graphite-reinforced epoxy has a manufacturing energy cost of 60% of a steel hood. In 5 years of use the graphite-epoxy hood will only reach 30% of the fuel consumption of a steel hood, and this figure includes the relatively high manufacturing energy. A one-piece glass-reinforced polyester hood would only reach 40% of the fuel consumption of a steel hood (Tables 15.12 and 15.13).
2. Specific strength and stiffness
3. Fatigue resistance
4. Damage tolerance
5. Corrosion resistance
6. Design flexibility

Examples of the weight saving possible in two transportation applications are (1) a total weight saving of 26% in a subsonic (commercial) aircraft and (2) a total weight saving of 315 lb per truck if heavy springs, made of graphite-polymer composite, were used instead of steel (Lewis and Brake, 1977).

Details of the engineering advantages of composites may be found in Kaiser (1978), Kelly and Hertzberg (1976), Payne (1978), Lewis and Brake (1977), Thomas (1978), Nardone (1964), and Theberge et al. (1976), Rabenborst (1975).

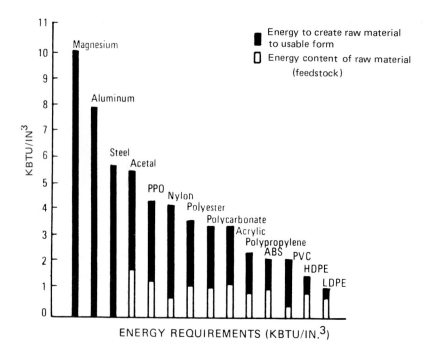

Fig. 15.19 Energy requirements of common metals and polymers. (After Thomas, 1978.)

Table 15.11 Energy Requirements for Producing Four Intermediate-Size Hoods

Hood	Weight (lb)	Material energy needs (Btu/lb)	Total energy to produce (kBtu/hood)	Energy savings over steel (kBtu/hood)
Steel	75	28,000	2100	Base
Aluminum	36.7	108,300	3975	−1875
2-Piece polyester composite	47	40,100	1885	215
1-Piece polyester composite	34	4,100	1363	737

Source: From Thomas (1978).

Table 15.12 Total Energy Savings During Manufacture and Lifetime

Hood	Production savings (kBTU)	Energy savings due to light weight (kBTU)	Total energy savings (kBTU)
Steel	Base	Base	Base
Aluminum	−1875	4388	2513
2-Piece polyester composite	215	3208	3423
1-Piece polyester composite	737	4697	5434

Source: From Thomas (1978).

D. Organic Matrix Composites

A study of a sampling of the available literature indicates that the principal likely candidates for present and future exploration are

> Glass-fiber reinforced polyester resins
> Glass-fiber reinforced epoxy resin
> Glass-fiber reinforced nylon resin
> Graphite-fiber reinforced epoxy resin
> Graphite-fiber reinforced polyimide resin
> Graphite-fiber reinforced phenolic resin

Details of the combinations are presented here with notes on usage and literature references.

Epoxy-boron (on tungsten or graphite) fiber reinforced can be used at service temperatures up to 225°C, where maximum stiffness is needed. Successful applications are in aircraft sections, helicopter blades, and gears and bearings. Boron fibers are very expensive with little prospect of early reduction in price. Hence usage will be limited to highly special designs (Clauser, 1973).

Epoxy-graphite fiber has an excellent combination of properties and for the price offers great stiffness and strength. Impact strength can be improved by use of hybrid glass-graphite fiber (June and Walter, 1978). This composite has excellent promise for automobile applications, particularly in drive trains, springs, and other critical parts (Lewis and Brake, 1977). It is currently being designed into airplanes (Witham et al., 1978) and will find many other uses in machinery where strength and stiffness combined with low density are required. There is continuing advantage to using this material since

Table 15.13 Component Lifetime Energy Savings with Composites (Hood)

Steel	Base (100% energy consumption)
Aluminum	25% (75% energy consumption)
Graphite-epoxy	65% (35% energy consumption)

Note: Savings are based on computing sum of manufacturing energy and fuel energy.
Source: From Thomas (1978).

energy consumption to overcome inertial losses will be minimized for the life of a machine. Other planned uses include energy storage flywheels (although glass-reinforced composites may be superior) (Rolston, 1977).

Epoxy-glass fiber (also glass microballoons and flakes) takes advantage of the high tensile strength of glass as well as low cost to bid to become widely used for structural parts. The moisture sensitivity of some E-glass composites has been shown by Thomas (1978) to be capable of being overcome by choice of resin system and ratio of curing agent. Epoxy-glass has been the low-cost common reinforcement. S-glass has been developed for higher performance composites. Problem areas are characterization and standardization of resin, fiber, and fiber surface treatment and standardization of processing.

Epoxy-aramid fiber is perhaps best used when the aramid is part of a hybrid reinforcement (Anon., Modern Plastics, May 1978). The outstanding contribution of the aramid is impact resistance. The aramids can be hybridized with graphite, glass, and fillers, such as mica flake. It is estimated that as much as 30% of the marine market will use hybrid reinforced composites by the late 1980s to take advantage of the large potential weight saving. The great stiffening contribution of mica flake to a composite should be noted.

Epoxy-basalt fiber composite: It has been shown that continuous mineral fibers made from basalt are technically and economically attractive in epoxy matrix composites (Subramanian et al., 1977). The composite properties are superior to those of glass-fiber reinforced epoxy composites in terms of bond of resin to fiber and hence of stress transfer from matrix to fiber. As in glass, silane treatment of the fiber improves the efficiency of the reinforcement. The usual problem areas of characterization and process control exist. There is no shortage of basaltic raw material in the Pacific Northwest region.

Phenolic-graphite fiber composite: The phenolic resins, as a class, include char-formers that act to slow fire spread. The low cost of the matrix and resistance to creep make these composites attractive. Slow fabrication speeds will militate against their use in mass production industries, like the automobile industry.

Polyimide-boron fiber composite will be limited to special high-temperature, high-strength, high-modulus applications because of the high price of its components. Problems of characterization, specification, and processing standards are present. Polyimide resin chemistry and a good bibliography are given by Lorenson (1975). A discussion of boron- and graphite-reinforced polyimide resin is given by Kollmansberger and Birchfield (1973).

Polyimide-graphite fiber composite: The comments on the preceding composite apply here except that graphite fiber is two orders of magnitude less costly than the boron fiber. The composite can compete with metals strengthwise and achieve a significant weight saving at service temperatures up to 600°F. The same problem areas exist as noted above. The work of Harper et al. (1977) on production of graphite-polyimide prepregs is noteworthy for paying attention to the need for processing parameters.

Polyimide-aluminum powder: Aluminum powder added to a polyimide resin used as an adhesive for titanium and for polyimide-graphite composites enhances the bond strength and permits service temperatures up to 315°C (St. Clair and St. Clair, 1976, 1978).

Polystyrene-glass fiber composite: Although highly regarded in the mid-1960s, this composite has been almost completely supplanted by polyester composites.

Polyester-glass fiber, microballoon, flake composites is the workhorse of the composites industry. The product can be enormously improved by the addition of graphite fiber in small quantities. Uses of polyester composites in marine hulls, tankage, and industrial equipment is going up steadily. The glass fiber industry looks to a doubling of volume from 1977 to 1981. However, the problems of characterization, nondestructive testing, and process standards, previously noted, inhibit critical design usage. It should be noted that reinforced reaction injection molding is entering the production field and will affect polyesters as well as the polyurethanes (Wood, 1978; SPI, 1975, 1977; du Pont, 1978).

Polyurethane composites: The reinforcing agents being used are glass, graphite, and aramid fibers and mica flake. The use of small amounts of graphite fiber significantly enhances the mechanical properties. The use of the RIM (reaction injection molding) process permits mass production, which interests the automobile industry. General Motors and Chrysler are known to be interested. It is expected that 1978 consumption of 21,500 tons of RIM plastics will be 200,000 tons by 1985 and of this about a third will be a composite. Markets are principally automotive (two-thirds of the total), recreational vehicles, sporting goods, agricultural equipment, construction, furniture, and so on (Wood, 1978). Particularly but not exclusively for polyurethane foams it should be noted that addition of fibers, flakes, and powders reinforce foams as well as cast and extruded plastics.

Nylon-fiber, flake, powder, composites have the outstanding benefits of impact resistance, stiffness, and low warpage. They are used for small parts in automobiles and special machines (Anon, Modern Plastics, 1978). Other advantages lie in the recently developed capability for stamping parts from sheets (SPI, 1975).

Polyolefin foam-asbestos composites have interesting properties that fit them for underground structural parts (SPI, 1977). Such usage includes utility boxes, manholes, septic tanks, and cable conduits. The properties of interest are the flexural modulus coupled with flexural strength, elevated heat distortion temperature, good impact strength, environmental resistance (very important for underground usage), and the relatively low cost and ease of reprocessing scrap.

E. Technical Problem Areas for Organic Matrix Composites

Competition for raw material supplies between energy and petrochemical uses and among petrochemical uses creates a study area on what is the best usage of available raw materials from the point of view of current needs and future needs.

The variety of resins and sources for resins makes it imperative that organic materials used as a matrix for a composite be adequately characterized. Not only are analytical and physicochemical data needed, but a determination of what constitutes adequate characterization must be made in the national interest. In this connection the NMAB report (NMAB-332, 1977) should be consulted and ongoing work on organic matrix composite characterization should be noted.

The early efforts of military contractors (Carpenter, 1978) to find adequate characterization modes should be taken as a point of departure to both concerns and techniques.

The nature of a composite dictates a study of the work function at the interface of the resin matrix and the reinforcing material. Important physical and chemical problems of wetting, reaction, and compatibility and stress transfer modes at the interface need to be addressed to give a fundamental understanding of the physical phenomena leading to reliable extrapolations to the macroworld of design.

The various reinforcing materials need to be characterized no less than the resins. It is unconscionable that scientists and engineers describe the essential materials of design by maker and catalog number. The Navy has made a good beginning with the work of Carpenter at McDonnell Aircraft Co. It should not be left to the Department of Defense to generate the data that will benefit all American industry.

Test methods need to be correlated with actual stresses in full-scale structures. After correlation the test methods need to be standardized for use in commerce. Through its activities in the voluntary standards organizations and its in-house activities, the Center for Materials Science can give a mighty assist to understanding the properties and the rational use of composites.

In its problem-solving mode the Center for Materials Science can empirically utilize the present state of knowledge in the national interest.

F. Ceramic Matrix Composites

This study will address the following matrix materials in the light of their contribution to advanced design. This choice is made because of their probable use in energy conversion and critical structural design in machinery, transportation, and communication. The materials of interest are alumina, silica, silicon carbide, silicon nitride, and carbon.

Alumina fibers are now available from pilot production. An important projected usage of alumina fibers is in high-temperature batteries, notably the sodium-sulfur traction battery operated at 300°C (Battelle, 1976).

Silica has been widely used for nucleated compositions with a very low coefficient of thermal expansion. It has a special advantage in its transparency to electromagnetic radiation.

Silicon carbide has been widely explored since it was learned that the stoichiometric composition would retain its mechanical properties beyond 1400°C with no appreciable loss, as contrasted to the earlier abrasive grade of silicon carbide with 8-10% excess silicon that degraded severely at 300°C.

Silicon nitride is a strong competitor to silicon carbide but is suspect because of possible high-temperature fatigue due to slow high-temperature crack propagation.

The carbon matrix is essentially used only in combination with carbon filamentary reinforcements in graphitized form. The developmental emphasis is currently on their use in high-temperature nozzles and other surfaces in heat engines. Currently the carbon-carbon composite is widely used in airplane brakes.

The entire category of ceramic matrix composites represents an opportunity of small present commercial volume but with substantial future promise. Civilian usage will extend to heat engines, textile equipment, energy storage, ocean-bed mining, hot bearings, self-lubricated bearings, crucibles and retorts, cutting tools, and so on.

Fiber-reinforced ceramic composites, already noted in the case of carbon-carbon composites, extend to metal fiber-reinforced composites (e.g., using tungsten fiber) and ceramic fiber-reinforced composites (e.g., using silicon carbide fiber to reinforce silicon nitride). The field is in its infancy but is growing rapidly.

For the field of ceramic matrix composites to continue to grow rapidly, it is necessary to overcome the three major drawbacks to widespread usage of ceramic in structural design. These are poor predictability, inadequate processing knowledge, and designer reluctance (NATO, 1972). All these areas represent opportunities for the Center of Materials Science of the National Bureau of Standards.

Although the opportunities are seen to be developmental and realization is some 5-10 years in the future, the following societally important goals (incomplete list) are being attacked with ceramic composites:

Battery plates and separators
Laser components
Structural beams
Friction components
Pressure vessels
Electrical housings and substrates
Jet engine components
Aerodynamic structures and airplane brakes
Automotive engines
Deep-sea mining equipment
Geothermal well components

The specifics of the opportunities are characterization, test methods, impact of processing variables on product, and, very importantly, developing a credible accepted theory of fracture.

G. Metal Matrix Composites

This part of this study is largely based on work supported by the Department of Defense. The metals of interest are those that are already familiar in current applications. Three classes of metal matrix composites will be considered here: (1) metal plus refractory fiber, (2) directionally solidified eutectic mixture, and (3) dispersion-strengthened metal (Payne, 1978). Examples of each class are as follows.

1. Aluminum-boron (on tungsten or graphite fiber), aluminum-silicon carbide-coated boron (on substrate fiber), aluminum-graphite fiber, titanium-boron (on tungsten or graphite fiber), titanium-silicon carbide, titanium-beryllium, magnesium-graphite fiber, lead-graphite fiber, and copper-graphite fiber
2. Nickel-nickel aluminide-nickel niobide, nickel aluminide-nickel niobide, nickel cobalt chromium-tantalum carbide, and cobalt nickel chromium aluminum-chromium carbide
3. Tungsten-tungsten carbide, tungsten-silicon carbide, aluminum-aluminum oxide, and nickel-thoria

1. Advantages of Metal Matrix Composites (NMAB, 1974)

The strength to density and modulus to density of an aluminum composite is two and a half times that of titanium and stainless steel alloys.
The fatigue strength of composites is advantageous.
The designer can adjust or orient the fibers to the principal axis of stress, thus designing the material for the application.
By adjusting fiber orientation, harmful natural vibrations can be "tuned out."
Metal matrix composites offer high-temperature capability far above organic matrix composites.
Joining by conventional methods is easy.

2. Disadvantages of Metal Matrix Composites

The disadvantages are those of a material in early development (NMAB, 1974).

> High cost due to small volume.
> Lack of adequate characterization data.
> Lack of standardization of materials and processes.
> Lack of standardized fully defined test methods and, particularly, nondestructive
>   testing and accelerated testing.
> Lack of a service-developed data base.
> Lack of demonstration of reproducibility (related to the first three items)
> Lack of data on cost effectiveness and life-cycle costs.
> Need for improvement in foreign object damage resistance for high-speed rotary
>   components.
> Lack of accepted comprehensive theory of fracture for metal matrix composites.
> Lack of adequate environmental testing based on a satisfactory modeling of the
>   conditions to which the composite will be subjected.

Subsequent to the NMAB study, which has been liberally cited above, the Institute
for Defense Analyses held three workshops that updated our knowledge in the area of
metal-matrix composites (Institute for Defense Analysis, 1975, 1976). Table 15.14
lists metal matrix composites that are currently under study.

The problem areas, opportunities for the Center of Materials Science, are identified as

> Nonlinear matrix response
> Analysis of effects of processing and service use on thermal and chemical properties
> Analysis of effects of processing and service use on mechanical properties
> Improved fibers
> Parameters associated with chemical vapor deposition processes
> Selection and preparation of reference metal matrix composite materials
> Nondestructive testing
> Nondestructive inspection
> Characterization
> Collection, analysis, and publication of preliminary design data, perhaps by a
>   National Bureau of Standards composite information center

The *civilian* prospects for metal matrix composites are found in components or
systems for

> Gas turbines
> Launch vehicles and ships
> Superconductors
> Electrical brushes
> Batteries
> Moderate temperature aircraft
> Lightweight transmission cases
> Lightweight bridging and other special structures

Note: The projected usage of organic matrix composites in Table 15.10 was computed by multiplying the current usage by the ratio of projected 1980 and 1985 business

Table 15.14 Current Metal Matrix Composites of Interest That Are Under Study

Fiber (vol %)	Fiber diameter (mils)	Fiber coating	Matrix material	Current interest
Boron (50)	8.0	None	1100 Aluminum	High impact strength
Boron (50)	8.0	None	6061 Aluminum	High strength
Boron (50)	5.6	None	2024 Aluminum	High strength
T-300 graphite (35)	0.3	Boride	201 Aluminum	Modulus = titanium
T-50	0.25	Boride	202 Aluminum	24×10^6 psi modulus
FP alumina (35)	0.8	None	Aluminum plus 2% lithium	Modulus = titanium
FP alumina (55)	0.8	None	Aluminum plus 2% lithium	Modulus = steel
FP alumina (55)	0.8	None	Magnesium alloys	Low cost, high specific strength, modulus
T-50 graphite (40)	0.25	None	Magnesium alloys	Low cost, high specific strength, modulus
Boron (50)	5.7	Silicon carbide	Titanium-3Al-2.5Sn	1000°F service
Beryllium (40)	Elongated particles		Titanium powder	800°F service
Silicon carbide (50)	5.6	Carbide	Superalloy M-200	2000-2200°F sheets
Tungsten (50)	10-20	None	FeCrAlY	2000-2200°F sheets

levels computed by the U.S. Bureau of Labor Statistics to current business levels. See
Kutscher (1976), Bowman and Morland (1976), and Personick and Sylvester (1976).

BIBLIOGRAPHY (for Sections I-IX)

1. A. M. Lovelace, Advanced Composites, AIAA Paper No. 74-242, January 1974.
2. G. Lubin and S. Dastin, Boron Epoxy Composites in the F-14 Stabilizer, Society of
    Plastics Industry Conference, 1971.
3. W. Ludwig, H. Erbacher, and G. Lubin, Composite Horizontal Stabilizer for the B-1,
    Society of Plastics Industry Conference, 1977.
4. R. N. Hadcock, The Application of Advanced Composites to Military Aircraft,
    10th ICAS Congress, 1976.
5. I. G. Hedrick and J. B. Whiteside, Effects of Environment on Advanced Composite
    Structures, AIAA Paper No. 77-463, March 1977.
6. L. Lackman, *Advanced Composites Design Guide,* Third Edition, Air Force Ma-
    terials Lab, Ohio, January 1973.
7. M. Salkind and G. Holister, eds., *Applications of Composite Materials,* STP524,
    ASTM, Philadelphia, 1973.
8. *Composite Materials: Testing and Design,* STP460, ASTM, Philadelphia, 1969.
9. *Composite Materials: Testing and Design (Second Conference),* STP497, ASTM,
    Philadelphia, 1972.
10. Analysis of Test Methods for High Modulus Fibers and Composites, STP521, ASTM,
    Philadelphia, 1973.
11. *Composite Materials: Testing and Design (Third Conference),* STP546, ASTM,
    Philadelphia, 1974.
12. E. Scala, *Composite Materials for Combined Functions,* Hayden, Rockelle Park,
    New Jersey, 1973.
13. G. Sendeckyj, ed., *Mechanics of Composite Materials, Composite Materials,* Vol. 2,
    Academic Press, New York, 1974.
14. L. Broutman, ed., *Fracture and Fatigue, Composite Materials,* Vol. 5, Academic
    Press, New York, 1974.
15. C. Chamis, ed., *Structural Design and Analysis—Part I, Composite Materials,* Vol. 7,
    Academic Press, New York, 1974.
16. C. Chamis, ed., *Structural Design and Analysis—Part II, Composite Materials,* Vol. 8,
    Academic Press, New York, 1975.
17. M. Rich and R. Welge, AIAA Paper 72-392, presented at the 13th SDM Conference,
    San Antonio, April 1972.
18. *Proceedings of the Second Conference on Fibrous Composites in Flight Vehicle De-
    sign,* AFFDL-TR-74-103, Air Force Flight Dynamics Lab, Ohio, Sept. 1974.
19. *Materials and Processes for the 70's,* SAMPE, Azusa, California, 1973.
20. *Materials on the Move,* SAMPE, Azusa, California, 1974.
21. *New Industries and Applications for Advanced Materials Technology,* SAMPE,
    Azusa, California, 1974.
22. C. Rogers, Structural Design with Composites, in *Fundamental Aspects of Fiber
    Reinforced Plastics,* R. Schwartz and H. Schwartz, eds., Wiley, New York, 1968,
    pp. 141-160.
23. G. Lubin and S. Dastin, First Boron Composite Structural Production Part, *26th SPI
    Conference Proceedings,* Society for the Plastics Ind., New York, 1971, Sec. 17-C.
24. G. Lehman and A. Mano, AIAA Paper 72-358, 13th SDM Conference, San Antonio,
    April 1972.

25. D. Smillie and D. Purdy, Advanced Material Application to Subsonic Transport Aviation, Douglas Paper 6186, Presented at the Ninth Congress of the International Conference of Aeronautical Sciences, Haifa, August, 1974.

26. A. Tucci, Fabrication of Advanced Fibrous Composite Structures, Douglas Paper 6097, presented at the 1973 WESTEC Conference, Los Angeles, March 1973.

27. L. Lassiter, Applications and Concepts for the Incorporation of Composites in Large Military Transport Aircraft, Paper No. 17, Presented at the Royal Aero. Soc. Conference on Reinforced Plastics in Aerospace Applications, The Plastics Institute, London, April 1973.

28. R. Hadcock, The Application of Mixed Fiber Composites to Military Aircraft, presented at the 10th National ACS Symposium, Washington, D.C., June 1974.

29. E. Jarosch and A. Stepan, *Journal of the American Helicopter Society,* Vol. 15, 1970.

30. T. Scarpeti, R. Sandford, and R. Powell, The Heavy Lift Helicopter Rotor Blade, Preprint No. 710, 29th AHS Forum, American Helicopter Society, New York, May 1973.

31. M. Salkind, New Composite Helicopter Rotor Concepts, Proceedings of the Conference on Fibrous Composites in Flight Vehecle Design, AFFDL-TR-72-130, Air Force Flight Dynamics Lab, Ohio, September 1972.

32. M. Salkind, *Fibre Science and Technology,* Vol. 8, No. 2, 1975.

33. S. Sattar, H. Stargardter, and D. Randall, The Development of JT8D Turbofan Engine Composite Fan Blades, AIAA 69-465, American Institute of Aeronautics and Astronautics, New York, June 1969.

34. W. Schulz, J. Mangiapane, and H. Stargardter, Development of BORSIC–Aluminum Composite Fan Blades for Supersonic Turbofan Engines, ASME 71-GT-90, American Society for Mechanical Engineers, New York, March 28, 1971.

35. I. Petker and W. Holmes, Boron/Polyimide Fan Blades–A Fabrication Study, SAE 71-0772, Society of Automotive Engineers, New York, September, 1971.

36. D. Stansbarger, et al., Low Cost Manufacturing Concepts of Advanced Composite Primary Aircraft Structures, Second Quarterly Report, Contract F33615-74-C-5153, November 1974.

37. M. Rich, G. Ridgley, and D. Lowry, Application of Advanced Composites to Helicopter Airframe Structures, Preprint No. 880, 30th National Forum. American Helicopter Society, New York, 1974.

38. R. Hayes, Application of Advances in Structures and Materials to the Design of the YF-17 Airplane, Paper No. 730891, Society of Automotive Engineers, New York, October 1973.

39. J. DeVault, Commercial Applications for Graphite Reinforced Composites, *SAMPE Quarterly,* Vol. 5, No. 2, Jan. 1974.

40. K. Marshall, High Strength Composites for America Cup Defenders, *26th SPI Conference Proceedings,* Society for the Plastics Ind., New York, 1971.

41. G. Lubin, *Handbook of Fiberglass and Advanced Plastics Composites,* Van Nostrand Reinhold Co., New York, 1969.

42. B. W. Rosen, A Simple Procedure for Experimental Determination of the Longitudinal Shear Modulus of Unidirectional Composites, *Journal of Composite Materials,* Vol. 6, October 1972.

43. C. C. Chamis, et al., Ten Degree Off Axis Tensile Tests for Interlaminar Shear Characterization of Fiber Composites, *NASA Technical Note* TN-D-8215, April 1976.

44. V. F. Mazzio and R. L. Mehan, Effects of Thermal Cycling on the Properties of Graphite-Epoxy Composites, Composite Materials: Testing and Design (Fourth Conference), ASTM STP 617, American Society for Testing and Materials, 1977.

45. J. M. Carlyle, Acoustic Emission in Fiber Reinforced Composites, Naval Air Development Center Report NADC-75082-30, May 1975.

REFERENCES (for Section X)

Anon., Modern Plastics, *55*(3):48-50, March 1978.

Bowman, C. T., and Morlan, T. H., Revised Projections of the U.S. Economy to 1980 and 1985. Monthly Labor Review, pp. 9-21, Government Printing Office, Washington, D.C., March 1976.

Carpenter, J. F., Test Program Evaluation of 3501-6 Resin, (Contract No. N00019-77-C-0155, Naval Air System Command) McDonnell Aircraft Co., McDonnell Douglas Corp., St. Louis, Mo., May 1978.

Carpenter, J. F., Quality Control of Structural Nonmetallics, (Contract No. N00019-76-C-0138, Naval Air Systems Command) McDonnell Aircraft Do., McDonnell Douglas Corp., St. Louis, Mo., June 1977.

Clauser, H., Advanced Composite Materials, Scientific American, 229 (1):36-44 (Jul. 1973).

du Pont Co., de Pont Introduces Polyester Resins Having Unusually High Rigidity, du Pont Industry News, du Pont Co., Public Affairs Dept., Wilmington, Delaware, Oct. 2, 1978.

Harper, S. E., Stoops, W. E., and Wilson, M. L., NASA Tech Brief LAR-12266 Support Package, Simplified Systematic Production of Graphite Polyimide Prepreg, Langley Research Center, Hampton, Va., Winter 1977.

Institute for Defense Analyses, Proceedings of First MMC Workshop: July 9-10, 1975, Technological Development of Metal Matrix Composites for DOD Application Requirements, Part I: Identification of Technology and Application Issues, Ellis L. Foster, Jr., Program Chairman, September 1975. Proceeding of Third MMC Workshop, September 15-16, 1976, Recommended Program Plan.

June, R. R., and Walter, R. W., SAMPE Q. (3):26ff, April 1978.

Kaiser, R., Technology Assessment of Advanced Composite Materials, (Contract ERS 77-19467, National Science Foundation) Argos Associates Inc., Winchester, Massachusetts, April 1978.

Kelly, A. (Chap. 9), Hertzberg, R. W. (Chap. 12), *Frontiers in Materials Science* (ed. by L. E. Murr and C. Stein), Marcel Dekker, New York, 1976.

Kollmansberger, R., and Birchfield, E., SAMPE Q, 4(2):28 (Jan. 1973).

Kutscher, R. E., Revised Bureau of Labor Statistics Projections to 1980 and 1985: An Overview, Monthly Labor Review, pp. 3-8 inclusive, Government Printing Office, Washington, D.C., March 1976.

Lewis, R. W., and Brake, P. F., Polymer Matrix Composites for Automotive Applications, Symposium and Workshop-Polymeric Materials and Their Use in Transportation, Polytechnic Institute of New York, Brooklyn, April 27-29, 1977.

Lorenson, L. E., SAMPE Q., 6(2):1 (Jan. 1975).

Nardone, J., Plastic Gears, Plastic Report 16, Plastics Technical Evaluation Center, Picatinny Arsenal, Dover, N.H., July 1964. See pages 12 and 13.

National Materials Advisory Board, National Academy of Science—National Academy of Engineering, Organic Polymer Characterization, NRC Report NMAB-332, Washington, D.C., 1977.

National Materials Advisory Board, National Academy of Science, National Academy of Engineering, Characterization of Materials, NRC Report MAB-229-M, Washington, D.C., 1967.

National Materials Advisory Board, National Academy of Science, National Academy of Engineering, Metal-Matrix Composites: Status and Prospects, NMAB-313, Washington, D.C., December, 1974.

NATO, Study on Non-Metallic Materials, Document AC/243-D/168, Brussels, Belgium, May 1972.

Payne, W. H., New Directions in Composites Research, Report CM-78-1, Smithsonian Science Information Exchange, Inc., Washington, D.C., 1978.

Personick, V. A., and Sylvester, R. A. Evaluation of Bureau of Labor Statistics 1970 Economic and Employment Projections, Monthly Labor Review, pp. 13-24 inclusive, Government Printing Office, Washington, D.C., August 1976.

Rabenborst, D. W., SAMPE Q., 6(2):23, January 1975.

Rolston, J. A., SAMPE Q., 8(2):7, January 1977.

St. Clair, A. K., and St. Clair, T. L., Polymer Engineering and Science, 16 (5), May 1976; NASA Tech Briefs, LAR-12257, Polyimide Adhesives for Titanium and Composite Bonding, Langley Research Center, Hampton, Va., Spring 1978.

SPI (Society of the Plastics Industry), Proceedings from Reinforced Plastics/Composites Institute, 30th Anniversary Conference, Feb. 4-7, 1075, Washington, D.C.

SPI (Society of the Plastics Industry), Proceedings from Reinforced Plastics/Composites Institute, Feb. 8-11, 1977, Washington, D.C.

Subramanian, R. V., Wang, T. J. Y., and Austin, H. F., SAMPE Q., 8(4):1 (July 1977).

Theberge, J., Arkles, B., and Robinson, R., Ind. Eng. Chem., Prod. Research Deve. 15(2): 100ff (1976).

Thomas, G., Presentation, The Pentagon, July 24, 1978.

Witham, et al., Low Cost Composite Wiring/Fuselage Manufacturing, Report 1R-434-7(1), U.S. Air Force Materials Laboratory, Wright-Patterson A.F.B., Ohio, Feb. 1978.

Wood, A. S., Mod. Plastics, 55(2):34 (Feb. 1978).

16

Miscellaneous Materials*

Preceding chapters have covered the structural nature, the properties, and the applications of materials that comprise the bulk of the materials available to the design engineer. In this chapter will be found miscellaneous, generally nonmetallic materials that are sometimes used independently as structural elements or in conjunction with other materials.

I. CARBON AND GRAPHITE

A. Designations

The element carbon occurs in three principal forms: the cubic crystal diamond; the hexagonal layerlike crystal, graphite; and amorphous carbon. Diamond is crystalline, extremely anisotropic, and very strong in the plane of the layers. The properties of amorphous carbon vary widely according to the source and method of manufacture.

B. Source

Diamond is found in natural deposits and is made synthetically (e.g., General Electric Co.'s "Man-made Diamond"). Graphite occurs in natural deposits, such as in Madagascar, Ceylon, and Mexico, but is currently largely made artificially from amorphous carbon by graphitizing at about 2980°C. Amorphous carbon is obtained by removing the volatile matter from petroleum or anthracite coke, lampblack, or carbon black.

C. Production Processes

Synthetic diamonds are produced from graphite at extremely high pressures (up to 2×10^6 psi) and elevated temperatures (up to 2450°C). The industrial use of diamond is as an abrasive or, infrequently, as a semiconductor. As an abrasive, diamond is either used loose in a paste or bonded by metal or resin in cutting tools or grinding wheels.

Industrial graphite is usually formed into the shape of the desired part from a combination of a carbonaceous binder and amorphous carbon. This is compacted by molding or extrusion and slow baked at 700-1200°C to carbonize the pitch and remove the

---

*Contributed by ROBERT S. SHANE.

volatile matter. The part is now a carbon part that is frequently usable as is, or it may be graphitized by heat treating at about 2980°C. (With appropriate modifications the foregoing applies to carbon and graphite fibers and cloths.)

Carbon and graphite parts made this way are slightly permeable. They may be made impermeable by impregnation with chemically resistant synthetic resins. Suitable choice of resin results in an impermeable material resistant to practically all chemicals that will not attack the carbon or graphite.

Parts of controlled porosity may be made by the techniques of ceramic processing (see Chap. 28). After careful sizing of the particles, they are shaped and sintered to form a rigid and uniformly porous material. Commercial grades are available with apparent porosity of 12-28% within fairly close limits and permeability of 0.2-15 mdar.

"Carbo-spheres" (also known as microballoons) are filler material in the form of hollow thin-walled spheres of carbon. They are 5-150 μm in diameter (average 40) with an approximate wall thickness of 1-2 μm and a bulk density of 0.12-0.14 g/ml.

Carbon fibers for use as reinforcing agents in resin, ceramic, and metal matrices have been known since the 1950s. Initially they were produced from rayon precursors in fiber, felt, or cloth form. These materials were and are still being used. However, they had only moderate strength (100,000-150,000 psi, tensile) and low stiffness (8 $\times$ 10^6 psi, elastic modulus). By applying stress to the yarn during the graphitizing step, the strength was more than doubled and the elastic modulus increased by almost an order of magnitude. Using a polyacrylonitrile precursor resulted in an increase of elongation at break to 1.1%, which was a tremendous improvement for an essentially ceramic product. However, this material was expensive, and Union Carbide Corp. developed a less costly pitch-based line of carbon and graphite fiber products. The tensile strength and modulus of elasticity are satisfactorily high, although the elongation at break is 0.3%, still better by far than the performance of a true ceramic. Additional information in the use of these fibers in composites and their design properties will be found in Chaps. 15 and 28, and by writing to Carbon Products Division, Union Carbide Corporation (270 Park Ave., New York, N.Y. 10017).

D. Properties

Diamond, graphite, and amorphous carbon are chemically inert in practically all acids, alkalis, and organic compounds with the exception of those that are highly oxidizing in behavior. In air, diamond burns at 1100°C, graphite begins to oxidize at 400°C, and carbon begins to oxidize at 345°C. In a reducing or nonoxidizing atmosphere, temperatures to about 3600°C are resisted, although amorphous carbon will have transformed to graphite below this temperature. Above 3700°C all forms of carbon sublime at atmospheric pressure. Of course, the temperature resistance of impregnated parts is limited to that of the resin impregnant.

The thermal conductivity of carbon is high in comparison with other nonmetallic materials; that of graphite exceeds values for some metals (in pyrolytic graphite thermal conductivity with the grain approaches that of copper); that of diamond is even higher. Such high thermal conductivity combined with low coefficient of thermal expansion gives carbon products great resistance to thermal shock. The specific heats of carbon and graphite are approximately equal and change almost linearly from 0.15 at 0°C to 0.47 at 1000°C. The inertness of carbon and graphite leads to use as high-temperature insulation in inert atmosphere and vacuum electric furnaces. Powder is being replaced by

Table 16.1 Typical Thermal Conductivity of Graphite Felt (BTU/hr-ft^2-in.$^\circ$F^{-1})

Hot face temperature ($^\circ$F)	Atmosphere	
	Vacuum	Inert
1000	0.5	1.0
2000	0.8	1.8
3000	1.5	3.5
4000	2.5	7.3
5000	4.5	11.5

Source: Courtesy of Ultra Carbon Corporation, Bay City, Michigan.

felt because of the lower density and increased rigidity. Thermal properties and nominal weight are given in Table 16.1 and Table 16.2.

Although the electrical resistance of carbon products is high compared with metals, these products are still good conductors. Carbon and graphite products are not strong compared with metals and most plastics, but they have sufficient strength for many mechanical applications. Typical property values of interest are given in Table 16.3.

E. Applications

Carbon products are available for structural applications as fibers, felts, fabrics, squares, bricks, tiles, rods, pipe, tubes, and custom shapes that are made by extrusion, spinning, weaving, and molding with or without subsequent machining.

Usage of carbon and graphite products is based on one or more of the outstanding properties: mechanical strength, chemical inertness, resistance to high temperature and thermal shock, lubricity, thermal and electrical conductivity, low thermal neutron absorption cross section, and high scattering cross section. Since graphite is more expensive than carbon, its use is reserved for applications requiring its special properties. Some examples of the use of carbon and graphite are found in Table 16.4.

Table 16.2 Nominal Weight for Carbon and Graphite Felt

Nominal thickness (in.)	Graphite felt (lb/ft^2)	Carbon felt (lb/ft^2)
1/8	0.048	0.052
1/4	0.10	0.11
3/8	0.17	0.18
1/2	0.20	0.21
3/4	0.31	0.32
1	6.48	0.51

Source: Courtesy of Ultra Carbon Corporation, Bay City, Michigan.

Table 16.3 Typical Property Values for Carbon and Graphite Parts and Fibers[a]

Material	Density (g/ml)	Tensile strength (psi)	Flexural strength (psi)	Compressive strength (psi)	Modulus of elasticity (10^6 psi)	Electrical resistance (ohm-in. at 20°C)	Thermal conductivity (BtU/hr-ft-°F)
Diamond	3.52						
Carbon	1.57	200-900	400-2600	1700-1900	0.75-1.9	0.0011-0.03	3-8.7
Graphite	1.54-1.78	440-900	800-2800	1700-5000	0.8-1.4	0.0003-0.0005	71-97
Impregnated carbon	1.92	1800	4400	10,000	2.8	0.0016	3
Impregnated graphite	2.03	2500	4700	9,000	2.2	0.0003	86
Fine porous carbon	1.1	190	600	850	–	0.007	–
Coarse porous carbon	1.1	80	160	300	–	0.008	–
Fine porous graphite	1.1-1.77	110	250	500	–	0.0012	–
Coarse porous graphite	1.1-1.68	50	140	270	–	0.002	–
High-modulus Graphite fiber		300×10^3	–	–	120	–	–

[a]Wide variation may be expected according to source, size of part, and degree of calcination or graphitization and direction of measurement (with grain or against grain; the values in the table are the higher values of measurement with the grain).

Table 16.4 Typical Uses of Carbon and Graphite

Carbon	Graphite
Filters for corrosive liquids	Chemical processing plants for non-oxidizing corrosive conditions
Heat exchangers for corrosive liquids	
Conveying pipes	Molds and riser rods in steel mills; runout tables and guides for aluminum extrusions
Valve linings	
Components and linings for absorption, fractionization, and distillation towers	Fluxing tubes
	Thermocouple sheaths for aluminum
Crucibles for molten metal	Pit-ingot molds
Furnace linings	Molds for centrifugally cast nonferrous metals
Trough linings	
Mold plugs	Sintering boats and die assemblies for powder metallurgy
Core shapes	
Chill plates	Bearing seals, valve and pump packing
Electrodes for producing abrasives, carbides, ferroalloys, and phosphorus	Semiconductor manufacture
	Induction-heating crucibles and dies for continuous casting, crystal growth, epitaxial growth, and diffusion
Arc-welding electrodes	
Generator brushes	
Nuclear applications (moderator and reflector)	
	Molds for hot pressing refractory oxides, carbides, and diamond-tipped drill bits
	Foundry mold facing
	High-performance fibers in composites
	Lubricant
	Resistance heaters
	Electric furnace electrodes
	Cathodes and anodes for electrochemical processes
	Nuclear applications (fuel elements, shielding, moderator, and reflector)
	Rocket motor nozzle components

Source: Graphite data courtesy of Great Lakes Carbon Corporation, Graphite Products Division, New York.

II. CEMENT, MORTAR, AND CONCRETE

A. Cement

The Romans were probably the first to use cement. They found that a water mixture of pulverized volcanic ash would harden to a firm mass, even under water, and termed the material Pozzolan cement. The next important development occurred in 1756, when Sweaton used lime, made from a limestone containing clay, for an underwater lighthouse mortar foundation. The two substances, clay and limestone, were first artificially mixed and calcined to form a hydraulic cement in 1824. This material was termed portland cement because its structures had the same appearance as a limestone variety of English building stone known as portland stone.

The proper mixture of clay and limestone for portland cement is sometimes found in natural rocks. Such ore is calcined as it comes from the quarry and then is crushed and ground, forming a natural cement. This product is usable for applications having low strength requirements. Sometimes blast-furnace slag has the right combination of ingredients, so that it may be ground and used like Pozzolan cement, but most cement used today is a true portland cement.

The mixture used for portland cement has about 80% carbonate of lime (such as limestone, chalk, or marl) and about 20% clay (as clay, shale, or slag). The materials are ground, wet or dry, in a ball mill and then placed in round tunnel kilns. The kilns are slightly inclined and up to 250 ft long. They are slowly rotated so that the charge gradually tumbles from the entry at the high side to the exit at the low side. Heating is accomplished by gas or powdered coal burning at the center of the lower end of the kiln. During passage through the kiln the mixture of limestone and clay partly fuses and sinters together, forming a clinker. The clinker is often mixed with a small percentage of calcium sulfate after cooling. It is then ground to a fine powder. The chemical composition is not simple. It consists of at least two aluminates of calcium oxide, two silicates of calcium oxide, and lime (CaO).

When mixed with water and allowed to dry, portland cement hardens, probably by the formation of hydrated lime, which crystallizes and bonds the other materials together. First setting takes place within 24 hr, but hardening through the mass continues for as long as 28 days, depending on the amount of calcium sulfate added to retard the process. High percentages of alumina replacing silica result in quick hardening in 1 or 2 days. Hardening may also be speeded to some extent by careful control of composition, calcining, and fine grinding of typical portland compositions.

A high-alumina low-iron variety known as white cement is used for ornamental work because of its white color. Waterproofing varieties contain soaps, oils, or tannic acid added during the final grinding.

B. Lime

The variety of lime usually used in mortar, either with or without cement, is known as quicklime, or common lime. It is made by heating calcium carbonate in a vertical or a rotary horizontal kiln at high temperature. The temperature is usually 900°C or above so the equilibrium pressure of the released carbon dioxode is above atmospheric pressure. The commercial product consists almost entirely or calcium oxide, with small percentages of iron oxide, silica, and magnesium oxide.

When mixed with water, or slaked, heat is released, the volume and weight of material increase about two or three times, and a white powder, calcium hydroxide or slaked lime, is formed. If an excess of water is used, the material becomes puttylike. It will set and harden by drying of moisture, by crystallization of the hydrate, and by absorbing carbon dioxide from the air, forming calcium carbonate. The process of hardening is slow.

Lime is available from suppliers in hydrated form. This form avoids the heat and reaction of slaking and gives a uniform product free of unslaked areas. Most lime used for mortar is of this type.

A magnesium variety is also available that contains magnesium oxide up to 40%. It slakes slower than common lime, expands less, evolves less heat, and sets more rapidly.

Besides its use in mortar, lime is also employed in marking athletic fields, as an acid neutralizer for soil, for the manufacture of paper, for the dehairing of hides prior to tanning, for the purification of illuminating gas, as a bleaching powder, and as a refractory or a chemical agent in metallurgical and chemical processes. The water suspension is used as whitewash.

C. Mortar

Cement mortar is made by mixing cement and sand with water. The usual proportions are 1 part of cement to 2 or 3 parts of sand. Enough water is added to mix the premixed cement and sand to a smooth working consistency.

The strength of cement mortar generally increases with (1) the proportion of cement used, (2) the coarseness of clean sand used, and (3) the density of the solid mass. Small additions of hydrated lime up to 10% or increased amounts of portland cement are used to improve workability and water tightness. Unslaked lime must never be added, since its expansion will disintegrate the mortar. The applications for cement mortar are for laying stonework, building blocks, and brickwork and for stucco exterior plastering.

Lime mortar is similarly made by mixing 1 part lime and 2 or 3 parts of clean sand with sufficient water. An alternative method is to mix the lime with water, forming a putty that is later mixed with sand when needed. With either method, the mortar or the putty should stand for a period of about 10 days for slaking to be complete. Even when hydrated lime is used, the mixture should stand for 1-3 days prior to use.

Lime mortars are not used for thick walls because of the slow rate at which they harden, nor are they used in contact with wet soil or under water because the hardening process requires the loss of excess moisture as well as the removal of the moisture released during the conversion of lime to calcium carbonate by absorption of carbon dioxide. The principal application of lime mortar is for interior plaster. Fiber (formerly cattle hair) is mixed with the mortar for base coats or scratch coats directly on wood, lath, plasterboard, or wire meshing. The fiber is usually added to the lime putty, sand is then turned in, and additional water is added as the material is to be applied. Plasterboard and wire-meshing applications require less added fiber than wood or lath applications. In the finer work the base coat for the latter is given a second coat, called brown coat, containing less fiber. The finish or skim coat may be made from lime putty and white sand, but more usually it consists of plaster of Paris or a mixture of 1 part lime putty and 1 or 2 parts plaster of Paris. The base coat is usually 1/4 in. thick, second coats 3/8-1/2 in. thick, and finish coats 1/8-1/4 in. thick.

Plaster of Paris is prepared by carefully driving half the water of combination from gypsum (calcium sulfate dihydrate) by heat. Gypsum itself is used as a retarder for portland cement, as a filler in paper and paints, and as blackboard chalk. When water is added to form a stiff paste, some of the plaster of Paris redissolves, forming gypsum crystals. The crystals precipitate out, leaving residual water free to dissolve more plaster of Paris, and the process continues in this manner until practically all the plaster of Paris has been converted to a hard mass of interlacing crystals of gypsum. Besides the use of plaster of Paris in plaster, it is also employed for plaster casts of statues, for example, for interior stucco, for bandages for broken bones, and for plaster mold casting of nonferrous metals, polymers, and ceramics.

D. Concrete

Concrete is made by mixing cement with aggregates of hard, inert, stonelike particles. The aggregate may consist of coarse sand, broken stone, gravel, cinders, or broken brick or mixtures of these materials, depending on the application. Portland cement is used for most concrete today, particularly for concrete reinforced with soft iron bars or mesh, for concrete load-bearing members, and for underwater concrete. It may be mixed on the job by hand, or by continuous or batch-process rotary mixers, or it may be premixed and brought to the job.

The strength of concrete depends on the mixture of materials used. Strength increases with (1) the proportion of cement, (2) the reduction in the amount of water used, if enough water is added to provide thorough mixing, (3) the size of the coarsest aggregate, and (4) the proportions of fine and coarse aggregate of given sizes that result in the highest density. Sand in excess of the amount required to fill voids in the coarse aggregate reduces strength. Rounded stone in place of angular or crushed stone also yields a weaker concrete. Air-entraining cement is used to decrease density and increase resistance to cracking resulting from thermal and mechanical shock.

Typical mixtures of concrete range from 1 part cement, 1½ sand, and 3 parts coarse stone aggregate (1:1½:3) through 1:2:4 and 1:2½:5 to 1:3:6. The first mixture is considered rich in cement and is used for highest strength requirements, as in columns, or for exceptional water requirements. The 1:2:4 mixture is standard for reinforced floors, beams, columns, and machine foundations and for tanks, sewers, and watertight work. The 1:2½:5 mixture is used for ordinary machine foundations, walls, piers, floors, abutments, sidewalks, heavy sewers, and so on. The leanest mixture is used for unimportant heavy work in walls and foundations where stationary loads are involved.

Strength values range from 3000 lb/in.^{2} compressive strength after 28 days for rich medium-wet concrete to 1000 lb/in.^{2} for lean mixtures. Cinder aggregate gives low strength values, 800-400 lb/in.^{2}, for similar mixtures. Strength increases over the 28 day value with aging by as much as 33 1/3% for medium-wet consistency and up to 50% for wet consistency. The benefit of keeping concrete damp during cure has been amply demonstrated.

The transverse modulus of rupture of concrete is considerably less than the tensile strength, of the order of 1/5-1/6. The true tensile strength is approximately 1/12 the compressive strength. Accordingly, concrete is seldom relied upon to support tensile stress. For example, beams, when required, are made with steel reinforcement.

Direct shear strength is 60-80% of the compressive strength, but shear failure resulting from diagonal tensile loads may be only 5-10% of the compressive strength.

Objects made from wet mixtures of concrete are completely water proof if properly filled, and if the junctions are well wetted before covering with neat cement and adding the next section. Watertightness increases with cement content, age, thickness, and aggregate size, provided the mixture is homogeneous. Dispersions of polymers have also been used in concrete to retard water penetration.

Concrete resists external attack by mineral oils and, if hardened, by animal fats, vegetable oils, and dilute acids. Strong acids, electrolysis, and seawater tend to disintegrate it under certain conditions. Freezing retards setting and hardening and may cause a surface scale to crack from exposed surfaces placed in freezing weather. This can be avoided by artificial heating or by covering the surface with straw or other insulation during the setting period.

III. CORK

A. Properties

Cork is a material that appears in nature. It is the spongy outer bark of the cork oak tree, which grows principally in northern Africa, Spain, and Portugal.

Cork has a cellular structure, each cell surrounded by an impenetrable skin that encloses air. This structure accounts for the low specific gravity of 0.15-0.20 when the outer hard bark is removed. It accounts for high resistance to the penetration of liquids, since capillarity is not effective in drawing liquids through it. And it also accounts for the low thermal conductivity of cork.

The natural material is truly compressible. It will undergo compression with very little side flow as its Poisson's ratio is practically zero. Long subjection to pressure does not destroy the structure or its ability to return to original size when the load is removed.

The coefficient of friction of natural cork is higher than that of most materials used for friction application (usually 0.5 dry and 0.25 in oil). The presence of moisture and oil does not deteriorate the property. The principal limit is operating temperature. Cork will not bear sustained temperatures above 250°F, although it may withstand short surges to 300 or 250°F. Dry applications are therefore limited to light or heavy duty involving little slippage.

Chemically, cork does not discolor or deteriorate in oil, grease, gasoline, and other petroleum products. Alkalis, some acids, and some organic solvents have a destructive action. Discoloration occurs with prolonged heating at 250°F or above as a result of slow distillation. Higher temperatures result in burning, but only in the presence of an external flame. Combustion produces a carbonized coating that inhibits further burning. Temperatures under 250°F, humidity, and atmospheric conditions do not cause aging. However, cork insulation has been largely superseded by synthetic polymeric closed-cell foams.

B. Applications

Cork is sometimes used in the form of blocks or slabs as trimmed from the natural material. In this form the material is quite variable, contains structural flaws, and is irregular in size, thickness, and shape. These limitations are often overcome by laminating the corkwood to produce greater uniformity and structural strength.

Compositions are also used to take advantage of the properties of cork and to alleviate its shortcomings for particular applications. Finely ground cork is combined with various binders to provide for expansion or contraction, and for fabrication into sheets, disks, tubes, blocks, rings, strip, balls, wheels, rods, and irregularly shaped articles. Laminations with cloth or vulcanized fiber improve tensile strength. Natural or synthetic rubber composites provide normal side flow under compression when that is required. Typical applications are as follows.

1. Stoppers, gaskets, and packings for seals where compressibility, resilience, and resistance to many chemicals, particularly oil and water at temperatures below 120°C, are needed.
2. Thermal insulation, particularly in cold-storage plants, trucks, and railway cars, either in block form or granulated for filling between walls. This use is being supplanted by mineral or polymeric closed-cell foams.
3. Vibration insulation, where compressibility and resilience are employed in cushions for machine tools, and where the unique surface is used for reducing reflection for air-borne noise.
4. Friction applications, such as clutches, wheels, feed rolls, fishing rod grips, and nonskid bases for light equipment.
5. Floats for life jackets, gages, pumps, and carburetors that make use of the low specific gravity.
6. Floor coverings, either cork tile, from the natural product, or linoleum, made from cork and various fillers with drying oils, and reinforced on the bottom side with burlap.

The newer synthetic plastics are replacing cork in many applications as technology reduces cost and offers additional design or service advantages to plastics. The applications of cork are still important and valuable.

IV. FELT

A. Manufacture of Industrial Felt

The manufacture of felt follows the general operations of wool textile manufacture, except conventional spinning, weaving, and knitting are not involved. Wool fiber constitutes the basic material. It may be new fiber, reprocessed fiber, and/or reused wool. For some applications, these classes of wool may be blended or replaced with other animal, synthetic, such as polyester or polyimide fibers, or vegetable fibers, particularly cotton. The fibers are blended and picked, carded, and then felted by a process known as hardening and fulling. The interlocking structure characteristic of felt is obtained in this operation through a combination of mechanical work, chemical action, moisture, and heat. Cross laying of carded fiber webs is usually employed to develop balanced strength. The rough surface of the wool fiber results in a nonraveling, porous, and resilient material.

Various finishing operations are employed, depending upon the application for the felt. Scouring and neutralizing follow the felting operation, and the material is then dyed and dried. The surface is sheared and sanded for lubrication and filtering uses; a rough or fuzzy surface is obtained for polishing uses, and an ironed or pressed surface may be applied to make the felt water repellent, to give it high wettability, to give stiffness for clean cutting or low distortion, and to proof the felt for resistance to mildew, fungi, flame, moth, and vermin.

Felt is available commercially according to standards of the SAE, the ASTM, and the Felt Association. These specifications cover two forms, roll felt and sheet felt. Roll felt is usually made in lengths 15-60 yards long, 60 or 72 in. wide, and in thicknesses from 3/64 to 1 in. Sheet felts are customarily made in 36 in. squares from 1/8 to 3 in. thick. Roll-felt standards cover thickness tolerances, weight (usually pounds per square yard) tolerances, percentage of wool fiber, combined content soluble in water and carbon tetrachloride, ash content, tensile strength, and splitting resistance. It is made in soft, firm, extra firm, and back-check grades from almost 0 tensile strength to 500-600 psi. Sheet felts are made in extra soft, soft, medium, hard, and rock-hard grades to thickness and weight tolerances.

B. Properties

The properties of felt depend on the fiber blend, the construction, the density, and surface and chemical treatments, if any. In general, all felts have high formability because of their high elongation before rupture. They have high compressibility and resilience, with less than 5% permanent deformation after 30% load deflection in compression. Their coefficient of static friction is approximately 0.20 with dry steel, 0.26 with glass, and 0.37 with clean wood. Their wear resistance varies directly with density and fiber quality, and inversely with applied load. Thermal conductivity ranges from 0.36 BTU/hr-sq^2-°F in.$^{-1}$ for the heavier densities to 0.25 for the lighter densities. Combinations of wool with kapok fiber produce even lower conductivity and 0.80 sound absorption coefficient above 512 cycles. Absorption for oil is exceptionally high, ranging from 160-180% for the heavy densities to over 300% for the lighter densities. Higher viscosity oils are absorbed somewhat better than low-viscosity oils. Capillarity increases with density of the felt and in inverse proportion to oil viscosity. Temperatures from −60°F to 212°F do not affect felt in the absence of water. Prolonged exposure to water causes slight swelling, shrinkage in length, increased elongation, and a reduction in tensile rupture strength of 25%. Water-repellent treatments decrease these effects. Exposure to air, ozone, ultraviolet, oil, grease, organic solvents, or weak inorganic acids has negligible effect, but strong acids, strong bases, or alkalis are injurious to strength.

C. Applications

The unique properties of felt are utilized in numerous applications. The following list is typical of the more important uses.

1. *Sealing*: High-density roll type is used for oil retention, intermediate density for grease shields, lower densities for household weather stripping, for dust shields on instrument cases and automobile windows, and on confined oil or grease shields.
2. *Wicking*: For moistening devices and bearing lubrication, high- and intermediate-density roll-type felt is used. Sheet types are used for ink-transfer rolls and rubber-stamp pads.
3. *Filtering*: Low-density felt in the form of cloths or bags is used for clarifying of oils, paints, dopes, lacquers, enamels, and solvents and as air filters for respirators and air-conditioning ducts.
4. *Wiping and polishing*: Sheet-type felt is used. The soft varieties are used for metal buffing wheels, medium grades are used for mirror- and lens-polishing laps, hard grades are used for glass tumbler, granite, and marble-polishing laps, and the extrahard grades are used for heaviest duty polishing wheels.

5. *Vibration and shock isolation*: Both sheet and roll types in all ranges of density are used for isolation, the denser types being chosen for devices to support heavier loads. Pads are used as stops, dampening supports for buffer blocks, and as padding in shipping cases.

6. *Surface protection*: Soft, roll-type felt is used for the underside of telephones, lamps, and similar articles to avoid damage to finished surfaces, to provide non-skid support, and for cushioning and packaging.

7. *Thermal insulation*: The lighter densities are used as linings for cabins, containers, and appliances to reduce heat losses or to control surface temperatures.

8. *Sound insulation*: Lighter density types of roll felt are used to dampen instrument panels, dashboards, containers, and appliances and to absorb air-borne noise above 500 hertz.

V. FIBER

A. Sources of Fibers

The term "fiber" is applied to a filament or threadlike piece of any material. Sometimes a raw material that can be separated into threads is also known as a fiber. Under such broad definition, fibers may be drawn from most any animal, vegetable, mineral, or, indeed many polymers. The polymeric fibers are discussed in Chap. 11.

1. Mineral Fibers

The mineral fibers of industrial importance are asbestos, glass fiber, slag wool, and metal wool. The latter is used mainly for filters and cleansing applications. Slag wool is used as thermal insulation, although glass fiber and asbestos are described elsewhere in this text. The use of asbestos in the United States has been limited recently due to the carcinogenic effect.

2. Animal Fibers

There are two principal types of animal fibers of commercial importance. One is animal hair obtained primarily from sheep, goats, camels, and pigs. Pig bristles are used in paint brushes. Many types of hair are used in upholstery and surface finishes, but sheep wool is the most important of the hair fibers. Its length varies from 1 to 8 in. and its diameter from 0.0018 to 0.004 in. It is graded by size and length, the "tops" being used for worsted yarns, and the "noils" for carded yarns. Both are used for fabrics having many personal and household applications. Wool waste for packing glands, for example, is made from carpet yarn with fibers not less than 3 in. long. Wool fibers are also employed for felt, as discussed in Sec. IV.

The other source of animal fiber is silk produced by the mulberry silkworm or from the larvae of other moths. Silk from the mulberry silkworm is of high strength and elongation, and varieties from other moths are inferior in strength.

3. Vegetable Fibers

All vegetable fibers consist mainly of cellulose (see Chap. 10). They may be seed hairs, such as cotton, or the inner bark of plants (bast hairs), such as flax, hemp, jute, and ramie. Cotton and ramie fibers are white, flax fiber is a gray brown, and jute and hemp fibers are brown. Those with a high proportion of cellulose are flexible and elastic, whereas the woody types are stiff and brittle. Their strength varies considerably, but the relative strength in decreasing order usually is manila hemp, sisal hemp, jute, linen, cotton, and ramie.

B. Products Made from Vegetable Fibers

Cotton is used for making batting, guncotton, cloth, and rope. The seed hairs or linters are separated from the seeds by ginning. The natural spirallike twist of the fiber makes it especially suited to spinning of yarn. Very fine yarns are made from sea-island cotton grown on islands off the shore of the Carolinas. A variety known as mercerized cotton is obtained by treating the fibers with strong caustic while it is under tension. The strength of the fiber is substantially increased by this treatment. The fiber also takes on a high luster and high absorption for dyestuffs.

Linen is the fiber obtained from the flax plant by a process called "retting," which involves steeping in water. The fiber consists of nearly pure cellulose and varies from pale yellow-white to gray. It may be bleached white by a chlorine bleach, but some deterioration occurs. Linen fiber has luster, is stronger than cotton, conducts heat better, and absorbs about 6-8% moisture from air, about the same amount as is absorbed by cotton. Linen does not withstand boiling in alkalis, bleaching powders, or oxidizing agents as well as cotton. It is used primarily for fine cloth.

Linen rags were the primary source of fiber for fine paper, but today most paper is made from wood pulp. Better papers contain an appreciable fraction of reinforcing fiber, such as cotton or other longer staple fiber. Cheaper paper is made by using the portion of wood pulp that does not dissolve in water. Better grades are made by heating wood chips with solutions of acid calcium sulfite, sodium hydroxide, or sodium sulfide. These reagents dissolve the lignin of the wood and leave a nearly pure cellulose residue. The residual pulp is then bleached with chlorine, an insoluble filler such as clay or barium sulfate is added, and glue or a mixture of aluminum hydroxide and soap is applied as a size. The mass is run between rolls of the paper machine, forming a coherent sheet of cellulose fibers with interstices filled by the filler, and the surface coated with sizing.

Jute is obtained from the stalks of the jute plant by retting. The fiber obtained is 4-7 ft long and a pale yellow-brown. The short fibers are employed in papermaking. Longer ones are used in the manufacture of coarse woven fabrics, such as burlap, carpet bindings, as a substitute for hemp in twine and small ropes, as a filler in cable, and as an adulterant for other fibers. The chief defects are deterioration under dampness and disintegration of the fiber if bleached.

Hemp is the name applied to the fiber obtained by retting the hemp plant. Manila hemp from the *Musa textilis* is white, lustrous, light, very strong, and of great durability. It is used in the manufacture of highest strength cordage. A variety known as sisal hemp is found in Central America, the West Indies, and Florida. It is light yellowish, straight, and smooth and is used for papermaking and for cordage, second only to manila hemp in strength. Other grades of hemp are dark brown and contain a mixture of cellulose and lignocellulose. They are more hygroscopic than cotton, less affected by moisture than jute, and are disintegrated by bleaching. They are used principally for twine and rope, and are little used for woven fabrics because of harshness and stiffness.

In making cordage, the fiber is first run into a continuous ribbon, which is then twisted or spun into yarn. A strand is then formed by twisting a number of yarns together in a direction opposite to the twist of the harn. A plain rope is "laid" by twisting a number of strands, three, four, or six, together in the opposite direction of the twist of the strand. A cable or hawser rope is laid by twisting three plain-laid three-strand ropes together in a direction opposite to the twist in the tope. The strength of the cable is less than that of the composite ropes, but greater elasticity and resistance to abrasion are obtained.

VI. FIBERBOARD

Fiberboard is also called vulcanized fiber, hard fiber, and fiber. The base material is prepared by treating paper wood pulp or a textile fiber with a saturated solution of zinc chloride. This mass is consolidated under heavy pressure, and the soluble materials are then leached out by soaking in water. The fiber is then dried, forming a hornlike material. A pliable variety may be prepared by treating with glycerine or glucose. The hard form may be sawed, drilled, and worked like wood. The soft forms may be molded. In general, vulcanized fiber parts have a specific gravity of 1.3, electrical strength of about 120 V/mil thickness, a tensile strength of 5000 psi in thin or thick sections and about 8000 psi for ¼ in. thicknesses, and after seasoning they have enough flexibility to bend on a radius 10 times their thickness at room temperature and 50 times their thickness after baking for 12 hr at 250°F. Vulcanized fiber is available in sheet and tube forms. Decorative textures or colors may be applied for some uses. In Sec. IX other types of board (wood particle board, wood fiber board, and molded boards) will be discussed.

VII. LEATHER

The term "leather" is used to designate the tanned and dehaired skin of an animal, and for articles made from a tanned animal hide. In industry the only leather of importance is that from steer hides. Best quality is obtained from the section back of the shoulders and close to the backbone. The size of this section is about 24-32 in. wide and 40-54 in. long.

Tanning is accomplished by treating the gelatin in a dehaired and cleaned hide to cause hardening. In this state the gelatin is not liable to decay. The treating process consists of steeping the hide in a water mixture of oak bark, tannin (tannic acid), chromic oxide, or certain other reagents from vegetable substances. Oak-tanned leather can be obtained 3/16 in. thick with a tensile strength of 3000-5000 psi, elongation (1 hr of loading) of 13.5%, modulus of elasticity of 18,000-32,000 psi, proportional limit of 23,000 psi, and density of about 0.035 lb/in.^{3} Chrome-tanned leather has a higher tensile strength of 8500-13,000 psi.

The primary application for leather has been in belting. Single thickness of 3/10 in. or cemented double thickness of 3/8 in. are employed. Wide belting and long belting is made by cementing the necessary number of strips together. Belts are cleaned with a coarse cloth and are dressed with neatsfoot or castor oil during idle periods. The oak-tanned belting is not suitable where exposed to steam, dampness, or dripping machine oil. Rawhide and chrome-tanned belts are superior to oak-tanned for this service, and mineral-tanned belts and waterproofed belts are available for the worst service. In recent years leather belting has been almost completely displaced by synthetic elastomers, of which polyvinyl chloride is the most prominent example, particularly for conveyers.

Rawhide is not a tanned leather. It is obtained by rubbing the raw hide with oils or fats, and then twisting and stretching the hide until the moisture is removed and the oil thoroughly impregnated. It is very strong, tough, and waterproof and is used mainly for belt lacings.

Cotton belts, both plain and impregnated, are used in place of leather for light power duty, and for hot, dry locations. Balata belting, a canvas belt impregnated with balata gum, is waterproof and is unaffected by animal oil. Mineral oil and temperatures over 120°F affect it, however. Rubber belting, sometimes with cotton or nylon cord, is well adapted to damp locations, but synthetic varieties should be used in applications subjected to steam, heat, oil, or grease. Metal mesh, wire, or sheet belting is also used. The trend to supplying motive power at each machine or tool has decreased the uses for belting, but it is still employed to operate auxiliary elements in many types of equipment and for conveyors. As noted above, polyvinyl chloride belting has captured the bulk of the belt market for conveyors and other synthetic elastomers have captured the bulk of the market for power transmission.

VIII. LUBRICANTS

A. Types of Lubricants

A lubricant is a material that is purposely placed between two machine elements that move intermittently or continuously with respect to each other, to reduce friction, temperature, and/or wear. Liquid lubricants are usually preferred where they may be retained. Plastic and solid lubricants are used where liquid lubricants cannot survive or remain.

1. Liquid Lubricants

Mineral oil is the most common lubricant of this type. It consists of a complex mixture of paraffin, naphthalene, aromatic, and unsaturated hydrocarbons that may be distinguished from crude oil by their lower viscosity. Redistillation, refining, and blending produce lubricating oils that range from spindle oils of low viscosity, hardly distinguishable from kerosene, to highly viscous, heat-resistant oils for airplane engines and steam cylinders. Compounding with animal and vegetable oils, or with fatty acids, is employed to obtain properties for specific purposes. Small quantities of synthetic materials may be added to retard oxidation (antioxidant), to retard rusting (antirust), to provide wetting, to withstand localized high pressure (extreme pressure additive), or to control foaming and emulsification.

Animal, vegetable, and fish oils were among the first used. Today they are still used alone or in compounding with mineral or other oils. Typical of these oils are neatsfoot, lard, and tallow oils of animal fat type; castor, polin, rapeseed, jojoba, and peanut oils of vegetable type; and sperm oil from the sperm whale (now largely supplanted by jojoba oil).

Silicone oils of synthetic origin are growing in application. Their chief advantages are high-temperature resistance, to 350-375°F, a pour point of −100°F, and a very flat temperature-viscosity curve. Since their lubricating value differs from typical mineral oils, applications should be made in accordance with manufacturers' recommendations.

Synthetic lubricants made from polymerized ethylene glycol and other precursors are on the market. They exhibit great lubricity and very long service life.

Water, air, and chemical liquids are employed as lubricants in some special applications. Water is used in ship propeller bearings. Pressurized air is used for "flating" very high speed centrifuge bearings. Chemical liquids often become lubricants when moving elements must be encased or submerged in processing such liquids.

2. Plastic Lubricants

Grease is the major lubricant of this type. Usually a grease is a suspension of metallic soaps in a lubricating liquid. Calcium and sodium soaps are most used and aluminum, lead, and lithium soap are used to a lesser extent. Synthetic resins and rubber may be added to some types to promote adhesive qualities. Available greases vary from nearly liquid to hard blocks, depending on the kind and quantity of the soap used. They are usually selected for relatively slow speed machine elements, especially open types operated intermittently. Another application is for high-temperature components that do not retain liquid lubricants.

3. Solid Lubricants

Graphite is the most common of the solid lubricants (see earlier section of this chapter), either alone or mixed with oil or grease. Other types sometimes used are soapstone, talc, wax, disulfides of tungsten or molybdenum, and silver sulfate. All are usually limited to slow-speed service. The sulfur-bearing types are suited to extremely high pressure with steel or iron machine elements.

B. Properties of Lubricants

There are a great many factors to be considered in applying lubricants. The major ones are

1. Nature of the relative motion involved
2. Relative speed between members
3. Nature of the load between bearing surfaces
4. Magnitude of the load between bearing surfaces
5. Temperature range under which the lubricant must operate
6. Cooling required and method of cooling
7. Method of supplying lubricant to the bearing surfaces
8. Atmosphere in which the bearings will operate
9. Bearing materials involved
10. Design and finish of the bearing members
11. Life required
12. Friction losses allowed

Each of these factors may impose limiting restrictions on the choice of lubricant. Selections should therefore be made with the aid of oil chemists or the supplier's representatives.

1. Viscosity

The most important property of liquid lubricants in thick-film lubrication is their viscosity. This property is defined as the ratio of shearing stress to the rate of shear, $\mu = f/$

(du/dh), where f = the friction drag through the oil film divided by the oil film area and du = gradient or difference in velocity in an oil film thickness dh. The units of viscosity coefficient μ are poises. One poise equals 1 dyne-sec/cm^2 in cgs units. Since the values are usually low, the centipoise, which equals 0.01 poise, is more frequently used. The centipoise equals 1.45×10^{-7} lb-sec/in.^{2}, a unit called the Reyn.

Viscosities are usually measured industrially by the Saybolt universal viscosimeter. This instrument measures the time for 60 ml oil at specified temperature to flow through a given orifice under its own weight. The Saybolt time in seconds cannot be used in design calculations but is converted into kinematic viscosity for this purpose by use of the equation KV, stokes = $0.0022 - (1.80/t)$, where t = Saybolt universal, seconds. The unit of kinematic viscosity is the stoke. The centistoke equals 0.01 stoke. The value of kinematic viscosity in stokes is multiplied by the density in cgs units to equal absolute viscosity coefficient μ in poises. Similarly, multiplying centistokes by density equals centipoises. The Saybolt viscosimeter is insensitive for low-viscosity oils below 32 sec, and for such oils the Ostwald viscosimeter is used. It reads directly in kinematic viscosity.

Mineral oils decrease in viscosity with an increase in temperature at a rate of 0.6-6% per °F. The ASTM has developed charts for predicting viscosities for temperatures intermediate of two measured values. The SAE has developed a numeric system for expressing viscosity grades of lubricants. High pressures (5000-15,000 psi) also cause a considerable increase in viscosity (125-1600%).

2. Oiliness

Bearings designed for thick-film lubrication often, as on starting, operate with boundary lubrication. Other bearings often have boundary condition throughout their service. Oiliness is the property that may be of the greatest importance under these conditions. In this sense, oiliness refers to a combination of wettability, surface tension, and slipperiness, all these qualities being required. There is no absolute method of measuring this property or any substitute for measuring start-up force and steady-state operating force by some form of dynamometer.

3. Specific Heat

Lubricants are often called upon to serve as a coolant while providing lubrication. The specific heat is therefore important in calculating heat balances. Values for petroleum hydrocarbons are approximately equal to

$$C = \frac{0.388 - 0.00045t}{S} \; \text{BTU/lb-°F}$$

where t = temperature (°F) and S = specific gravity at 60°F compared to water at the same temperature.

4. Specific Gravity

This property has no relation to lubricating value but is useful in changing kinematic viscosity to absolute viscosity. The specific gravity is usually referred to water, at 60°F for both. An American Petroleum Institute "Gravity" (API) is also recognized. It is API = 141.5/sp. gr. − 131.5. The temperature coefficient of expansion varies with specific gravity of oil from 6.5×10^{-4} for 0.70 sp. gr. to 3.5×10^{-4} for 1.0 sp. gr.

5. Other Properties

Other properties of lubricants are determined primarily as specification tests to control product uniformity. They are oxidation resistance, thermal stability, acidity, saponification number (mg potassium hydrate to saponify 1 g oil), iodine number (percent iodine absorbed, which indicates adulteration), lowest pour temperature, cloud point (temperature at which cloudiness occurs), flock point (temperature at which wax begins to precipitate), sedimentation, flash, and fire points or temperatures, and emulsification. Unfortunately, these specifications cannot be used alone for the selection or substitution of lubricants. If one lubricant with given test values performs satisfactorily, a second may not prove equally satisfactory even though the test values are essentially the same. These tests are helpful in indicating properties and uniformity of product, but performance tests are also needed to selection purposes. See Table 16.5 for additional information on lubricants.

Table 16.5 Miscellaneous Information on Lubricants

Gear Oil		
SAE no. (SAE J. 306 b, 1974)	Viscosity range (Saybolt universal seconds)	
	Minimum	Maximum
75 W at 210°F (98.9°C)	40	
80 W at 210°F (98.9°C)	49	
85 W at 210°F (98.9°C)	63	
90 W at 210°F (98.9°C)	74	120
140 W at 210°F (98.9°C)	120	200
250 W at 210°F (98.9°C)	200	

Motor Oil		
SAE No. (SAE J., 300 b, 1973)	Minimum	Maximum
5 W at 0°F		6,000
10 W at 0°F	6,000	12,000
20 W at 0°F	12,000	48,000
20 W at 210°F	45	58
30 W at 210°F	58	70
40 W at 210°F	70	85
50 W at 210°F	85	110

Greases	
NLGI consistency no. (SAE J. 310a, 1975)	Penetration at 77°F (25°C), ASTM worked penetration (60 strokes) in 0.1 mm
000	445-475
00	400-430
0	355-385
1	310-340
2	265-295
3	220-250
4	175-205
5	130-160
6	85-115

Table 16.5 (continued)

Animal and Vegetable Oils

Oil	Sp. gr. 20/4°C	Viscosity			
		Centistokes at		Saybolt (universal) seconds at	
		100°F (7.8°C)	210°F (98.9°C)	100°F (°C)	210°F (°C)
Almond	0.9188	43.2	8.74	201	54.0
Castor	0.9619	293.4	20.08	1368	97.7
Chinawood	0.9375	115.2	15.46	531	78.5
Coconut	0.9226	28.58	5.83	135	44.5
Cottonseed	0.9187	38.88	8.39	181	52.7
Glycerol, 99% CP	1.2588	224.40	11.56	1040	64.9
Lard	0.8968	31.66	7.24	148	48.9
Linseed	0.9297	29.60	7.33	139	49.2
Menhaden	0.9355	28.75	7.26	135	48.9
Neatsfoot	0.9158	43.15	8.50	200	53.1
Olive	0.9158	46.68	9.09	216	55.2
Palm kernel	0.9190	30.92	6.50	145	46.5
Rapeseed	0.9106	50.91	10.36	235	59.5
Sardine	0.9276	28.22	7.14	133	48.6
Soybean	0.9228	28.49	7.60	134	50.1
Sunflower	0.9207	33.31	7.68	156	41.9
Turkey red (sulfonated castor oil)	0.8948	20.63	5.83	100	41.9

Source: Data courtesy of Union Carbide Corporation, New York.

IX. WOOD*

A. Nature and Types of Wood

Although the term "wood" is used to denote the fuel and to describe a small stand of growing trees, its industrial use refers collectively to the material or finishes form of forest products. The words *timber* and *lumber* are also used. Timber describes large standing trees or large planks and beams cut from such trees. Lumber is a more general term used for boards, planks, and other raw material forms used in construction. Both timber and lumber are wood products.

Wood material consists of a skeleton of cellulose cells permeated by a mixture of other organic substances collectively known as "lignin." About 50% of the wood is cellulose; the balance is lignin plus about 2-10% minerals, waxes, tannins, and oils. A small percentage of incombustible, inorganic (mineral) matter, or ash, is also present.

---

*Based on information drawn from U.S. Forest Products Laboratory *Wood Handbook: Wood as an Engineering Material* (USDA Agricultural Handbook 72, revised), Washington, D.C., 1974.

Each year the new growth of trees forms new cells in the sapwood, the zone of wood next to the bark. The cells formed in the spring of the year are larger and have thinner walls than those grown in the summer and early fall. In the coniferous trees, that is, the ones that retain their leaves at all seasons, the spring growth is usually lighter in color than that grown later. Some of the decidous trees, that is, the ones that shed their leaves each fall, have less spring growth and it is darker than occurs later. Because of the color differences in the cellular growth added during the growing season, each year's growth appears as an annular ring on the cross section of the tree.

Since only the outer sapwood rings participate in the growing process, the inner sapwood ultimately changes into a lifeless form known as heartwood. The heartwood is darker in color, stronger, and less subject to decay. It may constitute 60% of old trees but is, of course, less in young trees.

The wood derived from American trees is described as softwood or hardwood. This classification is not related to measured hardness but is according to the type of tree. Some of the so-called softwoods are actually harder than varieties of the hardwoods.

The softwoods are those obtained from coniferous trees, ones that retain their leaves. They grow in all major forest areas, the northern, southern, Pacific, Rocky Mountain, and tropical forests. Typical softwoods are aspen, cedar (red, white, and Port Oxford), cypress, Douglas fir, balsam fir, hemlock, pine (Eastern white, red, Ponderosa, southern yellow, sugas, and Western white), redwood, and spruce. Usually their annular growth forms parallel cells and straight grain, which is easily split.

The hardwoods are the deciduous types that lose their leaves at the end of the growing season. Typical types are red alder, basswood, ash (18 varieties), beech, birch (20 varieties), cottonwood (12 varieties), elm (three varieties), gum, hickory, locust, maple (sugar, black, silver, red, and bigleaf), oak (15 commercial varieties usually divided into red and white groups), yellow poplar, and black walnut. Both annular and radial, or ray, cells are formed in hardwoods. This gives them a cross grain, and imparts high resistance to splitting. Fine-grain types have small, thin-walled cells and few vessels. Coarse-grained types have many vessels and large cells. The annular rings and the streaks of vessels in coarse-grained hardwoods impart desirable appearance, depending on the manner in which boards are cut from the logs.

A number of types of tropical wood are imported. Typical of these and the properties of chief interest are

> Balsa: lightest wood, 3-25 lb/ft^3
> Ebony: valued for its dark color, turning, and finishing
> Lignum vitae: one of the hardest, heaviest, and most durable woods.
> Mahogany: easily workable, relatively little shrinkage. African variety (Khaya)
>   and Philippine variety (Lauan) are also used.
> Rosewood: valued for its finishing and rose odor.
> Sandlewood: from India, valued for odor.
> Teak: high resistance to decay and low shrinkage.

B. Properties of Wood

1. Dryness

Moisture is present in green wood in three forms (Table 16.6). It comprises over 90% of the protoplasmic content of the living cells, it saturates the walls of the cells, and it

Table 16.6 Typical Range of Properties, Green Wood

Properties	Values
Green moisture, %	40-120
Density, lb/ft^3	27-71
Elastic modulus, psi	800,000-2,000,000
Modulus of rupture, psi	5,000-14,000
Crushing strength, parallel to grain, psi	4,000-9,200
Compressive strength, perpendicular to grain, psi	370-2,100
Shear strength, parallel to grain, psi	600-2,400
Hardness, load to indent 0.44 in. ball ½ diameter into into side grain, lb	320-2,150

fills the center of the lifeless cells. Loss of this moisture causes twisting, checking, warping, and shrinking. The shrinkage is usually largest in a lateral or tangential direction and may amount to 7-20% from green to oven-dry state. It causes checks or cracking longitudinally and reduces shearing resistance parallel to the grain. Although some checking and cracking result during drying, most wood is dried before it is used. Either air drying or kiln drying are employed.

In air drying, called seasoning, the boards are piled in an open shed and spacers are placed between layers to allow natural air circulation. This method of drying requires 1 month or more to dry 1 in. pieces to 10-20% moisture by weight. It may take several years for larger pieces. Because of the relatively long time required for air drying, it is being replaced by kiln drying.

Boards are piled in a kiln. Heated air is circulated at about 300°F, and moisture content may be reduced to 6-12% by weight in a few days. Hardwoods should be dried slower than softwoods. This method is generally employed for all hardwoods to be used for interior finish, patterns, and cabinet work, and for softwoods to be used for construction.

Dried wood is hygroscopic and will absorb moisture from the atmosphere up to the saturation point of the cell walls, usually 25-30% moisture by weight. The reabsorption of water is accompanied by swelling and reduced strength and machinability, and the bonding strength with adhesives is impaired. Wood should be kept dry prior to working and painted or coated for protection afterward when these characteristics are important.

2. Strength and Stiffness

The strength of wood increases in bending and compression with its density. Dense, fine-grained woods with at least six annular rings per in. and one-third of each ring of summer growth are therefore selected for heavy load-bearing members. Open-grain wood, which is light in density, is used for light framing, studs, and so on.

Defects, such as large, loose knots, excess sapwood checks, and decaying matter, reduce strength, especially in the bottom, or tensile side, of beams. Wood with these defects should be limited to concrete forms, boarding, and light framing. Tight knots that interrupt the grain on tensile members seriously reduce strength, compression is affected less (15-20%), and elastic modulus very little.

Strength and stiffness are increased as moisture content is decreased. Advantage may be taken of the dry and seasoned strength on small components free from defects and for protected applications. It is customary, however, to assume the green strength and stiffness in applying large members for outside use, since checking may occur in such pieces and moisture may be reabsorbed in use.

The mechanical properties of green wood generally lie within the ranges shown in Table 16.6. Values for many types of wood available may be obtained from the U.S. Department of Agriculture *Wood Handbook*. The variation in strength between pieces of the same grade is high, ranging from ½ to 1½ times the average strength values.

3. Thermal Properties

The specific heat of practically all wood is about 0.327 when oven dry. The value increases, of course, with moisture content. The calorific value also depends on moisture content since the moisture must be vaporized by burning the wood. Typical values range from 2700 BTU/lb with 50% moisture to 6300 BTU/lb when kiln dry. Density also affects the calorific value, the more dense types of wood having higher values.

C. Applications for Wood

Perhaps the largest application for wood is its use in construction, where it serves temporary uses as forms and scaffolds, and permanent uses as interior finish, exterior finish, light framing, and heavy load-bearing members. Economy is the prime consideration in selection of type and grade of wood. The denser varieties of highest quality are required for heavy load-bearing members, but these grades are the most expensive, heaviest to handle, and hardest to cut. Less dense and lower grades are employed when possible.

Grading standards have been established by such groups as the National Lumber Manufacturing Association, Southern Pine Association, West Coast Lumberman's Association, and the Departments of Agriculture and Commerce. They specify classifications for lumber according to nature and extent of typical defects. Also specified by such groups are the standard sizes and the manner of cutting logs into boards. Municipal codes in some cities define permissible working stresses for classes of structures and grades of wood used.

Other large uses for wood are furniture, pattern making, and wood products, such as toys and handles. Many furniture and cabinet products employ veneers, which consist of a thin piece of expensive wood glued to a backing of cheaper wood. The expensive wood may be sawed into thin slices, but this is wasteful because of the thicknesses required and sawdust formed. Slicing by moving the wood to a stationary cutting blade is more economical. Rotary cutting obtained by rotating the wood against the blade is most frequently employed. The wood is soaked in water or steamed before cutting. The thin sheets are held flat during drying and are then glued to their backing.

Plywood is another popular form made by gluing three or more 1/8 in. veneer layers together, the grain of each being placed at 90° to the adjacent ones. The material has good impact resistance and finds application in drawer bottoms, boxes, crates, paneling, flooring, and so forth.

Cooperage (barrels, kegs, and tubs) was a large user of wood but is gradually being replaced by other types of containers. The slack type is used for dry contents and the tight types for wet material. The latter must be made from high-grade wood with small pores and is often coated with a film of wax.

Coatings or impregnants are used for protection. Dry wood is very stable at ordinary temperatures, but is attacked by bacteria and fungi under warm, moist conditions, as in contact with the ground. The purposes of preservation treatments are to prevent entry of moisture into the wood and to prevent fungi from feeding on the wood. Sapwood is more vulnerable in this respect than heartwood, but late cuts are more desirable than spring cuts.

Preservatives used are of two types. Organic chemicals, such as creosote, a distillate or heavy oil of coal tar obtained from high-temperature carbonization of bituminous coal, is often used, either alone, or mixed with coal tar, petroleum, or wax. The latter materials are employed to seal the pores against moisture. Creosote coatings are often used on power poles, crossarms, and railway ties. Another commonly used organic chemical is pentachlorphenol in a mineral spirits solution. Another type of coating is water-soluble metallic salts, such as chromated zinc chloride. This material is odorless, clean, and serves as a fire retardant but is toxic. It can be painted but is subject to leaching in wet locations. Other materials that are being used are acid copper chromate, ammoniacal copper arsenite, chromated copper arsenate, and fluor chrome arsenate phenol. Methods used for applying preservatives are: (1) brush, mop, or spray; (2) dip in hot material; (3) steep in heated material; (4) pressure impregnate, which results in filling wood cells; and (5) follow pressure impregnation with vacuum, which results in removal of free impregnant, leaving cell walls saturated.

X. SILICONE PRODUCTS

Silicone polymers are the fundamental building block of silicone products. The most widely used silicone is poly(dimethyl siloxane):

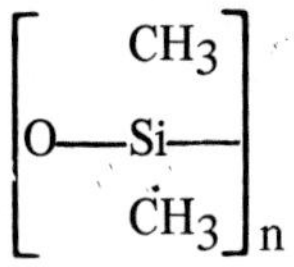

$$
\left[\begin{array}{c} CH_3 \\ O\!-\!Si\!- \\ CH_3 \end{array} \right]_n
$$

This is made by polymerization of a low-molecular-weight cyclic analog, octylmethyl cyclotetrasiloxane. The entire reaction from dimethyldichlorosilane to polymer is shown below.

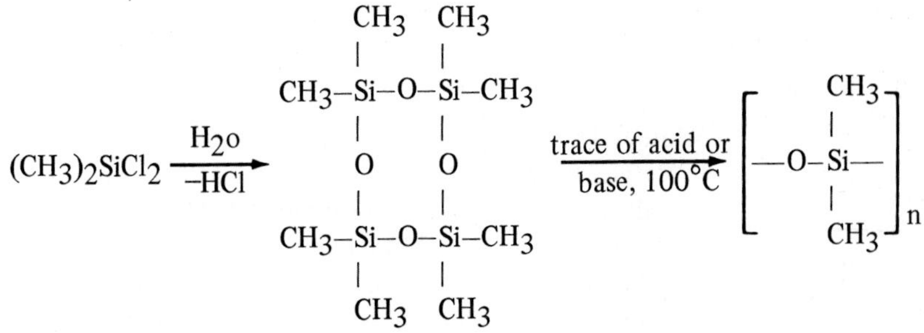

$$
(CH_3)_2SiCl_2 \xrightarrow[-HCl]{H_2O} \quad
\begin{array}{c}
CH_3 \quad CH_3 \\
| \qquad | \\
CH_3\!-\!Si\!-\!O\!-\!Si\!-\!CH_3 \\
| \qquad | \\
O \qquad O \\
| \qquad | \\
CH_3\!-\!Si\!-\!O\!-\!Si\!-\!CH_3 \\
| \qquad | \\
CH_3 \quad CH_3
\end{array}
\xrightarrow[\text{base, } 100^\circ C]{\text{trace of acid or}}
\left[\begin{array}{c} CH_3 \\ | \\ -O\!-\!Si\!- \\ | \\ CH_3 \end{array} \right]_n
$$

The chemistry of silicone polymers is discussed in Chap. 11. In this chapter we address the end uses of silicone polymers.

A. General Properties

As a class the following silicone properties are of importance to the designer.

> Flexibility and elasticity over a broad temperature range. The glass transition
> temperature is about −130°C; the temperature range of elasticity for sili-
> cone elastomers is −40-250°C.
> Water repellancy is high. This accounts for their use in polishes, antistick formula-
> tions, and biomedical devices.
> They can be designed to remain stable up to 300°C.
> They have interesting dielectric and lubricating properties.
> They adhere well to properly prepared polar surfaces.

B. Silicone Elastomers

These are vulcanizable either with heat or at room temperature, the rubberlike raw materials are compounded to achieve various degrees of stiffness, modulus, tear strength,
resilience, flame retardancy, and rate of cure. They are used for gaskets, seals, tubing,
hose, wire insulation, adhesion aids, mold making, adhesives, sealants, potting and en-
capsulating materials, component coating, and molded parts.

C. Silicone Liquids

Silicones are used in varnishes, surfactants, defoamers, vehicles, and carriers, in chemical
specialties, as dielectric medium, as heat transfer medium, as paint additives to improve
leveling and flow-out, reduce orange peel, and to improve mar resistance. They are also
used as release agents and lubricants.

D. Silicone Resins

The silicone resins are used in decorative and protective coatings to improve heat stability, increase water resistance, achieve desirable hardness in metallic paints, increase
chalk resistance, and to blend with organic resins to yield an improved combination
of properties.

XI. AMORPHOUS METAL ALLOYS

This is a class of substances that lacks crystalline structure but rather has much the makeup of a glass. The alloys are made from a melt by rapid cooling either by high-speed
cooling of very fine drops by a countercurrent gas stream, usually helium, or by casting
a melt as a thin ribbon on a rapidly moving cooled metal surface. The first method was
evolved by the Pratt and Whitney Division of United Technologies, Inc.; the second
method is practiced by the Metglas Products Department of Allied Chemical Corp.

    The supercooled metal alloys, although theoretically metastable, are in fact quite
stable at service temperatures, which may reach 2300°F in the case of the Arpalloys
made by Pratt and Whitney. The essential point is that alloys can be made that do not
conform to any rule that depends on an equilibrium between liquidus and solidus composition at solidification temperature. As a result exciting new properties become available
to the designer. In the case of Allied Chemical's Metglas alloys of iron, boron, silicon, and

carbon, new, attractive, high magnetic induction material is made. Other alloys contain cobalt, nickel, and molybdenum. In the case of Pratt and Whitney's Arpalloy (developed with the support of the U.S. Department of Defense Advanced Research Projects Agency), superior superalloys for use in jet engines have been made. The Allied Chemical material is available in ribbons and fabricated as such. The Pratt and Whitney material is produced as a powder and fabricated by powder metallurgy techniques. It has the special advantage of being fabricated to near net shape, thus conserving the scarce and expensive materials that enter into its composition.

XII. SPECIAL NONMETALLIC PLAIN BEARING MATERIALS*

Practically all engineering metals have been used in a variety of bearing applications at one time or another (see Chaps. 8, 9, and 19). In addition, many different types of plastics and other nonmetallic materials are used in special fields.

Plastic bearings offer many advantages in use, and their attributes generally include freedom from corrosion, quiet operation, and compatibility, which often minimizes or eliminates the need for lubrication. The range of molded plastics and impregnated or reinforced plastics available to industry make them useful for a number of specific applications. Phenolics, nylon, TFE fluorocarbons, and high-temperature injection-molded plastics (i.e., polycarbonates and acetal and polyimide types) are typical of the range of plastic material employed as bearings.

Many other nonmetallic materials find application where self-lubricating properties, high-temperature stability, solvent resistance, or low cost are considered. Carbon-graphite, wood, rubber, and ceramics are often used. These materials resist distortion at fairly elevated service temperatures, and generally their wear rates are low.

Operating limits for selected nonmetallic bearings are shown in Table 16.7. As with all bearing materials, careful consideration as to loads, speeds, contamination, service

---

*Contributed by JOSEPH B. LONG.

Table 16.7 Operating Limits for Selected Nonmetallic Bearing Materials

Bearing material	Load capacity (psi)	Maximum temperature (°F)	Speed (fpm)	PV limit p = psi load v = surface speed (fpm)
Phenolics	6,000	200	2,500	15,000
Nylon	1,000	200	1,000	3,000
TFE	500	500	50	1,000
Reinforced TFE	2,500	500	1,000	10,000
TFE fabric	60,000	500	50	25,000
Polycarbonate	1,000	220	1,000	3,000
Acetal	1,000	180	1,000	3,000
Carbon-graphite	600	750	2,500	15,000
Rubber	50	150	4,000	—
Wood	2,000	100	2,000	12,000

life, and operating temperatures, as well as the type and quantity of lubrication available, should be predominant in the selection of the proper bearing material.

XIII. MATERIALS FOR SPECIALIZED APPLICATIONS*

As new technologies emerge and old technologies advance, conventional materials become inadequate and new materials systems must be developed. In some instances, completely new materials must be developed. In other cases, fabrication methods are modified to improve materials properties. Finally, there are requirements that can be met only by combining new materials made by unconventional methods into materials systems with greatly enhanced performance. Three of the newer materials systems are described in this section.

A. Fiber-Reinforced Composites

Glass fiber-reinforced plastics have found wide application in many fields for many years. Their versatility and wide acceptance are due to the properties obtained by utilizing the high tensile strength of glass fibers (400,000 psi) in a tough resin matrix, such as polyester. However, such composites are limited to applications at ambient or moderately elevated temperatures.

Carbon or graphite fibers made by thermal decomposition of fibers, such as rayon, polyester, or acrylics, have tensile strengths equal to glass fibers combined with much higher stiffness and high temperature strength. Graphite-epoxy composites are finding wide acceptance in aircraft and automotive applications because of their excellent mechanical properties combined with very significant weight savings.

Among engineering materials, graphite, the crystalline form of carbon, is unique, since its tensile strength increases at elevated temperatures. At 5000°F, graphite is twice as strong as at room temperature and is ductile. Replacing the resin matrix with a carbon matrix produced a class of composites described as "carbon-carbon." These have found rapid acceptance for such applications as aircraft brakes and rocket motor components.

Fabrication of aircraft brakes starts with the production of graphite cloth from a rayon or polyacrylonitrile precursor. The precursor fabric is carbonized by controlled thermal decomposition in the absence of air, and the resulting carbon cloth is then heat treated to produce a graphite fabric. Layers of cloth cut to proper shape are bonded with a phenolic resin to produce the desired shape. This is then densified by either impregnating with pitch and carbonizing or by depositing carbon by a process called infiltration. In the latter process, methane is thermally decomposed at 1600°F to deposit carbon in the voids of the graphite cloth layup. The process is continued until a carbon-carbon composite of the desired density is obtained. Carbon-carbon aircraft brakes outperform conventional steel brakes by a factor of 20 while saving as much as 1500 lb in weight on large aircraft.

Carbon fibers can be converted into silicon carbide by a chemical vapor reaction called siliconizing. A silicon carbide matrix can be deposited around either graphite or silicon carbide fibers to produce erosion- and oxidation-resistant composites. This technology is growing rapidly, and new systems tailored to specific requirements are developed continually.

---

*Contributed by EDWIN T. MYSKOWSKI.

B. CVD Coatings

Chemical vapor deposited (CVD) coatings are produced by thermal decomposition of compounds and deposition of reaction products on suitable substrates. One such reaction uses methyl trichlorsilane to produce silicon carbide.

$$CH_3Cl_3Si + \Delta \rightarrow 3HCl + SiC$$

The reaction occurs at temperatures above 1600°F and deposits silicon carbide of about 100% density on such substrates as graphite or suitable metal. The deposit is very uniform and lacks a definite grain structure. The absence of grain boundaries greatly enhances erosion resistance compared with that of the conventional grades of silicon carbide, approaching that of diamond.

C. Free-Standing Shapes

For certain applications, free-standing parts of silicon carbide or other CVD materials are required. Tubes and other shapes can be fabricated using female mandrels made of graphite. These tubes can be used for extreme temperature heat exchangers and combustion chambers. The process is not limited to silicon carbide fabrication but can be used to process many other refractory compounds as well as metals. It is a very rapidly developing technology.